OPERATIONS RESEARCH

S.R. Yadav
Formerly, Professor
Department of Mathematics,
Birla Institute of Technology & Science (BITS)
&
B K Birla Institute of Engineering & Technology (BKBIET)
Pilani

A.K. Malik
Associate Professor
Department of Mathematics,
B K Birla Institute of Engineering & Technology (BKBIET)
Pilani

OXFORD
UNIVERSITY PRESS

OXFORD
UNIVERSITY PRESS

Oxford University Press is a department of the University of Oxford. It furthers the University's objective of excellence in research, scholarship, and education by publishing worldwide. Oxford is a registered trade mark of Oxford University Press in the UK and in certain other countries.

Published in India by
Oxford University Press
YMCA Library Building, 1 Jai Singh Road, New Delhi 110001, India

First published in 2014

ISBN-13: 978-0-19-809618-4
ISBN-10: 0-19-809618-6

Typeset in Times New Roman
by E-Edit Infotech Private Limited, Chennai
Printed in India by Tara Art Printers (P) Ltd, Noida

Preface

The depletion of various natural and physical resources and the possibility of complete exhaustion of these resources in the next few decades is a major area of concern in recent times. In order to deal with this threat, tremendous efforts are being made towards developing new technologies for effective usage and recycling of the available resources. Operations research is one branch of mathematics that has made this possible. It deals with the application of scientific methods and techniques to obtain optimal solutions for decision-making problems, and hence helps in efficient management of available resources. The formal initiation of operations research took place during the Second World War through a study of the effectiveness of military operations, primarily in the Royal Air Force.

With the growth of global markets and thus competition, the study of operations research has gained considerable importance. Every business now aims at providing good value products and services to its customers, which can be obtained from a fine combination of low cost, high quality, rapid availability of goods, and real-time information. This can be easily achieved by applying various techniques of operations research. In recent years, service organizations such as hospitals, airlines, banks, railways, and libraries have started recognizing the usefulness of operations research techniques in improving efficiency. These techniques also provide an effective base for decision-making.

ABOUT THE BOOK

Operations Research is a core subject for science, engineering, and allied disciplines in almost all universities, throughout the world. Although there are numerous books on this topic, this book deals with the subject in a new and refreshing way. The approach of this book is student friendly as it enables easy understanding of the fundamentals. It provides a variety of problems to encourage the reader to think through each formulation. A comprehensive coverage of both optimization techniques and operations management is provided.

The book introduces basic topics such as formulation and graphical solution of linear programming problems (LPP), simplex and revised simplex methods, duality and sensitivity analysis, transportation and assignment models. It then moves on to advanced topics, such as sequencing and scheduling (critical path method and program evaluation and review technique); dynamic, integer, and goal programming; game and decision theories; queuing and replacement models; simulation; inventory (deterministic and probabilistic) models; non-linear programming; and classical optimization techniques.

The book is designed to meet the needs of undergraduate, postgraduate, MPhil, and research students of science, mathematics, and operations research. The contents of this book have been chosen after careful perusal of the syllabi of the undergraduate (B Sc/B E/B Tech) and post-graduate (M Sc/M E/M Tech) courses. The book explains the fundamental principles of the various concepts with illustrations. The core material is interspersed with illustrative examples, theorems, and relevant discussions.

KEY FEATURES

- Written in a simple language with concepts introduced in a sequential manner
- Provides numerous figures, graphs, tables, and practical examples for easy comprehension of the concepts discussed
- Covers both optimization techniques and operations management in one volume
- Includes numerous fully solved examples within chapters for a better understanding of the topics discussed
- Provides multiple-choice questions, true or false questions, fill in the blanks, theoretical review questions, and numerical problems
- Includes a solved conceptual problem at the end of every chapter to help the reader evaluate their learning from the chapter
- Includes seven appendices, which discuss a few preliminary mathematical concepts in brief, and present a few tables that would be helpful in solving certain problems provided in the book

ONLINE RESOURCES

To aid the faculty and students in using this book, the companion website of this book http://oupinheonline.com/book/yadav-operations-research/9780198096184 provides the following resources:

For Faculty

- A solutions manual for the numerical problems given in the book
- A complete chapter-wise PowerPoint presentation to aid classroom teaching

For Students

- Online tests with multiple-choice questions, for self-evaluation and practice

CONTENTS AND COVERAGE

The first twelve chapters (1–12) of the book detail various optimization techniques used in operations research. The next nine chapters (13–21) discuss concepts of operations management. The last three chapters (22–24) explore a few advanced topics in optimization techniques. A chapter-wise scheme of the book is presented here.

Chapter 1 presents the basic definitions of operations research. It provides a historical overview of the development of the subject along with its scope and applications. The chapter further highlights various phases and models of operations research.

Chapter 2 deals with the formulation of an LPP. It provides insights into the limitations and applications of LPPs in businesses and industries.

Chapter 3 discusses the graphical solution of an LPP, some special cases, and limitations of solving an LPP using the graphical method.

Chapter 4 describes the simplex method, the most popular method to solve an LPP. It throws light on the criteria of optimality, special cases in simplex method, and artificial variable techniques required to obtain the solution to an LPP. The chapter further discusses determination of the solution to simultaneous linear equations as well as the inverse of a square matrix by using the simplex method in detail.

Chapter 5 elucidates the revised simplex method for solving an LPP.

Chapter 6 elaborates on the duality in linear programming. It deals with the symmetric form, primal–dual relationship, economic interpretation of duality, some important theorems on duality, and dual simplex method.

Chapter 7 focuses on post optimality (or sensitivity) analysis and parametric linear programming. It discusses the changes affecting feasibility and optimality, graphical sensitivity analysis, changes in the cost c_j (the coefficient of x_j) in the objective function, a_{ij}'s and b_i's (availabilities), addition of a new variable and constraint, and deletion of a constraint and variable.

Chapter 8 explains transportation problems. It provides insights into some of the important methods and techniques such as north-west corner method (NWCM), least cost entry method (LCM), Vogel's approximation method (VAM), modified distribution (MODI) method, and stepping stone method. It further discusses unbalanced and transshipment problems in detail.

Chapter 9 focuses on assignment problems. It discusses unbalanced and travelling salesman problems in detail.

Chapter 10 analyses sequencing problems—processing n jobs through two machines, n jobs through k machines, and 2-jobs through k machines. It helps in determining a proper sequence of jobs and machines to minimize the total elapsed time.

Chapter 11 deals with project scheduling, which is very important for any business organization. It discusses CPM, PERT, and optimum scheduling by CPM in detail.

Chapter 12 lays emphasis on dynamic programming. It encompasses the multi-decision process, Bellman's principle of optimality, deterministic and probabilistic dynamic programming, and applications of dynamic programming.

Chapter 13 helps in integer programming problem (IPP). It deals with Gomory's cutting plane method for all integer programming problems (AIPPs), mixed integer programming problems (MIPPs), branch and bound technique to obtain the solution to IPPs (Dankin's branching method) and zero-one IPP.

Chapter 14 explains the queuing theory. It discusses Markovian queues, probability distribution of n arrivals in time interval $(t, t+\Delta t)$ in a pure birth process, distribution of inter-arrivals time, distribution of departures in a pure death process, the birth and death process, and the various queuing models with their characteristic properties.

Chapter 15 elucidates goal programming. It discusses single and multiple-goal models, modified simplex method, and graphical solution of a goal programming problem.

Chapter 16 illustrates the concept of theory of games. It discusses lower and upper values of games, minimax principle with pure strategies, two-person zero–sum game with mixed strategies, and the dominance principle of game theory in detail.

Chapter 17 defines decision theory. It deals with decision-making under certainty, risk, uncertainty, and conflict. Aspects of decision theory such as posterior probability and Bayesian analysis are also discussed.

Chapter 18 helps us appreciate the applications of networking in operations research. It explains the shortest route, minimal spanning tree, and maximum flow problems.

Chapter 19 illustrates the concept of replacement models. It explains replacement policy under the following categories—when the value of money does not change with time, when the value of money changes with time, and when the items/units fail completely all of a sudden.

Chapter 20 defines simulation, random numbers, and pseudo-random numbers. It discusses the mid-square method of generating pseudo-random numbers, the multiplicative and mixed congruential method, the generation of random variates, the applications of simulation in queuing and inventory models, the advantages and disadvantages of simulation models, and the simulation languages.

Chapter 21 presents the inventory models. It further discusses the types of inventory, basic terminologies used in inventory, inventory control, inventory costs, inventory management and its benefits, the economic order quantity (EOQ), deterministic inventory models, EOQ problems with price break, probabilistic inventory models, and some important inventory control techniques.

Chapter 22 defines classical optimization techniques. It discusses single and multi-variable unconstrained optimization problems and explains multi-variable constrained optimization problems with equality and inequality constraints.

Chapter 23 helps us appreciate the applications of non-linear programming in optimization techniques. This chapter explains the search methods to solve unconstrained and constrained non-linear programming problems (NLPPs).

Chapter 24 discusses the solution to NLPP using quadratic and separable programming. It explains determination of the optimal solution of an NLPP using Kuhn–Tucker conditions.

At the end, we have a section on Appendices.

Appendix A provides the basic concepts of linear algebra. *Appendix B* deals with the matrices in brief. *Appendix C* presents some basics of calculus. *Appendix D* provides simple concepts of probability. *Appendix E* provides a table of 'area under the standard normal distribution (*Z*)'. It will help in solving certain problems discussed in Chapter 11. *Appendix F* provides a table of 'random numbers'. These will help in solving certain problems discussed in Chapter 20. *Appendix G* provides the table of 'Poisson probability distribution (*X*-table)'. It will aid in solving certain problems discussed in Chapter 14.

ACKNOWLEDGEMENTS

We owe our gratitude to Late Dr Birbal Singh, retired Professor of BITS, Pilani, who was a great source of inspiration. We are equally indebted to Dr S.R. Singh, Associate Professor, Mathematics, D.N. College, Meerut, for his constant motivation and guidance.

It gives us great pleasure to acknowledge the kind blessings and inspiration provided to us by Dr P.S. Bhatnagar, Director, BKBIET Pilani. We thank Prof. A.P. Girdhar, Prof. L. Solanki, Prof. S.B. Dandin, and Prof. S.K. Yadav for their valuable suggestions and continuous encouragement. We acknowledge the contributions of our former colleagues, Prof. P.H. Keskar, Prof. C.B. Gupta, and Prof. Chandra Shekhar (all from BITS Pilani) in the preparation of this book.

We also express our gratitude and appreciation for the valuable guidance provided by Dr S.K. Gupta, Associate Professor, Mathematics, D.N. College, Meerut; Prof. C.K. Jaggi, Head, Department of Operations Research, Delhi University, Delhi; and Prof. M.P. Singh, Head, Department of Mathematics and Statistics, Gurukula Kangri University, Haridwar.

We are extremely grateful to our families, who extended their unconditional support and graciously accepted our inability to attend to family chores during the course of writing this book.

Last, but not the least, we express our heartfelt thanks to the editorial team at Oxford University Press, India, who guided us throughout this project.

We eagerly look forward to suggestions and feedback from teachers and students. We can be reached at ajendermalik@gmail.com and sryadav1940@gmail.com.

S.R. Yadav
A.K. Malik

Features of the Book

After studying this chapter, the reader will be able to

- understand the symmetric form of a linear programming problem (LPP)
- write the dual form after reducing the LPP to the symmetric form
- identify the solution to the primal from the optimal solution of the dual
- understand the dual simplex method

Learning Objectives

Each chapter begins with an outline of the topics discussed in the chapter.

Solved Examples

Numerous solved examples are interspersed with the text to facilitate easier understanding of the concepts discussed.

Example 4.1 Write the standard form of the following LPP:

$$\text{Max } Z = x_1 + x_2 - 2x_3 \quad (4.1)$$

$$\text{Subject to} \quad x_1 + x_2 + x_3 \leq 15 \quad (4.2)$$

$$2x_1 - x_2 + x_3 \leq -10 \quad (4.3)$$

$$x_1 + 2x_3 = 10 \quad (4.4)$$

$x_1, x_2 \geq 0$, and x_3 is unrestricted in sign.

Solution Adding x_4 (slack variable) in Eq. (4.2), we get

$$x_1 + x_2 + x_3 + x_4 = 15 \quad (4.5)$$

Multiply Eq. (4.3) by (−1) to get $-2x_1 + x_2 - x_3 \geq 10$ (4.6)

Now subtract x_5 (surplus variable) from the left-hand side (LHS) of Eq. (4.6) to get

$$-2x_1 + x_2 - x_3 - x_5 = 10 \quad (4.7)$$

	Cost →			4	6	0	0	Minimum ratio X_B/α_i^j
	Variable →			x_1	x_2	x_3	x_4	
Table no.	C_B	Basic variable	X_B	α^1	α^2	α^3	α^4	
1	0	x_3	6	2	[3]	1	0	6/3 = 2→
	0	x_4	1	1	−1	0	1	...
		$c_j - Z_j$		4	6↑	0	0	–
2	6	x_2	2	2/3	1	1/3	0	–
	0	x_4	3	5/3	0	1/3	1	
		$c_j - Z_j$		0	0	−2	0	–

Tables

Provides numerous tables to illustrate various problem-solving techniques.

Recapitulation

Recapitulation at the end of each chapter provides a quick revision of the key concepts discussed in the chapter.

RECAPITULATION

1. Any maximization or minimization problem, where the objective function and constraints are linear, can be solved using the simplex method.
2. Every LPP can be converted into the canonical form, which is the requisite form for solving it using the application of the simplex method.
3. In the simplex method, we start with a readily available BFS, and move from one vertex (BFS) to another, skipping many intermediate vertices and arriving at the optimal solution rapidly.
4. For the simplex method, the leaving variable and entering variable rules have been devised in such a way that the next solution obtained is feasible and at the same time, the value of the objective function improves, ultimately leading to the optimal solution to the problem.

CONCEPTUAL PROBLEM

After conducting a market survey, a company decides to launch two new stylish leather belts, B_1 and B_2. The company takes into account all the aspects such as supply of raw material, production, advertising, and marketing of the product. In the beginning, the company arranges leather for the production of 800 belts (both B_1 and B_2). The B_1-type belts are fancy and cost more, whereas the B_2-type are ordinary and cost less. B_1 belts require twice as much time for manufacturing as B_2. Further, if all the belts are of type B_2, the company will be able to produce up to 1000 belts per day. The fancy buckles available for the B_1 belts are 400 per day and the ones for B_2 are 700 per day. The company expects to make a profit of ₹4 and ₹3 per belt, on types B_1 and B_2 respectively. What should the company do to earn maximum profit? To help the company, formulate the problem in an LP model and provide its solution using the revised simplex method.

Solution Let x_1 and x_2 be the quantity of leather belts B_1 and B_2, respectively. Then, the total profit is $4x_1 + 3x_2$.

Conceptual Problems

A case study is provided at the end of each chapter in the form of a conceptual problem, for better understanding of concepts discussed in the chapter

Exercises

End-chapter exercises in the form of multiple-choice questions, true or false questions, fill in the blanks, review questions, and numerical problems will help students test their skills and prepare for examinations.

Multiple-choice Questions

4.1 Which positive variables are added in the left-hand sides of the constraints of an LPP to convert them into equalities?
(a) Slack variable
(b) Surplus variab
4.2 Which positive va
convert them into
(a) Slack variable
(b) Surplus variab

State True or False

4.1 A BFS is said to be optimum, if it optimizes the objective function.
4.2 A basic variable can never assume the value zero.
4.3 A
4.4 A
4.5 A

Fill in the Blanks

4.1 The simplex method was developed by .... in 1947.
4.2 If a constraint in an LPP has the '≤' sign, then to convert it into an equation we use the
4.3 To co
sides

Review Questions

4.1 Write the procedure to solve an LPP using the simplex method.
4.2 Defin
(a)
(b)
(c)

Numerical Problems

4.1 Solve the following LPPs using the simplex method:
(a) A company produces two types of products: canned a profit, amounts, and availability of material, equipm produce each product are shown in the following table

A. Linear Algebra
B. Matrices
C. Calculus
D. Probability
E. Area Under the Standard Normal Distribution Z
F. Table of Random Numbers
G. Poisson Probability Distribution X Table

Appendices

Seven appendices on various mathematical topics have been provided at the end of the book. An understanding of these topics is a prerequisite for implementing various operations research techniques.

Brief Contents

Detailed Contents

1 Introduction to Operations Research

> After studying this chapter, the reader will be able to
>
> - understand the historical development of operations research
> - comprehend the various definitions of operations research
> - explain the various operations research models
> - appreciate the scope and applications of operations research

1.1 INTRODUCTION

Operations research is a branch of mathematics that deals with the application of scientific methods and techniques to obtain optimal solutions for decision-making problems. With the growth of the global market and thus the competition, operations research has gained much importance. In this era of competitiveness, every business aims at providing good value products and services to its customers. Good value results from a fine combination of low cost, high quality, rapid availability, and real-time information about goods and services. Operations research with its techniques provides an optimal solution to achieve business targets.

1.2 HISTORICAL DEVELOPMENT

The application of mathematical models for solving business problems first attracted the attention of the Russian mathematician, L.V. Kantorovich, who published a monograph, *Mathematical Methods in the Organization and Planning of Production*, in 1939. In this monograph, he pointed out that there are many classes of production problems that can be defined mathematically and can therefore, be solved numerically. (The formal initiation of operational research took place during the Second World War through a study of the effectiveness of military operations, primarily in the Royal Air Force.) Incidentally, the term 'operations research' was first coined by McClospy and Trefthen in the year 1940 in the course of their research on military operations during the Second World War.

In 1948, an operations research club known as the 'Operational Research Society of United Kingdom' was founded in England. It published a quarterly journal, the first issue of which appeared in 1950. Another big step in the field of operational research was the establishment of the 'Operations Research Society of America' (ORSA) in 1952 and the publication of its first journal 'Operations Research' the very next year.

In India, operations research came into existence in the year 1949 with the establishment of an operations research unit at the Regional Research Laboratory, Hyderabad, for the purpose of planning and organizing research. Of equal importance was the setting up of an operations research team at Defence Science Laboratory (later renamed Defence Science Centre) by Prof. R.S. Verma for the specific purpose of solving the problems of storing, planning, and purchasing. Operations Research received a further boost with the setting up of an operations research team in the Indian Statistical Institute, Kolkata, by Prof. P.C. Mahalanobis in 1953 for solving problems related to national planning and survey. Later, the 'Operations Research Society of India' (ORSI) was established in 1957. This society started publishing its journal OPSEARCH in 1964.

1.3 DEFINITIONS

It is very difficult to give a precise definition of operation research as it has a wide scope of applications. Here, we have provided the definitions given by some great scholars, who have defined the term based on their perspectives.

'*Operations research is a scientific approach to problem solving for executive management.*'

–H.M. Wagner

'*Operations research is the art of winning wars without actually fighting them.*'

–Aurther Clarke

'*Operations research is the application of scientific methods, techniques and tools to problems involving the operation of a system so as to provide those in control of the system with optimum solution to the problem.*'

–C. West Churchman, Russell L. Ackoff, and Leonard Arnoff

'*Operations research is the art of giving bad answers to problems which otherwise have worse answers.*'

–T.L. Saaty

'*Operations research may be described as a scientific approach to decision-making that involves the operations of organizational system.*'

–F.S. Hiller and G.J. Lieberman

'*Operations research is applied decision theory. It uses many scientific, mathematical or logical means to attempt to cope with the problems that confront the executive when he tries to achieve a thorough-going rationality in dealing with his decision problems.*'

–D.W. Miller and M.K. Starr

'*Operations research is a scientific knowledge through interdisciplinary team effort for the purpose of determining the best utilization of limited resources.*'

–H.A. Taha

'*Operations research is a scientific method of providing executive departments with a quantitative basic for decisions under their control.*'

–P.M. Morse and G.E. Kimball

1.4 MODELS

A model refers to an easy depiction of a problem. In an operations research model, the basic or the most important features of the problem under consideration are deliberated. A number of models are used for solving problems through the operations research techniques. The following are the various operations research models:

Mathematical programming models

1. Linear programming
2. Non-linear programming
3. Dynamic programming
4. Integer programming
5. Geometric programming
6. Multi-objective programming
7. Quadratic programming
8. Goal programming
9. Stochastic programming
10. Separable programming
11. Information theory
12. Sequencing theory
13. Game theory
14. Assignment methods
15. Transportation methods
16. Inventory control methods
17. Network scheduling methods
18. Differential calculus methods
19. Neural networks
20. Genetic algorithms
21. Simulated annealing
22. Calculus of variations

Stochastic process models

1. Queuing theory
2. Reliability theory
3. Statistical decision theory
4. Renewal theory
5. Simulation methods
6. Markov process

Statistical models

1. Correlation analysis
2. Regression analysis
3. Cluster analysis
4. Factor analysis
5. Design of experiments

1.5 SCOPE AND APPLICATIONS

In recent years, service organizations such as hospitals, airlines, banks, railways, and libraries have started recognizing the usefulness of operations research for improving efficiency. Operations research techniques also provide an effective base for taking management decisions. The importance of applications of operations research in solving various problems in the engineering field is now accepted universally. Operations research techniques are used in solving problems in the following fields:

1. Optimal inventory cost
2. Optimal designing of a control system
3. Optimal planning, scheduling, and controlling
4. Optimal designing of chemical processing equipment and plants

5. Optimal selection of a new site for an industry
6. Optimal designing of pipeline networks for process industries
7. Optimal designing of computer structure at minimum cost
8. Optimal designing of plastic structures
9. Optimal planning to obtain maximum profit in the presence of one or more competitors
10. Optimal designing of the electrical networks
11. Optimal designing of aircraft and aerospace structures of minimum weight
12. Optimal designing of earthquake structures of minimum weight
13. Optimal controlling of the waiting and idle times and queuing in production lines to reduce the costs
14. Optimal designing of civil engineering structures such as bridge, tower, dam, frames, and chimneys at minimum cost
15. Optimal designing of material handling equipment such as conveyors, trucks, and cranes at minimum cost
16. Optimal designing of water resources systems to achieve maximum benefit
17. Optimal designing of pumps, heat transfer equipment, and turbines to achieve maximum efficiency
18. Optimal designing of electrical machinery such as generators, transformers, and motors
19. Optimal designing of mechanical components such as gears, cams, and machine tools

1.6 PHASES

In general, various phases are involved in solving a problem through the operations research techniques. These phases are considered most important and relevant in operations research. Some of the main phases involved are as follows (refer to Fig. 1.1):

Decision phase In this phase, real-life problems comprising the selection of an appropriate aim, values of various variables with appropriate measurement, and formulation of an appropriate model for the problem are identified.

Research phase This phase is the longest phase and is extremely important. It involves observation and data collection, formulation of hypothesis, analysis of available information, and prediction of various results.

Planning and action phase This phase includes the implementation of the recommendations and the results predicted in the research phase.

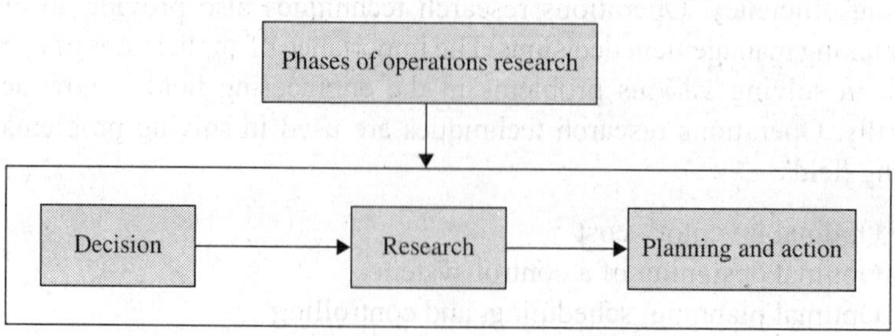

Figure 1.1 Main phases of operations research

RECAPITULATION

1. Operations research is a comprehensive scientific approach to provide the best possible solutions to complicated problems related to the management aspects of various disciplines.
2. The term operations research was first coined by McClosky and Trefthen in 1940 in UK.
3. Operations research came into existence during the Second World War for solving the strategical and tactical problems of the fighting forces.
4. In India, the first research unit of operations research was set up in 1949 at the Regional Research Laboratory in Hyderabad.
5. The Operations Research Society of India was established in 1957 and started publishing its journal OPSEARCH in 1964.

EXERCISES

Multiple-choice Questions

1.1 The first research laboratory was established in India in
(a) Delhi (b) Calcutta (c) Hyderabad (d) Mumbai

1.2 The definition of operations research 'Operations research is the art of winning wars without actually fighting them', was given by
(a) H.M. Wagner (b) H.A. Taha (c) R.S. Verma (d) Aurther Clarke

State True or False

1.1 The phases of operations research are decision, research, planning, and action.
1.2 The term operations research was first coined by H.M. Wagner.
1.3 The operations research unit of Indian Statistical Institute, Kolkata, was established in 1953.

Fill in the Blanks

1.1 Operations research came into existence during the ________ World War.
1.2 The Operations Research Society of India was established in the year ________.
1.3 The first operations research club was founded in ________ in the year 1948.

Review Questions

1.1 Write a short note on the historical development of operations research.
1.2 What is operations research? Give any five definitions of operations research.
1.3 Explain briefly the major phases of operations research.
1.4 Discuss the various models of operations research.
1.5 Write the scope and any ten applications of operations research.

2 Linear Programming Problem I—Formulation

After studying this chapter, the reader will be able to

- understand the formulation of a linear programming problem
- choose the appropriate decision variables for a real-life problem
- identify the objective of the given problem based on the chosen decision variables to be optimized
- understand the restrictions to be satisfied by the decision variables
- convert a real-life problem into a mathematical form consolidating the concepts learnt from the chapter

2.1 INTRODUCTION

Linear programming is an important optimization technique used in operations research. It was developed during the Second World War (1939–1945) by a group of British scientists for efficient utilization of depleting resources. The success of experiments with this technique attracted the attention of industries that were seeking methods to minimize their total costs and maximize their total profits. However, it was only in 1947 that the efforts of G.B. Dantzig led to the invention of the simplex method of linear programming. Later, new techniques in linear programming were developed and applied to various real-life problems.

Let us consider an example of a manufacturing company that produces personal computers and laptops. It follows the steps given here for the production of these goods:

1. The production department receives an order to produce a certain number of laptops and computers.
2. The material department ensures availability of the necessary raw materials for servicing this order.
3. After the production process is completed, the production department hands over the manufactured product to the sales department.

In each of these departments, various parameters can be optimized for gaining maximum profit—availability of machines and labour hours in the production department; availability of materials and storage capacity in the materials department; and the capacity of distributions and marketing in the sales department. The solution to such real-life problems can be effectively found using mathematical modelling.

2.2 LINEAR PROGRAMMING PROBLEM

Linear programming problem (LPP) is very useful for decision-making in business and other fields of day-to-day life for obtaining the optimum values of linear expressions subject to certain restrictions that are linear in nature. It is a powerful tool for dealing with problems of allocating limited resources among the activities necessary to attain the goals. In general, the problem is to maximize the net profit or minimize the total cost, taking into consideration the various restrictions imposed on the decision variables involved in the problem.

A general LPP can be described as follows:

$$\text{Optimize (maximize or minimize) } Z = c_1x_1 + c_2x_2 + \ldots + c_nx_n \tag{2.1}$$

Subject to conditions

$$\left.\begin{aligned} &a_{11}x_1 + a_{12}x_2 + \ldots + a_{1n}x_n \le \text{ or } = \text{ or } \ge b_1 \\ &a_{21}x_1 + a_{22}x_2 + \ldots + a_{2n}x_n \le \text{ or } = \text{ or } \ge b_2 \\ &\quad \cdot \qquad \cdot \qquad \cdot \\ &a_{m1}x_1 + a_{m2}x_2 + \ldots + a_{mn}x_n \le \text{ or } = \text{ or } \ge b_m \end{aligned}\right\} \tag{2.2}$$

$$\text{and} \quad x_1, x_2, \ldots, x_n \ge 0 \tag{2.3}$$

The function Z in Eq. (2.1) is called the *objective function*. The conditions (restrictions) given in Eq. (2.2) are called *constraints* and those given in Eq. (2.3) are called *non-negative restrictions*. The variables $x_1, x_2, \ldots, x_n$ are known as *decision variables*. The coefficients $c_1, c_2, \ldots, c_n$ in objective function are known as *costs*. The coefficients a_{ij} ($i = 1, 2, 3, \ldots, m$ and $j = 1, 2, 3, \ldots, n$) are known as *technological coefficients* and $b_1, b_2, \ldots, b_m$ are known as *availabilities*.

The aforementioned LPP can also be written in *matrix* notation as follows:

$$\text{Optimize (maximize or minimize) } Z = (c_1, c_2, \ldots, c_n)\begin{pmatrix} x_1 \\ x_2 \\ \cdots \\ x_n \end{pmatrix}$$

Subject to

$$\begin{pmatrix} a_{11} & a_{12} & \cdots & a_{1n} \\ a_{21} & a_{22} & \cdots & a_{2n} \\ \cdots & \cdots & \cdots & \cdots \\ a_{m1} & a_{m2} & \cdots & a_{mn} \end{pmatrix}\begin{pmatrix} x_1 \\ x_2 \\ \cdots \\ x_n \end{pmatrix} \le \text{or} = \text{or} \ge \begin{pmatrix} b_1 \\ b_2 \\ \cdots \\ b_m \end{pmatrix}$$

$$\text{and} \begin{pmatrix} x_1 \\ x_2 \\ \cdots \\ x_n \end{pmatrix} \geq 0 \quad \text{i.e., } (x_1 \geq 0, x_2 \geq 0, \ldots, x_n \geq 0)$$

or Optimize (maximize or minimize) $Z = C^T X$

Subject to $AX \leq \text{or} = \text{or} \geq b$

and $X \geq 0$

where $C = [c_1, c_2, \ldots, c_n]^T$, $b = [b_1, b_2, \ldots, b_m]^T$, $X = [x_1, x_2, \ldots, x_n]^T$, and $A = [a_{ij}]_{m \times n}$.

It has to be noted that C^T is a row matrix and both X and b are column matrices.

2.3 BASIC ASSUMPTIONS OF LINEAR PROGRAMMING PROBLEM

For an LPP, some basic assumptions are made. None of these assumptions should be violated; otherwise finding the correct solution is not possible. These assumptions are as follows:

1. The objective function of an LPP must have a linear function in decision variables.
2. The objective function must be to maximize or minimize under certain restrictions, which are again of '≤' or '≥' types.
3. The restrictions should not involve strict inequalities like '<' or '>'.
4. The decision variables must be non-negative.

2.4 FORMULATION OF LINEAR PROGRAMMING MODEL

The formulation of a linear programming model (LP model) means conversion of a descriptive problem into a mathematical model. Further, if some linear function is to be maximized or minimized under certain linear restrictions and the variables involved are required to be non-negative, then it is known as formulation of the problem in LP model. The following steps are required to formulate a problem in an LP model:

1. Identify the decision variables $x_1, x_2, \ldots, x_n$.
2. Identify the objective function and express it as a linear function of $x_1, x_2, \ldots, x_n$ and check whether it is to be maximized or minimized.
3. Identify the constraints as a linear function of $x_1, x_2, \ldots, x_n$ and express them in the form of inequalities.
4. Include proper restrictions for decision variables (i.e., x_i's≥ 0; $i = 1, 2, \ldots, n$).

2.5 LIMITATIONS OF LINEAR PROGRAMMING PROBLEM

The following are some of the limitations of LPPs:

1. In linear programming, there is only one objective function. However in practice, situations arise where multiple objectives are involved.
2. In an LP model, the objective function and constraints must be linear but these conditions are not satisfied in many practical problems.

3. Parameters such as costs, availabilities, and technological coefficients are assumed to be constant in an LP model, whereas they are neither constant nor deterministic in many practical problems.
4. An LP model does not always provide an integer solution. Therefore, when an integer solution is required, some other model should be used.

2.6 APPLICATIONS OF LINEAR PROGRAMMING PROBLEM IN BUSINESS AND INDUSTRIES

The following are some important applications of linear programming techniques or problems:

Assignment Linear programming technique is used to assign various jobs to available manpower to achieve maximum efficiency.

Transportation Transportation problems arise in businesses and industries where goods are required to be transported from the place of manufacturing or availability to other places of requirement. These problems require good application of the linear programming technique.

Product mix An LPP is used to determine the product mix for the production of a final product.

Diet An LPP is used in finding the best combination of daily requirement of nutrients that involve minimum cost.

Trim loss There are many situations where a commodity has to be prepared by cutting pieces from a metal sheet or a cloth bundle in some standard size. Linear programming technique provides the solution to problems where pieces have to be cut with minimum trim loss.

Blending By the application of the LP model, we can determine the most economical blend for a new product subject to the available ingredients.

Example 2.1 A pharmaceutical company produces three types of medicines M_1, M_2, and M_3. For manufacturing these medicines, three types of ingredients I_1, I_2, and I_3 are required. One M_1 requires 3 units of I_1 and 2 units of I_2; one M_2 requires 2 units of I_1, 3 units of I_2, and 4 units of I_3; and one M_3 requires 4 units of I_1 and 3 units of I_3. The company has a stock of 40 units of I_1, 30 units of I_2, and 45 units of I_3. The profit on selling one M_1 is ₹8, one M_2 is ₹13, and one M_3 is ₹10. If all the medicines produced are sold, formulate the problem as an LPP to maximize the profits.

Solution Tabulating the given data, we get

Ingredients ↓	Medicines			Total units available
	M_1	M_2	M_3	
I_1	3	2	4	40
I_2	2	3	0	30
I_3	0	4	3	45
Profit in ₹	8	13	10	

Let the company produce x_1, x_2, and x_3 number of medicines M_1, M_2, and M_3 respectively. The company would like to produce x_1, x_2, and x_3 such that the profit obtained as $(8x_1+13x_2+10x_3)$ is maximum. In producing x_1 units of M_1, x_2 units of M_2, and x_3 units of M_3, the required quantity of ingredient I_1 will be $(3x_1+2x_2+4x_3)$ units. Since the company has a stock of 40 units of ingredient I_1, we have $3x_1+2x_2+4x_3 \leq 40$. Similarly, for ingredient I_2, we have $2x_1+3x_2 \leq 30$, and for ingredient I_3, we have $4x_2+3x_3 \leq 45$. Since the units of the manufactured product cannot be negative, we have $x_1, x_2, x_3 \geq 0$.

Hence, the LPP is formulated as follows:

$$\begin{aligned} &\text{Maximize } Z = 8x_1+13x_2+10x_3 \\ &\text{Subject to} \quad 3x_1+2x_2+4x_3 \leq 40 \\ &\qquad\qquad\quad 2x_1+3x_2 \leq 30 \\ &\qquad\qquad\quad 4x_2+3x_3 \leq 45 \\ &\text{and} \qquad\qquad x_1, x_2, x_3 \geq 0 \end{aligned}$$

Example 2.2 A steel company manufactures three products P_1, P_2, and P_3. Each product has to pass through two machines M_1 and M_2. Each unit of P_1 requires 3 hours (h) of M_1 and 2 h of M_2; each unit of P_2 requires 2 h of M_1 and 5 h of M_2; and each unit of P_3 requires 2 h of M_1 and 3 h of M_2. The machines M_1 and M_2 are available for 30 h and 40 h respectively. The profit on each unit of products P_1, P_2, and P_3 is ₹4, ₹2, and ₹3 respectively. If all the manufactured products are sold, formulate the problem as an LPP to maximize the profit.

Solution Tabulating the given data, we get

Machines ↓	Products			Total time available (in h)
	P_1	P_2	P_3	
M_1	3	2	2	30
M_2	2	5	3	40
Profit in ₹	4	2	3	

Let the company produce x_1, x_2, and x_3 units of products P_1, P_2, and P_3 respectively. The total profit Z of the steel company is $4x_1+2x_2+3x_3$. The total time required on machine M_1 is $(3x_1+2x_2+2x_3)$ h. Since the machine M_1 is available for a maximum of 30 h, we have $3x_1+2x_2+2x_3 \leq 30$. Similarly, for the time restrictions on machine M_2, we have $2x_1+5x_2+3x_3 \leq 40$. Since the units of the manufactured product cannot be negative, we have $x_1, x_2, x_3 \geq 0$.

Hence, the LPP is formulated as follows:

$$\begin{aligned} &\text{Maximize } Z = 4x_1+2x_2+3x_3 \\ &\text{Subject to} \quad 3x_1+2x_2+2x_3 \leq 30 \\ &\qquad\qquad\quad 2x_1+5x_2+3x_3 \leq 40 \\ &\text{and} \qquad\qquad x_1, x_2, x_3 \geq 0 \end{aligned}$$

Example 2.3 A furniture company manufactures tables and chairs. Each table and chair is processed on two machines M_1 and M_2. One table requires 2 minutes (min) on M_1 and 4 min on M_2, and one chair requires 1 min on M_1 and 3 min on M_2. The machines M_1 and M_2 are available for 6 h and 4 h 10 min respectively, during any working day. The profit on each table and chair is ₹5 and ₹3 respectively. How many tables and chairs are manufactured to achieve maximum profit? Formulate the problem as an LPP, assuming that all the tables and chairs manufactured are sold.

Solution Tabulating the given data, we get

Machine	Time of processing		Time available (min)
	Table	Chair	
M_1	2	1	360
M_2	4	3	250
Profit in ₹	5	3	

Let the company manufacture x_1 number of tables and x_2 number of chairs. The total profit Z is $5x_1+3x_2$. The total time required on machine M_1 is $2x_1+x_2$ min. Since the machine M_1 is available for not more than 6 h, that is, 360 min, we have $2x_1+x_2\leq 360$. Similarly, for the total time required on machine M_2 and availability of machine M_2, we have $4x_1+3x_2\leq 250$. Since the units of the manufactured product cannot be negative, we have $x_1, x_2\geq 0$.

Hence, the LPP is formulated as follows:

$$\text{Maximize } Z=5x_1+3x_2$$

$$\begin{aligned}\text{Subject to}\quad & 2x_1+x_2\leq 360\\ & 4x_1+3x_2\leq 250\\ \text{and}\quad & x_1, x_2\geq 0\end{aligned}$$

Example 2.4 A publisher deals in two items, namely books and dictionaries. The publisher has ₹70,000 to invest and space for not more than 600 pieces (including both books and dictionaries). A book costs ₹500 and a dictionary, ₹700. The publisher can sell all the items that he buys, earning a profit of ₹50 for each book and ₹140 for each dictionary. Formulate the problem as an LPP to maximize the profit.

Solution Let the publisher buy x_1 books and x_2 dictionaries. As the publisher can store a maximum of 600 pieces of books and dictionaries, we have $x_1+x_2\leq 600$. The total cost incurred in purchasing x_1 books and x_2 dictionaries is $500x_1+700x_2$. However, since the publisher can invest only ₹70,000, we have $500x_1+700x_2\leq 70{,}000$ or $5x_1+7x_2\leq 700$. The total profit Z is $50x_1+140x_2$.

Hence, the LPP is formulated as follows:

$$\text{Maximize } Z=50x_1+140x_2$$

$$\begin{aligned}\text{Subject to}\quad & x_1+x_2\leq 600\\ & 5x_1+7x_2\leq 700\\ \text{and}\quad & x_1, x_2\geq 0\end{aligned}$$

Example 2.5 The supervisor of Reliance oil refinery takes a decision to combine two types of crudes, say C_1 and C_2, to produce fuel gas and petroleum using two processes—1 and 2. The inputs and outputs per production run are given in the following table:

Process	Input		Output	
	Crude C_1	Crude C_2	Petroleum	Fuel gas
1	5	3	5	7
2	4	5	3	4

The maximum units available for crudes C_1 and C_2 are 300 and 200 respectively. The market survey shows that at least 200 units of petroleum and 150 units of fuel gas must be produced. The profits per production run from processes 1 and 2 are ₹30 and ₹45 respectively. Formulate the problem as an LPP to maximize the profit, so that market demand is fulfilled.

Solution Let x_1 and x_2 be the number of production runs of processes 1 and 2 respectively. The amount of crudes C_1 and C_2 required are $5x_1+4x_2$ and $3x_1+5x_2$. As the maximum amounts of these crudes are 300 and 200 respectively, we have $5x_1+4x_2 \le 300$ and $3x_1+5x_2 \le 200$. The output of petroleum and fuel gas is $5x_1+3x_2$ and $7x_1+4x_2$, whereas market demands are at least 200 and 150 units respectively. Therefore, we have $5x_1+3x_2 \ge 200$ and $7x_1+4x_2 \ge 150$. The total profit is $30x_1+45x_2$.

Hence, the LPP is formulated as follows:

$$\begin{aligned} &\text{Maximize } Z = 30x_1 + 45x_2 \\ &\text{Subject to } \quad 5x_1+4x_2 \le 300 \\ &\qquad\qquad\quad 3x_1+5x_2 \le 200 \\ &\qquad\qquad\quad 5x_1+3x_2 \ge 200 \\ &\qquad\qquad\quad 7x_1+4x_2 \ge 150 \\ &\text{and} \qquad\qquad x_1 \ge 0,\ x_2 \ge 0 \end{aligned}$$

Example 2.6 A hotel operating 24 h has the following minimum requirement of servers:

Period	Time	Minimum number of servers required
1	7 a.m.–11 a.m.	6
2	11 a.m.–3 p.m.	12
3	3 p.m.–7 p.m.	8
4	7 p.m.–11 p.m.	16
5	11 p.m.–3 a.m.	5
6	3 a.m.–7 a.m.	3

A server reports to the hotel manager at the beginning of the period and continues to work for 8 h. The hotel manager wants to find the minimum number of servers available for each period. Formulate this as an LPP to minimize the total number of servers required.

Solution Let $x_1, x_2, \ldots, x_6$ be the number of servers reporting for duty at 7 a.m., 11 a.m., 3 p.m., 7 p.m., 11 p.m., and 3 a.m. respectively (i.e., at the beginning of the six periods). x_1 servers reporting at 7 a.m. will work up to 3 p.m. and x_2 servers reporting at 11 a.m. will work up to 7 p.m. Thus, the number of servers available between 11 a.m. and 3 p.m. is x_1+x_2. Similarly, x_2+x_3, x_3+x_4, x_5+x_6, and x_6+x_1 number of servers remain available between 11 a.m. to 3 p.m., 3 p.m. to 7 p.m., 7 p.m. to 11 p.m., 11 p.m. to 3 a.m., and 3 a.m. to 7 a.m. respectively. Therefore, the total number of servers reporting for duty in 24 h is $x_1+x_2+x_3+x_4+x_5+x_6$.

Hence, the LPP is formulated as follows:

$$\begin{aligned} &\text{Minimize } Z = x_1+x_2+x_3+x_4+x_5+x_6 \\ &\text{Subject to } \quad x_1+x_2 \ge 12 \\ &\qquad\qquad\quad x_2+x_3 \ge 8 \\ &\qquad\qquad\quad x_3+x_4 \ge 16 \\ &\qquad\qquad\quad x_4+x_5 \ge 5 \end{aligned}$$

$$x_5+x_6 \geq 3$$
$$x_6+x_1 \geq 6$$
and $\quad x_i \geq 0,\ i=1, 2, 3, \ldots, 6$

Example 2.7 A biscuit company can produce three products P_1, P_2, and P_3. Each of these products requires three different operations O_1, O_2, and O_3. One unit of P_1 requires 2 h of O_1, 2 h of O_2, and 3 h of O_3; one unit of P_2 requires 3 h of O_1, 1 h of O_2, and 2 h of O_3; and one unit of P_3 requires 4 h of O_1, 3 h of O_2, and 4 h of O_3. O_1, O_2, and O_3 can operate for a maximum of 40 h, 50 h, and 150 h respectively. The profit on each unit of P_1, P_2, and P_3 is ₹2, ₹4, and ₹3 respectively. If all the produce is sold, formulate the problem as an LPP to maximize the profit.

Solution Tabulating the given data, we get

Operations ↓	Product			Available operation time (in h)
	P_1	P_2	P_3	
O_1	2	3	4	40
O_2	2	1	3	50
O_3	3	2	4	150
Profit in ₹	2	4	3	

Let x_1, x_2, and x_3 be the units of products, P_1, P_2, and P_3 respectively, produced by the company. The total time required for operation O_1 to produce the products is $2x_1+3x_2+4x_3$. Since the capacity of operation O_1 is 40 h per unit, we have $2x_1+3x_2+4x_3 \leq 40$. Similarly, considering the times for O_2 and O_3, we have $2x_1+x_2+3x_3 \leq 50$ and $3x_1+2x_2+4x_3 \leq 150$. Further as the units produced cannot be negative, we have $x_1, x_2, x_3 \geq 0$. The profit (in ₹) is $2x_1+4x_2+3x_3$.

Hence the LPP is formulated as follows:

Maximize $Z=2x_1+4x_2+3x_3$

Subject to
$$2x_1+3x_2+4x_3 \leq 40$$
$$2x_1+x_2+3x_3 \leq 50$$
$$3x_1+2x_2+4x_3 \leq 150$$
and $\quad x_1, x_2, x_3 \geq 0$

Example 2.8 A dietician advises a wrestler to consume at least 90 grams of proteins, 100 grams of multivitamin, 150 grams of carbohydrates, and 70 grams of fats daily. The following table gives the analysis of the food items readily available in the market with their respective costs:

Food type	Food value (in grams) per 100 grams				Minimum daily requirement (grams)
	F_1	F_2	F_3	F_4	
Proteins	20	10	8	12	90
Multivitamins	10	16	18	12	100
Carbohydrates	5	15	25	10	150
Fats	12	10	20	14	70
Cost in ₹ (per kg)	5	4	8	6	

Formulate the problem as an LPP for an optimum diet.

Solution Let x_1, x_2, x_3, and x_4 kg of food F_1, F_2, F_3, and F_4 respectively be included in the daily diet. The total cost per day is $5x_1+4x_2+8x_3+6x_4$. The total amount of proteins in the daily diet is $(200x_1+100x_2+80x_3+120x_4)$. As the minimum daily requirement of proteins in the diet is 90 grams, we have $200x_1+100x_2+80x_3+120x_4 \geq 90$. Similarly, for the total amounts of multivitamins, carbohydrates, and fats in the diet, we have $100x_1+160x_2+180x_3+120x_4 \geq 100$, $50x_1+150x_2+250x_3+100x_4 \geq 150$, and $120x_1+100x_2+200x_3+140x_4 \geq 70$. Since the diet cannot contain negative quantity of any food items, we have $x_1, x_2, x_3, x_4 \geq 0$.

Hence, the LPP is formulated as follows:

Minimize $Z=5x_1+4x_2+8x_3+6x_4$

Subject to $200x_1+100x_2+80x_3+120x_4 \geq 90$
$100x_1+160x_2+180x_3+120x_4 \geq 100$
$50x_1+150x_2+250x_3+100x_4 \geq 150$
$120x_1+100x_2+200x_3+140x_4 \geq 70$

and $x_1, x_2, x_3, x_4 \geq 0$

Example 2.9 A company manufactures two products P_1 and P_2. Each product is processed on two machines M_1 and M_2. While P_1 requires 2 min on M_1 and 4 min on M_2, P_2 requires 3 min on M_1 and 3 min on M_2. The machines M_1 and M_2 are available for 1000 min and 1500 min respectively. The profit on each product P_1 and P_2 is ₹2 and ₹4 respectively to work. The company must manufacture 80 P_1's and 100 P_2's, but not more than 120 P_1's. Formulate the problem as an LPP to maximize the profit.

Solution Tabulating the given data, we get

Machine	Product		Total available time (in min)
	P_1	P_2	
M_1	2	3	1000
M_2	4	3	1500
Profit in ₹	2	4	

Let x_1 and x_2 be the number of P_1's and P_2's respectively produced by the company. The total time required to produce them on machines M_1 and M_2 are $2x_1+3x_2$ and $4x_1+3x_2$ min respectively. Since the time available with the machines M_1 and M_2 are 1000 and 1500 min respectively, we have $2x_1+3x_2 \leq 1000$ and $4x_1+3x_2 \leq 1500$. Moreover, the firm must manufacture 80 P_1's and 100 P_2's, but not more than 120 P_1's. Therefore, we have $80 \leq x_1 \leq 120$ and $x_2 \geq 100$. The profit (in ₹) is $2x_1+4x_2$.

Hence, the LPP is formulated as follows:

Maximize $Z=2x_1+4x_2$

Subject to $2x_1+3x_2 \leq 1000$
$4x_1+3x_2 \leq 1500$
$80 \leq x_1 \leq 120$

and $x_2 \geq 100$

Example 2.10 A doctor recommends two foods F_1 and F_2 to a patient for his diet, which includes at least 1000 units of vitamin, 850 units of protein, and 700 units of fat. While each unit of food F_1 contains 15 units of vitamin, 18 units of protein, and 12 units of fat, each unit of food F_2 contains 22 units of vitamin, 15 units of protein, and 16 units of fat. The cost of F_1 and F_2 is ₹5 and ₹8 respectively. Formulate the problem as an LPP to obtain the minimized cost for a diet.

Solution Tabulating the given data, we get

Food	F_1	F_2	Requirement (in units)
Vitamin	15	22	1000
Protein	18	15	850
Fat	12	16	700
Cost in ₹	5	8	

Let us consider x_1 and x_2 units of foods F_1 and F_2 respectively. In a diet, vitamin, protein, and fat units are $15x_1+22x_2$, $18x_1+15x_2$, and $12x_1+16x_2$. As the diet must have at least 1000, 850, and 700 units of vitamins, proteins, and fats respectively, we have $15x_1+22x_2 \geq 1000$, $18x_1+15x_2 \geq 850$, and $12x_1+16x_2 \geq 700$. Since the daily diet cannot contain negative quantity of any food items, we have $x_1, x_2 \geq 0$. The cost (in ₹) is $5x_1+8x_2$.

Hence, the LPP is formulated as follows:

$$\text{Minimize } Z = 5x_1 + 8x_2$$

$$\begin{aligned} \text{Subject to} \quad & 15x_1+22x_2 \geq 1000 \\ & 18x_1+15x_2 \geq 850 \\ & 12x_1+16x_2 \geq 700 \\ \text{and} \quad & x_1, x_2 \geq 0 \end{aligned}$$

RECAPITULATION

1. An optimization problem in which the objective function and constraints are linear and all decision variables are non-negative is called a linear programming problem (LPP).
2. No constraint should hold with strict inequalities such as '$<$' or '$>$' in any LPP.
3. In an LPP, there is only one objective function.
4. An LP model does not always provide an integer solution to the problem. To obtain an integer solution, we have to use other techniques like the integer programming technique.

EXERCISES

Multiple-choice Questions

2.1 Which of the following is required to be optimized in an LPP?
- (a) Objective function
- (b) Constraints
- (c) Decision variables
- (d) All of these

2.2 In an LPP, the restrictions under which the objective function is to be optimized are called
- (a) constraints
- (b) decision variables
- (c) objective function
- (d) none of these

2.3 Who invented the simplex method of linear programming?
- (a) George B. Dantzig
- (b) Cauchy
- (c) Gauss
- (d) Newton

2.4 A decision taken considering all the circumstances into account is called
(a) optimum (b) maximum (c) minimum (d) none of these

Review Questions

2.1 Define an LPP.
2.2 Write the basic assumptions needed in formulating a problem in LPP form.
2.3 What are the limitations of an LPP?
2.4 What are the applications of LPP in businesses and industries?

Numerical Problems

2.1 A newly established company has purchased sufficient amount of curtain cloth to furnish the staff offices. The curtain cloths are in pieces of length 20 feet each and can be cut in three sizes of length 3, 4, and 5 feet. The curtain requirement of the three sizes is given in the following table:

Curtain length (in feet)	Number required
3	1000
4	1500
5	3000

The company wishes to minimize the trim loss.
(a) Identify the patterns of cutting a piece of 20 feet length into requisite sizes.
(b) Identify the decision variables.
(c) Find the objective function.
(d) Write all the constraints.
(e) Formulate the problem as an LP model.

2.2 A cloth company manufactures T-shirts and pairs of jeans. Each cloth has to pass through three machines M_1, M_2, and M_3. One pair of jeans requires 3 h, 1 h, and 2 h on machines M_1, M_2, and M_3 respectively and one T-shirt requires 1 h each on machines M_1 and M_2 and 3 h on machine M_3. The profit on selling of one pair of jeans is ₹50, whereas for a T-shirt it is ₹55. The machines M_1, M_2, and M_3 are available for 60 h, 50 h, and 80 h per week respectively. How many pairs of jeans and T-shirts should be manufactured per week so as to maximize the profit? Formulate the problem as an LPP.

2.3 Asian Paints produces both exterior and interior paints from two raw materials P and Q as per the following table:

Raw material	Tonnes of raw material per tonne of		Maximum weekly availability (tonnes)
	Exterior paint	Interior paint	
P	7	5	35
Q	2	3	6
Profit per tonne (₹ in thousands)	6	4	

The market survey indicates that the maximum weekly demand of exterior paints is 3 tonnes and its weekly demand cannot exceed that of interior paints by more than 2 tonnes. The company wants to determine the best product mix of interior and exterior paints to maximize the total weekly profit. Answer the following with respect to the LP model of the aforementioned problem:
(a) What are the decision variables?

(b) The number of constraints are
(i) two (ii) three (iii) four (iv) five
(c) The objective function is ____________.
(d) The formulation is ____________.

2.4 A company manufactures two products A and B. Each product has to pass through two machines M_1 and M_2. The machines M_1 and M_2 are available for 100 h and 80 h per week respectively. To manufacture one unit of product A, 4 h of M_1 and 2 h of M_2 are required, whereas for one unit of product B, 3 h of M_1 and 1 h of M_2 are required. The profit on the sale of each unit of A and B is ₹15 and ₹10 respectively. If the manufacturer can sell all the items produced, how many of each should be produced to get maximum profit per week? Formulate the problem as an LPP.

2.5 A blacksmith who manufactures axes and knives takes 2 h to make an axe and 1 h to make a knife. It is assumed that he can work for a maximum of 15 h a day. The maximum number of axes and knives that he can handle per day is 20. While the profit on an axe is ₹15, the profit on a knife is ₹5. If all that is manufactured is sold, formulate the problem as an LPP to maximize the profit.

2.6 A pen manufacturing company manufactures two types of pens, ordinary and deluxe, say O and D. Each pen D takes twice as long to manufacture than one pen O. The company would have time to make a maximum of 500 pens per day if it produces only the ordinary version. The supply of plastic is sufficient to produce 400 pens per day (both O and D combined). The deluxe version requires fancy pens of which there are only 200 pieces per day available. The profit on the sale of each pen of O and D is ₹2 and ₹5 respectively. If all that is manufactured is sold, formulate the problem as an LPP to maximize the profit.

2.7 A company manufactures two types of cloths for curtains C_1 and C_2. Both curtains go through two machines M_1 and M_2. The details are given in the following table:

Types of machines	Processing time per piece (in h)		Total time required (in h)
	C_1	C_2	
M_1	3	1	100
M_2	1	4	80
Profit in ₹	5	8	

How many of each type of curtain should be manufactured per month to obtain the best returns? If all that is manufactured is sold, formulate the problem as an LPP.

2.8 A pharmaceutical company produces two medicines M_1 and M_2. The company has stocks of ingredients to make 30,000 bottles of M_1 and 40,000 bottles of M_2. However, there are only 50,000 bottles into which either of the medicines can be filled. Further, it takes 4 h to prepare enough material to fill 1000 bottles of medicine M_2 and 70 h are available for this operation. The profit on the sale of each bottle of M_1 and M_2 is ₹10 and ₹8 respectively. If all that is manufactured is sold, formulate the problem as an LPP to maximize the profit.

2.9 A leather factory manufactures two types of leather purses, L_1 (higher quality) and L_2 (lower quality). Each purse of type L_1 requires three times the time required for type L_2, and if all are of type L_2, the company could make 1200 purses per day. However, the supply of leather is sufficient for only 1000 purses per day. Purse L_1 requires a fancy buckle and only 600 buckles are available per day, whereas 900 buckles are available for purse L_2. The profit on each purse L_1 and L_2 is ₹5 and ₹2 respectively. What should the daily production of each type of purse be to get the maximum profit? Formulate the problem as an LPP.

2.10 A landlord owns 150 acres of land. He uses this land for growing grapes, coconuts, and apples. The price he can obtain is ₹1 per kg for apples. The average yield per acre is 2000 kg of grapes, 3000 kg of coconuts, and 1000 kg of apples. Fertilizer is available at ₹0.50 per kg and the amount required per acre is 100 kg each for grapes and coconuts and 50 kg for apples. Labour required for sowing, cultivating, and harvesting per acre is 5 man-days for grapes and apples, and 6 man-days for coconuts. A total of 500 man-days of labour are available at ₹20 per man-day. Formulate this problem as an LPP to maximize the profit.

2.11 A cloth manufacturing company produces three types of cloths C_1, C_2, and C_3, which require two types of wool, red and black. Assume one unit length of C_1 needs 1 yard of red wool and 2 yards of black wool, that of C_2 needs 3 yards of red wool and 4 yards of black wool, and that of C_3 needs 3 yards of red wool and 1 yard of black wool. The company has a stock of only 12 yards of red wool and 20 yards of black wool. The profit on the sale of each cloth C_1, C_2, and C_3 is ₹5, ₹10, and ₹15 respectively. If all that is manufactured is sold, formulate the problem as an LPP to maximize the profit.

2.12 A family doctor advises a minimum intake of 20, 25, and 15 units of vitamins V_1, V_2, and V_3 respectively. The doctor suggests two type of foods F_1 and F_2. F_1 contains 5, 4, and 3 units per gram and F_2 contains 3, 4, and 6 units per gram of vitamins V_1, V_2, and V_3 respectively. The costs of F_1 and F_2 are ₹4 and ₹5 per gram. How many grams of each food type should the family purchase every day to keep the food expense at a minimum? Formulate the problem as an LPP.

2.13 A computer company has two types of operators O_1 and O_2 to check the computer pieces produced by the company. It is required that at least 1000 pieces be inspected in an 8 h-day. Operator O_1 can check the pieces at the rate of 40 per h with an accuracy of 95% and operator O_2 can check pieces at the rate of 20 per h with an accuracy of 90%. The wage rate of operator O_1 is ₹5 per h and that of operator O_2 is ₹3 per h. Each time an error is made by an operator, the cost to company is ₹2. The company has 20 types of O_1 and 15 types of O_2 operators. Formulate the problem as an LPP to minimize the total cost of inspection.

2.14 A complete unit of a certain product consists of four and three units of components A and B respectively. A and B are manufactured from two different raw materials of which 100 and 200 units respectively are available with the company. Three departments are engaged in the production and they use different methods for manufacturing the component. The objective is to determine the number of production runs for each department that will maximize the total number of the final product. The following table gives the raw material requirement per production run and the resulting output of each component:

Department	Input per run (units)		Output per run (units)	
	Raw material		Component	
	R_1	R_2	A	B
1	7	5	6	4
2	4	8	5	8
3	2	7	7	3

Formulate the problem as an LPP.

2.15 A firm manufacturing two types of electrical items A and B can make a profit of ₹160 per unit of A and ₹240 per unit of B. Both A and B make use of two essential components: a motor and a transformer. While each unit of A requires 3 motors and 2 transformers, each unit of B requires 2 motors and 4 transformers. The total supply of components per month is restricted to 210 motors and 300 transformers. Further, type B is an export model requiring a voltage stabilizer, which has a supply restricted to 65 units per month. How many units each of A and B should the firm manufacture per month to maximize its profit? Formulate the problem as an LPP.

2.16 The data for two foods A and B is given in the following table:

Food value	One unit of food		Requirement
	Type A	Type B	
Calories	1000	2000	3000
Protein	25 grams	10 grams	100
Price (₹)	60	21	

Formulate the LPP for minimizing the expenditure on food satisfying the calorie and protein requirements.

CONCEPTUAL PROBLEM

A company wants to engage itself in the manufacturing of three kinds of paints. As a first step, the company has observed and analysed the working of other companies engaged in the manufacture and supply of these varieties of paints. The company has considered all the factors and wants to manufacture the best quality of these varieties at a competitive price. The company is in the process of developing a fortnight manufacturing programme to come out with three grades of paints P_1, P_2, and P_3. The plant works in three shifts and the following data is available:

Requirement	Grade			Availability
	P_1	P_2	P_3	
Special additive (per kilolitre)	0.25	0.2	0.70	700 tonnes
Packing (kilolitre per shift)	15	15	15	85 shifts

There are no restrictions on other resources. The particulars of the sale forecasts and the estimated profit are given in the following table:

Sales forecasts and estimated profit	Grade		
	P_1	P_2	P_3
Maximum possible sale per fortnight (kilolitre)	150	450	700
Profit (₹ per kilolitre)	4000	3600	2000

Due to prior commitments, a minimum of 200 kilolitres per fortnight of grade P_2 has to be supplied during the next year. The company plans to compare its products with other companies with respect to the quality and prices.

Formulate the LP model for developing the fortnight manufacturing programme to maximize the profit.

(*Contd*)

(*Contd*)

Solution Here the company is planning to produce three kinds of paints that can fetch different amounts of revenue for the company. Thus, the first task of the company is to find quantities of different grades of paints to be produced in such a way that the total revenue obtained is maximum. For this let x_1, x_2, and x_3 be the quantity in kilolitres of paint grades P_1, P_2, and P_3 respectively, which the company finally decides to manufacture fortnightly.

The total profit in producing these quantities of different grade paints is $4000x_1 + 3600x_2 + 2000x_3$.

The company would like to produce x_1, x_2, and x_3 such that the profit $4000x_1 + 3600x_2 + 2000x_3$ is maximum.

In producing x_1 units of P_1, x_2 units of P_2, and x_3 units of P_3, the required quantity of special additive will be $(0.25x_1 + 0.2x_2 + 0.7x_3)$ kilolitres.

Since the company has a stock of special additive of 700 tonnes, we have

$$0.25x_1 + 0.2x_2 + 0.7 \leq 700$$

Similarly, for packaging, we have $\frac{x_1}{15} + \frac{x_2}{15} + \frac{x_3}{15} \leq 85$

and $x_2 \geq 200$, $x_1 \leq 150$, $x_2 \leq 450$, $x_3 \leq 700$

Further, x_1, x_2, and x_3, being quantities in kilolitres of the paint, cannot be negative.

Thus, we have $x_1, x_2, x_3 \geq 0$.

Hence, the problem formulated is

$$\text{Maximize } Z = 4000x_1 + 3600x_2 + 2000x_3$$

$$\begin{aligned}
\text{Subject to} \quad & 5x_1 + 4x_2 + 14x_3 \leq 14000 \\
& \frac{x_1}{15} + \frac{x_2}{15} + \frac{x_3}{15} \leq 85 \\
& x_1 \leq 150 \\
& x_2 \leq 450 \\
& x_2 \geq 200 \\
& x_3 \leq 700 \\
\text{and} \quad & x_1, x_2, x_3 \geq 0
\end{aligned}$$

Linear Programming Problem II—Graphical Method

After studying this chapter, the reader will be able to

- understand the graphical method
- plot the inequalities
- find the common region called the set of feasible solutions (S_F), that is, a set of points where all inequalities are satisfied
- identify the vertices of S_F
- understand that the optimum of a linear programming problem (LPP) occurs at the vertices of S_F, provided S_F is bounded
- find the optimum solution to the LPP graphically

3.1 INTRODUCTION

The graphical technique provides a geometric picture of the solution to an LPP, and explains many other relevant aspects geometrically. It reveals that the maximization or minimization of an LPP occurs at the vertices of the set of feasible solutions (S_F), provided S_F is bounded. If the optimum occurs at more than one vertex of S_F, then it also occurs at every point on the line segment joining these vertices.

In the graphical method, we first identify the decision variables, then recognize the underlying objective of the problem, and ultimately obtain the given problem in the form of an LPP.

3.2 SOME DEFINITIONS

To solve an LPP using the graphical method, we use some basic terms as mentioned in this section.

Euclidean space Euclidean space in n dimensions denoted by E_n, is the set

$$E_n = \{(x_1, x_2, \ldots, x_n): \text{all } x_i \in R\}$$

1. $n = 1$, $E_1 = R_1 =$ Real line $= \{x: x \text{ is a real number}\}$
2. $n = 2$, $E_2 = R_2 = xy$-plane $= \{(x, y): x, y \in R\}$
3. $n = 3$, $E_3 = R_3 = xyz$-space $= \{(x, y, z): x, y, z \in R\}$

Neighbourhood The neighbourhood of a point X_O in E_n is the set S, $S = \{X \in E_n: |X - X_O| < \varepsilon, \varepsilon \text{ is a small positive number}\}$

1. It is an interval if $n = 1$.
2. It is the interior of a circle of radius ε with centre at X_O, if $n = 2$.
3. It is the interior of a sphere of radius ε with centre at X_O, if $n = 3$.

Boundary points A point X_O is called the boundary point of a set if its neighbourhood, however small, contains the points of the set as well as the points outside the set.

Closed set A set containing all its boundary points is called a closed set.

Open set A set that does not contain all its boundary points is called an open set.

Set bounded from below A subset $S \subset E_n$ is said to be bounded from below, if there exists a $Y \in E_n$ such that $Y \leq X$ for all $X \in S$, that is, every component y_j of $Y = (y_1, y_2, \ldots, y_n)^T$ is less than or equal to the corresponding component x_j of $X = (x_1, x_2, \ldots, x_n)^T$.

Set bounded from above A subset $S \subset E_n$ is said to be bounded from above, if there exists a $Y \in E_n$ such that $Y \geq X$ for all X in S.

Bounded set A subset S of E_n, which is bounded both from above and below, is a bounded set.

Example 3.1 Consider $S = \{(x, y): 0 \leq x \leq 1, 0 \leq y \leq 2\}$

Solution We can infer the following:

(a) S is bounded below by (0, 0).
(b) S is bounded above by (1, 2).

Thus, S, which is the set of all points on and inside the rectangle $0 \leq x \leq 1$, $0 \leq y \leq 2$ is a bounded set. Points on the rectangle are called boundary points of the set and points lying inside the set are called interior points. Points in E_2 outside the rectangle are called exterior points of S.

(c) Set S is a closed set as it contains all its boundary points.

Example 3.2 Consider the set $S = \{(x, y): x^2 + y^2 \leq 1\}$

Solution We can infer the following:

(a) S is a closed set.
(b) It has an infinite number of vertices.
(c) S is a bounded set.

Line Let $X_1, X_2 \in E^2$; then, the equation of line L through $X_1 = (x_1, y_1)$ and $X_2 = (x_2, y_2)$ is

$$\frac{y - y_1}{y_2 - y_1} = \frac{x - x_1}{x_2 - x_1} = \lambda \qquad \text{(say)}$$

Then, $x = x_1 + \lambda(x_2 - x_1) = \lambda x_2 + (1 - \lambda)x_1$ and $y = y_1 + \lambda(y_2 - y_1) = \lambda y_2 + (1 - \lambda)y_1$

Combining these in the form of a matrix notation, we get

$$\begin{pmatrix} x \\ y \end{pmatrix} = \lambda \begin{pmatrix} x_2 \\ y_2 \end{pmatrix} + (1-\lambda) \begin{pmatrix} x_1 \\ y_1 \end{pmatrix}$$

or

$$X = \lambda X_2 + (1 - \lambda) X_1$$

where $X = (x, y)^T$, $X_1 = (x_1, y_1)^T$, and $X_2 = (x_2, y_2)^T$.

Thus, a line L in plane E_2 passing through two points X_1 and X_2 can be defined as

$$L = \{X: X = \lambda X_2 + (1 - \lambda) X_1, \lambda \in R\} \qquad (3.1)$$

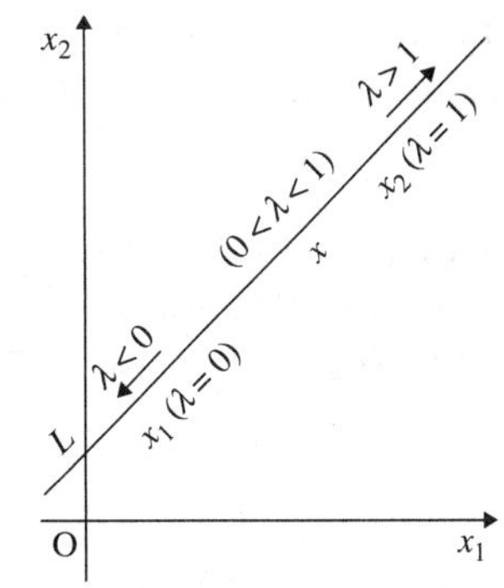

Figure 3.1 Convex linear combination of two points

1. If $0 \le \lambda \le 1$ in Eq. (3.1), we get the line segment joining X_1 and X_2 including points X_1 and X_2 (Fig. 3.1).
2. If $\lambda = 0$ in Eq. (3.1), we get a single point $X = X_1$ and if $\lambda = 1$, we get the point $X = X_2$.
3. For $\lambda < 0$, we get points on the arrow side from X_1, and for $\lambda > 1$, we get points on the arrow side away from X_2.

Hyper plane The set $S = \{X \in E_n: C^T X = \alpha\}$, where $C = (c_1, c_2, \ldots, c_n)^T$, $X = (x_1, x_2, \ldots, x_n)^T$, and α is a constant, is called a hyper plane in E_n. For $n = 3$, $S = \{(x_1, x_2, x_3) \in E_3: c_1x_1 + c_2x_2 + c_3x_3 = \alpha, c_1, c_2, c_3, \alpha \in R\}$ is a plane in E_3.

Half space The set $S = \{X \in E_n: C^T X \le \text{ or } \ge \alpha\}$, where $C = (c_1, c_2, \ldots, c_n)^T$, $X = (x_1, x_2, \ldots, x_n)^T$, and α is a constant, is called a half space in E_n. For $n = 2$, the set $S = \{(x_1, x_2): c_1x_1 + c_2x_2 \le \text{ or } \ge \alpha\}$ is also called a half plane in E_2.

Polytope and polyhedron The set of points satisfying all the constraints of an LPP with non-negative restrictions is called a polytope. A polytope may be bounded from above or below (Fig. 3.2b). A bounded polytope is known as a polyhedron (Fig. 3.2a).

Convex linear combination A point $X \in E_n$ is called the convex linear combination of two points X_1 and X_2 of E_n, if it can be written as $X = \lambda X_2 + (1 - \lambda) X_1$, $0 \le \lambda \le 1$.

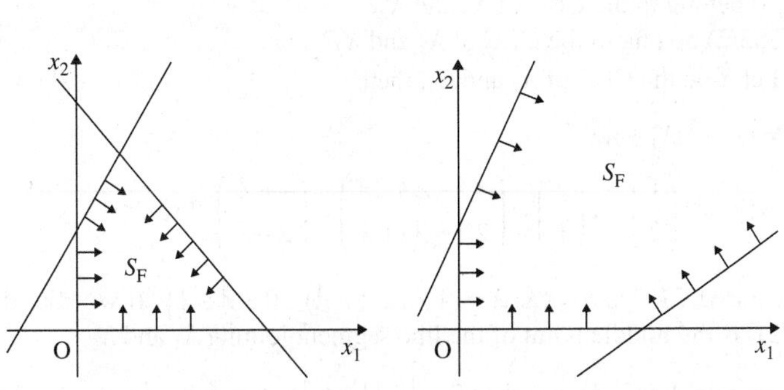

Figure 3.2 Set of points with non-negative restrictions (a) Bounded polytope (polyhedron) and (b) Polytope

In general, a point $X \in E_n$ is called the convex linear combination (CLC) of m points $X_1, X_2, \ldots, X_m$ in E_n, if there exist scalars $\lambda_i \geq 0$, $i = 1, 2, \ldots, m$ such that $X = \lambda_1 X_1 + \lambda_2 X_2 + \ldots + \lambda_m X_m$, where $\lambda_1 + \lambda_2 + \ldots + \lambda_m = 1$.

Convex set A subset $S \subset E_n$ is called a convex set (Fig. 3.3), if the CLC of every two points in S is also in S. That is, for all X_1 and X_2 in S, the point $X = (1 - \lambda) X_1 + \lambda X_2$, $0 \leq \lambda \leq 1$ is also in S.

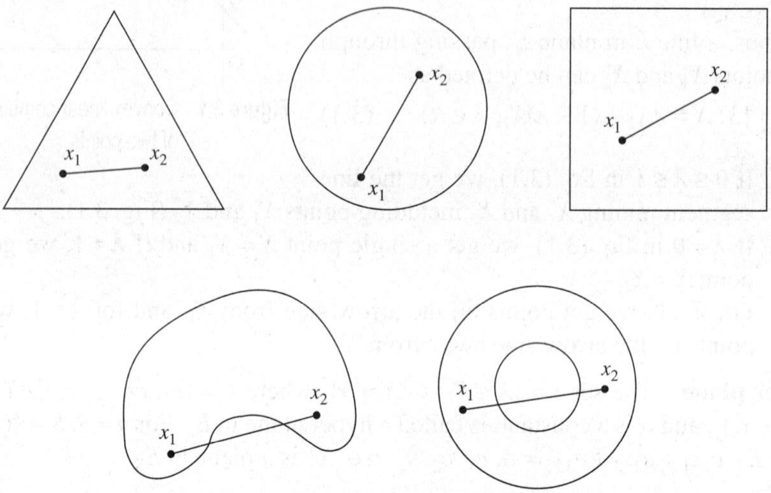

Figure 3.3 Convex and non-convex sets

The set S of all the CLCs of a finite number of points in E_n is called the convex polytope generated by these points. If we take two points X_1 and X_2, then S is the line segment joining X_1 and X_2. If X_1, X_2, and X_3 are three non-collinear points in E_3, then S is the set of points on and inside the triangle with X_1, X_2, and X_3 as its vertices.

Vertex A point X of a convex set S that cannot be expressed as a CLC of two distinct points X_1 and X_2 of S is called a vertex of S.

Example 3.3 Find the CLC of $X_1 = (1, 2)^T$ and $X_2 = (2, 1)^T$. Does

(a) (2, 3) belong to the CLC of X_1 and X_2?
(b) (3/2, 3/2) belong to the CLC of X_1 and X_2?

Solution Let X be the CLC of X_1 and X_2, then

$$X = (1 - \lambda) X_1 + \lambda X_2$$

$$= (1-\lambda)\begin{pmatrix} 1 \\ 2 \end{pmatrix} + \lambda \begin{pmatrix} 2 \\ 1 \end{pmatrix} = \begin{pmatrix} 1-\lambda+2\lambda \\ 2(1-\lambda)+\lambda \end{pmatrix} = \begin{pmatrix} 1+\lambda \\ 2-\lambda \end{pmatrix}, 0 \leq \lambda \leq 1$$

Thus, the required CLC is $S = \{X: X = (1 + \lambda, 2 - \lambda)^T, 0 \leq \lambda \leq 1\}$. If we take $\lambda = 1/2$, then $X = (3/2, 3/2)^T$ is the middle point of the line segment joining X_1 and X_2.

(a) If possible, let $(2, 3) \in S$, then $2 = 1 + \lambda \Rightarrow \lambda = 1$ and $2 - \lambda = 2 - 1 = 1 \neq 3$. Thus, (2, 3) does not belong to the CLC of X_1 and X_2.
(b) For $(3/2, 3/2) \in S$,

$$3/2 = 1 + \lambda \quad \Rightarrow \quad \lambda = 1/2 \tag{3.2}$$
$$\text{and } 2 - \lambda = 3/2 \quad \Rightarrow \quad \lambda = 1/2 \tag{3.3}$$

Eqs (3.2) and (3.3) are satisfied for $\lambda = 1/2$, which also satisfies the condition $0 \le \lambda \le 1$. Thus, (3/2, 3/2) belong to the CLC of X_1 and X_2.

Example 3.4 Let $X_1 = (-1, 0, 2)$, $X_2 = (0, 1, 2)$, and $X_3 = (0, 1, 1)$ be three points in E_3.

(a) Is $X = (1, 1, 1)$ a CLC of X_1, X_2, and X_3?
(b) Is $X = (-1/2, 1/2, 7/4)$ a CLC of X_1, X_2, and X_3?

Solution (a) For $X = (1, 1, 1)$ to be a CLC of X_1, X_2, and X_3

$$(1, 1, 1) = \lambda_1(-1, 0, 2) + \lambda_2 (0, 1, 2) + \lambda_3 (0, 1, 1),\ \lambda_i\text{'s} \ge 0 \text{ and } \lambda_1 + \lambda_2 + \lambda_3 = 1$$

or
$$(1, 1, 1) = (-\lambda_1, \lambda_2 + \lambda_3, 2\lambda_1 + 2\lambda_2 + \lambda_3)$$
$$\therefore -\lambda_1 = 1 \Rightarrow \lambda_1 = -1 \text{ i.e., } \lambda_1 < 0$$

Therefore, $X = (1, 1, 1)$ is not a CLC of X_1, X_2, and X_3.

(b) For $X = (-1/2, 1/2, 7/4)$ to be a CLC of X_1, X_2, and X_3,

$(-1/2, 1/2, 7/4) = \lambda_1(-1, 0, 2) + \lambda_2(0, 1, 2) + \lambda_3(0, 1, 1)$, $\lambda_1 + \lambda_2 + \lambda_3 = 1$ and λ_i's ≥ 0.

$$= (-\lambda_1, \lambda_2 + \lambda_3, 2\lambda_1 + 2\lambda_2 + \lambda_3)$$
$$\therefore -\lambda_1 = -1/2 \Rightarrow \lambda_1 = 1/2$$
$$\lambda_2 + \lambda_3 = 1/2 \Rightarrow \lambda_2 = (1/2) - \lambda_3$$
$$2\lambda_1 + 2\lambda_2 + \lambda_3 = 7/4 \Rightarrow \lambda_3 = (7/4) - 2\lambda_1 - 2\lambda_2 = (7/4) - 1 - 1 + 2\lambda_3 \Rightarrow \lambda_3 = 1/4$$

That is, $\lambda_1 = 1/2$, $\lambda_2 = 1/4$, and $\lambda_3 = 1/4$

Thus, $\lambda_i \ge 0$, $\forall\, i = 1, 2, 3$ and $\lambda_1 + \lambda_2 + \lambda_3 = (1/2) + (1/4) + (1/4) = 1$

Hence, (−1/2, 1/2, 7/4) is a CLC of X_1, X_2, and X_3.

3.3 SOME IMPORTANT THEOREMS

Theorem 3.1 A closed half space in E_n is a convex set.

Proof Let us consider a closed half space in E_n as

$S = \{X \in E_n: C^T X \ge \alpha\}$, where $C = (c_1, c_2, \ldots, c_n)^T$, $X = (x_1, x_2, \ldots, x_n)^T$ and α a constant. S is closed as all its boundary points are in S.

To prove that it is convex, consider any two points $X_1, X_2 \in S$. Then $X_1, X_2 \in S$ implies that

$$C^T X_1 \ge \alpha \text{ and } C^T X_2 \ge \alpha \tag{3.4}$$

Let $X = (1 - \lambda)X_1, + \lambda X_2$, $0 \le \lambda \le 1$ be a CLC of X_1 and X_2. Then

$$C^T X = C^T[(1 - \lambda)X_1, + \lambda X_2]$$
$$= (1 - \lambda)C^T X_1, + \lambda C^T X_2$$
$$\ge (1 - \lambda)\alpha + \lambda\alpha = \alpha$$

Thus, $C^T X \ge \alpha$ and hence $X \in S$.

Thus X, which is the CLC of any two points $X_1, X_2 \in S$ is again in S. Therefore, S is a convex set.

Theorem 3.2 The intersection of a finite number of convex sets in E_n is again a convex set.

Proof Let $S_1, S_2, \ldots, S_n$ be n convex sets in E_n and $S = S_1 \cap S_2 \cap S_3 \ldots \cap S_n$. Let $X_1, X_2 \in S$, then, $X_1, X_2 \in S_i$, $i = 1, 2, \ldots, n$.

Let $X = (1 - \lambda)X_1 + \lambda X_2$, $0 \le \lambda \le 1$ be a CLC of X_1 and X_2. As $X_1, X_2 \in S_i$ and S_i ($i = 1, 2, \ldots, n$) is a convex set, $X \in S_i$ ($i = 1, 2, \ldots, n$) and hence $X \in S$. Thus, we have proved that the CLC of any two points in S is also in S and hence S, which is an intersection of S_i ($i = 1, 2, \ldots, n$), is a convex set.

Note Let S be a set, which is non-empty, closed, convex, and bounded; then it has at least one vertex and every point of S is a CLC of its vertices.

Theorem 3.3 The optimum of the objective function $Z = f(X)$ of an LPP always occurs at some vertex of the set of feasible solutions (S_F), provided S_F is non-empty and bounded.

Proof Let us consider a minimization LPP. As an LPP has a finite number of constraints, the set of feasible solutions (S_F) is a bounded polytope with a finite number of vertices. Let $X_1, X_2, \ldots, X_n$ be the vertices of S_F. Let the objective function $Z = f(X)$ be minimum at some point $X_0 \in S_F$

$$\text{That is, } f(X_0) \le f(X) \text{ for all } X \in S_F \tag{3.5}$$

It is to be proved that $f(X_0) = f(X_i)$ for some vertex X_i ($i = 1, 2, \ldots, n$) of S_F. Here, X_0 may not be a vertex of S_F. We know that S_F is a non-empty, closed, convex, and bounded set, and hence any point belonging to S_F can be written as a CLC of its vertices. Therefore,

$$X_0 = \lambda_1 X_1 + \lambda_2 X_2 + \ldots + \lambda_n X_n, \quad \lambda_1 + \lambda_2 + \ldots + \lambda_n = 1, \quad \lambda_i \ge 0, \quad \forall i = 1, 2, \ldots, n$$

$$\therefore f(X_0) = f(\lambda_1 X_1 + \lambda_2 X_2 + \ldots + \lambda_n X_n)$$

$$\text{or } f(X_0) = \lambda_1 f(X_1) + \lambda_2 f(X_2) + \ldots + \lambda_n f(X_n), \text{ as } f \text{ is a linear function}$$

Let the least amongst $f(X_i)$'s, $i = 1, 2, \ldots, n$ be $f(X_k)$ for some $i = k$.

$$\text{Then, } f(X_0) \ge \lambda_1 f(X_k) + \lambda_2 f(X_k) + \ldots + \lambda_n f(X_k) = (\lambda_1 + \lambda_2 + \ldots + \lambda_n) f(X_k)$$

$$\text{or } f(X_0) \ge f(X_k) \tag{3.6}$$

In addition, from Eq. (3.5), we have

$$f(X_0) \le f(X_k), \tag{3.7}$$

as Eq. (3.5) is true for $X = X_k$ also

From Eqs (3.6) and (3.7) we have, $f(X_0) = f(X_k)$, which proves that the minimum of $f(X)$ occurs at some vertex $X_k \in S_F$.

3.4 GRAPHICAL METHOD

This section discusses how a linear programming model (LP model) in two variables is solved graphically. This method cannot be used when more than two variables are involved. Although LP models in two variables rarely occur in practice (where a typical LP model may involve many variables), the ideas revealed by the graphical procedure establish the foundation of the general solution technique (known as the simplex method), which we shall discuss in Chapter 4.

Any LP model includes three basic elements:

1. The non-negative decision variables whose values are to be determined
2. The objective function that we target to optimize
3. The restrictions (constraints) to be satisfied by the decision variables

The set of non-negative decision variables satisfying all the constraints is known as the set of *feasible solutions*. The feasible solution for which the value of the objective function is optimum (minimum in case of minimization problems and maximum in case of maximization problems) is called the *optimum feasible solution* of the LPP.

In the graphical method, we retain the inequality form of the constraints. An inequality constraint in two variables represents a half-plane, which is the set of points on one side of the line, including the points on the line. We sketch all these half planes created by all the constraints and take their intersection to get the set of feasible solutions (S_F). The following two steps are used in the graphical technique for solving an LP model in two variables:

1. Determination of the set of feasible solutions (solution space) that satisfies all the restrictions placed on the decision variables involved in the problem (including non-negativity restrictions).
2. Determination of the optimal solution(s) among all the points in the set of feasible solutions.

For obtaining the optimum solution in the graphical method, two methods are employed: (a) Corner point method and (b) Iso-profit method (isovalue line method).

3.4.1 Corner Point Method

We take the following steps to obtain the graphical solution to an LPP using the corner point method:

1. Sketch the inequalities (constraints).
2. Find the set of feasible solutions (S_F), by taking the intersection of inequalities.
3. Obtain the corner points (vertices) of S_F.
4. Select the optimum value of the objective function. This gives the value of the decision variables along with the optimum value of the objective function.

3.4.2 Iso-profit Method or Isovalue Line Method

In this method, the following steps are taken to obtain the optimal solution to an LPP:

1. Sketch all the constraints involved in the problem.
2. Identify the set of feasible solutions (S_F).
3. Draw parallel lines with different values of the objective function, moving in the direction in which the objective function is tending to increase or decrease, according to the maximization or minimization of the problem.
4. Continue to draw the parallel lines obtained by taking different values of the objective function in the desired direction, until one the following conditions holds good:
   (a) The last line passes through a vertex of S_F, where the optimum solution is obtained and any further movement of the parallel line pushes it out of S_F.
   (b) The parallel line surpasses the last vertex in the desired direction and still contains the points of S_F with no bounds on the objective function value. Hence, the solution is *unbounded*.

Note It is preferable to use the corner point method to obtain the optimal solution when the set of feasible solutions (S_F) is bounded. If S_F is not bounded, it would be more convenient to use the iso-profit method. In this case, the given problem may have finite, infinite, or unbounded solutions, depending on the form of the objective function. Further, in case the optimal solution exists at two vertices of S_F, then the optimal solution also exists at every point on the line segment joining these two vertices.

3.5 SPECIAL CASES IN GRAPHICAL METHOD

While solving an LPP using the graphical method, we come across some special situations where the problem has more than one optimal solution, no solution, or sometimes even an infinite number of solutions. In this section, we will discuss all such scenarios.

3.5.1 Alternate Optimal Solution

An LPP is said to have an alternate optimal solution, if it has an optimal solution at more than one vertex.

In Fig. 3.4, if the optimal solution exists at two vertices, say P_1 and P_2, then it will exist at every point X of the line segment joining the two vertices P_1 and P_2. Thus, the problem will have an infinite number of solutions.

3.5.2 No Feasible Solution

Consider the following LPP:

$$\text{Max } Z = 2x_1 + 5x_2$$

$$\text{Subject to } \quad x_1 + x_2 \le 1 \qquad (3.8)$$

$$3x_1 + 2x_2 \ge 6 \qquad (3.9)$$

$$\text{and} \quad x_1, x_2 \ge 0$$

As shown in Fig. 3.5, S_1 is the feasible solution space of Eq. (3.8) and S_2 is the solution space of Eq. (3.9). Therefore, the solution space of the problem as a whole is $S_F = S_1 \cap S_2 = \phi$. Thus, as the solution space is an empty set, the problem has no solution.

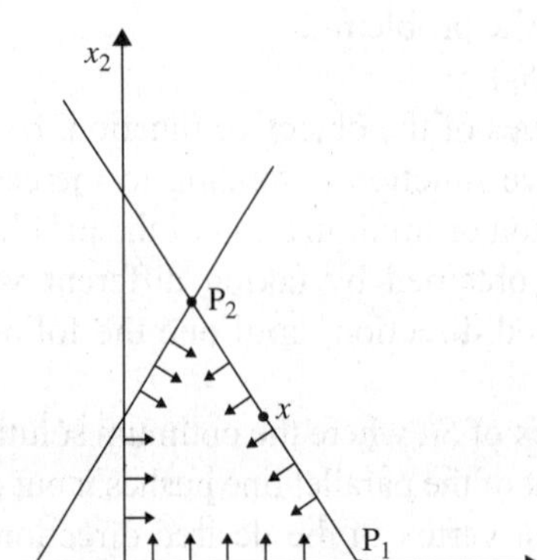

Figure 3.4 Optimal solution at two vertices P_1 and P_2

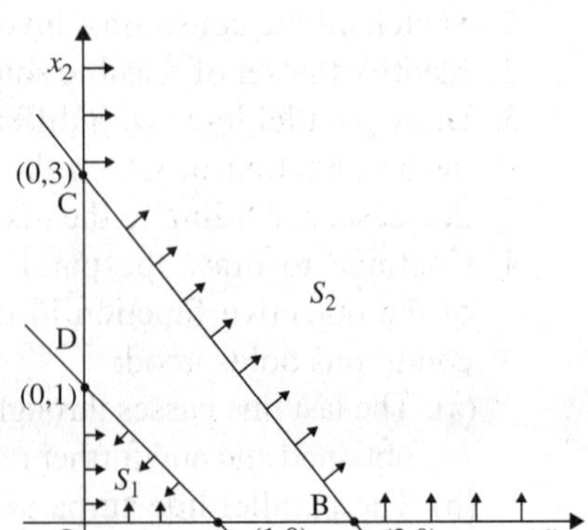

Figure 3.5 Empty solution space

3.5.3 Unbounded Solution Space but Bounded Optimal Solution

Consider the following LPP:

$$\text{Min } Z = x_1 + 2x_2$$

$$\begin{aligned} \text{Subject to} \quad & x_1 + x_2 \geq 3 && (3.10) \\ & x_1 - 3x_2 \leq 3 && (3.11) \\ \text{and} \quad & x_1, x_2 \geq 0 \end{aligned}$$

In Fig. 3.6, AB is the line $x_1 + x_2 = 3$, BC is the line $x_1 - 3x_2 = 3$, and S_F is the set of feasible solutions. Now, the problem is to find $(x_1, x_2) \in S_F$ that makes Z minimum. For this, consider the iso-profit parallel lines corresponding to the different values of Z. From the figure, it is clear that as we move in the direction of arrows, the Z value decreases. At the vertex B (3, 0), $Z = 3$. If we try to decrease Z further by moving in the direction of the arrow, the iso-profit line would not pass through S_F (we should remain in S_F) and hence, a further decrease in the Z value is not possible. Thus, Min $Z = 3$ at vertex B when $x_1 = 3$ and $x_2 = 0$.

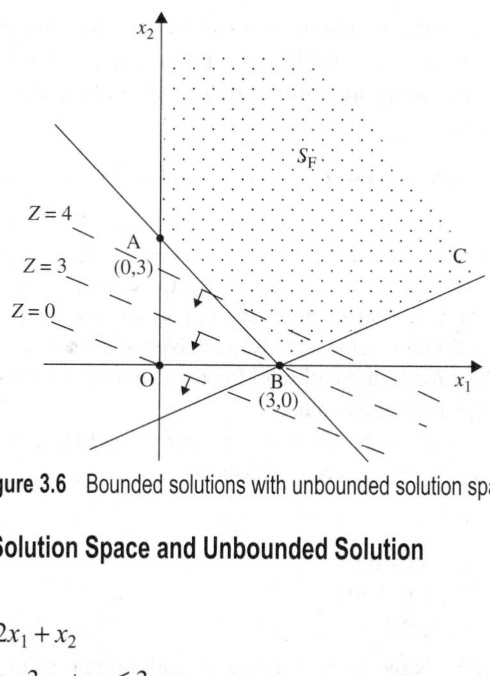

Figure 3.6 Bounded solutions with unbounded solution space

3.5.4 Unbounded Solution Space and Unbounded Solution

Consider the LPP

$$\text{Max } Z = 2x_1 + x_2$$

$$\begin{aligned} \text{Subject to} \quad & -3x_1 + x_2 \leq 3 && (3.12) \\ & x_1 + x_2 \geq 3 && (3.13) \\ & x_1 - 3x_2 \leq 3 && (3.14) \\ \text{and} \quad & x_1, x_2 \geq 0 \end{aligned}$$

In Fig. 3.7, line AB is $-3x_1 + x_2 = 3$, line BC is $x_1 + x_2 = 3$, line CD is $x_1 - 3x_2 = 3$, and S_F is the set of feasible solutions. Now the problem is to find point $(x_1, x_2) \in S_F$, which makes the Z value maximum. For this, consider the iso-profit parallel lines corresponding to the different values of Z. From the figure, it is clear that as we move in the direction of the arrows through the parallel iso-profit lines, the Z value is increasing. At the vertex B, $Z = 3$ and at C, $Z = 6$. If we move further in

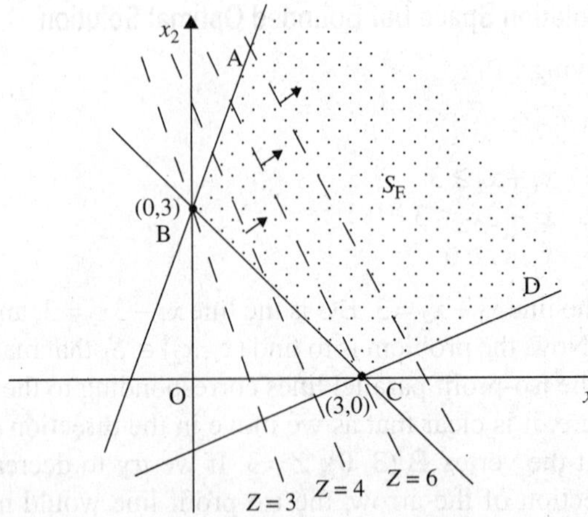

Figure 3.7 Unbounded solution with unbounded solution space

the direction of the arrows, the parallel iso-profit lines beyond vertex C would still pass across the set of feasible solutions (S_F), making the Z values larger with no upper bound. Thus, the solution is unbounded with unbounded solution space (as is clear from Fig. 3.7).

Example 3.5 A leather company manufactures two types of belts, B_1 and B_2. Each of these sell at a profit of ₹2 per belt. Each belt is processed on two machines, M_1 and M_2. The B_1 type belt requires one minute of processing time on machine M_1 and two minutes on machine M_2. The B_2 type belt requires one minute of processing time on each of the machines M_1 and M_2. While machine M_1 is available for a maximum of 6 h and 40 min, machine M_2 is available for 10 h on all working days. Formulate the given problem as an LPP and solve it using the graphical method.

Solution Let x_1, x_2 be the number of belts of types B_1 and B_2 to be manufactured. The LPP formulation of the given problem is as follows:

$$\text{Max } Z = 2x_1 + 2x_2 \tag{3.15}$$

$$\text{Subject to} \quad x_1 + x_2 \leq 400 \tag{3.16}$$

$$2x_1 + x_2 \leq 600 \tag{3.17}$$

$$\text{and} \quad x_1, x_2 \geq 0 \tag{3.18}$$

Corner point method Now the problem is to find a point $X = (x_1, x_2)^T$ for satisfying Eqs (3.16) and (3.17), where $Z = 2x_1 + 2x_2$ has the maximum value. We plot the inequalities in Eqs (3.16), (3.17), and (3.18) to get the set of feasible solutions S_F, as shown in Fig. 3.8.

Next, we find the coordinates of points O, A, B, and C, which are vertices (corner points) of S_F and calculate the Z values as given in the following table.

Thus, Max $Z = 800$, when $x_1 = 200$ and $x_2 = 200$. Further, Max $Z = 800$ when $x_1 = 0$ and $x_2 = 400$, which is another optimal solution called the alternate optimal solution of the problem. In fact, the optimum solution will occur at every point on the line segment joining the points B and C.

Points	Z
O(0, 0)	$Z=2\times 0+2\times 0=0$
A(300, 0)	$Z=2\times 300+2\times 0$ $=600$
B(200, 200)	$Z=2\times 200+2\times 200$ $=800$ (Max)
C(0, 400)	$Z=2\times 0+2\times 400$ $=800$ (Max)

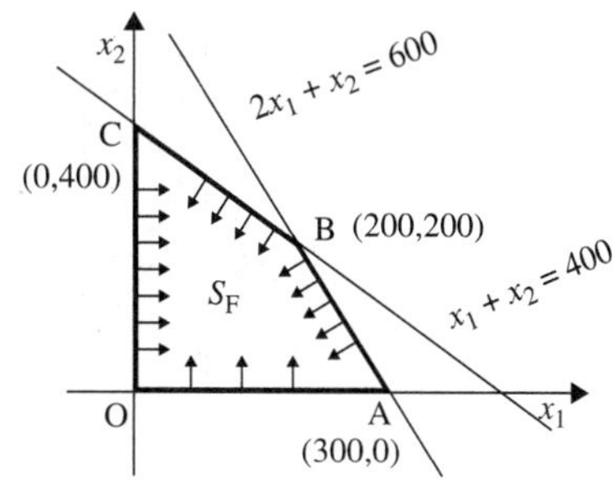

Figure 3.8 Alternate optimal solution at B and C

Example 3.6 An animal food company receives an order for producing at least 200 kg of a mixture per day, containing two ingredients A and B. The cost of the ingredients A and B is ₹2 per kg and ₹8 per kg respectively. A maximum of 80 kg of A and at least 60 kg of B must be used in preparing the food mixture. Formulate the LP model of the problem to find the amounts of the ingredients used, so that the cost of the food mixture comes to a minimum. Solve it using the graphical method.

Solution This problem can be considered as a decision-making one. The food company has to take the decision on producing the amount of the ingredients used in preparing the food mixture.

Let x_1 be the amount, in kg, of ingredient A used to produce the food mixture and x_2 be the amount, in kg, of ingredient B. The total amount of the food mixture produced is x_1+x_2. Thus, we have $x_1+x_2\geq 200$. The amount of ingredient A used is not more than 80 kg. Thus, we have $x_1\leq 80$. Similarly, $x_2\geq 60$ as at least 60 kg of ingredient B must be used.

Further, the cost involved in producing the food mixture is $2x_1+8x_2$. Our objective is to minimize the cost.

Thus, the problem formulated is

$$\text{Min } Z=2x_1+8x_2 \tag{3.19}$$

$$\text{Subject to} \quad x_1+x_2\geq 200 \tag{3.20}$$

$$x_1\leq 80 \tag{3.21}$$

$$x_2\geq 60 \tag{3.22}$$

$$\text{and} \quad x_1, x_2\geq 0$$

which is the LP model of formulation.

Corner point method Now, the problem is to find a point $X=(x_1, x_2)^T \in S_F$, satisfying Eqs (3.20), (3.21), and (3.22), where the value of $Z=2x_1+8x_2$ is minimum. We sketch the inequalities in (3.20), (3.21), and (3.22) to get the unbounded region S_F, the set of feasible solutions, as shown in Fig. 3.9.

We can determine the solution to this LPP using both the iso-profit method and the corner point method. Here, we solve the problem using the corner point method.

For this, we find the coordinates of points B(0, 200) and C(80, 120), which are the vertices (corner points) of S_F (Fig. 3.9) and calculate the value of Z there.

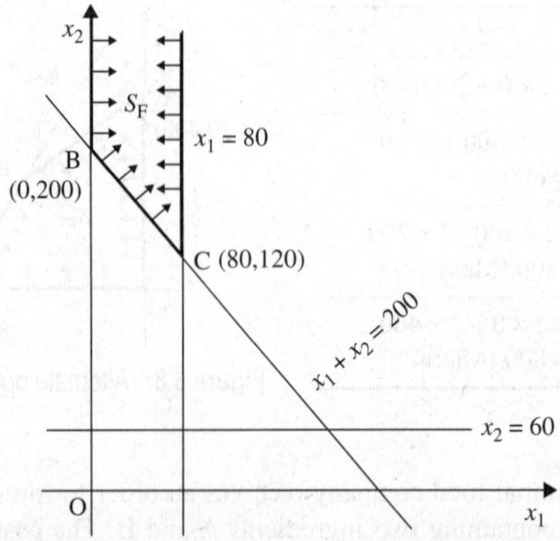

Figure 3.9 Bounded solution with unbounded solution space

Points	Z
B(0, 200)	$Z = 2 \times 0 + 8 \times 200 = 1600$
C(80, 120)	$Z = 2 \times 80 + 8 \times 120 = 1120$ (Min)

Thus, the optimum solution is Min $Z = 1120$, when $x_1 = 80$ and $x_2 = 120$.

Example 3.7 Solve the following LPP using the graphical method:

Maximize $Z = -x_1 + 2x_2$
Subject to $-x_1 + x_2 \leq 1$
$x_1 + x_2 \leq 2$
and $x_1, x_2 \geq 0$

Solution The given LPP is

Maximize $Z = -x_1 + 2x_2$ (3.23)
Subject to $-x_1 + x_2 \leq 1$ (3.24)
$x_1 + x_2 \leq 2$ (3.25)
and $x_1, x_2 \geq 0$ (3.26)

Now the problem is to find a point $X = (x_1, x_2)^T$ in the first quadrant, satisfying Eqs (3.24) and (3.25), where $Z = -x_1 + 2x_2$ has the maximum value. We sketch the inequalities in Eqs (3.24), (3.25), and (3.26) to get the region OABC (set of feasible solutions) as shown in Fig. 3.10. We shall determine the solution to the LPP using both the iso-profit and the corner point methods.

Iso-profit method Let us consider the parallel lines that correspond to different values of Z, say $Z = 0, 2, 5/2$. We see that Z increases as the movement of parallel lines is in the direction of the arrow, as shown in Fig. 3.10. The last such line passes through B(1/2, 3/2), with $Z = -(1/2) + 2(3/2) = 5/2$. Thus, the optimum solution is Max $Z = 5/2$, when $x_1 = 1/2$ and $x_2 = 3/2$.

Corner point method For this we find the coordinates of points O, A, B, and C, which are vertices (corner points) of S_F and calculate the Z values.

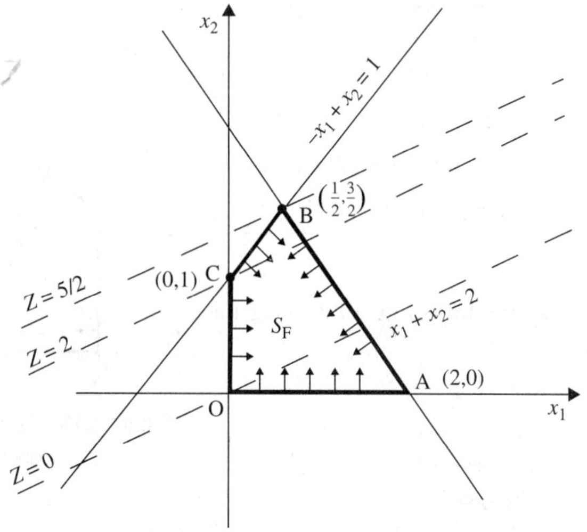

Figure 3.10 Bounded solution with bounded solution space

Points	Z
O(0, 0)	$Z = 0 + 2.0 = 0$
A(2, 0)	$Z = -2 + 2.0 = -2$
B(1/2, 3/2)	$Z = -(1/2) + 2(3/2) = 5/2$ (Max)
C(0, 1)	$Z = 0 + 2.1 = 2$

Thus optimum solution is Max $Z = 5/2$, when $x_1 = 1/2$ and $x_2 = 3/2$.

Example 3.8 Solve the following LPP using the graphical method.

$$\text{Min } Z = -5x_1 + 10x_2$$

$$\begin{aligned} \text{Subject to} \quad & -x_1 + x_2 \le 1 \\ & x_1 + x_2 \le 2 \\ \text{and} \quad & x_1, x_2 \ge 0 \end{aligned}$$

Solution The given LPP is

$$\text{Min } Z = -5x_1 + 10x_2 \tag{3.27}$$

$$\text{Subject to} \quad -x_1 + x_2 \le 1 \tag{3.28}$$

$$x_1 + x_2 \le 2 \tag{3.29}$$

$$\text{and} \quad x_1, x_2 \ge 0 \tag{3.30}$$

Corner point method The region of feasible solutions (S_F), which is the intersection of the sets, represented by Eqs (3.28), (3.29), and (3.30), is shown in Fig. 3.11.

We find the coordinates of the vertices O, A, B, and C of S_F as O(0, 0), A(2, 0), B(1/2, 3/2), and C(0, 1). We calculate the value of Z at these points as given in the following table.

Thus, the optimum solution is Min $Z = -10$ when $x_1 = 2$ and $x_2 = 0$.

Points	Z
O(0, 0)	$Z = -5(0) + 10(0) = 0$
A(2, 0)	$Z = -5(2) + 10(0) = -10$ (Min)
B(1/2, 3/2)	$Z = -5(1/2) + 10(3/2) = 25/2$
C(0, 1)	$Z = -5(0) + 10(1) = 10$

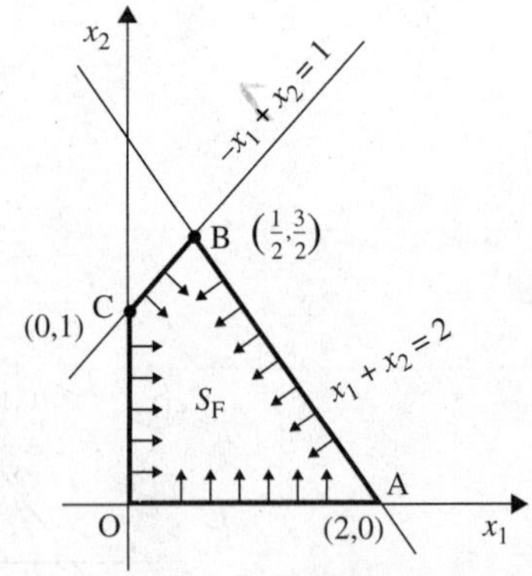

Figure 3.11 Bounded solution with bounded solution space

Example 3.9 Solve the following LPP using the graphical method:

$$\text{Max } Z = 10x_1 - 15x_2$$

$$\text{Subject to } \quad -x_1 + x_2 \leq 1$$
$$x_1 - 4x_2 \leq 2$$
$$\text{and } \quad x_1, x_2 \geq 0$$

Solution The given LPP is

$$\text{Max } Z = 10x_1 - 15x_2 \tag{3.31}$$

$$\text{Subject to } \quad -x_1 + x_2 \leq 1 \tag{3.32}$$
$$x_1 - 4x_2 \leq 2 \tag{3.33}$$
$$\text{and } \quad x_1, x_2 \geq 0 \tag{3.34}$$

Iso-profit method The region of feasible solutions (S_F), which is the intersection of the sets, represented by Eqs (3.32), (3.33), and (3.34), is shown in Fig. 3.12.

Here S_F is unbounded. We use the iso-profit technique to obtain the solution to the problem. For this, we draw parallel lines across S_F for various values of Z, say $Z = 0, 10, 20$. This shows that Z increases in the direction as shown by the arrow (→), having a value of 20 at vertex A, which is the last vertex in the direction of increasing Z. The further parallel movement of the line still passes across the feasible region and thus, there exists no upper bound on the Z value.

Thus, the given problem has an unbounded solution.

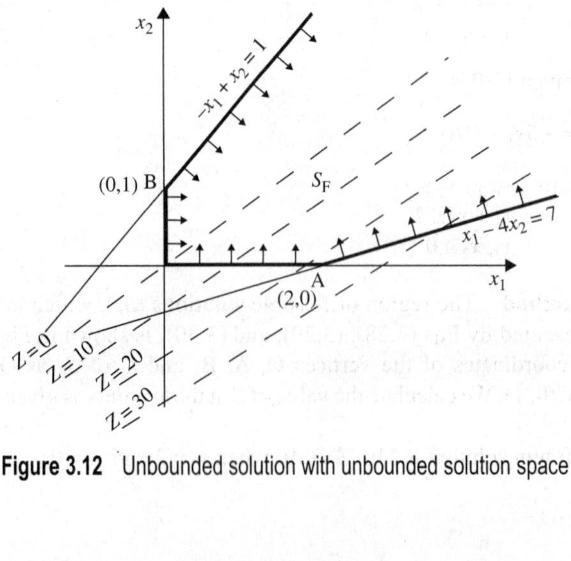

Figure 3.12 Unbounded solution with unbounded solution space

Example 3.10 Solve the following LPP using the graphical method:

$$\text{Max } Z = -2x_1 + 2x_2$$
$$\text{Subject to } \quad -x_1 + x_2 \le 1$$
$$x_1 + x_2 \le 2$$
$$\text{and} \quad x_1, x_2 \ge 0$$

Solution The given LPP is

$$\text{Max } Z = -2x_1 + 2x_2 \quad (3.35)$$
$$\text{Subject to } \quad -x_1 + x_2 \le 1 \quad (3.36)$$
$$x_1 + x_2 \le 2 \quad (3.37)$$
$$\text{and} \quad x_1, x_2 \ge 0 \quad (3.38)$$

Corner point method The region of feasible solutions (S_F), which is the intersection of the sets, represented by Eqs (3.36), (3.37), and (3.38), is shown in Fig. 3.13. We shall find the optimum solution using the corner point method.

We first find the coordinates of the vertices O, A, B, and C of S_F as O(0, 0), A(2, 0), B(1/2, 3/2), and C(0, 1), and then calculate the value of Z at these points as follows:

Points	Z
O(0, 0)	$Z = -2(0) + 2(0) = 0$
A(2, 0)	$Z = 2(2) + 2(0) = -4$
B(1/2, 3/2)	$Z = -2(1/2) + 2(3/2) = 2$ (Max)
C(0, 1)	$Z = -2(0) + 2(1) = 2$ (Max)

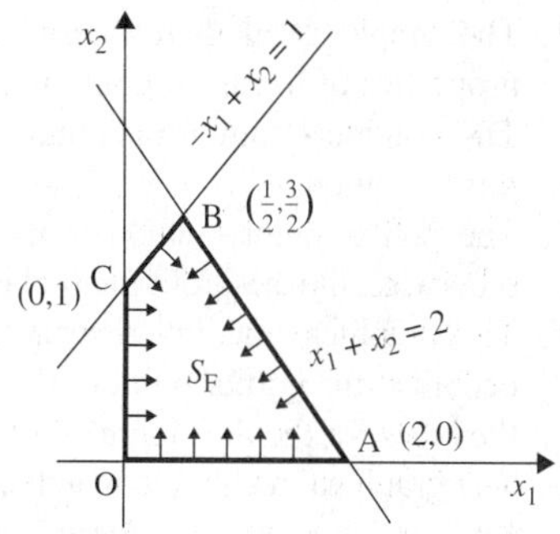

Figure 3.13 Bounded solution with bounded solution space

Thus (a) Max $Z=2$ when $x_1=1/2$ and $x_2=3/2$ and also (b) Max $Z=2$ when $x_1=0$ and $x_2=1$ is another optimal solution, called the alternate optimal solution of the problem. In fact, the optimum solution will occur at every point on the line segment joining the points BC.

Example 3.11 Solve the following LPP using the graphical method:

$$\text{Max } Z = 2x_1 + 3x_2$$
$$\text{Subject to } \quad x_1 + x_2 \le 3$$
$$5x_1 + 4x_2 \ge 20$$
$$\text{and} \quad x_1, x_2 \ge 0$$

Solution The given LPP is

$$\text{Max } Z = 2x_1 + 3x_2 \quad (3.39)$$
$$\text{Subject to } \quad x_1 + x_2 \le 3 \quad (3.40)$$
$$5x_1 + 4x_2 \ge 20 \quad (3.41)$$
$$\text{and} \quad x_1, x_2 \ge 0 \quad (3.42)$$

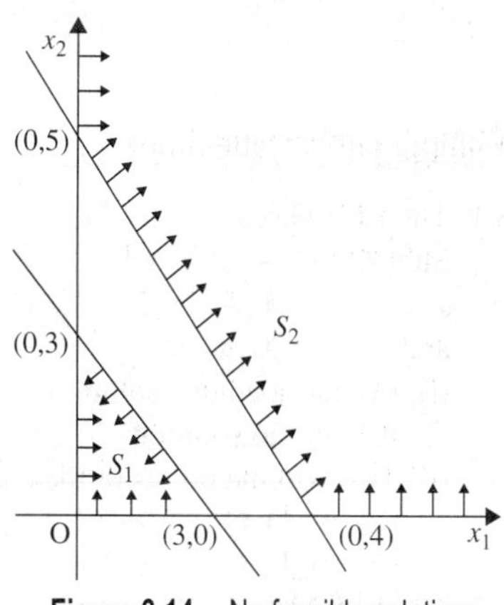

Figure 3.14 No feasible solution

Now, the problem is to find a point $X = (x_1, x_2)^T$ satisfying Eqs (3.40), (3.41), and (3.42), where the value of $Z = 2x_1 + 3x_2$ is maximum. We sketch the inequalities in Eqs (3.40), (3.41), and (3.42) to get the regions S_1 and S_2 as shown in Fig. 3.14.

Here $S_F = S_1 \cap S_2$ is empty. Therefore, the given problem has no feasible solution.

3.6 LIMITATIONS OF GRAPHICAL METHOD

The fundamental point in determining the solution to an LPP is that it occurs at some vertex of the set of feasible solutions (S_F). It is very difficult to visualize the vertices of S_F of an LPP involving three or more decision variables. However, when the number of decision variables is less than three, one can easily plot the constraints to find S_F and its vertices, and hence, the optimum solution to the given LPP. Thus, the graphical method works only when the number of decision variables in the given LPP is less than three. Hence, the biggest drawback of the graphical method is that it is not applicable to solve an LPP involving more than two variables.

In this method, the accurate plotting of the coordinates is of prime importance, but as the graphs are drawn manually, there is always scope for errors.

RECAPITULATION

1. The graphical method is used to draw some general conclusions regarding the properties of linear programming solutions.
2. The graphical solution is obtained either using the corner point method or the iso-profit method.
3. The corner point method is used when the set of feasible solutions is bounded; otherwise, the iso-profit method is used.
4. The graphical method reveals that the optimum value of the objective function occurs at the corner point of the set of feasible solutions. This key result is, in fact, the basis for the development of the simplex method.
5. The graphical technique to solve an LPP works only for two variables.

EXERCISES

Multiple-choice Questions

3.1 The LPP Max $Z = 2x_1 + 5x_2$

Subject to $-x_1 + x_2 \geq 1$

$x_1 + x_2 \leq 2$

and $x_1, x_2 \geq 0$

(i) (a) has a unique solution (c) has an unbounded solution

(b) has no solution (d) has an alternate optimum solution

(ii) The extreme points of the convex set of feasible solutions are

(a) (0, 1), (1/2, 3/2), (2, 1) (c) (0, 1), (0, 2), (1/2, 3/2)

(b) (0, 1), (0, 2), (2, 0) (d) none of these

3.2 The LPP Max $Z = 4x_1 + 4x_2$

Subject to $-x_1 + x_2 \geq 1$

$x_1 + x_2 \leq 2$

and $x_1, x_2 \geq 0$

The optimum solution of the convex set of feasible solutions is at the extreme points

(a) (0, 1), (0, 2) (c) (0, 1), (1/2, 3/2)

(b) (0, 2), (1/2, 3/2) (d) none of these

3.3 Let $S = \{(x_1, x_2): x_1^2 + x_2^2 < 1\}$. Then S has
(a) no vertex
(b) one vertex
(c) an infinite number of vertices
(d) none of these

3.4 Let $S = \{(x_1, x_2): x_1^2 + x_2^2 \le 2\}$. Then S has
(a) no vertex
(b) one vertex
(c) two vertices
(d) an infinite number of vertices

3.5 Let $S = \{(x_1, x_2): x_1^2 + x_2^2 \le 1\}$. Then every interior point of S is not a CLC of
(a) exactly one vertex of S
(b) exactly two vertices of S
(c) exactly three vertices of S
(d) more than three vertices of S

3.6 The intersection of two convex sets in E_n is
(a) a convex set
(b) not a convex set
(c) may or may not be a convex set, depending upon other conditions
(d) all of these

3.7 Let S be a non-empty closed convex set. Then S
(a) has a finite number of vertices
(b) has an infinite number of vertices
(c) may have finite or infinite number of vertices
(d) none of these

3.8 Let S_1 and S_2 be two convex sets. Then
(a) $S_1 \cup S_2$ is a convex set
(b) $S_1 \cap S_2$ is a convex set
(c) both (a) and (b) hold good
(d) none of these

3.9 The set $S = \{(x_1, x_2): x_1 + x_2 = 2\}$ has no vertex because it is
(a) not convex
(b) not closed
(c) bounded
(d) not bounded either from above or below

3.10 A CLC of $X_1 = (2, 3, -1)$ and $X_2 = (5, 3, 4)$ is
(a) $(-1, 0, 7)$ (b) $(0, 1, 3)$ (c) $(1, 7, 4)$ (d) none of these

3.11 A CLC of $X_1 = (1, 2, 3)$ and $X_2 = (-1, 2, 5)$ is
(a) $(2, 3, 4)$ (b) $(3, 4, 7)$ (c) both (a) and (b) (d) none of these

3.12 Let $S = \{(x_1, x_2): x_1^2 + x_2^2 \le 1\}$. Then S is a
(a) convex set (b) open set (c) non-convex set (d) all of these

3.13 In the LPP,
Min $Z = x_1 + 4x_2$
Subject to $-x_1 + x_2 \le 1$
$x_1 + x_2 \ge 2$
and $x_1, x_2 \ge 0$
(a) The feasible region is bounded but the solution space is unbounded.
(b) The feasible region as well as the solution space are bounded.
(c) The feasible region as well as the solution space are unbounded.
(d) The feasible region is unbounded but the solution space is bounded.

3.14 In the LPP,
Max $Z = 5x_1 + 2x_2$
Subject to $-x_1 + x_2 \le 1$
$x_1 + x_2 \ge 3$
and $x_1, x_2 \ge 0$
(i) (a) The feasible region is bounded but the solution space is unbounded.
(b) The feasible region as well as the solution space are bounded.
(c) The feasible region as well as the solution space are unbounded.
(d) The feasible region is unbounded but the solution space is bounded.
(ii) The extreme points of the convex set of feasible solutions are
(a) (3, 0), (1, 2)
(b) (0, 0), (3, 0), (1, 2)
(c) (0, 1), (0, 3), (1, 2)
(d) (0, 0), (0, 1), (3, 0), (1, 2)

3.15 For the LPP,
Min $Z = 10x_1 + 15x_2$
Subject to $x_1 + x_2 = 1$
$3x_1 + 2x_2 \leq 6$
and $x_1, x_2 \geq 0$

(i) The extreme points of the convex set of feasible solutions are
(a) (2, 0), (0, 2) (c) (0, 0), (0, 3)
(b) (2, 0), (0, 3) (d) (0, 2), (0, 3)

(ii) Optimum Z occurs at
(a) (0, 2) (c) (0, 2) and (2, 0)
(b) (2, 0) (d) none of these

State True or False

3.1 The set of feasible solutions (S_F) of an LPP is a convex set.
3.2 A set that is not convex can also have a vertex.
3.3 The vertex of the set of feasible solutions (S_F) is a boundary point of S_F.
3.4 Every boundary point of a convex set is its vertex.

Fill in the Blanks

3.1 If the value of the objective function (Z value) can be increased or decreased indefinitely, then such solutions are called __________ solutions.
3.2 A set of values of decision variables ($x_1, x_2, \ldots, x_n$) satisfying the constraints and non-negativity restrictions of a max LPP, for which the objective function has the largest value is called __________ solution.
3.3 A set of values of decision variables ($x_1, x_2, \ldots, x_n$) satisfying the constraints and non-negativity restrictions of an LPP is called a __________ solution.
3.4 An LPP with more than __________ variables cannot be solved graphically.

Review Questions

3.1 Define the following:
(a) Convex set
(b) Convex linear combination
(c) Vertex
(d) Half space in E_n
(e) Hyper plane in E_n
(f) Feasible solution of an LPP
(g) Optimal feasible solution of an LPP

3.2 How many methods are available to solve an LPP in two variables using the graphical method? Describe their operating process.
3.3 What are the limitations of an LPP? Mention at least five applications of an LPP.

Numerical Problems

3.1 Which of the following problems can be solved using the graphical method?

(a) Max $Z = 2x_1 + 3x_2$
Subject to $-x_1 + x_2 \leq 1$
$x_1 + x_2 \leq 2$
and $x_1, x_2 \geq 0$

(b) Min $Z = 2x_1 + 3x_2$
Subject to $-x_1 + x_2 \geq 1$
$x_1 + x_2 \leq 2$
and $x_1, x_2 \geq 0$

(c) Max $Z = 2x_1 - 4x_2 + x_3$
Subject to $-x_1 + x_2 + x_3 \leq 1$
$2x_1 - x_2 - 4x_3 \geq 2$
and $x_1, x_2, x_3 \geq 0$

3.2 Show that the set $S = \{(x_1, x_2, x_3): x_1^2 + x_2^2 + x_3^2 < 1\}$ is a convex set.

3.3 Find the convexity of the set $S = \{(x_1, x_2): 4x_1 + 3x_2 \le 6, x_1 + x_2 \ge 1\}$.

3.4 Which of the following sets are convex?

(a) $S = \{(x_1, x_2): 3x_1^2 + 2x_2^2 \le 6\}$
(b) $S = \{(x_1, x_2): x_1x_2 \le 1, x_1 \ge 0, x_2 \ge 0\}$
(c) $S = \{(x_1, x_2): x_1^2 + x_2^2 \le 1, x_1 + x_2 \ge 1\}$

3.5 Identify the vertices of the following sets:

(a) $S_1 = \{x \in R: 0 \le x < 1\}$
(b) $S_2 = \{x \in R: 0 < x \le 1\}$
(c) $S_3 = \{x \in R: 0 \le x \le 1\}$
(d) $S_4 = \{(x, y) \in E_2 : x^2+y^2 \le 1\}$

3.6 Prove that the set $S = \{X \in E_2: |X - X_O| \le 5, X_O$ is a fixed point in $E_2\}$ is a convex set.

3.7 Prove that the set $S = \{(x_1, x_2, \ldots, x_n) \in E_n: c_1x_1 + c_2x_2 + \ldots + c_nx_n = \alpha; c_1, c_2, \ldots, c_n, \alpha \in R\}$ is a convex set.

3.8 Solve the following using the graphical method:

(a) Min $Z = 20x_1+10x_2$
Subject to $x_1 + 2x_2 \le 40$
$3x_1 + x_2 \ge 30$
$4x_1 + 3x_2 \ge 60$
and $x_1, x_2 \ge 0$

(b) Max $Z = 3x_1+2x_2$
Subject to $x_1 - x_2 \le 1$
$x_1 + x_2 \ge 3$
and $x_1, x_2 \ge 0$

(c) Max $Z = x_1 + x_2$
Subject to $x_1 - x_2 \ge 0$
$-3x_1 + x_2 \ge 3$
and $x_1, x_2 \ge 0$

(d) Min $Z = x_1 + x_2$
Subject to $5x_1 + 10x_2 \le 50$
$x_1 + x_2 \ge 1$
$x_2 \le 4$
and $x_1, x_2 \ge 0$

(e) Max $Z = 6x_1-2x_2$
Subject to $2x_1 - x_2 \le 2$
$x_1 \le 3$
and $x_1, x_2 \ge 0$

3.9 A company manufactures two types of gift items: ordinary O and deluxe D. Each gift of type D takes twice as long to produce as that of type O, and the company would have time to make a maximum of 2000 gifts per day, if it produces only the ordinary items. The supply of the plastic is sufficient to produce 1500 gift items per day (both O and D). The deluxe gift item requires a fancy sheet, of which there are only 600 pieces available per day. If the company makes a profit of ₹3 and ₹5 per gift item respectively on items O and D, how many of each should be produced per day to maximize the profit?

3.10 A carpenter manufactures two wood items, A and B. The total number of items that the carpenter can handle per day at the most, is 24. It takes one hour to make item A and half an hour to make item B. It is assumed that the carpenter can work for a maximum of 16 hours a day. Further, the profit on item A is ₹300 and that on item B is ₹100. Formulate this LPP and also calculate the maximum number of both items that can be produced daily to maximize the profit.

3.11 A company manufactures trucks and tractors in a factory divided into two shops, S_1 and S_2. The shop S_1, which performs the basic assembly operation, must work five man-days on each tractor, but only two man-days on each truck. The shop S_2, which performs the finishing operation, must work three man-days on each truck or tractor that it produces. Due to men and machine limitations, while shop S_1 has 180 man-days per week available, shop S_2 has 135 man-days per week. If the manufacturer makes a profit of ₹300 on each tractor and ₹200 on each truck, how many of each should be produced to maximize the profit?

3.12 A family doctor recommends two types of food (F_1 and F_2) to a family for good health, which contains at least eight units of protein and 10 units of vitamin. Food F_1 contains two units/kg of protein and one unit/kg of vitamin, whereas food F_2 contains one unit/kg of protein and two units/kg of vitamin. F_1 costs ₹5 per kg and F_2 costs ₹7 per kg. Find the minimum cost for such a food combination.

CONCEPTUAL PROBLEM

Cairon India wants to enter the oil business. For this purpose, the company studies the working of those companies that are already engaged in the production of oil, with special references to planning, production, marketing, and profiting. After a proper study of these aspects and taking into consideration the availability of the raw material required, the company decides to produce two types of crudes C_1 and C_2. In producing these two types of crudes, materials of type M_1 and M_2 are required. To produce one unit of crude C_1, 4 units each of M_1 and M_2 are needed. To produce one unit of C_2, one unit of M_1 and two units of M_2 are needed. However, the constraint is that they cannot use more than 12 units of M_1 and 20 units of M_2. The profit per unit sale of C_1 and C_2 is ₹30 and ₹20 respectively. Let us assume all that is produced is sold. The company assigns the job of deciding the feasibility and quantum of production to a project engineer. The engineer has to determine how many units of each type of crude must be produced for earning the maximum revenue.

Solution Tabulating the given data, we get

Material used per unit of production	Crude		Availability
	C_1	C_2	
M_1	4	1	12
M_2	4	2	20
Profit	30	20	

Formulation of the problem Let x_1 and x_2 be the number of units of crudes C_1 and C_2 that are produced. The total profit is $30x_1 + 20x_2$. The company would like to determine x_1 and x_2 so that $30x_1 + 20x_2$ is maximum. The total number of units of material M_1 required to produce x_1 units of C_1 and x_2 units of C_2 is $4x_1 + 2x_2$. The availability restriction of M_1 gives $4x_1 + x_2 \leq 12$. Similarly for M_2, we have $4x_1 + 2x_2 \leq 20$. Further, x_1 and x_2 being the number of units of crudes produced, cannot be negative. Thus, we have $x_1, x_2 \geq 0$.

Thus, the formulated problem is

$$\text{Maximize } Z = 30x_1 + 20x_2 \tag{3.43}$$

$$\text{Subject to } \quad 4x_1 + x_2 \leq 12 \tag{3.44}$$

$$4x_1 + 2x_2 \leq 20 \tag{3.45}$$

$$\text{and} \quad x_1, x_2 \geq 0 \tag{3.46}$$

Graphical method Now the problem is to find a point $X=(x_1, y_2)^T$, satisfying Eqs (3.44), (3.45), and (3.46), where the value of $Z=30x_1+20x_2$ is maximum. We sketch the inequalities in Eqs (3.44), (3.45), and (3.46) to get the bounded region, S_F (set of feasible solutions), as shown in Fig. 3.15. We shall find the solution to this LPP using

(Contd)

(*Contd*)

both the iso-profit and the corner point methods. Here, we solve the problem using the corner point method. For this, we find the coordinates of points O(0, 0), A(3, 0), B(1, 8), and C(0, 10), which are the vertices (corner points) of S_F and calculate the value of Z.

Points	Z
O(0, 0)	$Z = 30 \times 0 + 20 \times 0 = 0$
A(3, 0)	$Z = 30 \times 3 + 20 \times 0 = 90$
B(1, 8)	$Z = 30 \times 1 + 20 \times 8 = 190$
C(0, 10)	$Z = 30 \times 0 + 20 \times 10 = 200$ (Max)

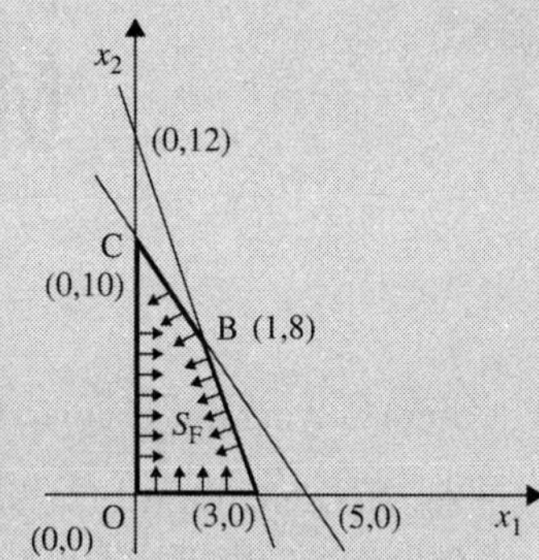

Figure 3.15 Bounded region

Thus, the optimum solution is Max $Z = 200$ when $x_1 = 0$ and $x_2 = 10$. From this table, we see that the company will earn a maximum profit of ₹200, if it produces only C_2 type crude amounting to 10 units.

4 Linear Programming Problem III—Simplex Method

After studying this chapter, the reader will be able to understand

- the standard form of a linear programming problem (LPP) and the conversion of an LPP in standard form
- the canonical form of an LPP and the conversion of an LPP in canonical form
- the basic terms and resolution used in an LPP for simplex method
- the criterion for optimality conditions
- the computational procedure for solving an LPP using the simplex method
- the use of artificial variable techniques (Big-M and two-phase methods)
- the solution of the simultaneous linear equations using the simplex method
- the inverse of a square matrix using the simplex method

4.1 INTRODUCTION

In Chapter 3, we discussed the graphical method for solving an LPP, which was limited to two-variable problems only. We observed that the optimum value of the objective function of an LPP occurs at the corner points of the feasible region (the region where the constraints and non-negativity restrictions hold good), provided the feasible region is bounded.

In the simplex method, we start with a vertex (corner point of the feasible region) and move to another vertex in a prescribed manner, so that the value of the objective function improves. In this way, we reach the vertex that gives the optimum value in a few steps. This method, developed by Dantzig, is such that we omit many vertices of S_F while going from one vertex to the other and then terminates in a few steps.

4.2 STANDARD FORM OF LINEAR PROGRAMMING PROBLEM

An LPP is said to be in standard form, if the following four conditions are satisfied:

1. The objective function must be either maximization or minimization.
2. All the constraints should hold with an equality sign.

3. The right-hand side (RHS) of all constraints is non-negative.
4. All the variables involved in the problem are non-negative.

Note Every LPP can be converted into the standard form.

4.3 SOME IMPORTANT TERMINOLOGIES

Before discussing the simplex method, we shall discuss certain necessary preliminaries.

Basic solution, basis matrix, and non-basic and basic variables In a system of m equations with n unknown variables ($n > m$), we equate ($n - m$) unknowns to zero. The solution of the resulting system is denoted by X_B and is called the *basic solution*, provided the coefficient matrix B of the resulting system is non-singular. Matrix B is called the *basis matrix*. The variables equated to zero are called *non-basic variables* and the variables constituting the basis matrix are called *basic variables*. Thus, we see that the basis matrix is the matrix of the columns of basic variables in a given system of equations.

Basic feasible solution and infeasible basic solution If the values of all the variables in the basic solution are non-negative, it is called a *basic feasible solution* (BFS). A basic solution having at least one variable with a negative value is called an *infeasible basic solution*.

Degenerate and non-degenerate basic feasible solutions A BFS with at least one basic variable assuming the value zero is called a *degenerate BFS*; otherwise it is called a *non-degenerate BFS*.

Slack and surplus variables A variable added to the '≤' type constraints to achieve equality is called a *slack variable*. A variable subtracted from the '≥' type constraints to achieve equality is called a *surplus variable*.

Artificial variable The variable added to the constraint to introduce the identity matrix as a submatrix of the coefficient matrix is called an *artificial variable*.

Redundancy In a system of m equations with n unknowns, we assume $n > m$, which is satisfied in almost all the cases after adding slack and surplus variables. We also assume that the rank of the coefficient matrix A is m and rank A = rank (A, b). This condition is a guarantee for consistency and for no constraint to be redundant. A constraint that does not form a boundary on the feasible region and has no effect on the solution of the problem is called a *redundant constraint*.

Canonical form of LPP A standard form of an LPP with the basis matrix as the unit matrix I (identity matrix) is called the canonical form of an LPP.

Note Every canonical form of an LPP is the standard form, but the converse may not hold good.

4.4 SOME IMPORTANT RESOLUTIONS USED IN LPP FOR SIMPLEX METHOD

To solve an LPP using the simplex method, it has to be brought to the requisite form using some important resolutions, which are discussed in this section.

Conversion of maximization problem to minimization problem and vice versa

$$\text{Max } Z = -\text{Min}(-Z), \quad \text{Min } Z = -\text{Max}(-Z)$$

In other words, if you want to find min Z, then find max$(-Z)$ and multiply the result with (-1).

Conversion of equality into inequality Sometimes, we need to convert an equality into an inequality. For this, we can write $\sum a_{ij}x_j = b_i$ in terms of two inequalities as $\sum a_{ij}x_j \geq b_i$ and $\sum a_{ij}x_j \leq b_i$.

Conversion of unrestricted variables in terms of non-negative variables If x_j is a variable that can assume any positive, zero, or negative value, then we call it an unrestricted variable. To convert it in terms of non-negative variables, we express it as the difference of two non-negative variables. For example, if a variable x_j is unrestricted in sign, it can be written as $x_j = x_j^+ - x_j^-$, where x_j^+, and x_j^- are non-negative.

Change of sense of inequality Many a time we need to convert an inequality with a '≤' sign into one with a '≥' sign and vice versa. For this, multiply both sides with (-1). For example, $\sum a_{ij}x_j \leq b_i$ can be written as $\sum(-a_{ij})x_j \geq (-b_i)$ and $\sum a_{ij}x_j \geq b_i$ can be written as $\sum(-a_{ij})x_j \leq (-b_i)$.

Procedure for reducing LPP to canonical form The following steps are to be followed for reducing a given LPP to the canonical form (see Table 4.1):

1. First, check if all the variables are non-negative. If not, then the following cases may arise:
   (a) If $x_i \leq 0$, then $-x_i = h_i \geq 0$. Now, replace x_i by $-h_i$.
   (b) If x_i is unrestricted in sign, then replace x_i by $(x_{i1} - x_{i2})$ where $x_{i1}, x_{i2} \geq 0$.
2. Check the RHS of all constraints. If it is negative for some constraint, multiply that constraint by (-1) and change the sign of the inequality.
3. For constraints of '≤' type, add slack variables.
4. For constraints of '≥' type, subtract the surplus variables and add an artificial variable.
5. For constraints with equality sign, add an artificial variable.

Table 4.1 Reducing a given LPP to canonical form

Type of constraints	Extra variable to be added	Coefficient of extra variable in the objective function		Presence of variables in the initial BFS (Yes/No)
		Max Z	Min Z	
Less than or equal to '≤'	Add only slack variable	0	0	Yes
Greater than or equal to '≥'	Subtract the surplus variable and add the artificial variable	0	0	No
		$-M$	$+M$	Yes
Equal to '='	Add the artificial variable	$-M$	$+M$	Yes

Note M is a large positive number.

Example 4.1 Write the standard form of the following LPP:

$$\text{Max } Z = x_1 + x_2 - 2x_3 \tag{4.1}$$

$$\text{Subject to} \quad x_1 + x_2 + x_3 \leq 15 \tag{4.2}$$

$$2x_1 - x_2 + x_3 \leq -10 \tag{4.3}$$

$$x_1 + 2x_3 = 10 \tag{4.4}$$

$x_1, x_2 \geq 0$, and x_3 is unrestricted in sign.

Solution Adding x_4 (slack variable) in Eq. (4.2), we get

$$x_1 + x_2 + x_3 + x_4 = 15 \tag{4.5}$$

Multiply Eq. (4.3) by (−1) to get $-2x_1 + x_2 - x_3 \geq 10$ (4.6)

Now subtract x_5 (surplus variable) from the left-hand side (LHS) of Eq. (4.6) to get

$$-2x_1 + x_2 - x_3 - x_5 = 10 \tag{4.7}$$

As x_3 is unrestricted in sign, it can be written as $x_3 = x_{31} - x_{32}$, where $x_{31}, x_{32} \geq 0$.

Now replacing x_3 by $x_{31} - x_{32}$ in Eqs (4.1), (4.4), (4.5), and (4.7), we get the following:

$$\text{Max } Z = x_1 + x_2 - 2(x_{31} - x_{32})$$

$$\begin{aligned} \text{Subject to} \quad & x_1 + x_2 + (x_{31} - x_{32}) + x_4 = 15 \\ & -2x_1 + x_2 - (x_{31} - x_{32}) - x_5 = 10 \\ & x_1 + 2(x_{31} - x_{32}) = 10 \\ & x_1, x_2, x_{31}, x_{32} \geq 0 \end{aligned}$$

Here, all the aforementioned four conditions of standard form are satisfied and hence, it is the required form.

Example 4.2 Write the canonical form of the following LPP:

$$\text{Max } Z = x_1 + x_2$$

$$\begin{aligned} \text{Subject to} \quad & -x_1 + x_2 \leq 5 \\ & x_1 + x_2 \geq 3 \\ \text{and} \quad & x_1, x_2 \geq 0 \end{aligned}$$

Solution The given LPP can be written in the standard form as follows:

$$\text{Max } Z = x_1 + x_2$$

$$\begin{aligned} \text{Subject to} \quad & -x_1 + x_2 + x_3 = 5 \\ & x_1 + x_2 - x_4 = 3 \\ \text{and} \quad & x_1, x_2, x_3, x_4 \geq 0 \end{aligned}$$

This can be written in matrix notation as follows:

$$\text{Max } Z = (1, 1)(x_1, x_2)^{\mathrm{T}}$$

$$\begin{pmatrix} -1 & 1 & 1 & 0 \\ 1 & 1 & 0 & -1 \end{pmatrix} \begin{pmatrix} x_1 \\ x_2 \\ x_3 \\ x_4 \end{pmatrix} = \begin{pmatrix} 1 \\ 3 \end{pmatrix}$$

The coefficient matrix A of the system is as follows:

$$A = \begin{pmatrix} -1 & 1 & 1 & 0 \\ 1 & 1 & 0 & -1 \end{pmatrix}$$

Here, the coefficient matrix A does not have the identity matrix as its submatrix. As a result, the simplex method cannot be applied to solve the problem. So we add a new variable x_5 with the coefficient 1 in the second equation to obtain the second identity column. This variable x_5 is called the artificial variable. The new system obtained is as follows:

$$\text{Max } Z = x_1 + x_2 + 0{\cdot}x_3 + 0{\cdot}x_4 - Mx_5$$
$$-x_1 + x_2 + x_3 = 5$$
$$-x_1 + x_2\, x_4 + x_5 = 3$$

This is the canonical form of the LPP with coefficient matrix A= $\begin{pmatrix} -1 & 1 & 1 & 0 & 0 \\ 1 & 1 & 0 & -1 & 1 \end{pmatrix}$ and identity matrix I= $\begin{pmatrix} 1 & 0 \\ 0 & 1 \end{pmatrix}$ as its submatrix.

Note The purpose of adding an artificial variable is to bring the identity matrix as the submatrix of the coefficient matrix.

4.5 SIMPLEX METHOD

Consider the following LPP in the standard form:

$$\text{Max } Z = f(X) = c_1x_1 + c_2x_2 + c_3x_3 + \ldots + c_nx_n$$
$$a_{11}x_1 + a_{12}x_2 + \ldots + a_{1n}x_1 = b_1$$
$$a_{21}x_1 + a_{22}x_2 + \ldots + a_{2n}x_2 = b_2$$
$$\ldots$$
$$a_{m1}x_1 + a_{m2}x_2 + \ldots + a_{mn}x_n = b_m$$

Where $X = (x_1, x_2, \ldots, x_n)^T \geq 0$ (4.8)

or $$\text{Max } Z = f(X) = (c_1, c_2, \ldots, c_n)\begin{pmatrix} x_1 \\ x_2 \\ \ldots \\ x_n \end{pmatrix}$$

$$\begin{pmatrix} a_{11} \\ a_{21} \\ \ldots \\ a_{m1} \end{pmatrix} x_1 + \begin{pmatrix} a_{12} \\ a_{22} \\ \ldots \\ a_{m2} \end{pmatrix} x_2 + \ldots + \begin{pmatrix} a_{1n} \\ a_{2n} \\ \ldots \\ a_{mn} \end{pmatrix} x_n = \begin{pmatrix} b_1 \\ b_2 \\ \ldots \\ b_m \end{pmatrix}$$

or Max $Z = C^TX$

Subject to $A_1x_1 + A_2x_2 + \ldots + A_nx_n = b$

$x_1, x_2, \ldots, x_n \geq 0$

where $A_1, A_2, \ldots, A_n$ are n columns of coefficient matrix $A = [a_{ij}]_{m\times n}$

or Max $Z = C^TX$

Subject to $AX = b$

$X \geq 0,\ b \geq 0$

where $C = (c_1, c_2, \ldots, c_n)^T$, $X = (x_1, x_2, \ldots, x_n)^T$, $b = (b_1, b_2, \ldots, b_m)^T$.

Now considering one basic solution obtained by putting $(n - m)$ unknowns, say $x_{m+1}, x_{m+2}, \ldots, x_n$ equal to zero in Eq. (4.8), the resulting system can be written as follows:

$$\text{Max } Z = f(X) = c_1x_1 + c_2x_2 + c_3x_3 + \ldots + c_mx_m$$

$$\begin{aligned} &a_{11}x_1 + a_{12}x_2 + \ldots + a_{1m}x_m = b_1 \\ &a_{21}x_1 + a_{22}x_2 + \ldots + a_{2m}x_m = b_2 \\ &\ldots \\ &a_{m1}x_1 + a_{m2}x_2 + \ldots + a_{mm}x_m = b_m \\ &x_1, x_2, \ldots, x_m \geq 0 \end{aligned}$$

or $\text{Max } Z = f(X) = c_1x_1 + c_2x_2 + c_3x_3 + \ldots + c_mx_m$

$$\begin{pmatrix} a_{11} & a_{12} & \ldots & a_{1m} \\ a_{21} & a_{22} & \ldots & a_{2m} \\ \ldots & \ldots & \ldots & \ldots \\ a_{m1} & a_{m2} & \ldots & a_{mn} \end{pmatrix} \begin{pmatrix} x_1 \\ x_2 \\ \ldots \\ x_m \end{pmatrix} = \begin{pmatrix} b_1 \\ b_2 \\ \ldots \\ b_m \end{pmatrix} \text{ or } BX_B = b \Rightarrow X_B \geq 0$$

or $\text{Max } Z = C_B^{\mathrm{T}} X_B$ (4.9)

Subject to $BX_B = b$ (4.10)

$X_B \geq 0$ (4.11)

$$\text{where } B = (A_1, A_2, \ldots, A_m) = \begin{pmatrix} a_{11} & a_{12} & \ldots & a_{1m} \\ a_{21} & a_{22} & \ldots & a_{2m} \\ \ldots & \ldots & \ldots & \ldots \\ a_{m1} & a_{m2} & \ldots & a_{mn} \end{pmatrix}, \; X_B = \begin{pmatrix} x_1 \\ x_2 \\ \ldots \\ x_m \end{pmatrix}, \; b = \begin{pmatrix} b_1 \\ b_2 \\ \ldots \\ b_m \end{pmatrix},$$

$$C_B = \begin{pmatrix} c_1 \\ c_2 \\ \ldots \\ c_m \end{pmatrix}$$

if B is non-singular. Further from Eq. (4.10), we have

$$X_B = B^{-1}b \tag{4.12}$$

Note Vector X_B stands for the basic variables $(x_1, x_2, \ldots, x_m)$ as its components. X_B (as in Eq. 4.12) is also used for the basic solution.

B is called the basis matrix and X_B is called the corresponding BFS. Columns of B form a basis for E_m. Thus, any of A_j $(j = m+1, m+2, \ldots, n)$ can be written as

$$\left.\begin{aligned} A_{m+1} &= \alpha_1^{m+1}A_1 + \alpha_2^{m+1}A_2 + \cdots + \alpha_m^{m+1}A_m \\ A_{m+2} &= \alpha_1^{m+2}A_1 + \alpha_2^{m+2}A_2 + \cdots + \alpha_m^{m+2}A_m \\ &\ldots \\ A_n &= \alpha_1^{n}A_1 + \alpha_2^{n}A_2 + \cdots + \alpha_m^{n}A_m \end{aligned}\right\} \tag{4.13}$$

Combining all the equations in (4.13), we can write as follows:

$$A_j = \alpha_1^j A_1 + \alpha_2^j A_2 + \ldots + \alpha_m^j A_m$$

$$= (A_1, A_2, \ldots, A_m) \begin{pmatrix} \alpha_1^j \\ \alpha_2^j \\ \cdots \\ \alpha_m^j \end{pmatrix} = B\alpha^j \text{ or } \alpha^j = B^{-1}A_j \qquad (4.14)$$

where $\alpha^j = \begin{pmatrix} \alpha_1^j \\ \alpha_2^j \\ \cdots \\ \alpha_m^j \end{pmatrix} = (\alpha_1^j, \alpha_2^j, \ldots, \alpha_m^j)^{\mathrm{T}}$ and $\alpha_1^j, \alpha_2^j, \ldots, \alpha_m^j$ are m components of the column vector α^j.

Note In the aforementioned discussion, X_B has been used as a vector with components $x_1, x_2, \ldots, x_m$.

In the simplex method, we move from one BFS to the other, maintaining the feasibility of the solution in such a way that the value of the objective function improves. For this, we should have another set of m basic variables, which can be obtained by replacing one of the existing basic variables by some non-basic variable. This amounts to the same thing as replacing a basis column of the existing basic matrix B by one of the non-basic columns A_j, $j = m + 1, m + 2, \ldots, n$, to have a new basis matrix, say B_1.

$$B = (A_1, A_2, \ldots, A_{r-1}, \overset{\uparrow}{A_r}, A_{r+1}, \ldots, A_m),\ A = (A_1, A_2, \ldots, A_{r-1}, \overset{\downarrow}{A_j}, A_{r+1}, \ldots, A_m)$$

Thus $B_1 = (A_1, A_2, \ldots, A_{r-1}, A_j, A_{r+1}, \ldots, A_m)$, where A_r of B is replaced by A_j of A, A_j is one of $A_{m+1}, A_{m+2}, \ldots, A_n$. Since B is the basis matrix in E_m,

$$A_j = \alpha_1^j A_1 + \alpha_2^j A_2 + \ldots + \alpha_r^j A_r + \ldots + \alpha_m^j A_m$$

If $\alpha_r^j \neq 0$, we get

$$A_r = \frac{1}{\alpha_r^j}(-\alpha_1^j A_1 - \alpha_2^j A_2 -, \ldots, -\alpha_{r-1}^j A_{r-1} + A_j - \alpha_{r+1}^j A_r, \ldots, -\alpha_m^j A_m) \qquad (4.15)$$

The original basis matrix B gave

$$x_1A_1 + x_2A_2 + \ldots + x_rA_r + \ldots + x_mA_m = b \qquad (4.16)$$

Substituting the value of A_r from Eq. (4.15) in Eq. (4.16), we get

$$\left(x_1 - \alpha_1^j \frac{x_r}{\alpha_r^j}\right)A_1 + \ldots + \left(x_{r-1} - \alpha_{r-1}^j \frac{x_r}{\alpha_r^j}\right)A_{r-1} + \frac{x_r}{\alpha_r^j}A_j$$
$$+\left(x_{r+1} - \alpha_{r+1}^j \frac{x_r}{\alpha_r^j}\right)A_{r+1} + \ldots + \left(x_m - \alpha_m^j \frac{x_r}{\alpha_r^j}\right)A_m = b \qquad (4.17)$$

Eq. (4.17) shows that when A_r leaves and A_j enters, the new basic solution is x_1', x_2', ..., x_r', ..., x_m', where

$$x_1' = \left(x_1 - \alpha_1^j \frac{x_r}{\alpha_r^j}\right), \ldots, x_{r-1}' = \left(x_{r-1} - \alpha_{r-1}^j \frac{x_r}{\alpha_r^j}\right), \; x_r' = \frac{x_r}{\alpha_r^j},$$
$$x_{r+1}' = \left(x_{r+1} - \alpha_{r+1}^j \frac{x_r}{\alpha_r^j}\right), \ldots, x_m' = \left(x_m - \alpha_m^j \frac{x_r}{\alpha_r^j}\right) \tag{4.18}$$

With m basic variables and $(n-m)$ non-basic variables satisfying the system $AX=b$, we should choose the leaving column A_r in such a way that the basic solution given in Eq. (4.18) becomes feasible and hence, a BFS. To ensure this, we choose r such that

$$\frac{x_{B_r}}{\alpha_r^j} = \min_i \left(\frac{x_{B_i}}{\alpha_i^j}, \alpha_i^j \succ 0 \right) = \theta_j \text{ (say)} \tag{4.19}$$

This determines r, that is, which column A_r (variable x_r) should leave so that column A_j (i.e., variable x_j) enters the basis (j is yet to be determined). Eq. (4.19) the *leaving variable rule*, which ensures the feasibility of the new basic solution.

Thus, Eq. (4.18), along with Eq. (4.19) is a BFS. The new value Z' of the objective function corresponding to this BFS is given in Eq. (4.20).

$$Z' = c_1x_1' + c_2x_2' + c_3x_3' + \ldots + c_jx_r' + \ldots + c_mx_m'. \tag{4.20}$$

Here c_r is changed to c_j. Change in the value of the objective function is

$$Z' - Z = (c_1x_1' + c_2x_2' + \ldots + c_jx_r' + \ldots + c_mx_m') - (c_1x_1 + c_2x_2 + \ldots + c_rx_r + \ldots + c_mx_m)$$
$$Z' - Z = c_1(x_1 - \theta_j\alpha_1^j) + c_2(x_2 - \theta_j\alpha_2^j) + \ldots + c_{r-1}(x_{r-1} - \theta_j\alpha_{r-1}^j) + c_j\theta_j + c_{r+1}(x_{r+1} - \theta_j\alpha_{r+1}^j) + \ldots + c_m(x_m - \theta_j\alpha_m^j) - (c_1x_1 + c_2x_2 + \ldots + c_rx_r + \ldots + c_mx_m)$$
$$Z' - Z = -\theta_j(c_1\alpha_1^j + c_2\alpha_2^j + \ldots + c_{r-1}\alpha_{r-1}^j + c_r\alpha_r^j + c_{r+1}\alpha_{r+1}^j + \ldots + c_m\alpha_m^j) + c_j\theta_j$$
$$Z' - Z = \theta_j(c_j - Z_j)$$

where $Z_j = c_1\alpha_1^j + c_2\alpha_2^j + \ldots + c_{r-1}\alpha_{r-1}^j + c_r\alpha_r^j + c_{r+1}\alpha_{r+1}^j + \ldots + c_m\alpha_m^j = C_B^T\alpha^j$ and $\alpha^j = (\alpha_1^j, \alpha_2^j, \ldots, \alpha_m^j)^T$

For a maximization problem, $\theta_j(c_j - Z_j)$ must be most positive. Thus for maximum change, we select that column A_j of the coefficient matrix A for which $\theta_j(c_j - Z_j)$ is most positive. However, this rule is not convenient as we have to calculate θ_j for all the non-basic columns. To simplify this, we select A_j as the entering column for which $c_j - Z_j$ is the most positive. The rule, based on the $c_j - Z_j$ value, may increase one or two simplex iterations, but saves a lot of calculations later on. Thus, the variable x_j with the most positive $c_j - Z_j$ value should enter the basis for the improvement of the value of the objective function. This may be referred to as the *entering variable rule*.

4.6 SIMPLEX TABLE

In order to solve an LPP using the simplex method in the most economically organized manner, it is written in the form of a table known as the simplex table, as depicted in Table 4.2.

Table 4.2 Simplex table

Costs of all variables involved in the problem →				c_1	c_2	c_3	...	c_j	...	c_n	Minimum Ratio $\frac{X_{B_i}}{\alpha_i^j}$
All variables involved in the problem →				x_1	x_2	x_3	...	x_j	...	x_n	
Table no.	Costs of basic variables C_B	Basic variable	Value of basic variable X_B	α^1	α^2	α^3	...	α^j	...	α^n	
1	C_{B1}	x_1	X_{B1}	α_1^1	α_1^2	α_1^3	...	α_1^j	...	α_1^n	...
	C_{B2}	x_2	X_{B2}	α_2^1	α_2^2	α_2^3	...	α_2^j	...	α_2^n	...
	...	...	...	...	...	...	...	...	...	...	...
	C_{Bm}	x_m	X_{Bm}	α_m^1	α_m^2	α_m^3	...	α_m^j	...	α_m^n	...
	$c_j - Z_j$			Δ_1	Δ_2	Δ_3	...	Δ_j	...	Δ_m	

The following points are noted in this table:

1. $\Delta_j = c_j - Z_j = c_j - C_B^{\mathrm{T}}\alpha^j$, Δ_j is the relative cost of variable x_j. In particular, $\Delta_1 = c_1 - Z_1 = c_1 - C_B^{\mathrm{T}}\alpha^1$
2. In the C_B column, the costs of basic variables are written.
3. In the basic variable column, all the basic variables are written in a definite order. In this column, x_i stands for the ith basic variable. Its column will have the ith entry as 1 and all other entries as zero.
4. In the X_B column, we write the values of basic variables, that is, $X_B = B^{-1}b$.
5. $\alpha^j = (\alpha_1^j, \alpha_2^j, \ldots, \alpha_m^j)^{\mathrm{T}}, j = 1, 2, \ldots, n$
6. In the initial simplex table, α^j is equal to the jth column A_j of the coefficient matrix A (i.e., $\alpha^j = A_j$).

4.7 CRITERIA OF OPTIMALITY

The criterion of optimality is based on the following theorem:

Theorem 4.1 Consider the following LPP:

$$\text{Max } Z = C^{\mathrm{T}}X \tag{4.21}$$

$$\text{Subject to } \quad AX = b, X \geq 0$$

where $A = [a_{ij}]_{m\times n}$, $X = (x_1, x_2, \ldots, x_n)^{\mathrm{T}}$, $C = (c_1, c_2, \ldots, c_n)^{\mathrm{T}}$, $b = (b_1, b_2, \ldots, b_m)^{\mathrm{T}}$.

If for a BFS, X_B of an LPP in Eq. (4.21), $c_j - Z_j \leq 0$ for all j ($j = 1, 2, \ldots, n$), then X_B is the optimal BFS of Eq. (4.21).

4.8 COMPUTATIONAL OR ITERATIVE PROCEDURE FOR SOLVING LINEAR PROGRAMMING PROBLEM USING SIMPLEX METHOD

The following steps list the procedure for solving an LPP using the simplex method:

1. First, write the given LPP in the canonical form using slack, surplus, and artificial variables. In case the artificial variable(s) is (are) used, employ the artificial variable technique to solve the problem, which will be discussed in Section 4.10.
2. Identify the simplest BFS.
3. Write the resulting problem obtained in step 2 in the form of a simplex table (Table 4.2).
4. To start the simplex iteration, ensure that the following conditions are satisfied:
   (a) Identity matrix is the submatrix of the coefficient matrix.
   (b) For basic variables, the value of $c_j - Z_j$ is zero.

If these conditions are satisfied in Table 4.2, we can start the simplex iteration to obtain a new BFS. If all Δ_j are non-positive (i.e., $\Delta_j \leq 0$), then the starting simplex table provides the optimal solution. Otherwise, go to step 5.

5. To get a new BFS, one of the non-basic variables should enter the basis and one of the existing basic variables should leave the basis. In the simplex method, we first decide the entering variable.

Entering variable rule The variable x_j with the most positive $c_j - Z_j$ value will enter the basis.

Leaving variable rule The entering variable rule fixes the column α^j. Now for the leaving variable, calculate the ratio of the basic variable values x_{Bi} to the corresponding entries α_i^j, that is, $x_{Bi}/\alpha_i^j, \alpha_i^j > 0$ and take the minimum ratio. Let this minimum ratio be for $i=r$, then x_r will leave the basis. Thus, when x_j enters, x_r leaves the basis.

6. Locate the entry at the intersection of the entering variable column and the leaving variable row. This entry is called the *pivot* (key) *entry* and the row (column) containing the pivot entry is called the *pivot row* (column). Enrectangle the pivot entry. Now apply the *pivot reduction*. In other words, to prepare the next simplex table, use the appropriate elementary row operations to bring 1 in the position corresponding to the pivot entry and zeros above and below this 1 (with the help of the pivot row).
7. Repeat steps 5 and 6 until $\Delta_j > 0$ and one of the following conditions is satisfied:
   (a) If all the entries in the entering column α^j are non-positive, but the entry Δ_j is positive, then the solution to the problem is unbounded and the procedure ends.
   (b) If $\Delta_j \leq 0$ for all j, then the solution is optimum. The table thus obtained will be the optimal table and the simplex iterations will stop. Further, in the optimum table for some non-basic variable x_j if
      (i) $\Delta_j = 0$, then there exists an alternate optimal solution.
      (ii) $\Delta_j < 0$ for all j, then the optimal solution is unique.

Notes 1. We place the starting basic variables in all problems towards the end in a simplex table.
2. Three elementary row operations that are frequently used in simplex iterations include (a) division or multiplication of a row by a non-zero number, (b) subtraction or addition of a constant multiple of a row in another row, and (c) interchange of any two rows.

4.9 SPECIAL CASES IN SIMPLEX METHOD

While solving an LPP using the simplex method, we come across some special cases, such as an LPP having an infeasible solution, unboundedness, degeneracy, alternate optimal solution, and cycling. These cases are discussed in the following subsections.

4.9.1 Infeasibility

Infeasibility may occur in the solution to an LPP only when artificial variables are involved, in converting it to the canonical form to apply the simplex method. The presence of one or more artificial variables with a positive value in the optimal table of (a) phase-I of the two-phase method or (b) the Big-M method are the indications of the existence of an infeasible solution. In case an infeasible solution exists, the LP model is either improperly formulated or some of the constraints involved in the problem are incompatible. As a result, the LP model should be reformulated.

4.9.2 Unboundedness

Unboundedness occurs under the following two circumstances:

1. When in a maximization problem, all entries in the entering variable column (pivotal column) are negative—this implies that as we increase the value of the entering variable, the value of the objective function (Z-value) increases with no bound and thus, the solution is unbounded with unbounded solution space.
2. In a column of non-basic variables of an optimal table in case of a minimization problem, all entries are non-positive—thus, the solution obtained is optimal, but the solution space is unbounded.

4.9.3 Degeneracy

Degeneracy in a transportation problem can occur at any stage while performing a simplex iteration, when the mini ratio rule gives the same minimum value for two rows of the existing basic variables. As a result, any one of these variables leaves the basis (and become non-basic), while the other variable continues as a basic variable with zero value. Thus, the solution obtained is degenerate. Degeneracy can also occur, when in the optimal table, at least one of the decision variables is present with zero value.

4.9.4 Alternate or More Than One Optimal Solution

Sometimes an LPP has more than one optimum solution. Such solutions are called the alternate optimal solutions of the given LPP. The presence of a zero in $(c_j - Z_j)$ row below the non-basic variables in the optimal simplex table is an indication of the existence of an alternate optimal BFS. A BFS corresponds to a vertex of a set of feasible solutions (S_F) of an LPP. This means that in case of an alternate optimum solution, the optimal solution occurs at more than one vertex of S_F. As a result, the optimum occurs at every convex linear combination (CLC) of these vertices and hence, the problem will have an infinite number of optimal solutions. The solutions obtained by these CLC's are non-basic.

4.9.5 Cycling

In case there is a cycle of entering and leaving variables (while applying simplex method rules), that is, variables are repeated without any improvement in the value of the objective function, we say that cycling has occurred. Such cases are rare in practice. Certain techniques are available to avoid cycling.

Example 4.3 A company manufactures two types of products, P_1 and P_2. Product P_1 is of lower quality and product P_2 is of standard quality. The profit on P_1 is ₹30 and that on P_2 is ₹40. The raw material (in kg) required per unit of the product P_1 and P_2 is 60 and 120 respectively. The machinery (in h) needed per unit of producing P_1 and P_2 are 8 and 5 respectively. The man-hours for the assembly per unit of P_1 and P_2 are 3 and 4 respectively. The total availability of raw material (in kg), machinery (in h), and assembly (in h) are 12,000, 600, and 500 respectively. Solve the problem using the simplex method for maximum overall profit.

Solution Let the company manufacture x_1 and x_2 units of products P_1 and P_2 respectively. The overall profit is $30x_1 + 40x_2$.

The problem formulated is as follows:

$$\begin{aligned} &\text{Maximize} && Z = 30x_1 + 40x_2 \\ &\text{Subject to} && 60x_1 + 120x_2 \le 12{,}000 \\ &&& 8x_1 + 5x_2 \le 600 \\ &&& 3x_1 + 4x_2 \le 500 \\ &\text{and} && x_1, x_2 \ge 0 \end{aligned}$$

The problem can be written in the canonical form as follows:

$$\begin{aligned} &\text{Max } Z = 30x_1 + 40x_2 + 0{\cdot}x_3 + 0{\cdot}x_4 + 0{\cdot}x_5 \\ &\text{Subject to} \quad 6x_1 + 12x_2 + x_3 = 1200 \\ &\qquad\qquad\quad 8x_1 + 5x_2 + x_4 = 600, \\ &\qquad\qquad\quad 3x_1 + 4x_2 + x_5 = 500, \\ &\text{and} \qquad\quad x_1, x_2, x_3, x_4, x_5 \ge 0,\ x_3,\ x_4, \text{ and } x_5 \text{ are slack variables.} \end{aligned}$$

The starting BFS is obtained by equating $x_1 = x_2 = 0$ so that $x_3 = 1200$, $x_4 = 600$, and $x_5 = 500$ with x_3, x_4, and x_5 as basic variables and x_1 and x_2 as non-basic variables. Thus, the starting simplex table is as follows:

Cost →				30	40	0	0	0	Minimum ratio $\frac{X_{B_i}}{\alpha_i^j}$
Variable →				x_1	x_2	x_3	x_4	x_5	
Table no.	C_B	Basic variable	X_B	α^1	α^2	α^3	α^4	α^5	
1	0	x_3	1200	6	12	1	0	0	
	0	x_4	600	8	5	0	1	0	
	0	x_5	500	3	4	0	0	1	
	$c_j - Z_j$			30	40	0	0	0	

In the following table, we see that x_2 is the variable with the most positive entry $\Delta_2 = 40$ in the $c_j - Z_j$ row and hence will enter the basis. This fixes the α^2 column. For the leaving variable, the minimum ratio x_{Bi}/α_i^j is for x_3 row and hence x_3 will leave the basis. Thus, when x_2 enters, x_3 leaves the basis (table 1).

		Cost →		30	40	0	0	0	Minimum ratio
		Variable →		x_1	x_2	x_3	x_4	x_5	$\frac{X_{B_i}}{\alpha_i^j}$
Table no.	C_B	Basic variable	X_B	α^1	α^2	α^3	α^4	α^5	
1	0	x_3	1200	6	[12]	1	0	0	1200/12 = 100 →
	0	x_4	600	8	5	0	1	0	600/5 = 120
	0	x_5	500	3	4	0	0	1	500/4 = 125
		$c_j - Z_j$		30	40↑	0	0	0	

Now, apply the pivot reduction—to prepare the next simplex table, use appropriate elementary row operations to bring 1 in the position corresponding to the pivot entry and zeros below this 1, using elementary row operations with the help of the pivot row. We get the following simplex table (use the iterations until the optimum simplex table is obtained):

		Cost →		30	40	0	0	0	Minimum ratio
		Variable →		x_1	x_2	x_3	x_4	x_5	$\frac{X_{B_i}}{\alpha_i^j}$
Table no.	C_B	Basic variable	X_B	α^1	α^2	α^3	α^4	α^5	
2	40	x_2	100	1/2	1	1/12	0	0	$\frac{100}{1/2} = 200$
	0	x_4	100	[11/2]	0	−5/12	1	0	$\frac{100}{11/2} = \frac{200}{11}$ →
	0	x_5	100	1	0	−1/3	0	1	100/1 = 100
		$c_j - Z_j$		10↑	0	−10/3	0	0	
3	40	x_2	100/11	0	1	4/33	−1/11	0	
	30	x_1	200/11	1	0	−5/66	2/11	0	
	0	x_5	900/11	0	0	−17/66	−2/11	1	
		$c_j - Z_j$		0	0	−85/33	−20/11	0	

In this table, $\Delta_j \le 0$ for all j, therefore the solution is optimal. Hence, the optimal solution is $x_1 = 200/11$, $x_2 = 1000/11$, and Max $Z = 46{,}000/11$.

Example 4.4 Solve the following LPP using the simplex method:

$$\text{Max } Z = 4x_1 + 6x_2$$
$$\text{Subject to } \quad 2x_1 + 3x_2 \le 6$$
$$x_1 - x_2 \le 1$$
$$\text{and } x_1, x_2 \ge 0$$

Further, determine the alternate optimal solution if it exists.

Solution The problem can be written in the canonical form as follows:

$$\text{Max } Z = 4x_1 + 6x_2 + 0\cdot x_3 + 0\cdot x_4$$
$$\text{Subject to } \quad 2x_1 + 3x_2 + x_3 + 0\cdot x_4 = 6$$

$$x_1 - x_2 + 0 \cdot x_3 + x_4 = 1$$

and $x_1, x_2, x_3, x_4 \geq 0$, x_3 and x_4 are slack variables.

By taking x_1 and x_2 as non-basic variables (i.e., $x_1=0$ and $x_2=0$), we write the corresponding basic solution with $x_3=6$ and $x_4=1$ in the form of the following simplex table and thereafter use simplex iterations following the entering and leaving variable rules till we obtain an optimal solution:

Table no.	C_B	Basic variable	X_B	4	6	0	0	Minimum ratio $\frac{X_{B_i}}{\alpha_j^i}$
		Variable →		x_1	x_2	x_3	x_4	
				α^1	α^2	α^3	α^4	
1	0	x_3	6	2	[3]	1	0	$6/3 = 2$→
	0	x_4	1	1	−1	0	1	…
		c_j-Z_j		4	6↑	0	0	
2	6	x_2	2	2/3	1	1/3	0	
	0	x_4	3	5/3	0	1/3	1	
		c_j-Z_j		0	0	−2	0	

In this table, $\Delta_j \leq 0$ for all j, therefore the solution is optimal. Hence, the optimal solution is $x_1 = 0$, $x_2 = 2$, and Max $Z = 12.4$.

Now we see that in the aforegiven table, zero entry occurs in the $c_j - Z_j$ row under the non-basic variable x_1 (zero entry in the $c_j - Z_j$ row below a non-basic variable indicates the existence of an alternate optimal solution of the problem). So if x_1 is made to enter the basis, the new solution obtained will again be an optimal solution.

Table no.	C_B	Basic variable	X_B	4	6	0	0	Minimum ratio $\frac{X_{B_i}}{\alpha_j^i}$
		Variable →		x_1	x_2	x_3	x_4	
				α^1	α^2	α^3	α^4	
2	6	x_2	2	2/3	1	1/3	0	$\frac{2}{2/3} = 3$
	0	x_4	3	[5/3]	0	1/3	1	$\frac{3}{5/3} = 9/5$→
		$c_j - Z_j$		0↑	0	−2	0	
3	6	x_2	4/5	0	1	1/5	−2/5	
	4	x_1	9/5	1	0	1/5	3/5	
		$c_j - Z_j$		0	0	−2	0	

In this table, the alternate optimal solution is $x_1 = 9/5$, $x_2 = 4/5$, and Max $Z = 12$.

Example 4.5 Solve the following LPP using the simplex method:

Max $Z = 3x_1 + 2x_2$

Subject to $x_1 + x_2 \leq 4$

$$x_1 - x_2 \le 2$$

and $\quad x_1, x_2 \ge 0$

Solution The problem can be written in the canonical form as follows:

Max $Z = 3x_1 + 2x_2 + 0x_3 + 0x_4$

Subject to $\quad x_1 + x_2 + x_3 + 0x_4 = 4$

$\quad x_1 - x_2 + 0x_3 + x_4 = 2$

and $\quad x_1, x_2, x_3, x_4 \ge 0$, x_3 and x_4 are slack variables.

By taking x_1 and x_2 as non-basic variables (i.e., $x_1 = 0$ and $x_2 = 0$), we write the corresponding basic solution with $x_3 = 4$ and $x_4 = 2$ in the form of the following simplex table and thereafter, use simplex iterations following the entering and leaving variable rules till we obtain an optimal solution:

		Cost →		3	2	0	0	Minimum ratio
		Variable →		x_1	x_2	x_3	x_4	$\frac{X_{B_i}}{\alpha_i^j}$
Table no.	C_B	**Basic variable**	X_B	α^1	α^2	α^3	α^4	
1	0	x_3	4	1	1	1	0	4/1 = 4
	0	x_4	2	[1]	−1	0	1	2/1 = 2→
		$c_j - Z_j$		3↑	2	0	0	
2	0	x_3	2	0	[2]	1	−1	2/2 = 1→
	3	x_1	2	1	−1	0	1	…
		$c_j - Z_j$		0	5↑	0	−3	
3	2	x_2	1	0	1	1/2	−1/2	
	3	x_1	3	1	0	1/2	1/2	
		$c_j - Z_j$		0	0	−5/2	−1/2	

In table 3, $\Delta_j \le 0$ for all j, therefore the solution is optimal. Hence, the optimal solution is $x_1 = 3$, $x_2 = 1$, and Max $Z = 11$.

Example 4.6 Solve the following LPP using the simplex method:

Max $Z = -4x_1 + x_2$

Subject to $\quad -x_1 + x_2 \le 1$

$\quad x_1 - 2x_2 \le 2$

and $\quad x_1, x_2 \ge 0$

Solution The problem can be written in the canonical form as follows:

Max $Z = -4x_1 + x_2 + 0 \cdot x_3 + 0 \cdot x_4$

Subject to $\quad -x_1 + x_2 + x_3 + 0 \cdot x_4 = 1$

$\quad x_1 - 2x_2 + 0 \cdot x_3 + x_4 = 2$

and $\quad x_1, x_2, x_3, x_4 \ge 0$, x_3 and x_4 are slack variables.

By taking x_1 and x_2 as non-basic variables (i.e., $x_1 = 0$ and $x_2 = 0$), we write the corresponding basic solution with $x_3 = 1$ and $x_4 = 2$ in the form of the following simplex table and thereafter, use simplex iterations following the entering and leaving variable rules till we obtain an optimal solution:

		Cost →		−4	1	0	0	Minimum ratio
		Variable →		x_1	x_2	x_3	x_4	
Table no.	C_B	Basic variable	X_B	α^1	α^2	α^3	α^4	$\frac{X_{B_i}}{\alpha_i^j}$
1	0	x_3	1	−1	[1]	1	0	1/1 = 1→
	0	x_4	2	1	−2	0	1	…
		$c_j - Z_j$		−4	1↑	0	0	
2	1	x_2	1	−1	1	1	0	
	0	x_4	4	−1	0	2	1	
		$c_j - Z_j$		−3	0	−1	0	

In table 2, $\Delta_j \le 0$ for all j, therefore the solution is optimal. Hence, the optimal solution is $x_1 = 0, x_2 = 1$, and Max $Z = 1$.

Example 4.7 Solve the following LPP using the simplex method:

$$\begin{aligned}&\text{Max } Z = -x_1 + 4x_2\\&\text{Subject to } \quad x_1 - x_2 \le 1\\&\qquad\qquad -2x_1 + x_2 \le 2\\&\text{and} \qquad\quad x_1, x_2 \ge 0\end{aligned}$$

Solution The problem can be written in the canonical form as follows:

$$\begin{aligned}&\text{Max } Z = -x_1 + 4x_2 + 0\cdot x_3 + 0\cdot x_4\\&\text{Subject to } \quad x_1 - x_2 + x_3 + 0\cdot x_4 = 1\\&\qquad\qquad -2x_1 + x_2 + 0\cdot x_3 + x_4 = 2\\&\text{and} \qquad\quad x_1, x_2, x_3, x_4 \ge 0,\ x_3 \text{ and } x_4 \text{ are slack variables.}\end{aligned}$$

By taking x_1 and x_2 as non-basic variables (i.e., $x_1 = 0$ and $x_2 = 0$), we write the corresponding basic solution with $x_3 = 1$ and $x_4 = 2$ in the form of the following simplex table and thereafter, use simplex iterations following the entering and leaving variable rules till we obtain an optimal solution.

		Cost →		−1	4	0	0	Minimum ratio
		Variable →		x_1	x_2	x_3	x_4	
Table no.	C_B	Basic variable	X_B	α^1	α^2	α^3	α^4	$\frac{X_{B_i}}{\alpha_i^j}$
1	0	x_3	1	1	−1	1	0	…
	0	x_4	2	−2	[1]	0	1	2/1 = 1→
		$c_j - Z_j$		−1	4↑	0	0	
2	0	x_3	3	−1	0	1	1	
	4	x_2	2	−2	1	0	1	
		$c_j - Z_j$		7↑	0	0	−4	

In this table, x_1 enters as $\Delta_1 = 7$, but there is no variable to leave the basis as a consequence of all entries in α^1 column being negative. Hence, the solution is unbounded.

Example 4.8 Solve the following LPP using the simplex method:

$$\text{Max } Z = 2x_1 - x_2 + x_3$$
$$\text{Subject to } \quad 3x_1 + x_2 + x_3 \le 60$$
$$x_1 - x_2 + 2x_3 \le 10$$
$$-x_1 - x_2 + x_3 \ge -20$$
$$x_1, x_2, x_3 \ge 0$$

Solution Multiplying $-x_1 - x_2 + x_3 \ge -20$ with (–1) to make its RHS positive and adding slack variables x_4, x_5, and x_6, we get

$$\text{Max } Z = 2x_1 - x_2 + x_3 + 0{\cdot}x_4 + 0{\cdot}x_5 + 0{\cdot}x_6$$
$$\text{Subject to } \quad 3x_1 + x_2 + x_3 + x_4 + 0{\cdot}x_5 + 0{\cdot}x_6 = 60$$
$$x_1 - x_2 + 2x_3 + 0{\cdot}x_4 + x_5 + 0{\cdot}x_6 = 10$$
$$x_1 + x_2 - x_3 + 0{\cdot}x_4 + 0{\cdot}x_5 + x_6 = 20$$
$$x_1, x_2, x_3, x_4, x_5, x_6 \ge 0$$

which is the canonical form of the given LPP.

The starting BFS is obtained by equating $x_1 = x_2 = x_3 = 0$ so that $x_4 = 60$, $x_5 = 10$, and $x_6 = 20$ with x_4, x_5, and x_6 as basic variables and x_1, x_2, and x_3 as non-basic variables. Thus, the starting simplex table is

Cost →				2	–1	1	0	0	0	Minimum ratio $\frac{X_{B_i}}{\alpha_i^j}$
Variable →				x_1	x_2	x_3	x_4	x_5	x_6	
Table no.	C_B	Basic variable	X_B	α^1	α^2	α^3	α^4	α^5	α^6	
1	0	x_4	60	3	1	1	1	0	0	
	0	x_5	10	1	–1	2	0	1	0	
	0	x_6	20	1	1	–1	0	0	1	
	$c_j - Z_j$			2	–1	1	0	0	0	

In the following table, we see that x_1 is the variable with the most positive entry $\Delta_1 = 2$ in $c_j - Z_j$ row and hence will enter the basis. This fixes the α^1 column. For the leaving variable, the minimum ratio x_{Bi}/α_i^j is for x_5 row and hence, x_5 will leave the basis. Thus when x_1 enters, x_5 leaves the basis (see the following table 1).

Cost →				2	–1	1	0	0	0	Minimum ratio $\frac{X_{B_i}}{\alpha_i^j}$
Variable →				x_1	x_2	x_3	x_4	x_5	x_6	
Table no.	C_B	Basic variable	X_B	α^1	α^2	α^3	α^4	α^5	α^6	
1	0	x_4	60	3	1	1	1	0	0	60/3 = 20
	0	x_5	10	[1]	–1	2	0	1	0	10/1 = 10→
	0	x_6	20	1	1	–1	0	0	1	20/1 = 20
	$c_j - Z_j$			2↑	–1	1	0	0	0	

Now, apply the pivot reduction—to prepare the next simplex table using appropriate elementary row operations to bring 1 in the position corresponding to the pivot entry and zeros above and below this 1, using elementary row operations with the help of the pivot row. We get the following simplex table (repeat the iterations until the optimum simplex table is obtained):

		Cost →		2	−1	1	0	0	0	Minimum ratio
		Variable →		x_1	x_2	x_3	x_4	x_5	x_6	$\frac{X_{B_i}}{\alpha_i^j}$
Table no.	C_B	Basic variable	X_B	α^1	α^2	α^3	α^4	α^5	α^6	
2	0	x_4	30	0	4	−5	1	−3	0	30/4 = 15/2
	2	x_1	10	1	−1	2	0	1	0	–
	0	x_6	10	0	[2]	−3	0	−1	1	10/2 = 5→
		$c_j - Z_j$		0	1↑	−3	0	−2	0	
3	0	x_4	10	0	0	1	1	−1	−2	
	2	x_1	15	1	0	1/2	0	1/2	1/2	
	−1	x_2	5	0	1	−3/2	0	−1/2	1/2	
		$c_j - Z_j$		0	0	−3/2	0	−3/2	−1/2	

In table 3, $\Delta_j \le 0$ for all j, therefore the solution is optimal. Hence, the optimal solution is $x_1 = 15, x_2 = 5, x_3 = 0$, and Max $Z = 25$.

Example 4.9 Solve the following LPP using the simplex method:

$$\text{Min } Z = x_1 - 3x_2 + 2x_3$$

$$\text{Subject to } \quad 3x_1 - x_2 + 3x_3 \le 7$$
$$-2x_1 + 4x_2 \le 12$$
$$-4x_1 + 3x_2 + 8x_3 \le 10$$
$$\text{and} \quad x_1, x_2, x_3 \ge 0$$

Solution Converting the minimization problem into a maximization problem, we have

$$\text{Max } Z' = -x_1 + 3x_2 - 2x_3, \text{ where } Z' = -Z$$

The problem can be written in the canonical form as follows:

$$\text{Max } Z' = -x_1 + 3x_2 - 2x_3$$

$$\text{Subject to } \quad 3x_1 - x_2 + 3x_3 + x_4 = 7$$
$$-2x_1 + 4x_2 + x_5 = 12$$
$$-4x_1 + 3x_2 + 8x_3 + x_6 = 10$$
$$\text{and} \quad x_1, x_2, x_3, x_4, x_5, x_6 \ge 0, \; x_4, x_5, \text{ and } x_6 \text{ are slack variables.}$$

By taking x_1, x_2, and x_3 as non-basic variables (i.e., $x_1 = 0$, $x_2 = 0$, and $x_3 = 0$), we write the corresponding basic solution with $x_4 = 7$, $x_5 = 12$, and $x_6 = 10$ in the form of the following simplex table and thereafter, use simplex iterations following the entering and leaving variable rules until we obtain an optimal solution:

		Cost →		−1	3	−2	0	0	0	Minimum ratio
		Variable →		x_1	x_2	x_3	x_4	x_5	x_6	$\frac{X_{B_i}}{\alpha_i^j}$
Table no.	C_B	Basic variable	X_B	α^1	α^2	α^3	α^4	α^5	α^6	
1	0	x_4	7	3	−1	3	1	0	0	–
	0	x_5	12	−2	[4]	0	0	1	0	12/4 = 3→
	0	x_6	10	−4	3	8	0	0	1	10/3 = 3.33
		$c_j - Z_j$		−1	3↑	−2	0	0	0	–

(Contd)

(Contd)

Table no.	C_B	Basic variable	X_B	Cost → −1 Variable → x_1 α^1	3 x_2 α^2	−2 x_3 α^3	0 x_4 α^4	0 x_5 α^5	0 x_6 α^6	Minimum ratio $\frac{X_{B_i}}{\alpha_i^j}$
2	0	x_4	10	[5/2]	0	3	1	1/4	0	$\frac{10}{5/2}=4\rightarrow$
	3	x_2	3	−1/2	1	0	0	1/4	0	–
	0	x_6	1	−5/2	0	8	0	−3/4	1	–
		$c_j - Z_j$		1/2↑	0	−2	0	−3/4	0	
3	−1	x_1	4	1	0	6/5	2/5	1/10	0	
	3	x_2	5	0	1	3/5	1/5	3/10	0	
	0	x_6	11	0	0	11	1	−1/2	1	
		$c_j - Z_j$		0	0	−13/5	−5/2	−4/5	0	

In this table, $\Delta_j \leq 0$ for all j, therefore the solution is optimal. Thus, the optimal solution is $x_1 = 4$, $x_2 = 5$, $x_3 = 0$, and Min $Z = -Z' = -11$.

4.10 ARTIFICIAL VARIABLE TECHNIQUE FOR SOLVING LINEAR PROGRAMMING PROBLEMS

To solve an LPP using the simplex method, one starts with a BFS, where the basis matrix is in the form of an identity matrix. In the earlier sections, we discussed the simplex method technique to solve an LPP, where all the constraints were of the '≤' type. In such cases, the basis matrix is readily available in the form of an identity matrix; however, there are many instances where all the constraints are not of the '≤' type. Some of them may be of the '≥' or '=' types. In such cases, the basis matrix is not readily available in the form of an identity matrix; as a result, we cannot start simplex iterations. To avoid this difficulty, we introduce artificial variables in the problem to easily obtain a basis matrix in the form of an identity matrix. This facilitates the use of the simplex procedure initially. Thus, the purpose of adding artificial variables in the problem is to obtain the identity matrix as the submatrix of the coefficient matrix, so as to have the starting BFS of the LPP with the basis matrix in the form of an identity matrix. Here, we shall discuss two methods to solve an LPP involving artificial variables—the Big-M method and the two-phase method.

Note While solving an LPP by the simplex method, if the canonical form is not obtained, we use artificial variables to get it and solve the LPP using the Big-M or two-phase method.

4.10.1 Big-M Method

In the Big-M method, the objective function of the problem is modified by adding ($-M$) times the sum of the artificial variables in case of a maximization problem and ($+M$) times the sum of the artificial variables in case of a minimization problem, where M is a large positive number.

The following steps are useful in solving an LPP:

1. Reduce the given LPP to the canonical form using slack, surplus, and artificial variables.
2. In case artificial variables are used, modify the objective function by adding ($-M$) times the sum of the artificial variables involved in the problem if it is a maximization problem.
3. Write the simplex table and use simplex iterations after bringing it to the requisite format, until the optimal solution is obtained.

Notes
1. The large positive value of M in the Big-M method will tend to push the artificial variables out of the basis in simplex iterations, if the solution of the problem exists. However, if the artificial variable continues with the positive value in the optimal simplex table, the given problem has *no solution.*
2. A no solution situation for an LPP will arise in the artificial variable technique only when it is solved by the simplex method.
3. Whenever there is a tie among the leaving variables with an artificial variable, then the artificial variable is preferred, because our aim is to bring it to zero level.
4. An artificial variable is never considered for re-entry into the basis once it leaves the basis. We drop its column and the column entries are not calculated in subsequent simplex tables unless B^{-1} is needed in the optimal table.

Example 4.10 Per unit of finished product P weighing 150 gm is manufactured by a company, using two raw materials, R_1 and R_2 at a cost of ₹2 and ₹8 per unit respectively. In the production stage, at least 14 units of R_2 and a maximum of 20 units of R_1 must be used. Each unit of R_1 and R_2 weighs 5 and 10 gm respectively. How much of R_1 and R_2 must be used for each unit of P in order to minimize the cost.

Solution Let x_1 and x_2 be the units of raw materials R_1 and R_2 respectively required to minimize the cost. The total minimum cost is $2x_1 + 8x_2$.

The formulation of the LPP is as follows:

$$\begin{aligned} &\text{Min } Z = 2x_1 + 8x_2 \\ &\text{Subject to} \quad 5x_1 + 10x_2 = 150 \\ &\qquad\qquad\qquad x_1 \le 20 \\ &\qquad\qquad\qquad x_2 \ge 14 \\ &\text{and} \qquad\qquad x_1, x_2 \ge 0 \end{aligned}$$

Converting the minimization problem into maximization, we have

$$\text{Max } Z' = -2x_1 - 8x_2$$

The problem can be written in the canonical form (after modifying the objective function) as follows:

$$\begin{aligned} &\text{Max } Z' = -2x_1 - 8x_2 - M(x_5 + x_6) \\ &\text{Subject to} \quad 5x_1 + 10x_2 + x_5 = 150 \\ &\qquad\qquad\qquad x_1 + x_3 = 20 \\ &\qquad\qquad\qquad x_2 - x_4 + x_6 = 14 \end{aligned}$$

where $x_1, x_2, x_3, x_4, x_5, x_6 \ge 0$, x_3 slack, x_4 surplus, x_5 and x_6 are artificial variables and M is a large positive number.

By taking x_1, x_2, and x_4 as non-basic variables (i.e., $x_1 = 0, x_2 = 0$, and $x_4 = 0$), we write the corresponding basic solution with $x_3 = 20, x_5 = 150$, and $x_6 = 14$ in the form of the following simplex table and thereafter use simplex iterations following the entering and leaving variable rules until we obtain an optimal solution:

		Cost →		−2	−8	0	0	−M	−M	Minimum ratio
		Variable →		x_1	x_2	x_3	x_4	x_5	x_6	$\frac{X_{B_i}}{\alpha_i^j}$
Table no.	C_B	Basic variable	X_B	α^1	α^2	α^3	α^4	α^5	α^6	
1	0	x_3	20	1	0	1	0	0	0	–
	−M	x_5	150	5	10	0	0	1	0	150/10 = 15
	−M	x_6	14	0	[1]	0	−1	0	1	14/1 = 14→
		$c_j - Z_j$		5M−2	11M−8↑	0	−M	0	0	
2	0	x_3	20	1	0	1	0	0		–
	−M	x_5	10	5	0	0	[10]	1		10/10 = 1→
	−8	x_2	14	0	1	0	−1	0		–
		$c_j - Z_j$		5M−2	0	0	10M−8↑	0		
3	0	x_3	20	1	0	1	0			20/1 = 20
	0	x_4	1	[1/2]	0	0	1			$\frac{1}{1/2} = 2$ →
	−8	x_2	15	1/2	1	0	0			$\frac{15}{1/2} = 30$
		$c_j - Z_j$		2↑	0	0	0			
4	0	x_3	18	0	0	1	−2			
	−2	x_1	2	1	0	0	2			
	−8	x_2	14	0	1	0	−1			
		$c_j - Z_j$		0	0	0	−4			

In table 4, $\Delta_j \leq 0$ for all j, therefore, the solution is optimal. Thus, the optimal solution is $x_1 = 2$, $x_2 = 14$, and Min $Z = -$Max $Z' = 116$.

Example 4.11 A dietician recommends two type of food, F_1 and F_2 to a person. F_1 contains 15 units of carbohydrates and 15 units of proteins per gram. F_2 contains 20 units of carbohydrates and 30 units of proteins per gram. The daily minimum human requirement of carbohydrates and proteins are 900 and 1200 units respectively. The cost per unit for F_1 and F_2 is ₹5 and ₹8. How should the person plan to meet the food requirements to minimize the food cost? Identify the decision variables. Formulate the LPP model of the problem and solve it using both graphical method and simplex method.

Solution The problem can be tabulated as follows:

Type of Food	Units per gram		Profit
	Carbohydrates	Proteins	
F_1	15	15	5
F_2	20	30	8
Availability	900	1200	–

Let x_1 grams of F_1 and x_2 grams of F_2 be included in the food to be taken to minimize the food cost. Now, the person would like to choose the food amounts x_1 and x_2 in such a way that the total food cost $5x_1 + 8x_2$ is minimum. Here x_1 and x_2 are the decision variables. A person's minimum requirement of carbohydrates and proteins can be given as $15x_1 + 20x_2 \geq 900$ and $15x_1 + 30x_2 \geq 1200$ respectively. Further, the food amount x_1 of F_1 and x_2 of F_2 cannot be negative, so we have $x_1, x_2 \geq 0$.

Thus, the problem formulated as an LP model can be given as follows:

$$\text{Min } Z = 5x_1 + 8x_2 \tag{4.22}$$

$$\text{Subject to} \quad 15x_1 + 20x_2 \geq 900 \tag{4.23}$$

$$15x_1 + 30x_2 \geq 1200 \tag{4.24}$$

$$\text{and} \quad x_1, x_2 \geq 0 \tag{4.25}$$

Graphical method Now the problem is to find a point $X = (x_1, x_2)^T$ satisfying Eqs (4.23), (4.24), and (4.25), where the value of $Z(=5x_1 + 8x_2)$ is minimum. We sketch the inequalities in Eqs (4.23), (4.24), and (4.25) to obtain the unbounded region S_F (set of feasible solutions) as shown in Fig. 4.1.

Here, we solve the problem by the corner point method. For this, we find the coordinates of points A(0, 45), B(20, 30), and C(80, 0), which are the vertices (corner points) of S_F and calculate the value of Z.

Points	Z
A(0, 45)	$Z = 5 \times 0 + 8 \times 45 = 360$
B(20, 30)	$Z = 5 \times 20 + 8 \times 30 = 340$ (Min)
C(80, 0)	$Z = 5 \times 80 + 8 \times 0 = 400$

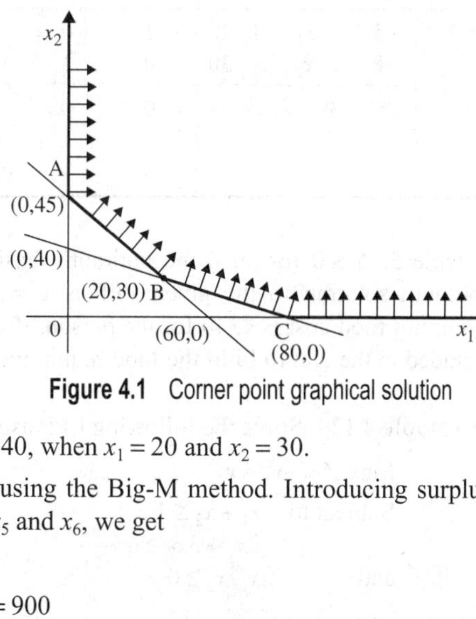

Figure 4.1 Corner point graphical solution

Thus, the optimum solution is Min $Z = 340$, when $x_1 = 20$ and $x_2 = 30$.

Simplex method We solve the problem using the Big-M method. Introducing surplus variables x_3 and x_4, and artificial variables x_5 and x_6, we get

$$\begin{aligned} &\text{Min } Z = 5x_1 + 8x_2 \\ &\text{Subject to} \quad 15x_1 + 20x_2 - x_3 + x_5 = 900 \\ &\qquad\qquad\quad 15x_1 + 30x_2 - x_4 + x_6 = 1200 \\ &\text{and} \qquad\quad x_1, x_2, x_3, x_4, x_5, x_6 \geq 0 \end{aligned}$$

Converting the problem into maximization and modifying the resulting objective function by subtracting $-M(x_5 + x_6)$, we get

$$\begin{aligned} &\text{Max } Z' = -5x_1 - 8x_2 - Mx_5 - Mx_6 \\ &\text{Subject to} \quad 15x_1 + 20x_2 - x_3 + x_5 = 900 \\ &\qquad\qquad\quad 15x_1 + 30x_2 - x_4 + x_6 = 1200 \\ &\text{and} \qquad\quad x_1, x_2, x_3, x_4, x_5, x_6 \geq 0, M \text{ is a large positive number.} \end{aligned}$$

By taking x_1, x_2, x_3, and x_4 as non-basic variables (i.e., $x_1 = 0$, $x_2 = 0$, $x_3 = 0$ and $x_4 = 0$), we write the corresponding basic solution with $x_5 = 900$ and $x_6 = 1200$ in the form of the following simplex table and thereafter, use simplex iterations following the entering and leaving variable rules until we obtain an optimal solution:

	Cost →			−5	−8	0	0	−M	−M	Minimum ratio
	Variable →			x_1	x_2	x_3	x_5	x_4	x_6	$\frac{X_{B_i}}{\alpha_i^j}$
Table no.	C_B	Basic variable	X_B	α^1	α^2	α^3	α^4	α^5	α^6	
1	−M	x_5	900	15	20	−1	0	1	0	900/20 = 45
	−M	x_6	1200	15	[30]	0	−1	0	1	1200/30 = 40→
	$c_j - Z_j$			30M− 5	50M− 8↑	−M	−M	0	0	
2	M	x_5	100	[5]	0	−1	2/3	1	−2/3	100/5 = 20→
	−8	x_2	40	1/2	1	0	−1/30	0	1/30	$\frac{40}{1/2} = 80$
	$c_j - Z_j$			5M− 1↑	0	−M	(2/3) M− (8/30)	0	(2/3) M− (8/30)	
3	−5	x_1	20	1	0	−1/5	2/15	1/5	−2/15	
	−8	x_2	30	0	1	1/10	−1/10	−1/5	1/10	
	$c_j - Z_j$			0	0	−12 M+ (9/5)	−2/15	0	0	

In table 3, $\Delta_j \leq 0$ for all j, the optimum feasible solution has been obtained. Thus, the optimum solution to the given LPP is $x_1 = 20$, $x_2 = 30$, and Min $Z = 340$. Hence, the minimum food cost is ₹340 for the person, if at least 20 units of F_1 and 30 units of F_2 are included in the diet to fulfil the food requirements.

Example 4.12 Solve the following LPP using the Big-M method:

Max $Z = -x_1 + x_2$

Subject to $x_1 + x_2 \leq 1$

$2x_1 + 3x_2 \geq 6$

and $x_1, x_2 \geq 0$

Solution The problem can be written in the canonical form as follows:

$$\left.\begin{aligned} &\text{Max } Z = -x_1 + x_2 + 0{\cdot}x_3 + 0{\cdot}x_4 \\ &\text{Subject to } x_1 + x_2 + x_3 = 1 \\ &\qquad 2x_1 + 3x_2 - x_4 = 6 \end{aligned}\right\} \quad (4.26)$$

and $x_1, x_2, x_3, x_4 \geq 0$, x_3 is the slack variable, and x_4 is the surplus variable.

The modified problem using the Big-M method is as follows:

$$\left.\begin{aligned} &\text{Max } Z = -x_1 + x_2 + 0{\cdot}x_3 + 0{\cdot}x_4 - Mx_5 \\ &\text{Subject to } x_1 + x_2 + x_3 = 1 \\ &\qquad 2x_1 + 3x_2 - x_4 + x_5 = 6 \\ &\text{and } x_1, x_2, x_3, x_4, x_5 \geq 0,\ x_5 \text{ is the artificial variable.} \end{aligned}\right\} \quad (4.27)$$

The optimal solution of (4.27) with $x_5 = 0$ is also an optimal solution of (4.26). By taking x_1, x_2, and x_4 as non-basic variables (i.e., $x_1 = 0, x_2 = 0$, and $x_4 = 0$), we write the corresponding basic solution with $x_3 = 1$ and $x_5 = 6$ in the form of the following simplex table 1 and thereafter, use simplex iterations following the entering and leaving variable rules until we obtain an optimal solution as given in table 2:

Cost →				−1	1	0	0	−M	Minimum ratio $\frac{X_{B_i}}{\alpha_i^j}$
Variable →				x_1	x_2	x_3	x_4	x_5	
Table no.	C_B	Basic variable	X_B	α^1	α^2	α^3	α^4	α^5	
1	0	x_3	1	1	[1]	0	1	0	1/1 = 1→
	−M	x_5	6	2	3	−1	0	1	6/3 = 2
		$c_j - Z_j$		−1	1↑	−M	0	0	
2	1	x_2	1	1	1	0	1	1	
	−M	x_5	3	−1	0	−1	0	1	
		$c_j - Z_j$		−2−M	0	−M	−1−3M	0	

This is the optimal table in which the artificial variable x_5 is present with positive value $x_5 = 3$. Thus, the problem has no solution.

Note When artificial variable is present in the optimal table with positive value while using Big M-method, then we conclude that the given problem has no solution.

Example 4.13 Solve the following LPP using the Big-M method:

$$\text{Max } Z = 4x_1 + 5x_2 - 3x_3$$

$$\begin{aligned} \text{Subject to } \quad & x_1 + x_2 + x_3 = 10 \\ & x_1 - x_2 \geq 1 \\ & 2x_1 + 3x_2 + x_3 \leq 40 \\ \text{and} \quad & x_1, x_2, x_3 \geq 0 \end{aligned}$$

Solution The problem can be written in the standard form as follows:

$$\left.\begin{aligned} & \text{Max } Z = 4x_1 + 5x_2 - 3x_3 + 0 \cdot x_5 + 0 \cdot x_7 \\ & \text{Subject to } \quad x_1 + x_2 + x_3 = 10 \\ & \qquad x_1 - x_2 - x_5 = 1 \\ & \qquad 2x_1 + 3x_2 + x_3 + x_7 = 40 \end{aligned}\right\} \quad (4.28)$$

and $x_1, x_2, x_3, x_5, x_7 \geq 0$, x_5 is the slack variable, and x_7 is the surplus variable. Stating the problem in canonical form by introducing artificial variables x_4 and x_6 and modifying the objective function to solve the problem using the Big-M method:

$$\left.\begin{aligned} & \text{Max } Z = 4x_1 + 5x_2 - 3x_3 - Mx_4 + 0 \cdot x_5 - Mx_6 + 0 \cdot x_7 \\ & \text{Subject to } \quad x_1 + x_2 + x_3 + x_4 = 10 \\ & \qquad x_1 - x_2 - x_5 + x_6 = 1 \\ & \qquad 2x_1 + 3x_2 + x_3 + x_7 = 40 \end{aligned}\right\} \quad (4.29)$$

and $\quad x_1, x_2, x_3, x_4, x_5, x_6, x_7 \geq 0$, x_4 and x_6 are artificial variables.

The optimal solution of (4.29) with $x_4 = 0$, $x_6 = 0$ is also an optimal solution of (4.28). By taking x_1, x_2, x_3, and x_5 as non-basic variables (i.e., $x_1 = 0, x_2 = 0, x_3 = 0$ and $x_5 = 0$), we write the corresponding basic solution with $x_4 = 10$, $x_6 = 1$, and $x_7 = 40$ in the form of the following simplex table 1. Thereafter, using simplex iterations following the entering and leaving variable rules until we obtain the simplex tabels 2 and 3:

Cost →				**4**	**5**	**−3**	**0**	**−*M***	**−*M***	**0**	**Minimum ratio**
Variable →				x_1	x_2	x_3	x_5	x_4	x_6	x_7	$\frac{X_{B_i}}{\alpha_i^j}$
Table no.	C_B	**Basic variable**	X_B	α^1	α^2	α^3	α^5	α^4	α^6	α^7	
1	$-M$	x_4	10	1	1	1	0	1	0	0	$10/1 = 10$
	$-M$	x_6	1	[1]	−1	0	−1	0	1	0	$1/1 = 1\rightarrow$
	0	x_7	40	2	3	1	0	0	0	1	$40/2 = 20$
		$c_j - Z_j$		$4 + 2M\uparrow$	5	$-3 + M$	$-M$	0	0	0	
2	$-M$	x_4	9	0	[2]	1	1	1		0	$9/2 = 4.5\rightarrow$
	4	x_1	1	1	−1	0	−1	0		0	–
	0	x_7	38	0	5	1	2	0		1	$38/5 = 7.6$
		$c_j - Z_j$		0	$9 + 2M\uparrow$	$-3 + M$	$M + 4$	0		0	
3	5	x_2	9/2	0	1	1/2	1/2	0			
	4	x_1	11/2	1	0	1/2	−1/2	0			
	0	x_7	31/2	0	0	−3/2	−1/2	1			
		$c_j - Z_j$		0	0	−15/2	−1/2	0			

In table 3, $\Delta_j \leq 0$ for all j, therefore, the solution is optimal. Thus, the optimal solution is $x_1 = 11/2, x_2 = 9/2, x_3 = 0$, and Max $Z = 89/2$.

4.10.2 Two-phase Method

In this method of the artificial variable technique, the LPP is solved in two phases and thus, the method is called the two-phase method.

Phase I We first reduce the LPP to its canonical form by introducing artificial variables. Introduce an additional objective function in place of the given objective function and retain the constraints and non-negativity restrictions as such. The additional objective function, called the auxiliary objective function is as follows:

$$\text{Min } Z_0 = R_1 + R_2 + \ldots + R_p \text{ (the sum of artificial variables)}$$

Irrespective of the original problem being a maximization or minimization problem, the phase I problem will always be a minimization problem. This is called the LPP of phase I and is solved using simplex method rules to get the optimal table of phase I.

Notes
1. If the artificial variable is present with a non-zero value in the optimal table of phase I, then the given problem has no solution.
2. The purpose of using phase I is to get the starting BFS for phase II.

Phase II In phase II, the optimal solution of phase I is the starting point. Include the original objective function of the given problem and form the simplex table. This gives the initial simplex table of phase II. Now apply simplex iterations until the optimality conditions are satisfied to get the optimal table of phase II. The optimal solution given by this optimal table is the optimal solution of the given problem.

4.10.3 Comparison between Big-M and Two-phase Methods

The following points list a comparison between the two methods:

1. The Big-M method solves an LPP in a single phase, whereas the two-phase method has two phases.
2. In the Big-M method, the objective function is modified by introducing a large coefficient of the artificial variable whereas in the two-phase method, the auxiliary objective is introduced in phase 1.
3. The Big-M method is not suitable for computers because of M being large and not a definite quantity, whereas the two-phase method is adaptable to computers.

Example 4.14 Solve the following LPP using the two-phase method:

$$\text{Max } Z = 3x_1 - x_2$$

$$\begin{aligned} \text{Subject to} \quad & 2x_1 + x_2 \geq 2 \\ & x_1 + 3x_2 \leq 2 \\ & x_2 \leq 4 \\ \text{and} \quad & x_1, x_2 \geq 0 \end{aligned}$$

Solution The problem can be written in the standard form as follows:

$$\text{Max } Z = 3x_1 - x_2 + 0{\cdot}x_3 + 0{\cdot}x_5 + 0{\cdot}x_6$$

$$\begin{aligned} \text{Subject to} \quad & 2x_1 + x_2 - x_3 = 2 \\ & x_1 + 3x_2 + x_5 = 2 \\ & x_2 + x_6 = 4 \end{aligned}$$

and $x_1, x_2, x_3, x_4, x_5, x_6 \geq 0$, x_3 is the surplus variable, x_5 and x_6 are the slack variables.

The canonical form is as follows:

$$\text{Max } Z = 3x_1 - x_2 + 0{\cdot}x_3 + 0{\cdot}x_5 + 0{\cdot}x_6 - x_4$$

$$\begin{aligned} \text{Subject to} \quad & 2x_1 + x_2 - x_3 + x_4 = 2 \\ & x_1 + 3x_2 + x_5 = 2 \\ & x_2 + x_6 = 4 \end{aligned}$$

and $x_1, x_2, x_3, x_4, x_5, x_6 \geq 0$, x_4 is the artificial variable.

Phase I: In phase I, we introduce an auxiliary objective function, which is to be minimized.

$$\text{Min } Z' = x_4$$

$$\begin{aligned} \text{Subject to} \quad & 2x_1 + x_2 - x_3 + x_4 = 2 \\ & x_1 + 3x_2 + x_5 = 2 \\ & x_2 + x_6 = 4 \\ \text{and} \quad & x_1, x_2, x_3, x_4, x_5, x_6 \geq 0 \end{aligned}$$

Converting the minimization problem into maximization, we have

$$\text{Max } Z'' = -Z' = -x_4$$

$$\text{Subject to} \quad 2x_1 + x_2 - x_3 + x_4 = 2$$
$$x_1 + 3x_2 + x_5 = 2$$
$$x_2 + x_6 = 4$$
$$\text{and} \quad x_1, x_2, x_3, x_4, x_5, x_6 \geq 0$$

By taking x_1, x_2, and x_3 as non-basic variables (i.e., $x_1 = 0$, $x_2 = 0$ and $x_3 = 0$), we write the corresponding basic solution with $x_4 = 2$, $x_5 = 2$, and $x_6 = 4$ in the form of the following simplex table 1 and thereafter, use simplex iterations following the entering and leaving variable rules until we obtain an optimal solution:

	Cost →			0	0	0	−1	0	0	Minimum ratio
	Variable →			x_1	x_2	x_3	x_4	x_5	x_6	$\frac{X_{B_i}}{\alpha_i^j}$
Table no.	C_B	Basic variable	X_B	α^1	α^2	α^3	α^4	α^5	α^6	
1	−1	x_4	2	[2]	1	−1	1	0	0	2/2 = 1→
	0	x_5	2	1	3	0	0	1	0	2/1 = 2
	0	x_6	4	0	1	0	0	0	1	…
	$c_j - Z_j$			2↑	1	−1	0	0	0	
2	0	x_1	1	1	1/2	−1/2		0	0	
	0	x_5	1	0	5/2	1/2		1	0	
	0	x_6	4	0	1	0		0	1	
	$c_j - Z_j$			0	0	0		0	0	

In table 2, $\Delta_j \leq 0$ for all j, and hence it is the optimal table of phase I.

Phase II: In phase II, consider the original objective function of the given problem, omit the artificial variable column, and use the optimal table of phase I to obtain the following table 3:

Thereafter, use simplex method rules of leaving and entering variable to obtain table 4.

	Cost →			3	−1	0	0	0	Minimum ratio
	Variable →			x_1	x_2	x_3	x_5	x_6	$\frac{X_{B_i}}{\alpha_i^j}$
Table no.	C_B	Basic variable	X_B	α^1	α^2	α^3	α^5	α^6	
3	3	x_1	1	1	1/2	−1/2	0	0	…
	0	x_5	1	0	5/2	[1/2]	1	0	$\frac{1}{1/2} = 2$→
	0	x_6	4	0	1	0	0	1	…
	$c_j - Z_j$			0	−5/2	3/2↑	0	0	
4	3	x_1	2	1	3	0	1	0	
	0	x_3	2	0	5	1	2	0	
	0	x_6	4	0	1	0	0	1	
	$c_j - Z_j$			0	−10	0	−3	0	

In table 4, $\Delta_j \leq 0$ for all j, therefore, the solution is optimal. Thus, an optimal solution is $x_1 = 2$, $x_2 = 0$, and Max $Z = 6$.

Example 4.15 Solve the following LPP using the two-phase method:

$$\text{Max } Z = 3x_1 + 2x_2 - 3x_3$$
$$\text{Subject to } \quad 2x_1 + x_2 + x_3 \leq 2$$
$$3x_1 + 4x_2 - x_3 \geq 12$$
$$\text{and } \quad x_1, x_2, x_3 \geq 0$$

Solution The problem can be written in the standard form as follows:

$$\text{Max } Z = 3x_1 + 2x_2 - 3x_3 + 0 \cdot x_4 + 0 \cdot x_5$$
$$\text{Subject to } \quad 2x_1 + x_2 + x_3 + x_4 = 2$$
$$3x_1 + 4x_2 - x_3 - x_5 = 12$$

and $x_1, x_2, x_3, x_4, x_5 \geq 0$, x_4 is the slack, and x_5 is the surplus variable.

The canonical form is

$$\text{Max } Z = 3x_1 + 2x_2 - 3x_3 + 0 \cdot x_4 + 0 \cdot x_5 - x_6$$
$$\text{Subject to } \quad 2x_1 + x_2 + x_3 + x_4 = 2$$
$$3x_1 + 4x_2 - x_3 - x_5 + x_6 = 12$$

and $x_1, x_2, x_3, x_4, x_5, x_6 \geq 0$, x_6 is the artificial variable.

Phase I: In phase I, we introduce an auxiliary objective function, which is to be minimized and retain the remaining equations as they are to obtain the following:

$$\text{Min } Z' = x_6 \quad \text{(sum of artificial variables)}$$
$$\text{Subject to } \quad 2x_1 + x_2 + x_3 + x_4 = 2$$
$$3x_1 + 4x_2 - x_3 - x_5 + x_6 = 12$$
$$\text{and } \quad x_1, x_2, x_3, x_4, x_5, x_6 \geq 0$$

Converting the minimization problem into maximization, we have the following:

$$\text{Max } Z'' = -Z' = -x_6$$
$$\text{Subject to } \quad 2x_1 + x_2 + x_3 + x_4 = 2$$
$$3x_1 + 4x_2 - x_3 - x_5 + x_6 = 12$$
$$\text{and } \quad x_1, x_2, x_3, x_4, x_5, x_6 \geq 0$$

By taking x_1, x_2, x_3, and x_5 as non-basic variables (i.e., $x_1 = 0$, $x_2 = 0$, $x_3 = 0$, and $x_5 = 0$), we write the corresponding basic solution with $x_4 = 2$ and $x_6 = 12$ in the form of the following simplex table 1 and thereafter, use simplex iterations following the entering and leaving variable rules until we obtain an optimal table 2:

		Cost →		0	0	0	0	0	−1	Minimum ratio
		Variable →		x_1	x_2	x_3	x_5	x_4	x_6	$\frac{X_{B_i}}{\alpha_i^j}$
Table no.	C_B	**Basic variable**	X_B	α^1	α^2	α^3	α^5	α^4	α^6	
1	0	x_4	2	2	[1]	1	0	1	0	2/1 = 2→
	−1	x_6	12	3	4	−1	−1	0	1	12/4 = 3
		$c_j - Z_j$		3	4↑	−1	−1	0	0	

(*Contd*)

(*Contd*)

	Cost →			0	0	0	0	0	−1	Minimum ratio
	Variable →			x_1	x_2	x_3	x_5	x_4	x_6	$\frac{X_{B_i}}{\alpha_i^j}$
Table no.	C_B	Basic variable	X_B	α^1	α^2	α^3	α^5	α^4	α^6	
2	0	x_2	2	2	1	1	0	1	0	
	−1	x_6	4	−5	0	−5	−1	−4	1	
	$c_j - Z_j$			−5	0	−5	−1	−4	0	

In table 2, $\Delta_j \leq 0$ for all j, but the artificial variable x_6 is present with a positive value ($x_6 = 4$). Hence, the given problem has no solution.

4.11 SOLVING SIMULTANEOUS LINEAR EQUATIONS USING SIMPLEX METHOD

The following steps are used to solve a given system of simultaneous linear equations using the simplex method:

1. Add an artificial variable to each equation of the system.
2. Introduce a dummy objective function, say, Max $Z = 0 \cdot X - 1 \cdot X_a$, where X_a are artificial variables and X are unrestricted in sign.

Example 4.16 Solve the following system of linear equations using the simplex method:

$$2x_1 + 3x_2 = 2 \text{ and } x_1 - x_2 = 1,\ x_1, x_2 \geq 0$$

Solution We place the given system in the LPP format of phase 1 of the two-phase mehod as follows:

$$\text{Max } Z = 0 \cdot x_1 + 0 \cdot x_2 - x_3 - x_4$$

Subject to $2x_1 + 3x_2 + x_3 = 2$

$x_1 - x_2 + x_4 = 1$

and $x_1, x_2, x_3, x_4 \geq 0$, x_3 and x_4 are artificial variables.

By taking x_1 and x_2 as non-basic variables (i.e., $x_1 = 0$ and $x_2 = 0$), we write the corresponding basic solution with $x_3 = 2$ and $x_4 = 4$ in the form of the following simplex table 1 and thereafter, use simplex iterations following the entering and leaving variable rules to obtain tables 2 and 3 [Here $X_a = (x_3, x_4)^{\mathrm{T}}$ and $X = (x_1, x_2)^{\mathrm{T}}$]:

	Cost →			0	0	−1	−1	Minimum ratio
	Variable →			x_1	x_2	x_3	x_4	$\frac{X_{B_i}}{\alpha_i^j}$
Table no.	C_B	Basic variable	X_B	α^1	α^2	α^3	α^4	
1	−1	x_3	2	2	3	1	0	2/2 = 1
	−1	x_4	1	1	−1	0	1	1/1 = 1
	$c_j - Z_j$			3↑	2	0	0	

In table 1, a tie exists in the leaving variables. Break it arbitrarily. Let x_4 leave the basis when x_1 enters. Now, apply simplex iterations to get the following tables:

Cost →				0	0	−1	−1	Minimum ratio
Variable →				x_1	x_2	x_3	x_4	
Table no.	C_B	Basic variable	X_B	α^1	α^2	α^3	α^4	$\frac{X_{B_i}}{\alpha_i^j}$
1	−1	x_3	2	2	3	1	0	2/2 = 1
	−1	x_4	1	[1]	−1	0	1	1/1 = 1→
		$c_j - Z_j$		3↑	2	0	0	
2	−1	x_3	0	0	[5]	1	−2	0/5 = 0→
	0	x_1	1	1	−1	0	1	…
		$c_j - Z_j$		0	5↑	0	−3	
3	0	x_2	0	0	1	1/5	−2/5	
	0	x_1	1	1	0	1/5	3/5	
		$c_j - Z_j$		0	0	−1	1	

In this table, $\Delta_j \leq 0$ for all j, therefore, the solution is optimal. Thus, the obtained optimal solution is $x_1 = 1$ and $x_2 = 0$.

4.12 FINDING INVERSE OF SQUARE MATRIX USING SIMPLEX METHOD

The following steps are used to find the inverse of a square matrix using the simplex method:

1. Take an arbitrary RHS column vector b so that system is obtained in the form $AX = b$, with A as the given matrix.
2. Add an artificial variable to each equation of the system obtained in step 1.
3. To apply the simplex method, introduce an objective function of maximization form with cost zero for the given variables and (−1) for the artificial variables.
4. Now apply the simplex method to solve the problem as an LPP, to get the optimal table with an identity matrix below the given variables. The matrix under the artificial variables in this optimal table is the inverse of the given matrix.

However, in case the identity matrix is not obtained, continue the simplex iterations until all columns below the original variables of the problem become identity columns (the resulting solution may be infeasible).

Example 4.17 Find the inverse of matrix $A = \begin{bmatrix} 2 & 3 \\ 1 & -1 \end{bmatrix}$ using the simplex method.

Solution Taking $b = \begin{bmatrix} 1 \\ 2 \end{bmatrix}$, the system of equation is $\begin{bmatrix} 2 & 3 \\ 1 & -1 \end{bmatrix} \begin{bmatrix} x_1 \\ x_2 \end{bmatrix} = \begin{bmatrix} 1 \\ 2 \end{bmatrix}$

To solve this, the system of equations can be written as follows:

$$\text{Max } Z = 0 \cdot x_1 + 0 \cdot x_2 - x_3 - x_4$$

$$\text{Subject to} \quad 2x_1 + 3x_2 + x_3 = 1$$

$$x_1 - x_2 + x_4 = 2$$

$$\text{and} \quad x_1, x_2, x_3, x_4 \geq 0, x_3 \text{ and } x_4 \text{ are artificial variables}$$

By taking x_1 and x_2 as non-basic variables (i.e., $x_1 = 0$ and $x_2 = 0$), we write the corresponding basic solution with $x_3 = 1$ and $x_4 = 2$ in the form of the following simplex table and thereafter, use simplex iterations following the entering and leaving variable rules until we obtain an optimal solution:

Cost →				0	0	−1	−1	Minimum ratio
Variable →				x_1	x_2	x_3	x_4	
Table no.	C_B	Basic variable	X_B	α^1	α^2	α^3	α^4	$\frac{X_{B_i}}{\alpha_i^j}$
1	−1	x_3	1	[2]	3	1	0	1/2→
	−1	x_4	2	1	−1	0	1	2/1 = 2
		$c_j - Z_j$		3↑	2	0	0	
2	0	x_1	1/2	1	3/2	1/2	0	
	−1	x_4	3/2	0	[−5/2]	−1/2	1	
		$c_j - Z_j$		0	−5/2↑	−3/2	0	
3	0	x_1	−2/5	1	0	1/5	3/5	
	0	x_2	−3/5	0	1	1/5	−2/5	
		$c_j - Z_j$		0	0	2/5	−1	

Here, the given matrix A is converted into a unit matrix, so A^{-1} is given by the x_3 and x_4 columns in simplex table 3, that is, $A^{-1} = \begin{bmatrix} 1/5 & 3/5 \\ 1/5 & -2/5 \end{bmatrix}$

Note In the aforegiven table 2, an optimum solution is obtained. However, matrix A has not yet been converted into a unit matrix. To get unit matrix, introduce x_2 in the basis and remove x_4 from the basis to obtain table 3.

RECAPITULATION

1. Any maximization or minimization problem, where the objective function and constraints are linear, can be solved using the simplex method.
2. Every LPP can be converted into the canonical form, which is the requisite form for solving it using the application of the simplex method.
3. In the simplex method, we start with a readily available BFS, and move from one vertex (BFS) to another, skipping many intermediate vertices and arriving at the optimal solution rapidly.
4. For the simplex method, the leaving variable and entering variable rules have been devised in such a way that the next solution obtained is feasible and at the same time, the value of the objective function improves, ultimately leading to the optimal solution to the problem.

(Contd)

(*Contd*)

5. In case the starting BFS for an LPP is not readily available, it is obtained by introducing fictitious variables (called artificial variables). However, in that case, the solution to the resulting problem need not be a solution to the original problem. However, the solution of the resulting problem with artificial variables equal to zero in the optimal table is also an optimal solution to the original problem.
6. When artificial variables are introduced in the LPP, we solve it using either the Big-M method or the two-phase method.
7. The simplex method can also be used to solve simultaneous linear equations and to obtain the inverse of a non-singular square matrix.

EXERCISES

Multiple-choice Questions

4.1 Which positive variables are added in the left-hand sides of the constraints of an LPP to convert them into equalities?
(a) Slack variables
(b) Surplus variables
(c) Artificial variables
(d) Basic variables

4.2 Which positive variables are subtracted from the left-hand sides of the constraints to convert them into equalities?
(a) Slack variables
(b) Surplus variables
(c) Artificial variables
(d) Basic variables

4.3 The LPP, Max $Z = f(X)$, Subject to $AX = b$, $X \geq 0$, with a set of feasible solutions S_F has no solution. This means that S_F is
(a) unbounded
(b) empty
(c) not closed
(d) none of these

4.4 If the feasible region of an LPP is an empty set, then we say that the problem has
(a) an unbounded solution
(b) an infinite number of solutions
(c) no solution
(d) none of these

4.5 The following set of equations is given: $x_1 + 4x_2 - x_3 = 3$, $5x_1 + 2x_2 + 3x_3 = 4$. The BFS involving x_1 and x_2 is
(a) (5/9, 11/18, 0)
(b) (4/9, 5/9, 0)
(c) (4/9, 11/18, 0)
(d) none of these

4.6 The maximum number of basic solutions to a set of m simultaneous equations in n unknowns ($n \geq m$) is
(a) m
(b) $n - m$
(c) ${}^nc_{n-m}$
(d) none of these

4.7 If m is the number of basic solutions of an LPP and n is the number of extreme points of the set of feasible solutions of the LPP, then
(a) $m < n$
(b) $m > n$
(c) $m \geq n$
(d) none of these

4.8 The presence of an artificial variable with a non-zero value in the optimal table of phase I of the two-phase method indicates that the problem has
(a) no solution
(b) unbounded solution
(c) alternate optimal solution
(d) none of these

4.9 In the simplex table, at any iterative step, the presence of all negative α^j entries in the entering variable column indicate that the problem has
(a) no solution (c) alternate optimal solution
(b) unbounded solution (d) none of these

10. If in the minimum ratio rule, the minimum ratio is equal for two rows, then the next basic solution will be
(a) non-degenerate (c) unbounded
(b) degenerate (d) none of these

State True or False

4.1 A BFS is said to be optimum, if it optimizes the objective function.
4.2 A basic variable can never assume the value zero.
4.3 A BFS is degenerate if one or more basic variables are valued zero.
4.4 A basic solution is always feasible.
4.5 A BFS is the optimal solution to the LPP.
4.6 The LPP Min $Z = 2x_1 + 3x_2$, subject to $-x_1 + x_2 \le 1$, $-2x_2 + x_1 \ge 2$, and $x_1, x_2 \ge 0$ has a feasible solution.
4.7 The LPP Max $Z = 2x_1 + 3x_2$, subject to $x_1 + x_2 \le 2$, $3x_1 + x_2 \ge 6$, and $x_1, x_2 \ge 0$ has a single point as its feasible region.
4.8 The LPP Min $Z = x_1 + x_2$, subject to $x_1 - x_2 \ge 0$, $3x_1 - x_2 \le -3$, and $x_1, x_2 \ge 0$ has a feasible solution.
4.9 In the simplex method, the leaving variable rule ensures that the next solution remains feasible.
4.10 In the simplex method, if the entering variable rule is not followed, then the next solution may not improve.
4.11 In the simplex method, first we fix the leaving variable and thereafter the entering variable.
4.12 In the simplex method, we start with the infeasible solution and using the simplex rule, we proceed to improve the feasibility of the solution.
4.13 In the simplex method, we start with the feasible solution and proceed to improve the value of the objective function.
4.14 The purpose of phase I in the two-phase method is to obtain the basic feasible solution.
4.15 If the phase I optimal table contains an artificial variable even with zero value, then the LPP has no solution.

Fill in the Blanks

4.1 The simplex method was developed by ________ in 1947.
4.2 If a constraint in an LPP has the '≤' sign, then to convert it into an equation we use the ________ variable and if it has the '≥' sign, we use the ________ variable.
4.3 To convert a minimization problem into a maximization problem, we multiply both sides by ________
4.4 The basis matrix is the submatrix constituted by the columns of ________ variables in the coefficient matrix A, obtained after converting the problem to the canonical form.
4.5 The number of basic variables in any LPP equals the number of ________ in the LPP.
4.6 The non-basic variables in a basic solution are ________ valued variables.

Review Questions

4.1 Write the procedure to solve an LPP using the simplex method.

4.2 Define the following terms:
   (a) Basic solution
   (b) Basic feasible solution
   (c) Slack and surplus variables
   (d) Artificial variables
   (e) Redundancy in simplex method
   (f) Conversion of equality into inequality

4.3 Write the conditions of optimality in the simplex method.

4.4 Explain the various artificial variable techniques used in the simplex method.

Numerical Problems

4.1 Solve the following LPPs using the simplex method:

(a) A company produces two types of products: canned apple and canned juice. The profit, amounts, and availability of material, equipment, and labour required to produce each product are shown in the following table:

	Canned juice	**Canned apple**	**Availability**
Material	1	1.4	4.9
Equipment (Machine h)	1	2.3	6.9
Labour (Man h)	3	2	12
Profit (in ₹)	2	1	

Find the product mix that will maximize the profit.

(b) A dietician suggests two types of foods F_1 and F_2. One unit of food F_1 contains 1 mg of vitamins, 100 mg of proteins, and 10 mg of carbohydrates, and one unit of food F_2 contains 1 mg of vitamins, 10 mg of proteins, and 100 mg of carbohydrates. The daily human requirements of vitamins, proteins, and carbohydrates are 1 mg, 50 mg, and 10 mg respectively. The food cost of one unit of F_1 and F_2 is ₹1 and ₹1.5 respectively. Determine the minimum cost to prepare food types F_1 and F_2 such that it supplies the body with atleast the minimum requirements of vitamins, proteins, and carbohydrates.

4.2 Solve the following LPPs using the simplex method:

(a) Max $Z = 7x_1 + 5x_2$
Subject to $-x_1 - 2x_2 \geq -6$
$4x_1 + 3x_2 \leq 12$
and $x_1, x_2 \geq 0$

(b) Max $Z = 5x_1 + 3x_2$
Subject to $3x_1 + 5x_2 \leq 15$
$5x_1 + 2x_2 \leq 10$
and $x_1, x_2 \geq 0$

(c) Max $Z = 5x_1 + 7x_2$
Subject to $x_1 + x_2 \leq 4$
$3x_1 - 8x_2 \leq 24$
$10x_1 + 7x_2 \leq 35$
and $x_1, x_2 \geq 0$

(g) Min $Z = x_1 + x_2 + 3x_3$
Subject to $3x_1 + 2x_2 + x_3 \leq 3$
$2x_1 + x_2 + 2x_3 \leq 2$
and $x_1, x_2, x_3 \geq 0$

(h) Max $Z = 3x_1 + 5x_2 + 4x_3$
Subject to $2x_1 + 3x_2 \leq 8$
$2x_2 + 5x_3 \leq 10$
$3x_1 + 2x_2 + 4x_3 \leq 15$
and $x_1, x_2, x_3 \geq 0$

(i) Max $Z = 3x_1 + 2x_2 + 5x_3$
Subject to $x_1 + 2x_2 + x_3 \leq 430$
$3x_1 + 2x_3 \leq 460$
$x_1 + 4x_2 \leq 420$
and $x_1, x_2, x_3 \geq 0$

(d) Max $Z = 2x_1 + x_2$
Subject to $x_1 + 2x_2 \le 10$
$x_1 + x_2 \le 6$
$x_1 - x_2 \le 2$
$x_1 - 2x_2 \le 1$
and $x_1, x_2 \ge 0$

(e) Max $Z = 3x_1 + 4x_2$
Subject to $x_1 + 3x_2 \le 9$
$2x_1 - x_2 \le 8$
$x_1 + x_2 \le 5$
and $x_1, x_2 \ge 0$

(f) Max $Z = 7x_1 + x_2 + 2x_3$
Subject to $x_1 + x_2 - 2x_3 \le 10$
$4x_1 + x_2 + x_3 \le 20$
and $x_1, x_2, x_3 \ge 0$

(j) Max $Z = 3x_1 + 4x_2$
Subject to $x_1 - x_2 \le 1$
$-x_1 + x_2 \le 2$
and $x_1, x_2 \ge 0$

(k) Max $Z = 4x_1 + 10x_2$
Subject to $2x_1 + x_2 \le 50$
$2x_1 + 5x_2 \le 100$
$2x_1 + 3x_2 \le 90$
and $x_1, x_2 \ge 0$

4.3 Solve the following LPPs using the Big-M method:

(a) A family doctor suggests two types of foods F_1 and F_2. Per gm of F_1 contains 20 units of vitamin and 40 units of protein and per gm of F_2 contains 30 units of each, vitamin and protein. The daily human requirements of vitamins and proteins are 900 units and 1200 units respectively. The food cost per gm of F_1 and F_2 is ₹0.60 and ₹0.80 respectively. How many gm each of F_1 and F_2 should be consumed, to minimize the food cost?

(b) Min $Z = 2x_1 + 3x_2$
Subject to $x_1 + x_2 \ge 5$
$x_1 + 2x_2 \ge 6$
and $x_1, x_2 \ge 0$

(c) Min $Z = 4x_1 + 8x_2 + 3x_3$
Subject to $x_1 + x_2 \ge 2$
$2x_1 + x_3 \ge 5$
and $x_1, x_2, x_3 \ge 0$

(d) Max $Z = x_1 + 2x_2 + 3x_3 - x_4$
Subject to $x_1 + 2x_2 + 3x_3 = 15$
$2x_1 + x_2 + 5x_3 = 20$
$x_1 + 2x_2 + x_3 + x_4 = 10$
and $x_1, x_2, x_3, x_4 \ge 0$

(e) Max $Z = -x_1 - x_2$
Subject to $3x_1 + 2x_2 \ge 30$
$-2x_1 + 3x_2 \le -30$
$x_1 + x_2 \le 5$
and $x_1, x_2 \ge 0$

4.4 Solve the following LPPs using the two-phase method:

(a) A company produces two types of products P_1 and P_2. To manufacture these products, the company needs three types of chemicals, C_1, C_2, and C_3 whose minimum requirements are 10, 12, and 12 units respectively. Product P_1 contains 5, 2, and 1 unit of C_1, C_2, and C_3 per packet and P_2 contains 1, 2, and 4 units of C_1, C_2, and C_3 per packet respectively. Production per packet of products P_1 and P_2 costs ₹3 and ₹2 respectively to the company. How many packets of each product be produced to minimize the cost and obtain the requirements.

(b) Min $Z = x_1 + x_2$
Subject to $2x_1 + x_2 \ge 4$
$x_1 + 7x_2 \ge 7$
and $x_1, x_2 \ge 0$

(c) Max $Z = 5x_1 + 8x_2$
Subject to $3x_1 + 2x_2 \ge 3$
$x_1 + 4x_2 \ge 4$
$x_1 + x_2 \le 5$
and $x_1, x_2 \ge 0$

(d) Min $Z = x_1 - 2x_2 - 3x_3$
Subject to $-2x_1 + x_2 + 3x_3 = 2$
$2x_1 + 3x_2 + 4x_3 = 1$
and $x_1, x_2, x_3 \ge 0$

4.5 Solve the following system of linear equations using the simplex method:

(a) $3x_1 + 2x_2 = 4$ and $4x_1 - x_2 = 6$ (b) $x_1 + x_2 = 1$ and $2x_1 + x_2 = 3$

(c) Find the inverse of $\begin{bmatrix} 1 & 2 \\ 3 & 2 \end{bmatrix}$ using the simplex method.

CONCEPTUAL PROBLEM

'Libra' is a well-known company in Kanpur, which is engaged in the manufacturing of leather goods such as footwear, bags, caps, and jackets. The company wants to increase its activities by manufacturing some more leather goods. It is especially interested in the manufacturing of leather belts, which have a good market. The company intends to manufacture various types of belts, some fancy and some affordable by all. After doing a market survey, the company has decided to produce three kinds of belts A_1, A_2, and A_3. Both A_2 and A_3 require equal manufacturing time, whereas A_1 requires double the time of A_2 (or A_3). The supply of leather is sufficient for only 2000 belts per month (A_1, A_2, and A_3 combined). If all belts were of type A_2, the company could produce 3000 belts per month. The buckles used for type A_1, A_2, and A_3 are of different types and 600, 700, and 800 respectively, of these are available per month. The profit per belt is ₹6, ₹4, and ₹3, respectively. Now, the company faces a problem regarding the number of belts of types A_1, A_2, and A_3 that have to be produced so that it earns the maximum revenue, assuming that all that is produced is sold. Formulate a mathematical model to solve this problem. Further, provide the physical interpretation of the result obtained.

Solution Let x_1, x_2, and x_3 number of leather belts of types A_1, A_2, and A_3 respectively be produced each month. Here, x_1, x_2, and x_3 are the decision variables. The total profit on production is $6x_1 + 4x_2 + 3x_3$. The company would like to choose x_1, x_2, and x_3 in such a way that the total profit $6x_1 + 4x_2 + 3x_3$ is maximum. The restriction on the availability of leather is $x_1 + x_2 + x_3 \leq 2000$. The company can produce 3000 belts if only type A_2 is manufactured. However, it wants to produce all of them, that is, A_1, A_2, and A_3 in x_1, x_2, and x_3 number so as to have the maximum profit. Thus, we have $2x_1 + x_2 + x_3 \leq 3000$. The restrictions on the number of buckles available for each type of belt yields $x_1 \leq 600$, $x_2 \leq 700$, and $x_3 \leq 800$. Further, the production of belts, cannot be negative. Hence, so we have $x_1, x_2, x_3 \geq 0$. Thus, the problem is formulated as an LP model as follows:

$$\begin{aligned} &\text{Max } Z = 6x_1 + 4x_2 + 3x_3 \\ &\text{Subject to} \quad 2x_1 + x_2 + x_3 \leq 3000 \\ &\qquad x_1 + x_2 + x_3 \leq 2000 \\ &\qquad x_1 \leq 600 \\ &\qquad x_2 \leq 700 \\ &\qquad x_3 \leq 800 \\ &\text{and} \quad x_1, x_2, x_3 \geq 0 \end{aligned}$$

The canonical form of the LPP is (convert equalities into equalities using slack variables s_i, $i = 1, 2, 3, 4, 5$).

$$\begin{aligned} &\text{Max } Z = 6x_1 + 4x_2 + 3x_3 \\ &\text{Subject to} \quad 2x_1 + x_2 + x_3 + x_4 = 3000 \\ &\qquad x_1 + x_2 + x_3 + x_5 = 2000 \\ &\qquad x_1 + x_6 = 600 \end{aligned}$$

(*Contd*)

(Contd)

$$x_2 + x_7 = 700$$
$$x_3 + x_8 = 800$$

and $\quad x_1, x_2, x_3 \geq 0$, x_4, x_5, x_6, x_7, and x_8 are slack variables

By taking x_1 and x_2 as non-basic variables (i.e., $x_1 = 0$, $x_2 = 0$ and $x_3 = 0$), we write the corresponding basic solution with $x_4 = 3000$, $x_5 = 2000$, $x_6 = 600$, $x_7 = 700$, and $x_8 = 800$ in the form of the following simplex table 1 and thereafter, use simplex iterations following the entering and leaving variable rules until we obtain an optimal solution:

		Cost →		6	4	3	0	0	0	0	0	Minimum ratio
		Variable →		x_1	x_2	x_3	x_4	x_5	x_6	x_7	x_8	$\frac{X_{B_i}}{\alpha_i^j}$
Table no.	C_B	**Basic variable**	X_B	α^1	α^2	α^3	α^4	α^5	α^6	α^7	α^8	
1	0	x_4	3000	2	1	1	1	0	0	0	0	3000/2 = 1500
	0	x_5	2000	1	1	1	0	1	0	0	0	2000
	0	x_6	600	[1]	0	0	0	0	1	0	0	600→
	0	x_7	700	0	0	0	0	0	0	1	0	…
	0	x_8	800	0	0	0	0	0	0	0	1	…
		$c_j - Z_j$		6↑	4	3	0	0	0	0	0	
2	0	x_4	1800	0	1	1	1	0	−2	0	0	1800
	0	x_5	1400	0	1	1	0	1	−1	0	0	1400
	6	x_1	600	1	0	0	0	0	1	0	0	…
	0	x_7	700	0	[1]	0	0	0	0	1	0	700→
	0	x_8	800	0	0	1	0	0	0	0	1	…
		$c_j - Z_j$		0	4↑	3	0	0	−6	0	0	
3	0	x_4	1100	0	0	1	1	0	−2	−1	0	1100
	0	x_5	700	0	0	[1]	0	1	−1	−1	0	700→
	6	x_1	600	1	0	0	0	0	1	0	0	…
	4	x_2	700	0	1	0	0	0	0	1	0	…
	0	x_8	800	0	0	1	0	0	0	0	1	800
		$c_j - Z_j$		0	0	3↑	0	0	−6	−4	0	
4	0	x_4	400	0	0	0	1	−1	−1	0	0	
	3	x_3	700	0	0	1	0	1	−1	−1	0	
	6	x_1	600	1	0	0	0	0	1	0	0	
	4	x_2	700	0	1	0	0	0	0	1	0	
	0	x_8	100	0	0	0	0	−1	1	1	1	
		$c_j - Z_j$		0	0	0	0	−3	−3	−1	0	

Since all $c_j - Z_j \leq 0$, the optimum feasible solution has been obtained. Thus, the optimum solution to the given LPP is $x_1 = 600$, $x_2 = 700$, $x_3 = 700$, and Max $Z = 8500$. Hence, the company should produce 600 belts of type A_1 and 700 belts each of type A_2 and A_3 to achieve the maximum profit of ₹8500.

5 Linear Programming Problem IV—Revised Simplex Method

After studying this chapter, the reader will be able to

- identify the basics of the revised simplex method
- understand the computational procedure for solving a linear programming problem (LPP) using the revised simplex method
- find B^{-1} for the next table and check the optimality of the solution at any stage

5.1 INTRODUCTION

When an LPP involves a large number of variables, it becomes a cumbersome task to solve the problem using the simplex method, either manually or using a computer, as the variables and constraints occupy a large storage space in the computer's memory. To overcome this problem, the revised simplex method is used, where the number of calculations required is small. In this chapter, we will discuss the revised simplex method in detail.

5.2 REVISED SIMPLEX METHOD

The revised simplex method is an improvisation of the simplex method. In the simplex method, all the entries are calculated to complete a simplex table, whereas in the revised simplex method, we don't need to develop the entire simplex table. In this method, only the necessary information required for completing the current simplex iteration is updated and stored. The revised simplex method is a powerful tool to solve many problems that involve a large number of decision variables and constraints, such as inventory, diet, product mix, and trim loss problems.

5.3 COMPUTATIONAL PROCEDURE FOR SOLVING LPP USING REVISED SIMPLEX METHOD

The following steps are used to solve an LPP using the revised simplex method:

1. Bring the given LPP to the canonical form, such that the identity matrix (I) is the submatrix of the coefficient matrix A.

2. Identify the basic variables $x_1, x_2, \ldots, x_m$; the corresponding costs C_B^T; the basic solution X_B; and the basis matrix B. Find B^{-1}. Since, initially $B = I$, therefore, $B^{-1} = I$.
3. Tabulate the identified data from step 2 as shown here:

Table no.	C_B	Basic variable	$X_B = C_B^T B^{-1}$	B^{-1}	Non-basic variables $x_{m+1}, x_{m+2}, x_j, \ldots, x_n$	Minimum ratio $\frac{X_{B_i}}{\alpha_i^j}$
1	C_{B1}	x_1	X_{B1}		α_i^j	
	…	…	…		…	
	C_{Br}	x_r	X_{Br}	–	α_r^j	–
	…	…	…		…	
	C_{Bm}	x_m	X_{Bm}		α_m^j	
	$c_j - Z_j$			–	–	–

4. To check the optimality of the solution obtained in step 3, calculate Δ_j for non-basic variables x_j (where $j = m+1, m+2, \ldots, n$), using $\Delta_j = c_j - Z_j = c_j - C_B^T B^{-1} A_j$. $C_B^T B^{-1}$ is called the *simplex multiplier*. If $\Delta_j \leq 0$ for all $j = m+1, m+2, \ldots, n$; then the solution obtained is the optimal solution. If at least, one $\Delta_j > 0$, then the variable x_j with the most positive Δ_j value enters the basis. Calculate column α^j to decide the leaving variable x_r (x_r is one of the basic variables $x_1, x_2, \ldots, x_m$). Find $\min_i(x_i/\alpha_i^j, \alpha_i^j > 0)$ for $i = 1, 2, \ldots, m$ and say is equal to x_r/α_r^j for row $i = r$. The variable x_r then leaves the basis. Thus, when x_j enters, the variable x_r leaves the basis.

Notes 1. 'Variable x_j enters the basis' means it becomes a basic variable.
2. 'Variable x_r leaves the basis' means it becomes a non-basic variable.
3. The word 'basis' refers to the list of basic variables.

5. Update the simplex table of step 3 using the data calculated in step 4. The entry at the intersection of the entering and leaving variables is called the pivot entry. Enrectangle the pivot entry. The leaving variable row is called the pivot row and the entering variable column, the pivot column. To obtain new B^{-1} and X_B, interchange the positions of the leaving and entering variables in the updated simplex table. Use the same elementary row operations on the existing B^{-1} and X_B columns that are required to bring '1' to the position of the pivot entry and '0' below and above the entry 1. These computations result in new basic variables, C_B^T, X_B, and B^{-1}.
6. Prepare the next simplex iteration table incorporating computations done in step 5 with the new basic variables, C_B^T, X_B, and B^{-1} and check the optimality of the solution.
7. To obtain the optimal solution, repeat steps 4–6 until all Δ_j become non-positive ($\Delta_j \leq 0$) for non-basic variables.

Note Step 5 provides a method to find the new X_B and B^{-1} for the next simplex table.

Example 5.1 Solve the following LPP using the revised simplex method:

$$\text{Max } Z = -x_1 + 4x_2$$

Subject to $x_1 - x_2 \le 1$

$-2x_1 + x_2 \le 2$

and $x_1, x_2 \ge 0$

Solution The problem can be written in the standard form as

$$\text{Max } Z = -x_1 + 4x_2 + 0\cdot x_3 + 0\cdot x_4$$

Subject to $x_1 - x_2 + x_3 + 0\cdot x_4 = 1$

$-2x_1 + x_2 + 0\cdot x_3 + x_4 = 2$

and $x_1, x_2, x_3, x_4 \ge 0$, x_3 and x_4 are slack variables.

Here, the coefficient matrix $A = (A_1, A_2, A_3, A_4) = \begin{pmatrix} 1 & -1 & 1 & 0 \\ -2 & 1 & 0 & 1 \end{pmatrix}$.

The basic variables are in the order x_3 and x_4. The basis matrix $B = \begin{pmatrix} 1 & 0 \\ 0 & 1 \end{pmatrix}$,

$B^{-1} = \begin{pmatrix} 1 & 0 \\ 0 & 1 \end{pmatrix}$, $b = \begin{pmatrix} 1 \\ 2 \end{pmatrix}$, and $C^{\mathrm{T}} = (-1, 4, 0, 0)$.

The BFS $X_B = B^{-1}b = (1, 2)$.

Now to check the optimality, calculate Δ_j for non-basic variables (x_1 and x_2).

$$\Delta_j = c_j - Z_j = c_j - C_B^{\mathrm{T}} B^{-1} A_j$$

$$C_B^{\mathrm{T}} B^{-1} = (0, 0)\begin{pmatrix} 1 & 0 \\ 0 & 1 \end{pmatrix} = (0, 0)$$

$$C_B^{\mathrm{T}} B^{-1} A_j = (0, 0)\begin{pmatrix} 1 & -1 \\ -2 & 1 \end{pmatrix} = (0, 0)$$

$$\Delta_j = c_j - Z_j = (-1, 4) - (0, 0) = (-1, 4)$$

The solution is not optimal.

Here, $\Delta_j = 4$ is the most positive for the x_2 column, so x_2 will be the entering variable. For the leaving variable, calculate α^2 (the pivotal column).

$$\alpha^2 = B^{-1}A_2 = \begin{pmatrix} 1 & 0 \\ 0 & 1 \end{pmatrix}\begin{pmatrix} -1 \\ 1 \end{pmatrix} = \begin{pmatrix} -1 \\ 1 \end{pmatrix}$$

Tabulating the aforementioned computations, we get:

Table no.	C_B	Basic variable	X_B	B^{-1}		Non-basic variables x_1	x_2	Minimum ratio $\frac{X_{B_i}}{\alpha_i^j}$
1	0	x_3	1	1	0		−1	…
	0	x_4	2	0	1		[1]	2/1 = 2→
	$c_j - Z_j$			–		−1	4↑	–

Now use the same elementary row operations on columns X_B and B^{-1} of the aforegiven table, which are needed to bring zero above the pivot entry 1. Incorporating these computations and interchanging the position of the entering variable x_2 and the leaving variable x_4, we get the following table:

Table no.	C_B	Basic variable	X_B	B^{-1}		Non-basic variables x_1	x_4	Minimum ratio $\frac{X_{B_i}}{\alpha_i^j}$
2	0 4	x_3 x_2	3 2	1 0	1 1			
	$c_j - Z_j$			–		–		–

Now, to check the optimality of the solution of this table, we calculate Δ_j for non-basic variables (x_1, x_4).

$$\Delta_j = c_j - Z_j = c_j - C_B^{\mathrm{T}} B^{-1} A_j,\ j = 1, 4$$

$$C_B^{\mathrm{T}} B^{-1} = (0, 4)\begin{pmatrix} 1 & 1 \\ 0 & 1 \end{pmatrix} = (0, 4)$$

$$C_B^{\mathrm{T}} B^{-1} A_j = (0, 4)\begin{pmatrix} 1 & 0 \\ -2 & 1 \end{pmatrix} = (-8, 4)$$

$$\Delta_j = c_j - Z_j = (-1, 0) - (-8, 4) = (7, -4)$$

The solution is not optimal.

Here, $\Delta_j = 7$ is the most positive for the x_1 column, so x_1 will be the entering variable. For the leaving variable, calculate α^1 (the pivotal column).

$$\alpha^1 = B^{-1} A_1 = \begin{pmatrix} 1 & 1 \\ 0 & 1 \end{pmatrix}\begin{pmatrix} 1 \\ -2 \end{pmatrix} = \begin{pmatrix} -1 \\ -2 \end{pmatrix}$$

Tabulating the aforementioned computations, we get the following table:

Table no.	C_B	Basic variable	X_B	B^{-1}		Non-basic variables x_1	x_4	Minimum ratio $\frac{X_{B_i}}{\alpha_i^j}$
3	0 4	x_3 x_2	3 2	1 0	1 1	–1 –2		–
	$c_j - Z_j$			–		7↑		–

In the aforegiven table, x_1 enters as $\Delta_1 = 7$, but there is no variable to leave the basis because of all entries in the α^1 columnt being negative. Hence, the solution is unbounded.

Example 5.2 Solve the following LPP using the revised simplex method:

$$\text{Max } Z = -2x_1 - x_2$$
$$3x_1 + x_2 = 3$$
$$4x_1 + 3x_2 \geq 6$$
$$x_1 + 2x_2 \leq 4$$
$$x_1, x_2 \geq 0$$

Solution LPP can be written in the standard form as follows:

$$\text{Max } Z = -2x_1 - x_2 + 0{\cdot}x_3 + 0{\cdot}x_4$$
$$\text{Subject to} \quad 3x_1 + x_2 = 3$$
$$4x_1 + 3x_2 - x_3 = 6$$

$$x_1 + 2x_2 + x_4 = 4$$
$$x_1, x_2, x_3, x_4 \geq 0, x_3 \text{ is the surplus, and } x_4 \text{ is the slack variable}$$

Stating the problem in canonical form by introducing artificial variables and modifying the objective function by introducing M, a large positive number, we have,

$$\text{Max } Z = -2x_1 - x_2 + 0{\cdot}x_3 + 0{\cdot}x_4 - M{\cdot}x_5 - M{\cdot}x_6$$
$$\text{Subject to } 3x_1 + x_2 + x_5 = 3$$
$$4x_1 + 3x_2 - x_3 + x_6 = 6$$
$$x_1 + 2x_2 + x_4 = 4$$

and $x_1, x_2, x_3, x_4, x_5, x_6 \geq 0$, x_3 is the surplus, x_4 is the slack, and x_5 and x_6 are the artificial variables.

Here, the coefficient matrix $A = (A_1, A_2, A_3, A_5, A_6, A_4) = \begin{pmatrix} 3 & 1 & 0 & 1 & 0 & 0 \\ 4 & 3 & -1 & 0 & 1 & 0 \\ 1 & 2 & 0 & 0 & 0 & 1 \end{pmatrix}$.

The basic variables are in the order x_5, x_6, and x_4. The basic matrix $B = \begin{pmatrix} 1 & 0 & 0 \\ 0 & 1 & 0 \\ 0 & 0 & 1 \end{pmatrix}$,

$B^{-1} = \begin{pmatrix} 1 & 0 & 0 \\ 0 & 1 & 0 \\ 0 & 0 & 1 \end{pmatrix}$, $b = \begin{pmatrix} 3 \\ 6 \\ 4 \end{pmatrix}$, and $C^{\mathrm{T}} = (-2, -1, 0, -M, -M, 0)$. The basic feasible solution (BFS) $X_B = B^{-1}b = (3, 6, 4)$. Now to check the optimality, calculate Δ_j for non-basic variables (x_1, x_2, x_3).

$$\Delta_j = c_j - Z_j = c_j - C_B^{\mathrm{T}} B^{-1} A_j,\ j = 1, 2, 3$$

$$C_B^{\mathrm{T}} B^{-1} = (-M, -M, 0) \begin{pmatrix} 1 & 0 & 0 \\ 0 & 1 & 0 \\ 0 & 0 & 1 \end{pmatrix} = (-M, -M, 0)$$

$$C_B^{\mathrm{T}} B^{-1} A_j = (-M, -M, 0) \begin{pmatrix} 3 & 1 & 0 \\ 4 & 3 & -1 \\ 1 & 2 & 0 \end{pmatrix} = (-7M, -4M, M)$$

$$\Delta_j = c_j - Z_j = (-2, -1, 0) - (-7M, -4M, M) = (-2 + 7M, -1 + 4M, -M)$$

The solution is not optimal.

Here $\Delta_j = -2 + 7M$ is the most positive for the x_1 column, so x_1 will be the entering variable. For the leaving variable, calculate α^1 (the pivotal column).

$$\alpha^1 = B^{-1}A_1 = \begin{pmatrix} 1 & 0 & 0 \\ 0 & 1 & 0 \\ 0 & 0 & 1 \end{pmatrix} \begin{pmatrix} 3 \\ 4 \\ 1 \end{pmatrix} = \begin{pmatrix} 3 \\ 4 \\ 1 \end{pmatrix}$$

Tabulating the aforementioned computations, we get

Table no.	C_B	Basic variable	X_B	B^{-1}			Non-basic variables x_1	x_2	x_3	Minimum ratio $\frac{X_{B_i}}{\alpha_i^j}$
1	$-M$	x_5	3	1	0	0	[3]			$3/3 = 1 \rightarrow$
	$-M$	x_6	6	0	1	0	4			$6/4 = 1.5$
	0	x_4	4	0	0	1	1			$4/1 = 4$
	$c_j - Z_j$			–			$-2 + 7M \uparrow$	$-1 + 4M$	$-M$	–

Now, use the same elementary row operations on the X_B and B^{-1} columns, which are needed to bring 1 to the pivotal position $\boxed{3}$ and zeros below 1, thus obtained. Incorporating these computations and interchanging the position of the entering variable x_1 and the leaving variable x_5, we get the next simplex table:

Table no.	C_B	Basic variable	X_B	B^{-1}			Non-basic variables x_5	x_2	x_3	Minimum ratio $\frac{X_{B_i}}{\alpha_i^j}$
2	−2	x_1	1	1/3	0	0				
	−M	x_6	2	−4/3	1	0		–		–
	0	x_4	3	−1/3	0	1				
	$c_j - Z_j$							–		–

Now, to check the optimality of the solution of the aforegiven table, we calculate Δ_j for non-basic variables x_2 and x_3 (omitting calculations for x_5 being an artificial variable).

$$\Delta_j = c_j - Z_j = c_j - C_B^T B^{-1} A_j,\ j = 2, 3$$

$$C_B^T B^{-1} = (-2, -M, 0)\begin{pmatrix} 1/3 & 0 & 0 \\ -4/3 & 1 & 0 \\ -1/3 & 0 & 1 \end{pmatrix} = \left(-\frac{2}{3} + \frac{4M}{3}, -M, 0\right)$$

$$C_B^T B^{-1} A_j = \left(-\frac{2}{3} + \frac{4M}{3}, -M, 0\right)\begin{pmatrix} 1 & 0 \\ 3 & -1 \\ 2 & 0 \end{pmatrix} = \left(-\frac{2}{3} - \frac{5M}{3}, M\right)$$

$$\Delta_j = c_j - Z_j = (-1, 0) - \left(-\frac{2}{3} - \frac{5M}{3}, M\right) = \left(-\frac{1}{3} + \frac{5M}{3}, -M\right)$$

Therefore, the solution is not optimal.

Here $\Delta_j = -1/3 + 5M/3$ is the most positive for the x_2 column, so x_2 will be the entering variable. For the leaving variable, calculate α^2 (the pivotal column).

$$\alpha^2 = B^{-1} A_2 = \begin{pmatrix} 1/3 & 0 & 0 \\ -4/3 & 1 & 0 \\ -1/3 & 0 & 1 \end{pmatrix}\begin{pmatrix} 1 \\ 3 \\ 2 \end{pmatrix} = \begin{pmatrix} 1/3 \\ 5/3 \\ 5/3 \end{pmatrix}$$

Tabulating the aforementioned calculations, we get

Table no.	C_B	Basic variable	X_B	B^{-1}			Non-basic variables x_5	x_2	x_3	Minimum ratio $\frac{X_{B_i}}{\alpha_i^j}$
3	−2	x_1	1	1/3	0	0		1/3		$\frac{1}{1/3} = 3$
	−M	x_6	2	−4/3	1	0		$\boxed{5/3}$		$\frac{2}{5/3} = \frac{6}{5} \rightarrow$
	0	x_4	3	−1/3	0	1		5/3		$\frac{3}{5/3} = \frac{9}{5}$
	$c_j - Z_j$						–	−1/3+5M/3↑	−3	–

Now, use the same elementary row operations on columns X_B and B^{-1} of this table, which are needed to bring 1 to the position of the pivot entry $\boxed{5/3}$ and zeros below and above this 1, thus obtained. Incorporating these computations and interchanging the position of the entering variable x_2 and the leaving variable x_6, we get the following table:

Table no.	C_B	Basic variable	X_B	B^{-1}			Non-basic variables x_5 x_6 x_3	Minimum ratio $\frac{X_{B_i}}{\alpha_i^j}$
4	−2	x_1	3/5	3/5	−1/5	0		
	−1	x_2	6/5	−4/5	3/5	0	–	–
	0	x_4	1	1	−1	1		
		$c_j - Z_j$		–			–	–

Now to check the optimality of the solution of this table, we calculate Δ_j for the non-basic variable x_3 (omitting calculations for both x_5 and x_6 being artificial variables).

$$\Delta_j = c_j - Z_j = c_j - C_B^T B^{-1} A_j$$

$$C_B^T B^{-1} = (-2, -1, 0)\begin{pmatrix} 3/5 & -1/5 & 0 \\ -4/5 & 3/5 & 0 \\ 1 & -1 & 1 \end{pmatrix} = \left(-\frac{2}{5}, -\frac{1}{5}, 0\right)$$

$$C_B^T B^{-1} A_j = \left(-\frac{2}{5}, -\frac{1}{5}, 0\right)\begin{pmatrix} 0 \\ -1 \\ 0 \end{pmatrix} = \left(\frac{1}{5}\right)$$

$$\Delta_j = c_j - Z_j = (0) - \left(\frac{1}{5}\right) = \left(-\frac{1}{5}\right)$$

The solution is optimal as $\Delta_j \le 0$, for the non-basic variable x_3.
Hence, the optimal solution is $x_1 = 3/5$, $x_2 = 6/5$, and Max $Z = -12/5$.

Example 5.3 Solve the following LPP using the revised simplex method:

Max $Z = 2x_1 - x_2 + x_3$

Subject to $3x_1 + x_2 + x_3 \le 60$

$x_1 - x_2 + 2x_3 \le 10$

$x_1 + x_2 - x_3 \le 20$

and $x_1, x_2, x_3 \ge 0$

Solution The problem can be written in the standard form as follows:

Max $Z = 2x_1 - x_2 + x_3 + 0{\cdot}x_4 + 0{\cdot}x_5 + 0{\cdot}x_6$

Subject to $3x_1 + x_2 + x_3 + x_4 + 0{\cdot}x_5 + 0{\cdot}x_6 = 60$

$x_1 - x_2 + 2x_3 + 0{\cdot}x_4 + x_5 + 0{\cdot}x_6 = 10$

$x_1 + x_2 - x_3 + 0{\cdot}x_4 + 0{\cdot}x_5 + x_6 = 20$

$x_1, x_2, x_3, x_4, x_5, x_6 \ge 0$, x_4, x_5, and x_6 are slack variables.

Here, the coefficient matrix $A = (A_1, A_2, A_3, A_4, A_5, A_6) = \begin{pmatrix} 3 & 1 & 1 & 1 & 0 & 0 \\ 1 & -1 & 2 & 0 & 1 & 0 \\ 1 & 1 & -1 & 0 & 0 & 1 \end{pmatrix}$

The basic variables are x_4, x_5, and x_6. $B = \begin{pmatrix} 1 & 0 & 0 \\ 0 & 1 & 0 \\ 0 & 0 & 1 \end{pmatrix}$, $B^{-1} = \begin{pmatrix} 1 & 0 & 0 \\ 0 & 1 & 0 \\ 0 & 0 & 1 \end{pmatrix}$, $b = \begin{pmatrix} 60 \\ 10 \\ 20 \end{pmatrix}$, and

$$C_B^{\mathrm{T}} = (0, 0, 0).$$

The BFS $X_B = B^{-1}b = (60, 10, 20)$.

Now to check the optimality, calculate Δ_j for the non-basic variables (x_1, x_2, x_3).

$$\Delta_j = c_j - Z_j = c_j - C_B^{\mathrm{T}}B^{-1}A_j,\ j = 1, 2, 3$$

$$C_B^{\mathrm{T}}B^{-1} = (0, 0, 0)\begin{pmatrix} 1 & 0 & 0 \\ 0 & 1 & 0 \\ 0 & 0 & 1 \end{pmatrix} = (0, 0, 0)$$

$$C_B^{\mathrm{T}}B^{-1}A_j = (0, 0, 0)\begin{pmatrix} 3 & 1 & 1 \\ 1 & -1 & 2 \\ 1 & 1 & -1 \end{pmatrix} = (0, 0, 0)$$

$$\Delta_j = c_j - Z_j = (2, -1, 1) - (0, 0, 0) = (2, -1, 1)$$

The solution is not optimal.

Here $\Delta_j = 2$ is the most positive for the x_1 column, so x_1 will be the entering variable. For the leaving variable, calculate α^1 (the pivotal column).

$$\alpha^1 = B^{-1}A_1 = \begin{pmatrix} 1 & 0 & 0 \\ 0 & 1 & 0 \\ 0 & 0 & 1 \end{pmatrix}\begin{pmatrix} 3 \\ 1 \\ 1 \end{pmatrix} = \begin{pmatrix} 3 \\ 1 \\ 1 \end{pmatrix}$$

Tabulating the aforementioned computations, we get

Table no.	C_B	Basic variable	X_B	B^{-1}			Non-basic variables x_1	x_2	x_3	Minimum ratio $\frac{X_{B_i}}{\alpha_i^j}$
1	0	x_4	60	1	0	0	3			60/3 = 20
	0	x_5	10	0	1	0	[1]			10/1 = 10 →
	0	x_6	20	0	0	1	1			20/1 = 20
		$c_j - Z_j$			–		2↑	−1	1	–

Now, use the same elementary row operations on columns X_B and B^{-1} of the aforegiven table, which are needed to bring zeros below and above the pivot entry 1 and interchange the position of the entering variable x_1 and the leaving variable x_5. Then, we get the following table:

Table no.	C_B	Basic variable	X_B	B^{-1}			Non-basic variables x_5	x_2	x_3	Minimum ratio $\frac{X_{B_i}}{\alpha_i^j}$
2	0	x_4	30	1	−3	0				
	2	x_1	10	0	1	0		–		–
	0	x_6	10	0	−1	1				
		$c_j - Z_j$			–			–		–

Now, to check the optimality of the solution of this table again, we calculate Δ_j for the non-basic variables (x_5, x_2, x_3).

$$\Delta_j = c_j - Z_j = c_j - C_B^T B^{-1} A_j,\ j = 5, 2, 3$$

$$C_B^T B^{-1} = (0, 2, 0) \begin{pmatrix} 1 & -3 & 0 \\ 0 & 1 & 0 \\ 0 & -1 & 1 \end{pmatrix} = (0,\ 2,\ 0)$$

$$C_B^T B^{-1} A_j = (0, 2, 0) \begin{pmatrix} 0 & 1 & 1 \\ 1 & -1 & 2 \\ 0 & 1 & -1 \end{pmatrix} = (2, -2, 4)$$

$$\Delta_j = c_j - Z_j = (0, -1, 1) - (2, -2, 4) = (-2, 1, -3)$$

The solution is not optimal.

Here, $\Delta_j = 1$ is the most positive for the x_2 column, so x_2 will be the entering variable. For the leaving variable, calculate α^2 (the pivotal column).

$$\alpha^2 = B^{-1} A_2 = \begin{pmatrix} 1 & -3 & 0 \\ 0 & 1 & 0 \\ 0 & -1 & 1 \end{pmatrix} \begin{pmatrix} 1 \\ -1 \\ 1 \end{pmatrix} = \begin{pmatrix} 4 \\ -1 \\ 2 \end{pmatrix}$$

Tabulating the aforementioned computations, we get

Table no.	C_B	Basic variable	X_B	B^{-1}			Non-basic variables			Minimum ratio $\frac{X_{B_i}}{\alpha_i^j}$
							x_5	x_2	x_3	
3	0	x_4	30	1	−3	0		4		30/4 = 7.5
	2	x_1	10	0	1	0		−1		...
	0	x_6	10	0	−1	1		[2]		10/2 = 5 →
	$c_j - Z_j$			–			−2	1↑	−3	–

Now, use the same elementary row operations on columns X_B and B^{-1} of the aforegiven table, which are needed to bring 1 to the position of the pivot entry [2] and zeros below and above the pivot entry 1, thus obtained and incorporating and interchanging the position of the entering variable x_2 and the leaving variable x_6. Then, we get the following table:

Table no.	C_B	Basic variable	X_B	B^{-1}			Non-basic variables			Minimum ratio $\frac{X_{B_i}}{\alpha_i^j}$
							x_5	x_6	x_3	
4	0	x_4	10	1	−1	−2				
	2	x_1	15	0	1/2	1/2		–		–
	−1	x_2	5	0	−1/2	1/2				
	$c_j - Z_j$			–				–		–

Now, to check the optimality of the solution of the aforegiven table again, we calculate Δ_j for the non-basic variables (x_5, x_6, x_3).

$$\Delta_j = c_j - Z_j = c_j - C_B^T B^{-1} A_j$$

$$C_B^T B^{-1} = (0, 2, -1)\begin{pmatrix} 1 & -1 & -2 \\ 0 & \tfrac{1}{2} & \tfrac{1}{2} \\ 0 & -\tfrac{1}{2} & \tfrac{1}{2} \end{pmatrix} = (0, 3/2, 1/2)$$

$$C_B^T B^{-1} A_j = (0, 3/2, 1/2)\begin{pmatrix} 0 & 0 & 1 \\ 1 & 0 & 2 \\ 0 & 1 & -1 \end{pmatrix} = (3/2, 1/2, 5/2)$$

$\Delta_j = c_j - Z_j = (0, 0, 1) - (3/2, 1/2, 5/2) = (-3/2, -1/2, -3/2)$. The solution is optimal as $\Delta_j \le 0$ for all non-basic variables x_j, $j = 5, 6, 3$.
Hence, the optimal solution is $x_1 = 15$, $x_2 = 5$, $x_3 = 0$, and Max $Z = 25$.

Example 5.4 A firm manufactures two types of electric equipment E_1 and E_2. Profit per unit of E_1 and E_2 is ₹150 and ₹200, respectively. Each unit of E_1 requires 3 motors and 2 transformers, whereas each unit of E_2 requires 2 motors and 4 transformers. The equipment E_1 is of a special type and is equipped with a voltage stabilizer. The monthly availability of motors, transformers, and stabilizers is not more than 210, 300, and 60 respectively. What quantity of E_1 and E_2 should be manufactured, so that the profit earned is the maximum?

(a) Identify the decision variables.
(b) Construct the objective function.
(c) Identify the constraints.
(d) Formulate the problem in the linear programming (LP) model.
(e) Write the formulated problem in the standard form, identifying the slack (or surplus) variables used.
(f) Sketch the graph of the set of feasible solutions (S_F) and obtain the optimal solution using the graphical method.
(g) Solve it using the revised simplex method.

Solution (a) Let x_1 and x_2 number of units of equipment E_1 and E_2 respectively be manufactured every month. Then, x_1 and x_2 are the decision variables.

(b) In producing x_1 and x_2 units of E_1 and E_2 respectively, the total profit received will be ₹$(150x_1 + 200x_2)$. The firm's objective is to find x_1 and x_2 in such a way that $150x_1 + 200x_2$ is maximum.

Thus, the objective function of the problem is

$$\text{Max } Z = 150x_1 + 200x_2 \tag{5.1}$$

(c) In Eq. (5.1), x_1 and x_2 are restricted on account of availabilities of resources (limited availability of motors, transformers, and stabilizers).

As only 210 motors are available, the total number of motors required for x_1 and x_2 units of E_1 and E_2, that is, $3x_1 + 2x_2$ cannot exceed that number. Thus, we have

	$3x_1 + 2x_2 \le 210$	(motor availability constraints)
Similarly,	$2x_1 + 4x_2 \le 300$	(transformer availability constraints)
and	$x_1 \le 60$	(stabilizer availability constraints)

Further, the number of units of E_1 and E_2 produced cannot be negative, so we have $x_1, x_2 \ge 0$.

(d) The formulation of the problem in the LPP model is as follows:

$$\text{Max } Z = 150x_1 + 200x_2$$

$$\text{Subject to } 3x_1 + 2x_2 \leq 210$$
$$2x_1 + 4x_2 \leq 300$$
$$x_1 \leq 60$$
$$\text{and } x_1, x_2 \geq 0$$

(e) The standard form of the aforementioned LPP is as follows:

$$\text{Max } Z = 150x_1 + 200x_2 \quad (5.2)$$

$$\text{Subject to } 3x_1 + 2x_2 + x_3 = 210 \quad (5.3)$$
$$2x_1 + 4x_2 + x_4 = 300 \quad (5.4)$$
$$x_1 + x_5 = 60 \quad (5.5)$$

and $x_1, x_2, x_3, x_4, x_5 \geq 0$, (where x_3, x_4, and x_5 are slack variables)

(f) Graphical method: We will use the corner point method to obtain the graphical solution. To find a point $X=(x_1, x_2)^T$ satisfying Eqs (5.3), (5.4), and (5.5) such that the value of $Z=150x_1+200x_2$ is maximum, let us sketch on a graph the inequalities given in Eqs (5.3), (5.4), and (5.5) to get the bounded region S_F (the set of feasible solutions) as shown Fig. 5.1. The coordinates of points O(0, 0), A(60, 0), B(60, 15), C(30, 60), and D(0, 75), which form the vertices (corner points) of S_F, and the value of Z is calculated from this bounded region.

Points	Z
O(0, 0)	$Z = 150 \times 0 + 200 \times 0 = 0$
A(60, 0)	$Z = 150 \times 60 + 200 \times 0 = 9000$
B(60, 15)	$Z = 150 \times 60 + 200 \times 15 = 12,000$
C(30, 60)	$Z = 150 \times 30 + 200 \times 60 = 16,500$ (Max)
D(0, 75)	$Z = 150 \times 0 + 200 \times 75 = 15,000$

Among the Z values given by the aforegiven table, ₹16,500 is the maximum. Thus, the maximum profit is ₹16,500. A maximum of 30 E_1 and 60 E_2 should be manufactured to achieve this profit.

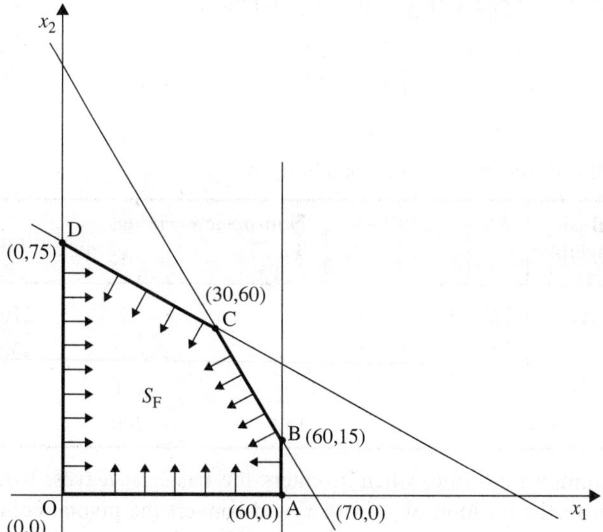

Figure 5.1 Corner point graphical solution

(g) Revised simplex method:

Here, the coefficient matrix is
$$A = \begin{pmatrix} x_1 & x_2 & x_3 & x_4 & x_5 \\ 3 & 2 & 1 & 0 & 0 \\ 2 & 4 & 0 & 1 & 0 \\ 1 & 0 & 0 & 0 & 1 \\ A_1 & A_2 & A_3 & A_4 & A_5 \end{pmatrix}$$

Basic variables are $(x_3, x_4, \text{and} x_5)$ and the basic matrix is $B = \begin{pmatrix} 1 & 0 & 0 \\ 0 & 1 & 0 \\ 0 & 0 & 1 \end{pmatrix}$, $B^{-1} = \begin{pmatrix} 1 & 0 & 0 \\ 0 & 1 & 0 \\ 0 & 0 & 1 \end{pmatrix}$,

$b = \begin{pmatrix} 210 \\ 300 \\ 60 \end{pmatrix}$, and $C^{\mathrm{T}} = (150, 200, 0, 0, 0)$. The BFS $X_B = B^{-1}b = (210, 300, 60)$.

Now to check the optimality, calculate Δ_j for the non-basic variables (x_1, x_2).

$$\Delta_j = c_j - Z_j = c_j - C_B^{\mathrm{T}} B^{-1} A_j$$

$$C_B^{\mathrm{T}} B^{-1} = (0, 0, 0) \begin{pmatrix} 1 & 0 & 0 \\ 0 & 1 & 0 \\ 0 & 0 & 1 \end{pmatrix} = (0, 0, 0)$$

$$C_B^{\mathrm{T}} B^{-1} A_j = (0, 0, 0) \begin{pmatrix} 3 & 2 \\ 2 & 4 \\ 1 & 0 \end{pmatrix} = (0, 0)$$

$$\Delta_j = c_j - Z_j = (150, 200) - (0, 0) = (150, 200)$$

The solution is not optimal.

Here $\Delta_j = 200$ is the most positive for the x_2 column, so x_2 will be the entering variable. For the leaving variable, calculate α^2 (the pivotal column).

$$\alpha^2 = B^{-1} A_2 = \begin{pmatrix} 1 & 0 & 0 \\ 0 & 1 & 0 \\ 0 & 0 & 1 \end{pmatrix} \begin{pmatrix} 2 \\ 4 \\ 0 \end{pmatrix} = \begin{pmatrix} 2 \\ 4 \\ 0 \end{pmatrix}$$

Tabulating the aforementioned computations, we get

Table no.	C_B	Basic variable	X_B	B^{-1}			Non-basic variables x_1	x_2	Minimum ratio $\frac{X_{B_i}}{\alpha_i^j}$
1	0	x_3	210	1	0	0		2	210/2 = 105
	0	$\leftarrow x_4$	300	0	1	0		[4]	300/4 = 75
	0	x_5	60	0	0	1		0	
		$c_j - Z_j$	–				150	200↑	–

By the minimum ratio rule, when x_2 enters the basis, x_4 leaves. Now for the next table, interchange the position of x_4 with x_2 and convert the pivotal column α^2 into an identity column by applying the appropriate row operations on the solution column and B^{-1} entries only.

For this, the new x_2 row is obtained by dividing the old x_4 row by 4 as (75; 0, 1/4, 0). The new x_3 row is (210; 1, 0, 0) – 2(75; 0, 1/4, 0) = (60; 0, –1/2, 0). The new x_5 row = (60; 0, 0, 1) – 0(75; 0, 1/4, 0) = (60; 0, 0, 1). Using these computations, the next BFS is given by the following simplex table:

Table no.	C_B	Basic variable	X_B	B^{-1}			Non-basic variables x_1 x_4	Minimum ratio $\frac{X_{B_i}}{\alpha_i^j}$
2	0	x_3	60	1	–1/2	0		
	200	x_2	75	0	1/4	0	–	–
	0	x_5	60	0	0	1		
	$c_j - Z_j$			–			–	–

Now again to check the optimality of the solution of the aforegiven table, we calculate Δ_j for non-basic variables x_1 and x_4.

$$C_B^T B^{-1} = (0, 200, 0)\begin{pmatrix} 1 & -\frac{1}{2} & 0 \\ 0 & \frac{1}{4} & 0 \\ 0 & 0 & 1 \end{pmatrix} = (0,\ 50,\ 0)$$

$$C_B^T B^{-1} A_j = (0, 50, 0)\begin{pmatrix} 3 & 0 \\ 2 & 1 \\ 1 & 0 \end{pmatrix} = (100,\ 50)$$

$$\Delta_j = c_j - Z_j = (150, 0) - (100, 50) = (50, -50)$$

The solution is not optimal.

Here $\Delta_j = 50$ is the most positive for the x_1 column, so x_1 will be the entering variable. For the leaving variable, calculate α^1 (the pivotal column).

$$\alpha^1 = B^{-1}A_1 = \begin{pmatrix} 1 & -\frac{1}{2} & 0 \\ 0 & \frac{1}{4} & 0 \\ 0 & 0 & 1 \end{pmatrix}\begin{pmatrix} 3 \\ 2 \\ 1 \end{pmatrix} = \begin{pmatrix} 2 \\ \frac{1}{2} \\ 1 \end{pmatrix}$$

Tabulating the aforementioned computations, we get

Table no.	C_B	Basic variable	X_B	B^{-1}			Non-basic variables x_1	x_4	Minimum ratio $\frac{X_{B_i}}{\alpha_i^j}$
3	0	$\leftarrow x_3$	60	1	–1/2	0	[2]		60/2 = 30
	200	x_2	75	0	1/4	0	1/2		75/(1/2) = 150
	0	x_5	60	0	0	1	1		60/1 = 60
	$c_j - Z_j$			–			50↑	–50	–

By the minimum ratio rule, when x_1 enters, x_3 leaves the basis. For the next table, interchange x_3 with x_1. To get the new x_1 row, divide the old x_3 row by the pivot entry 2 to get (30; 1/2, –1/4, 0).

The new x_2 row is (75, 0, 1/4, 0) – (1/2)(30, 1/2, –1/4, 0) = (60, –1/4, 3/8, 0) and the new x_5 row is (60, 0, 0, 1) – (30, 1/2, –1/4, 0) = (30, –1/2, 1/4, 1).

Using these computations, the next BFS is given by the following simplex table:

Table no.	C_B	Basic variable	X_B	B^{-1}			Non-basic variables x_3	x_4	Minimum ratio $\frac{X_{B_i}}{\alpha_i^j}$
4	150	x_1	30	1/2	−1/4	0			
	200	x_2	60	−1/4	3/8	0	–		–
	0	x_5	30	−1/2	1/4	1			
	$c_j - Z_j$			–			–		–

Now again to check the optimality of the solution of table 4, we calculate Δ_j for non-basic variables x_3 and x_4.

$$C_B^{\mathrm{T}}B^{-1} = (150, 200, 0)\begin{pmatrix} \frac{1}{2} & -\frac{1}{4} & 0 \\ -\frac{1}{4} & \frac{3}{8} & 0 \\ -\frac{1}{2} & \frac{1}{4} & 1 \end{pmatrix} = (25, 37.5, 0)$$

$$C_B^{\mathrm{T}}B^{-1}A_j = (25, 37.5, 0)\begin{pmatrix} 1 & 0 \\ 0 & 1 \\ 0 & 0 \end{pmatrix} = (25, 37.5)$$

$$\Delta_j = c_j - Z_j = (0, 0) - (25, 37.5) = (-25, -37.5)$$

The solution is optimal as $\Delta_j \leq 0$ for all non-basic variables x_j.

Hence, the optimal solution is $x_1 = 30$, $x_2 = 60$, and Max $Z = 16{,}500$.

Example 5.5 A company manufactures two types of wood items, W_1 and W_2, which have to go through machines M_1, M_2, and M_3 to attain their final shape. Each W_1 item requires 4 h on machine M_1, 5 h on machine M_2, and 3 h on machine M_3. Each W_2 item requires 5 h on machine M_1, 2 h on machine M_2, and 8 h on machine M_3. The total time available per day on machine M_1 is at least 10 h, on M_2 is 10 h, and on M_3 is 12 h. The profit per unit of wood items W_1 and W_2 is ₹5 and ₹3 respectively. Formulate the problem in the LP model to maximize the profit and solve it using the revised simplex method.

Solution Let x_1 and x_2 be the number of items of W_1 and W_2 to be manufactured, respectively, to maximize the profit. We have,

$$Z = 5x_1 + 3x_2$$

Under the given restrictions

$$4x_1 + 5x_2 \geq 10 \text{ (constraint on machine } M_1)$$
$$5x_1 + 2x_2 \leq 10 \text{ (constraint on machine } M_2)$$
$$3x_1 + 8x_2 \leq 12 \text{ (constraint on machine } M_3)$$

Again as the number of wood items cannot be negative, we have

$$x_1, x_2 \geq 0$$

The formulation of the LP model is as follows:

$$\begin{aligned} &\text{Max } Z = 5x_1 + 3x_2 \\ &\text{Subject to } \quad 4x_1 + 5x_2 \geq 10 \\ &\qquad\qquad\quad 5x_1 + 2x_2 \leq 10 \\ &\qquad\qquad\quad 3x_1 + 8x_2 \leq 12 \\ &\text{and} \qquad\quad\; x_1, x_2 \geq 0 \end{aligned}$$

Revised simplex method: The standard form of this LPP is as follows:

Max $Z = 5x_1 + 3x_2 + 0{\cdot}x_3 + 0{\cdot}x_4 + 0{\cdot}x_5 - Mx_6$

Subject to $4x_1 + 5x_2 - x_3 + x_6 = 10$

$5x_1 + 2x_2 + x_4 = 10$

$3x_1 + 8x_2 + x_5 = 12$

and $x_1, x_2, x_3, x_4, x_5, x_6 \geq 0$, x_3 is the surplus variable; x_4 and x_5 are slack variables, and x_6 is the artificial variable

Here, the coefficient matrix is $A = \begin{pmatrix} x_1 & x_2 & x_3 & x_6 & x_4 & x_5 \\ 4 & 5 & -1 & 1 & 0 & 0 \\ 5 & 2 & 0 & 0 & 1 & 0 \\ 3 & 8 & 0 & 0 & 0 & 1 \\ A_1 & A_2 & A_3 & A_6 & A_4 & A_5 \end{pmatrix}$

The basic variables are (x_6, x_4, x_5) and the basic matrix is $B = \begin{pmatrix} 1 & 0 & 0 \\ 0 & 1 & 0 \\ 0 & 0 & 1 \end{pmatrix}$, $B^{-1} = \begin{pmatrix} 1 & 0 & 0 \\ 0 & 1 & 0 \\ 0 & 0 & 1 \end{pmatrix}$,

$b = \begin{pmatrix} 10 \\ 10 \\ 12 \end{pmatrix}$ and $C_B^{\mathrm{T}} = (-M, 0, 0)$ and non-basic variables are (x_1, x_2, x_3).

The BFS, $X_B = B^{-1}b = (10, 10, 12)$.

Now to check the optimality, calculate Δ_j for the non-basic variables (x_1, x_2, x_3).

$$\Delta_j = c_j - Z_j = c_j - C_B^{\mathrm{T}} B^{-1} A_j$$

$$C_B^{\mathrm{T}} B^{-1} = (-M, 0, 0) \begin{pmatrix} 1 & 0 & 0 \\ 0 & 1 & 0 \\ 0 & 0 & 1 \end{pmatrix} = (-M, 0, 0)$$

$$C_B^{\mathrm{T}} B^{-1} A_j = (-M, 0, 0) \begin{pmatrix} 4 & 5 & -1 \\ 5 & 2 & 0 \\ 3 & 8 & 0 \end{pmatrix} = (-4M, -5M, M)$$

$$\Delta_j = -c_j - Z_j = (5, 3, 0) - (-4M, -5M, M) = (5 + 4M, 3 + 5M, -M).$$

The solution is not optimal. Here $\Delta_j = 3 + 5M$ is the most positive for the x_2 column, so x_2 will be the entering variable. For the leaving variable, calculate α^2 (the pivotal column).

$$\alpha^2 = B^{-1}A_2 = \begin{pmatrix} 1 & 0 & 0 \\ 0 & 1 & 0 \\ 0 & 0 & 1 \end{pmatrix} \begin{pmatrix} 5 \\ 2 \\ 8 \end{pmatrix} = \begin{pmatrix} 5 \\ 2 \\ 8 \end{pmatrix}$$

Tabulating the aforementioned computations, we get

Table no.	C_B	Basic variable	X_B	B^{-1}			Non-basic variables x_1	x_2	x_3	Minimum ratio $\frac{X_{B_i}}{\alpha_i^j}$
1	$-M$	x_6	10	1	0	0		5		$10/5 = 2$
	0	x_4	10	0	1	0		2		$10/2 = 5$
	0	$\leftarrow x_5$	12	0	0	1		[8]		$12/8 = 1.5$
		$c_j - Z_j$		–			$5 + 4M$	$3 + 5M\uparrow$	$-M$	–

Thus when x_2 becomes the basic variable, x_5 leaves. Here, the enrectangled entry 8 is the pivot entry. Using the requisite pivot operations (elementary row operations) as used in the simplex method to get the next simplex table and interchanging the positions of x_5 and x_2, we get the following table:

Table no.	C_B	Basic variable	X_B	B^{-1}			Non-basic variables x_1	x_5	x_3	Minimum ratio $\frac{X_{B_i}}{\alpha_i^j}$
2	$-M$	x_6	5/2	1	0	−5/8				
	0	x_4	7	0	1	−1/4		–		–
	3	x_2	3/2	0	0	1/8				
		$c_j - Z_j$						–		–

Now again to check the optimality of the solution of this table, we calculate Δ_j for the non-basic variables x_1, x_5, and x_3.

$$C_B^{\mathrm{T}}B^{-1} = (-M, 0, 3)\begin{pmatrix} 1 & 0 & -\frac{5}{8} \\ 0 & 1 & -\frac{1}{4} \\ 0 & 0 & \frac{1}{8} \end{pmatrix} = \left(-M, 0, \frac{5}{8}M + \frac{3}{8}\right)$$

$$C_B^{\mathrm{T}}B^{-1}A_j = \left(-M, 0, \frac{5}{8}M + \frac{3}{8}\right)\begin{pmatrix} 4 & 0 & -1 \\ 5 & 1 & 0 \\ 3 & 0 & 0 \end{pmatrix} = \left(\frac{17}{8}M + \frac{9}{8}, \frac{5}{8}M + \frac{3}{8}, M\right)$$

$$\Delta_j = c_j - Z_j = (5, 0, 0) - \left(\frac{17}{8}M + \frac{9}{8}, \frac{5}{8}M + \frac{3}{8}, M\right) = \left(\frac{17M + 31}{8}, -\frac{5}{8}M - \frac{3}{8}, -M\right)$$

The solution is not optimal.

Here $\Delta_j = (17M+31)/8$ is the most positive for the x_1 column, so x_1 will be the entering variable. For the leaving variable, calculate α^1 (the pivotal column).

$$\alpha^1 = B^{-1}A_1 = \begin{pmatrix} 1 & 0 & -\frac{5}{8} \\ 0 & 1 & -\frac{1}{4} \\ 0 & 0 & \frac{1}{8} \end{pmatrix}\begin{pmatrix} 4 \\ 5 \\ 3 \end{pmatrix} = \begin{pmatrix} \frac{17}{8} \\ \frac{17}{4} \\ \frac{3}{8} \end{pmatrix}$$

Tabulating the aforementioned computations, we get

Table no.	C_B	Basic variable	X_B	B^{-1}			Non-basic variables x_1	x_5	x_3	Minimum ratio $\frac{X_{B_i}}{\alpha_i^j}$
3	$-M$	$\leftarrow x_6$	5/2	1	0	−5/8	**17/8** (boxed)			(5/2)/(17/8) = 20/17→
	0	x_4	7	0	1	−1/4	17/4			7/(17/4) = 28/17
	3	x_2	3/2	0	0	1/8	3/8			(3/2)/(3/8) = 4
		$c_j - Z_j$					$\frac{17M+31}{8}\uparrow$	$-\frac{5}{8}M - \frac{3}{8}$	$-M$	

Thus, when x_1 enters the basis, x_6 leaves. Entry 17/8, which is enrectangled in the aforegiven table, is the pivot entry. Using pivot operations, we get the following table:

Table no.	C_B	Basic variable	X_B	B^{-1}			Non-basic variables x_6 x_5 x_4	Minimum ratio $\frac{X_{B_i}}{\alpha_i^j}$
4	5	x_1	20/17	8/17	0	−5/17		
	0	x_4	2	−2	1	1	–	–
	3	x_2	18/17	−3/17	0	4/17		
	$c_j - Z_j$			–			–	–

Now again to check the optimality of the solution of this table, we calculate Δ_j for all non-basic variables x_5 and x_3 (omit calculations for x_6 as it is an artificial variable).

$$C_B^T B^{-1} = (5, 0, 3)\begin{pmatrix} 8/17 & 0 & -5/17 \\ -2 & 1 & 1 \\ -3/17 & 0 & 4/17 \end{pmatrix} = \left(\frac{31}{17}, 0, -\frac{13}{17}\right)$$

$$C_B^T B^{-1} A_j = \left(\frac{31}{17}, 0, -\frac{13}{17}\right)\begin{pmatrix} 0 & -1 \\ 0 & 0 \\ 1 & 0 \end{pmatrix} = \left(-\frac{13}{17}, -\frac{31}{17}\right)$$

$$\Delta_j = c_j - Z_j = (0, 0) - \left(-\frac{13}{17}, -\frac{31}{17}\right) = \left(\frac{13}{17}, \frac{31}{17}\right)$$

The solution is not optimal.
Here, $\Delta_j = 31/17$ is the most positive for the x_3 column, so x_3 will be the entering variable. For the leaving variable, calculate α^3 (the pivotal column).

$$\alpha^3 = B^{-1}A_3 = \begin{pmatrix} 8/17 & 0 & -5/17 \\ -2 & 1 & 1 \\ -3/17 & 0 & 4/17 \end{pmatrix}\begin{pmatrix} -1 \\ 0 \\ 0 \end{pmatrix} = \begin{pmatrix} -8/17 \\ 2 \\ 3/17 \end{pmatrix}$$

Tabulating the aforementioned computations, we get

Table no.	C_B	Basic variable	X_B	B^{-1}	Non-basic variables x_6	x_5	x_3	Minimum ratio $\frac{X_{B_i}}{\alpha_i^j}$
5	5	x_1	20/17	8/17 0 −5/17			−8/17	…
	0	←x_4	2	−2 1 1			[2]	2/2 = 1→
	3	x_2	18/17	−3/17 0 4/17			3/17	(18/7)/(3/7) = 6
	$c_j - Z_j$			–	−M − 31/17	13/17	31/17↑	–

Thus when x_3 enters the basis, x_4 leave the basis. Entry 2, which is enrectangled in the aforegiven table, is the pivot entry. Using the same pivot operations on the X_B and B^{-1} columns of this table, which are needed to bring 1 to the pivotal position and zeros below and above 1 thus obtained, we get the following table:

Table no.	C_B	Basic variable	X_B	B^{-1}			Non-basic variables x_6 x_5 x_4	Minimum ratio $\frac{X_{B_i}}{\alpha_i^j}$
6	5	x_1	28/17	0	4/17	−1/17		
	0	x_3	1	−1	1/2	1/2	–	–
	3	x_2	15/17	0	−3/34	5/34		
	$c_j - Z_j$			–			–	–

Now again to check the optimality of the solution of the aforegiven table, we calculate Δ_j for the non-basic variables x_5 and x_3.

$$C_B^T B^{-1} = (5, 0, 3)\begin{pmatrix} 0 & 4/17 & -1/17 \\ -1 & 1/2 & 1/2 \\ 0 & -3/34 & 5/34 \end{pmatrix} = \left(0, \frac{31}{34}, \frac{5}{34}\right)$$

$$C_B^T B^{-1} A_j = \left(0, \frac{31}{34}, \frac{5}{34}\right)\begin{pmatrix} 0 & 0 \\ 0 & 1 \\ 1 & 0 \end{pmatrix} = \left(\frac{5}{34}, \frac{31}{34}\right)$$

$$\Delta_j = c_j - Z_j = (0, 0) - \left(\frac{5}{34}, \frac{31}{34}\right) = \left(-\frac{5}{34}, -\frac{31}{34}\right).$$

The solution is optimal as $\Delta_j \le 0$ for all the non-basic variables x_j.

Hence, the optimal solution is $x_1 = 28/17$, $x_2 = 15/17$, and Max $Z = 185/17$.

RECAPITULATION

1. The revised simplex method is extremely useful for obtaining an optimal solution to an LPP when a large number of decision variables are involved.
2. In the revised simplex method, only computations required for the next simplex table are carried out.
3. The revised simplex method uses the same procedure to solve an LPP as we use in simplex method.
4. In the revised simplex method, B^{-1} at any stage is obtained using the same elementary row operations on the rows of B^{-1} of the earlier table, which convert the pivot column α^j into the identity column. The B^{-1}, thus obtained is used for the next simplex table, and is required to determine the optimality conditions for the solution to that table.

EXERCISES

Multiple-choice Questions

5.1 To start with the simplex iteration, the starting basis matrix B is
   (a) I
   (b) $-I$
   (c) zero matrix
   (d) none of these

5.2 The simplex multiplier is
   (a) C_B^T
   (b) $C_B^T B$
   (c) $C_B^T B^{-1}$
   (d) none of these

State True or False

5.1 When an LPP is solved by the revised simplex method,
   (a) the optimal solution may not be a BFS.
   (b) the optimal solution may not be a feasible solution.
   (c) there may exist more than one optimal solution.
   (d) there may exist an unbounded solution.
   (e) there may exist no solution.

Fill in the Blanks

5.1 The basic solution procedure of the revised simplex method and ________ method is the same.
5.2 The revised simplex method requires ________ computer storage and time than the simplex method.
5.3 In the ________ method, it is not necessary to calculate all the entries in the simplex table for each iteration.
5.4 The revised simplex method is more useful in solving an LPP involving ________ number of variables.
5.5 The elementary ________ operations can be used in both simplex and revised simplex methods.

Review Questions

5.1 How is the revised simplex method more useful than the simplex method?
5.2 What is a simplex multiplier?
5.3 What are the uses of the simplex multiplier?
5.4 Explain the method of finding the inverse of the current basis using the inverse of the previous basis in the revised simplex method.
5.5 Write the procedure for solving an LPP using the revised simplex method.

Numerical Problems

5.1 Consider the following LPP:

Max $Z = 6x_1 - 2x_2 + 3x_3$

Subject to $2x_1 - x_2 + 2x_3 \leq 2$

$x_1 + 4x_3 \leq 4$

and $x_1, x_2, x_3 \geq 0$

Using x_4 and x_5 as slack variables and solving the LPP using the revised simplex method, we arrive at the following simplex table after some iterations:

Table no.	C_B	Basic variable	X_B	B^{-1}		Non-basic variables x_4	x_5	x_3	Minimum ratio $\frac{X_{B_i}}{\alpha_i^j}$
…	6	x_1	4	0	1		–		–
	−2	x_2	6	−1	2				
		$c_j - Z_j$		–		…	…	…	–

Answer the following questions:

(a) The simplex multiplier is ________

(b) Δ_j for the non-basic variables $(x_4, x_5, x_3) = (\_\_\_\_, \_\_\_\_, \_\_\_\_)$.

(c) This table gives ________

(i) an optimal solution
(ii) an infeasible solution
(iii) an alternate optimal solution
(iv) no solution

(d) Find the optimal solution to the LPP, if it exists.

5.2 Solve the following LPPs using the revised simplex method:

(a) Max $Z = 7x_1 + 5x_2$

Subject to $-x_1 - 2x_2 \geq -6$

$4x_1 + 3x_2 \leq 12$

and $x_1, x_2 \geq 0$

(b) Max $Z = 3x_1 + 4x_2$

Subject to $x_1 - x_2 \leq 1$

$-x_1 + x_2 \leq 2$

and $x_1, x_2 \geq 0$

(c) Max $Z = 2x_1 + x_2$

Subject to $x_1 + 2x_2 \le 10$

$x_1 + x_2 \le 6$

$x_1 - x_2 \le 2$

$x_1 - 2x_2 \le 1$

and $x_1, x_2 \ge 0$

(d) Max $Z = 3x_1 + 5x_2 + 4x_3$

Subject to $2x_1 + 3x_2 \le 8$

$2x_2 + 5x_3 \le 10$

$3x_1 + 2x_2 + 4x_3 \le 15$

and $x_1, x_2, x_3 \ge 0$

(e) Max $Z = 3x_1 + 2x_2 + 5x_3$

Subject to $x_1 + 2x_2 + x_3 \le 430$

$3x_1 + 2x_3 \le 460$

$x_1 + 4x_2 \le 420$

and $x_1, x_2, x_3 \ge 0$

(f) Max $Z = 4x_1 + 10x_2$

Subject to $2x_1 + x_2 \le 50$

$2x_1 + 5x_2 \le 100$

$2x_1 + 3x_2 \le 90$

and $x_1, x_2 \ge 0$

(g) Min $Z = 4x_1 + 8x_2 + 3x_3$

Subject to $x_1 + x_2 \ge 2$

$2x_1 + x_3 \ge 5$

and $x_1, x_2, x_3 \ge 0$

(h) Max $Z = x_1 + 2x_2 + 3x_3 - x_4$

Subject to $x_1 + 2x_2 + 3x_3 = 15$

$2x_1 + x_2 + 5x_3 = 20$

$x_1 + 2x_2 + x_3 + x_4 = 10$

and $x_1, x_2, x_3, x_4 \ge 0$

(i) Max $Z = -x_1 - x_2$

Subject to $3x_1 + 2x_2 \ge 30$

$-2x_1 + 3x_2 \le -30$

$x_1 + x_2 \le 5$

and $x_1, x_2 \ge 0$

(j) Max $Z = 5x_1 + 8x_2$

Subject to $3x_1 + 2x_2 \ge 3$

$x_1 + 4x_2 \ge 4$

$x_1 + x_2 \le 5$

and $x_1, x_2 \ge 0$

CONCEPTUAL PROBLEM

After conducting a market survey, a company decides to launch two new stylish leather belts, B_1 and B_2. The company takes into account all the aspects such as supply of raw material, production, advertising, and marketing of the product. In the beginning, the company arranges leather for the production of 800 belts (both B_1 and B_2). The B_1-type belts are fancy and cost more, whereas the B_2-type are ordinary and cost less. B_1 belts require twice as much time for manufacturing as B_2. Further, if all the belts are of type B_2, the company will be able to produce up to 1000 belts per day. The fancy buckles available for the B_1 belts are 400 per day and the ones for B_2 are 700 per day. The company expects to make a profit of ₹4 and ₹3 per belt, on types B_1 and B_2 respectively. What should the company do to earn maximum profit? To help the company, formulate the problem in an LP model and provide its solution using the revised simplex method.

Solution Let x_1 and x_2 be the quantity of leather belts B_1 and B_2, respectively. Then, the total profit is $4x_1 + 3x_2$.

Further x_1 and x_2 being the quantities of the belts, cannot be negative. Thus, we have $x_1, x_2 \ge 0$

The problem formulated is as follows:

Max $Z = 4x_1 + 3x_2$

Subject to $2x_1 + x_2 \le 1000$

$x_1 + x_2 \le 800$

$x_1 \le 400$

(Contd)

(*Contd*)

$x_2 \le 700$

and $x_1, x_2 \ge 0$

The problem can be written in the standard form as follows:

Max $Z = 4x_1 + 3x_2 + 0{\cdot}x_3 + 0{\cdot}x_4 + 0{\cdot}x_5 + 0{\cdot}x_6$

Subject to $2x_1 + x_2 + x_3 = 1000$

$x_1 + x_2 + x_4 = 800$

$x_1 + x_5 = 400$

$x_2 + x_6 = 700$

$x_1, x_2, x_3, x_4, x_5, x_6 \ge 0$ and x_3, x_4, x_5, and x_6 are slack variables.

Here, the coefficient matrix $A = (A_1, A_2, A_3, A_4, A_5, A_6) = \begin{pmatrix} 2 & 1 & 1 & 0 & 0 & 0 \\ 1 & 1 & 0 & 1 & 0 & 0 \\ 1 & 0 & 0 & 0 & 1 & 0 \\ 0 & 1 & 0 & 0 & 0 & 1 \end{pmatrix}$.

The basic variables are x_3, x_4, x_5, and x_6.

$$B = \begin{pmatrix} 1 & 0 & 0 & 0 \\ 0 & 1 & 1 & 0 \\ 0 & 0 & 1 & 0 \\ 0 & 0 & 0 & 1 \end{pmatrix},\ B^{-1} = \begin{pmatrix} 1 & 0 & 0 & 0 \\ 0 & 1 & 1 & 0 \\ 0 & 0 & 1 & 0 \\ 0 & 0 & 0 & 1 \end{pmatrix},\ b = \begin{pmatrix} 1000 \\ 800 \\ 400 \\ 700 \end{pmatrix}$$ and $C^{\mathrm{T}} = (4, 3, 0, 0,$

$0, 0)$. The BFS $X_B = B^{-1}b = (1000, 800, 400, 700)$.

To check the optimality, calculate Δ_j for the non-basic variables (x_1, x_2).

$$\Delta_j = c_j - Z_j = c_j - C_B^{\mathrm{T}} B^{-1} A_j$$

$$C_B^{\mathrm{T}} B^{-1} = (0, 0, 0, 0) \begin{pmatrix} 1 & 0 & 0 & 0 \\ 0 & 1 & 1 & 0 \\ 0 & 0 & 1 & 0 \\ 0 & 0 & 0 & 1 \end{pmatrix} = (0, 0, 0, 0)$$

$$C_B^{\mathrm{T}} B^{-1} A_j = (0, 0, 0, 0) \begin{pmatrix} 2 & 1 \\ 1 & 1 \\ 1 & 0 \\ 0 & 1 \end{pmatrix} = (0, 0)$$

$$\Delta_j = c_j - Z_j = (4, 3) - (0, 0) = (4, 3).$$

The solution is not optimal.

Here, $\Delta_j = 4$ is the most positive for the x_1 column, so x_1 will be the entering variable. For the leaving variable, calculate α^1 (the pivotal column).

$$\alpha^1 = B^{-1}A_1 = \begin{pmatrix} 1 & 0 & 0 & 0 \\ 0 & 1 & 1 & 0 \\ 0 & 0 & 1 & 0 \\ 0 & 0 & 0 & 1 \end{pmatrix} \begin{pmatrix} 2 \\ 1 \\ 1 \\ 0 \end{pmatrix} = \begin{pmatrix} 2 \\ 1 \\ 1 \\ 0 \end{pmatrix}$$

(*Contd*)

(*Contd*)

Tabulating these computations, we get

Table no.	C_B	Basic variable	X_B	B^{-1}				Non-basic variables x_1	x_2	Minimum ratio $\frac{X_{B_i}}{\alpha_i^j}$
1	0	x_3	1000	1	0	0	0	2		1000/2 = 500
	0	x_4	800	0	1	0	0	1		800/1 = 800
	0	x_5	400	0	0	1	0	[1]		400/1 = 400 →
	0	x_6	700	0	0	0	1	0		
	$c_j - Z_j$			–				4↑	3	–

Now use the same elementary row operations on columns X_B and B^{-1} of the aforegiven table, which are needed to bring zeros below and above the pivot entry 1. Interchange the position of the entering variable x_1 and leaving variable x_5, to get the following table:

Table no.	C_B	Basic variable	X_B	B^{-1}				Non-basic variables x_5	x_2	Minimum ratio $\frac{X_{B_i}}{\alpha_i^j}$
2	0	x_3	200	1	0	−2	0			
	0	x_4	400	0	1	−1	0	–		–
	4	x_1	400	0	0	1	0			
	0	x_6	700	0	0	0	1			
	$c_j - Z_j$			–				–		–

Now to check the optimality of the solution of the aforegiven table, we calculate Δ_j for the non-basic variables (x_5, x_2).

$$\Delta_j = c_j - Z_j = c_j - C_B^T B^{-1} A_j$$

$$C_B^T B^{-1} = (0, 0, 4, 0)\begin{pmatrix} 1 & 0 & -2 & 0 \\ 0 & 1 & -1 & 0 \\ 0 & 0 & 1 & 0 \\ 0 & 0 & 0 & 1 \end{pmatrix} = (0, 0, 4, 0)$$

$$C_B^T B^{-1} A_j = (0, 0, 4, 0)\begin{pmatrix} 0 & 1 \\ 0 & 1 \\ 1 & 0 \\ 0 & 1 \end{pmatrix} = (4, 0)$$

$$\Delta_j = c_j - Z_j = (0, 3) - (4, 0) = (-4, 3)$$

The solution is not optimal.

Here $\Delta_j = 3$ is the most positive for the x_2 column, so x_2 will be the entering variable. For the leaving variable, calculate α^2 (the pivotal column).

$$\alpha^2 = B^{-1} A_2 = \begin{pmatrix} 1 & 0 & -2 & 0 \\ 0 & 1 & -1 & 0 \\ 0 & 0 & 1 & 0 \\ 0 & 0 & 0 & 1 \end{pmatrix}\begin{pmatrix} 1 \\ 1 \\ 0 \\ 1 \end{pmatrix} = \begin{pmatrix} 1 \\ 1 \\ 0 \\ 1 \end{pmatrix}$$

(*Contd*)

(*Contd*)

Tabulating these computations, we get

Table no.	C_B	Basic variable	X_B	B^{-1}				Non-basic variables		Minimum ratio $\frac{X_{B_i}}{\alpha_i^j}$
								x_5	x_2	
3	0	x_3	200	1	0	−2	0		1 (boxed)	200
	0	x_4	400	0	1	−1	0		1	400
	4	x_1	400	0	0	1	0		0	…
	0	x_6	700	0	0	0	1		1	700
	$c_j - Z_j$			–				−4	3↑	–

Now use the same elementary row operations on columns X_B and B^{-1} of this table, which are needed to bring 1 to the position of pivot entry and zeros below the pivot entry 1. Interchanging the position of entering variable x_2 and leaving variable x_3, we get the following table:

Table no.	C_B	Basic variable	X_B	B^{-1}				Non-basic variables		Minimum ratio $\frac{X_{B_i}}{\alpha_i^j}$
								x_5	x_3	
4	3	x_2	200	1	0	−2	0	–		–
	0	x_4	200	−1	1	−1	0			
	4	x_1	400	0	0	1	0			
	0	x_6	500	−1	0	2	1			
	$c_j - Z_j$			–				–		–

Now to check the optimality of the solution of this table, we calculate Δ_j for the non-basic variables (x_5, x_3).

$$\Delta_j = c_j - Z_j = c_j - C_B^{\mathrm{T}} B^{-1} A_j$$

$$C_B^{\mathrm{T}} B^{-1} = (3, 0, 4, 0)\begin{pmatrix} 1 & 0 & -2 & 0 \\ -1 & 1 & -1 & 0 \\ 0 & 0 & 1 & 0 \\ -1 & 0 & 2 & 1 \end{pmatrix} = (3, 0, -2, 0)$$

$$C_B^{\mathrm{T}} B^{-1} A_j = (3, 0, -2, 0)\begin{pmatrix} 0 & 1 \\ 0 & 0 \\ 1 & 0 \\ 0 & 0 \end{pmatrix} = (-2, 3)$$

$$\Delta_j = c_j - Z_j = (0, 0) - (-2, 3) = (2, -3)$$

The solution is not optimal.

Here $\Delta_j = 2$ is the most positive for the x_5 column, so x_5 will be the entering variable. For the leaving variable, calculate α^5 (the pivotal column).

(*Contd*)

(*Contd*)

$$\alpha^5 = B^{-1}A_5 = \begin{pmatrix} 1 & 0 & -2 & 0 \\ -1 & 1 & -1 & 0 \\ 0 & 0 & 1 & 0 \\ -1 & 0 & 2 & 1 \end{pmatrix}\begin{pmatrix} 0 \\ 0 \\ 1 \\ 0 \end{pmatrix} = \begin{pmatrix} -2 \\ 1 \\ 1 \\ 2 \end{pmatrix}$$

Tabulating these computations, we get

Table no.	C_B	Basic variable	X_B	B^{-1}				Non-basic variables x_5	x_3	Minimum ratio $\frac{X_{B_i}}{\alpha_i^j}$
5	3	x_2	200	1	0	−2	0	−2		…
	0	x_4	200	−1	1	−1	0	[1]		200
	4	x_1	400	0	0	1	0	1		400
	0	x_6	500	−1	0	2	1	2		250
	$c_j - Z_j$			–				2↑	−3	–

Now use the same elementary row operations on columns X_B and B^{-1} of this table, which are needed to bring 1 to the position of pivot entry, zero below and above the pivot entry 1, and interchange the positions of the entering variable x_5 and the leaving variable x_4, to get the following table:

Table no.	C_B	Basic variable	X_B	B^{-1}				Non-basic variables x_4 x_3	Minimum ratio $\frac{X_{B_i}}{\alpha_i^j}$
6	3	x_2	600	−1	2	−4	0		
	0	x_5	200	−1	1	1	0	–	–
	4	x_1	200	1	−1	2	0		
	0	x_6	100	1	−2	4	1		
	$c_j - Z_j$			–				–	–

Now again to check the optimality of the solution of this table, we calculate Δ_j for the non-basic variables (x_4, x_3).

$$\Delta_j = c_j - Z_j = c_j - C_B^{\mathrm{T}}B^{-1}A_j$$

$$C_B^{\mathrm{T}}B^{-1} = (3, 0, 4, 0)\begin{pmatrix} 1 & 2 & 0 & 0 \\ -1 & 1 & 1 & 0 \\ 1 & -1 & 0 & 0 \\ 1 & -2 & 0 & 1 \end{pmatrix} = (1, 2, 0, 0)$$

$$C_B^{\mathrm{T}}B^{-1}A_j = (1, 2, 0, 0)\begin{pmatrix} 0 & 1 \\ 1 & 0 \\ 0 & 0 \\ 0 & 0 \end{pmatrix} = (2, 1)$$

$$\Delta_j = c_j - Z_j = (0, 0) - (1, 2) = (-1, -2)$$

The solution is optimal as $\Delta_j \le 0$ for non-basic variables x_j.

(*Contd*)

(*Contd*)

Thus, the company should produce 200 belts of type B_1 and 600 belts of type B_2 per day. Then, this plan will fetch ₹2600, which will be the maximum profit. Graphically, we can explain the solution as shown in Fig. 5.2.

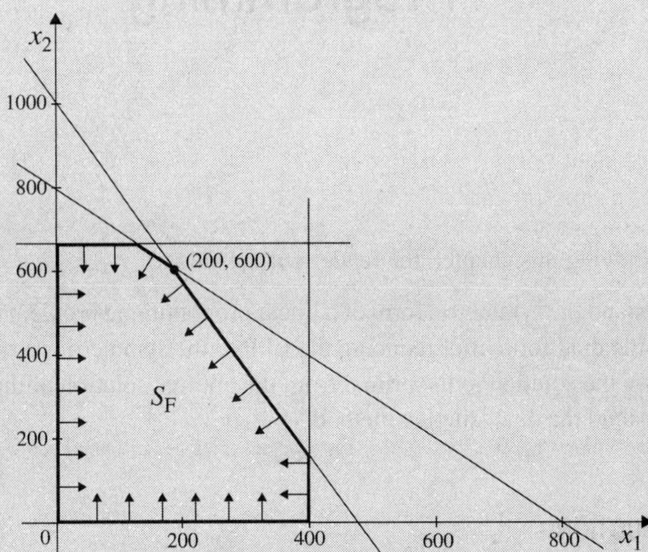

Figure 5.2 Graphical representation of optimal solution

6 Duality in Linear Programming

After studying this chapter, the reader will be able to

- understand the symmetric form of a linear programming problem (LPP)
- write the dual form after reducing the LPP to the symmetric form
- identify the solution to the primal from the optimal solution of the dual
- understand the dual simplex method

6.1 INTRODUCTION

For any given linear programming problem (LPP), known as the 'primal', there exists an associated LPP called the 'dual'. The solutions to these two problems are closely related to each other. The solution to any one of the problems can be obtained from the solution of the other. Solving a problem having a lesser number of constraints is easier. Thus, duality gives us the freedom to choose the LPP—primal or dual—that has a smaller number of constraints, identify solutions, and then, if required, obtain the solution to the other problem.

The graphical method of solving an LPP in two variables is more suitable when the constraints are given in equality or inequality forms. To solve the same problem using the simplex method, it has to be reduced to its canonical form. Further, to solve the same problem using duality, it must be reduced to its symmetric form. In a nutshell, to solve an LPP by different methods, it has to be reduced to different forms such as the inequality, canonical, and symmetric. We have already discussed the first two forms in Chapters 3 and 4. In this chapter, the symmetric form of an LPP, definition of its dual, and steps for finding the solution to the original problem from the solution of its dual are discussed.

6.2 SYMMETRIC FORM

An LPP is said to be in the symmetric form if

1. it is a maximization (minimization) problem where all constraints are of the '$\leq$' ('$\geq$') type; and
2. all the decision variables are ≥ 0.

Note In the symmetric form of an LPP, there is no restriction on the sign of b_i's (RHS). Any LPP can be reduced to the symmetric form for writing its dual.

The following are some of the examples of the symmetric form of an LPP:

Example 6.1 Max $Z = c_1x_1 + c_2x_2 + \ldots + c_nx_n = C^{T}X$

Subject to $AX \leq b$

and $X \geq 0$

where $C = (c_1, c_2, \ldots, c_n)^{T}$, $X = (x_1, x_2, \ldots, x_n)^{T}$, $A = [a_{ij}]_{m \times n}$, and $b = (b_1, b_2, \ldots, b_m)^{T}$.

Example 6.2 Min $Z = c_1x_1 + c_2x_2 + \ldots + c_nx_n = C^{T}X$

Subject to $AX \geq b$

and $X \geq 0$

6.3 DEFINITION OF DUAL OF LINEAR PROGRAMMING PROBLEM

Let us consider the following LPP:

$$\text{Max } Z_P = C^{T}X \tag{6.1}$$

$$\text{Subject to } \quad AX \leq b \tag{6.2}$$

$$\text{and } \quad X \geq 0 \tag{6.3}$$

where $C=(c_1, c_2, \ldots, c_n)^{T}$, $X=(x_1, x_2, \ldots, x_n)^{T}$, $A=[a_{ij}]_{m \times n}$, and $b=(b_1, b_2, \ldots, b_m)^{T}$.

The dual of the aforementioned LPP is defined as follows:

$$\text{Min } Z_D = b^{T}Y \tag{6.4}$$

$$\text{Subject to } \quad A^{T}Y \geq C \tag{6.5}$$

$$\text{and } \quad (y_1, y_2, \ldots, y_m)^{T} = Y \geq 0 \tag{6.6}$$

where $Y \geq 0$ means, $y_i \geq 0$ for all $i = 1, 2, \ldots, m$. The LPP given in Eqs (6.1)–(6.3) is called the primal problem and the m constraints given in Eq. (6.5) are called the dual constraints. The variable y_i is called the dual variable associated with the ith primal constraints.

Notes

1. The dual of a maximization problem becomes a minimization problem and vice versa.
2. If in the primal problem, some constraints are given in the equality form, then the corresponding variables in the dual problem will be unrestricted in sign (i.e., they may be positive or negative). Conversely, if the primal variable is unrestricted in sign, then the corresponding dual constraint will hold with the equality sign.

6.4 PRIMAL–DUAL RELATIONSHIP

The relationship between a primal problem and its dual can be explained as follows:

1. If the primal problem is a maximization problem, then its dual will be a minimization problem and vice versa.

2. The requirement vector (b) in the primal becomes the price-vector (C^T) in its dual and vice versa.
3. The primal coefficient matrix A becomes A^T in its dual.
4. The constraints with '≤' sign in the primal have the '≥' sign in their dual and vice versa.
5. A primal with n decision variables and m constraints becomes a dual with m dual variables and n constraints and vice versa.
6. If the ith constraints in primal hold with equality, then the ith dual variable will be unrestricted in sign.
7. If the jth primal variable is unrestricted in sign, then the jth dual constraint will hold with the equality sign.
8. If the ith slack or surplus variable in a primal problem is positive in the optimal table, then the ith dual variable will be zero in the optimal table of the dual and vice versa.
9. If the primal problem has a solution, then its dual also has a solution.
10. If the primal has no solution, then its dual also has no solution.
11. If the primal has an unbounded solution, then its dual has either no solution or an unbounded solution.

Example 6.3 Find the dual of the following LPP:

$$\text{Max } Z = x_1 - x_2 + 3x_3$$

$$\text{Subject to } \quad x_1 + x_2 + x_3 \le 10$$

$$2x_1 - x_3 \le 3$$

$$2x_1 - 2x_2 + 3x_3 \le 2$$

$$\text{and } \quad x_1, x_2, x_3 \ge 0$$

Solution The given problem is in a symmetric form. It can be written in the matrix form as follows:

$$\text{Max } Z = C^T X = (1, -1, 3)\,(x_1, x_2, x_3)^T$$

$$\text{Subject to } \quad \begin{pmatrix} 1 & 1 & 1 \\ 2 & 0 & -1 \\ 2 & -2 & 3 \end{pmatrix} \begin{pmatrix} x_1 \\ x_2 \\ x_3 \end{pmatrix} \le \begin{pmatrix} 10 \\ 3 \\ 2 \end{pmatrix}$$

$$\text{and } \quad x_1, x_2, x_3 \ge 0$$

$$\text{where } \quad C = (1, -1, 3)^T, X = (x_1, x_2, x_3)^T, b = (10, 3, 2)^T$$

The dual of the aforementioned LPP is as follows:

$$\text{Min } Z_D = (10, 3, 2)(y_1, y_2, y_3)^T = 10y_1 + 3y_2 + 2y_3$$

$$\text{Subject to } \quad \begin{pmatrix} 1 & 2 & 2 \\ 1 & 0 & -2 \\ 1 & -1 & 3 \end{pmatrix} \begin{pmatrix} y_1 \\ y_2 \\ y_3 \end{pmatrix} \ge \begin{pmatrix} 1 \\ -1 \\ 3 \end{pmatrix}$$

$$\text{or } \quad y_1 + 2y_2 + 2y_3 \ge 1$$

$$y_1 - 2y_3 \ge -1$$

$$y_1 - y_2 + 3y_3 \ge 3$$

$$\text{and } \quad y_1, y_2, y_3 \ge 0$$

Short cut method The given LPP is in symmetric form having three constraints. Let us associate three dual variables y_1, y_2, and y_3 with these constraints. We have,

$$\begin{aligned}
&\text{Max } Z = x_1 - x_2 + 3x_3 \\
&\text{Subject to} \quad x_1 + x_2 + x_3 \le 10, \qquad y_1 \\
&\qquad\qquad\quad 2x_1 + 0 \cdot x_2 - x_3 \le 3, \qquad y_2 \\
&\qquad\qquad\quad 2x_1 - 2x_2 + 3x_3 \le 2, \qquad y_3 \\
&\text{and} \qquad\quad x_1, x_2, x_3 \ge 0
\end{aligned}$$

Then the dual is as follows:

$$\begin{aligned}
&\text{Min } Z_D = 10y_1 + 3y_2 + 2y_3 \\
&\text{Subject to} \quad y_1 + 2y_2 + 2y_3 \ge 1 \\
&\qquad\qquad\quad y_1 + 0 \cdot y_2 - 2y_3 \ge -1 \\
&\qquad\qquad\quad y_1 - y_2 + 3y_3 \ge 3 \\
&\text{and} \qquad\quad y_1, y_2, y_3 \ge 0
\end{aligned}$$

From this equation, it can be observed that to get the objective function of the dual, we multiply the right-hand side (RHS) of the primal problem with the corresponding y_i's. Max Z is replaced by Min Z_D. The coefficients of x_1, x_2, and x_3 columns are multiplied with the corresponding y_i's to get the constraints of the dual problem. Further, '$\le$' signs become '$\ge$' signs and x_i's ≥ 0 are replaced by y_i's ≥ 0.

Example 6.4 Find the dual of the following LPP:

$$\begin{aligned}
&\text{Min } Z = 3x_1 + 4x_2 \\
&\text{Subject to} \quad 2x_1 + 6x_2 \ge 16 \\
&\qquad\qquad\quad 5x_1 + 2x_2 \ge 20 \\
&\text{and} \qquad\quad x_1, x_2 \ge 0
\end{aligned}$$

Solution The given LPP is in symmetric form having two constraints. Let us associate two dual variables y_1 and y_2 with these constraints.

$$\begin{aligned}
&\text{Min } Z = 3x_1 + 4x_2 \\
&\text{Subject to} \quad 2x_1 + 6x_2 \ge 16, \quad y_1 \\
&\qquad\qquad\quad 5x_1 + 2x_2 \ge 20, \quad y_2 \\
&\text{and} \qquad\quad x_1, x_2 \ge 0
\end{aligned}$$

The dual of this problem is as follows:

$$\begin{aligned}
&\text{Max } Z_D = 16y_1 + 20y_2 \\
&\text{Subject to} \quad 2y_1 + 5y_2 \le 3 \\
&\qquad\qquad\quad 6y_1 + 2y_2 \le 4 \\
&\text{and} \qquad\quad y_1, y_2 \ge 0
\end{aligned}$$

It can be observed here that to attain the objective function of the dual, the RHS of its primal problem is multiplied with the corresponding y_i's. Min Z is replaced by Max Z_D. The coefficients of the x_1 and x_2 columns are multiplied with the corresponding y_i's and added to obtain the constraints of the dual problem. The '$\ge$' signs are replaced by the '$\le$' signs, respectively, and x_i's ≥ 0 are replaced by y_i's ≥ 0.

Example 6.5 Find the dual of the following LPP:

$$\begin{aligned}
&\text{Max } Z = 2x_1 + 3x_2 \\
&\text{Subject to} \quad x_1 + 2x_2 = 2 \\
&\qquad\qquad\quad 5x_1 + 4x_2 \le 20 \\
&\qquad\qquad\quad x_1 \ge 0 \text{ and } x_2 \text{ is unrestricted in sign}
\end{aligned}$$

Solution The given LPP is

$$\left.\begin{array}{ll} \text{Max } Z = 2x_1 + 3x_2 & \\ \text{Subject to} & x_1 + 2x_2 = 2 \\ & 5x_1 + 4x_2 \le 20 \\ & x_1 \ge 0 \text{ and } x_2 \text{ is unrestricted in sign} \end{array}\right\} \quad (6.7)$$

Here, the constraint $x_1 + 2x_2 = 2$ is equivalent to the two inequality constraints $x_1 + 2x_2 \ge 2$ and $x_1 + 2x_2 \le 2$. Further, x_2, which is unrestricted in sign can be expressed as the difference of two positive variables x_2^+ and x_2^- as $x_2 = x_2^+ - x_2^-$. Thus, the problem in Eq. (6.7) can be converted to the symmetric form as follows:

$$\left.\begin{array}{ll} \text{Max } Z = 2x_1 + 3(x_2^+ - x_2^-) & \\ \text{Subject to} & x_1 + 2(x_2^+ - x_2^-) \le 2 \\ & -x_1 - 2(x_2^+ - x_2^-) \le -2 \\ & 5x_1 + 4(x_2^+ - x_2^-) \le 20 \\ \text{and} & x_1, x_2^+, x_2^- \ge 0 \end{array}\right\} \quad (6.8)$$

or

$$\begin{array}{ll} \text{Max } Z = C^T X & \\ \text{Subject to} & AX \le b \\ \text{and} & X \ge 0 \end{array} \quad (6.9)$$

where $C = (2, 3, -3)^T$, $b = (2, -2, 20)^T$, $X = (x_1, x_2^+, x_2^-)^T$ and $A = \begin{pmatrix} 1 & 2 & -2 \\ -1 & -2 & 2 \\ 5 & 4 & -4 \end{pmatrix}$.

The dual of Eq. (6.9) is as follows:

$$\text{Min } Z = (2, -2, 20)\begin{pmatrix} y_1 \\ y_2 \\ y_3 \end{pmatrix}$$

$$\text{Subject to} \quad \begin{pmatrix} 1 & -1 & 5 \\ 2 & -2 & 4 \\ -2 & 2 & -4 \end{pmatrix}\begin{pmatrix} y_1 \\ y_2 \\ y_3 \end{pmatrix} \ge \begin{pmatrix} 2 \\ 3 \\ -3 \end{pmatrix}$$

$$\text{and} \quad y_1, y_2, y_3 \ge 0$$

or

$$\begin{array}{ll} \text{Min } Z = 2y_1 - 2y_2 + 20y_3 & \\ \text{Subject to} & y_1 - y_2 + 5y_3 \ge 2 \\ & 2y_1 - 2y_2 + 4y_3 \ge 3 \\ & -2y_1 + 2y_2 - 4y_3 \ge -3 \\ \text{and} & y_1, y_2, y_3 \ge 0 \end{array}$$

or

$$\begin{array}{ll} \text{Min } Z = 2(y_1 - y_2) + 20y_3 & \\ \text{Subject to} & (y_1 - y_2) + 5y_3 \ge 2 \\ & 2(y_1 - y_2) + 4y_3 \ge 3 \\ & -2(y_1 - y_2) - 4y_3 \ge -3 \\ \text{and} & y_1, y_2, y_3 \ge 0 \end{array}$$

In the aforementioned equation, replacing the group $(y_1 - y_2)$ by Y_1, y_3 by Y_2, and combining the last two inequalities in the form of an equation, we get the following:

$$\begin{array}{ll} \text{Min } Z = 2Y_1 + 20Y_2 & \\ \text{Subject to} & Y_1 + 5Y_2 \ge 2 \\ & 2Y_1 + 4Y_2 = 3 \\ & Y_2 \ge 0 \text{ and } Y_1 \text{ is unrestricted in sign.} \end{array}$$

Here we see that while writing the dual, a constraint with an equality sign can be treated to be of the '≤' sign in maximization problems and similarly of the '≥' sign in minimization problems for recognition of the problem in its symmetric form. The corresponding dual variable will be unrestricted in sign. Further, if a primal variable is unrestricted in sign, then the corresponding dual constraint will hold with the equality sign.

Example 6.6 Find the dual of the following LPP:

$$\text{Min } Z = 5x_1 + 12x_2 + 4x_3$$
$$\text{Subject to } \quad x_1 + 2x_2 + x_3 \le 5$$
$$2x_1 - x_2 + 3x_3 = 2$$
$$x_1 + 4x_2 - x_3 \ge 1$$
$$x_1, x_2 \ge 0, \text{ and } x_3 \text{ is unrestricted in sign}$$

Solution The given LPP has three constraints. Multiplying the first constraint with (–1) and associating three dual variables y_1, y_2, and y_3 with these constraints, we get Eq. (6.10).

$$\left.\begin{aligned} &\text{Min } Z = 5x_1 + 12x_2 + 4x_3 \\ &\text{Subject to } \quad -x_1 - 2x_2 - x_3 \ge -5, \quad y_1 \\ &\qquad 2x_1 - x_2 + 3x_3 = 2, \quad y_2 \\ &\qquad x_1 + 4x_2 - x_3 \ge 1, \quad y_3 \\ &\qquad x_1, x_2 \ge 0, \text{ and } x_3 \text{ is unrestricted in sign} \end{aligned}\right\} \quad (6.10)$$

Now, treat the second constraint in Eq. (6.10) of the '≥' type for writing the dual (the corresponding dual variable will become unrestricted in sign), then the dual of Eq. (6.10) is formulated as Eq. (6.11).

$$\left.\begin{aligned} &\text{Max } Z_D = -5y_1 + 2y_2 + y_3 \\ &\text{Subject to } \quad -y_1 + 2y_2 + y_3 \le 5 \\ &\qquad -2y_1 - y_2 + 4y_3 \le 12 \\ &\qquad -y_1 + 3y_2 - y_3 = 4 \\ &\qquad y_1, y_3 \ge 0, \text{ and } y_2 \text{ is unrestricted in sign} \end{aligned}\right\} \quad (6.11)$$

Here in Eq. (6.11), the third constraint holds good with the equality sign as the third primal variable x_3 in Eq. (6.10) is unrestricted in sign.

Here, it can be observed that to get the objective function of the dual, the RHS of the primal problem given in Eq. (6.10) is multiplied with the corresponding y_i's. Min Z is replaced by Max Z_D. The coefficients of x_1, x_2, and x_3 columns are multiplied with the corresponding y_i's and added to get the constraints of the dual problem given in Eq. (6.11).

6.5 ECONOMIC INTERPRETATION OF DUALITY

Consider the primal given here:

$$\left.\begin{aligned} &\text{Max } Z = \sum_{j=1}^{n} C_j x_j \\ &\text{Subject to } \quad \sum_{j-1}^{n} a_{ij} x_j \le b_i, \; i = 1, 2, 3, \ldots, m \\ &\qquad x_j \ge 0, \; j = 1, 2, 3, \ldots, n \end{aligned}\right\} \quad (6.12)$$

Then, its dual is formulated as follows:

$$\left.\begin{aligned} &\text{Min } Z = \sum_{i=1}^{m} b_i y_i \\ &\text{Subject to } \quad \sum_{i=1}^{m} a_{ji} y_i \ge C_j, \; j = 1, 2, 3, \ldots, n \\ &\qquad y_i \le 0, \; i = 1, 2, 3, \ldots, m \end{aligned}\right\} \quad (6.13)$$

The LPP given in Eq. (6.12) can be viewed as a resource allocation model wherein the objective function can be considered to maximize the revenue or profit by utilizing the limited resources b_i's available. According to this model, the primal problem in Eq. (6.12) has n economic activities and m resources. The coefficient C_j in Eq. (6.12) represents the profit per unit on undertaking the activity j. The resources are represented by i, whose maximum availability b_i is consumed at the rate of a_{ij} units per unit of activity j. Thus, the LHS of the ith constraint gives the total amount consumed by the activities j ($j=1, 2, \ldots, n$), which must be less than or equal to maximum availability b_i of the ith source. Looking at the problem from this viewpoint, the associated dual problem in Eq. (6.13) offers an interesting economic interpretation to minimize the linear programming resource availability rate under certain restrictions. This can be explained with the help of Example 6.7.

Example 6.7 A cement manufacturing company A is facing a problem of minimizing cost of transporting cement bags from its two warehouses W_1 and W_2 to its two shops S_1 and S_2 situated in different locations. The stock available in the warehouses W_1 and W_2 are 6000 and 4000 bags respectively. The requirements for the shops S_1 and S_2 are 3000 and 7000 bags respectively. The unit transportation costs from the warehouses to the shops are given in the adjoining table:

Shops → Warehouses ↓	S_1	S_2
W_1	20	25
W_2	26	22

Formulate a problem to determine the most economic cost transporting schedule to fulfil the demands at the shops from the stock available in the warehouses. Write the dual of the problem and give the economic interpretation of this dual.

Solution Let x_{ij} be the quantity transported from warehouse W_i to shop S_j ($i, j=1, 2$). The total cost of transportation is $20x_{11}+25x_{12}+26x_{21}+22x_{22}$. The company would like to determine x_{ij}'s such that this cost is minimum. Thus, the objective function is Min $Z=20x_{11}+25x_{12}+26x_{21}+22x_{22}$. Further taking into consideration, the availabilities at the warehouses and requirements at the shops, we have

$$\begin{aligned} x_{11}+x_{12} &\le 6000 \\ x_{21}+x_{22} &\le 4000 \\ x_{11}+x_{21} &\ge 3000 \\ x_{12}+x_{22} &\ge 7000 \end{aligned}$$

and $x_{ij} \ge 0, \forall\, i, j=1, 2$ (as the units transported is non-negative).

Thus, the most economic transporting schedule is determined using Eq. (6.14).

$$\left.\begin{aligned} &\text{Min } Z=20x_{11}+25x_{12}+26x_{21}+22x_{22} \\ &\text{Subject to } \quad x_{11}+x_{12} \le 6000 \\ &\qquad\qquad\quad x_{21}+x_{22} \le 4000 \\ &\qquad\qquad\quad x_{11}+x_{21} \ge 3000 \\ &\qquad\qquad\quad x_{12}+x_{22} \ge 7000 \\ &\qquad\qquad\quad x_{ij} \ge 0, \forall\, i, j=1, 2 \end{aligned}\right\} \tag{6.14}$$

The symmetric form of Eq. (6.14) is Eq. (6.15).

$$\left.\begin{aligned} &\text{Min } Z=20x_{11}+25x_{12}+26x_{21}+22x_{22} \\ &\text{Subject to } \quad -x_{11}-x_{12} \ge -6000, \qquad y_1 \\ &\qquad\qquad\quad -x_{21}-x_{22} \ge -4000, \qquad y_2 \\ &\qquad\qquad\quad x_{11}+x_{21} \ge 3000, \qquad y_3 \\ &\qquad\qquad\quad x_{12}+x_{22} \ge 7000, \qquad y_4 \\ &\qquad\qquad\quad x_{ij} \ge 0, \forall\, i, j=1, 2 \end{aligned}\right\} \tag{6.15}$$

Thus, the dual of Eq. (6.15), and hence, of Eq. (6.14) is Eq. (6.16).

$$\left.\begin{array}{lll} \text{Max } Z_D = -6000y_1 - 4000y_2 + 3000y_3 + 7000y_4 & & (6.16a) \\ \text{Subject to } \; -y_1 + y_3 \le 20 & & (6.16b) \\ \qquad -y_1 + y_4 \le 25 & & (6.16c) \\ \qquad -y_2 + y_3 \le 26 & & (6.16d) \\ \qquad -y_2 + y_4 \le 22 & & (6.16e) \\ \qquad y_i\text{'s} \ge 0, \text{ and } i = 1, 2, 3, 4 & & \end{array}\right\} \quad (6.16)$$

where y_i's are the dual variable associated with the four constraints.

Economic interpretation Let us imagine that another company B approaches company A with a proposal to purchase all the 6000 units available at W_1 at ₹y_1 per unit and all the 4000 units at W_2 at ₹y_2 per unit. It will then supply 3000 and 7000 units to shops S_1 and S_2, respectively, at a unit price of ₹y_3 and ₹y_4. Now, B may follow dual constraints in Eq. (6.16) for the benefit of company A (the manufacturing company) wherein B's net cost of selling and purchasing is less than or equal to A's cost of transporting the goods on their own. For example, consider Eq. (6.16b). Suppose the manufacturer's unit cost of transporting the goods from W_1 to S_1 is ₹20 and that of company B is $y_3 - y_1$, which is less than or equal to ₹20 as shown in Eq. (6.16b). Similarly, each dual constraint in Eq. (6.16) states that the unit transportation cost of company B is cheaper than that of A. Therefore, company A will accept the proposal put forward by company B in the hope that it will reduce its transportation costs. On the other hand, company B will fix the values of y_1, y_2, y_3, and y_4 in such a manner that dual constraints in Eq. (6.16) are satisfied and its net profit in Eq. (6.16a) is maximized.

6.6 IMPORTANT THEOREMS

We shall now prove some results that usually hold true for the primal and dual problems. These results show how closely the primal and dual problems are related to each other.

Let us consider the following primal:

$$\left.\begin{array}{lll} \text{Min } Z = C^T X & & (a) \\ \text{Subject to } \; AX \ge b & & (b) \\ \text{and } \qquad X \ge 0 & & (c) \end{array}\right\} \quad (P)$$

Then, its dual is as follows:

$$\left.\begin{array}{lll} \text{Max } Z_D = b^T Y & & (a) \\ \text{Subject to } \; A^T Y \le C & & (b) \\ \qquad Y \ge 0 & & (c) \end{array}\right\} \quad (D)$$

Theorem 6.1 The dual of a dual is primal.

Proof Consider the following LPP:

$$\left.\begin{array}{ll} \text{Max } Z = C^T X & \\ \text{Subject to } \; AX \le b & \\ \text{and } \qquad X \ge 0 & \\ \text{where } \qquad A = [a_{ij}]_{m \times n} & \end{array}\right\} \quad (6.17)$$

The dual of Eq. (6.17) is Eq. (6.18).

$$\left.\begin{aligned}&\text{Min } Z' = b^{\mathrm{T}}Y\\&\text{Subject to}\quad A^{\mathrm{T}}Y \ge C\\&Y = (y_1, y_2, \ldots, y_n)^{\mathrm{T}} \ge 0,\ y_i\text{'s are the dual variables of Eq. (6.17)}\end{aligned}\right\}\quad(6.18)$$

The dual of Eq. (6.18) is as follows:

$$\left.\begin{aligned}&\text{Max } Z'' = C^{\mathrm{T}}W && \text{or} && \text{Max } Z'' = C^{\mathrm{T}}W\\&\text{Subject to}\quad (A^{\mathrm{T}})^{\mathrm{T}}W \le (b^{\mathrm{T}})^{\mathrm{T}} && && \text{Subject to}\quad AW \le b\\&\text{and}\quad W \ge 0 && && \text{and}\quad W \ge 0\end{aligned}\right\}\quad(6.19)$$

The problem in Eq. (6.19) is the same as that of Eq. (6.17).

Example 6.8 Verify that the dual of dual is primal for the following LPP:

$$\begin{aligned}&\text{Max } Z = 2x_1 - 3x_2\\&\text{Subject to}\quad -x_1 + x_2 \le 1\\&\qquad x_1 + x_2 \le 2\\&\text{and}\quad x_1, x_2 \ge 0\end{aligned}$$

Solution The given problem can be written as Eq. (6.20).

$$\left.\begin{aligned}&\text{Max } Z = (2, -3)\,(x_1, x_2)^{\mathrm{T}}\\&\text{Subject to}\quad \begin{pmatrix}-1 & 1\\ 1 & 1\end{pmatrix}\begin{pmatrix}x_1\\ x_2\end{pmatrix} \le \begin{pmatrix}1\\ 2\end{pmatrix}\\&\text{and}\quad x_1, x_2 \ge 0\end{aligned}\right\}\quad(6.20)$$

$$\text{where } C = (2, -3)^{\mathrm{T}},\ X = (x_1, x_2)^{\mathrm{T}},\ A = \begin{pmatrix}-1 & 1\\ 1 & 1\end{pmatrix},\ b = (1, 2)^{\mathrm{T}}$$

We know that the dual of the primal problem is formulated as follows:

$$\begin{aligned}&\text{Min } Z_{\mathrm{D}} = b^{\mathrm{T}}Y\\&\text{Subject to}\quad A^{\mathrm{T}}Y \ge (C^{\mathrm{T}})^{\mathrm{T}} (=C)\\&\text{and}\quad Y \ge 0\end{aligned}$$

Therefore, the dual of the problem in Eq. (6.20) is as follows:

$$\begin{aligned}&\text{Min } Z_{\mathrm{D}} = (1, 2)(y_1, y_2)^{\mathrm{T}} = y_1 + 2y_2\\&\text{Subject to}\quad \begin{pmatrix}-1 & 1\\ 1 & 1\end{pmatrix}\begin{pmatrix}y_1\\ y_2\end{pmatrix} \ge \begin{pmatrix}2\\ -3\end{pmatrix}\\&\text{and}\quad y_1, y_2 \ge 0\end{aligned}$$

$$\text{where } (C^{\mathrm{T}})^{\mathrm{T}} = (2, -3)^{\mathrm{T}},\ Y = (y_1, y_2)^{\mathrm{T}},\ A^{T} = \begin{pmatrix}-1 & 1\\ 1 & 1\end{pmatrix},\ b^{\mathrm{T}} = (1, 2)$$

or

$$\left.\begin{aligned}&\text{Min } Z_{\mathrm{D}} = y_1 + 2y_2\\&\text{Subject to}\quad -y_1 + y_2 \ge 2\\&\qquad y_1 + y_2 \ge -3\\&\text{and}\quad y_1, y_2 \ge 0;\ y_i\text{'s are the dual variables}\end{aligned}\right\}\quad(6.21)$$

The dual of Eq. (6.21) is as follows:

$$\begin{aligned}&\text{Max } Z = C^{\mathrm{T}}W\\&\text{Subject to}\quad (A^{\mathrm{T}})^{\mathrm{T}}W \le (b^{\mathrm{T}})^{\mathrm{T}}\\&\text{and}\quad W \ge 0\end{aligned}$$

or $\quad \text{Max } Z = (2, -3)(w_1, w_2)^{\mathrm{T}} = 2w_1 - 3w_3$

$$\text{Subject to} \quad \begin{pmatrix} -1 & 1 \\ 1 & 1 \end{pmatrix}\begin{pmatrix} w_1 \\ w_2 \end{pmatrix} \le \begin{pmatrix} 1 \\ 2 \end{pmatrix}$$

and $\quad w_1, w_2 \ge 0$; w_i's are the dual variables of Eq. (6.21)

where $\quad C^{\mathrm{T}} = (2, -3),\ W = (w_1, w_2)^{\mathrm{T}},\ (A^{\mathrm{T}})^{\mathrm{T}} = \begin{pmatrix} -1 & 1 \\ 1 & 1 \end{pmatrix},\ (b^{\mathrm{T}})^{\mathrm{T}} = (1, 2)$

or $\quad \text{Max } Z = 2w_1 - 3w_2$

$$\text{Subject to} \quad -w_1 + w_2 \le 1$$
$$w_1 + w_2 \le 2$$

and $\quad w_1, w_2 \ge 0$

which is same as the given problem as in Eq. (6.20) and thus verifies that the dual of dual is primal

---

Theorem 6.2 If X is a feasible solution of the primal problem (P) and Y is a feasible solution of the dual problem (D), then $Z \ge Z_{\mathrm{D}}$, that is, $C^{\mathrm{T}}X \ge b^{\mathrm{T}}Y$.

Proof As X is a feasible solution of Eq. (P), it satisfies

$$AX \ge b$$

Multiplying the aforementioned equation by Y^{T}, we get Eq. (6.22)

$$Y^{\mathrm{T}}AX \ge Y^{\mathrm{T}}b \quad (\text{as } Y \ge 0) \tag{6.22}$$

Further, since Y is given as a feasible solution of (D), from part (b) of Eq. (D), we get

$$A^{\mathrm{T}}Y \le C \quad \text{or} \quad Y^{\mathrm{T}}A \le C^{\mathrm{T}} \tag{6.23}$$

Multiplying Eq. (6.23) by X, we get $Y^{\mathrm{T}}AX \le C^{\mathrm{T}}X$ (6.24)

From Eqs (6.22) and (6.24), we have $Y^{\mathrm{T}}b \le C^{\mathrm{T}}X$ or $Z_{\mathrm{D}} \le Z$

Theorem 6.3 If X^* is a feasible solution of the primal problem (P) and Y^* is a feasible solution of its dual problem (D) such that the value of the objective function f of the primal at X^* equals the value of the objective function ϕ of the dual at Y^* that is, $f(X^*) = \phi(Y^*)$, then X^* and Y^* are the optimal solutions of (P) and (D) respectively.

Proof As $X^*$and $Y^*$are the feasible solutions of (P) and (D), respectively, we have [from Eqs (Pb) and (Db)],

$$AX^* \ge b \text{ and } A^{\mathrm{T}}Y^* \le C \tag{6.25}$$

$$\text{Given } f(X^*) = \phi(Y^*) \text{ that is, } C^{\mathrm{T}}X^* = b^{\mathrm{T}}Y^* \tag{6.26}$$

Now, let X be any feasible solution of (P). Since Y^* is a feasible solution of (D), from Theorem 6.2, we have $C^{\mathrm{T}}X \ge b^{\mathrm{T}}Y^* = C^{\mathrm{T}}X^*$ [using Eq. (6.26)].

This shows that $C^{\mathrm{T}}X \ge C^{\mathrm{T}}X^*$ for all feasible solutions X of (P) and thus proves that X^* is the optimal solution of (P).

Next let Y be any feasible solution of (D). Since X^* is given as a feasible solution of (P), from Theorem 6.2, we have

$b^{\mathrm{T}}Y \le C^{\mathrm{T}}X^* = b^{\mathrm{T}}Y^*$, that is, $b^{\mathrm{T}}Y \le b^{\mathrm{T}}Y^*$ for all feasible solutions Y of (D) and hence Y^* is the optimal solution of (D).

Theorem 6.4 *Fundamental theorem of duality*: If the primal problem (P) has an optimal solution, then its dual problem (D) also has an optimal solution and vice versa.

Proof (a) $b > 0$

Let the primal problem have an optimal solution optimal X. This means artificial variables if any are at zero level and

$$c_j - Z_j = c_j - C_B^{\mathrm{T}}B^{-1}A_j \geq 0 \text{ for all } j = 1, 2, \ldots, n$$

or $\quad C_B^{\mathrm{T}}B^{-1}A \leq C^{\mathrm{T}} \quad \text{or} \quad Y^{\mathrm{T}}A \leq C^{\mathrm{T}} \quad [C_B^{\mathrm{T}}B^{-1} = Y^{\mathrm{T}}]$

or $\quad YA^{\mathrm{T}} \leq C$

This shows that Y (defined) is a solution of the dual problem (D).

Now, if we can show that $Y \geq 0$, then it will become a feasible solution for the dual problem. For this, $c_j - Z_j \geq 0$, for all surplus variables. Thus,

$$\begin{aligned} c_j - Z_j &= c_j - C_B^{\mathrm{T}}\alpha^j \\ &= c_j - C_B^{\mathrm{T}}B^{-1}(-e_j), \text{ for } e_j \text{ is the unit vector with its } j\text{th component } 1 \\ &= C_B^{\mathrm{T}}B^{-1}I, \quad [c_j = 0 \text{ for all surplus variables}] \end{aligned}$$

Therefore, $c_j - Z_j = C_B^{\mathrm{T}}B^{-1} \geq 0$ or $Y^{\mathrm{T}} \geq 0$

Thus, the solution Y of dual is its feasible solution. Now, we have $b^{\mathrm{T}}Y = Y^{\mathrm{T}}b = C_B^{\mathrm{T}}B^{-1}b = C_B^{\mathrm{T}}X_B$.

Thus, if X is a feasible solution of the primal and Y is a feasible solution of its dual, then

$$Z_{\mathrm{D}} = b^{\mathrm{T}}Y = Y^{\mathrm{T}}b = C_B^{\mathrm{T}}X_B = Z$$

Hence from Theorem 6.3, Y^{T} is an optimal solution of the dual problem.

(b) If $b_i < 0$ in the primal, we avoid the proof.

Hint Multiply the ith constraint with (−1) and use the simplex method to determine the optimal solution of the primal. Then, we calculate $Y^{\mathrm{T}} = C_B^{\mathrm{T}}B^{-1}$ for the optimal primal table. The optimal solution of this is Y^{T}, with the sign of y_i changed.

Notes
1. Any given minimization or maximization problem can be considered as primal.
2. Theorem 6.4 provides a method to obtain the optimal solution to the primal from the optimal solution of its dual. $C_B^{\mathrm{T}}B^{-1}$ is called the simplex multiplier, which is calculated from the optimal table of the dual to determine the optimal solution of the primal.
3. In any simplex table, the matrix under the initial basic variables (excluding $c_j - Z_j$ row) is B^{-1} of the basis matrix B corresponding to the simplex table.

Theorem 6.5 If the slack or surplus variable associated with the kth primal constraint appears as a basic variable in the optimal table, then the dual variable y_k corresponding to the kth constraint is zero in the optimal table of the dual problem.

Proof Let P(min) be the primal problem and D(max) be the its dual problem. Then, the optimal table of primal $c_j - Z_j \geq 0$ for all columns and $c_j - Z_j = 0$ for all basic columns. Since by hypothesis x_k is the basic variable in the optimal solution of the primal, we have

$$c_k - Z_k = 0$$

or $\quad Z_k = 0$, because cost of x_k is zero

or $\quad C_B^{\mathrm{T}}B^{-1}(e_k) = 0$ if x_k is a slack variable

or $\quad C_B^{\mathrm{T}}B^{-1}(-e_k) = 0$ if x_k is a surplus variable,

where $e_k(-e_k)$ is the unit vector whose kth component is 1(−1) and all others are zero.

Hence,

$$C_B^{\mathrm{T}}B^{-1}e_k = 0 \quad \text{or} \quad Y^{\mathrm{T}}e_k = 0 \text{ where } Y^{\mathrm{T}} = C_B^{\mathrm{T}}B^{-1}$$

or $\quad y_k = 0 \quad (\because Y^{\mathrm{T}}e_k = y_k)$

Theorem 6.6 If the variable x_k of the primal problem appears as a basic variable in the optimal solution, then the kth dual constraint will hold with equality in the optimal dual solution (i.e., the corresponding slack or surplus variable is zero in the optimal dual solution).

Proof Since x_k is a basic variable in the optimal solution of the primal, we have

$$c_k - Z_k = c_k - C_B^{\mathrm{T}}\alpha^k = c_k - C_B^{\mathrm{T}}B^{-1}A_k = 0$$
$$C_B^{\mathrm{T}}B^{-1}A_k = c_k \quad \text{or} \quad Y^{\mathrm{T}}A_k = c_k$$
$$[Y^{\mathrm{T}} = (y_1, y_2, \ldots, y_k, \ldots, y_m)^{\mathrm{T}} \text{ and } A_k = (A_{1k}, \ldots, A_{mk})]$$
$$y_1a_{1k} + y_2a_{2k} + \ldots + y_ma_{mk} = c_k.$$

The kth dual constraint $Y^{\mathrm{T}}A_k = c_k$ holds with the equality sign.

Theorem 6.7 *Existence theorem*: If the primal problem has an unbounded solution, then the dual has either no solution or an unbounded solution.

Proof Let S_{FP} and S_{FD} denote the set of feasible solutions for a primal problem P(min problem) and its dual problem D(max problem) respectively. Since the primal problem has an unbounded solution, there exists an $X \in S_{\mathrm{FP}}$ such that $Z = C^{\mathrm{T}}X < m$, however small m may be, that is, Z has no lower bound. Now, let us assume that the dual problem has a feasible solution, say $Y \in S_{\mathrm{FD}}$. Then using Theorem 6.2, we have $Z_{\mathrm{D}} = b^{\mathrm{T}}Y \le C^{\mathrm{T}}X = Z$, that is, $b^{\mathrm{T}}Y$ is the lower bound for Z.

This contradicts the fact that Z has no lower bound. This means, our assumption that the dual problem has a feasible solution is wrong. Thus, if the primal problem has an unbounded solution, then its dual has no feasible solution.

Note Theorems 6.5 and 6.6 are called the *complementary slackness theorems* and establish a closer relationship between the primal and dual problems.

Example 6.9 A family doctor suggests two types of food items F_1 and F_2 for a child. A unit of food F_1 contains 1 gm each of vitamins B and C, and a unit of food F_2 contains 2 gms of vitamin B and 3 gms of vitamin C. The daily minimum requirement of vitamins B and C is 1 and 2 gms respectively. A unit of F_1 and F_2 costs ₹3 and ₹1 respectively. Formulate the problem in a linear programming (LP) model to minimize the food cost. Write the dual and then from its solution, determine the solution to the primal or original problem.

Solution The problem is formulated in the LP model as follows:

$$\text{Min } Z = 3x_1 + x_2$$
$$\text{Subject to} \quad x_1 + x_2 \ge 1 \qquad y_1$$
$$2x_1 + 3x_2 \ge 2 \qquad y_2$$
$$\text{and} \quad x_1, x_2 \ge 0$$

The given problem is in a symmetric form, having two constraints. Let us associate two dual variables y_1 and y_2 with these constraints. The dual of this problem is as follows:

$$\text{Max } Z = y_1 + 2y_2$$
$$\text{Subject to} \quad y_1 + 2y_2 \le 3$$
$$y_1 + 3y_2 \le 1$$
$$\text{and} \quad y_1, y_2 \ge 0$$

The given problem can be written in the canonical form as follows:

$$\text{Max } Z = y_1 + 2y_2$$
$$\text{Subject to} \quad y_1 + 2y_2 + y_3 = 3$$
$$y_1 + 3y_2 + y_4 = 1$$
$$\text{and} \quad y_1, y_2 \ge 0, y_3 \text{ and } y_4 \text{ are slack variables}$$

The initial simplex table is as follows:

		Cost →		1	2	0	0	Minimum ratio
		Variable →		y_1	y_2	y_3	y_4	
Table no.	C_B	Basic variable	X_B	α^1	α^2	α^3	α^4	$\frac{X_{B_i}}{\alpha_i^j}$
1	0	y_3	3	1	2	1	0	3/2
	0	$\leftarrow y_4$	1	1	[3]	0	1	1/3
		$c_j - Z_j$		1	2↑	0	0	
2	0	y_3	7/3	1/3	0	1	−2/3	7
	2	$\leftarrow y_2$	1/3	[1/3]	1	0	1/3	1
		$c_j - Z_j$		1/3↑	0	0	−2/3	
3	0	y_3	2	0	−1	1	−1	
	1	y_1	1	1	3	0	1	
		$c_j - Z_j$		0	−1	0	−1	

In table 3 of this simplex table, all the basic variables become non-negative. Thus, the optimum solution is $y_1 = 1$, $y_2 = 0$, and Max $Z = 1$.

Now, the solution of the primal problem is obtained using a simplex multiplier, that is,

$$(x_1, x_2) = C_B^{\mathrm{T}} B^{-1} = (0,\ 1)\begin{pmatrix} 1 & -1 \\ 0 & 1 \end{pmatrix} = (0,\ 1)$$

Hence, $x_1 = 0$, $x_2 = 1$, and Min $Z = 1$.

Example 6.10 Write the dual of the following LPP:

$$\begin{aligned} &\text{Min } Z = 2y_1 + 3y_2 + 5y_3 \\ &\text{Subject to} \quad y_1 + 2y_2 + 3y_3 \geq 2 \\ &\qquad\qquad\quad 2y_1 - y_2 + y_3 \geq 3 \\ &\text{and} \qquad\quad\ y_1, y_2, y_3 \geq 0 \end{aligned}$$

From the solution of the dual, determine the solution to the given (primal) problem.

Solution The given problem is in the symmetric form. Its dual is Eq. (6.27)

$$\left.\begin{aligned} &\text{Max } Z_D = 2x_1 + 3x_2 \\ &\text{Subject to} \quad x_1 + 2x_2 \leq 2 \\ &\qquad\qquad\quad 2x_1 - x_2 \leq 3 \\ &\qquad\qquad\quad 3x_1 + x_2 \leq 5 \\ &\text{and} \qquad\quad\ x_1, x_2 \geq 0 \end{aligned}\right\} \tag{6.27}$$

Substituting Eq. (6.27) in the canonical form of an LPP using slack variables x_3, x_4, and x_5, we get the following

$$\begin{aligned} &\text{Max } Z_D = 2x_1 + 3x_2 \\ &\text{Subject to} \quad x_1 + 2x_2 + x_3 = 2 \\ &\qquad\qquad\quad 2x_1 - x_2 + x_4 = 3 \\ &\qquad\qquad\quad 3x_1 + x_2 + x_5 = 5 \\ &\text{and} \qquad\quad\ x_1, x_2, x_3, x_4, x_5 \geq 0 \end{aligned}$$

The initial simplex table is as follows:

Cost →				2	3	0	0	0	Minimum ratio $\frac{X_{B_i}}{\alpha_i^j}$
Variable →				x_1	x_2	x_3	x_4	x_5	
Table no.	C_B	Basic variable	X_B	α^1	α^2	α^3	α^4	α^5	
1	0	x_3	2	1	[2]	1	0	0	2/2 = 1 min←
	0	x_4	3	2	−1	0	1	0	…
	0	x_5	5	3	1	0	0	1	5/1 = 5
		$c_j - Z_j$		2	3↑	0	0	0	
2	3	x_2	1	1/2	1	1/2	0	0	1/(1/2) = 2
	0	x_4	4	[5/2]	0	1/2	1	0	4/(5/2) = 8/5→
	0	x_5	4	5/2	0	−1/2	0	1	4/(5/2) = 8/5
		$c_j - Z_j$		1/2↑	0	−3/2	0	0	
3	3	x_2	1/5	0	1	2/5	−1/5	0	
	2	x_1	8/5	1	0	1/5	2/5	0	
	0	x_5	0	0	0	−1	−1	1	
		$c_j - Z_j$		0	0	−8/5	−1/5	0	

In table 3, $c_j - Z_j \le 0$ for all the non-basic variables. The optimum solution is $x_1 = 8/5, x_2 = 1/5$, and Max $Z_D = 19/5$. The solution of the primal is given by the simplex multiplier, that is,

$$(y_1, y_2, y_3) = C_B^T B^{-1} = (3, 2, 0)\begin{pmatrix} 2/5 & -1/5 & 0 \\ 1/5 & 2/5 & 0 \\ -1 & -1 & 1 \end{pmatrix} = (8/5,\ 1/5,\ 0)$$

Hence, $y_1 = 8/5$, $y_2 = 1/5$, $y_3 = 0$ and Min $Z = 19/5$.

6.7 DUAL SIMPLEX METHOD

Sometimes in a simplex table, a particular basic solution is infeasible, that is, one or more basic variables are negative but the optimality criterion is fulfilled. Such situations usually arise in sensitivity analysis. In these situations, we use a method called the dual simplex method. In the simplex method, feasibility conditions are satisfied in the initial simplex table, that is, we start with a basic feasible solution and move towards the optimal solution. However, in the dual simplex method, we proceed the other way, that is, the optimality criterion is satisfied in the initial simplex table, but the basic solution is infeasible. Starting with this infeasible solution, we move towards a feasible solution maintaining the optimality criterion.

6.7.1 Procedure for Solving Linear Programming Problems

The following steps are used to solve an LPP using the dual simplex method:

1. Formulate the given LPP in the following format:

   Max or Min $Z = f(X) = C^T X$

   Subject to $AX = b$

   $$X = (x_1, x_2, \ldots, x_n)^T \ge 0$$

where $A = [a_{ij}]_{m \times n}$ is the coefficient matrix containing the identity matrix as its submatrix and one or more b_i's are negative.

2. Prepare the simplex table and check that optimality conditions ($c_j - Z_j \le 0$ of the maximization problem for all j) are satisfied, but the basic solution of the table is infeasible.
3. Restore feasibility using the following rules of leaving and entering variables:
   (a) *Leaving variable rule*: The variable x_r of the row r, which is the most negative among the basic variables, leaves the basis. Therefore, the leaving variable $x_r = \min_i \{x_i ; x_i < 0\}$. This rule ensures that the existing non-feasible solution will move towards feasibility.
   (b) *Entering variable rule*: Locate the negative entries in the row of the leaving variable, calculate their ratio with the corresponding entries in the $c_j - Z_j$ row, and select the entering variable x_k of the column for which this ratio is minimum in an absolute sign, that is,

$$\operatorname*{Min}_{j}\left\{\left|\frac{c_j - Z_j}{\alpha_r^j}\right|, \alpha_r^j < 0\right\} = \theta, \text{ say for } j = k$$

   This rule ensures that optimality conditions continue to hold good. If all α_r^j are non-negative, then the given problem has no feasible solution.
4. Continue step 3 until all the basic variables become non-negative. The table thus obtained is the optimal table and the solution given by this table is the optimal solution.

Note In the dual simplex method, we first determine the leaving variable and then the entering variable.

Example 6.11 Solve the following LPP by dual simplex method:

$$\text{Max } Z = -4x_1 - 6x_2 - 18x_3$$

Subject to $x_1 + 3x_3 \ge 3$

$x_2 + 2x_3 \ge 5$

and $x_1, x_2, x_3 \ge 0.$

Solution The given problem can also be written as follows:

Max $Z = -4x_1 - 6x_2 - 18x_3$

Subject to $-x_1 - 3x_3 \le -3$

$-x_2 - 2x_3 \le -5$

and $x_1, x_2, x_3 \ge 0$

or

Max $Z = -4x_1 - 6x_2 - 18x_3 + 0x_4 + 0x_5$

Subject to $-x_1 - 3x_3 + x_4 = -3$

$-x_2 - 2x_3 + x_5 = -5$

and $x_1, x_2, x_3, x_4, x_5 \ge 0$, x_4 and x_5 are slack variables

The initial dual simplex table is as follows:

	Cost →			−4	−6	−18	0	0	Minimum ratio
	Variable →			x_1	x_2	x_3	x_4	x_5	$\min_j \left\lvert\frac{c_j - Z_j}{\alpha_r^j}\right\rvert$
Table no.	C_B	**Basic variable**	X_B	α^1	α^2	α^3	α^4	α^5	
1	0	x_4	−3	−1	0	−3	1	0	…
	0	← x_5	−5	0	[−1]	−2	0	1	$\min\left(\left\lvert\frac{-6}{-1}\right\rvert, \left\lvert\frac{-18}{-2}\right\rvert\right) = 6$ for x_2
		$c_j - Z_j$		−4	−6↑	−18	0	0	

In this table, the optimality conditions are satisfied, but the feasibility conditions are not satisfied (as basic variables x_4 and x_5 are assuming negative values). In such situations, we use the dual simplex method to restore the feasibility conditions. Here, x_5 being the most negative will leave the basis. This fixes the row x_5. Locate the negative entries in this row and take the ratios of the corresponding $c_j - Z_j$ with these negative entries with absolute values. The variable of the column for which this ratio is minimum will enter the basis. Here when x_5 leaves, x_2 will enter the basis. The next table is presented here:

	Cost →			−4	−6	−18	0	0	Minimum ratio
	Variable →			x_1	x_2	x_3	x_4	x_5	$\min_j \left\lvert \frac{c_j - Z_j}{\alpha_r^j} \right\rvert$
Table no.	C_B	**Basic variable**	X_B	α^1	α^2	α^3	α^4	α^5	
2	0	$\leftarrow x_4$	−3	−1	0	[−3]	1	0	$\min\left(\left\lvert\frac{-4}{-1}\right\rvert, \left\lvert\frac{-6}{-3}\right\rvert\right) = 2$ for x_3
	−6	x_2	5	0	1	2	0	−1	…
		$c_j - Z_j$		−4	0	−6↑	0	−6	
3	−18	x_3	1	1/3	0	1	−1/3	0	
	−6	x_2	3	−2/3	1	0	2/3	−1	
		$c_j - Z_j$		−2	0	0	−2	0	

In table 3 of this simplex table, all the basic variables become non-negative. The optimum solution is $x_1 = 0$, $x_2 = 3$, $x_3 = 1$, and Max $Z = -36$.

Example 6.12 Solve the following LPP using the dual simplex method:

Min $Z = 2x_1 + 3x_2 + 5x_3$

Subject to $x_1 + 2x_2 + 3x_3 \geq 2$

$2x_1 - x_2 + x_3 \geq 3$

and $x_1, x_2, x_3 \geq 0$

Solution Converting the minimization problem into maximization, we have

Max $Z' = -2x_1 - 3x_2 - 5x_3$ where $Z' = -Z$

The problem can also be written as follows:

Max $Z' = -2x_1 - 3x_2 - 5x_3$

Subject to $-x_1 - 2x_2 - 3x_3 + x_4 = -2$

$-2x_1 - x_2 - x_3 + x_5 = -3$

and $x_1, x_2, x_3, x_4, x_5 \geq 0$, x_4 and x_5 are slack variables.

The initial dual simplex table is as follows:

	Cost →			−2	−3	−5	0	0	Minimum ratio
	Variable →			x_1	x_2	x_3	x_4	x_5	$\min_j \left\lvert \frac{c_j - Z_j}{\alpha_r^j} \right\rvert$
Table no.	C_B	**Basic variable**	X_B	α^1	α^2	α^3	α^4	α^5	
1	0	x_4	−2	−1	−2	−3	1	0	…
	0	$\leftarrow x_5$	−3	[−2]	1	−1	0	1	$\min\left(\left\lvert\frac{-2}{-2}\right\rvert, \left\lvert\frac{-5}{-1}\right\rvert\right) = 1$ for x_1
		$c_j - Z_j$		−2↑	−3	−5	0	0	

(*Contd*)

(Contd)

Table no.	C_B	Basic variable	X_B	x_1 (α^1), Cost −2	x_2 (α^2), Cost −3	x_3 (α^3), Cost −5	x_4 (α^4), Cost 0	x_5 (α^5), Cost 0	Minimum ratio $\min_j \left\| \frac{c_j - Z_j}{\alpha_r^j} \right\|$
2	0	$\leftarrow x_4$	−1/2	0	[−5/2]	−5/2	1	−1/2	$\min\left(\left\|\frac{-4}{-5/2}\right\|, \left\|\frac{-4}{-5/2}\right\| \left\|\frac{-1}{-1/2}\right\|\right) = \frac{8}{5}$ for x_2
	−2	x_1	3/2	1	−1/2	1/2	0	−1/2	…
		$c_j - Z_j$		0	−4↑	−4	0	−1	
3	−3	x_2	1/5	0	1	1	−2/5	1/5	
	−2	x_1	8/5	1	0	1	−1/5	−2/5	
		$c_j - Z_j$		−2	0	0	−2	0	

In table number 3 of this simplex table, all the basic variables become non-negative. The optimum solution is $x_1 = 8/5$, $x_2 = 1/5$, $x_3 = 0$, and Min $Z = 19/5$.

RECAPITULATION

1. To solve an LPP using duality, it has to be first reduced to its symmetric form.
2. Every LPP can be reduced to the symmetric form.
3. With any LPP, we can associate another LPP.
4. A given problem in a linear form is called the primal and its associated problem is called the dual.
5. An LPP with lesser number of constraints is easier to solve.
6. The optimal solution of an LPP is obtained from the optimal solution of its dual using the simplex multiplier ($C_B^T B^{-1}$).
7. When the basic solution of an LPP is infeasible but fulfils the optimality conditions, the dual simplex method is used to restore feasibility while maintaining the optimality conditions.
8. In the dual simplex method, though the rules of entering and leaving variable are derived from the dual problem, they are used in the primal problem.

EXERCISES

Multiple-choice Questions

6.1 If the primal variable x_j is unrestricted in sign, then the corresponding dual constraint will hold with the
(a) '≤' sign, if the primal problem is a minimization problem
(b) '≥' sign, if the primal problem is a maximization problem
(c) '=' sign
(d) none of these

6.2 If the given problem has no solution then its dual has
(a) no solution
(b) a bounded solution
(c) no solution or has an unbounded solution
(d) none of these

6.3 If the given problem has unbounded solution then its dual has
(a) no solution
(b) a bounded solution
(c) an unbounded solution
(d) none of these

6.4 Let x_j be the basic variable in some simplex iteration of dual simplex method with a negative value and all entries in the x_j row are ≥ 0. Then,
(a) the set of feasible solutions is an unbounded solution
(b) the set of feasible solutions is empty
(c) the given problem has an alternate optimal solution
(d) none of these

6.5 If the given problem has an alternate optimal solution, then its dual has
(a) to degenerate optimal solution
(b) no solution
(c) an alternate optimal solution
(d) none of these

6.6 If the given problem has a degenerate optimal solution, then its dual
(a) also has a degenerate optimal solution
(b) has no solution
(c) has an alternate optimal solution
(d) none of these

6.7 While using the dual simplex method for a given problem, if the rule for entering a variable is not followed, then
(a) the optimality will be disturbed
(b) the feasibility will deteriorate further
(c) the optimality will never be disturbed
(d) none of these

6.8 In the dual simplex method, the following conditions must be satisfied to start a simplex iteration:
(a) Optimality
(b) Feasibility
(c) Both optimality and feasibility
(d) None of these

State True or False

6.1 In a dual simplex method, iterations
(a) move towards feasibility maintaining optimality.
(b) move towards optimality maintaining feasibility.
(c) maintain both feasibility and optimality.

6.2 The dual simplex method is applicable when in the initial table
(a) the optimality condition is satisfied but the feasibility condition is not satisfied.
(b) the optimality condition is not satisfied.
(c) the feasibility condition is satisfied, but the optimality condition is not satisfied.
(d) the feasibility condition is satisfied.
(e) both feasibility and optimality are satisfied.
(f) neither feasibility nor optimality is satisfied.

Fill in the Blanks

6.1 In a primal LPP, if the jth constraint holds with '=' sign, then the corresponding dual variable will be of __________ sign.

6.2 In duality, if the given problem has an optimal solution then its dual has __________ solution.

6.3 In writing the dual of an LPP after reducing it to a symmetric form, the RHS (availability vector b) has __________.

6.4 The dual of the __________ of an LPP is primal.

6.5 The dual of an LPP, Max $Z = C^T X$, $AX \leq b$, $X \geq 0$ is __________.

6.6 If the LPP is given in a symmetric form with $\geq$ sign in all constraints and satisfies the non-negativity restriction, then
  (a) the problem is a ____________ problem
  (b) its dual constraints will be with ____________ sign
  (c) its dual problem is a ____________ problem

Review Questions

6.1 State and prove the duality theorem.
6.2 Define the dual of a given LPP.
6.3 Prove that if any variable in a primal problem is unrestricted in sign, then the corresponding constraint in the dual problem will hold with the equality sign.
6.4 Explain the economic interpretation of duality.
6.5 Write the procedure for solving an LPP through the dual simplex method.

Numerical Problems

6.1 Find the dual of the following LPPs:

(a) Max $Z = 3x_1 + 5x_2 + 4x_3$
Subject to $2x_1 + 3x_2 \leq 8$
$2x_2 + 5x_3 \leq 10$
$3x_1 + 2x_2 + 4x_3 \leq 15$
and $x_1, x_2, x_3 \geq 0$

(b) Min $Z = 3x_1 + 2x_2$
Subject to $2x_1 + 3x_2 \geq 2$
$x_1 + x_2 \geq 1$
and $x_1, x_2 \geq 0$

(c) Min $Z = 10x_1 + 20x_2$
Subject to $3x_1 + 2x_2 \geq 18$
$x_1 + 3x_2 \geq 8$
$2x_1 - x_2 \leq 6$
and $x_1, x_2 \geq 0$

(d) Min $Z = 7x_1 + 3x_2 + 8x_3$
Subject to $8x_1 + 2x_2 + x_3 \geq 3$
$3x_1 + 6x_2 + 4x_3 \geq 4$
$4x_1 + x_2 + 5x_3 \geq 1$
$x_1 + 5x_2 + 2x_3 \geq 7$
and $x_1, x_2, x_3 \geq 0$

(e) Min $Z = 2x_1 + 2x_2 + 4x_3$
Subject to $2x_1 + 3x_2 + 5x_3 \geq 2$
$3x_1 + x_2 + 4x_3 \leq 3$
$x_1 + 4x_2 + 6x_3 \leq 5$
and $x_1, x_2, x_3 \geq 0$

(f) Max $Z = 3x_1 + x_2 + 4x_3 + x_4 + 9x_5$
Subject to $4x_1 - 5x_2 - 9x_3 + x_4 - 2x_5 \leq 6$
$2x_1 + 3x_2 + 4x_3 - 5x_4 + x_5 \leq 9$
$x_1 + x_2 - 5x_3 - 7x_4 + 11x_5 \leq 10$
and $x_1, x_2, x_3, x_4, x_5 \geq 0$

(g) Min $Z = 3x_1 - 2x_2 + 4x_3$
Subject to $3x_1 + 5x_2 + 4x_3 \geq 7$
$6x_1 + x_2 + 3x_3 \geq 4$
$7x_1 - 2x_2 - x_3 \leq 10$
$x_1 - 2x_2 + 5x_3 \geq 3$
$4x_1 + 7x_2 - 2x_3 \geq 2$
and $x_1, x_2, x_3 \geq 0$

(h) Max $Z = 2x_1 + 3x_2 + x_3$
Subject to $4x_1 + 3x_2 + x_3 = 6$
$x_1 + 2x_2 + 5x_3 = 4$
and $x_1, x_2, x_3 \geq 0$

(i) Max $Z = x_1 + 3x_2$
Subject to $3x_1 + 2x_2 \leq 6$
$3x_1 + x_2 = 4$
and $x_1, x_2 \geq 0$

(j) Min $Z = x_1 + x_2 + x_3$
Subject to $x_1 - 3x_2 + 4x_3 = 5$
$x_1 - 2x_2 \leq 3$
$2x_2 - x_3 \geq 4$
and $x_1, x_3 \geq 0$, x_3 is unrestricted in sign

6.2 Solve the following LPPs using the dual simplex method:

(a) Min $Z = 2x_1 + 2x_3 + 2x_5$
Subject to $3x_1 + 2x_2 + x_3 \geq 1500$
$-x_2 + x_4 + 2x_5 \geq 1000$
$x_3 - x_4 \geq 3000$
and $x_1, x_2, x_3, x_4, x_5 \geq 0$

(b) Min $Z = x_1 + 4x_2 + 3x_4$
Subject to $x_1 + 2x_2 - x_3 + x_4 \geq 3$
$-2x_1 - x_2 + 4x_3 + x_4 \geq 2$
and $x_1, x_2, x_3, x_4 \geq 0$

(c) Min $Z = 4x_1 + 6x_2 + 18x_3$
Subject to $x_1 + 3x_3 \geq 3$
$x_2 + 2x_3 \geq 5$
and $x_1, x_2, x_3 \geq 0$

(d) Max $Z = -x_1 - 4x_2 - 3x_4$
Subject to $x_1 + 2x_2 - x_3 + x_4 \geq 3$
$-2x_1 - x_2 + 4x_3 + x_4 \geq 2$
and $x_1, x_2, x_3, x_4 \geq 0$

CONCEPTUAL PROBLEM

There is an increase in demand for food grains in the country today as the population is increasing. This, in turn, is leading to a rise in the production of food grains, which depends largely on availability of proper irrigation facilities. For proper irrigation, good quality pipes are required. Manufacturing of plastic pipes is attracting the attention of many companies, as the cost of producing them is less than that of the steel pipes. A company has decided to manufacture two types of plastic pipes, P_1 and P_2, for irrigation purposes. The three resources needed for manufacturing such pipes are extrusion hours, packaging hours, and a special kind of additive to provide strength to the plastic raw material. The production of unit length each (one unit length = 100 m) of type P_1 and P_2 requires 5 h and 10 h of extrusion time, 4 h and 2 h of packing time, and 1 kg of additive mix. Usually, companies prepare a production plan for one week. The maximum resource time available per week for extrusion and packing is 50 h and 20 h, respectively. Further, the maximum additive mix available per week is 6 kg. The profit per unit length of type P_1 and P_2 is ₹300 and ₹450 respectively. Now, the company faces the problem of formulating a mathematical model of the problem, that is, to determine the unit lengths of P_1 and P_2 to be manufactured to earn maximum profit under the given restrictions in terms of availability of resources.

Solution The company expert proceeds to solve the given problem using the dual simplex method. Here, the profit earned depends on the unit lengths of P_1 and P_2. So for decision variables, let x_1 and x_2 be the unit lengths, respectively, of pipes P_1 and P_2 to be manufactured. Then, the total profit is $300x_1 + 450x_2$. The restriction on extrusion hours available results in a constraint $5x_1 + 10x_2 \leq 50$. Similarly, the restriction on available packing hours results in a constraint $4x_1 + 2x_2 \leq 20$. Further, the restriction on special additive results in another constraint $x_1 + x_2 \leq 6$. As x_1 and x_2 are the unit lengths of pipes to be manufactured and thus cannot be negative, we have $x_1, x_2 \geq 0$. Hence, the formulated LP model is as follows:

Max $Z = 300x_1 + 450x_2$ (6.23)
Subject to $5x_1 + 10x_2 \leq 50$ (6.24)
$4x_1 + 2x_2 \leq 20$ (6.25)
$x_1 + x_2 \leq 6$ (6.26)
and $x_1, x_2 \geq 0$ (6.27)

It is a maximization problem with all '≤' constraints and all variables involved are '≥' zero. Hence, it is in its symmetric form for writing its dual. Let y_1, y_2, and y_3 be the dual variables associated with the constraints given in Eqs (6.24), (6.25), and (6.26), respectively.

Then dual is as follows:

Min $Z_D = 50y_1 + 20y_2 + 6y_3$
Subject to $5y_1 + 4y_2 + y_3 \geq 300$
$10y_1 + 2y_2 + y_3 \geq 450$
and $y_1, y_2, y_3 \geq 0$

(Contd)

(*Contd*)

Now, let us convert the minimization problem into a maximization problem. Introducing surplus and artificial variables and modifying the objective function by adding $-M$ times the sum of artificial variables, we get the following:

$$\text{Max } Z = -50y_1 - 20y_2 - 6y_3 - My_6 - My_7$$

$$\text{Subject to} \quad 5y_1 + 4y_2 + y_3 - y_4 + y_6 = 300$$

$$10y_1 + 2y_2 + y_3 - y_5 + y_7 = 450$$

$$\text{and} \quad y_1, y_2, y_3, y_4, y_5, y_6, y_7 \geq 0$$

where y_4 and y_5 are surplus variables, y_6 and y_7 are artificial variables, and M is a large positive number.

The initial simplex table is as follows:

Cost →				−50	−6	−18	0	0	−M	−M	Minimum ratio
Variable →				y_1	y_2	y_3	y_4	y_5	y_6	y_7	$\frac{X_{B_i}}{\alpha_i^j}$
Table no.	C_B	Basic variable	X_B	α^1	α^2	α^3	α^4	α^5	α^6	α^7	
1	$-M$	y_6	300	5	4	1	−1	0	1	0	300/5 = 60
	$-M$	$\leftarrow y_7$	450	[10]	2	1	0	−1	0	1	450/10 = 45 min
		$c_j - Z_j$		$-50 + 15M\uparrow$	$-20 + 6M$	$-6 + 2M$	$-M$	$-M$	0	0	–
2	$-M$	$\leftarrow y_6$	75	0	[3]	1/2	−1	1/2	1	−1/2	75/3=25 min
	−50	y_1	45	1	1/5	1/10	0	−1/10	0	1/10	$\frac{45}{1/5} = 225$
		$c_j - Z_j$		0	$3M + 10\uparrow$	$(M/2) - 1$	$-M$	$(M/2) - 5$	0	$-(3/2)M + 5$	–
3	−20	$\leftarrow y_2$	25	0	1	1/6	−1/3	1/6	1/3	−1/6	150 min
	−50	y_1	40	1	0	1/15	1/15	−2/5	−1/15	2/15	600
		$c_j - Z_j$		0	0	$2/3\uparrow$	−10/3	−10/3	$-M + (10/3)$	$-M + (10/3)$	–
4	−6	y_3	150	0	6	1	−2	1	2	−1	–
	−50	y_1	30	1	−2/5	0	1/3	−1/5	−1/5	−1/5	
		$c_j - Z_j$		0	−4	0	−2	−4	$-M + 2$	$-M + 4$	–

In table 4 of this simplex table, all the basic variables become non-negative. The optimum solution is $y_1 = 30$, $y_2 = 0$, $y_3 = 150$, and Min $Z_D = 1500 + 900 = 2400$.

Now, the solution of the primal problem is obtained using the simplex multiplier, that is,

$$(-x_1, -x_2) = C_B^T B^{-1} = (-6, -50)\begin{pmatrix} 2 & -1 \\ -1/5 & 1/5 \end{pmatrix} = (-12 + 10, 6 - 10) = (-2, -4)$$

Thus, $x_1 = 2$, $x_2 = 4$, and profit $Z = 300x_1 + 450x_2 = 2400$

Hence, the company can earn a maximum profit of ₹2400 if it manufactures 2 and 4 unit lengths of P_1 and P_2 respectively.

Post-optimality Analysis or Sensitivity Analysis

After studying this chapter, the reader will be able to

- understand sensitivity analysis
- understand graphical sensitivity analysis
- find the solution to the linear programming problem when certain changes, such as addition or deletion of a variable/constraint, are made in the problem after obtaining its optimal solution
- understand parametric linear programming

7.1 INTRODUCTION

While formulating a linear programming problem (LPP), the values of all the variables are assumed to be known. Since variables such as resource availability, their consumption, market demand, and profit do not remain constant, the optimal solution may also change. Thus, it is always desirable to study the effect of such changes on the optimal solutions. *Sensitivity analysis* or *post-optimal analysis* refers to the study of such effects on the optimal solutions. In such analyses, the optimal solution of a problem with respect to the changing variables is determined. In other words, sensitivity analysis is the determination of maximum variation in any parameter such that the solution remains optimal. Sensitivity analysis is used only when discrete changes are made in parameters. However, in real-life problems, when continuous variations are required in the parameters of the problem, then a technique called *parametric linear programming* is used.

Let us consider the following LPP:

$$\text{Max or Min } Z = C^{\mathrm{T}}X$$

$$\text{Subject to} \quad AX = b$$
$$X \geq 0$$

where $A=[a_{ij}]_{m\times n}$, $C=(c_1, c_2, \ldots, c_n)^{\mathrm{T}}$ and $X=(x_1, x_2, \ldots, x_n)^{\mathrm{T}}$. In this chapter, we will analyse how sensitive a solution of an LPP is with respect to the following changes made in the given problem, after its optimal solution has been obtained:

1. Change in c_j's (the coefficients of x_j's in the objective function)

2. Change in b_i's (the right-hand side of constraints) after bringing the problem to its standard form
3. Change in a_{ij}'s, the columns of the coefficient matrix $A = [a_{ij}]_{m\times n}$
4. Deletion and addition of constraints
5. Deletion and addition of decision variables

7.2 CHANGES AFFECTING FEASIBILITY AND OPTIMALITY

Let us consider the following LPP

$$\text{Max } Z = C^{\mathrm{T}}X \tag{7.1}$$

$$\text{Subject to } \quad AX = b$$

where $X = (x_1, x_2, \ldots, x_n)^{\mathrm{T}} \geq 0$, $C = (c_1, c_2, \ldots, c_n)^{\mathrm{T}}$, $b = (b_1, b_2, \ldots, b_m)^{\mathrm{T}}$, and $A = [a_{ij}]_{m\times n} = (A_1, A_2, \ldots, A_n)$.

Let $x_1, x_2, \ldots, x_m$ be the basic variables of a simplex table in any simplex iteration, with the basis matrix $B = (A_1, A_2, \ldots, A_m)$, then

$$Z = C_B^{\mathrm{T}} X_B \tag{7.2}$$

$$\text{and} \quad X_B = B^{-1}b \tag{7.3}$$

$$Z_j = C_B^{\mathrm{T}}\alpha^j, \qquad \alpha^j = (\alpha_1^j, \alpha_2^j, \ldots, \alpha_m^j) = B^{-1}A_j$$

$$= C_B^{\mathrm{T}}B^{-1}A_j$$

$$\Delta_j = c_j - Z_j = c_j - C_B^{\mathrm{T}}B^{-1}A_j, \qquad j = 1, 2, \ldots, n \tag{7.4}$$

where $\alpha^j = (\alpha_1^j, \alpha_2^j, \ldots, \alpha_m^j)^{\mathrm{T}} = B^{-1}A_j$ and $C_B^{\mathrm{T}}B^{-1}$ is called the simplex multiplier. In the initial simplex table, $\alpha^j = A_j$. In the sensitivity analysis, Eq. (7.3) is used to check the feasibility condition and Eq. (7.4) is used to check the optimality condition. This is indicated by the following simplex table:

		Cost →		c_1 c_2 … … c_j … … c_n	Minimum ratio
		Variables →		x_1 x_2 … … x_j … … x_n	$\frac{X_{B_i}}{\alpha_i^j}, \alpha_i^j > 0$
Table no.	C_B	**Basic variables**	X_B	α^1 α^2 … … α^j … … α^n	$i = 1, 2, \ldots, m$
1	C_{B1}	X_{B1}		α_1^1 α_1^2 … … α_1^j … … α_1^n	
	C_{B2}	X_{B2}	↑	α_2^1 α_2^2 … … α_2^j … … α_2^n	
	…	…	$X_B = B^{-1}b$	… … … … … … … …	–
	…	…	↓	… … … … … … … …	
	C_{Bm}	X_{Bm}		α_m^1 α_m^2 … … α_m^j … … α_m^n	
	$\Delta_j = c_j - Z_j$ for $j = 1, 2, \ldots, n$			Δ_1 Δ_2 … … Δ_j … … Δ_n	–

Any changes made in an LP problem after obtaining its optimal solution may affect either the optimality condition (change in the value of Δ_j) or the feasibility condition (change in the value of X_B). If the optimality of a problem is affected, the simplex method can be used to restore it. In case the feasibility is affected, the dual simplex method can be used to restore the feasibility conditions. Equations (7.2), (7.3), and (7.4) indicate the changes in the problem that affect the feasibility and optimality

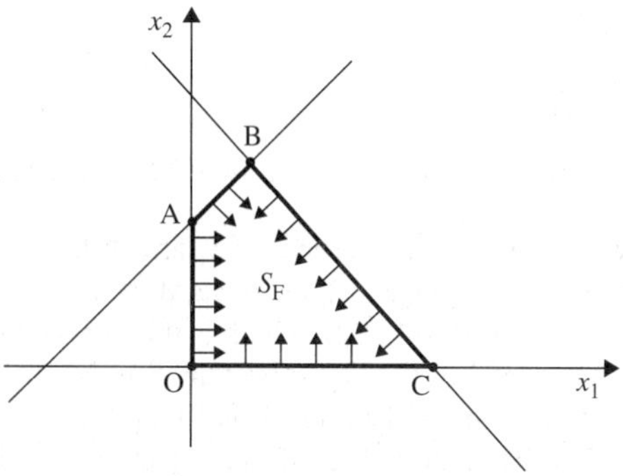

Figure 7.1 Optimum at point B

conditions. If any change is made to c_j's, then from Eq. (7.4), it can be seen that Δ_j changes, and hence, optimality of the solution may be affected. If any of the b_i's are changed, then from Eq. (7.3), vector b changes and hence, $X_B = B^{-1}b$ changes. As a result, the feasibility may be affected.

In all such cases, $Z = C_B^T X_B$ [as given in Eq. (7.2)] is calculated to get the optimum value of the objective function. Changes affecting either optimality or feasibility are considered at one point of time. For the changes in a problem where both the conditions (feasibility and optimality) are affected simultaneously, there is no method to find the solution.

7.3 GRAPHICAL SENSITIVITY ANALYSIS

Graphical sensitivity analysis is useful if an LPP is considered in two dimensions (as we can illustrate the problem by a graph in two dimensions). In an LPP, the maximum or minimum of the objective function always occurs at the vertex, provided the set of feasible solutions (S_F) is bounded.

Let us consider an LPP in two variables, x_1 and x_2, having two constraints with boundary lines AB and BC, where the existing coefficients of x_1 and x_2 in the objective function are changed to new coefficients c_1 and c_2, respectively. The new objective function thus becomes $c_1x_1 + c_2x_2$.

Let the lines AB and BC intersect at a point B (Fig. 7.1). Assume the maximum of the objective function $c_1x_1 + c_2x_2$ of the LPP occurs at B. Now, any change in the values of c_1 and c_2 can change the position of the optimum vertex B. Our aim is to find the range of variation for both c_1 and c_2, so that the current optimum does not change. More specifically, the range of optimality for the ratio c_1/c_2 (or c_2/c_1) that will keep the existing optimum solution of the given LPP unchanged has to be determined.

Example 7.1 For the solution of the following problem to remain optimal, find the range of variation of

(a) the ratio (c_1/c_2), when in the following LPP, the coefficient 1 of x_1 and coefficient 2 of x_2 in the objective function are changed to c_1 and c_2 respectively:

Max $Z = x_1 + 2x_2$

Subject to $-x_1 + x_2 \leq 1$ (7.5)

$x_1 + x_2 \leq 2$ (7.6)

and $x_1, x_2 \geq 0$

(b) b_1 (where the RHS 1 of the constraint in Eq. (7.5) is changed to b_1).

Solution (a) As shown in Fig. 7.2, the vertices of S_F are O(0, 0), A(0, 1), B(1/2, 3/2), and C(2, 0). At O(0, 0), $Z = 0$; at A(0, 1), $Z = 2$; at B(1/2, 3/2), $Z = 7/2$; and at C(2, 0), $Z = 2$. The maximum $Z = 7/2$ occurs at B(1/2, 3/2), which is the intersection of constraints given in Eqs (7.5) and (7.6). The slope of the objective function is –1/2. The slope of the line BC is –1 [Eq. (7.6)] and the slope of the line AB is 1 [Eq. (7.5)].

In Fig. 7.2, we see that the slope (–1/2) of the objective function varies between the slopes –1 and 1 of the two constraints given in Eqs (7.6) and (7.5) respectively, intersecting each other at the optimal point *B*, without affecting the optimality of the existing problem.

Now, if we change the coefficients (1 of x_1 to c_1 and 2 of x_2 to c_2), then the slope $(-c_1/c_2)$ of the new objective function (Max $Z = c_1x_1 + c_2x_2$) must lie between the slopes of the two lines, namely $x_1 + x_2 = 2$ and $-x_1 + x_2 = 1$, intersecting at *B*, so that the existing solution at *B* remains optimal when $x_1 = 1/2$ and $x_2 = 3/2$. We can express this relationship algebraically as

$$-\frac{1}{1} \leq \frac{-c_1}{c_2} \leq -\left(\frac{-1}{1}\right)$$

In other words, the range of variation of the ratio $(-c_1/c_2)$ is between –1 and 1.

(b) As seen in (a), the optimal solution is Max $Z = 7/2$ when $x_1 = 1/2$ and $x_2 = 3/2$.

Now, let the RHS of the constraint given in Eq. (7.5) be changed to b_1, that is, 1 is changed to b_1 so that the first constraint becomes $-x_1+x_2=b_1$. Now, as b_1 changes (between –2 to 2), the position of the optimal point *B* slides along the line $x_1+x_2=2$ from D(0, 2) to C(2, 0) for the constraint $-x_1+x_2 \leq b_1$ to remain in the feasible region as shown in Fig. 7.3. The solution at D(0, 2) (i.e., $x_1=0$, $x_2=2$) is $Z=0+4=4$ and the solution at C(2, 0) (i.e., $x_1=2$, $x_2=0$) is $Z=2$. Thus, the *Z* value continuously decreases from 4 to 2 as b_1 decreases from 2 to –2, staying in the feasible region. Thus, the range of b_1 is $-2 \leq b_1 \leq 2$ (Fig. 7.3).

This method has limited applications as problems with changes in only two variables can be solved through the graphical method.

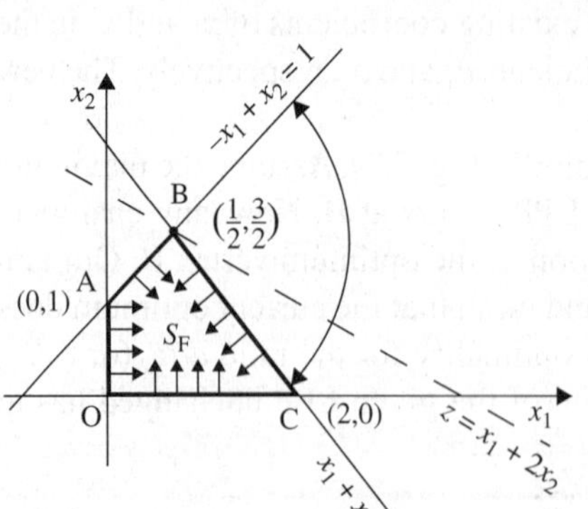

Figure 7.2 Range of variation of ratio $(-c_1/c_2)$

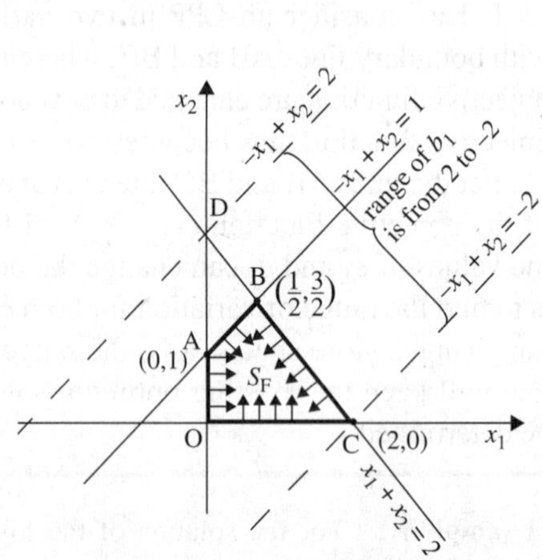

Figure 7.3 Range of variation of b_1

Let us now discuss sensitivity analysis in general.

7.4 CHANGES IN COST c_j IN OBJECTIVE FUNCTION

Sometimes, a change in the cost or profit variable leads to a change in the coefficient of the objective function, which in turn affects the optimal solution.

Let us consider the following LPP:

$$\left.\begin{array}{ll} \text{Max } Z = C^{\mathrm{T}}X & \\ \text{Subject to} & AX = b \\ \text{and} & X \geq 0 \end{array}\right\} \tag{7.7}$$

where $X = (x_1, x_2, \ldots, x_n)^{\mathrm{T}} \geq 0$, $C = (c_1, c_2, \ldots, c_n)^{\mathrm{T}}$, $b = (b_1, b_2, \ldots, b_m)^{\mathrm{T}}$ and $A = [a_{ij}]_{m \times n}$. Here, m is the number of constraints and n is the number of decision variables.

Let $X_B = B^{-1}b$ be the optimal basic feasible solution (BFS) of Eq. (7.7), with B as the basis matrix. The optimality condition $c_j - Z_j \leq 0$, $\forall j$ is satisfied for this optimal BFS. After getting the optimal solution of Eq. (7.7), if a change is made in cost c_j (the coefficient of variable x_j), then $X_B = B^{-1}b$ remains unaffected, as it does not involve the cost c_j; however, the value of $c_j - Z_j$ changes, which may affect the optimality of the solution. The cost c_j is either associated with a basic or a non–basic variable and thus, the following two cases arise:

1. Let the cost c_j in which the change is made be the cost of the non-basic variable, that is, $c_j \notin C_B^{\mathrm{T}}$. Then, calculate the new $c_j - Z_j$ for the column x_j only. If the new $c_j - Z_j \leq 0$, the same solution remains optimal and if the new $c_j - Z_j > 0$, then use the simplex method to restore the optimality condition and also find optimum $Z = C_B^{\mathrm{T}} X_B$.
2. Let the cost c_j in which the change is made be the cost of the basic variable, that is, $c_j \in C_B^{\mathrm{T}}$, and is the coefficient of the rth basic variable x_r. Let the new cost c_j be $C_{Br} + \Delta C_{Br}$. Now calculate $c_j - Z_j$ for all non-basic variables x_j using the new cost of the basic variable.

Thus, for non-basic variables

$$c_j - (\text{new } Z_j) = c_j - \left\{ \sum_{i=1, i\neq r}^{m} C_{B_i}\alpha_i^j + (C_{B_r} + \Delta C_{B_r})\alpha_r^j \right\}$$

$$= c_j - \sum_{i=1}^{m} C_{B_i}\alpha_i^j - \Delta C_{B_r}\alpha_r^j$$

$$= \left(c_j - \sum_{i=1}^{m} C_{B_i}\alpha_i^j \right) - \Delta C_{B_r}\alpha_r^j = (c_j - Z_j) - \Delta C_{B_r}\alpha_r^j$$

Now for the same solution to remain optimal,

$$(c_j - Z_j) - \Delta C_{B_r}\alpha_r^j \leq 0 \tag{7.8}$$

$$\text{or} \quad \Delta C_{B_r} \geq \frac{c_j - Z_j}{\alpha_r^j} \text{ for } \alpha_r^j > 0 \tag{7.9}$$

$$\text{and} \quad \Delta C_{B_r} \leq \frac{c_j - Z_j}{\alpha_r^j} \text{ for } \alpha_r^j < 0 \tag{7.10}$$

Combining Eqs (7.9) and (7.10), we have

$$\max_j \left\{ \frac{c_j - Z_j}{\alpha_r^j}, \alpha_r^j > 0 \right\} \le \Delta C_{B_r} \le \min_j \left\{ \frac{c_j - Z_j}{\alpha_r^j}, \alpha_r^j < 0 \right\}. \tag{7.11}$$

Example 7.2 Consider the following LPP:

Max $Z = 2x_1 - x_2$

Subject to $x_1 + x_2 \le 2$

$x_1 - x_2 \le 1$

and $x_1, x_2 \ge 0$

The optimal table of the LPP is

		Cost →		2	1	0	0
		Variable →		x_1	x_2	x_3	x_4
Table no.	C_B	**Basic variable**	X_B	α^1	α^2	α^3	α^4
1	−1	x_2	1/2	0	1	1/2	−1/2
	2	x_1	3/2	1	0	1/2	1/2
		$c_j - Z_j$		0	0	−1/2	−3/2

Find the optimal solution of the aforementioned LPP, using sensitivity analysis for the following changes:

(a) Cost c_1 (the coefficient of x_1 in the objective function) is changed from 2 to 3
(b) Cost c_1 (the coefficient of x_1 in the objective function) is changed from 2 to −1
(c) How much can c_1 (=2) be changed so that the optimality is not disturbed and the aforegiven table remains the optimal table?

Solution The given LPP is

Max $Z = 2x_1 - x_2$

Subject to $x_1 + x_2 \le 2$

$x_1 - x_2 \le 1$

and $x_1, x_2 \ge 0$

(a) c_1 is changed from 2 to 3. Further, $c_1 \in C_B^T$, so we have to calculate the values of entries $c_j - Z_j$ for all columns of the non-basic variables x_j.

$$c_j - Z_j = c_j - C_B^T B^{-1} A_j \tag{7.12}$$

Here, new $C_B^T = (-1, 3)$, $B^{-1} = \begin{pmatrix} 1/2 & -1/2 \\ 1/2 & 1/2 \end{pmatrix}$, $c_j = (0, 0)$ is the cost matrix for the non-basic variables x_3 and x_4

and

$$A = (A_1, A_2, A_3, A_4) = \begin{pmatrix} 1 & 1 & 1 & 0 \\ 1 & -1 & 0 & 1 \end{pmatrix}$$

$$\therefore C_B^T B^{-1} = (-1, 3) \begin{pmatrix} 1/2 & -1/2 \\ 1/2 & 1/2 \end{pmatrix} = (1, 2)$$

and $C_B^T B^{-1} A_j = (1, 2) \begin{pmatrix} 1 & 0 \\ 0 & 1 \end{pmatrix} = (1, 2)$, $[(A_j)_{j=3,4} = (A_3, A_4)]$ (7.13)

From Eqs (7.12) and (7.13), we have

$$c_j - Z_j = c_j - C_B^{\mathrm{T}} B^{-1} A_j = (0, 0) - (1, 2) = (-1, -2), j = 3, 4$$

Hence, $c_3 - z_3 = -1 \leq 0$ and $c_4 - z_4 = -2 \leq 0$.

Thus, for the given change, optimality conditions are satisfied. The feasibility conditions are independent of c_j.

Thus, the BFS of the changed problem is

$$(x_2, x_1)^{\mathrm{T}} = X_B = B^{-1}b = \begin{pmatrix} 1/2 & -1/2 \\ 1/2 & 1/2 \end{pmatrix}\begin{pmatrix} 2 \\ 1 \end{pmatrix} = (1/2, 3/2)^{\mathrm{T}} \Rightarrow x_2 = 1/2, x_1 = 3/2$$

Max Z can be obtained by substituting $x_1 = 3/2$ and $x_2 = 1/2$ directly in the objective function of the changed problem, that is, $3x_1 - x_2 = 3 \times (3/2) - (1/2) = 4$.

The optimum solution of the changed problem is $x_1 = 3/2$, $x_2 = 1/2$, and Max $Z = 4$.

(b) c_1 is changed from 2 to -1. $c_1 \in C_B^{\mathrm{T}}$, so we have to calculate $c_j - Z_j$ for the non-basic variables x_3 and x_4.

Here, new $C_B^{\mathrm{T}} = (-1, -1)$, $B^{-1} = \begin{pmatrix} 1/2 & -1/2 \\ 1/2 & 1/2 \end{pmatrix}$

$c_j = (0, 0)$ is the cost matrix for the non-basic variables x_3 and x_4

$$A = (A_1, A_2, A_3, A_4) = \begin{pmatrix} 1 & 1 & 1 & 0 \\ 1 & -1 & 0 & 1 \end{pmatrix}$$

$$C_B^{\mathrm{T}} B^{-1} = (-1, -1)\begin{pmatrix} 1/2 & -1/2 \\ 1/2 & 1/2 \end{pmatrix} = (-1, 0)$$

$$C_B^{\mathrm{T}} B^{-1} A_j = (-1, 0)\begin{pmatrix} 1 & 0 \\ 0 & 1 \end{pmatrix} = (-1, 0), \ [(A_j)_{j=3,4} = (A_3, A_4)]$$

$$\therefore c_j - Z_j = c_j - C_B^{\mathrm{T}} B^{-1} A_j = (0, 0) - (-1, 0) = (1, 0), j = 3, 4$$

So $c_3 - z_3 = 1$ and $c_4 - z_4 = 0$. Hence, optimality is disturbed. Applying simplex iterations on the resulting simplex table 2, we get the following optimal table:

Cost→				−1	−1	0	0	Minimum ratio
Variable→				x_1	x_2	x_3	x_4	$\frac{X_{B_i}}{\alpha_i^j}, \alpha_i^j > 0$
Table no.	C_B	Basic variable	X_B	α^1	α^2	α^3	α^4	
2	−1	$\leftarrow x_2$	1/2	0	1	[1/2]	−1/2	$\frac{1/2}{1/2} = 1$
	−1	x_1	3/2	1	0	1/2	1/2	$\frac{3/2}{1/2} = 3$
	$c_j - Z_j$			0	0	1↑	0	–
3	0	x_3	1	0	2	1	−1	…
	−1	$\leftarrow x_1$	1	1	−1	0	[1]	1/1 = 1
	$c_j - Z_j$			0	−2	0	1↑	–
4	0	x_3	2	1	1	1	0	–
	0	x_4	1	1	−1	0	1	
	$c_j - Z_j$			−1	−1	0	0	–

The optimal solution is $x_1 = 0, x_2 = 0$, and Max $Z = 0$

(c) We know that

$$\max_j \left\{ \frac{c_j - Z_j}{\alpha_r^j}, \alpha_r^j > 0 \right\} \le \Delta C_{B_r} \le \min_j \left\{ \frac{c_j - Z_j}{\alpha_r^j}, \alpha_r^j < 0 \right\}.$$

Here $r = 2$ ($C_{B2} = c_1 = 2$ as x_1 is the second basic variable and hence $\Delta C_{Br} = \Delta c_1$)

x_3 and x_4 are non-basic variables.

$$\alpha_2^3 = 1/2 > 0, \alpha_2^4 = 1/2 > 0, c_3 - Z_3 = -1/2 \text{ and } c_4 - Z_4 = -3/2$$

$$\text{Thus, } \max\left\{ \frac{c_3 - Z_3}{\alpha_2^3}, \frac{c_4 - Z_4}{\alpha_2^4} \right\} = \max\left\{ \frac{-1/2}{1/2}, \frac{-3/2}{1/2} \right\} = \max\{-1, -3\} = -1.$$

Hence, $-1 \le \Delta c_1 < \infty$

As there is no α_r^j, which is negative, the upper limit of Δc_1 is ∞.

$$-1 + 2 < c_1 + \Delta c_1 < 2 + \infty \quad \Rightarrow \quad 1 \le \text{New } c_1 < \infty \quad \Rightarrow \quad 1 \le c_1^* < \infty$$

Thus when $c_1 = 2$ is changed to c_1^* (=new c_1), then that c_1^* can vary from 1 to ∞ leaving the existing simplex table optimal.

Alternate method We can also find the range of variation of c_1 by the ab-initio method (i.e., using the inequality in Eq. 7.8 directly). For this, let c_1 be changed to $c_1' = 2 + \Delta c_1$.

$$\text{Then } C_B^T = (-1, c_1'), B^{-1} = \begin{pmatrix} 1/2 & -1/2 \\ 1/2 & 1/2 \end{pmatrix}$$

$c_j = (0, 0)$ is the cost matrix for the non-basic variables x_3 and x_4

$$A = (A_1, A_2, A_3, A_4) = \begin{pmatrix} 1 & 1 & 1 & 0 \\ 1 & -1 & 0 & 1 \end{pmatrix}, A_j = \begin{pmatrix} 1 & 0 \\ 0 & 1 \end{pmatrix}, j = 3, 4$$

$$C_B^T B^{-1} = (-1, c_1') \begin{pmatrix} 1/2 & -1/2 \\ 1/2 & 1/2 \end{pmatrix} = \left(-\frac{1}{2} + \frac{c_1'}{2}, \frac{1}{2} + \frac{c_1'}{2} \right)$$

$$C_B^T B^{-1} A_j = \left(-\frac{1}{2} + \frac{c_1'}{2}, \frac{1}{2} + \frac{c_1'}{2} \right) \begin{pmatrix} 1 & 0 \\ 0 & 1 \end{pmatrix} = \left(-\frac{1}{2} + \frac{c_1'}{2}, \frac{1}{2} + \frac{c_1'}{2} \right), [(A_j)_{j=3,4} = (A_3, A_4)]$$

$$\therefore c_j - Z_j = c_j - C_B^T B^{-1} A_j = (0, 0) - \left(-\frac{1}{2} + \frac{c_1'}{2}, \frac{1}{2} + \frac{c_1'}{2} \right) = \left(\frac{1}{2} - \frac{c_1'}{2}, -\frac{1}{2} - \frac{c_1'}{2} \right)$$

$$\text{Thus, } c_3 - z_3 = \frac{1}{2} - \frac{c_1'}{2} \text{ and } c_4 - z_4 = -\frac{1}{2} - \frac{c_1'}{2}.$$

For optimality conditions to be satisfied, we have

$$\frac{1}{2} - \frac{c_1'}{2} \le 0 \Rightarrow c_1' \ge 1 \text{ or } 2 + \Delta c_1 \ge 1 \Rightarrow \Delta c_1 \ge -1 \; [c_1' = 2 + \Delta c_1] \tag{7.14}$$

$$\text{and } -\frac{1}{2} - \frac{c_1'}{2} \le 0 \Rightarrow c_1' \ge -1 \text{ or } 2 + \Delta c_1 \ge -1 \Rightarrow \Delta c_1 \ge -3 \tag{7.15}$$

From Eqs (7.14) and (7.15), we have $\Delta c_1 \ge -1$. This implies that we can decrease c_1 by 1 and there is no bound for its increase. Thus, c_1 can vary from 1 to ∞ or $1 \le c_1 < \infty$.

7.5 CHANGES IN b_i's AVAILABILITIES

We know that $X_B = B^{-1}b$ and $Z = C_B^T X_B$

The optimality conditions $(c_j - Z_j = c_j - C_B^T B^{-1} A_j \leq 0)$ remain unaffected for any change in b_i's (as they are independent of b). However, any change in the value of b_i's (the availability of resources) may affect X_B and Z and as a result, the new basic solution X_B may become infeasible. Thus, for any change in the availability vector b, first calculate $X_B^* = B^{-1}b$ (with the changed value of b).

The following two cases may arise:

1. If $X_B^* = B^{-1}$ (new b) ≥ 0, then the new solution is feasible and is given by $X_B^* = B^{-1}$ (new b) and Max $Z = C_B^T X_B^*$.
2. If $X_B^* = B^{-1}$ (new b) < 0, then the new solution becomes infeasible. In that case, the dual simplex method is applied to restore feasibility and the solution is found from the optimal simplex table thus obtained.

Example 7.3 A company produces two type of products, P_1 and P_2, using two raw materials, M_1 and M_2, as per the adjacent table:

Raw material	Product (tons)		Maximum daily availability
	P_1	P_2	
M_1	8	4	16
M_2	1	2	4
Profit per ton	5	4	

As per the market survey, the daily demand of P_2 cannot exceed that of P_1 by more than one ton.

(a) Formulate the problem in the LP model.

(b) Solve it using the simplex method.

(c) Starting from the optimal table obtained in (b) and using sensitivity analysis, find the range in which the availability of raw materials M_1 and M_2 can vary so that the optimal simplex table of (b) remains optimal (treat (c) independently).

Solution (a) Let x_1 and x_2 tons of products P_1 and P_2 be produced daily. The company would like to produce products P_1 and P_2 such that the daily profit $5x_1 + 4x_2$ is maximum. Thus, the objective is Max $Z = 5x_1 + 4x_2$. The usage of material $M_1 = 8x_1 + 4x_2$ tons and the usage of material $M_2 = x_1 + 2x_2$ tons. Since the daily availabilities of the raw materials M_1 and M_2 are limited to 16 and 4 tons respectively, we have $8x_1 + 4x_2 \leq 16$ and $x_1 + 2x_2 \leq 4$. Further, the demand of P_2 cannot exceed that of P_1 by more than 1 ton, we have $x_2 \leq x_1 + 1$, that is, $-x_1 + x_2 \leq 1$. Thus, the complete LP model of the problem can be written as follows:

$$
\begin{aligned}
&\text{Max } Z = 5x_1 + 4x_2\\
&\text{Subject to } \quad 8x_1 + 4x_2 \leq 16\\
&\qquad\qquad\qquad x_1 + 2x_2 \leq 4\\
&\qquad\qquad\qquad -x_1 + x_2 \leq 1\\
&\qquad\qquad\qquad x_1, x_2 \geq 0
\end{aligned}
$$

(b) The problem can be written in the canonical form as follows:

$$
\begin{aligned}
&\text{Max } Z = 5x_1 + 4x_2 + 0{\cdot}x_3 + 0{\cdot}x_4 + 0{\cdot}x_5\\
&\text{Subject to } \quad 8x_1 + 4x_2 + x_3 = 16\\
&\qquad\qquad\qquad x_1 + 2x_2 + x_4 = 4
\end{aligned}
$$

$$-x_1 + x_2 + x_5 = 1$$

and $x_1, x_2, x_3, x_4, x_5 \geq 0$, x_3, x_4, and x_5 are slack variables.

The simplest BFS is obtained by equating $x_1 = x_2 = 0$ so that $x_3 = 16$, $x_4 = 4$, and $x_5 = 1$ will be the BFS, with x_3, x_4, and x_5 as basic variables and x_1 and x_2 as non-basic variables.

Thus, the initial simplex table is as follows:

Cost →				5	4	0	0	0	Minimum ratio
Variable →				x_1	x_2	x_3	x_4	x_5	$\frac{X_{B_i}}{\alpha_i^j}, \alpha_i^j > 0$
Table no.	C_B	Basic variable	X_B	α^1	α^2	α^3	α^4	α^5	
1	0	$\leftarrow x_3$	16	[8]	4	1	0	0	$16/8 = 2 \rightarrow$
	0	x_4	4	1	2	0	1	0	$4/1 = 4$
	0	x_5	1	−1	1	0	0	1	...
		$c_j - Z_j$		5↑	4	0	0	0	
2	5	x_1	2	1	1/2	1/8	0	0	$\frac{2}{1/2} = 4$
	0	$\leftarrow x_4$	2	0	[3/2]	−1/8	1	0	$\frac{2}{3/2} = 4/3 \rightarrow$
	0	x_5	3	0	3/2	1/8	0	1	$\frac{3}{3/2} = 2$
		$c_j - Z_j$		0	3/2↑	−5/8	0	0	
3	5	x_1	4/3	1	0	1/12	−1/3	0	
	4	x_2	4/3	0	1	−1/12	2/3	0	
	0	x_5	1	0	0	−3/12	−1	1	
		$c_j - Z_j$		0	0	−1/4	−1	0	

In the aforegiven table 3, $\Delta_j \leq 0$ for all j, therefore the solution is optimal. Hence, the optimal solution is $x_1 = 4/3$, $x_2 = 4/3$, and Max $Z = 12$.

(c) (i) To find the range of availability of raw material M_1, let the present availability 16 be changed to b_1. Then,

$$\text{new } X_B = B^{-1}(\text{new } b) = \begin{pmatrix} 1/12 & -1/3 & 0 \\ -1/12 & 2/3 & 0 \\ 3/12 & -1 & 1 \end{pmatrix} \begin{pmatrix} b_1 \\ 4 \\ 1 \end{pmatrix} = \begin{pmatrix} \frac{b_1}{12} - \frac{4}{3} \\ -\frac{b_1}{12} + \frac{8}{3} \\ \frac{3}{12} b_1 - 4 + 1 \end{pmatrix} \geq 0,$$

for feasibility to be satisfied.

Thus $\frac{b_1}{12} - \frac{4}{3} \geq 0$, i.e., $b_1 \geq 16$ (7.16)

$-\frac{b_1}{12} + \frac{8}{3} \geq 0$, i.e., $b_1 \leq 32$ (7.17)

$\frac{3}{12} b_1 - 4 + 1 \geq 0$, i.e., $b_1 \geq 12$ (7.18)

Now from Eqs (7.16), (7.17), and (7.18), we get $16 \le b_1 \le 32$, which is the range of variability of b_1.

To find the range of availability of raw material M_2, let the present availability 4 be changed to b_2.

Then

$$\text{New } X_B = B^{-1}(\text{new } b) = \begin{pmatrix} 1/12 & -1/3 & 0 \\ -1/12 & 2/3 & 0 \\ 3/12 & -1 & 1 \end{pmatrix} \begin{pmatrix} 16 \\ b_2 \\ 1 \end{pmatrix} = \begin{pmatrix} \frac{16}{12} - \frac{b_2}{3} \\ -\frac{16}{12} + \frac{2}{3} b_2 \\ \frac{3 \times 16}{12} - b_2 + 1 \end{pmatrix} \ge 0$$

for feasibility to be satisfied.

Thus $\frac{16}{12} - \frac{b_2}{3} \ge 0$, i.e., $b_2 \le 4$ (7.19)

$-\frac{16}{12} + \frac{2}{3} b_2 \ge 0$, i.e., $b_2 \ge 2$ (7.20)

$\frac{3 \times 16}{12} - b_2 + 1 \ge 0$, i.e., $b_2 \le 5$ (7.21)

Now from Eqs (7.19), (7.20), and (7.21), we get $2 \le b_2 \le 4$, which is the range of variability of b_2.

Example 7.4 Consider the following LPP:

Min $Z = 2x_1 + x_2$

Subject to $3x_1 + x_2 = 3$

$4x_1 + 3x_2 \ge 6$

$x_1 + 2x_2 \le 3$

and $x_1, x_2 \ge 0$

Writing this LPP in the canonical form and after modifying the objective function, we get

Max $Z' = -2x_1 - x_2 - M(x_3 + x_4)$

Subject to $3x_1 + x_2 + x_4 = 3$

$4x_1 + 3x_2 - x_3 + x_5 = 6$

$x_1 + 2x_2 + x_6 = 3$

$x_1, x_2, x_3, x_4, x_5, x_6 \ge 0$, x_4 and x_5 are artificial variables, x_6 is a slack variable, x_3 is a surplus variable, and M is a large positive number.

The optimum solution of this LPP is given by the following table:

Cost →				−2	−1	0	−M	−M	0
Variable →				x_1	x_2	x_3	x_4	x_5	x_6
Table no.	C_B	Basic variable	X_B	α^1	α^2	α^3	α^4	α^5	α^6
1	−2	x_1	3/5	1	0	1/5	3/5	−1/5	0
	−1	x_2	6/5	0	1	−3/5	−4/5	3/5	0
	0	x_6	0	0	0	1	1	−1	1
		$c_j - Z_j$		0	0	−1/5	$-M+(2/5)$	$-M+(1/5)$	0

with $x_1 = 3/5$, $x_2 = 6/5$ and Min $Z = 12/5$, Max $Z' = -12/5$.

Using sensitivity analysis, find the solution if the following changes are made:

(a) $b = (3, 6, 3)^T$ is changed to $(4, 7, 3)^T$

(b) $b = (3, 6, 3)^T$ is changed to $(5, 5, 3)^T$

Solution (a) If we change $b = (3, 6, 3)^T$ to $(4, 7, 3)^T$, then the new solution is

$$X_B' = B^{-1}b_{\text{new}} = \begin{pmatrix} 3/5 & -1/5 & 0 \\ -4/5 & 3/5 & 0 \\ 1 & -1 & 1 \end{pmatrix}\begin{pmatrix} 4 \\ 7 \\ 3 \end{pmatrix} = \begin{pmatrix} 1 \\ 1 \\ 0 \end{pmatrix} \geq 0, \text{ feasibility is not disturbed.}$$

Thus, the solution of the changed problem is Min $Z = 2 + 1 = 3$ or Max $Z' = -3$ when $x_1 = 1, x_2 = 1$.

(b) When $b = (3, 6, 3)^T$ is changed to $(5, 5, 3)^T$, then the new solution is

$$X_B' = B^{-1}b_{\text{new}} = \begin{pmatrix} 3/5 & -1/5 & 0 \\ -4/5 & 3/5 & 0 \\ 1 & -1 & 1 \end{pmatrix}\begin{pmatrix} 5 \\ 5 \\ 3 \end{pmatrix} = \begin{pmatrix} 2 \\ -1 \\ 3 \end{pmatrix} \text{ which is infeasible.}$$

We now use the dual simplex method to restore the feasibility in table 1 after incorporating the change. The new simplex table obtained is as follows:

	Cost →			−2	−1	0	−M	−M	0	Minimum ratio
	Variable →			x_1	x_2	x_3	x_4	x_5	x_6	
Table no.	C_B	**Basic variable**	X_B	α^1	α^2	α^3	α^4	α^5	α^6	
1	−2	x_1	2	1	0	1/5	3/5	−1/5	0	x_4 is not considered as being artificial
	−1	$\leftarrow x_2$	−1	0	1	−3/5	−4/5	3/5	0	
	0	x_6	3	0	0	1	1	−1	1	
		$c_j - Z_j$		0	0	−1/5↑	−M+(2/5)	−M+(1/5)	0	
2	−2	x_1	5/3	1	1/3	0	1/3	0	0	
	0	x_3	5/3	0	−5/3	1	4/3	−1	0	
	0	x_6	4/3	0	5/3	0	−1/3	0	1	
		$c_j - Z_j$		0	−1/3	0	−M+(2/3)	−M	0	

Thus in table 1, when x_2 leaves, x_3 enters. The optimal solution (given by table 2) of the changed problem is $x_1 = 5/3$, $x_2 = 0$ and Max $Z' = -10/3$ or Min $Z = 10/3$.

7.6 ADDITION OF NEW VARIABLES

Addition of a new variable, say x_j, results in adding one more term $c_j x_j$ in the objective function and a new column A_j in the coefficient matrix. This addition, in turn, leads to the addition of a column α^j of a new variable in the optimal table of the given problem. The value $c_j - Z_j$ is calculated for the newly added column. This may affect the optimality of the existing solution in the optimal table. In case the optimality is disturbed, the simplex method is used to restore it.

Let a variable x_j with cost c_j and column A_j be added to a given problem. Now since C_B^T, B^{-1}, A_j, and c_j are known, $C_B^T B^{-1}$, $C_B^T B^{-1} A_j$ and $c_j - Z_j (= c_j - C_B^T B^{-1} A_j)$ can be calculated. Two cases arise:

1. If $c_j - Z_j \leq 0$, then the solution of the original problem remains optimal for the changed problem.

2. If $c_j - Z_j > 0$ for the column of the added variable, then the optimality is disturbed. To restore the optimality, simplex iterations are applied till the optimal solution of the changed problem is obtained.

Example 7.5 The optimal table of the LPP

$$\text{Max } Z = 2x_1 + 3x_2$$
$$\text{Subject to } \quad x_1 + 2x_2 \le 2$$
$$2x_1 - x_2 \le 3$$
$$3x_1 + x_2 \le 5$$
$$\text{and} \quad x_1, x_2 \ge 0$$

with slack variables x_3, x_4, and x_5 added to the first, second, and third constraints respectively is given in the adjacent table.

	Cost →			2	3	0	0	0
	Variable →			x_1	x_2	x_3	x_4	x_5
Table no.	C_B	Basic variable	X_B	α^1	α^2	α^3	α^4	α^5
0	3	x_2	1/5	0	1	2/5	−1/5	0
	2	x_1	8/5	1	0	1/5	2/5	0
	0	x_5	0	0	0	−1	−1	1
	$c_j - Z_j$			0	0	−8/5	−1/5	0

Using sensitivity analysis, find the solution of the new problem obtained with the following changes in the given problem:

(a) Add a variable x_a with cost 5 (coefficient of x_a in the objective function) and column $(5, 10, 15)^{\mathrm{T}}$ in the coefficient matrix.

(b) Add a variable x_b with cost 4 (coefficient of x_b in the objective function) and column $(0, 5, 4)^{\mathrm{T}}$ in the coefficient matrix.

Solution The changed problem, when written in the standard form of the LPP is

$$\text{Max } Z = 2x_1 + 3x_2 + 5x_a$$
$$\text{Subject to } \quad x_1 + 2x_2 + 5x_a + x_3 = 2$$
$$2x_1 - x_2 + 10x_a + x_4 = 3$$
$$3x_1 + x_2 + 15x_a + x_5 = 5$$
$$\text{and} \quad x_1, x_2, x_3, x_4, x_5, x_a \ge 0$$

(a) The variable added is x_a with cost $c_a = 5$ and column $(5, 10, 15)^{\mathrm{T}} = A_a$. We first calculate $c_j - Z_j$ for the column of this variable. From the given optimal table we have $C_B^{\mathrm{T}} = (3, 2, 0)$

$$B^{-1} = \begin{pmatrix} 2/5 & -1/5 & 0 \\ 1/5 & 2/5 & 0 \\ -1 & -1 & 1 \end{pmatrix}; \; C_B^{\mathrm{T}} B^{-1} = (3, 2, 0) \begin{pmatrix} 2/5 & -1/5 & 0 \\ 1/5 & 2/5 & 0 \\ -1 & -1 & 1 \end{pmatrix} = (8/5, 1/5, 0)$$

the following are also given: $c_a = 5$ and $A_a = (5, 10, 15)^{\mathrm{T}}$

$$\therefore c_a - Z_j = c_a - C_B^{\mathrm{T}} B^{-1} A_a = 5 - (8/5,\ 1/5,\ 0)(5, 10, 15)^{\mathrm{T}}$$

$$= 5 - 10 = -5 < 0$$

Thus, the optimality condition is not disturbed, and the optimal solution of the original problem remains the optimal solution of the changed problem. Hence, the optimal solution of the changed problem is Max $Z = 19/5$ with $x_1 = 8/5$, $x_2 = 1/5$, and $x_a = 0$.

(b) Let the variable added be x_b with cost $c_b = 4$ and column $A_b = (0, 5, 4)^{\mathrm{T}}$. From the given table, simplex multiplier $C_B^{\mathrm{T}} B^{-1}$ is

$$C_B^{\mathrm{T}} B^{-1} = (3, 2, 0) \begin{pmatrix} 2/5 & -1/5 & 0 \\ 1/5 & 2/5 & 0 \\ -1 & -1 & 1 \end{pmatrix} = \left(\frac{8}{5}, \frac{1}{5}, 0 \right)$$

the following are also given: $c_b = 4$ and $A_b = (0, 5, 4)^T$

$$c_b - Z_j = c_b - C_B^T B^{-1} A_b = 4 - \left(\frac{8}{5}, \frac{1}{5}, 0\right)(0, 5, 4)^T = 4 - 1 = 3 > 0$$

Hence, the optimality condition is disturbed. To prepare the simplex table for the changed problem, we calculate the new column vector α^b as

$$\alpha^b = B^{-1}A_b = \begin{pmatrix} 2/5 & -1/5 & 0 \\ 1/5 & 2/5 & 0 \\ -1 & -1 & 1 \end{pmatrix}\begin{pmatrix} 0 \\ 5 \\ 4 \end{pmatrix} = \begin{pmatrix} -1 \\ 2 \\ -1 \end{pmatrix}$$

The simplex table for the changed problem obtained from the original problem by adding the column of vector α^b in the given optimal table and $c_b - Z_b$ entry in $c_j - Z_j$ row is table 1. Thereafter using the simplex method, we get table 2.

Cost →				2	3	4	0	0	0	Minimum ratio $\frac{X_{B_i}}{\alpha_i^j}$
Variable →				x_1	x_2	x_b	x_3	x_4	x_5	
Table no.	C_B	Basic variable	X_B	α^1	α^2	α^b	α^3	α^4	α^5	
1	3	x_2	1/5	0	1	−1	2/5	−1/5	0	…
	2	x_1	8/5	1	0	[2]	1/5	2/5	0	$\frac{8/5}{2} = 4/5$
	0	x_5	0	0	0	−1	−1	−1	1	…
		$c_j - Z_j$		0	0	3↑	−8/5	−1/5	0	
2	3	x_2	1	1/2	1	0	1/2	0	0	
	4	x_b	4/5	1/2	0	1	1/10	1/5	0	
	0	x_5	4/5	1/2	0	0	−9/10	−4/5	1	
		$c_j - Z_j$		−3/2	0	0	−19/10	−4/5	0	

The optimum solution is Max $Z = 31/5$, $x_1 = 0$, $x_2 = 1$, and $x_b = 4/5$.

7.7 DELETION OF CONSTRAINTS

To see the effect of the deletion of a constraint, let us consider the following LPP in two variables with three constraints:

$$\begin{aligned} &C_1: a_1x_1 + b_1x_2 \le a \\ &C_2: a_2x_1 + b_2x_2 \le b \\ &C_3: a_3x_1 + b_3x_2 \le c \end{aligned} \quad \text{or} \quad \begin{aligned} &C_1: a_1x_1 + b_1x_2 + x_3 = a \\ &C_2: a_2x_1 + b_2x_2 + x_4 = b \\ &C_3: a_3x_1 + b_3x_2 + x_5 = c \end{aligned}$$

with corresponding line representations C_1, C_2, and C_3 respectively, as shown in Fig. 7.4. Let us assume the optimum occurs at the point C. Now, if the third constraint (C_3) is deleted, it has no effect on the solution of the existing optimal solution, as its boundary lines do not pass through the optimal point C. However, if C_1 or C_2 is deleted, then the existing optimal solution of the given problem is affected, as they pass through the optimal point C. On deletion of one of these constraints, point C does not remain the optimal point anymore. As a result, the optimal solution of the problem changes. Now delete c_1 and let us add a new variable ($-x_3'$) to the given constraint C_1, so that it becomes $a_1x_1 + b_1x_2 + x_3 - x_3' = a$. This form of the constraint is satisfied, if in the optimal solution of the new problem, $x_3 = 0$ and $x_3' > 0$ or $x_3 > 0$ and $x_3' = 0$. As a result, the points satisfying the optimal solution can be on either side of C_1. Thus, C_1 becomes ineffective and is considered as deleted.

Note Deletion of a constraint can affect optimality conditions, as its deletion amounts to the addition of a variable. This addition of a variable can affect the optimality condition.

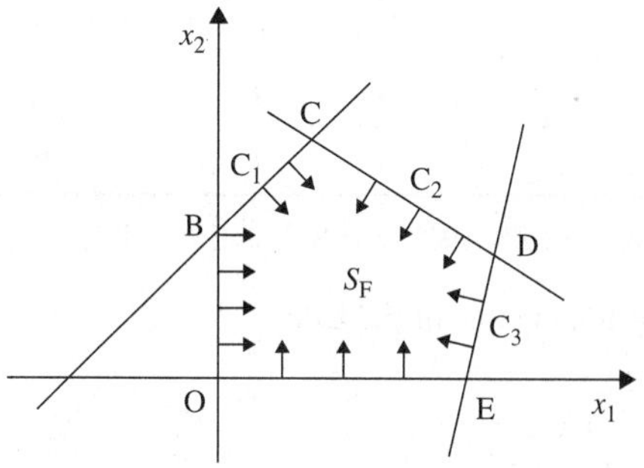

Figure 7.4 Graph of constraints C_1, C_2, and C_3

Example 7.6 The optimal table of the LPP

Max $Z = 2x_1 + 3x_2$

Subject to $x_1 + 2x_2 \le 2$

$2x_1 - x_2 \le 3$

$3x_1 + x_2 \le 5$

and $x_1, x_2 \ge 0$

or

Max $Z = 2x_1 + 3x_2 + 0 \cdot x_3 + 0 \cdot x_4$

Subject to $x_1 + 2x_2 + x_3 = 2$

$2x_1 - x_2 + x_4 = 3$

$3x_1 + x_2 + x_5 = 5$

Cost →				2	3	0	0	0
Variable →				x_1	x_2	x_3	x_4	x_5
Table no.	C_B	Basic variable	X_B	α^1	α^2	α^3	α^4	α^5
0	3	x_2	1/5	0	1	2/5	−1/5	0
	2	x_1	8/5	1	0	1/5	2/5	0
	0	x_5	0	0	0	−1	−1	1
		$c_j - Z_j$		0	0	−8/5	−1/5	0

$x_1, x_2, x_3, x_4, x_5 \ge 0$; x_3, x_4, x_5 are slack variables, is given in the aforegiven table: Using sensitivity analysis, find the solution to the new problem obtained by deleting the third constraint $3x_1 + x_2 \le 5$.

Solution The new third constraint after adding the variable ($-x_5'$) is as follows:

$$3x_1 + x_2 + x_5 - x_5' = 5, x_5' > 0$$

To calculate the new column for the added variable x_5', we calculate the α^j column and the $c_j - Z_j$ entry for the new simplex table of the changed problem.

$$\alpha^j = B^{-1}A_j = \begin{pmatrix} 2/5 & -2/5 & 0 \\ 1/5 & 2/5 & 0 \\ -1 & -1 & 1 \end{pmatrix}\begin{pmatrix} 0 \\ 0 \\ -1 \end{pmatrix} = \begin{pmatrix} 0 \\ 0 \\ -1 \end{pmatrix}$$

$$C_B^{\mathrm{T}}B^{-1} = (3, 2, 0)\begin{pmatrix} 2/5 & -1/5 & 0 \\ 1/5 & 2/5 & 0 \\ -1 & -1 & 1 \end{pmatrix} = \left(\frac{8}{5}, \frac{1}{5}, 0\right)$$

$c_j - Z_j = c_j - C_B^T B^{-1} A_j = 0 - (8/5, 1/5, 0)(0, 0, -1)^T = 0 - 0 = 0$

The simplex table for the new (changed) problem is as follows:

		Cost →		2	3	0	0	0	0
		Variable →		x_1	x_2	$x_5^!$	x_3	x_4	x_5
Table no.	C_B	**Basic variable**	X_B	α^1	α^2	α^j	α^3	α^4	α^5
1	3	x_2	1/5	0	1	0	2/5	−1/5	0
	2	x_1	8/5	1	0	0	1/5	2/5	0
	0	x_5	0	0	0	−1	−1	−1	1
		$c_j - Z_j$		0	0	0	−8/5	−1/5	0

The optimum solution is Max $Z = 19/5$, $x_1 = 8/5$, and $x_2 = 1/5$.

Example 7.7 The optimum table of the LPP:

Max $Z = x_1 + 2x_2$

Subject to $-x_1 + x_2 \le 1$ (7.22)

$x_1 + x_2 \le 2$ (7.23)

$x_1, x_2 \ge 0$, is given as follows:

		Cost →		1	2	0	0
		Variable →		x_1	x_2	x_3	x_4
Table no.	C_B	**Basic variable**	X_B	α^1	α^2	α^3	α^4
0	2	x_2	3/2	0	1	1/2	1/2
	1	x_1	1/2	1	0	−1/2	1/2
		$c_j - Z_j$		0	0	−1/2	−3/2

where x_3 and x_4 are the slack variables added to constraints given in Eqs (7.22) and (7.23) respectively. Using sensitivity analysis, find the optimal solution to the problem with the following changes:

(a) Constraint in Eq. (7.22) is deleted

(b) Constraint in Eq. (7.23) is deleted

Solution (a) The graphical representation of the given LPP is shown in Fig. 7.5. The optimal solution of the given problem is $x_1 = 1/2, x_2 = 3/2$. If we place these values in the constraint in Eq. (7.22), it is satisfied as exact equality and hence if we delete it, the optimal solution will change. To get the solution of the changed problem, we write the constraint in Eq. (7.22) as $-x_1 + x_2 + x_3 - x_3^+ = 1$. This results in the deletion of the constraint in Eq. (7.22).

Thus, the deletion of constraint Eq. 7.22 means the addition of a new variable $x_3^+ \ge 0$ with the cost 0 and column $(-1, 0)^T$ in the coefficient matrix of the original problem. Calculate

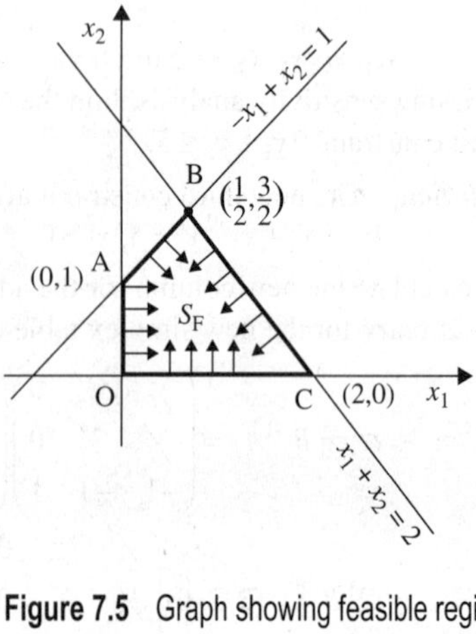

Figure 7.5 Graph showing feasible region

$$\alpha^j = B^{-1}A_j = \begin{pmatrix} \tfrac{1}{2} & \tfrac{1}{2} \\ -\tfrac{1}{2} & \tfrac{1}{2} \end{pmatrix}\begin{pmatrix} -1 \\ 0 \end{pmatrix} = \begin{pmatrix} -\tfrac{1}{2} \\ \tfrac{1}{2} \end{pmatrix} \tag{7.24}$$

$$c_j - Z_j = c_j - C_B^{\mathrm{T}} B^{-1} A_j = 0 - (2,\ 1)(-1/2,\ 1/2)^{\mathrm{T}} = 0 - (-1/2) = 1/2$$

The optimality is affected. We restore it using the simplex method.

Incorporating these computations in the given optimal simplex table, we get the following simplex table:

		Cost →		1	2	0	0	0
		Variable →		x_1	x_2	x_3^+	x_3	x_4
Table no.	C_B	**Basic variable**	X_B	α^1	α^2	α^{3+}	α^3	α^4
1	2	x_2	3/2	0	1	−1/2	1/2	1/2
	1	$\leftarrow x_1$	1/2	1	0	1/2	−1/2	1/2
		$c_j - Z_j$		0	0	1/2↑	−1/2	−3/2
2	2	x_2	2	1	1	0	0	1
	0	x_3^+	1	2	0	1	−1	1
		$c_j - Z_j$		−1	0	0	0	−2

This is the optimal table. The optimal solution is $x_1 = 0$, $x_2 = 2$, and Max $Z = 4$.

(b) Given: $x_1 = 1/2$ and $x_2 = 3/2$ as the optimal solution of the original problem. The constraint to be deleted is $x_1 + x_2 \le 2$, which is satisfied as strict equality by the optimal solution of the original problem. Hence, the deletion of the constraint will change the solution. We write the constraint $x_1 + x_2 \le 2$ to be deleted as $x_1 + x_2 + x_4 - x_4^+ = 2$. Thus, a new variable x_4^+ with the cost 0 and column $A_4^+ = (0,\ -1)^{\mathrm{T}}$ is added in the problem. For this

$$\alpha^j = B^{-1}A_j = \begin{pmatrix} \tfrac{1}{2} & \tfrac{1}{2} \\ -\tfrac{1}{2} & \tfrac{1}{2} \end{pmatrix}\begin{pmatrix} 0 \\ -1 \end{pmatrix} = \begin{pmatrix} -\tfrac{1}{2} \\ -\tfrac{1}{2} \end{pmatrix} \tag{7.25}$$

$$c_j - Z_j = c_j - C_B^{\mathrm{T}} B^{-1} A_j = 0 - (2,\ 1)(-1/2,\ -1/2)^{\mathrm{T}} = 0 - (-3/2) = 3/2$$

The optimality is disturbed. We use the simplex method to restore it. The simplex table for the new problem obtained from the optimal table of the original problem by adding an additional column $(-1/2,\ -1/2)^{\mathrm{T}}$ of the variable x_4^+ and the entry $c_j - Z_j = 3/2$ in the $c_j - Z_j$ row is given as follows:

		Cost →		1	2	0	0	0
		Variable →		x_1	x_2	x_4^+	x_3	x_4
Table no.	C_B	**Basic variable**	X_B	α^1	α^2	α^{4+}	α^3	α^4
1	2	x_2	3/2	0	1	−1/2	1/2	1/2
	1	x_1	1/2	1	0	−1/2	−1/2	1/2
		$c_j - Z_j$		0	0	3/2	−1/2	−3/2

Thus optimality is disturbed as the entry 3/2 in the x_4^+ column and the $c_j - Z_j$ row is positive. Now, variable x_4^+ can enter the basis but all entries in the x_4^+ column are negative and hence, there is no variable to leave the basis. Therefore, the solution of the changed problem is unbounded.

7.8 DELETION OF VARIABLES

Sometimes, an organization may decide to stop using a certain resource variable, either due to its non-availability or it becoming obsolete. In this section, we discuss the effect of such a situation on the optimal solution of a problem.

Suppose we have obtained the optimal solution of an LPP and after that we decide to delete (drop) one of the variables of the problem. In such a situation, the variable to be deleted can either be (a) a non-basic variable or (b) a basic variable. In the first case, the non-basic variable assumes the zero value in the optimal solution. Its deletion, therefore, has no effect on the existing solution and the same solution remains optimal for the new problem also. However in the second case, when the variable to be deleted is basic, the column of this variable is deleted from the given simplex table and its row is multiplied with (–1) (to make the constraints ineffective). As a result, the solution becomes infeasible (optimality conditions remain unaffected) and the dual simplex method is used to restore the feasibility of the solution.

Example 7.8 The optimal table of the LPP

$$\text{Max } Z = -x_1 + 3x_2 - 2x_3$$

$$\text{Subject to } \quad 3x_1 - x_2 + 2x_3 \le 7$$
$$-2x_1 + 4x_2 \le 12$$
$$-4x_1 + 3x_2 + 8x_3 \le 10$$
$$x_1, x_2, x_3 \ge 0, \text{ is}$$

	Cost →			2	3	–2	0	0	0
	Variable →			x_1	x_2	x_3	x_4	x_5	x_6
Table no.	C_B	Basic variable	X_B	α^1	α^2	α^3	α^4	α^5	α^6
1	–1	x_1	4	1	0	4/5	2/5	1/10	0
	3	x_2	5	0	1	2/5	1/5	3/10	0
	0	x_6	11	0	0	10	1	–1/2	1
		$c_j - Z_j$		0	0	–12/5	–1/5	–4/5	0

Using sensitivity analysis, find the optimal solution of the problem obtained after making the following changes in the given problem: (a) delete the variable x_3 (b) delete the variable x_1.

Solution (a) The variable x_3 to be deleted is a non-basic variable and hence $x_3 = 0$ in the given optimal solution. Hence, on deleting the x_3 column from the given optimal table, we get the same solution as the optimal solution of the new problem, that is, $x_1 = 4$, $x_2 = 5$, and Max $Z = 11$.

(b) In this case, x_1 is a basic variable. Hence, delete the x_1 column from the given optimal table and multiply the x_1 row by (–1) to get the simplex table of the new problem. Using the dual simplex method to restore the feasibility, we get the optimal table of the new problem.

Cost →				3	−2	0	0	0	Minimum ratio
Variable →				x_2	x_3	x_4	x_5	x_6	
Table no.	C_B	Basic variable	X_B	α^2	α^3	α^4	α^5	α^6	
1	−1	$\leftarrow x_1$	−4	0	−4/5	[−2/5]	−1/10	0	$\min\left\{\left\|\frac{-12/5}{-4/5}\right\|, \left\|\frac{-1/5}{-2/5}\right\|, \left\|\frac{-4/5}{-1/10}\right\|\right\}$
	3	x_2	5	1	2/5	1/5	3/10	0	
	0	x_6	11	0	10	1	−1/2	1	$=1/2$ for x_4 column, when x_1 leaves x_4 enters.
	$c_j - Z_j$			0	−12/5	−1/5↑	−4/5	0	
2	0	x_4	10	0	2	1	1/4	0	
	3	x_2	3	1	0	0	1/4	0	
	0	x_6	1	0	8	0	3/4	0	
	$c_j - Z_j$			0	−2	0	−3/4	0	

The optimal solution is $x_2 = 3$, $x_3 = 0$, and Max $Z = 9$.

7.9 ADDITION OF CONSTRAINTS

Any simplex table has as many basic variables (hence as many rows) as the number of constraints involved in the problem. The addition of a constraint in the given problem adds one more variable and a row as well to the given optimal simplex table, which may affect the feasibility of the solution. While adding the new constraint, it is first ascertained if the given optimal solution satisfies the constraint. If it is satisfied, then the same solution remains optimal for the new problem also. If the optimal solution does not satisfy the constraint, then first, the given constraint is expressed in the '≤' form (if it is not in this form) and a slack variable is added to it. Now, introduce a new row of this constraint and a unit column of added slack variable in the given optimal table. This disturbs the standard format of the simplex table. To bring back the resulting simplex table in the standard format, elementary row operations are used. If feasibility is not disturbed, the same solution remains optimal. If the basic solution thus obtained is infeasible, the dual simplex method is applied till the optimal BFS is obtained. Note that in adding a constraint, an artificial variable is never used. Thus, if a constraint to be added is of the '≥' type, it is first multiplied with (−1), and the slack variable is added to convert it to the equality form.

Example 7.9 Max $Z = 12x_1 + 15x_2$

Subject to $4x_1 + 3x_2 \leq 12$

$2x_1 + 5x_2 \leq 10$

$x_1, x_2 \geq 0$

The optimal table of the aforementioned LPP with slack variables x_3 and x_4 is given as follows:

Cost →				12	15	0	0
Variable →				x_1	x_2	x_3	x_4
Table no.	C_B	Basic variable	X_B	α^1	α^2	α^3	α^4
1	12	x_1	15/7	1	0	5/14	−3/14
	15	x_2	8/7	0	1	−1/7	2/7
	$c_j - Z_j$			0	0	−15/7	−12/7

Using sensitivity analysis, find the solution of this problem when constraint (a) $x_1 + x_2 \geq 2$ is added (b) $x_1 - 2x_2 \geq 2$ is added.

Solution (a) The constraint $x_1 + x_2 \geq 2$ is satisfied by the optimal solution $x_1 = 15/7$ and $x_2 = 8/7$ of the existing problem. Therefore, the same solution remains optimal for the changed problem as well. The optimal solution of the changed problem is Max $Z = 300/7$ when $x_1 = 15/7$ and $x_2 = 8/7$.

(b) When we put $x_1 = 15/7$ and $x_2 = 8/7$ in the constraint $x_1 - 2x_2 \geq 2$, it is not satisfied. So the optimal solution of the changed problem may be different. To find it, we first multiply the given constraint with (–1) and then add a slack variable, say x_5, to get $-x_1 + 2x_2 + x_5 = -2$.

Add this as a row in the given optimal table with x_5 as the third basic variable. In addition, add a column of x_5 with the cost '0' and the column vector $(0, 0, 1)^T$. The new simplex table is as follows:

		Cost →		12	15	0	0	0
		Variable →		x_1	x_2	x_3	x_4	x_5
Table no.	C_B	Basic variable	X_B	α^1	α^2	α^3	α^4	α^5
1	12	x_1	15/7	1	0	5/14	–3/14	0
	15	x_2	8/7	0	1	–1/7	2/7	0
	0	x_5	–2	–1	2	0	0	1
		$c_j - Z_j$		0	0	–15/7	–12/7	0

The simplex table is not in its standard format, as the x_1 column must be (1, 0, 0, 0) instead of (1, 0, –1; 0). Similarly, the x_2 column is (0, 1, 2, 0), which must be (0, 1, 0, 0). To restore the standard format, add x_1 row and (–2) times of the x_2 row in the corresponding entries of the x_5 row to get the following simplex table:

		Cost →		12	15	0	0	0
		Variable →		x_1	x_2	x_3	x_4	x_5
Table no.	C_B	Basic variable	X_B	α^1	α^2	α^3	α^4	α^5
1	12	x_1	15/7	1	0	5/14	–3/14	0
	15	x_2	8/7	0	1	–1/7	2/7	0
	0	$\leftarrow x_5$	–15/7	0	0	9/14	–11/14	1
		$c_j - Z_j$		0	0	–15/7	–12/7↑	0
2	12	x_1	30/11	1	0	331/308	0	–3/11
	15	x_2	4/11	0	1	1/11	0	4/11
	0	x_4	30/11	0	0	–9/11	1	–14/11
		$c_j - Z_j$		0	0	–39/7	0	–24/11

The optimum solution is Max $Z = 420/11$, $x_1 = 30/11$, and $x_2 = 4/11$.

7.10 CHANGE IN COLUMN A_j OF COEFFICIENT MATRIX A

Change in one of the columns A_j of the coefficient matrix $A=[a_{ij}]_{m\times n}=(A_1, A_2, \ldots, A_n)$ may affect the condition of optimality, $c_j-Z_j=c_j-C_B^TB^{-1}A_j$. As a result, two cases may arise:

1. The column A_j may be of a non-basic variable.
   (a) If the optimality conditions are satisfied after the change is made, then the same solution will remain optimal.
   (b) If the optimality conditions are not satisfied, then the simplex method is used to restore the optimality condition.
2. The column A_j may be of a basic variable. In this case, we use the following technique. Replace the existing cost c_j of the basic variable x_j by large $M>0$. Now, add a new variable x_j' with the existing cost, c_j in the objective function and a new column A_j' in the coefficient matrix A and a new column α_j' in the optimal table. Now, use the simplex method, treating x_j as an artificial variable.

Example 7.10 The optimal table of the LPP

Max $Z=3x_1-2x_2+4x_3$

Subject to $x_1+2x_2+2x_3\le 430$

$3x_1+2x_2\le 460$

$x_1+4x_2\le 420$

$x_1, x_2, x_3\ge 0$, with x_4, x_5, and x_6 as slack variables is as follows:

		Cost →		3	−2	4	0	0	0
		Variable →		x_1	x_2	x_3	x_4	x_5	x_6
Table no.	C_B	Basic variable	X_B	α^1	α^2	α^3	α^4	α^5	α^6
1	0	x_4	200	−1/2	2	0	1	−1/2	0
	4	x_3	230	3/2	0	1	0	1/2	0
	0	x_6	420	1	4	0	0	0	1
		c_j-Z_j		−3	−2	0	0	−2	0

Using sensitivity analysis, find the solution of the LPP with respect to the following changes:

(a) $A_1=(1, 3, 1)^T$ of the coefficient matrix is changed to $(3, 3, 4)^T$.
(b) $A_1=(1, 3, 1)^T$ is changed to $(2, 1, 3)^T$.

Solution (a) Change in the column (the coefficients of x_1) may affect only the optimality condition (c_1-Z_1).

$$c_1-Z_1=c_1-C_B^TB^{-1}A_1$$

$$C_B^TB^{-1}=(0,4,0)\begin{pmatrix}1 & -\frac{1}{2} & 0\\ 0 & \frac{1}{2} & 0\\ 0 & 0 & 1\end{pmatrix}=(0,2,0)$$

Given, the new $A_1=(3, 3, 4)^T$

Thus, $C_B^T B^{-1} A_I = (0, 2, 0)\ (3, 3, 4)^T = 6$

Therefore, the new $(c_1 - Z_1) = 3 - 6 = -3$

The optimality condition is not disturbed. Hence, the original solution remains optimal for the changed problem as well, that is, $x_1 = x_2 = 0$ and $x_3 = 230$, Max $Z = 920$.

(b) Here again, $C_B^T B^{-1} = (0, 2, 0)$ and the new $A_1 = (2, 1, 3)^T$. Therefore, the new $(c_1 - Z_1) = 3 - (0, 2, 0)(2, 1, 3)^T = 3 - 2 = 1 > 0$.

Hence, the optimality condition is disturbed. To restore optimality, we use the regular simplex method.

The new α^1 is given as follows:

$$\alpha^1 = B^{-1} \times \text{new } A_1 = \begin{pmatrix} 1 & -1/2 & 0 \\ 0 & 1/2 & 0 \\ 0 & 0 & 1 \end{pmatrix} \begin{pmatrix} 2 \\ 1 \\ 3 \end{pmatrix} = (3/2, 1/2, 3)^T$$

Thus, after incorporating these changes, the resulting simplex table is as follows:

Cost →				3	−2	4	0	0	0	Minimum ratio
Variable →				x_1	x_2	x_3	x_4	x_5	x_6	
Table no.	C_B	Basic variable	X_B	α^1	α^2	α^3	α^4	α^5	α^6	
1	0	←x_4	200	[3/2]	2	0	1	−1/2	0	400/3 (min)
	4	x_3	230	1/2	0	1	0	1/2	0	460
	0	x_6	420	3	4	0	0	0	1	140
	$c_j - Z_j$			1↑	−2	0	0	−2	0	
2	0	x_1	400/3	1	4/3	0	2/3	−1/3	0	
	4	x_3	490/3	0	−2/3	1	−1/3	2/3	0	
	0	x_6	20	0	0	0	−2	1	1	
	$c_j - Z_j$			0	−10/3	0	−2/3	−5/3	0	

The optimal solution is $x_1 = 400/3$, $x_2 = 0$, $x_3 = 490/3$, and Max $Z = 3160/3$.

7.11 PARAMETRIC LINEAR PROGRAMMING

Parametric linear programming is a part of sensitivity analysis. It investigates the changes in the optimum LP solution that result from predetermined continuous variations in the (a) coefficients C of the objective function and (b) the RHS b of the constraints of the LP, defined as

$$\text{Max } Z = C^T X$$

$$\text{Subject to } \sum_{j=1}^{n} A_j x_j = b \ (m \text{ constraints in } n \text{ variables})$$

where $X = (x_1, x_2, \ldots, x_n)^T \geq 0$, $C = (c_1, c_2, \ldots, c_n)^T$, and $b = (b_1, b_2, \ldots, b_m)^T \geq 0$.

The typical parametric function is $C(t) = c + tc'$, in case of variations in the coefficients of the objective function and $b(t) = b + tb'$, in case of variation in RHS of constraints, where t is the parameter of variation, and c' and b' are the given vectors. For convenience, we assume $t \geq 0$.

Note The variation of the parameter in parametric linear programming may be linear or non-linear. Computations in a non-linear case are complicated.

7.11.1 Parametric Changes in Cost Vector *c*

The main steps involved in parametric analysis for parametric changes in c are as follows:

1. Find the optimal solution of the given LP (if not given) at $t = 0$.
2. Consider the given parametric changes in c. Using the optimality conditions of the simplex method, determine the range $0 \le t \le t_1$, for which the solution obtained at $t = 0$ remains optimal. In this case, the value t_1 of t is called the critical value. This process continues by determining the successive critical values t_i's and their corresponding optimal feasible solutions, till there is an indication that the last solution will remain unchanged for $t > t_i$, or that no feasible solution exists beyond the point t_i.

Example 7.11 For $t \ge 0$, determine the critical values of t for which the solution of the following parametric LPP remains optimal basic feasible.

$$\text{Max } Z = (3 - 6t)x_1 + (2 - 2t)x_2 + (5 + 5t)x_3$$

Subject to $x_1 + 2x_2 + x_3 \le 40$

$3x_1 + 2x_3 \le 60$

$x_1 + 4x_2 \le 30$

and $x_1, x_2, x_3 \ge 0$

Solution The standard form of the given LPP is

$$\text{Max } Z = (3 - 6t)x_1 + (2 - 2t)x_2 + (5 + 5t)x_3 + 0{\cdot}x_4 + 0{\cdot}x_5 + 0{\cdot}x_6$$

Subject to $x_1 + 2x_2 + x_3 + x_4 = 40$

$3x_1 + 2x_3 + x_5 = 60$

$x_1 + 4x_2 + x_6 = 30$

and $x_1, x_2, x_3, x_4, x_5, x_6 \ge 0$; x_4, x_5, and x_6 are slack variables

Here $c(t) = (3 - 6t, 2 - 2t, 5 + 5t)^T$. The optimal solution of the given LPP at $t = 0$ is

	Cost →		$3-6t$	$2-2t$	$5+5t$	0	0	0
	Variable →		x_1	x_2	x_3	x_4	x_5	x_6
C_{B_0}	Basic variable	X_{B_0}	α^1	α^2	α^3	α^4	α^5	α^6
$2-2t$	x_2	5	−1/4	1	0	1/2	−1/4	0
$5+5t$	x_3	30	3/2	0	1	0	1/2	0
0	x_6	10	2	0	0	−2	1	1

Here $X_{B_0} = (x_2, x_3, x_6)^T = (5, 30, 10)^T, C_{B_0}(t) = (2 - 2t, 5 + 5t, 0)^T$, and $B_0^{-1} = \begin{pmatrix} 1/2 & -1/4 & 0 \\ 0 & 1/2 & 0 \\ -2 & 1 & 1 \end{pmatrix}$

The optimality conditions for the non-basic variables x_1, x_4, and x_5 in the aforegiven table are

$$\left(c_j(t) - Z_j(t)\right)_{j=1,4,5} = \left(c_j(t) - C_{B_0}(t)B_0^{-1}A_j\right)_{j=1,4,5}$$

$$= (3-6t, 0, 0) - (2-2t, 5+5t, 0)\begin{pmatrix} \frac{1}{2} & -\frac{1}{4} & 0 \\ 0 & \frac{1}{2} & 0 \\ -2 & 1 & 1 \end{pmatrix}\begin{pmatrix} 1 & 1 & 0 \\ 3 & 0 & 1 \\ 1 & 0 & 0 \end{pmatrix}$$

$$= (-4-14t, -1+t, -2-3t) \le 0$$

Thus X_{B_0} remains optimal as long as the following conditions are satisfied: $-4-14t \le 0$, $-1+t \le 0$, and $-2-3t \le 0$.

Since $t \ge 0$, the first and third inequalities are satisfied, and the second inequality is satisfied if $t \le 1$. Thus, X_{B_0} remains optimal if $0 \le t \le 1$, and $t_1 = 1$ is the first critical value. The optimal solution is $x_1 = 0$, $x_2 = 5$, $x_3 = 30$, and $Z = 160 + 140t$, $0 \le t \le 1$.

At $t = 1$, $c_4(t) - Z_4(t) = -1 + t$ becomes zero and is positive for $t > 1$. Hence, for $t > 1$, the optimality condition $-1 + t \le 0$ is not satisfied, and x_4 becomes the entering basic variable. To find the leaving variable, use the minimum ratio rule and the revised simplex method; x_2 leaves the basis when x_4 enters the basis. The revised simplex table is

C_{B_0}	Basic variable	X_{B_0}	B_0^{-1}			Non-basic variables x_1 x_4 x_5	Minimum ratio $\frac{X_{B_i}}{\alpha_i^j}$
$2-2t$	$\leftarrow x_2$	5	1/2	−1/4	0	[1/2]	$\frac{5}{1/2} = 10$
$5+5t$	x_3	30	0	1/2	0	0	…
0	x_6	10	−2	1	1	−2	…
$c_j - Z_j$						$t-1\uparrow$	

Using the same elementary row operations on column of X_{B_0} and B_0^{-1}, which will reduce the pivot entry [1/2] to 1 and all entries below '1', thus obtained to zeros, we get the new inverse (B_1^{-1}) for the next simplex table. The new revised simplex table is as follows:

C_{B_1}	Basic variable	X_{B_1}	B_0^{-1}			Non-basic variables x_1 x_2 x_5	Minimum ratio $\frac{X_{B_i}}{\alpha_i^j}$
0	x_4	10	1	−1/2	0		
$5+5t$	x_3	30	0	1/2	0		
0	x_6	30	0	0	1		
$c_j - Z_j$							

Here, $X_{B_1} = (x_4, x_3, x_6)^{\mathrm{T}} = (10, 30, 30)^{\mathrm{T}}$, $C_{B_1}(t) = (0, 5+5t, 0)^{\mathrm{T}}$.

The optimality conditions are

$$\left(c_j(t) - Z_j(t)\right)_{j=1,2,5} = \left(c_j(t) - C_{B_0}(t)B_0^{-1}A_j\right)_{j=1,2,5}$$

$$= (3-6t, 2-2t, 0) - (0, 5+5t, 0)\begin{pmatrix} 1 & -\frac{1}{2} & 0 \\ 0 & \frac{1}{2} & 0 \\ 0 & 0 & 1 \end{pmatrix}\begin{pmatrix} 1 & 2 & 0 \\ 3 & 0 & 1 \\ 1 & 4 & 0 \end{pmatrix}$$

$$= \left(-\frac{1+27t}{2}, 2-2t, -\frac{5+5t}{2}\right) \le 0 \text{ is satisfied for } t > 1.$$

Thus, X_{B_1} remains optimal for $t \geq 1$. This means the second critical value is $t_2 = \infty$ and the process ends. Thus, the overall optimal solution is

$$x_1 = 0, x_2 = 5, x_3 = 30, \text{ and } Z = 160 + 140t, \text{ for } 0 \leq t \leq 1$$

$$x_1 = 0, x_2 = 0, x_3 = 30, \text{ and } Z = 150 + 150t, \text{ for } t \geq 1$$

7.11.2 Parametric Changes in Requirement Vector *b*

Consider the LPP

$$\text{Max } Z = C^T X$$
$$\text{Subject to } \quad AX = b$$
$$X \geq 0$$

Let the requirement vector b of the LPP be replaced by $b(\lambda)=b+\lambda b'$, where b' represents the predetermined variation in b, and λ is a parameter. We take $\lambda \geq 0$. Here again, the given LPP is initially solved for $\lambda=0$ using the simplex method technique. Now, any changes in the requirement vector $b(\lambda)$ can only affect the feasibility of the existing optimal solution obtained at $\lambda = 0$. The critical values of λ are thus determined by applying the conditions of feasibility, that is, $X_{B_0}(\lambda) = B_0^{-1}(b + \lambda b') \geq 0$. B_0 is the basis matrix of the basic variables in the optimal solution of the LPP corresponding to $\lambda = 0$. Similarly, we shall use the notation B_i for the basis matrix corresponding to the critical value λ_i. This condition also determines the range of λ in which the solution remains optimal and feasible.

Example 7.12 Determine the critical values (range) of λ for which the solution of the following parametric LPP remains optimal and basic feasible.

$$\text{Max } Z = 3x_1 + 2x_2 + 5x_3$$
$$\text{Subject to } \quad x_1 + 2x_2 + x_3 \leq 430 + \lambda$$
$$3x_1 + 2x_3 \leq 460 - 4\lambda$$
$$x_1 + 4x_2 \leq 420 - 4\lambda$$
$$\text{and } \quad x_1, x_2, x_3 \geq 0, \lambda \text{ is a non-negative parameter}$$

Solution Here $b=(430, 460, 420)^T$, $b'=(1, -4, -4)^T$. Taking $\lambda = 0$ and x_4, x_5, and x_6 as slack variables, the given LPP is solved using the simplex method. The optimum BFS obtained is given by Table 7.1.

Table 7.1 Optimum BFS for Example for 7.12

	Cost →		3	2	5	0	0	0
	Variable →		x_1	x_2	x_3	x_4	x_5	x_6
C_{B_0}	Basic variable	X_{B_0}	α^1	α^2	α^3	α^4	α^5	α^6
2	x_2	100	–1/4	1	0	1/2	–1/4	0
5	x_3	230	3/2	0	1	0	1/2	0
0	x_6	20	2	0	0	–2	1	1
	$c_j - Z_j$		–4	0	0	–1	–2	0

Here $X_{B_0}=(x_2,x_3,x_6)^{\mathrm{T}}=(100,230,20)^{\mathrm{T}}, C_{B_0}(t)=(2,5,0)^{\mathrm{T}}$ and

$$B_0^{-1}=\begin{pmatrix} 1/2 & -1/4 & 0 \\ 0 & 1/2 & 0 \\ -2 & 1 & 1 \end{pmatrix}$$

Predetermined variation in $b=b'=(1,-4,-4)^{\mathrm{T}}$.

To find the range over which the current optimal solution remains feasible, we first calculate

$$X'_{B_n}=B_0^{-1}b'=\begin{pmatrix} 1/2 & -1/4 & 0 \\ 0 & 1/2 & 0 \\ -2 & 1 & 1 \end{pmatrix}\begin{pmatrix} 1 \\ -4 \\ -4 \end{pmatrix}=\begin{pmatrix} 3/2 \\ -2 \\ -10 \end{pmatrix} \tag{7.26}$$

$$X_{B_n}=B_0^{-1}b=(100,230,20)^{\mathrm{T}} \tag{7.27}$$

For the current solution $X_{B_0}(\lambda)$ to remain feasible,

$$X_{B_0}(\lambda)=B_0^{-1}(b+\lambda\, b')=B_0^{-1}b+\lambda B_0^{-1}b'\geq 0$$

$$X_{B_0}(\lambda)=(100,230,20)^{\mathrm{T}}+\lambda(3/2,-2,-10)^{\mathrm{T}}\geq 0 \quad \text{[Using Eqs (7.26) and (7.27)]}$$

or $\quad X_{B_0}(\lambda)=\left(100+\frac{3}{2}\lambda, 230-2\lambda, 20-10\lambda\right)^{\mathrm{T}}\geq 0$

$\Rightarrow 100+\frac{3}{2}\lambda\geq 0, 230-2\lambda\geq 0$ and $20-10\lambda\geq 0$

$\Rightarrow \lambda\leq 2$

Thus, $\lambda=2$ is the first critical value and the existing solution of Table 7.1 remains optimal and feasible in the range $0\leq\lambda\leq 2$.

The objective function value in this range is

$$Z(\lambda)=C_B^{\mathrm{T}}X_{B_0}(\lambda)=(2,5,0)\left(100+\frac{3}{2}\lambda, 230-2\lambda, 20-10\lambda\right)^{\mathrm{T}}=1350-7\lambda$$

At $\lambda=2, x_2=103, x_3=226, x_6=0$ and the optimum simplex table will be as shown in Table 7.2.

Table 7.2 Optimum simplex table for Example for 7.12

	Cost →		**3**	**2**	**5**	**0**	**0**	**0**
	Variable →		x_1	x_2	x_3	x_4	x_5	x_6
C_{B_0}	**Basic variable**	X_{B_0}	α^1	α^2	α^3	α^4	α^5	α^6
2	x_2	103	−1/4	1	0	1/2	−1/4	0
5	x_3	226	3/2	0	1	0	1/2	0
0	x_6	0	2	0	0	−2	1	1
	c_j-Z_j		−4	0	0	−1	−2	0

When $\lambda>2$, x_6 $(=20-10\lambda)$ becomes negative and therefore by the dual simplex method, x_6 leaves and x_4 (enters) becomes the basic variable. As a result, the improved simplex table becomes

	Cost →		3	2	5	0	0	0
	Variable →		x_1	x_2	x_3	x_4	x_5	x_6
C_{B_1}	Basic variable	X_{B_1}	α^1	α^2	α^3	α^4	α^5	α^6
2	x_2	103	1/4	1	0	0	0	1/4
5	x_3	226	3/2	0	1	0	1/2	0
0	x_4	0	−1	0	0	1	−1/2	−1/2
	$c_j - Z_j$		−5	0	0	0	−5/2	−1/2

For the present table, we first calculate $B_1^{-1}b' = X'_{B_1}$:

$$X'_{B_1} = B_1^{-1}b' = \begin{pmatrix} 0 & 0 & \frac{1}{4} \\ 0 & \frac{1}{2} & 0 \\ 1 & -\frac{1}{2} & -\frac{1}{2} \end{pmatrix} \begin{pmatrix} 1 \\ -4 \\ -4 \end{pmatrix} = \begin{pmatrix} -1 \\ -2 \\ 5 \end{pmatrix}$$

For the current solution $X_B(\lambda)$ to remain feasible

$$X_B(\lambda) = (x_2, x_3, x_4)^{\mathrm{T}} = B_1^{-1}(b + \lambda b') = X_{B_1} + \lambda X'_{B_1} \geq 0$$

$$= (103, 226, 0)^{\mathrm{T}} + \lambda(-1, -2, 5)^{\mathrm{T}} \geq 0$$

$$= (103 - \lambda, 226 - 2\lambda, 5\lambda)^{\mathrm{T}} \geq 0 \quad \Rightarrow \quad \lambda \leq 103$$

Thus, the second critical value of λ is $\lambda = 103$ and the existing solution $x_1 = 0, x_2 = 103 - \lambda$, and $x_3 = 226 - 2\lambda$ is optimal and feasible in the range $2 \leq \lambda \leq 103$. The objective function value in this range is

$$Z(\lambda) = C_B^{\mathrm{T}} X_{B_1}(\lambda) = (2, 5, 0)(103 - \lambda, 226 - 2\lambda, 5\lambda)^{\mathrm{T}} = 1336 - 12\lambda$$

When $\lambda > 103$, the basic variable x_2 becomes negative and the solution becomes infeasible. However, all entries in the row corresponding to this basic variable are positive. As a result, the infeasible solution cannot be improved by the dual simplex method and no solution exists for $\lambda > 103$. Summarizing the details, the ranges of λ for which the solution remains optimum and feasible are as follows:

λ	x_1	x_2	x_3	Z
$0 \leq \lambda \leq 2$	0	$100 + \frac{3}{2}\lambda$	$230 - 2\lambda$	$1350 - 7\lambda$
$2 \leq \lambda \leq 103$	0	$103 - \lambda$	$226 - 2\lambda$	$1336 - 12\lambda$

7.12 DIFFERENCE BETWEEN SENSITIVITY ANALYSIS AND PARAMETRIC LINEAR PROGRAMMING

Sensitivity analysis is used to understand the effect of discrete changes in the parameters after obtaining the optimal solution of the LPP, whereas parametric linear programming is used to understand the effect of continuous changes in the parameters after obtaining the optimal solution of the LPP.

RECAPITULATION

1. Sensitivity analysis or post-optimal analysis is used to find the optimal solution, when the different parameters of the LP model (such as the coefficients of a decision variable column in constraints, RHS of constraints, coefficients of decision variables in the objective function) are changed.
2. The main purpose of the post-optimality or sensitivity analysis is to determine the new optimal feasible solution, if it exists, in the most effective and efficient way, when changes in the problem are made after getting its optimal solution.
3. In sensitivity analysis, if any changes made in the problem disturb the optimality conditions, then the simplex method is used to restore the optimality.
4. The dual simplex method is used if the feasibility conditions are disturbed.
5. If any changes made in the problem disturb both the feasibility and optimality conditions, then sensitivity or post-optimality analysis fails to give the solution for the changed problem.
6. The addition of a new variable may or may not affect the optimality conditions. If the optimality conditions are disturbed, use the simplex method to restore them. The same solution remains optimal, if the optimal conditions are not disturbed.
7. The deletion of a constraint means the addition of a new variable, and hence can affect optimality conditions.
8. The deletion of a variable and the addition of a new constraint may affect feasibility.
9. Parametric linear programming is applied to both linear and non-linear parametric variations. Generally, parametric linear programming is used in continuous parametric changes in the cost vector C and the RHS of b of constraints.

EXERCISES

Multiple-choice Questions

Consider the following LPP:

$$\left.\begin{aligned} &\text{Max } Z = 3x_1 - 2x_2 + 4x_3 \\ &\text{Subject to} \quad x_1 + 2x_2 + x_3 \le 430 \\ &\qquad\qquad\quad 3x_1 + 2x_3 \le 460 \\ &\qquad\qquad\quad x_1 + 4x_2 \le 420 \\ &\text{and} \qquad\quad \text{all } x_i' \ge 0 \end{aligned}\right\} \tag{7.28}$$

The optimal table is as follows:

		Cost →		3	−2	4	0	0	0
		Variable →		x_1	x_2	x_3	x_4	x_5	x_6
Table no.	C_B	**Basic variable**	X_B	α^1	α^2	α^3	α^4	α^5	α^6
1	0	x_4	200	−1/2	2	0	1	−1/2	0
	4	x_3	230	3/2	0	1	0	1/2	0
	0	x_6	420	1	4	0	0	0	1
		$c_j - Z_j$		−3	−2	0	0	−2	0

7.1 If the first constraint of the given LPP is deleted, then the optimal solution of the changed problem is
(a) $x_1 = 0, x_2 = 0, x_3 = 230$
(b) $x_1 = 100, x_2 = 80, x_3 = 80$
(c) $x_1 = 0, x_2 = 230, x_3 = 0$
(d) none of these

7.2 If in Eq. (7.28), b is changed to $(400, 400, 400)^T$, then the optimal solution of the changed problem is
(a) $x_1 = 0, x_2 = 0, x_3 = 230$
(b) $x_1 = 0, x_2 = 0, x_3 = 200$
(c) $x_1 = 200, x_2 = 400, x_3 = 230$
(d) none of these

7.3 If the constraint $x_1 - x_2 + x_3 \le 400$ is added to Eq. (7.28), then the optimal solution of the changed problem is
(a) $x_1 = 0, x_2 = 0, x_3 = 230$
(b) $x_1 = 100, x_2 = 80, x_3 = 80$
(c) $x_1 = 100, x_2 = 0, x_3 = 230$
(d) none of these

7.4 If the constraint $2x_1 + x_2 + x_3 \le 200$ is added to Eq. (7.28), then the optimal solution of the changed problem is
(a) $x_1 = 0, x_2 = 0, x_3 = 0$
(b) $x_1 = 0, x_2 = 0, x_3 = -30$
(c) $x_1 = 200, x_2 = 400, x_3 = 200$
(d) none of these

7.5 Which is the largest interval for c_1, provided in Eq. (7.28), for which the optimal solution does not change?
(a) $0 \le c_1 \le 3$
(b) $-\infty < c_1 \le 6$
(c) $0 \le c_1 \le 6$
(d) $-\infty \le c_1 \le \infty$

7.6 If x_3 is deleted from Eq. (7.28), then the optimal solution of the changed problem is
(a) $x_1 = 460/3, x_2 = 0, x_3 = 0$
(b) $x_1 = 0, x_2 = 0, x_3 = 230$
(c) $x_1 = 460/3, x_2 = 0$
(d) no feasible solution

7.7 Parametric programming is used when the variations made in the parameter are
(a) continuous
(b) discrete
(c) may be continuous or discrete
(d) none of these

7.8 Any change in costs and availabilities of an LPP simultaneously may disturb
(a) only optimality (b) only feasibility (c) both optimality and feasibility

State True or False

7.1 The addition of a new variable may affect only the optimality condition.
7.2 A change in vector b may affect the feasibility of the solution.
7.3 The addition of a new constraint can affect only the feasibility of the optimal solution of the original problem.
7.4 In sensitivity analysis, when the feasibility of an LPP is disturbed, we use the dual simplex method to restore it.
7.5 In sensitivity analysis, when the optimality of an LPP is disturbed, we use the simplex method to restore it.
7.6 In sensitivity analysis, while replacing an existing constraint with a new constraint, we first delete the existing constraint, and then add the new constraint.
7.7 In sensitivity analysis, while replacing an existing constraint with a new constraint, we first add the new constraint and then delete the existing constraint.
7.8 The parametric LPP is only applicable to linear parametric variations.
7.9 The parametric variation in the parametric LP can be non-linear in nature.

Fill in the Blanks

7.1 The simultaneous change in the availability vector and the addition of a new constraint to an LPP may disturb ___________.

7.2 The simultaneous addition of a new variable and deletion of an existing variable in an LPP may disturb ___________.

7.3 The simultaneous addition of a new constraint and deletion of an existing constraint in an LPP may disturb ___________.

7.4 Suppose S_F of an LPP is non-empty and bounded, and a constraint is deleted from this LPP, then the new S_F may become ___________.

7.5 Suppose S_F of an LPP is non-empty and bounded, and a new constraint is added to this LPP, then the new S_F may become ___________.

7.6 The computational work is comparatively easier if we consider ___________ parametric variations in the parametric LPP.

Review Questions

7.1 Write a short note on graphical sensitivity analysis.

7.2 Explain what changes in sensitivity analysis affect (a) feasibility and (b) optimality.

7.3 Can you use sensitivity analysis when both the feasibility and optimality conditions are disturbed simultaneously?

7.4 Write a short note on parametric linear programming.

Numerical Problems

7.1 Consider the following LPP:

Max $Z = 3x_1 + 5x_2$

Subject to $x_1 + x_2 \leq 1$

$2x_1 + 3x_2 \leq 1$

and $x_1, x_2 \geq 0$

Find the range at which (a) c_1 can vary (b) c_2 can vary without affecting the optimality conditions of the optimal solution for the given problem.

7.2 The given LPP is

Max $Z = 3x_1 + 5x_2$

Subject to $x_1 \leq 4$

$x_2 \leq 6$

$3x_1 + 2x_2 \leq 18$

and $x_1, x_2 \geq 0$

Find the solution of the new problem using sensitivity analysis, if c_2 is changed from 5 to 1.

7.3 Find the range within which the RHS of the constraints (taken one by one) of the following LPP can vary without disturbing the feasibility:

Max $Z = -x_1 + 2x_2 - x_3$

Subject to $3x_1 + x_2 - x_3 \leq 10$

$-x_1 + 4x_2 + x_3 \geq 6$

$x_2 + x_3 \leq 4$

and $x_1, x_2, x_3 \geq 0$

7.4 Consider the following LPP:
Max $Z = 3x_1 + 5x_2$
Subject to $x_1 + x_3 = 4$
$3x_1 + 2x_2 + x_4 = 18$
and $x_1, x_2, x_3, x_4 \geq 0$
Find the solution of the new problem using sensitivity analysis, if we add a new variable x_5 with the cost 7 and column vector $(1, 2)^T$ to the aforementioned LPP.

7.5 The optimal table of the LPP
Max $Z = x_1 - x_2 - x_3$
Subject to $-2x_1 + x_2 + x_3 \geq 2$
$x_1 - 2x_2 + 2x_3 = 2$
$x_1, x_2, x_3 \geq 0$, is

	Cost →			1	−1	−1	0	$-M$	$-M$
	Variable →			x_1	x_2	x_3	x_4	x_5	x_6
Table no.	C_B	Basic variable	X_B	α^1	α^2	α^3	α^4	α^5	α^6
–	−1	x_2	1/2	−5/4	1	0	−1/2	1/2	−1/4
	−1	x_3	3/2	−3/4	0	1	−1/2	1/2	1/4
		$c_j - Z_j$		−1	0	0	−1	$1-M$	$-M$

x_4 is the surplus variable, and x_5, x_6 are the artificial variables associated with the first and second constraints respectively. Using sensitivity analysis, find the optimal solution of the LPP, obtained after the following changes are made in the given LPP (all the parts are independent):

(a) Change the RHS of the second constraint: (i) from 2 to 4 (ii) from 2 to 8.
(b) Change the first constraint from '≥' to '≤'.
(c) Add a new constraint $x_1 + x_2 + x_3 \leq 1$.
(d) Change the cost c_1 (the coefficient of x_1 in the objective function) from 1 to 3.
(e) Add a new constraint $x_1 + x_2 + x_3 \geq 4$.
(f) Add a new variable x_7 with the cost '2' and column $(2, -1)^T$.
(g) Find the range of variation of c_2 so that the optimality of the existing simplex table is not disturbed.
(h) Find the range of variation of b_1 (= 2) so that the feasibility of the existing simplex table is not disturbed.

7.6 For $t \geq 0$, determine the critical values of t for which the solution of the following parametric LPP remains optimal basic feasible.
Max $Z = (3 - 6t)x_1 + (2 - 2t)x_2 + (5 + 5t)x_3$
Subject to $x_1 + 2x_2 + x_3 \leq 430$
$3x_1 + 2x_3 \leq 460$
$x_1 + 4x_2 \leq 420$
and $x_1, x_2, x_3 \geq 0$

7.7 Determine the critical values (range) of λ for which the solution of the following parametric LPP remains optimal basic feasible.
Max $Z = 3x_1 + 2x_2 + 5x_3$
Subject to $x_1 + 2x_2 + x_3 \leq 40 - \lambda$
$3x_1 + 2x_3 \leq 60 + 2\lambda$
$x_1 + 4x_2 \leq 30 - 7\lambda$
and $x_1, x_2, x_3 \geq 0$, λ is a parameter.

CONCEPTUAL PROBLEM

An electric infrastructure company engaged in manufacturing electronic items is planning to expand its business. After conducting a market survey, the company decides to manufacture electrical toys, which may be beneficial and are in demand. The company decides to manufacture two models of electrical toys, M_1 and M_2. To manufacture these models, three resources—resistors, capacitors, and chips are required as raw materials and the company manager contacts various sources for their supply. The following table summarizes the data of the situation:

Resource	Unit resource requirements		Maximum availability (units)
	Model M_1 (units)	Model M_2 (units)	
Resistor	2	3	1200
Capacitor	2	1	1000
Chips	0	4	800
Unit profit (₹ in hundreds)	3	4	–

Now, considering the limited availability of raw materials, the company faces a problem in estimating the number of models of M_1 and M_2 to be manufactured, in order to earn maximum profit. The company wants solutions for the following:

1. What is the status of each resource at the optimal level?
2. After getting the optimal solution, will it be advantageous for the company, if the availability of
   (a) resistors is increased to 1300
   (b) chips is reduced to 350

Solution Let x_1 and x_2 be the amounts produced of models M_1 and M_2 respectively. The following are the details of the LP model and its associated optimal simplex table (obtained using the simplex method):

Max $Z = 3x_1 + 4x_2$

Subject to $2x_1 + 3x_2 \leq 1200$ (Resistor constraint)

$2x_1 + x_2 \leq 1000$ (Capacitor constraint)

$0x_1 + 4x_2 \leq 800$ (Chips constraint)

Cost →				3	4	0	0	0
Variable →				x_1	x_2	x_3	x_4	x_5
Table no.	C_B	Basic variable	X_B	α^1	α^2	α^3	α^4	α^5
...	3	x_1	450	1	0	–1/4	3/4	0
	0	x_5	400	0	0	–2	2	1
	4	x_2	100	0	1	1/2	–1/2	0
		$c_j - Z_j$		0	0	–5/4	–1/4	0

x_3, x_4, and x_5 are slack variables added to the three constraints. The optimal solution is $x_1 = 450$ and $x_2 = 100$, and max profit $= 3x_1 + 4x_2 = 1750$.

(Contd)

(*Contd*)

1. (a) Status of resources:
   The solution is $x_1 = 450$, $x_2 = 100$, and Max $Z = 3x_1 + 4x_2 = 1750$.

Status of first resource Requirement of first resource $= 2x_1 + 3x_2 = 2 \times 450 + 3 \times 100 = 900 + 300 = 1200$

Availability of the first resource = 1200

Thus, the first resource is completely utilized to get maximum profit.

Status of second resource Requirement of second resource $= 2x_1 + x_2 = 2 \times 450 + 100 = 1000$

Availability of second resource = 1000

Thus, the second resource is also fully utilized.

Status of third resource Requirement of third resource $= 4x_2 = 2 \times 100 = 400$

Availability of third resource = 400

Thus, the third resource (chips) is not completely used and is in excess of $800 - 400 = 400$ units.

2. (a) With the increase in the number of resistors to 1300 (from 1200), the RHS of the constraints is changed and the resultant is

$$\begin{pmatrix} x_1 \\ x_5 \\ x_2 \end{pmatrix} = B^{-1} \times (\text{new RHS}) = \begin{pmatrix} -1/4 & 3/4 & 0 \\ -2 & 2 & 1 \\ 1/2 & -1/2 & 0 \end{pmatrix} \begin{pmatrix} 1300 \\ 1000 \\ 800 \end{pmatrix} = \begin{pmatrix} 425 \\ 200 \\ 150 \end{pmatrix}$$

Thus, the current basic variables x_1, x_5, and x_2 remain feasible at the new values $x_1 = 425$, $x_5 = 200$, and $x_2 = 150$. The associated maximum profit for the resulting problem is 1875. The maximum profit in the given problem is 1750. Thus, the increase in the number of resistors to 1300 (from 1200) is advantageous, and for this, the company should manufacture 425 units of model M_1 and 150 units of model M_2 electrical toys.

(b) With the decrease in the number of chips to 350 (from 800), the RHS of the constraints is changed to $(1200, 1000, 350)^T$ and the resultant is

$$\begin{pmatrix} x_1 \\ x_5 \\ x_2 \end{pmatrix} = B^{-1}b = \begin{pmatrix} -1/4 & 3/4 & 0 \\ -2 & 2 & 1 \\ 1/2 & -1/2 & 0 \end{pmatrix} \begin{pmatrix} 1200 \\ 1000 \\ 350 \end{pmatrix} = \begin{pmatrix} 450 \\ -50 \\ 100 \end{pmatrix}$$

The resultant solution is infeasible because $x_5 = -50$. We apply the dual simplex method to restore the feasibility. Modifying the existing table by incorporating the resulting infeasible solution, we get the following table:

		Cost →		3	4	0	0	0
		Variable →		x_1	x_2	x_3	x_4	x_5
Table no.	C_B	Basic variable	X_B	α^1	α^2	α^3	α^4	α^5
...	3	x_1	450	1	0	−1/4	3/4	0
	0	$\leftarrow x_5$	−50	0	0	−2	2	1
	4	x_2	100	0	1	1/2	−1/2	0
		$c_j - Z_j$		0	0	−5/4↑	−1/4	0

(*Contd*)

(Contd)

From the dual simplex method, x_5 leaves and x_3 enters, and we get the following table:

		Cost →		3	4	0	0	0
		Variable →		x_1	x_2	x_3	x_4	x_5
Table no.	C_B	**Basic variable**	X_B	α^1	α^2	α^3	α^4	α^5
...	3	x_1	1825/4	1	0	0	1/2	−1/8
	0	x_3	25	0	0	1	−1	−1/2
	4	x_2	175/2	0	1	0	1	−1/4
		$c_j - Z_j$		0	0	0	−3/2	−5/8

This is the optimal table. The optimal solution of the resulting problem is $x_1 = 1825/4$, $x_2 = 175/2$, and profit $= 3x_1 + 4x_2 = 1718.75$. The maximum profit of the given problem is ₹1750. Hence, the reduction in the number of chips to 350 is not advantageous to the company.

Transportation Problems

> After studying this chapter, the reader will be able to
>
> - determine the initial basic feasible solution of the transportation problem using the north-west corner, least cost entry, and Vogel's approximation methods
> - solve the balanced and unbalanced transportation problems
> - obtain the optimal solution to the transportation problem using the modified distribution and stepping stone methods
> - understand the transshipment problem and obtain its solution

8.1 INTRODUCTION

In previous chapters, we have studied in detail about the linear programming problem (LPP) and solved it using various methods like the simplex method. A transportation problem is an LPP that deals with identifying an optimal solution for the transportation and allocation of resources to various destinations and from one site to another, while keeping the expenditure to a minimum. The coefficient matrix used in a transportation problem is of a special nature, and if the simplex method is used, the solution obtained is highly degenerate. Moreover, due to a large number of constraints and variables involved in the problem, a large number of computations are required to obtain the solution. To avoid these hassles, the simplex method has been modified and is used to develop an algorithm to obtain solutions to all transportation problems.

8.2 FORMULATION OF TRANSPORTATION PROBLEM

Let us consider a problem of transporting a specific commodity from m sources S_i (the places of manufacturing or availability of the commodity) $i = 1, 2, \ldots, m$ to n destinations D_j (the places of requirement of the commodity) $j = 1, 2, \ldots, n$ as shown graphically in Fig. 8.1. Further let

a_i = the quantity manufactured (or available) at source S_i; $i = 1, 2, \ldots, m$
b_j = the quantity required at destination D_j; $j = 1, 2, \ldots, n$
c_{ij} = the cost of transportation of one unit of the commodity (item) from S_i to D_j; $i = 1, 2, \ldots, m$; $j = 1, 2, \ldots, n$

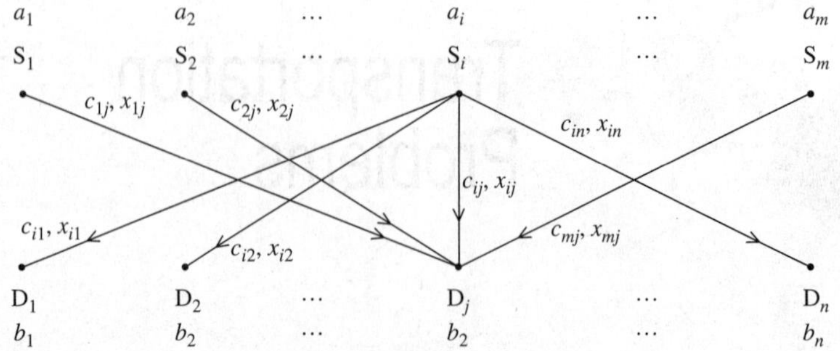

Figure 8.1 Transportation model

Now, the problem is how much of the commodity can be transported from what source to which destination to minimize the total cost of transportation? For this let x_{ij} = units of the commodity to be transported from S_i to D_j; $i = 1, 2, \ldots, m; j = 1, 2, \ldots, n$.

Then, the total cost of transportation is

$$(c_{11}x_{11} + c_{12}x_{12} + \ldots + c_{1n}x_{1n}) + (c_{21}x_{21} + c_{22}x_{22} + \ldots + c_{2n}x_{2n}) + \ldots + (c_{m1}x_{m1} + c_{m2}x_{m2} + \ldots + c_{mn}x_{mn})$$

Our objective is to minimize this total cost, that is,

$$\text{Min } Z = (c_{11}x_{11} + c_{12}x_{12} + \ldots + c_{1n}x_{1n}) + (c_{21}x_{21} + c_{22}x_{22} + \ldots + c_{2n}x_{2n}) + \ldots + (c_{m1}x_{m1} + c_{m2}x_{m2} + \ldots + c_{mn}x_{mn}) \quad (8.1)$$

$$\text{Subject to} \quad x_{i1} + x_{i2} + \ldots + x_{in} = a_i, \quad a_i > 0, i = 1, 2, \ldots, m \quad (8.2)$$

$$x_{1j} + x_{2j} + \ldots + x_{mj} = b_j, \quad b_j > 0, j = 1, 2, \ldots, n \quad (8.3)$$

$$x_{ij} \geq 0, \forall i, j,\ i = 1, 2, \ldots, m; j = 1, 2, \ldots, n \quad (8.4)$$

From Eqs (8.2) and (8.3), we have

$$\sum_{i=1}^{m} a_i = \sum_{j=1}^{n} b_j \quad (8.5)$$

The problem satisfying the condition in Eq. (8.5) is called the balanced transportation problem. From Eq. (8.5), the number of basic variables is $(m + n - 1)$ as one of the a_i's or b_j's can be expressed in terms of the remaining a_i's or b_j's.

8.3 DEVELOPMENT OF TRANSPORTATION ALGORITHM

The problem presented in Section 8.2 is an LPP with a large number of constraints, and variables and a coefficient matrix having a large number of zero entries. Thus, the usual simplex method is not suitable to solve such problems and its modified form called the transportation algorithm is used. The first step towards solving such problems is to obtain the initial basic feasible solution (BFS) as done in the simplex method and then determine the optimal solution using the transportation algorithm. There are many methods available to obtain the initial BFS. In this chapter, we will discuss the following three most popular methods:

1. North-west corner method (NWCM)
2. Least cost entry method (LCM)
3. Vogel's approximation method (VAM)

Table 8.1 Tabular form of transportation problem

Destination → Source ↓	D_1	D_2	D_3	D_4	D_5	Availability ↓
S_1	c_{11} (x_{11})	c_{12} (x_{12})	c_{13} (x_{13})	c_{14} (x_{14})	c_{15} (x_{15})	a_1
S_2	c_{21} (x_{21})	c_{22} (x_{22})	c_{23} (x_{23})	c_{24} (x_{24})	c_{25} (x_{25})	a_2
S_3	c_{31} (x_{31})	c_{32} (x_{32})	c_{33} (x_{33})	c_{34} (x_{34})	c_{35} (x_{35})	a_3
Requirement →	b_1	b_2	b_3	b_4	b_5	$\sum a_i=\sum b_j$

To begin with, let us consider a transportation problem with three sources S_1, S_2, and S_3 and five destinations D_1, D_2, D_3, D_4, and D_5, that is, put $m = 3$ and $n = 5$ in Eqs (8.1)–(8.5). Tabulating the given values (which are different from the simplex table), we get Table 8.1.

In Table 8.1, the rectangle obtained at the intersection of a row and column may be called a cell. While the cells in which allocation has been made are called the basic cells, the cells where allocation has not been made are called the non-basic cells. In any balanced transportation problem with m sources and n destinations, $m+n-1$ basic cells should be present. The allocation made in the basic cells gives us the BFS.

Now, we shall illustrate the three methods, namely the NWCM, LCM, and VAM to obtain the initial BFS. In the following text, we shall note that it is easier to get the initial BFS from the first two methods, that is, NWCM and LCM, but one has to perform many iterations after that to obtain the optimal solution of the problem. However, VAM gives either an optimal solution or a solution very close to the optimal solution. We, therefore, use VAM to obtain the initial BFS unless it is mentioned in the problem to use a specific method.

8.4 SOLUTION OF TRANSPORTATION PROBLEM

The following steps are involved in solving a transportation problem:

1. Formulation of the transportation problem.
2. Tabulating the formulated problem.
3. If the given transportation problem is unbalanced, then first balance it by introducing an additional row or column called the fictitious source or destination, respectively.

For example, consider the adjoining transportation problem given in Table 8.2.

Table 8.2 Unbalanced transportation problem

	D_1	D_2	D_3	Availability
S_1	4	3	2	10
S_2	2	5	0	13
S_3	3	8	6	12
Req.	8	5	4	–

Here, $a_1 = 10$, $a_2 = 13$, $a_3 = 12$; $b_1 = 8$, $b_2 = 5$, $b_3 = 4$. Such that

$$\sum_{i=1}^{3} a_i = 35, \sum_{j=1}^{3} b_j = 17$$

$$\sum_{i=1}^{3} a_i \neq \sum_{j=1}^{3} b_j \text{ and } \sum_{i=1}^{3} a_i - \sum_{j=1}^{3} b_j = 18, \text{ so it is unbalanced.}$$

First, we need to balance the problem. So, we introduce an additional column of D_F (fictitious destination) and enter zeros in this column in the position of costs and '18' in the requirement row. This balances the problem as given in Table 8.3 (balanced transportation problem).

Table 8.3 Balanced transportation problem

	D_1	D_2	D_3	D_F	Availability
S_1	4	3	2	0	10
S_2	2	5	0	0	13
S_3	3	8	6	0	12
Req.	8	5	4	18	35

4. If the problem is given as a maximization problem, then first convert it into a minimization problem by multiplying each entry in the matrix with (–1).
5. Find the initial BFS using one of the following methods:
   (a) NWC method
   (b) LCM or matrix minima method
   (c) VAM
6. Find the optimal solution starting from the solution obtained in step 5 by using either the (a) stepping stone method or (b) modified distribution (MODI) method

Notes 1. Step 5 is applicable to the balanced transportation problem only when it is in minimization form.

2. In most cases, the transportation problem is given in the tabular form. In such cases, we need not consider the first two steps for solving the problem.

Let us now discuss the NWCM, LCM, and VAM in detail.

8.4.1 North-west Corner Method

In this method, the following steps are involved:

1. Allocate the maximum possible units $\{x_{11} = \min(a_1, b_1)\}$ in the extreme north-west corner of the table and then neglect the row or column that is satisfied.
2. The following three cases arise:
   (a) If $b_1 < a_1$, we have $x_{11} = b_1$ and the first column is satisfied. As the quantity $(a_1 - b_1)$ is left unallocated in the first row, prepare the next table after omitting the first column. Now, move to the right-hand side along the first row in the table thus obtained and make the second allocation of the units $x_{12} = \min(a_1 - x_{11}, b_2)$ to the extreme north-west cell (1, 2) and so on.
   (b) If $b_1 > a_1$, we have $x_{11} = a_1$ and the first row is satisfied. As the quantity $(b_1 - a_1)$ is left unallocated in the first column, prepare the next table after omitting the first row. Now, move vertically downward in the first column of the table thus obtained and make the second allocation of $x_{21} = \min(a_2, b_1 - x_{11})$ to the extreme north-west cell (2, 1) and so on.

(c) If $b_1 = a_1$ then $x_{11} = a_1 = b_1$. Thus, both the first row and the first column are satisfied simultaneously. This is the case of degeneracy. In this case, we arbitrarily assume one to be satisfied and other not to be satisfied (Write the balance zero. This zero against the retaining satisfied row or column will be allocated somewhere and give rise to a degenerate solution).

3. Repeat steps 1–3 till the allocation is complete.

Example 8.1 Find the initial BFS of the following transportation problem using the NWC method:

	D_1	D_2	D_3	Available
S_1	2	7	4	5
S_2	3	3	1	8
S_3	5	4	7	7
S_4	1	6	2	14
Requirement	7	9	18	34

Solution Consider the transportation problem of the given table. In this table, the north-west corner is the x_{11} position. The maximum possible allocation in the x_{11} position $= \min(a_1, b_1) = \min(5, 7) = 5$. We enter 5 in x_{11} position and subtract 5 from a_1 and b_1 to obtain Table 8.4.

Table 8.4 Iteration 1

	D_1	D_2	D_3	
S_1	2(5)	7	4	~~5~~
S_2	3	3	1	8
S_3	5	4	7	7
S_4	1	6	2	14
	~~7~~ 2	9	18	

Here, S_1 row is satisfied and is removed to get Table 8.5.

Now, we make the maximum possible allocation in the x_{21} position. The units allocated is $x_{21} = \min(2, 8) = 2$. We enter 2 in the x_{21} position and subtract 2 from 2 and 8 to get Table 8.6.

Here, D_1 column is satisfied and is removed to get Table 8.7.

Again, we allocate the maximum possible units, which is equal to $\min(6, 9) = 6$ and enter 6 in the x_{22} position. Now, subtract 6 from 6 and 9 to get Table 8.8.

Here, on removing the satisfied S_2 row, we get the Table 8.9.

Now, we allocate the maximum possible units, which is $\min(3, 7) = 3$ and enter 3 in the x_{32} position. Subtract 3 from 3 and 7 to get Table 8.10.

Here, the satisfied D_2 column is removed to get Table 8.11.

We allocate 4 and 14 in x_{33} and x_{43} positions to respectively (see Table 8.12) to complete the overall allocation so that both the row and column totals are satisfied.

Thus, the resulting allocation made is shown in Table 8.13.

Table 8.5 Iteration 2

	D_1	D_2	D_3	
S_2	3	3	1	8
S_3	5	4	7	7
S_4	1	6	2	14
	2	9	18	

Table 8.6 Iteration 3

	D_1	D_2	D_3	
S_2	3(2)	3	1	~~8~~6
S_3	5	4	7	7
S_4	1	6	2	14
	~~2~~	9	18	

Table 8.7 Iteration 4

	D_2	D_3	
S_2	3	1	6
S_3	4	7	7
S_4	6	2	14
	9	18	

Table 8.8 Iteration 5

	D_2	D_3	
S_2	3(6)	1	~~6~~
S_3	4	7	7
S_4	6	2	14
	~~9~~ 3	18	

Table 8.9 Iteration 6

	D_2	D_3	
S_3	4	7	7
S_4	6	2	14
	3	18	

Table 8.10 Iteration 7

	D_2	D_3	
S_3	4(3)	7	~~7~~4
S_4	6	2	14
	~~3~~	18	

Table 8.11 Iteration 8

	D_3	
S_3	7	4
S_4	2	14
	18	

Table 8.12 Iteration 9

	D_3	
S_3	7(4)	4
S_4	2(14)	~~14~~
	~~18~~	

Table 8.13 Iteration 10

	D_1	D_2	D_3	Available
S_1	2(5)	7	4	5
S_2	3(2)	3(6)	1	8
S_3	5	4(3)	7(4)	7
S_4	1	6	2(14)	14
Requirement	7	9	18	34

The BFS is $x_{11} = 5, x_{21} = 2, x_{22} = 6, x_{32} = 3, x_{33} = 4, x_{43} = 14$.

The total cost of transportation $= 5 \times 2 + 2 \times 3 + 6 \times 3 + 3 \times 4 + 4 \times 7 + 14 \times 2 = 102$.

Example 8.2 Use NWCM to determine an initial BFS to the following transportation problem:

		Destination				
		D_1	D_2	D_3	D_4	Supply
Origin	O_1	6	4	1	5	14
	O_2	8	9	2	7	16
	O_3	4	3	6	2	5
Demand		6	10	15	4	35

Solution In this table, the north-west corner is the x_{11} position. The maximum possible allocation in the x_{11} position $= \min(a_1, b_1) = \min(14, 6) = 6$. We enter 6 in the x_{11} position and subtract 6 from 14 and 6 to obtain Table 8.14.

Here, D_1 column is satisfied and is removed to get Table 8.15.

Now, we make the maximum possible allocation in the x_{12} position. Thus, $x_{12} = \min(8, 10) = 8$ units are allocated. We enter 8 in the x_{12} position and subtract 8 from 8 and 10 to get Table 8.16.

Here, O_1 row is satisfied and is removed to get Table 8.17.

Again, we allocate the maximum possible units equal to min(2, 16) = 2 and enter 2 in the x_{22} position. Now, subtract 2 from 2 and 16 to get Table 8.18.

Here, removing D_2 column, which is satisfied, we get Table 8.19.

Now, we allocate the maximum possible units, that is, min(14, 15) = 14 and enter 14 in the x_{23} position. Subtract 14 from 14 and 15 to get Table 8.20.

Here, the satisfied O_2 row is removed to get Table 8.21.

We allocate 1 and 4 in x_{33} and x_{34} positions respectively to complete the overall allocation (Table 8.22) so that both the row and column totals are satisfied.

Thus, the overall allocation is shown in Table 8.23.

Table 8.14 Iteration 1

	D_1	D_2	D_3	D_4	
O_1	6(6)	4	1	5	~~14~~ 8
O_2	8	9	2	7	16
O_3	4	3	6	2	5
	~~6~~	10	15	4	

Table 8.15 Iteration 2

	D_2	D_3	D_4	
O_1	4	1	5	8
O_2	9	2	7	16
O_3	3	6	2	5
	10	15	4	

Table 8.16 Iteration 3

	D_2	D_3	D_4	
O_1	4(8)	1	5	~~8~~
O_2	9	2	7	16
O_3	3	6	2	5
	~~10~~ 2	15	4	

Table 8.17 Iteration 4

	D_2	D_3	D_4	
O_2	9	2	7	16
O_3	3	6	2	5
	2	15	4	

Table 8.18 Iteration 5

	D_2	D_3	D_4	
O_2	9(2)	2	7	~~16~~ 14
O_3	3	6	2	5
	~~2~~	15	4	

Table 8.19 Iteration 6

	D_3	D_4	
O_2	2	7	14
O_3	6	2	5
	15	4	

Table 8.20 Iteration 7

	D_3	D_4	
O_2	2(14)	7	~~14~~
O_3	6	2	5
	~~15~~ 1	4	

Table 8.21 Iteration 8

	D_3	D_4	
O_3	6	2	5
	1	4	

Table 8.22 Iteration 9

	D_3	D_4	
O_3	6(1)	2(4)	~~5~~
	~~1~~	~~4~~	

Table 8.23 Iteration 10

		Destination				
		D_1	D_2	D_3	D_4	Supply
	O_1	6(6)	4(8)	1	5	14
Origin	O_2	8	9(2)	2(14)	7	16
	O_3	4	3	6(1)	2(4)	5
Demand		6	10	15	4	35

The BFS is $x_{11} = 6$, $x_{12} = 8$, $x_{22} = 2$, $x_{23} = 14$, $x_{33} = 1$, $x_{34} = 4$.

8.4.2 Least Cost Entry or Matrix Minima Method

In this method, we make the maximum possible allocation to the cell with the least cost (c_{ij}). If the cell with the least cost is not unique, then make the choice among cells of the least cost where more allocation can be made. Again, if the cell is not unique then an arbitrary choice for the cell can be made for allocation. In this process of allocation, neglect the row or column that is satisfied and there is no scope for any more allocation. Continue the process till all the a_i's and b_j's are exhausted.

Example 8.3 Find the initial BFS of the following transportation problem using LCM:

	D_1	D_2	D_3	Available
S_1	2	7	4	5
S_2	3	3	1	8
S_3	5	4	7	7
S_4	1	6	2	14
Requirement	7	9	18	34

Solution As we can see in the cost matrix, the lowest cost is 1 in cells (2, 3) and (4, 1). The maximum units that we can allocate in cell (2, 3) is min(8, 18)=8 and that in cell (4, 1) is min(14, 7)=7. Since more units can be allocated in cell (2, 3), so we allocate 8 units in cell (2, 3) and x_{23}=8. This exhausts the availability from S_2 and therefore, the S_2 row is neglected for further allocation.

	D_1	D_2	D_3	
S_1	2	7	4	5
S_3	5	4	7	7
S_4	1	6	2	14
	7	9	10	

In the aforegiven reduced transportation table, we find that the least cost is 1 in cell (4, 1) and allocate min(14, 7)=7 units in it. Thus, x_{41}=7. This satisfies the requirement at D_1. Leaving the D_1 column, we get the adjacent table:

	D_2	D_3	
S_1	7	4	5
S_3	4	7	7
S_4	6	2	7
	9	10	

Now, we see that in aforegiven table the lowest cost is 2 in cell (4, 3). Here, we allocate min(7, 10) = 7 units in it. Thus, x_{43} = 7. This satisfies the requirement at S_4. Leaving the S_4 row, we get the adjacent table:

	D_2	D_3	
S_1	7	4	5
S_3	4	7	7
	9	3	

In this reduced transportation table, the lowest cost is 4 in cells (1, 3) and (3, 2). We prefer cell (3, 2) since more units can be allocated. The maximum units that we can place in cell (3, 2) is min(7, 9) = 7. So, we allocate 7 units in cell (3, 2). This exhausts the availability from S_3 and it is neglected for further allocation. This gives x_{32} = 7.

	D_2	D_3	
S_1	7	4	5
	2	3	

We allocate 2 in x_{12} and 3 in x_{13} to complete the overall allocation. Thus, the resulting allocation made is shown in the following table:

	D_1	D_2	D_3	Available
S_1	2	7(2)	4(3)	5
S_2	3	3	1(8)	8
S_3	5	4(7)	7	7
S_4	1(7)	6	2(7)	14
Requirement	7	9	18	34

The BFS is $x_{12} = 2, x_{13} = 3, x_{23} = 8, x_{32} = 7, x_{41} = 7, x_{43} = 7$.
The total cost of transportation $= 7 \times 2 + 4 \times 3 + 1 \times 8 + 4 \times 7 + 1 \times 7 + 2 \times 7 = 83$.

Example 8.4 Use LCM to determine the initial BFS to the following transportation problem:

	Destination				
	D_1	D_2	D_3	D_4	Capacity
O_1	1	2	3	4	6
O_2	4	3	2	0	8
O_3	0	2	2	1	10
Demand	4	6	8	6	24

Solution As we can see in the cost matrix, the lowest cost is 0 in cells (2, 4) and (3, 1). The maximum units that we can allocate in cell (2, 4) is min(8, 6) = 6. So, we allocate 6 units in cell (2, 4). This satisfies the demand at D_4 and this is negelcted for further allocation. This gives $x_{24} = 6$.

	D_1	D_2	D_3	
O_1	1	2	3	6
O_2	4	3	2	2
O_3	0	2	2	10
	4	6	8	

In the aforegiven reduced transportation table, we find that there is the least cost 0 in cell (3, 1). Here, we allocate the maximum possible units, that is, 4. Thus, $x_{31} = 4$. This satisfies the demand at D_1. Leaving the D_1 column, we get the adjacent table:

	D_2	D_3	
O_1	2	3	6
O_2	3	2	2
O_3	2	2	6
	6	8	

Now in the aforegiven table, the lowest cost is 2 in cells (1, 2), (2, 3), (3, 2), and (3, 3). Here, we can allocate the maximum possible units, that is, 6 in x_{12}, x_{32}, and x_{33} and arbitrarily choose x_{12} for allocation. Thus, $x_{12} = 6$. This satisfies the requirement at O_1 and D_2. Leaving the O_1 row, we assume D_2 column not to be satisfied. Thus, replacing 6 by 0, we get the adjacent table:

	D_2	D_3	
O_2	3	2	2
O_3	2	2	6
	0	8	

In this reduced transportation table, the lowest cost is 2 in cells (2, 3), (3, 2), and (3, 3). Now, by allocating the maximum possible units, that is, 6 in x_{33}, we get the adjacent table:

	D_2	D_3	
O_2	3	2	2
	0	2	

In this reduced transportation table, we allocate 2 in x_{23} and 0 in x_{22} to complete the overall allocation. Thus, the resulting allocation made is shown in the following table:

	Destination				
	D_1	D_2	D_3	D_4	Capacity
O_1	1	2(6)	3	4	6
O_2	4	3(0)	2(2)	0(6)	8
O_3	0(4)	2	2(6)	1	10
Demand	4	6	8	6	24

The BFS is $x_{12} = 6, x_2 = 0, x_{23} = 2, x_{24} = 6, x_{31} = 4, x_{33} = 6$.

8.4.3 Vogel's Approximation Method

Although Vogel's approximation method (VAM) takes more time to get the initial BFS, the time consumed in obtaining the optimal solution is reduced. Thus, it is advised to obtain the initial BFS by VAM using the following procedure:

1. Calculate the difference between the lowest and the next higher (i.e., the next lowest cost) cost for each row and column. These differences are called row and column penalties, respectively. Write these differences alongside the respective row or column.
2. Choose the row/column with the highest penalty. Allocate the maximum possible units to the cell with the lowest cost in the chosen row/column. If the largest penalty is not unique, then choose from among the penalties tied, the row/column having a cell with the lowest cost. If the problem persists, then select the row or column with the minimum cost from the cells where more allocation can be made. If the problem is still not resolved, then choose arbitrarily.
3. Omit the satisfied row or column from future considerations.
4. Repeat steps 1–3 till all a_i's and b_j's are exhausted.

Example 8.5 Find the initial BFS of the following transportation problem using VAM:

	D_1	D_2	D_3	Available
S_1	2	7	4	5
S_2	3	3	1	8
S_3	5	4	7	7
S_4	1	6	2	14
Requirement	7	9	18	34

Solution First, compute the penalties for all rows and column. Tabulating the given data, we get

	D_1	D_2	D_3	Available	Penalty
S_1	2	7	4	5	(2)
S_2	3	3	1	8	(2)
S_3	5	4	7	7	(1)
S_4	1	6	2	14	(1)
Requirement	7	9	18	34	
Penalty	(1)	(1)	(1)		

We find that maximum penalty (2) is for rows S_1 and S_2. However, S_2 has a cell with the minimum cost 1 as compared to S_1 that has a cell with the smaller cost 2. So, we prefer to select S_2 for allocation and allocate maximum possible units in cell (2, 3). Thus $x_{23} = \min(8, 18) = 8$. This exhausts the availability from S_2. We omit S_2 for further allocation.

The reduced cost matrix is given here:

	D_1	D_2	D_3	Available	Penalty
S_1	2	7	4	5	(2)
S_3	5	4	7	7	(1)
S_4	1	6	2	14	(1)
Requirement	7	9	10		
Penalty	(1)	(2)	(2)		

We note that the units still needed at destination D_3 is 10. Again, we calculate the row and column penalties and enter them alongside the rows and columns. The maximum penalty is 2 for columns D_2 and D_3 and also for row S_1. We choose those rows or columns that have a cell with the lowest cost. (1, 1) and (4, 3) are two such cells. However, as we can allocate more units in (4, 3), we select it for allocation. The units allocated in (4, 3) is min(10, 14) = 10, thus $x_{43} = 10$. As a result, the requirement at D_3 is fulfilled and we omit it for further allocation.

The reduced cost matrix is as follows:

	D_1	D_2	Available	Penalty
S_1	2	7	5	(5)
S_3	5	4	7	(1)
S_4	1	6	4	(5)
Requirement	7	9		
Penalty	(1)	(2)		

Now, we again calculate the row and column penalties and enter them alongside the rows and columns as in the aforegiven reduced cost matrix table. The maximum penalty is 5 for row S_1 and S_4. We choose rows having a cell with the lowest cost. (1, 1) and (4, 1) are two such cells. However as we can allocate more in (4, 1), we select it for allocation.

The units allocated in (4, 1) is min(4, 7) = 4, thus $x_{41} = 4$. As a result, the requirement at S_4 is fulfilled and we omit S_4 row for further allocation. The reduced cost matrix is as follows:

	D_1	D_2	Available	Penalty
S_1	2	7	5	(5)
S_3	5	4	7	(1)
Requirement	3	9		
Penalty	(3)	(3)		

	D_2	Available
S_1	7	2
S_3	4	7
Req.	9	

Here, we calculate the row and column penalties and enter them alongside the rows and columns as in the previous table. The maximum penalty is 5 for row S_1. We choose a column having a cell with the lowest cost. (1, 1) is such a cell. The units allocated in (1, 1) is min(3, 5) = 3, thus $x_{11} = 3$. As a result, the requirement at D_1 is fulfilled and we omit D_1 column for further allocation. The reduced cost matrix is shown in the aforegiven table.

We allocate 2 in x_{12} and 7 in x_{32} to complete the overall allocation. Thus, the resulting allocation made is shown in the following table:

	D_1	D_2	D_3	Available
S_1	2(3)	7(2)	4	5
S_2	3	3	1(8)	8
S_3	5	4(7)	7	7
S_4	1(4)	6	2(10)	14
Requirement	7	9	18	34

The BFS is $x_{11} = 3$, $x_{12} = 2$, $x_{23} = 8$, $x_{32} = 7$, $x_{41} = 4$, $x_{43} = 10$.

The total cost of transportation $= 2 \times 3 + 7 \times 2 + 1 \times 8 + 4 \times 7 + 1 \times 4 + 2 \times 10 = 80$.

Example 8.6 Determine an initial BFS to the following transportation table using VAM:

		To				
		I	II	III	IV	Supply
From	A	5	1	3	3	34
	B	3	3	5	4	15
	C	6	4	4	3	12
	D	4	1	4	2	19
Requirement		21	25	17	17	80

Solution Compute the penalties for all rows and columns. Tabulating the given data, we get

	I	II	III	IV	Available	Penalty
A	5	1	3	3	34	(2)
B	3	3	5	4	15	(1)
C	6	4	4	3	12	(1)
D	4	1	4	2	19	(1)
Requirement	21	25	17	17		
Penalty	(1)	(2)	(1)	(1)		

We find that the maximum penalty (2) is for row A and column II. We choose a row or column having a cell with the lowest cost. (1, 2) and (4, 2) are such cells. However, as we can allocate more in (1, 2), we select it for allocation. The units allocated in (1, 2) is min (34, 25) = 25, thus $x_{12} = 25$. As a result, the requirement at II is fulfilled and we omit column II for further allocation. The reduced cost matrix is as follows:

	I	III	IV	Available	Penalty
A	5	3	3	9	(2)
B	3	5	4	15	(1)
C	6	4	3	12	(1)
D	4	4	2	19	(2)
Requirement	21	17	17		
Penalty	(1)	(1)	(1)		

We note that the units still needed to requirement A is 9. Again, we calculate the row and column penalties and enter them alongside the rows and columns as in the aforegiven cost and requirement matrix table. The maximum penalty is 2 for rows A and D. We choose the row having a cell with the lowest cost which is in cell (4, 4) for row D. So, we select it for allocation. The units allocated in (4, 4) is min(19, 17) = 17, thus $x_{44} = 17$. As a result, the requirement at column IV is fulfilled and we omit this column for further allocation. The reduced cost matrix is as follows:

	I	III	Available	Penalty
A	5	3	9	(2)
B	3	5	15	(2)
C	6	4	12	(2)
D	4	4	2	(0)
Requirement	21	17		
Penalty	(1)	(1)		

Now, we again calculate the row and column penalties and enter them alongside the rows and columns. The maximum penalty is 2 for rows A, B, and C. We choose the rows having a cell with the lowest cost. Such are the cells (1, 3) and (2, 1). However, more allocation can be made in (2, 1). Therefore, we selected (2, 1) and allocate min(15, 21) = 15 units. Thus, $x_{21} = 15$. As a result, the availability at row B is fulfilled and we omit this row for further allocation. The reduced cost matrix is as follows:

	I	III	Available	Penalty
A	5	3	9	(2)
C	6	4	12	(2)
D	4	4	2	(0)
Requirement	6	17		
Penalty	(1)	(1)		

Here, we calculate the row and column penalties and enter them alongside the rows and columns. The maximum penalty is 2 for rows A and C. We choose the row having a cell with the lowest cost. (1, 3) is one such cell. The units allocated in (1, 3) is min(9, 17) = 9, thus $x_{13} = 9$. As a result, the requirement at B is fulfilled and we omit this row for further allocation. The reduced cost matrix is as follows:

	I	III	Available	Penalty
C	6	4	12	(2)
D	4	4	2	(0)
Requirement	6	8		
Penalty	(2)	(0)		

Now, we calculate the row and column penalties and enter them alongside the rows and columns. The maximum penalty is 2 for row C and column I. We choose that row or column having a cell with the lowest cost. Such cells are (3, 3) and (4, 1). However, more allocation can be made in cell (3, 3). Thus, we select cell (3, 3) and allocate min(12, 8) = 8 units, thus $x_{33} = 8$. As a result, the requirement at column III is fulfilled and we omit this column for further allocation. The reduced cost matrix is as follows:

	I	Available
C	6	4
D	4	2
Requirement	6	

We allocate 4 in x_{31} and 2 in x_{41} positions to complete the overall allocation. Thus, the resulting allocation made is shown in the following table:

	I	II	III	IV	Available
A	5	1(25)	3(9)	3	34
B	3(15)	3	5	4	15
C	6(4)	4	4(8)	3	12
D	4(2)	1	4	2(17)	19
Requirement	21	25	17	17	

The BFS is $x_{12} = 25, x_{13} = 9, x_{21} = 15, x_{31} = 4, x_{33} = 8, x_{41} = 2, x_{44} = 17$.

8.5 TEST OF OPTIMALITY

As mentioned in earlier sections, to solve any transportation problem, the first step is to obtain an initial BFS. The initial BFS can be obtained by any of the three methods mentioned in Section 8.4. At times, one obtains the initial BFS and assumes that this is the final solution. This is incorrect. To solve a transportation problem is to obtain its optimal solution. However, the initial BFS obtained need not be the optimal solution. We therefore, have to check whether the BFS obtained is optimal or not. There are two methods available to check the optimality of the initial BFS:

1. Modified distribution
2. Stepping stone

8.5.1 Modified Distribution Method

The following steps are used to obtain the optimal solution using the MODI method:

1. Construct a transportation table by entering the costs c_{ij} at the upper right corner of the cells. Write the availabilities a_i's at different sources against each row on the extreme right of the source row and requirements b_j's at the bottom of each destination column. Enter the initial BFS obtained by one of the three methods in appropriate cells within braces. The cells where allocation is made are called the basic cells and the remaining ones are called the non-basic cells.
2. Associate dual variables u_i's to rows and v_j's to columns such that $u_i + v_j = c_{ij}$ holds good for the basic cells. These will be $(m + n - 1)$. Equations in $(m + n)$ unknowns where m is the number of rows (sources) and n is the number of columns (destinations). Solve these equations to find u_i's and v_j's. To get the unique solution, we take one of the u_i's or v_j's to be zero so that the number of variables is equal to the number of equations. It is convenient to take the dual variable that corresponds to the row or column that has maximum number of entries x_{ij} (i.e., the maximum number of basic cells) as zero. Write u_i's in rows on extreme right and v_j's in columns at the bottom of the table.

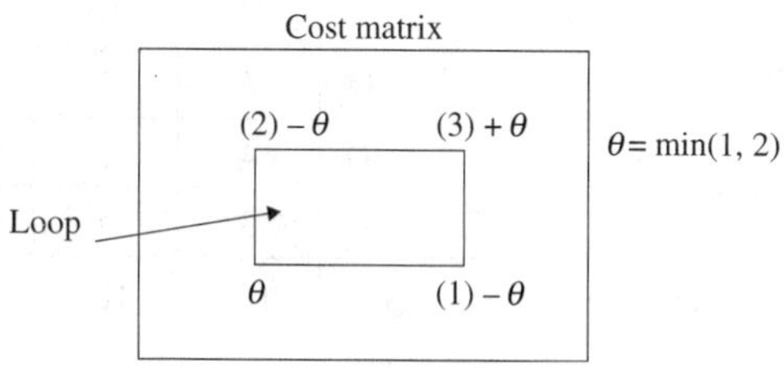

Figure 8.2 Loop for allocation in a non-basic cell

3. Once u_i's and v_j's are known, calculate $u_i + v_j - c_{ij} = d_{ij}$ (say) for each non-basic cell and enter this number in the south-west corner of the cell. Examine d_{ij} for each non-basic cell.
   (a) If all $d_{ij} \leq 0$, then the existing BFS is optimal. Write the optimal solution. If at least one $d_{ij} = 0$, then it is the indication of the existence of an alternate optimal solution.
   (b) If at least one $d_{ij} > 0$, then the solution is not optimal. In such cases, go to step 4.
4. (a) *Entering variable rule*: Select the variable with the largest positive d_{ij} in the south-west corner to enter the basis. To maintain the total number $(m + n - 1)$ of the basic variables, one of the existing basic variables has to leave the basis.
   (b) *Leaving variable rule*: For this we construct a loop as shown in Fig. 8.2. Start from the cell of the entering variable, move horizontally and vertically, take turns only at a basic cell, and return back to the starting cell. The closed loop thus obtained is unique and will have its edges at the basic cells.
   (c) Let θ be the maximum units that can be allocated at the entering variable cell. Subtract and add θ from the row edges and column edges respectively maintaining row and column totals (see Fig. 8.2). The value of θ is the minimum of the allocations from which θ units have

been subtracted. This will be a new BFS. Now repeat steps 2–4 until there is no positive entry in the south-west corner of any non-basic cell. The solution thus obtained will be the optimal solution.

5. Write the solution of the given problem, the value of the basic variables, and the minimum cost.

Example 8.7 Determine the optimum BFS to the following transportation problem using the MODI method:

	D_1	D_2	D_3	D_4	$a_i \downarrow$
O_1	5	3	6	2	19
O_2	4	7	9	1	37
O_3	3	4	7	5	34
$b_j \rightarrow$	16	18	31	25	

Solution The following steps are used to determine the optimum BFS by the MODI method:

1. First, we find the initial BFS using VAM, obtained as follows:

	D_1	D_2	D_3	D_4	
O_1	5	3 (18)	6 (1)	2	19
O_2	4 (12)	7	9	1 (25)	37
O_3	3 (4)	4	7 (30)	5	34
	16	18	31	25	

Now, we check the optimality of the solution.

2. Calculate u_i's and v_j's taking $u_i + v_j = c_{ij}$ for the basic cells

   $u_1 + v_2 = 3$, $u_1 + v_3 = 6$, $u_2 + v_1 = 4$, $u_2 + v_4 = 1$, $u_3 + v_1 = 3$, and $u_3 + v_3 = 7$.

   Solving these with $u_1 = 0$, we get $u_1 = 0, u_2 = 2, u_3 = 1, v_1 = 2, v_2 = 3$, $v_3 = 6$, and $v_4 = -1$ as shown in the adjacent table:

					$u_i \downarrow$
	5	3 (18)	6 (1)	2	0
	4 (12)	7	9	1 (25)	2
	3 (4)	4	7 (30)	5	1
$v_j \rightarrow$	2	3	6	−1	

3. Now, calculate $d_{ij} = u_i + v_j - c_{ij}$ for the non-basic cells and enter this number in the south-west corner of the cell as follows:

   $d_{11} = u_1 + v_1 - c_{11} = 0 + 2 - 5 = -3$

   $d_{14} = u_1 + v_4 - c_{14} = 0 - 1 - 2 = -3$

$d_{22} = u_2 + v_2 - c_{22} =$
$2 + 3 - 7 = -2$
$d_{23} = u_2 + v_3 - c_{23} =$
$2 + 6 - 9 = -1$
$d_{32} = u_3 + v_2 - c_{32} =$
$1 + 3 - 4 = 0$
$d_{34} = u_3 + v_4 - c_{34} =$
$1 - 1 - 5 = -5$

				$u_i \downarrow$
5 −3	3 (18)	6 (1)	2 −3	0
4 (12)	7 −2	9 −1	1 (25)	2
3 (4)	4 0	7 (30)	5 −5	1
$v_j \rightarrow$ 2	3	6	−1	

Here, all d_{ij}'s are non-positive, that is, $d_{ij} \leq 0$. Thus, the existing BFS is optimal.

Hence, the solution of the problem is $x_{12} = 18$, $x_{13} = 1$, $x_{21} = 12$, $x_{24} = 25$, $x_{31} = 4$, and $x_{33} = 30$. The minimum transportation cost $= 18 \times 3 + 1 \times 6 + 12 \times 4 + 25 \times 1 + 4 \times 3 + 30 \times 7 =$ ₹355.

Note In this problem, one $d_{ij} = 0$ ($d_{32} = 0$) in the optimal table, which is the indication of the existence of an alternate optimum solution. The alternate optimal solution can be obtained (if needed) by entering the variable of the cell, containing zero entry, into the basis.

Example 8.8 Solve the following transportation problem using the MODI method:

	To			Available
From	2	7	4	5
	3	3	1	8
	5	4	7	7
	1	6	2	14
Required	7	9	18	

Solution The steps involved in solving this transportation problem are as follows:

1. First, we find the initial BFS using VAM, obtained as follows:

2 (5)	7	4	5
3	3	1 (8)	7
5	4 (7)	7	8
1 (2)	6 (2)	2 (10)	14
7	9	18	

Now, we check the optimality of the solution.

2. Calculate u_i's and v_j's taking $u_i + v_j = c_{ij}$ for the basic cells
$u_1+v_1=2,\quad u_2+v_3=1,\quad u_3+v_2=4,$
$u_4+v_1=1, u_4+v_2=6,$ and $u_4+v_3=2.$
Solving these with $u_4=0$, we get $u_1=1$, $u_2=-1$, $u_3=-2$, $u_4=0$, $v_1=1$, $v_2=6$, and $v_3=2$ as shown in the adjacent table:

			$u_i \downarrow$
2 (5)	7	4	1
3	3	1 (8)	−1
5	4 (7)	7	−2
1 (2)	6 (2)	2 (10)	0
$v_j \rightarrow$ 1	6	2	

3. Now calculate $d_{ij}=u_i+v_j-c_{ij}$ for the non-basic cells and enter this number in the south-west corner of the cell
$d_{12}=u_1+v_2-c_{12}=1+6-7=0$
$d_{13}=u_1+v_3-c_{13}=1+2-4=-1$
$d_{21}=u_2+v_1-c_{21}=-1+1-3=-3$
$d_{22}=u_2+v_2-c_{22}=-1+6-3=2$
$d_{31}=u_3+v_1-c_{31}=-2+1-5=-6$
$d_{33}=u_3+v_3-c_{33}=-2+2-7=-7$
Here, all the d_{ij}'s are not negative. Thus, the existing BFS is not optimal. This solution can be improved as shown in the next step.

			$u_i \downarrow$
2 (5)	7 0	4 −1	1
3 −3	3 2	1 (8)	−1
5 −6	4 (7)	7 −7	−2
1 (2)	6 (2)	2 (10)	0
$v_j \rightarrow$ 1	6	2	

4. Here, $d_{22}=2$ is the most positive, so we select the variable x_{22} to enter the basis. Now, to maintain the number of basic variables, one of the existing basic variables has to leave the basis. For this, we construct a closed loop starting from the cell of the entering variable as depicted here.

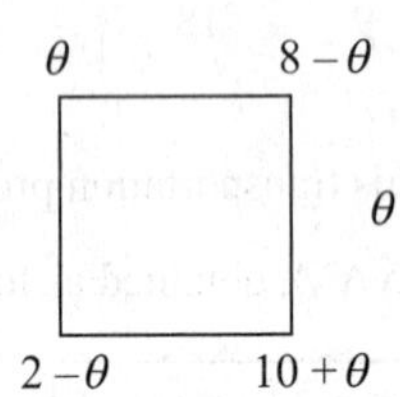

$\theta = \min(2, 8) = 2$

			$u_i \downarrow$
2 (5)	7 0	4 −1	1
3 −3	3 (2)	1 (6)	−1
5 −6	4 (7)	7 −7	−2
1 (2)	6	2 (12)	0
$v_j \rightarrow$ 1	6	2	

The transportation cost $= 2 \times 5 + 3 \times 2 + 1 \times 6 + 4 \times 7 + 1 \times 2 + 2 \times 12 =$ ₹76.

Now, again we shall test the optimality. Proceeding through steps 2–4, we get the following table:

			$u_i \downarrow$
2 (5)	7 0	4 −1	1
3 −3	3 (2)	1 (6)	−1
5 −6	4 (7)	7 −7	0
1 (2)	6 0	2 (12)	0
$v_j \rightarrow$ 1	4	2	

Here, all the d_{ij}'s are non-positive, that is, $d_{ij} \leq 0$. Thus, the existing BFS is optimal.

Hence, the solution of the problem is $x_{11} = 5$, $x_{22} = 2$, $x_{23} = 6$, $x_{32} = 7$, $x_{41} = 2$, and $x_{43} = 12$.

The minimum transportation cost $= 2 \times 5 + 3 \times 2 + 1 \times 6 + 4 \times 7 + 1 \times 2 + 2 \times 12 =$ ₹76.

Example 8.9 A manufacturer wants to ship 8 load units of his product as shown in the following table:

	D_1	D_2	D_3	Available
O_1	50	30	220	1
O_2	90	45	170	3
O_3	250	200	50	4
Required	4	2	2	

The matrix gives the mileage from origin O_i's to destination D_j's. The shipping costs are ₹10 per load unit per mile. What is the optimal shipping schedule?

Solution The following steps are involved in determining the optimal shipping schedule for this problem:

1. First, we find the initial BFS using VAM, obtained as follows:

	D_1	D_2	D_3	
O_1	50 (1)	30	220	1
O_2	90 (3)	45 (0)	170	3
O_3	250	200 (2)	50 (2)	4
	4	2	2	

Another alternate solution is

	D_1	D_2	D_3	
O_1	50 (1)	30	220	1
O_2	90 (3)	45	170	3
O_3	250 (0)	200 (2)	50 (2)	4
	4	2	2	

Here, one of the initial basic variables is zero, so the initial BFS is degenerate.

Let us check the optimality of the BFS. For this, we go to steps 2 and 3.

2. Calculate u_i's and v_j's taking $u_i + v_j = c_{ij}$ for the basic cells. We have $u_1 + v_1 = 50$, $u_2 + v_1 = 90$, $u_2 + v_2 = 45$, $u_3 + v_2 = 200$, and $u_3 + v_3 = 50$. Solving these with $u_3 = 0$, we get $u_1 = -195$, $u_2 = -155$, $u_3 = 0$, $v_1 = 245$, $v_2 = 200$, and $v_3 = 50$ as shown in the following table.

				$u_i \downarrow$
	50 (1)	30	220	−195
	90 (3)	45 (0)	170	−155
	250	200 (2)	50 (2)	0
$v_j \rightarrow$	245	200	50	

3. Now, calculate $d_{ij} = u_i + v_j - c_{ij}$ for the non-basic cells and enter this number in the south-west corner of the cells as follows:

				$u_i \downarrow$
	50 (1)	30 −25	220 −365	−195
	90 (3)	45 (0)	170 −275	−155
	250 −5	200 (2)	50 (2)	0
$v_j \rightarrow$	245	200	50	

Here, all the d_{ij}'s are non-positive, that is, $d_{ij} \leq 0$. Thus, the existing BFS is optimal.

Hence, the solution of the problem is $x_{11} = 1, x_{21} = 3, x_{22} = 0, x_{32} = 2$, and $x_{33} = 2$.
The minimum transportation cost $= 1 \times 50 + 3 \times 90 + 0 \times 45 + 2 \times 200 + 2 \times 50 =$ 820, that is, ₹8200.

8.5.2 Stepping Stone Method

This method is used to determine the optimal solution to the transportation problem. For the application of this method, first we find the initial BFS of the transportation problem by using one of VAM, LCM, and NWCM or any other known methods. After obtaining the initial BFS, we use the following steps to find the optimal solution:

1. Choose a non-basic cell of the transportation matrix. Prepare a closed loop starting from the chosen cell, move along the horizontal and vertical paths, take turns only at the basic cells, and come back to the initial non-basic cell. Such a loop is unique. Tentatively allocate 1 unit of the commodity in the chosen cell and add or subtract 1 unit to or from the allocations made at the basic cells (corner points of the loop) located at the turning points of the loop, such that both the column and row totals remain intact.
2. As there are changes in allocations resulting from step 1, calculate the net change in the transportation cost (i.e., the total additions in the transportation cost – the total reductions in the transportation cost) and enter this in the south-west corner of the chosen non-basic cell.
3. Repeat steps 1 and 2 for all the non-basic cells. The following two cases arise:
   (a) If all the entries for the net change in transportation cost as calculated in step 2 for the non-basic cells are non-negative, then the existing BFS is optimal.
   (b) If all entries are not ≥ 0, then choose the non-basic cell with the most negative entry in the south-west corner for the maximum possible allocation adjusting the overall allocation.
4. Repeat steps 1–3 until no negative net change in the transportation cost is obtained. The table satisfying the conditions of step 4 will be the optimal table and its solution will be the optimal solution to the transportation problem.

Example 8.10 Using VAM, find the initial BFS of the following transportation problem given in the standard notation:

		D_1	D_2	D_3	D_4	Available
	S_1	5	2	4	3	60
Sources	S_2	6	4	9	5	60
	S_3	2	3	8	1	90
Requirement		50	65	65	30	210

Further, use the stepping stone method to find the optimal solution if the BFS obtained is not optimal.

Solution Compute the penalties for all the rows and columns to get the data in the form of following table:

		D_1	D_2	D_3	D_4	Available	Penalty
	S_1	5	2	(60)4	3	~~60~~	1
Sources	S_2	6	4	9	5	60	1
	S_3	2	3	8	1	90	1
Requirement		50	65	~~65~~ 5	30		
Penalty		3	1	4	2		

We select column D_3 having the maximum penalty 4 and choose cell (S_1, D_3) of this column for allocation of maximum possible units 60 [min(60, 65)]. This satisfies row S_1 and we subtract 60 from the row availability (60) and column requirement (65). This satisfies row S_1. Now, omit row S_1 for further considerations and again calculate the row and column penalties for the resulting table to get the following table:

	D_1	D_2	D_3	D_4	Available	Penalty
S_2	6	4	9	5	60	1
S_3	2	3	8	1	90	1
Requirement	50	65	5	30		
Penalty	4	1	1	4		

As per the aforementioned table, the maximum penalty is for columns D_1 and D_4. As there is a tie, we select a column comprising a cell with the lowest cost. Thus, we select cell (3, 4) to allocate the maximum possible units 30 [min(30, 90)].

Repeating this process till the available row units at sources and requirement column units for destinations are exhausted we obtain the following table:

	D_1	D_2	D_3	D_4
S_1	5	2	4 (60)	3
S_2	6	4 (55)	9 (5)	5
S_3	2 (50)	3 (10)	8	1 (30)

Let us now use the stepping stone method to check the optimality of this solution. For this, calculate the net change in the transportation cost for the non-basic cells (1, 1), (1, 2), (1, 4), (2, 1), (2, 4), and (3, 3).

1. Consider the non-basic cell (1, 1) and tentatively allocate 1 unit of the commodity to get the following table:

5 +1	2	4 60 – 1	3
6	4 55 – 1	9 5 + 1	5
2 50 – 1	3 10 + 1	8	1 30

The loop formed is as shown in the adjoining figure:
Addition in the transportation cost = 5 + 9 + 3 = 17
Reduction in the transportation cost = 4 + 4 + 2 = 10
Net change in the transportation cost = 17 – 10 = 7

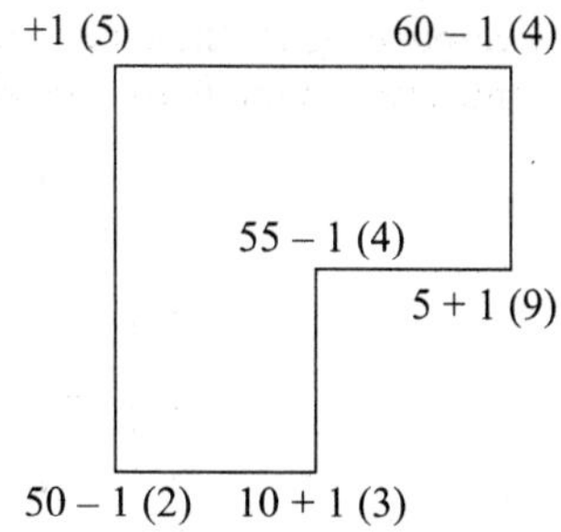

2. For cell (1, 2):
Addition in the transportation cost = 2 + 9 = 11
Reduction in the transportation cost = 4 + 4 = 8
Net change in the transportation cost = 11 – 8 = 3

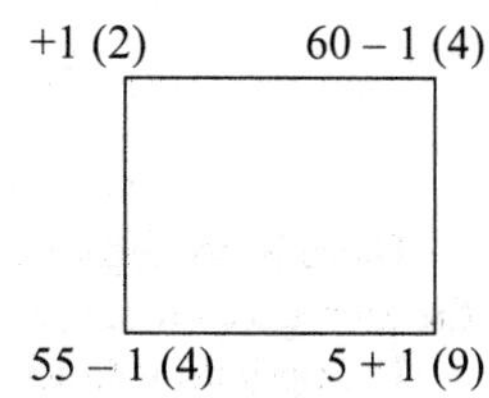

3. For cell (1, 4):
Addition in the transportation cost = 3 + 3 + 9 = 15
Reduction in the transportation cost = 1 + 4 = 8
Net change in the transportation cost = 15 – 9 = 6

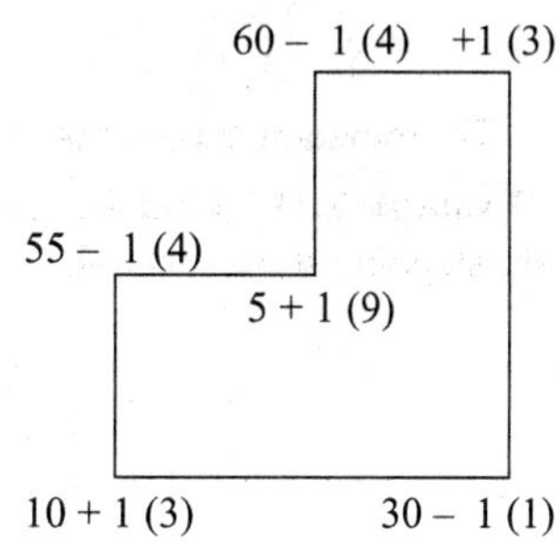

4. For cell (2, 1):
Addition in the transportation cost = 6 + 3 = 9
Reduction in the transportation cost = 4 + 2 = 6
Net change in the transportation cost = 9 – 6 = 3

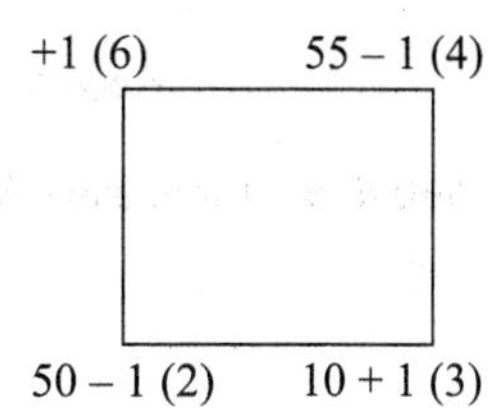

5. For cell (2, 4):
Addition in the transportation cost = 5 + 3 = 8
Reduction in the transportation cost = 1 + 4 = 5
Net change in the transportation cost = 8 – 5 = 3

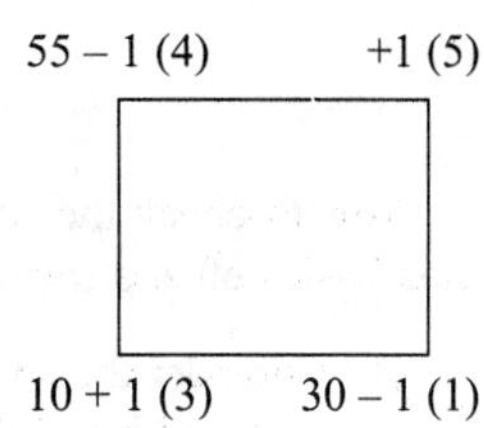

6. For cell (3, 1):
Addition in the transportation cost = 4 + 8 = 12
Reduction in the transportation cost = 9 + 3 = 12
Net change in the transportation cost = 12 – 12 = 0

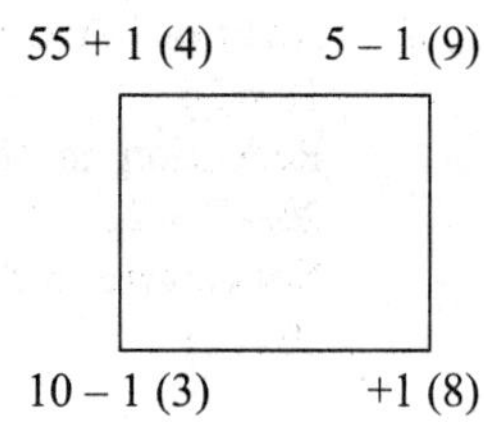

Incorporating all these net changes in the transportation cost table –1 in the south-west corner, we get the following table:

	D_1	D_2	D_3	D_4
S_1	5 7	2 3	4 (60)	3 6
S_2	6 3	4 (55)	9 (5)	5 3
S_3	2 (50)	3 (10)	8 0	1 (30)

There is no negative entry in the south-west corner of the allocation matrix and hence, the aforegiven table gives an optimal solution.

The optimal solution is

$$S_1 \xrightarrow{x_{13}=60} D_3,\quad S_2 \xrightarrow{x_{22}=55} D_2,\quad S_2 \xrightarrow{x_{23}=5} D_3,\quad S_3 \xrightarrow{x_{31}=50} D_1,\quad S_3 \xrightarrow{x_{32}=10} D_2,\quad S_3 \xrightarrow{x_{34}=30} D_4$$

The minimum transportation cost $= 60 \times 4 + 55 \times 4 + 5 \times 9 + 50 \times 2 + 10 \times 3 + 30 \times 1 = 665.$

Example 8.11 Find the optimal solution to the following transportation problem using the stepping stone method.

	D_1	D_2	D_3	D_4	Available
S_1	10	0	20	11	15
S_2	12	7	9	20	25
S_3	0	14	16	18	5
Requirement	5	15	15	10	

Solution First, using VAM, obtain the following allocation table of initial BFS:

	D_1	D_2	D_3	D_4
S_1	10	0 (15)	20	11
S_2	12	7 (0)	9 (15)	20 (10)
S_3	0 (5)	14	16	18 (0)

Now to check the optimality of the BFS in the aforegiven table, allocate one unit in a non-basic cell and consider all the non-basic cells one by one.

1. Consider the non-basic cell (1, 1). The loop obtained is as follows:

   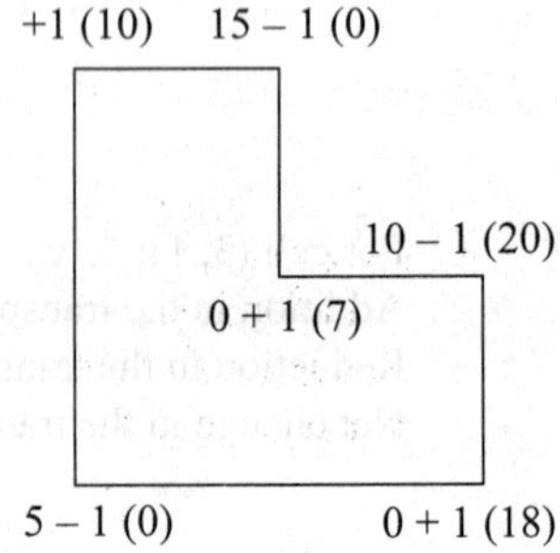

   Addition in the transportation cost = 10 + 7 + 18 = 35

   Reduction in the transportation cost = 0 + 20 + 0 = 20

   Net change in the transportation cost = 35 – 20 = 15

Note The entry in the brackets is the unit cost of the cell.

2. For cell (1, 3):
   Addition in the transportation cost = 20 + 7 = 27
   Reduction in transportation cost = 0 + 9 = 9
   Net change in transportation cost = 27 – 9 = 18

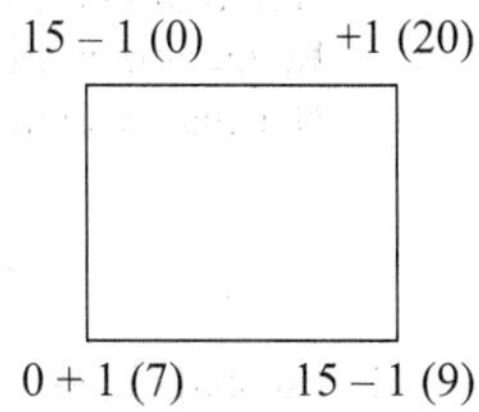

3. For cell (1, 4):
   Addition in the transportation cost = 11 + 7 = 18
   Reduction in the transportation cost = 0 + 20 = 20
   Net change in the transportation cost = 18 – 20 = –2

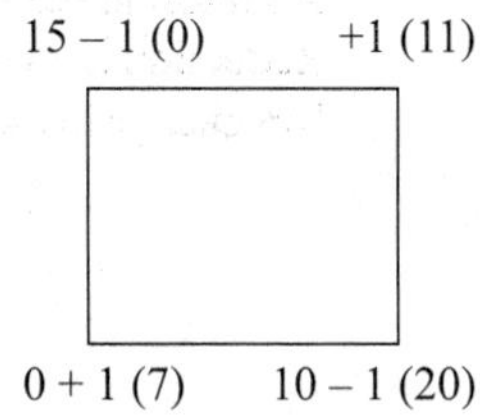

4. For cell (2, 1):
   Addition in the transportation cost = 12 + 18 = 30
   Reduction in the transportation cost = 20 + 0 = 20
   Net change in the transportation cost = 30 – 20 = 10

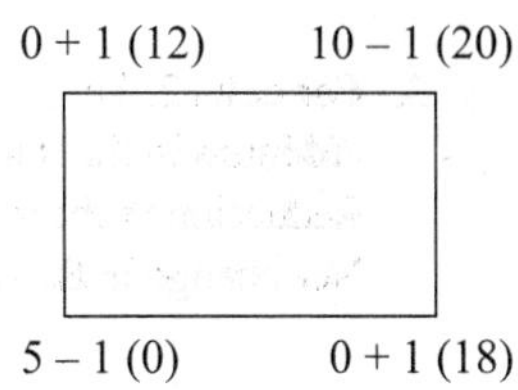

5. For cell (3, 2):
   Addition in the transportation cost = 14 + 20 = 34
   Reduction in the transportation cost = 18 + 7 = 25
   Net change in the transportation cost = 34 – 25 = 9

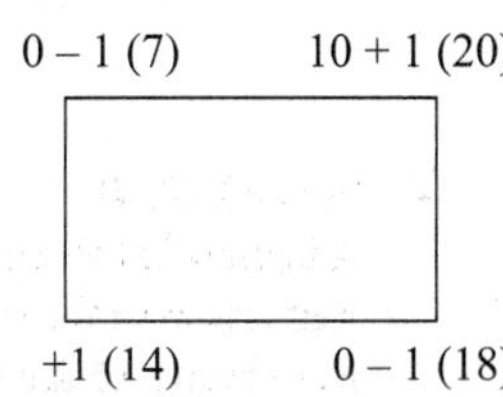

6. For cell (3, 3):
   Addition in the transportation cost = 16 + 20 = 36
   Reduction in the transportation cost = 18 + 9 = 27
   Net change in the transportation cost = 36 – 27 = 9

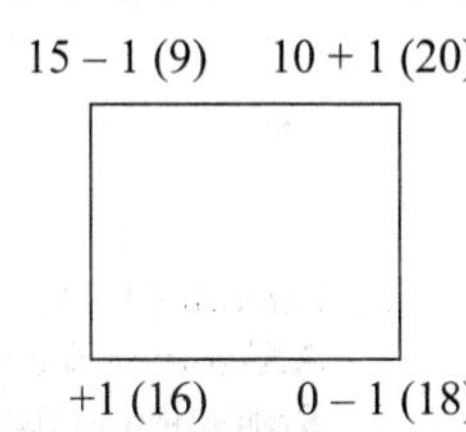

From these calculations, we observe that the net change in the transportation cost is negative in case of cell (1, 4), indicating that the solution given by the initial BFS is not optimal. Now, prepare the next allocation table by allocating the maximum possible units min(10, 15) = 10 in the non-basic cell (1, 4) and adjusting the allocations in the existing basic cells. The resulting allocation table is as follows:

D_1	D_2	D_3	D_4
10	0 (5)	20	11 (10)
12	7 (10)	9 (15)	20
0 (5)	14	16	18 (0)

To check the optimality of the solution, consider again all non-basic cells one by one.

1. For cell (1, 1)
   Addition in the transportation cost = 10 + 18 = 28
   Reduction in the transportation cost = 11 + 0 = 11
   Net change in the transportation cost = 28 – 11 = 17

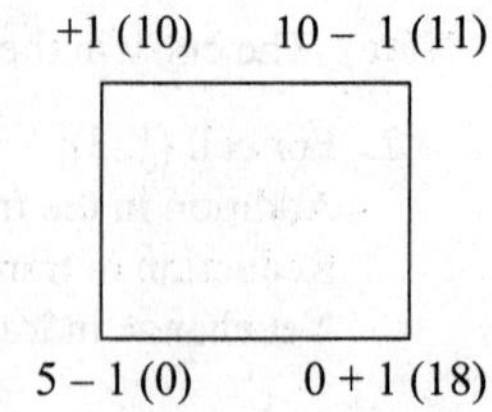

2. For cell (1, 3):
   Addition in the transportation cost = 7 + 20 = 27
   Reduction in the transportation cost = 9 + 0 = 9
   Net change in the transportation cost = 27 – 9 = 18

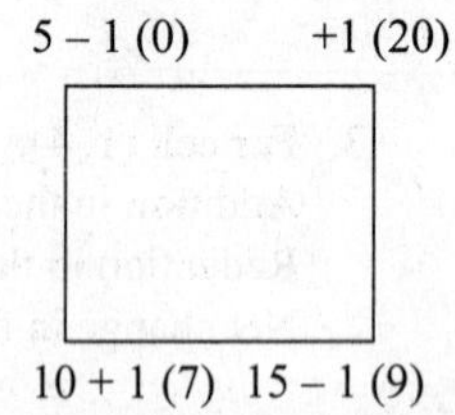

3. For cell (2, 1):
   Addition in the transportation cost = 12 + 0 + 18 = 30
   Reduction in the transportation cost = 7 + 11 + 0 = 18
   Net change in the transportation cost = 30 – 18 = 12

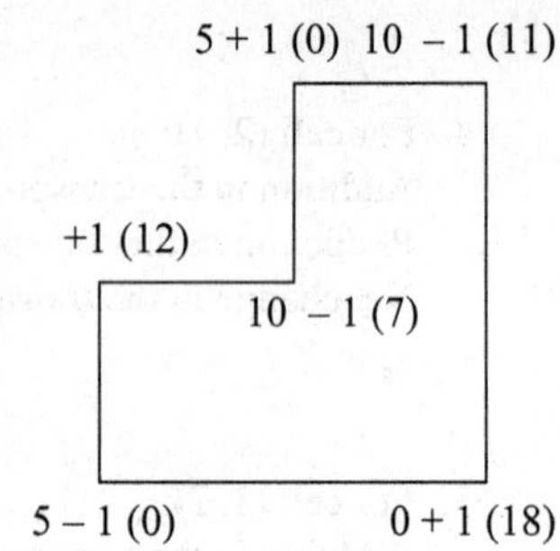

4. For cell (2, 4):
   Addition in the transportation cost = 20 + 0 = 20
   Reduction in the transportation cost = 11 + 7 = 18
   Net change in the transportation cost = 20 – 18 = 2

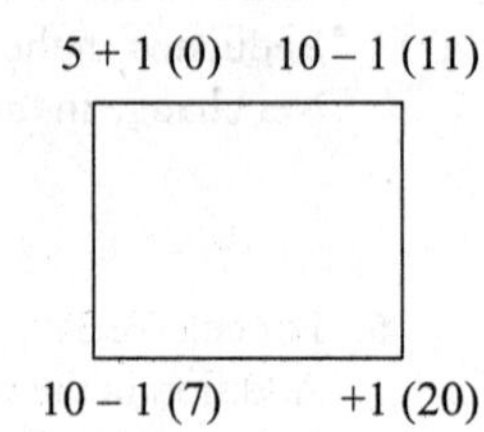

5. For cell (3, 2):
   Addition in the transportation cost = 14 + 11 = 25
   Reduction in the transportation cost = 18 + 0 = 18
   Net change in the transportation cost = 25 – 18 = 7

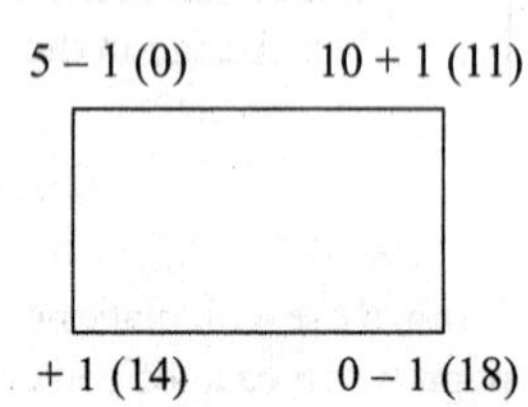

6. For cell (3, 3):
   Addition in the transportation cost = 16 + 11 + 7 = 34
   Reduction in the transportation cost = 0 + 9 + 18 = 27
   Net change in the transportation cost = 34 – 27 = 7

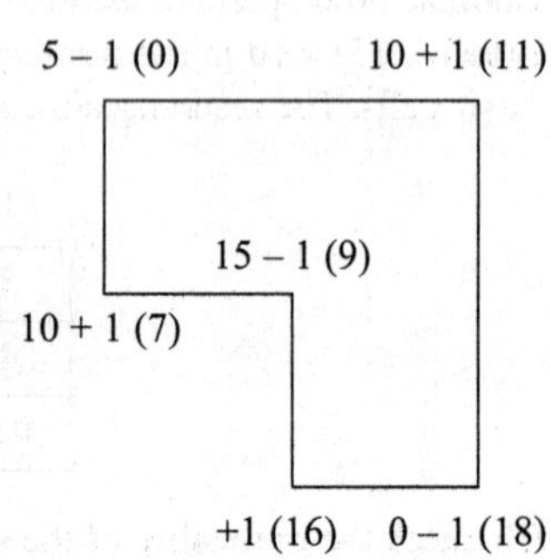

The resulting allocation table is as follows:

	D_1	D_2	D_3	D_4
S_1	10 17	0 (15)	20 18	11 (10)
S_2	12 12	7 (10)	9 (15)	20 2
S_3	0 (5)	14 7	16 7	18 (0)

There is no negative entry in the south-west corner of this allocation matrix and hence the aforegiven table gives an optimal solution. The optimal solution is

$$S_1 \xrightarrow{x_{12}=5} D_2,\quad S_1 \xrightarrow{x_{14}=10} D_4,\quad S_2 \xrightarrow{x_{22}=10} D_2,\quad S_2 \xrightarrow{x_{23}=15} D_3,\quad S_3 \xrightarrow{x_{31}=5} D_1,\quad S_3 \xrightarrow{x_{34}=0} D_4$$

The minimum transportation cost $= 0 \times 5 + 11 \times 10 + 7 \times 10 + 9 \times 15 + 0 \times 5 + 18 \times 0 = 315$.

8.6 DEGENERACY IN TRANSPORTATION PROBLEM

The degeneracy in a transportation problem can occur in the following two ways:

1. Usually, when we determine the initial BFS to a transportation problem, either a row or a column is satisfied at a time. The satisfied row or column is omitted for any further allocation. However, sometimes we come across situations when both row and column are satisfied simultaneously. Now, if we omit both the satisfied row and column for further allocation, then the number of basic variables will be less than $m + n - 1$. However, the total number has to be $m + n - 1$ in a transportation problem having m sources and n destinations. To overcome this difficulty whenever a row and column are satisfied together, we assume only one to be satisfied and insert zero against the other (row or column) treating that it is not satisfied and proceed as usual. This zero will be allocated somewhere and the solution thus obtained will be degenerate. (See Example 8.12).
2. Sometimes, a degenerate BFS is also obtained while determining the optimal solution when the minimum allocation occurs at more than one corner of a loop from which θ is substracted. Here, θ is the allocation, which is the minimum of the corner values from which it is subtracted. The cell with the minimum allocation becomes non-basic. However, if there are more than one such basic cell, then any one of them may be considered as a non-basic cell and zero is allocated in the remaining basic cells. The BFS thus obtained will be degenerate. Figures 8.3 and 8.4 illustrate the examples of having non-degenerate and degenerate BFS respectively, while forming a loop. Thus after placing the value of θ if we get zero in more than one place, then the only one cell is kept empty and zero is entered in other basic cells, and the BFS thus obtained will be degenerate.

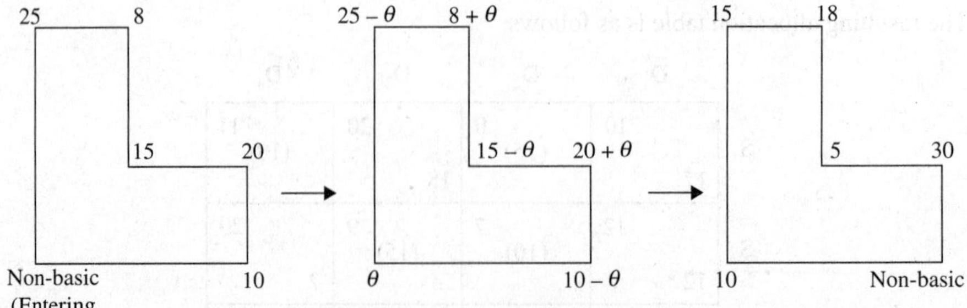

Here $\theta = \min(10, 15, 25) = 10$.

Figure 8.3 Giving non-degenerate BFS

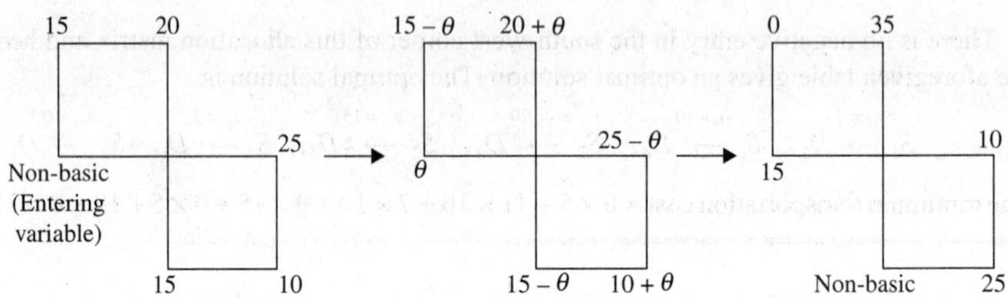

Here $\theta = \min(15, 25, 15) = 15$.

Figure 8.4 Giving degenerate BFS

We consider Example 8.12, where degeneracy occurs while obtaining the initial BFS using the NWC method.

Example 8.12 Use the NWC method to determine the initial BFS to the following transportation problem:

	Destination				
	D_1	D_2	D_3	D_4	Capacity (Availability)
O_1	10	0	20	11	5
O_2	12	7	9	20	25
O_3	0	14	16	18	15
Demand (Requirement)	5	15	15	10	45

Solution In this table, the north-west corner is the x_{11} position. The maximum possible allocation in x_{11} position = $\min(a_1, b_1) = \min(5, 5) = 5$. We enter 5 in the x_{11} position and subtract 5 from a_1 and b_1. Here, the first row and the first column are satisfied simultaneously, and is a case of degeneracy. In this case, assume arbitrarily that the first row is satisfied and replace entry 5 against the first column by 0

	D_1	D_2	D_3	D_4	
O_2	12	7	9	20	25
O_3	0	14	16	18	15
	0	15	15	10	

(assuming that the first column is not satisfied). Thus neglecting the first row, we get the adjacent table.

Now, we make the maximum possible allocation in the x_{21} position. The units allocated are $x_{21} = \min(0, 25) = 0$. We enter 0 in the x_{21} position and omit the first column to get the following table:

	D_2	D_3	D_4	
O_2	7	9	20	25
O_3	14	16	18	15
	15	15	10	

Now, we make the maximum possible allocation in the x_{22} position. The units allocated are $x_{22} = \min(15, 25) = 15$. We enter 15 in the x_{22} position, and subtract 15 from 25 and omit the D_2 column to get the adjacent table:

	D_3	D_4	
O_2	9	20	10
O_3	16	18	15
	15	10	

Now, we make the maximum possible allocation in the x_{23} position. The units allocated is $x_{23} = \min(15, 10) = 10$. We enter 10 in the x_{23} position and subtract 10 from 15 and omit second row to get the adjacent table.

	D_3	D_4	
O_3	16	18	15
	5	10	

We allocate 5 in x_{33} and 10 in the x_{34} position to complete the overall allocation to get the adjacent table.

	D_3	D_4	
O_3	16(5)	18(10)	~~15~~
	~~5~~	~~10~~	

Thus, the overall allocation made is shown in the following table:

	Destination				
	D_1	D_2	D_3	D_4	Supply
O_1	10(5)	0	20	11	5
Origin O_2	12(0)	7(15)	9(10)	20	25
O_3	0	14	16(5)	18(10)	15
Demand	5	15	15	10	45

The BFS is $x_{11} = 5, x_{21} = 0, x_{22} = 15, x_{23} = 10, x_{33} = 5, x_{34} = 10$ and the minimum transportation cost $= 10 \times 5 + 12 \times 0 + 7 \times 15 + 9 \times 10 + 16 \times 5 + 18 \times 10 =$ ₹505. Here degeneracy occurs while finding the initial BFS.

Example 8.13 Solve the following transportation problem given in the usual notation:

	D_1	D_2	D_3	D_4	$a_i \downarrow$
O_1	10	0	20	11	20
O_2	12	7	9	20	30
O_3	0	14	1	18	10
$b_j \rightarrow$	5	15	15	25	

Solution The steps involved in solving this transportation problem are as follows:

1. First we find the initial BFS using VAM, which is obtained as follows:

	D_1	D_2	D_3	D_4	
O_1	10	0 (15)	20	11 (5)	20
O_2	12	7	9 (15)	20 (15)	30
O_3	0 (5)	14	16	18 (5)	10
	5	15	15	25	

Now, we check the optimality of the solution.

2. Calculate u_i's and v_j's taking $u_i + v_j = c_{ij}$ for basic cells. We have

   $u_1 + v_2 = 0$, $u_1 + v_4 = 11$, $u_2 + v_3 = 9$, $u_2 + v_4 = 20$, $u_3 + v_1 = 0$, and $u_3 + v_4 = 18$.

   Solving these with $v_4 = 0$, we get $u_1 = 11, u_2 = 20, u_3 = 18, v_1 = -18, v_2 = -11, v_3 = -11$, and $v_4 = 0$

3. Now, calculate $d_{ij} = u_i + v_j - c_{ij}$ for the non-basic cells and enter this number in the south-west corner of the cell.

$d_{11} = u_1 + v_1 - c_{11} = 11 - 18 - 10 = -17$

$d_{13} = u_1 + v_3 - c_{13} = 11 - 11 - 20 = -20$

$d_{21} = u_2 + v_1 - c_{21} = 20 - 18 - 12 = -10$

$d_{22} = u_2 + v_2 - c_{22} = 20 - 11 - 7 = 2$

$d_{32} = u_3 + v_2 - c_{32} = 18 - 11 - 14 = -7$

$d_{33} = u_3 + v_3 - c_{33} = 18 - 11 - 16 = -9$

					$u_i \downarrow$
	10 / −17	0 (15)	20 / −20	11 (5)	11
	12 / −10	7 / 2	9 (15)	20 (15)	20
	0 (5)	14 / −7	16 / −9	18 (5)	18
$v_j \rightarrow$	−18	−11	−11	0	

Using these computations, we get the adjacent transportation table:

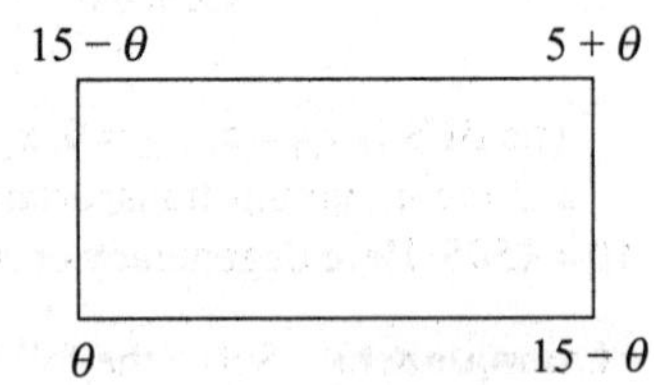

Here, $d_{22} = 2$ is the only positive entry among d_{ij}'s. Therefore, variable x_{22} enters the basis. To decide about the leaving variable, we prepare a loop starting from cell x_{22} and move horizontally and vertically taking turns at the basic cells and coming back to the initial cell to get the adjacent closed loop:

Here, $\theta = \min(15, 15) = 15$. This will make two variables x_{12} and x_{24} as non-basic variables. However, one variable should become non-basic at a time. So we arbitrarily retain x_{24} in the basis with value zero and obtain a degenerate BFS as given in the adjacent table:

10	0	20	11 (20)
12	7 (15)	9 (15)	20 (0)
0 (5)	14	16	18 (5)

Starting from the aforegiven BFS, we proceed as usual and ultimately the following optimal table is obtained:

10 −17	0 −2	20 −20	11 (20)
12 −10	7 (15)	9 (15)	20 (0)
0 (5)	14 −9	16 −9	18 (5)

Hence, the optimal solution to the problem is $x_{14}=20$, $x_{22}=15$, $x_{23}=15$, $x_{24}=0$, $x_{31}=5$, $x_{35}=5$ and the minimum transportation cost$=11\times20+7\times15+9\times15+20\times0+0\times5+18\times5=$₹550. Here degeneracy occurs while finding the optimal solution.

8.7 UNBALANCED TRANSPORTATION PROBLEM

A transportation problem is called unbalanced if the sum of all the available quantities is not equal to the sum of all the requirements, that is, $\sum_{i=1}^{m} a_i \neq \sum_{j=1}^{n} b_j$, m and n being the number of sources and destinations, respectively. There are two different situations when one can get an unbalanced transportation problem:

1. *Excess of availabilities* (*i.e.*, $\sum_{i=1}^{m} a_i > \sum_{j=1}^{n} b_j$): To convert this type of unbalanced transportation problem into a balanced one, we introduce a column of dummy (fictitious) destination denoted by D_F with demand $(\sum_{i=1}^{m} a_i - \sum_{j=1}^{n} b_j)$ and the unit transportation costs in column cells are taken to be zeros. The units transported from a source to a dummy destination will remain unused at the source.

Example 8.14 Balance the following transportation problem:

	D_1	D_2	D_3	$a_i \downarrow$
S_1	5	1	0	8
S_2	3	2	4	7
S_3	7	5	2	15
S_4	9	6	4	15
$b_j \rightarrow$	5	15	10	

Solution Here, $\sum_{i=1}^{4} a_i = 8+7+15+15=45$ and $\sum_{j=1}^{3} b_j = 5+15+15+10=30$ where $\sum_{i=1}^{4} a_i > \sum_{j=1}^{3} b_j$ and is an unbalanced transportation problem. To make this a balanced transportation problem, we introduce a column of dummy destination D_F (fictitious destination) with demand $b_4 = \sum_{i=1}^{m} a_i - \sum_{j=1}^{n} b_j = 45 - 30 = 15$ and enter zeros as costs. The balanced problem thus obtained is as follows:

	D_1	D_2	D_3	D_F	$a_i \downarrow$
S_1	5	1	0	0	8
S_2	3	2	4	0	7
S_3	7	5	2	0	15
S_4	9	6	4	0	15
$b_j \rightarrow$	5	15	10	15	

2. *Shortage in availabilities* $\left(i.e., \sum_{i=1}^{m} a_i < \sum_{j=1}^{n} b_j\right)$: Here, demand is more than the availability.

To make the problem balanced, add a row of fictitious (dummy) source (i.e., a row) with availability equal to $\left(\sum_{j=1}^{n} b_j - \sum_{i=1}^{m} a_i\right)$: and zero costs. The units transported from this dummy source to a destination is the shortage at that destination.

Example 8.15 Balance the following transportation problem:

	D_1	D_2	D_3	D_4	$a_i \downarrow$
S_1	10	0	20	11	10
S_2	12	7	94	20	8
S_3	0	14	15	18	7
$b_j \rightarrow$	5	15	10	15	

Solution This is a transportation problem with three sources S_1, S_2, S_3 and four destinations D_1, D_2, D_3, and D_4. Here, $\sum_{i=1}^{3} a_i = 10 + 8 + 7 = 25$ and $\sum_{j=1}^{4} b_j = 5 + 15 + 15 + 10 = 45$. Thus, $\sum_{i=1}^{3} a_i < \sum_{j=1}^{4} b_j$, demand is more than the requirement. We add a row of fictitious (dummy) source S_F with availability $a_4 = \sum_{j=1}^{4} b_j - \sum_{i=1}^{3} a_i = 20$ and zero costs to get the following balanced transportation problem:

	D_1	D_2	D_3	D_4	$a_i \downarrow$
S_1	10	0	20	11	10
S_2	12	7	94	20	8
S_3	0	14	15	18	7
S_4	0	0	0	0	20
$b_j \rightarrow$	5	15	15	10	

Example 8.16 Solve the following unbalanced transportation problem given in usual notations:

	D_1	D_2	D_3	$a_i \downarrow$
O_1	4	3	2	10
O_2	2	5	0	13
O_3	3	8	6	12
$b_j \rightarrow$	8	5	4	

Solution In the given problem,

$$\sum_{i=1}^{3} a_i = 10 + 13 + 12 = 35 \text{ and } \sum_{j=1}^{3} b_j = 8 + 5 + 4 = 17$$

Since $\sum_{i=1}^{3} a_i > \sum_{j=1}^{4} b_j$, the problem is unbalanced. To convert this problem into a balanced one, we introduce a column of fictitious destination D_4 with the requirement $35 - 17 = 18$ having all the transportation costs equal to zero. We obtain the adjacent table for the balanced transportation problem.

	D_1	D_2	D_3	D_4	$a_i \downarrow$
O_1	4	3	2	0	10
O_2	2	5	0	0	13
O_3	3	8	6	0	12
$b_j \rightarrow$	8	5	4	18	

1. First, we find the initial BFS using VAM, which is obtained as follows:

	D_1	D_2	D_3	D_4	$a_i \downarrow$
	4	3 (5)	2	0 (5)	10
	2 (8)	5	0 (4)	0 (1)	13
	3	8	6	0 (12)	12
$b_j \rightarrow$	8	5	4	18	

Now, we check the optimality of the solution.

2. Calculate u_i's and v_j's taking $u_i + v_j = c_{ij}$ for the basic cells. We have

$$u_1 + v_2 = 3,\ u_1 + v_4 = 0,\ u_2 + v_1 = 2,\ u_2 + v_3 = 0,\ u_2 + v_4 = 0, \text{ and } u_3 + v_4 = 0$$

Solving these with $u_2 = 0$, we get $u_1 = 0$, $u_2 = 0$, $u_3 = 0$, $v_1 = 2$, $v_2 = 3$, $v_3 = 0$, and $v_4 = 0$

					$u_i \downarrow$
	4	3 (5)	2	0 (5)	0
	2 (8)	5	0 (4)	0 (1)	0
	3	8	6	0 (12)	0
$v_j \rightarrow$	2	3	0	0	

3. Now, calculate $d_{ij} = u_i + v_j - c_{ij}$ for the non-basic cells and enter this number in the south-west corner of the cell as follows:

	1	2	3	4	$u_i \downarrow$
	4 / −2	3 / (5)	2 / −2	0 / (5)	0
	5 / (8)	5 / −2	0 / (4)	0 / (1)	0
	3 / −1	8 / −5	6 / −6	0 / (12)	0
$v_j \rightarrow$	2	3	0	0	

Here, all the d_{ij}'s are non-positive, that is, $d_{ij} \leq 0$. Thus, the existing BFS is optimal.

Hence, the solution of the problem is $x_{12}=5$, $x_{14}=5$, $x_{21}=8$, $x_{23}=4$, $x_{24}=1$, and $x_{34}=12$ and the total transportation cost$=3\times5+0\times5+2\times8+0\times4+0\times1+0\times12=$₹31.

Example 8.17 Solve the following unbalanced transportation problem if the penalty costs for unit storage at destinations 1, 2, and 3 are 5, 3, and 2 respectively.

		To 1	2	3	Supply
	1	5	1	7	10
Source	2	6	4	6	80
	3	3	2	5	15
Demand		75	20	50	

Solution The given problem is an unbalanced transportation problem because, demand = 75 + 20 + 50 = 145 and supply = 10 + 80 + 15 = 105 are not equal. Therefore we introduce a dummy source 4 with supply 145 – 105 = 40 and costs 5, 3, and 2, respectively.

1. First, we find the initial BFS using VAM, which is obtained as follows:

1	2	3	
5	1 / (10)	7	10
6 / (60)	4 / (10)	6 / (10)	80
3 / (15)	2	5	15
5	3	2 / (40)	40

Here, the number of allocations = 6 = $m + n - 1 = 4 + 3 - 1$ and they are at independent positions.

Now, we check the optimality of the solution.

2. Calculate $\underline{u}_i$'s and v_j's taking $u_i + v_j = c_{ij}$ for the basic cells. We have

$$u_1 + v_2 = 1,\ u_2 + v_1 = 6,\ u_2 + v_2 = 4,\ u_2 + v_3 = 6,\ u_3 + v_1 = 3, \text{ and } u_4 + v_3 = 2.$$

Solving these with $u_2 = 0$, we get $u_1 = -3$, $u_2 = 0$, $u_3 = -3$, $u_4 = -4$, $v_1 = 6$, $v_2 = 4$, and $v_3 = 6$.

	1	2	3	$u_i \downarrow$
	5	1 (10)	7	−3
	6 (60)	4 (10)	6 (10)	0
	3 (15)	2	5	−3
	5	3	2 (40)	−4
$v_j \rightarrow$	6	4	6	

3. Now calculate $d_{ij} = u_i + v_j - c_{ij}$ for the non-basic cells and enter this number in the south-west corner of the cell as follows:

	1	2	3	$u_i \downarrow$
	5 −2	1 (10)	7 −4	−3
	6 (60)	4 (10)	6 (10)	0
	3 (15)	2 −1	5 −2	−3
	5 −3	3 −3	2 (40)	−4
$v_j \rightarrow$	6	4	6	

Here, all the d_{ij}'s are non-positive, that is, $d_{ij} \leq 0$. Thus, the existing BFS is optimal.

Hence, the optimal solution is $x_{12} = 10$, $x_{21} = 60$, $x_{22} = 10$, $x_{23} = 10$, $x_{31} = 15$ and unfilled demand is $x_{43} = 40$, that is, the demand at the destination is not fully satisfied and it is unfilled by 40 units.

Hence, the total transportation cost $= 10 \times 1 + 60 \times 6 + 10 \times 4 + 10 \times 6 + 15 \times 3 + 40 \times 2 =$ ₹595.

8.8 TRANSSHIPMENT PROBLEM

A transportation problem where a commodity moves from one source to another source or destination before arriving at the final destination is called a transhipment problem. In real life, it may be cheaper to transport a commodity to an intermediate (or transient) node before it reaches the final place of requirement. This concept is more general than the one applied in the transportation model where only direct transportation from the source to a destination is allowed. Let us consider two plants P_1 and P_2. P_1 and P_2 supply luxury cars to three centers C_1, C_2, and C_3, which are again connected with four dealers D_1, D_2, D_3, and D_4. The supply (available units) at P_1 and P_2 are 500 and 2000 luxury cars respectively and the requirements of dealers D_1, D_2, D_3, and D_4 are 800, 950, 1050, and 1200 luxury cars, respectively.

The transportation cost per luxury car (in hundreds of rupees) between the pairs of node is shown on the connecting lines in Fig. 8.5. This is a case of transshipment as the total supply of $1500 + 2500 = 4000$ luxury cars at nodes P_1 and P_2 passes through several other nodes before reaching the final destinations D_1, D_2, D_3, and D_4. This transshipment model can be converted into a regular transportation model with eight sources P_1, P_2, C_1, C_2, C_3, D_1, D_2, and D_3 and seven destinations C_1, C_2, C_3, D_1, D_2, D_3, and D_4. Here, P_1 and P_2 are the pure supply nodes, D_4 is a pure demand node, and the nodes C_1, C_2, C_3, D_1, D_2, and D_3 act as both sources and destinations (with both inputs and outputs) and are referred to as transshipment nodes. The units of supply and demand at different nodes are calculated as follows:

1. Supply at a pure supply node = Original supply
2. Supply at a transshipment node = Original supply + Buffer units
3. Demand at a pure demand node = Original demand
4. Demand at a transshipment node = Original demand + Buffer units

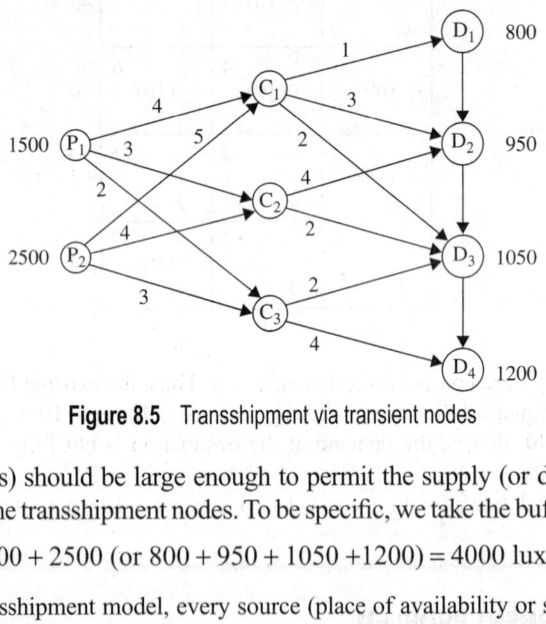

Figure 8.5 Transshipment via transient nodes

The buffer (units) should be large enough to permit the supply (or demand) of units through any of the transshipment nodes. To be specific, we take the buffer B as follows:

$$B = 1500 + 2500 \text{ (or } 800 + 950 + 1050 + 1200) = 4000 \text{ luxury cars}$$

Note In a transshipment model, every source (place of availability or supply) and every destination (place of requirement or demand) is considered as a transshipment node.

Example 8.18 Two show rooms S_1 and S_2 of a scooter manufacturing company have a stock of 200 and 300 scooters respectively. Out of these, 200, 150, and 150 scooters are supplied to three dealers D_1, D_2, and D_3 respectively. The unit transportation costs matrices are given as follows:

	D_1	D_2	D_3	
S_1	2	5	3	200
S_2	4	2	6	300
	200	150	150	

	S_1	S_2
S_1	0	2
S_2	2	0

	D_1	D_2	D_3
D_1	0	1	2
D_2	1	0	4
D_3	2	4	0

How to supply scooters to three dealers such that their requirement is fulfilled and the cost of transportation is minimum? Solve the problem using the transshipment model.

Solution Treating all S_1, S_2, D_1, D_2, and D_3 as transshipment nodes and buffer $B = 200 + 300 = 500$, the given transhipment problem can be designed in the following transportation model:

	S_1	S_2	D_1	D_2	D_3	
S_1	0	2	2	5	3	700
S_2	2	0	4	2	6	800
D_1	2	4	0	1	2	500
D_2	5	2	1	0	4	500
D_3	3	6	2	4	0	500
	500	500	700	650	650	

Using VAM, we get the following allocation matrix:

	S_1	S_2	D_1	D_2	D_3	
S_1	0 (500)	2	2 (50)	5	3 (150)	700
S_2	2	0 (500)	4	2 (300)	6	800
D_1	2	4	0 (500)	1	2	500
D_2	5	2	1 (150)	0 (350)	4	500
D_3	3	6	2	4	0 (500)	500
	500	500	700	650	650	

Now, we check the optimality of the solution using the MODI method:

1. Calculate u_i's and v_j's taking $u_i + v_j = c_{ij}$ for the basic cells. We have
   $u_1 + v_1 = 0$, $u_1 + v_3 = 2$, $u_1 + v_5 = 3$, $u_2 + v_2 = 0$, $u_2 + v_4 = 2$, $u_3 + v_3 = 0$, $u_4 + v_3 = 1$, $u_4 + v_4 = 0$, and $u_5 + v_5 = 0$.

Solving these with $u_1 = 0$, we get $u_1 = 0$, $u_2 = 1$, $u_3 = -2$, $u_4 = -1$, $u_5 = -3$, $v_1 = 0$, $v_2 = -1$, $v_3 = 2$, $v_4 = 1$, and $v_5 = 3$

2. Now, calculate $d_{ij} = u_i + v_j - c_{ij}$ for the non-basic cells.

$d_{12} = u_1 + v_2 - c_{12} = 0 - 1 - 2 = -3$
$d_{14} = u_1 + v_4 - c_{14} = 0 + 1 - 5 = -4$
$d_{21} = u_2 + v_1 - c_{21} = 1 + 0 - 2 = -1$
$d_{23} = u_2 + v_3 - c_{23} = 1 + 2 - 4 = -1$
$d_{25} = u_2 + v_5 - c_{25} = 1 + 3 - 6 = -2$
$d_{31} = u_3 + v_1 - c_{31} = -2 + 0 - 2 = -4$
$d_{32} = u_3 + v_2 - c_{32} = -2 - 1 - 4 = -7$
$d_{34} = u_3 + v_4 - c_{34} = -2 + 1 - 1 = -2$
$d_{35} = u_3 + v_5 - c_{35} = -2 + 3 - 2 = -1$
$d_{41} = u_4 + v_1 - c_{41} = -1 + 0 - 5 = -6$
$d_{42} = u_4 + v_2 - c_{42} = -1 - 1 - 2 = -4$
$d_{45} = u_4 + v_5 - c_{45} = -1 + 3 - 4 = -2$
$d_{51} = u_5 + v_1 - c_{51} = -3 + 0 - 3 = -6$
$d_{52} = u_5 + v_2 - c_{52} = -3 - 1 - 6 = -10$
$d_{53} = u_5 + v_3 - c_{53} = -3 + 2 - 2 = -3$
$d_{54} = u_5 + v_4 - c_{54} = -3 + 1 - 4 = -6$

Here, all the d_{ij}'s are non-positive, that is, $d_{ij} \leq 0$. Thus, the existing BFS is optimal.

Hence, the optimum solution of the problem is $x_{12} = 500$, $x_{13} = 50$, $x_{15} = 150$, $x_{22} = 500$, $x_{24} = 300$, $x_{33} = 500$, $x_{43} = 150$, $x_{44} = 350$, and $x_{55} = 500$.

The minimum transportation cost is $= 500 \times 0 + 50 \times 2 + 150 \times 3 + 500 \times 0 + 300 \times 2 + 500 \times 0 + 150 \times 1 + 350 \times 0 + 500 \times 0 =$ ₹1300.

RECAPITULATION

1. A transportation problem is a linear programming problem.
2. The solution of a transportation problem is highly degenerate and hence the simplex method is not suitable to solve it.
3. To solve a transportation problem, start with a basic feasible solution (BFS).
4. There are many methods to get the initial feasible solution to a transportation problem. A few of them are the (a) north-west corner method (NWCM), (b) least cost entry method (LCM), and (c) Vogel's approximation method (VAM).
5. VAM is the most efficient method as it provides a solution, which is either optimal or requires only a few iterations to get an optimal solution.
6. The MODI or stepping stone methods are used to get an optimal solution starting from the BFS obtained by using one of the aforementioned methods.
7. The transportation model is used when the problem is a balanced minimization problem.
8. When we formulate a real-life problem in a mathematical model, it sometimes turns out to be an unbalanced maximization problem. Such problems after converting into balanced minimization problem, can be solved using the transportation algorithm.
9. In some cases, it may be cheaper to transport a commodity to a final destination via another source or destination or via a combination of both instead of transporting it directly to the final destination. For such problems, we use the transshipment model.

EXERCISES

Multiple-choice Questions

8.1 While determining the initial BFS in a transportation problem using any of the given methods, if a row and a column are satisfied simultaneously, then the
(a) transportation problem has no solution.
(b) transportation problem has a degenerate BFS.
(c) transportation problem has a non-basic solution.
(d) none of these

8.2 In a transportation problem, the dual variables u_i and v_j are
(a) non-negative
(b) non-positive
(c) unrestricted in sign
(d) none of these

8.3 The number of independent constraints in an unbalanced LPP with m sources and n destinations is
(a) $m+n+1$ (b) $m+n$ (c) $m+n-1$ (d) none of these

8.4 In a balanced transportation problem with m sources and n destinations, the number of basic variables is
(a) $m+n$ (b) $m+n+1$ (c) $m+n-1$ (d) none of these

8.5 If an LPP with m sources and n destinations is not balanced, then the number of basic variables in its BFS is
(a) $m+n-1$ (b) $m+n+1$ (c) $m+n$ (d) none of these

8.6 The transportation problem is
(a) a non-linear programming problem (c) a quadratic programming problem
(b) an LPP (d) none of these

8.7 The transportation algorithm has been developed as
(a) a maximization problem (c) both maximization and minimization problem
(b) a minimization problem (d) none of these

State True or False

8.1 In a transportation problem while obtaining the initial BFS using any of the methods, every time exactly one row or one column is satisfied and not both.

8.2 In a balanced transportation problem with m sources and n destinations, there are $(m+n-1)$ basic variables because one of the constraints is redundant.

8.3 In a transportation problem while obtaining the initial BFS if a row and a column are satisfied simultaneously then BFS thus obtained will be degenerate.

8.4 We cannot use VAM, NWCM, and LCM for obtaining the inital BFS unless the transportation problem is balanced.

8.5 A balanced transportation problem always has a unique solution.

8.6 Amongst VAM, NWCM, and LCM, VAM is the most efficient method to get the initial BFS to a transportation problem.

8.7 If VAM method is used to get the initial BFS to a transportation problem, then no iteration or only a few more iterations are needed to get the optimal solution.

8.8 In solving a transportation problem, degeneracy may occur while obtaining its initial BFS and sometimes during the process of obtaining its optimal solution.

Fill in the Blanks

8.1 Amongst the VAM, LCM, and NWCM to obtain the initial BFS, the most efficient method is the __________.

8.2 All constraints in a transportation problem hold with __________ sign.

8.3 A transportation model is a __________.

8.4 The transportation problem is an LPP, but __________ method is not used to solved it.

8.5 Consider the transportation problem having m sources with availabilities a_i, $i = 1, 2, \ldots, m$ and n destinations, D_j with requirements b_j, $j = 1, 2, \ldots, n$. With reference to this transportation problem, fill in the following blanks:
(a) It is balanced if __________
(b) It is unbalanced if __________
(c) If $\sum a_i = \sum b_j$ then it is called __________
(d) If $\sum a_i \neq \sum b_j$ then it is called __________
(e) If not balanced, then we have to __________ it before starting to obtain the initial BFS.
(f) If it is unbalanced, then the number of BFS, that is, __________.

(g) If it is balanced, then the number of BFS, that is, ________.
(h) If it is formulated as a maximization problem, then we have to reduce it into a ________ problem before applying a transportation algorithm.
(i) A transportation problem algorithm is developed as a ________ problem.
(j) The number of constraints is ________.

Review Questions

8.1 Write the general mathematical formulation of a transportation problem.
8.2 Write a short note on the following transportation problem methods to obtain the initial BFS:
(a) NWCM
(b) LCM or matrix minima method
(c) VAM
8.3 Explain the MODI method in detail.
8.4 Write a short note on the stepping stone method.
8.5 Explain about the occurrence of degeneracy in a transportation problem.
8.6 What is a transshipment problem?

Numerical Problems

8.1 Use NWCM to obtain an initial BFS to the following transportation problems:

(a)

		To				
		D_1	D_2	D_3	D_4	Supply
	A	13	11	15	20	2
From	B	17	14	12	13	6
	C	18	18	15	12	7
Demand		3	3	4	5	15

(b)

		Destination					
		I	II	III	IV	V	Supply
	A	2	11	10	30	7	4
Origin	B	1	4	7	2	1	8
	C	3	9	4	8	12	9
Demand		3	3	4	5	6	21

8.2 Use LCM to obtain the initial BFS to the following transportation problem:

	Destination				
	D_1	D_2	D_3	D_4	Supply
I	1	5	3	3	34
II	3	3	1	2	15
III	0	2	2	3	12
IV	2	7	2	4	19
Demand	21	25	17	17	80

8.3 Determine an initial BFS to the following transportation table using (a) LCM (b) VAM:

	Destination				
	D_1	D_2	D_3	D_4	Supply
O_1	1	2	1	4	30
Origin O_2	3	3	2	1	50
O_3	4	2	5	9	20
Demand	20	40	30	10	100

8.4 Determine an initial BFS of the following transportation table using (a) NWCM (b) LCM (c) VAM:

	Warehouse				
	W_1	W_2	W_3	W_4	Capacity
F_1	19	30	50	10	7
F_2	70	30	40	60	9
F_3	40	8	70	20	18
Requirement	5	8	7	14	34

8.5 Solve the following transportation problems:

(a)

	D_1	D_2	D_3	D_4	Supply
O_1	23	27	16	18	30
O_2	12	17	20	51	40
O_3	22	28	12	32	53
Demand	22	35	25	41	

(b)

	D_1	D_2	D_3	D_4	Availability
O_1	1	2	1	4	30
O_2	3	3	2	1	50
O_3	4	2	5	9	20
Required	20	40	30	10	100

8.6 A company has four warehouses W_1, W_2, W_3, and W_4 from which it supplies to three markets M_1, M_2, and M_3. Determine the optimal transportation cost from the following data:

	W_1	W_2	W_3	W_4	Required
M_1	19	14	23	11	11
M_2	15	16	12	21	13
M_3	30	25	16	39	19
Available	6	10	12	15	

8.7 A company has three warehouses W_1, W_2, and W_3 with availabilities of 14, 12, and 5 tons respectively of a certain kind of goods. The goods have to be supplied to three retailers A, B, and C with requirements of 6, 10, and 15 tons respectively. The cost matrix is as follows:

	A	B	C
W_1	6	8	4
W_2	4	9	3
W_3	1	2	6

Find the transportation schedule or policy which minimizes the cost.

8.8 Obtain an initial BFS to the following transportation problem. Is this solution an optimal solution? If not, obtain the optimal solution.

	W_1	W_2	W_3	W_4	$a_i \downarrow$
F_1	19	30	50	10	7
F_2	70	30	40	60	9
F_3	40	8	70	20	18
$b_j \rightarrow$	5	8	7	14	34

8.9 Solve the following transportation problems:

(a)

	D_1	D_2	D_3	D_4	D_5	$a_i \downarrow$
O_1	73	40	9	79	20	8
O_2	62	93	96	8	13	7
O_3	96	65	80	50	65	9
O_4	57	58	29	12	87	3
O_5	56	23	87	18	12	5
$b_j \rightarrow$	6	8	10	4	4	

(b)

		To				
		I	II	III	IV	
	A	15	10	17	18	2
From	B	16	13	12	13	6
	C	12	17	20	11	7
		3	3	4	5	

(c)

	To				Supply
	21	16	25	13	11
From	17	18	14	23	13
	32	27	18	41	19
Demand	6	10	12	15	

(d)

	D_1	D_2	D_3	D_4	D_5	Availability
S_1	5	7	6	8	7	13
Sources S_2	7	6	8	9	8	6
S_3	6	4	7	7	9	8
Requirement	6	7	3	5	6	27

8.10 Obtain an optimum BFS to the following transportation problem:

		To		Available
	7	3	4	2
From	2	1	3	3
	3	4	6	5
Demand	4	1	5	

8.11 Solve the following transportation problems:

(a)

	D_1	D_2	D_3	D_4	D_5	Available
O_1	5	5	6	4	2	9
O_2	6	9	7	8	5	13
O_3	5	6	4	6	3	9
Requirement	3	7	8	5	8	27

(b)

	D_1	D_2	D_3	D_4	D_5	D_6	Available
O_1	9	12	9	6	9	10	5
O_2	7	3	7	7	5	5	6
O_3	6	5	9	11	3	11	2
O_4	6	8	11	2	2	10	9
Required	4	4	6	2	4	4	22

8.12 The costs of transporting material from supply points A, B, C, and D to demand points E, F, G, H, and J are given in the following table.

		To					
		E	F	G	H	J	
	A	8	10	12	17	15	100
From	B	15	13	18	11	9	10
	C	14	20	6	10	13	170
	D	13	19	7	5	12	280
		90	20	50	210	190	

Find the optimum solution.

8.13 Is $x_{13} = 50$, $x_{14} = 20$, $x_{21} = 55$, $x_{31} = 30$, $x_{32} = 35$, $x_{34} = 25$ an optimum solution to the following transportation problem?

					Available units
	6	1	9	3	70
	11	5	2	8	55
	10	12	4	7	90
Required units	85	35	50	45	

If not, modify it to obtain its optimum BFS.

8.14 Solve the following transportation problem:

		To			
		D_1	D_2	D_3	$a_i \downarrow$
From	O_1	7	4	0	5
	O_2	6	8	0	15
	O_3	3	9	0	9
	$b_j \rightarrow$	15	6	8	

8.15 Solve the following cost-minimizing transportation problem:

	D_1	D_2	D_3	D_4	Available
O_1	5	3	6	5	15
O_2	10	7	12	4	11
O_3	7	5	8	4	13
Demand	8	12	13	6	

8.16 Solve the following transportation problem:

							Available
A	5	3	7	3	8	5	3
B	5	6	12	5	7	11	4
C	2	1	2	4	8	2	2
D	9	6	10	5	10	9	8
Required	3	3	6	2	1	2	

8.17 Determine an optimum BFS to the following transportation problem:

	D_1	D_2	D_3	D_4	Capacity
O_1	1	2	3	4	6
O_2	4	3	2	0	8
O_3	0	2	2	1	10
Demand	4	6	8	6	24

where O_i and D_j denote ith origin and jth destination respectively.

8.18 Solve the following transportation problem:

	D_1	D_2	D_3
S_1	8	7	3
S_2	3	8	9
S_3	11	3	5

8.19 The unit cost array with supplies a_i ($i = 1, 2, 3$) and demand b_j ($j = 1, 2, 3, 4$) are given in the following table:

					Available units
	6	1	9	3	70
	11	5	2	8	55
	10	12	4	7	90
Required units	85	35	50	45	

Find the optimum solution to this problem.

8.20 Solve the following transportation problems:

(a)

	W_1	W_2	W_3	W_4	W_5	Available
F_1	4	3	1	2	6	40
F_2	5	2	3	4	5	30
F_3	3	5	6	3	2	20
F_4	2	4	4	5	3	10
Required	30	30	15	20	5	

(b)

	M_1	M_2	M_3	M_4	M_5	Available
F_1	4	2	3	2	6	8
F_2	5	4	5	2	1	12
F_3	6	5	4	7	3	14
Required	4	4	6	8	8	

8.21 A company has three plants P_1, P_2, and P_3 which supply to warehouses at W_1, W_2, W_3, and W_4. The monthly plant capacities are 160, 150, and 190 units and the warehouse requirements are 80, 90, 110, and 160 units respectively. The unit transportation costs (in ₹) are as follows:

	To			
	W_1	W_2	W_3	W_4
F_1	42	48	38	37
F_2	40	49	52	51
F_3	39	38	40	43

Solve this problem to minimize transportation costs.

CONCEPTUAL PROBLEM

Four villages (V_1, V_2, V_3, and V_4) in Bihar state of India are severely affected by the heavy rains in the state, triggering a flood-like situation. The road link to these villages is completely disrupted. The government has decided to deploy the paramilitary forces for supplying the food grains to these villages by three aircraft A_1, A_2, and A_3. As the four villages vary in terms of population and are also spread apart from the sources of food grain supply, the aircraft can make only a limited number of trips in a day and carry limited units of food grains per trip to these villages. Further, the aircraft differ in their load carrying capacity and in respect of sources from where they take off to these different villages. The units of the food grains that aircraft A_1, A_2, and A_3 can each carry in one trip to villages V_1, V_2, V_3, and V_4 are, respectively, 10, 8, 6, 3; 8, 10, 5, 3; and 2, 5, 9, 10. The total number of trips required per day to villages V_1, V_2, V_3, and V_4 are respectively 30, 50, 60, and 40. However, the total numbers of trips that aircraft A_1, A_2, and A_3 can make in one day to these villages are 60, 40, and 50 respectively. The commander of the paramilitary group deployed to supply the food grains has to prepare a plan to calculate the total number of trips that aircraft A_i, ($i = 1, 2, 3$) have to make to villages V_j, ($j = 1, 2, 3, 4$) such that the total quantity of food grains dropped in a day is maximum. Further, the commander has to report to the government as to which of the villages are in short supply of trips and also submit the full analysis of the food supply operation. Can you help to solve this problem?

Solution The following is the formulation of this problem:

The commander of the group faces the problem of determining the number of trips that aircraft A_i, $i = 1, 2, 3$ have to make to villages V_j, $j = 1, 2, 3, 4$ to deliver maximum units of food grains.

Let x_{ij} = the number of trips per day that aircraft A_i makes to village V_j, $i = 1, 2, 3$, and $j = 1, 2, 3, 4$. The total number of trips in a day made by aircraft $A_1 = x_{11} + x_{12} + x_{13} + x_{14}$.

Thus,

$$x_{11} + x_{12} + x_{13} + x_{14} = 60 \quad (8.6)$$

Similarly, $$x_{21} + x_{22} + x_{23} + x_{24} = 40 \quad (8.7)$$

and $$x_{31} + x_{32} + x_{33} + x_{34} = 50 \quad (8.8)$$

The total number of trips required per day to village $V_1 = x_{11} + x_{21} + x_{31} + x_{41}$

Thus, $$x_{11} + x_{21} + x_{31} + x_{41} = 30 \quad (8.9)$$

Similarly, $$x_{12} + x_{22} + x_{32} + x_{42} = 50 \quad (8.10)$$

$$x_{13} + x_{23} + x_{33} + x_{43} = 60 \quad (8.11)$$

and $$x_{14} + x_{24} + x_{34} + x_{44} = 40 \quad (8.12)$$

Z = The units of food grains delivered in one day by three aircraft to four villages

$$= 10x_{11} + 8x_{12} + 6x_{13} + 3x_{14} + 8x_{21} + 10x_{22} + 5x_{23} + 3x_{24} + 2x_{31} + 5x_{32} + 9x_{33} + 10x_3 \quad (8.13)$$

Therefore, the problem before the group commander is to determine x_{ij}, $i = 1, 2, 3$; $j = 1, 2, 3, 4$ so as to maximize Z in (8.13) with restrictions (8.6) to (8.12). The problem is an LPP and to be very specific it is a transportation type problem. Hence, we can make an attempt to solve it using the transportation algorithm. Tabulating the given data, we get the following table:

(*Contd*)

(*Contd*)

	V_1	V_2	V_3	V_4	a_i
A_1	10	8	6	3	60
A_2	8	10	5	3	40
A_3	2	5	9	10	50
b_j	30	50	60	40	

where column a_i provides the total number of trips that aircraft A_i can make in one day, row b_j displays the number of trips required to villages V_j, and the entries within the rectangle denote the amount of food grains that aircraft A_i can carry to village V_j in one trip. This is a maximization and unbalanced problem. To solve it using the transportation algorithm, convert it into a minimization problem and balance it. Then, use VAM to get the initial BFS. The minimized and balanced problem thus obtained is as follows:

	V_1	V_2	V_3	V_4	Available
A_1	−10	−8	−6	−3	60
A_2	−8	−10	−5	−3	40
A_3	−2	−5	−9	−10	50
A_4	0	0	0	0	30
Requirement	30	50	60	40	

The initial BFS using VAM, is given as follows:

	V_1	V_2	V_3	V_4	
A_1	−10 (30)	−8 (10)	−6 (20)	−3	60
A_2	−8	−10 (40)	−5	−3	40
A_3	−2	−5	−9 (10)	−10 (40)	50
A_4	0	0	0 (30)	0	30
	30	50	60	40	

Now, we check the optimality of the solution using the MODI method.

Calculate u_i's and v_j's taking $u_i + v_j = c_{ij}$ for the basic cells and $d_{ij} = u_i + v_j - c_{ij}$ for the non-basic cells and enter this number in the south-west corner of the cell.

$$u_1 + v_1 = -10$$
$$u_1 + v_2 = -8$$

(*Contd*)

(*Contd*)

$u_1 + v_3 = -6$
$u_2 + v_2 = -10$
$u_3 + v_3 = 9$
$u_3 + v_4 = -10$
and $u_4 + v_3 = 0$

Solving these with $u_1 = 0$, we get $u_1 = 0$, $u_2 = -2$, $u_3 = -3$, $u_4 = 6$, $v_1 = -10$, $v_2 = -8$, $v_3 = -6$, and $v_4 = -7$

Now, $d_{14} = u_1 + v_4 - c_{14} = 0 - 7 - (-3) = -4$
$d_{21} = u_2 + v_1 - c_{21} = -2 - 10 - (-8) = -4$
$d_{23} = u_2 + v_3 - c_{23} = -2 - 6 - (-5) = -3$
$d_{24} = u_2 + v_4 - c_{24} = -2 - 7 - (-3) = -6$
$d_{31} = u_3 + v_1 - c_{31} = -3 - 10 - (-2) = -11$
$d_{32} = u_3 + v_2 - c_{32} = -3 - 8 - (-5) = -6$
$d_{41} = u_4 + v_1 - c_{41} = 6 - 10 - 0 = -4$
$d_{42} = u_4 + v_2 - c_{42} = 6 - 8 - 0 = -2$
$d_{44} = u_4 + v_4 - c_{44} = 6 - 7 - 0 = -1$

	$v_1 = -10$	$v_2 = -8$	$v_3 = -6$	$v_4 = -7$	
$u_1 = 0$	−10 (30)	−8 (10)	−6 (20)	−3 −4	60
$u_2 = -2$	−8 −4	−10 (40)	−5 −3	−3 −6	40
$u_3 = -3$	−2 −11	−5 −6	−9 (10)	−10 (40)	50
$u_4 = 6$	0 −4	0 −2	0 (30)	0 −1	30
	30	50	60	40	

Here, all the d_{ij}'s are non-positive, that is, $d_{ij} \leq 0$. Thus, the existing BFS is optimal.

Hence, the solution of the problem is $x_{11} = 30$, $x_{12} = 10$, $x_{13} = 20$, $x_{22} = 40$, $x_{33} = 10$, and $x_{34} = 40$.

Thus, the aircraft A_1 makes 30, 10, and 20 trips to villages V_1, V_2, and V_3 respectively. Aircraft A_2 makes 40 trips to village V_2. Aircraft A_3 makes 10 and 40 trips to villages V_3 and V_4 respectively. Village V_3 falls short of 30 trips.

The maximum amount of food grains dropped in one day $= 30 \times 10 + 10 \times 8 + 20 \times 6 + 40 \times 10 + 10 \times 9 + 40 \times 10 = 1390$ units.

9 Assignment Problems

After studying this chapter, the reader will be able to

- understand the Hungarian algorithm
- solve the travelling salesman problem using the Hungarian algorithm
- deal with the situation when the Hungarian algorithm does not work
- solve the assignment problem when it is balanced or unbalanced

9.1 INTRODUCTION

The assignment of jobs (activities) to different people (facilities) can be done if there exists a one-to-one correspondence between them and the efficiency of every person on the job is known. Our objective here is to create an assignment that maximizes the efficiency or minimizes the inefficiency. To accomplish this objective, the Hungarian algorithm for assignment is used.

The assignment problem is a special case of the linear programming problem (LPP). We can also say that it is a special case of the transportation problem with $m = n$ and $a_i = b_i = 1, i = 1, 2, \ldots, n$. The aim of an assignment problem is to estimate how a job involving manpower can be solved most efficiently with minimal cost. Assume n jobs (tasks) are assigned to n persons (facilities). Each person can perform every job but with varying efficiencies. Here, a problem arises as to how the most optimum assignment can be made, that is, which person can do what job most efficiently. Let us now define the assignment problem mathematically:

$$x_{ij} = \begin{cases} 1, & \text{if } i\text{th person is assigned } j\text{th job} \\ 0, & \text{if } i\text{th person is not assigned } j\text{th job} \end{cases}$$

$$c_{ij} = \text{inefficiency of } i\text{th person in doing the } j\text{th job}$$

Our objective is to assign the jobs to people in a way that the overall inefficiency of doing the jobs is minimized. Mathematically, it can be represented as

$$\text{Minimize } Z = \sum_{j=1}^{n} \sum_{i=1}^{n} c_{ij} x_{ij}$$

Subject to the condition

$$\sum_{i=1}^{n} x_{ij} = 1, \; j = 1, 2, ..., n, \text{ that is, } i\text{th person is assigned only one job.}$$

$$\sum_{j=1}^{n} x_{ij} = 1, \; i = 1, 2, ..., n, \text{ that is, } j\text{th job is assigned to only one person.}$$

The assignment correspondence is one-to-one between the person and the job. The assignment problem can be represented in the form of a matrix called effectiveness (cost) matrix:

		Jobs				
		J_1	J_2	J_3	...	J_n
	P_1	c_{11}	c_{12}	c_{13}	...	c_{1n}
	P_2	c_{21}	c_{22}	c_{23}	...	c_{2n}
Persons	P_3	c_{31}	c_{32}	c_{33}	...	c_{3n}
	...	...	...	...	...	...
	P_n	c_{n1}	c_{n2}	c_{n3}	...	c_{nn}

9.2 SOLVING ASSIGNMENT PROBLEMS USING HUNGARIAN METHOD

The following are the steps of the Hungarian algorithm:

1. Locate the minimum entry in each row of the cost matrix, and subtract it from each entry of that row. The matrix thus obtained is called the reduced matrix, having at least one zero entry in each row.
2. Now subtract the smallest element of each column from every element of that column. We get the second reduced matrix containing at least one zero in every column. Thus, steps 1 and 2 bring at least one zero in every row and column of the reduced matrix.
3. Now, examine the first row of the matrix obtained in step 2. If there is only one zero, then surround it by a square as '□'and cross (X) all other zeros in the column of the surrounded zero. Repeat it for all the rows. If there is more than one zero in any row, skip it and move to the next row. Repeat the procedure for all the columns as well.
4. If there is exactly one surrounded zero in every row and column, then the complete optimal assignment is done. The mark '□' represents the assignment of the job to the respective person and we stop at this stage. If the total number of surrounded zeros is not exactly n, we go to step 5.
5. Draw (by trial) the minimum number of vertical or horizontal lines to cover all the zeros. Select the smallest element, say x, among all the elements uncovered by these lines. Subtract the value x from all the elements not covered by the lines, and add x to all those elements that lie at the intersection of the lines. Repeat steps 3 and 4 to get the optimum assignment.

Note To draw the minimum number of lines, the procedure mentioned here can be followed. If a zero is surrounded while going around a row (column), then draw a vertical (horizontal) line through this zero. When this is done for all the surrounded zeros, it will give the minimum number of lines covering all (surrounded or crossed) zeros.

9.3 MINIMAL ASSIGNMENT PROBLEM

The Hungarian algorithm discussed in Section 9.2 is used to solve minimal assignment problems. The following examples help in understanding the method better.

Example 9.1 Solve the following assignment problem:

Job

Person		J_1	J_2	J_3	J_4
	A	10	14	22	12
	B	16	10	18	12
	C	8	14	8	14
	D	20	8	16	6

Solution

1. Locate the minimum entry in each row of the cost matrix and subtract it from each entry of that row. The matrix thus obtained is called the reduced matrix, having at least one zero entry in each row.

	J_1	J_2	J_3	J_4
A	0	4	12	2
B	6	0	8	2
C	0	6	0	6
D	14	2	10	0

2. Now, subtract the smallest element of each column from every element of that column. We get the second reduced matrix containing at least one zero in every column. Thus, steps 1 and 2 will bring at least one zero in every row and column of the reduced matrix.

	J_1	J_2	J_3	J_4
A	0	4	12	2
B	6	0	8	2
C	0	6	0	6
D	14	2	10	0

3. Now, examine the first row of the matrix obtained in step 2. There is one zero in this row, so surround it by a square and cross all the other zeros in the column of this surrounded zero. In the second row there is only one zero, so surround it by a square. Repeat it for all the rows. If there is more than one zero in any row, then skip it and pass on to the next row. Repeat this procedure for the columns as well. By doing so, we get the following table:

	J_1	J_2	J_3	J_4
A	[0]	4	12	2
B	6	[0]	8	2

(Contd)

(*Contd*)

	J_1	J_2	J_3	J_4
C	~~0~~	6	[0]	6
D	14	2	10	[0]

In the aforegiven table, we see that all the zeros are either marked by □ or cancelled, and there is an assignment □ in every row and column. Hence, we get the solution of the given problem. The optimum solution is A → J_1, B → J_2, C → J_3, and D → J_4.

Example 9.2 Solve the following assignment problem:

		Job		
		J_1	J_2	J_3
	P	20	15	18
Person	Q	25	15	12
	R	30	20	20

Solution

1. Locate the minimum entry in each row of the cost matrix and subtract it from each entry of that row. The matrix thus obtained is as given here.

	J_1	J_2	J_3
P	5	0	3
Q	13	3	0
R	10	0	0

2. Now subtract the smallest element of each column from every element of that column. We get the following as the second reduced matrix. Thus, steps 1 and 2 will bring at least one zero in every row and column of the reduced matrix.

	J_1	J_2	J_3
P	0	0	3
Q	8	3	0
R	5	0	0

3. Now examine the first row of the matrix obtained in step 2. There are two zeros in this row, and hence we skip it and pass on to the next row. In the second row, there is only one zero, so surround this by a square and cross all the other zeros in the column of surrounded zero. Repeat it for all the rows. If there is more than one zero in any row, skip it and pass on to the next row. Repeat this procedure for the columns as well. By doing so, we get the following table:

	J_1	J_2	J_3
P	[0]	~~0~~	3
Q	8	3	[0]
R	5	[0]	~~0~~

In the aforegiven table, we see that all the zeros are either marked by □ or cancelled, and there is an assignment □ in every row and column. The optimum solution is P → J_1, Q → J_3, and R → J_2.

Example 9.3 Solve the following assignment problem:

		Job				
		1	2	3	4	5
Person	A	8	4	2	6	1
	B	0	9	5	5	4
	C	3	8	9	2	6
	D	4	3	1	0	3
	E	9	5	8	9	5

Solution

1. Locate the minimum entry in each row of the cost matrix and subtract it from each entry of that row. The following is the matrix thus obtained.

	1	2	3	4	5
A	7	3	1	5	0
B	0	9	5	5	4
C	1	6	7	0	4
D	4	3	1	0	3
E	4	0	3	4	0

2. Now subtract the smallest element of each column from every element of that column. We get the following as the second reduced matrix. Thus, steps 1 and 2 will bring at least one zero in every row and column of the reduced matrix.

	1	2	3	4	5
A	7	3	0	5	0
B	0	9	4	5	4
C	1	6	6	0	4
D	4	3	0	0	3
E	4	0	2	4	0

3. Now examine the first row of the matrix obtained in step 2. There are two zeros in this row, and hence we skip it and pass on to the next row. In the second row, there is only one zero, so surround this by a square and cross all the zeros in the column of this surrounded zero. Repeat it for all the rows. If there is more than one zero in any row, then skip it and pass on to the next row. Repeat this procedure for the columns as well. By doing so and after making the assignment, we get the following table:

7	3	~~0~~	5	[0]
[0]	9	4	5	4
1	6	6	[0]	4
4	3	[0]	~~0~~	3
4	[0]	2	4	~~0~~

In the aforegiven table, we see that all the zeros are either marked by □ or cancelled, and there is an assignment □ in every row and column. Hence, we get the solution of the given problem. The optimum solution is A → 5, B → 1, C → 4, D → 3, and E → 2.

Example 9.4 The principal of a college has four peons, and four tasks have to be assigned to them. The peons differ in efficiency and the tasks differ in their intrinsic difficulty. The time each peon would take to perform each task is given in the following effectiveness matrix. How should the tasks be allocated, one to each peon, so as to minimize the total peon hours?

		Subordinates			
		I	II	III	IV
Tasks	A	8	26	17	11
	B	13	28	4	26
	C	38	19	18	15
	D	19	26	24	10

Solution

1. Locate the minimum entry in each row of the cost matrix and subtract it from each entry of that row. The following is the matrix thus obtained.

	I	II	III	IV
A	0	18	9	3
B	9	24	0	22
C	23	4	3	0
D	9	16	14	0

2. Now subtract the smallest element of each column from every element of that column. We get the following second reduced matrix. Steps 1 and 2 will bring at least one zero in every row and column of the reduced matrix.

	I	II	III	IV
A	0	14	9	3
B	9	20	0	22
C	23	0	3	0
D	9	12	14	0

3. Now, examine the first row of the matrix obtained in step 2. If there is only one zero, surround it by a square and cross all the other zeros in the column. If there is more than one zero in any row, skip it and pass on to the next row. Repeat it for all the rows and columns as well. By doing so, we get the following table:

	I	II	III	IV
A	[0]	14	9	3
B	9	20	[0]	22
C	23	[0]	3	~~0~~
D	9	12	14	[0]

In the aforegiven table, we see that all the zeros are either marked by □ or cancelled, and there is an assignment □ in every row and column. Hence, we get the solution of the given problem. The optimum solution is A → I, B → III, C → II, and D → IV.

Example 9.5 Solve the following assignment problem:

	1	2	3	4
A	10	12	19	11
B	5	10	7	8
C	12	14	13	11
D	8	15	11	9

Solution

1. Locate the minimum entry in each row of the cost matrix and subtract it from each entry of that row. The matrix thus obtained is called the reduced matrix, having at least one zero entry in each row.

	1	2	3	4
A	0	2	9	1
B	0	5	2	3
C	1	3	2	0
D	0	7	3	1

2. Now, subtract the smallest element of each column from every element of that column. We get the second reduced matrix containing at least one zero in every column. Thus steps 1 and 2 will bring at least one zero in every row and column of the reduced matrix.

	1	2	3	4
A	0	0	7	1
B	0	3	0	3
C	1	1	0	0
D	0	5	1	1

3. Now, examine the first row of the matrix obtained in step 2. There are two zeros in this row. Hence, skip it and pass on to the next row. If there is only one zero, then surround this by a square and cross all the other zeros in the column of the surrounded zero. If there is more than one zero in any row, skip it and pass on to the next row. Repeat it for all the rows and columns as well. By doing so, we get the following table:

~~0~~	[0]	7	1
~~0~~	3	[0]	3
1	1	~~0~~	[0]
[0]	5	1	1

In the aforegiven table, we see that all the zeros are either marked by □ or cancelled, and there is an assignment □ in every row and column. Hence, we get the solution of the given problem. The optimum solution is A → 2, B → 3, C → 4, and D → 1.

Example 9.6 Solve the following assignment problem:

	J_1	J_2	J_3	J_4	J_5
A	40	35	25	20	15
B	16	26	48	50	24
C	22	12	32	15	17
D	16	18	12	15	36
E	45	20	36	24	18

Solution

1. Locate the minimum entry in each row of the cost matrix and subtract it from each entry of that row. The following matrix is thus obtained.

	J_1	J_2	J_3	J_4	J_5
A	25	20	10	5	0
B	0	10	32	34	8
C	10	0	20	3	5
D	4	6	0	3	24
E	27	2	18	6	0

2. Now subtract the smallest element of each column from every element of that column. We get the following as the second reduced matrix containing at least one zero in every column. Thus steps 1 and 2 will bring at least one zero in every row and column of the reduced matrix.

	J_1	J_2	J_3	J_4	J_5
A	25	20	10	2	0
B	0	10	32	31	8
C	10	0	20	0	5
D	4	6	0	0	24
E	27	2	18	3	0

3. Now, examine the first row of the matrix obtained in step 2. In the first row, there is only one zero, so surround it by a square and cross all the other zeros in the column. In the second row, there is only one zero, so surround it by a square. If there is more than one zero in any row, skip it and pass on to the next row. Repeat it for all the rows and columns as well. By doing so and after making the assignment, we get the following table:

	J_1	J_2	J_3	J_4	J_5
A	25	20	10	2	[0]
B	[0]	10	32	31	8

(Contd)

(*Contd*)

C	10	[0]	20	~~0~~	5
D	4	6	[0]	~~0~~	24
E	27	2	18	3	~~0~~

4. In the aforegiven table, we see that row 5 and column 4 have no assignment, and hence we go to the next step.
5. Draw (by trial) the minimum number of vertical or horizontal lines to cover all the zeros.

	J_1	J_2	J_3	J_4	J_5
A	25	20	10	2	[0]
B	[0]	10	32	31	8
C	10	[0]	20	~~0~~	5
D	4	6	[0]	~~0~~	24
E	27	2	18	3	~~0~~

Select the smallest element among all the elements uncovered by lines, that is, 2. Subtract this value 2 from all the elements not covered by the lines, and add 2 to all those elements that lie at the intersection of the lines. We get the following matrix after making the assignment.

	J_1	J_2	J_3	J_4	J_5
A	25	18	8	0	0
B	[0]	8	30	29	8
C	12	0	20	0	7
D	6	6	[0]	~~0~~	26
E	27	0	16	1	0

Here, all the zeros are not crossed or enrectangled, so the assignment is not achieved. Now, we proceed to make arbitrary assignments. In the first row there are two zeros, we arbitrarily enrectangle the zero lying in the first row and fourth column, and cross out the remaining zeros of the first row and the fourth column.

	J_1	J_2	J_3	J_4	J_5
A	25	18	8	[0]	~~0~~
B	[0]	8	30	29	8
C	12	[0]	20	~~0~~	7
D	6	6	[0]	~~0~~	26
E	27	~~0~~	16	1	[0]

In the aforegiven table, we see that all the zeros are either marked by ☐ or cancelled, and there is an assignment ☐ in every row and column. Hence, we get the solution of the given problem. The optimum solution is $A \rightarrow J_4$, $B \rightarrow J_1$, $C \rightarrow J_2$, $D \rightarrow J_3$, and $E \rightarrow J_5$.

We could have enrectangled the zero lying in the first row and the fifth column; in that case the assignment table would be as given here:

	J_1	J_2	J_3	J_4	J_5
A	25	18	8	~~0~~	[0]
B	[0]	8	30	29	8
C	12	~~0~~	20	[0]	7
D	6	6	[0]	~~0~~	26
E	27	[0]	16	1	~~0~~

The alternate optimum solution is $A \rightarrow J_5$, $B \rightarrow J_1$, $C \rightarrow J_4$, $D \rightarrow J_3$, and $E \rightarrow J_2$.

Example 9.7 Solve the following assignment problem:

	B_1	B_2	B_3	B_4
A_1	3	2	3	6
A_2	1	4	2	2
A_3	6	3	4	6
A_4	5	6	5	4

Solution

1. Locate the minimum entry in each row of the cost matrix and subtract it from each entry of that row. The following matrix is thus obtained, having at least one zero entry in each row.

	B_1	B_2	B_3	B_4
A_1	1	0	1	4
A_2	0	3	1	1
A_3	3	0	1	3
A_4	1	2	1	0

2. Now subtract the smallest element of each column from every element of that column. We get the following as the second reduced matrix containing at least one zero in every column. Thus steps 1 and 2 will bring at least one zero in every row and column of the reduced matrix.

	B_1	B_2	B_3	B_4
A_1	1	0	0	4
A_2	0	3	0	1
A_3	3	0	0	3
A_4	1	2	0	0

3. Now, examine the first row of the matrix obtained in step 2. There are two zeros in this row. Hence, we can skip it and pass on to the next row. If there is more than one zero in any row, skip it and pass on to the next row. Repeat it for all the

rows and columns as well. By doing so and after making the assignment, we get the following table:

	B_1	B_2	B_3	B_4
A_1	1	0	0	4
A_2	[0]	3	~~0~~	1
A_3	3	0	0	3
A_4	1	2	~~0~~	[0]

Here, all the zeros are not crossed or enrectangled, so the assignment is not achieved. Now, we proceed to make arbitrary assignments. In the first row there are two zeros. We arbitrarily enrectangle the zero lying in the first row and second column, and cross out the remaining zeros of the first row and second column.

	B_1	B_2	B_3	B_4
A_1	1	[0]	~~0~~	4
A_2	[0]	3	~~0~~	1
A_3	3	~~0~~	[0]	3
A_4	1	2	~~0~~	[0]

In the aforegiven table, we see that all the zeros are either marked by □ or cancelled, and there is an assignment □ in every row and column. Hence, we get the solution of the given problem. The optimum solution is $A_1 \rightarrow B_2$, $A_2 \rightarrow B_1$, $A_3 \rightarrow B_3$, and $A_4 \rightarrow B_4$.

We could have enrectangled the zero lying in the first row and third column; in that case the assignment table would be as given here:

	B_1	B_2	B_3	B_4
A_1	1	~~0~~	[0]	4
A_2	[0]	3	~~0~~	1
A_3	3	[0]	~~0~~	3
A_4	1	2	~~0~~	[0]

The alternate optimum solution is $A_1 \rightarrow B_3$, $A_2 \rightarrow B_1$, $A_3 \rightarrow B_2$, and $A_4 \rightarrow B_4$.

Example 9.8 Air India, which operates seven days a week, has the following timetable:

Delhi–Dehradun			Dehradun–Delhi		
Flight no.	**Departure**	**Arrival**	**Flight no.**	**Departure**	**Arrival**
1	8.00 a.m.	10.00 a.m.	101	10.00 a.m.	12.00 noon
2	10.00 a.m.	12.00 noon	102	11.00 a.m.	1.00 p.m.
3	2.30 p.m.	4.30 p.m.	103	4.30 p.m.	6.30 p.m.
4	8.30 p.m.	10.30 p.m.	104	9.00 p.m.	11.00 p.m.

The crew must have a minimum layover of 6h between flights. Obtain the pairing of flights that minimizes the layover time away from home. For any given pairing, the crew will be based in the city that results in the smaller layover. In addition, mention the town where the crew should be based for each pair.

Solution First, we prepare tables for the layover times between the flights. Suppose we pair flight no. 101 when the crew is based in Delhi. The time of stay in Dehradun will be the layover time away from home. Now flight no. 1, which reaches Dehradun at 10.00 a.m., cannot fly at 10.00 a.m. on the same day from Dehradun, as the minimum layover time is 6h. So, it will depart Dehradun the next day at 10.00 a.m., which will result in a layover time of 24h. Similarly, the other layover times (in hours) can be calculated as shown in the following tables:

Crew based in Delhi

	101	102	103	104
1	24	25	6.5	11
2	22	23	28.5	9
3	17.5	18.5	24	28.5
4	11.5	12.5	18	22.5

Crew based in Dehradun

	1	2	3	4
101	20	22	26.5	8.5
102	19	21	25.5	7.5
103	13.5	15.5	20	26
104	9	11	27.5	21.5

Here, we assume the layover times in terms of quarter hour (0.25 h or 15 minutes) as one unit of time to avoid fractions. Thus, multiplying by 4 in each table, we get the following tables:

Crew based in Delhi

	101	102	103	104
1	96	100	26	44
2	88	92	114	36
3	70	74	96	114
4	46	50	72	90

Crew based in Dehradun

	1	2	3	4
101	80	88	106	34
102	76	84	102	30
103	54	62	80	104
104	36	44	110	86

For combining the aforegiven tables, we choose the base that gives lesser layover time for each pairing. The layover times marked with '*' denote that the crew is based in Dehradun; else the crew is based in Delhi.

	101	102	103	104
1	80*	76*	26	36*
2	88*	84*	62*	36
3	70	74	80*	110*
4	34*	30*	72	86*

1. Locate the minimum entry in each row of the cost matrix and subtract it from each entry of that row. The following matrix thus obtained has at least one zero entry in each row.

	101	102	103	104
1	54*	50*	0	10*
2	52*	58*	26*	0
3	0	4	10*	40*
4	4*	0*	42	56*

2. Now subtract the smallest element of each column from every element of that column. We get the following matrix containing at least one zero in every column. Thus, steps 1 and 2 will bring at least one zero in every row and column of the reduced matrix.

	101	102	103	104
1	54*	50*	0	10*
2	52*	58*	26*	0
3	0	4	10*	40*
4	4*	0*	42	56*

3. Now, examine the first row of the matrix obtained in step 2. There is only one zero; surround it by a square and cross all the other zeros in the column. Repeat it for all the rows. If there is more than one zero in any row, skip it and pass on to the next row. Repeat this procedure for the columns as well. By doing so, we get the following table:

	101	102	103	104
1	54*	50*	[0]	10*
2	52*	58*	26*	[0]
3	[0]	4	10*	40*
4	4*	[0]*	42	56*

In the aforegiven table, we see that all zeros are either marked by □ and there is an assignment in every row and column. Hence, we get the solution of the given problem. The optimum solution is 1 → 103, 2 → 104, 3 → 101, 4 → 103, that is, the crew is based in Delhi-Delhi-Delhi-Dehradun.

9.4 MAXIMAL ASSIGNMENT PROBLEM

The Hungarian algorithm is applicable to the minimal assignment problem. However, sometimes in assigning jobs to different persons, the information data is in terms of sales, efficiency, or effectiveness of different jobs or persons. In such a situation, one would like to maximize the sales, efficiency, or effectiveness and thus, the problem becomes a maximization problem. In order to apply the Hungarian algorithm to solve it, we have to convert it into a minimization problem. This can be achieved using any of the following two methods:

1. Add a negative sign before every entry in the assignment matrix and follow step 1 onwards of the Hungarian algorithm.

2. Locate the largest entry in the assignment matrix, replace each entry with the result of its subtraction from the largest entry, and then follow step 1 onwards of the Hungarian algorithm.

Example 9.9 A company has 5 salesmen and 5 sales districts. Considering the capabilities of the salesmen and the nature of the districts, the company manager estimates that sales per month (in ₹ hundred) for each salesman in each district would be as follows:

		Districts				
		A	B	C	D	E
	1	32	38	40	28	40
	2	40	24	28	21	36
Person	3	41	27	33	30	37
	4	22	38	41	36	36
	5	29	33	40	35	39

Find the assignment of salesmen to districts that will result in the maximization of sales.

Solution First, we convert the maximization problem into a minimization problem by multiplying each element by (–1). The table thus obtained is given here:

–32	–38	–40	–28	–40
–40	–24	–28	–21	–36
–41	–27	–33	–30	–37
–22	–38	–41	–36	–36
–29	–33	–40	–35	–39

1. Locate the minimum entry in each row of the cost matrix and subtract it from each entry of that row. The following matrix thus obtained has at least one zero entry in each row.

8	2	0	12	0
0	16	12	19	4
0	14	8	11	4
19	3	0	5	5
11	7	0	5	1

2. Now subtract the smallest element of each column from every element of that column. We get the following second reduced matrix containing at least one zero in every column. Thus steps 1 and 2 will bring at least one zero in every row and column of the reduced matrix.

8	0	0	7	0
0	14	12	14	4
0	12	8	6	4
19	1	0	0	5
11	5	0	0	1

3. Now, examine the first row of the matrix obtained in step 2. There are three zeros in this row. Hence, skip it and pass on to the next row. In the second row, there is only one zero, so surround it by a square and cross all the other zeros in the column. If there is only one zero, then surround it by a square and cross all the other zeros in the column. If there is more than one zero in any row, skip it and pass on to the next row. Repeat it for all the rows and columns as well. By doing so, we get the following table:

8	[0]	~~0~~	7	~~0~~
[0]	14	12	14	4
~~0~~	12	8	6	4
19	1	[0]	~~0~~	5
11	5	~~0~~	[0]	1

4. Here, we see that in the aforementioned table, row 3 and column 5 have no assignment; therefore, we go to the next step.
5. Draw (by trial) the minimum number of vertical or horizontal lines to cover all zeros.

8	[0]	~~0~~	7	~~0~~
[0]	14	12	14	4
~~0~~	12	8	6	4
19	1	[0]	~~0~~	5
11	5	~~0~~	[0]	1

Select the smallest element, among all elements uncovered by the lines, that is, 4. Subtract the value 4 from all elements not covered by the lines and add 4 to all those elements that lie at the intersection of the lines.

12	0	0	7	0
0	10	8	10	0
0	8	4	2	0
23	1	0	0	5
15	5	0	0	1

Repeat steps 3 and 4 to get the following optimum assignment after successively selecting cells (2, 1) and (4, 3) arbitrarily for assignment.

12	[0]	~~0~~	7	~~0~~
[0]	10	8	10	~~0~~
~~0~~	8	4	2	[0]
23	1	[0]	~~0~~	5
15	5	~~0~~	[0]	1

In the aforegiven table, we see that all zeros are either marked by □ or cancelled, and there is an assignment □ in every row and column. Hence, we get the solution of the given problem. The optimum solution is 1 → B, 2 → A, 3 → E, 4 → C, and 5 → D. The maximum sales = 38 + 40 + 37 + 41 + 35 = 191 = 19,100.

Example 9.10 Solve the following maximal assignment problem:

	A	B	C	D
J_1	18	33	9	31
J_2	23	30	28	14
J_3	12	30	21	15
J_4	44	25	24	21

Solution First, we convert the maximization problem into a minimization problem by locating the largest entry in the assignment matrix, which is 44 in cell (4, 1) and replacing each entry after subtracting it from the largest entry. The table thus obtained is as given here.

	A	B	C	D
J_1	26	11	35	13
J_2	21	14	16	30
J_3	32	14	23	29
J_4	0	19	20	23

1. Locate the minimum entry in each row of the cost matrix and subtract it from each entry of that row. The following matrix thus obtained has at least one zero entry in each row.

	A	B	C	D
J_1	15	0	24	2
J_2	7	0	2	16
J_3	18	0	9	15
J_4	0	19	20	23

2. Now subtract the smallest element of each column from every element of that column. We get the second reduced matrix containing at least one zero in every column. Thus, steps 1 and 2 will bring at least one zero in every row and column of the reduced matrix.

	A	B	C	D
J_1	15	0	22	0
J_2	7	0	0	14
J_3	18	0	7	13
J_4	0	19	18	21

3. Now, examine the first row of the matrix obtained in step 2. There are two zeros in this row. Hence, skip it and pass on to the next row. If there is only one zero, then surround it by a square and cross all the other zeros in the column. If there is more

than one zero in any row, skip it and pass on to the next row. Repeat it for all the rows and columns as well. By doing so, we get the following table:

	A	B	C	D
J_1	15	~~0~~	22	$\boxed{0}$
J_2	7	~~0~~	$\boxed{0}$	14
J_3	18	$\boxed{0}$	7	13
J_4	$\boxed{0}$	19	18	21

In the aforegiven table, we see that all the zeros are either marked by □ or cancelled, and there is an assignment □ in every row and column. Hence, we get the solution of the given problem. The optimum solution is $J_1 \rightarrow D$, $J_2 \rightarrow C$, $J_3 \rightarrow B$, and $J_4 \rightarrow A$.

9.5 UNBALANCED ASSIGNMENT PROBLEM

In any assignment problem, when the number of jobs (tasks) is not equal to the number of persons (facilities), it is called an unbalanced assignment problem. Such unbalanced problems cannot be solved using the Hungarian algorithm, as it has been developed to solve only balanced assignment problems. Thus, to solve an unbalanced assignment problem, first the problem is balanced by adding a dummy (fictitious) row or column with zero costs in the new cells. Now, the problem can be solved as usual by using the Hungarian algorithm.

Example 9.11 The head of the department of mathematics has to assign four jobs to three peons. The peons differ in efficiency. The estimates of the time each peon would take to perform a task are given in the following matrix. How should the head allocate the jobs, one to each peon, so as to minimize the total man-hours?

		Peons		
		1	2	3
Jobs	I	9	26	15
	II	13	27	6
	III	35	20	15
	IV	18	30	20

Solution The given problem is unbalanced. To solve an unbalanced assignment problem, we first balance the problem by adding a dummy (fictitious) column with zero costs in the newly created cells.

	1	2	3	4
I	9	26	15	0
II	13	27	6	0
III	35	20	15	0
IV	18	30	20	0

1. Locate the minimum entry in each row of the cost matrix and subtract it from each entry of that row. The following matrix thus obtained has at least one zero entry in each row.

	1	2	3	4
I	9	26	15	0
II	13	27	6	0
III	35	20	15	0
IV	18	30	20	0

2. Now subtract the smallest element of each column from every element of that column. The following is the second reduced matrix containing at least one zero in every column. Thus, steps 1 and 2 will bring at least one zero in every row and column of the reduced matrix.

	1	2	3	4
I	0	6	9	0
II	4	7	0	0
III	26	0	9	0
IV	9	10	14	0

3. Now, examine the first row of the matrix obtained in step 2. There are two zeros in this row. Hence, skip it and pass on to the next row. In the fourth row, there is only one zero, so surround it by a square. Repeat this procedure for the columns as well. By doing so, we get the following table:

	1	2	3	4
I	[0]	6	9	~~0~~
II	4	7	[0]	~~0~~
III	26	[0]	9	~~0~~
IV	9	10	14	[0]

In the aforegiven table, we see that all the zeros are either marked by □ or cancelled, and there is an assignment □ in every row and column. Hence, we get the solution of the given problem.

The optimum solution is I → 1, II → 3, III → 2, and task IV remains unassigned. The total time (man-hours) = 9 + 6 + 20 = 35.

Example 9.12 Solve the following assignment problem:

		Job			
		J_1	J_2	J_3	J_4
Person	A	15	18	20	5
	B	10	5	15	7
	C	20	8	16	6

Solution The given problem is unbalanced. To solve an unbalanced assignment problem, we first balance the problem by adding a dummy (fictitious) row (say D) with zero costs in the newly created cells to get the following assignment matrix.

	J_1	J_2	J_3	J_4
A	15	18	20	5
B	10	5	15	7
C	2	8	9	17
D	0	0	0	0

1. Locate the minimum entry in each row of the cost matrix and subtract it from each entry of that row. The following matrix is thus obtained has at least one zero entry in each row.

	J_1	J_2	J_3	J_4
A	10	13	15	0
B	5	0	10	2
C	0	6	7	15
D	0	0	0	0

2. Now subtract the smallest element of each column from every element of that column. We get the second reduced matrix containing at least one zero in every column. Thus steps 1 and 2 will bring at least one zero in every row and column of the reduced matrix.

	J_1	J_2	J_3	J_4
A	10	13	15	0
B	5	0	10	2
C	0	6	7	15
D	0	0	0	0

3. Now, examine the first row of the matrix obtained in step 2. There is one zero in this row, so surround it by a square and cross all the other zeros in the column. In the second row, there is only one zero, so surround it by a square and cross all the other zeros in the column. Repeat it for all the rows. If there is more than one zero in any row, skip it and pass on to the next row. Repeat this procedure for the columns as well. By doing so, we get the following table:

	J_1	J_2	J_3	J_4
A	10	13	15	[0]
B	5	[0]	10	2
C	[0]	6	7	15
D	~~0~~	~~0~~	[0]	~~0~~

In the aforegiven table, we see that all the zeros are either marked by □ or cancelled, and there is an assignment □ in every row and column. Hence, we get the solution of the given problem. The optimum solution is A→J_4, B→J_2, C→J_1, and D→J_3. However, as reality person D does not exist and hence job J_3 remains unassigned.

9.6 ASSIGNMENT PROBLEMS UNDER CERTAIN RESTRICTIONS

Sometimes due to unavoidable reasons, a particular job is not assigned to a particular person, who would have been the best suited for it. In such situations, to find an optimal solution, we assign a very high cost, say ∞, in the corresponding cell. The problem is now solved by the usual method using the Hungarian algorithm. For example, if job J_i is not suitable to person P_j, replace c_{ij} by ∞.

Given cost matrix (Rows: Jobs; Columns: Persons)

Jobs	P_1	…	…	P_j	…	P_n
J_1	…	…	…	c_{1j}	…	…
…	…	…	…	c_{2j}	…	…
…	…	…	…	…	…	…
J_i	c_{i1}	c_{i2}	…	c_{ij}	…	c_{in}
…	…	…	…	…	…	…
J_n	…	…	…	c_{nj}	…	…

→

Modified cost matrix (Rows: Jobs; Columns: Persons)

Jobs	P_1	…	…	P_j	…	P_n
J_1	…	…	…	c_{1j}	…	…
…	…	…	…	c_{2j}	…	…
…	…	…	…	…	…	…
J_i	c_{i1}	c_{i2}	…	∞	…	c_{in}
…	…	…	…	…	…	…
J_n	…	…	…	c_{nj}	…	…

In particular,

Given cost matrix

Jobs	1	2	3
J_1	2	3	5
J_2	6	4	Not suitable
J_3	3	2	8

→

Modified cost matrix

Jobs	1	2	3
J_1	2	3	5
J_2	6	4	∞
J_3	3	2	8

A situation may also arise when a particular person P_j is assigned only a specific job J_i. In such cases, we assign a very high cost, say ∞, in all the P_j column cells except the cell (i, j). In that cell, the price is kept unchanged. For example,

Given cost matrix (Rows: Jobs; Columns: Persons)

Jobs	P_1	…	…	P_j	…	P_n
J_1	…	…	…	c_{1j}	…	…
…	…	…	…	c_{2j}	…	…
…	…	…	…	…	…	…
J_i	c_{i1}	c_{i2}	…	c_{ij}	…	c_{in}
…	…	…	…	…	…	…
J_n	…	…	…	c_{nj}	…	…

→

Modified cost matrix (Rows: Jobs; Columns: Persons)

Jobs	P_1	…	…	P_j	…	P_n
J_1	…	…	…	∞	…	…
…	…	…	…	∞	…	…
…	…	…	…	…	…	…
J_i	c_{i1}	c_{i2}	…	c_{ij}	…	c_{in}
…	…	…	…	…	…	…
J_n	…	…	…	∞	…	…

In particular,

Given cost matrix

Jobs	1	2	3
J_1	2	3	Not suitable
J_2	6	4	4
J_3	3	2	Not suitable

→

Modified cost matrix

Jobs	1	2	3
J_1	2	3	∞
J_2	6	4	4
J_3	3	2	∞

Example 9.13 Four engineering students E_1, E_2, E_3, and E_4 of an institute are to design four projects P_1, P_2, P_3, and P_4 respectively. E_2 is not competent enough to design the project P_2. Given the following time estimates needed for each engineering student to design a given project, determine how each student should be assigned to the projects so as to minimize the total design time of the four projects.

		Projects			
		P_1	P_2	P_3	P_4
Engg student	E_1	12	10	10	8
	E_2	14	Not suitable	15	11
	E_3	6	10	16	4
	E_4	8	10	9	7

Solution Sometimes due to some unavoidable reasons a particular job is not assigned to a particular person. To tackle such a problem, we assign a very high cost, say ∞, in the corresponding cell. Thus the given problem of the aforegiven table is reduced to

12	10	10	8
14	∞	15	11
6	10	16	4
8	10	9	7

1. Locate the minimum entry in each row of the cost matrix and subtract it from each entry of that row. The following matrix thus obtained has at least one zero entry in each row.

4	2	2	0
3	∞	4	0
2	6	12	0
1	3	2	0

2. Now subtract the smallest element of each column from every element of that column. We get the following as the second reduced matrix containing at least one zero in every column. Thus steps 1 and 2 will bring at least one zero in every row and column of the reduced matrix.

3	0	0	0
2	∞	2	0
1	4	10	0
0	1	0	0

3. Now, examine the first row of the matrix obtained in step 2. There are two zeros in this row. Hence, skip it and pass on to the next row. In the second row, there is only one zero, so surround it by a square and cross all the other zeros in the column. If there is only one zero, then surround it by a square and cross all the other zeros in the column. If there is more than one zero in any row, skip it and pass on to the

next row. Repeat it for all the rows and columns as well. By doing so, we get the following table:

3	[0]	~~0~~	~~0~~
2	∞	2	[0]
1	4	10	0
[0]	1	~~0~~	~~0~~

4. Here, we see that in the aforegiven table, row 3 and column 3 have no assignment; therefore, we go to the next step.
5. Draw (by trial) the minimum number of vertical or horizontal lines to cover all the zeros.

3	[0]	~~0~~	~~0~~
2	∞	2	[0]
1	4	10	0
[0]	1	~~0~~	~~0~~

Select the smallest element, among all the elements uncovered by the lines; it is 1 in this case. Subtract the value 1 from all the elements not covered by the lines, and add 1 to all those elements that lie at the intersection of the lines to get the following matrix:

3	0	0	1
1	∞	1	0
0	3	9	0
0	1	0	1

Repeat steps 3 and 4 to get the optimum assignment.

3	[0]	~~0~~	1
1	∞	1	[0]
[0]	3	9	~~0~~
~~0~~	1	[0]	1

In the aforegiven table, we see that all the zeros are either marked by □ or cancelled, and there is an assignment □ in every row and column. Hence, we get the solution of the given problem. The optimum solution is 1 → B, 2 → D, 3 → A, and 4 → C. The total minimum time = 10 + 11 + 6 + 9 = 36.

Example 9.14 Solve the following assignment problem:

		Jobs			
		A_1	A_2	A_3	A_4
Person	P_1	30	40	20	25
	P_2	25	35	Not suitable	20
	P_3	40	25	30	45
	P_4	30	25	15	15

Solution To tackle this problem, we assign a very high cost, say ∞, in the (2, 3) cell. Thus, the given problem of the aforegiven table is reduced to

30	40	20	25
25	35	∞	20
40	25	30	45
30	25	15	15

1. Locate the minimum entry in each row of the cost matrix and subtract it from each entry of that row. The matrix thus obtained is called the reduced matrix, having at least one zero entry in each row.

10	20	0	5
5	15	∞	0
15	0	5	20
15	10	0	0

2. Now subtract the smallest element of each column from every element of that column. We get the second reduced matrix containing at least one zero in every column. Thus steps 1 and 2 will bring at least one zero in every row and column of the reduced matrix.

5	20	0	5
0	15	∞	0
10	0	5	20
10	10	0	0

3. Now, examine the first row of the matrix obtained in step 2. In the first row, there is only one zero, so surround it by a square and cross all the other zeros in the column. In the second row, there are two zeros. Hence, skip it and pass on to the next row. If there is only one zero, then surround it by a square and cross all the other zeros in the column. If there is more than one zero in any row, skip it and pass on to the next row. Repeat it for all the rows and columns as well. By doing so, we get the following table:

5	20	[0]	5
[0]	15	∞	~~0~~
10	[0]	5	20
10	10	~~0~~	[0]

In the aforegiven table, we see that all the zeros are either marked by □ or cancelled, and there is an assignment □ in every row and column. Hence, we get the solution of the given problem.

The optimum solution is $P_1 \rightarrow A_3$, $P_2 \rightarrow A_1$, $P_3 \rightarrow A_2$, and $P_4 \rightarrow A_4$.

9.7 TRAVELLING SALESMAN PROBLEM

Suppose a travelling salesman wants to visit a certain number of cities. He already knows the costs (distances or time) involved in going from one city to the other for all the city pairs. Now, the problem is to choose a route for visiting all the cities once, and returning to the city he started from, with the minimum cost (distance or time) involved. This problem can be treated as a kind of assignment problem, and can be solved using the Hungarian algorithm. Sometimes, when the assignment is not possible, we use the method of numeration by assigning the next minimum element of the matrix in the place of zero.

The travelling salesman problem can be expressed in the LP form as follows:

$$\text{Min } Z = \sum_{i=1}^{n} \sum_{j=1}^{n} c_{ij} x_{ij}$$

$$\sum_{j=1}^{n} x_{ij} = 1, i = 1, 2, \ldots, n$$

$$\sum_{i=1}^{n} x_{ij} = 1, j = 1, 2, \ldots, n$$

$$x_{ij} = \begin{cases} 1, \text{ if salesman travels from city } i \text{ to } j \\ 0, \text{ otherwise} \end{cases}$$

$$c_{ii} = \infty, \forall i = 1, 2, \ldots, n.$$

Where c_{ij} is the distance (or cost or time) of travelling from city i to j.

Note that $c_{ij} \neq c_{ji}$, that is, the cost of travelling from city i to j may not always be equal to the cost of travelling from city j to i.

Example 9.15 Solve the following travelling salesman problem:

	A	B	C	D
A	∞	15	20	10
B	25	∞	35	15
C	5	25	∞	20
D	10	20	15	∞

Solution

1. Locate the minimum entry in each row of the cost matrix and subtract it from each entry of that row. The following matrix thus obtained has at least one zero entry in each row.

	A	B	C	D
A	∞	5	10	0
B	10	∞	20	0
C	0	20	∞	15
D	0	10	5	∞

2. Now subtract the smallest element of each column from every element of that column. We get the second reduced matrix containing at least one zero in every column. Thus steps 1 and 2 will bring at least one zero in every row and column of the reduced matrix.

	A	B	C	D
A	∞	0	5	0
B	10	∞	15	0
C	0	15	∞	15
D	0	5	0	∞

3. Now, examine the first row of the matrix obtained in step 2. If there is only one zero, then surround it by a square and cross all the other zeros in the column. If there is more than one zero in any row, skip it and pass on to the next row. Repeat it for all the rows and columns as well. By doing so, we get the following table:

	A	B	C	D
A	∞	[0]	5	~~0~~
B	10	∞	15	[0]
C	[0]	15	∞	15
D	~~0~~	5	[0]	∞

In the aforegiven table, we see that all the zeros are either marked by □ or cancelled, and there is an assignment □ in every row and column. Hence, we get the solution of the given problem. Hence the optimum solution is A → B → D → C → A.

Example 9.16 Solve the travelling salesman problem with the following data:

$$c_{12} = 4,\ c_{13} = 7,\ c_{14} = 3,\ c_{23} = 6,\ c_{24} = 3,\ c_{34} = 7,\ \text{and } c_{ij} = c_{ji}.$$

Solution From the given data, the following cost matrix can be derived:

	1	2	3	4
1	∞	4	7	3
2	4	∞	6	3
3	7	6	∞	7
4	3	6	7	∞

1. Locate the minimum entry in each row of the cost matrix and subtract it from each entry of that row. The following matrix thus obtained has at least one zero entry in each row.

	1	2	3	4
1	∞	1	4	0
2	1	∞	3	0
3	1	0	∞	1
4	0	3	4	∞

2. Now, subtract the smallest element of each column from every element of that column. We get the second reduced matrix containing at least one zero in every column. Thus steps 1 and 2 will bring at least one zero in every row and column of the reduced matrix.

	1	2	3	4
1	∞	1	1	0
2	1	∞	0	0
3	1	0	∞	1
4	0	3	1	∞

3. Now, examine the first row of the matrix obtained in step 2. If there is only one zero, surround it by a square and cross all the other zeros in the column. If there is more than one zero in any row, skip it and pass on to the next row. Repeat it for all the rows and columns as well. By doing so, we get the following table:

	1	2	3	4
1	∞	1	1	[0]
2	1	∞	[0]	~~0~~
3	1	[0]	∞	1
4	[0]	3	1	∞

In the aforegiven table, we see that all the zeros are either marked by □ or cancelled, and there is an assignment □ in every row and column. Hence, we get the solution of the given problem. The optimum solution is $1 \rightarrow 4$, $2 \rightarrow 3$, $3 \rightarrow 2$, and $4 \rightarrow 1$, that is, $1 \rightarrow 4 \rightarrow 1$ and $2 \rightarrow 3 \rightarrow 2$.

It is not an acceptable solution of the travelling salesman problem. Now, we try to get the next best solution which satisfies the extra condition of visiting all the cities. The smallest element in the table other than zero is 1. So we try to bring 1 into the solution. Since 1 occurs at 6 places, we consider all cases and determine the acceptable solution.

Case 1 Shifting the assignment from cell (1, 4) to cell (1, 3), and then using step 3, we get

	1	2	3	4
1	∞	1	[1]	~~0~~
2	1	∞	~~0~~	[0]
3	1	[0]	∞	1
4	[0]	3	1	∞

In the aforegiven table, we see that all the zeros are either marked by □ or cancelled, and there is an assignment □ in every row and column. Hence, we get the solution of the given problem. The optimum solution is $1 \rightarrow 3 \rightarrow 2 \rightarrow 4 \rightarrow 1$, and the total minimum cost $= 7 + 6 + 3 + 3 = 19$.

Case 2 If we shift the assignment from cell (1, 4) to cell (1, 2), then using step 3, we get

	1	2	3	4
1	∞	[1]	1	~~0~~
2	1	∞	[0]	~~0~~
3	1	~~0~~	∞	[1]
4	[0]	3	1	∞

In the aforegiven table, we see that all the zeros are either marked by □ or cancelled, and there is an assignment □ in every row and column. Hence, we get the solution of the given problem. The optimum solution is $1 \to 2 \to 3 \to 4 \to 1$, and the total minimum cost $= 4 + 6 + 7 + 3 = 20$.

It can be observed that in all other cases, the cost in the reduced matrix will be more than that obtained in case 1. Hence the optimum solution is $1 \to 3 \to 2 \to 4 \to 1$, and the total optimum cost $= 7 + 6 + 3 + 3 = 19$.

9.8 DIFFERENCE BETWEEN ASSIGNMENT AND TRANSPORTATION PROBLEMS

The following are the main differences between the assignment and transportation problems:

Assignment problem	Transportation problem
The formulation of the assignment problem is Min $Z = \sum_{j=1}^{n} \sum_{i=1}^{n} c_{ij} x_{ij}$ Subject to $\sum_{i=1}^{n} x_{ij} = 1, \quad j = 1, 2, \ldots, n$, *i.e.*, the ith person is assigned only one job. $\sum_{j=1}^{n} x_{ij} = 1, \quad i = 1, 2, \ldots, n$, *i.e.*, the jth job is assigned to only one person. $x_{ij} = 0$ or 1 where c_{ij} is the inefficiency of the ith person doing the jth job.	The formulation of the transportation problem is Min $Z = \sum_{j=1}^{m} \sum_{i=1}^{n} c_{ij} x_{ij}$ Subject to $\sum_{j=1}^{n} x_{ij} = a_i, \quad j = 1, 2, \ldots, n, \quad a_i > 0.$ $\sum_{i=1}^{m} x_{ij} = b_j, \quad i = 1, 2, \ldots, m, \quad b_j > 0.$ $\sum_{i=1}^{m} a_i = \sum_{j=1}^{n} b_j, \qquad x_{ij} \geq 0, \forall\, i, j.$ where c_{ij} is the per unit transportation cost of the commodity from the ith source to the jth destination, x_{ij} is the amount of the commodity to be transported from the ith source to the jth destination, a_i is the amount available at the ith source, and b_j is the amount required at the jth destination.
In the assignment problem, the number of jobs is equal to the number of persons.	In the transportation problem, the number of sources (m) may not be equal to number of destinations (n).
In the optimal table of the assignment problem, each row and column has exactly one entry.	In the optimal table of the transportation problem, each row and column has at least one entry.

RECAPITULATION

1. A transportation problem with m sources and n destinations can still be reduced to a simpler form known as the assignment problem with $m = n$ and $a_i = b_i = 1, i = 1, 2, \ldots, n$.
2. In an assignment problem with n jobs to be assigned to n persons, there are n^2 variables with $2n$ constraints.
3. An assignment problem has a highly degenerate solution. Hence, a special algorithm called the Hungarian algorithm, and not the transportation algorithm, is used to solve it.

(Contd)

(*Contd*)

4. When all zeros are either surrounded or crossed, but there is no surrounded zero in each row and column, an assignment cannot be made. In such situations, a minimum number of horizontal and vertical lines are drawn to cover all zeros. For this, the procedure is as follows: if a zero is surrounded while going along a row (column), then a vertical (horizontal) line is drawn through this zero. Doing this for all surrounded zeros gives the minimum number of lines covering all zeros.

EXERCISES

Multiple-choice Questions

9.1 An assignment problem of assigning n jobs to n persons has
   (a) n non-zero basic variables in the optimum assignment
   (b) $n - 1$ basic variables in the optimum assignment
   (c) $2n - 1$ non-zero basic variables in the optimum assignment
   (d) none of these

9.2 To solve the assignment problem, we use
   (a) simplex algorithm
   (b) Hungarian algorithm
   (c) transportation algorithm
   (d) none of these

9.3 An assignment problem is a
   (a) non-linear programming problem
   (b) linear programming problem
   (c) quadratic programming problem
   (d) none of these

9.4 The assignment problem is a special case of
   (a) a non-linear programming problem
   (b) a transportation problem
   (c) a quadratic programming problem
   (d) none of these

State True or False

9.1 An assignment problem can be solved using the simplex method.

9.2 An assignment problem is not a linear programming problem.

9.3 In the assignment algorithm, we first bring a zero in every row and column to get a (row or column) reduced matrix. If this process is reversed (i.e., first get the column and then the row reduced matrix), then again we shall get the same reduced matrix.

9.4 In the assignment algorithm, we first bring a zero in every row and then in every column, or first bring a zero in every column and then in every row. Then, the reduced matrices in the two processes are different, but the ultimate solution of the problem is the same.

9.5 In an assignment problem with m jobs to be assigned to m persons, the solution obtained is a degenerate basic feasible solution.

Fill in the Blanks

9.1 An assignment problem is a special case of ________.

9.2 The method to solve an assignment problem is called ________.

9.3 The assignment problem is an ________ problem.

9.4 Consider the assignment problem of assigning n jobs to n persons:
   (a) The number of constraints is ________.
   (b) The number of assignment pairs is ________.
   (c) The number of basic variables at zero level is ________.

Review Questions

9.1 Write a short note on the assignment algorithm (Hungarian assignment method).

9.2 Write a short note on the assignment problem (a) with certain restrictions and (b) to solve the travelling salesman problem.

9.3 What is an unbalanced assignment problem?

9.4 Write a short note on the maximal assignment problem.

9.5 What are the differences between assignment and transportation problems?

Numerical Problems

9.1 The head of the mathematics department in an educational institute has three expert faculty members (FM) in the department who can prepare teaching material for three new courses. The time taken (in days) to prepare the material for different courses is as per the adjoining table:

Find the optimum assignment.

FM \ Courses	A	B	C
1	120	100	80
2	70	90	110
3	110	140	120

9.2 Solve the following assignment problems.

(a)

Jobs \ Persons	1	2	3	4	5
I	12	8	7	15	4
II	7	9	17	14	10
III	9	6	12	6	7
IV	7	6	14	6	10
V	9	6	12	10	6

(b)

	I	II	III	IV
A	2	3	4	5
B	4	5	6	7
C	7	8	9	8
D	3	5	8	4

(c)

	a	b	c	d	e
A	160	130	175	190	200
B	135	120	130	160	175
C	140	110	155	170	185
D	50	50	80	80	110
E	55	35	70	80	105

(d)

	I	II	III	IV	V	VI
A	9	22	58	11	19	27
B	43	78	72	50	63	48
C	41	28	91	37	45	33
D	74	42	27	49	39	32
E	36	11	57	22	25	18
F	3	56	53	31	17	28

(e)

	I	II	III	IV	V
A	11	17	8	16	20
B	9	7	12	6	15
C	13	16	15	12	16
D	21	24	17	28	26
E	14	10	12	11	15

(f)

	I	II	III	IV
A	5	3	1	8
B	7	9	2	6
C	6	4	5	7
D	5	7	7	6

(g)

		Jobs				
		I	II	III	IV	V
Machines	A	5	11	10	12	4
	B	2	4	6	3	5
	C	3	12	5	14	6
	D	6	14	4	11	7
	E	7	9	8	12	5

(h)

		Persons				
		I	II	III	IV	V
Jobs	A	1	3	2	3	6
	B	2	4	3	1	5
	C	5	6	3	4	6
	D	3	1	4	2	2
	E	1	5	6	5	4

(i)

	7	8	9	10	11	12
1	41	72	39	52	25	51
2	22	29	49	65	81	50
3	27	39	60	51	32	32
4	45	50	48	52	37	43
5	29	40	39	26	30	33
6	82	40	40	60	51	30

(j)

From \ To	7	8	9	10	11	12
1	31	62	29	42	15	41
2	12	19	39	55	71	40
3	17	29	50	41	22	22
4	35	40	38	42	27	33
5	19	30	39	16	20	23
6	72	30	30	50	41	20

(k)

	I	II	III	IV	V
1	10	5	9	18	11
2	13	9	6	12	14
3	3	2	4	4	5
4	18	9	12	17	15
5	11	6	14	19	10

(l)

	I	II	III	IV	V
1	–2	–4	–8	–6	–1
2	0	–9	–5	–5	–4
3	–3	–8	0	–2	–6
4	–4	–3	–1	0	–3
5	–9	–5	–9	–9	–5

9.3 In a company, three jobs J_1, J_2, and J_3 are to be assigned to three machines M_1, M_2, and M_3. The processing costs (in ₹) are as given in the following matrix:

Find the allocation that will minimize the overall processing cost.

Jobs \ Machines	M_1	M_2	M_3
J_1	19	28	31
J_2	11	17	16
J_3	12	15	13

9.4 Air India operates seven days a week, as per the following timetable:

Delhi–Agra			Agra–Delhi		
Flight no.	**Departure**	**Arrival**	**Flight no.**	**Departure**	**Arrival**
1	8.00 a.m.	9.00 a.m.	101	9.00 a.m.	10.15 a.m.
2	9.00 a.m.	10.00 a.m.	102	9.30 a.m.	10.45 a.m.
3	2.30 p.m.	3.30 p.m.	103	1.00 p.m.	2.15 p.m.
4	7.30 p.m.	8.30 p.m.	104	6.30 p.m.	7.45 p.m.

The crew must have a minimum layover of 5 h between flights. Obtain the pair of flights that minimizes the layover time away from home. For any given pair, the crew will be based in the city that results in the smallest layover. For each pair, mention the town where the crew should be based.

9.5 A small company wants five jobs to be done. The following matrix shows the return (in ₹) on assigning the ith ($i = 1, 2, 3, 4, 5$) person to the jth job ($j = J_1, J_2, J_3, J_4, J_5$). Assign the five jobs to the five persons, so as to maximize the total expected return.

		Jobs				
		J_1	J_2	J_3	J_4	J_5
Persons	1	5	11	10	12	4
	2	2	4	6	3	5
	3	3	12	5	14	6
	4	6	14	4	11	7
	5	7	9	8	12	5

9.6 A company has four plants A, B, C, and D, each of which can manufacture any of the four products P_1, P_2, P_3, and P_4. The production costs differ from plant to plant, as does the sales revenue. From the following data, obtain which product each plant should produce to maximize profit.

Sales revenue (₹1000)

	Product			
Plant ↓	P_1	P_2	P_3	P_4
A	50	68	49	62
B	60	70	51	74
C	55	67	53	70
D	58	65	54	69

Product cost (₹1000)

	Product			
Plant ↓	P_1	P_2	P_3	P_4
A	49	60	45	61
B	55	63	45	69
C	52	62	49	68
D	55	64	48	66

9.7 In a company, four salesmen are to be assigned to four districts. The estimates of the sales revenue (in ₹ hundreds) are as follows:

		Districts			
		A	B	C	D
Salesmen	1	320	350	400	280
	2	400	250	300	220
	3	420	270	340	300
	4	250	390	410	350

Find the optimum assignment to maximize total revenue.

9.8 A TV company has four areas open for sales and four salesmen available for the assignment. The areas are not equally rich in their sales potential; it is estimated that a typical salesman operating in each area would bring in the following annual sales:

Area	I	II	III	IV
Annual sales(₹)	60,000	50,000	40,000	30,000

The four salesmen are also considered to differ in their ability, and it is estimated that working under the same conditions, their yearly sales would vary proportionately as follows:

Salesman	A	B	C	D
Proportion	7	5	5	4

If the objective is the expected maximum total sales, then the intuitive answer is to assign the best salesman to the richest area, the next best salesman to the second richest, and so on. Verify this answer by the assignment technique.

9.9 A department head has four clerks to do three jobs. Each job can be assigned to only one clerk. The cost of each job with each clerk is given in the following table:

Job	Clerks: W	X	Y	Z
A	18	24	28	32
B	8	13	17	19
C	10	15	19	22

What are the job assignments which will minimize the cost?

9.10 In a company, three persons, P_1, P_2 and P_3 are to be assigned five jobs J_1, J_2, J_3, J_4, and J_5. Each person can do only one job. Using the Hungarian method, find which two jobs should be left undone in the following cost minimizing assignment problem:

	J_1	J_2	J_3	J_4	J_5
P_1	7	8	6	5	9
P_2	9	6	7	6	10
P_3	8	7	9	5	6

9.11 Solve the following assignment problem for minimizing the cost:

Machines	Jobs: 1	2	3	4	5
1	2.5	5	1	6	1
2	2	5	1.5	7	3
3	3	6.5	2	8	3
4	3.5	7	2	9	4.5
5	4	7	3	9	6
6	6	9	5	10	6

9.12 Solve the following assignment problem for maximizing the profit:

Jobs \ Machines	A	B	C	D
1	3	6	2	6
2	7	1	4	4
3	3	8	5	8
4	6	4	3	7
5	5	2	4	3
6	5	7	6	4

9.13 The manager of a company has four persons available to whom jobs can be assigned for the day. Five jobs are offered with the expected profit (in ₹) for every person for each job being as follows:

Persons \ Jobs	A	B	C	D	E
1	6.20	7.80	5.00	10.10	8.20
2	7.10	8.40	6.10	7.30	5.90
3	8.70	9.20	11.10	7.10	8.10
4	4.80	6.40	8.70	7.70	8.00

Find the assignment of persons to jobs that will result in the maximum profit. Which job should be declined?

9.14 Solve the following assignment problem:

Operators \ Machines	I	II	III	IV	V
A	5	5	...	2	6
B	7	4	2	3	4
C	9	3	5	...	3
D	7	2	6	7	2
E	6	5	7	9	1

9.15 Solve the following travelling salesman problems:

(a)

	1	2	3	4	5
1	∞	6	12	6	4
2	6	∞	10	5	4
3	8	7	∞	11	3
4	5	4	11	∞	5
5	5	2	7	8	∞

(b)

From \ To	A_1	A_2	A_3	A_4	A_5
A_1	∞	2	5	7	1
A_2	6	∞	3	8	2
A_3	8	7	∞	4	7
A_4	12	4	6	∞	5
A_5	1	3	2	8	∞

(c)

From \ To	1	2	3	4	5	6
1	∞	20	23	27	29	34
2	21	∞	19	26	31	24
3	26	28	∞	15	36	26
4	25	16	25	∞	23	18
5	23	40	23	31	∞	10
6	27	18	12	35	16	∞

(d) $c_{12} = 20$, $c_{13} = 4$, $c_{14} = 10$, $c_{23} = 5$, $c_{34} = 6$, $c_{25} = 10$, $c_{35} = 6$, and $c_{45} = 20$, where $c_{ij} = c_{ji}$ and there is no route between i and j if the value for c_{ij} is not given.

(e) A travelling salesman has to visit five cities A, B, C, D, and E. The distances (in hundred miles) between the five cities are as follows:

From \ To	A	B	C	D	E
A	...	7	6	8	4
B	7	...	8	5	6
C	6	8	...	9	7
D	8	5	9	...	8
E	4	6	7	8	...

If the salesman starts from city A and has to come back to city A, which route should he choose, so that the total distance travelled is minimum.

CONCEPTUAL PROBLEM

Indian Airlines is a big player in the field of aviation and has a large network of flights. It intends to run regular flights on all the seven days of the week. At present, the company is operating flights only on international routes all over the world. Now, it wants to start service on domestic routes as well. To ensure its success, a thorough study of the existing flights of some domestic airlines is necessary. It is proposed to analyse the schedules of some other airlines in relation with issues such as class, fair, and other

(Contd)

(*Contd*)

facilities. The timings of the flights have been fixed, keeping in mind the needs of the passengers, so that they reach their destination early in the morning, whether they are direct flights or with a break in the journey, and return in the evening. After due consideration, Indian Airlines decides to start three up and down flights from Delhi to Mumbai, with their departure and arrival times as per the following table:

Delhi–Mumbai			Mumbai–Delhi		
Flight no.	**Departure**	**Arrival**	**Flight no.**	**Departure**	**Arrival**
101	7.00 a.m.	9.00 a.m.	201	8.00 a.m.	9.30 a.m.
102	2.00 p.m.	3.30 p.m.	202	1.00 p.m.	3.00 p.m.
103	7.00 p.m.	9.00 p.m.	203	8.00 p.m.	10.00 p.m.

As a facility, the company has decided to provide a minimum 6 h layover (rest) time to the crew between flights. X, the chief executive of the company, faces the problem of preparing a plan of pairing flights that minimizes the total layover time away from home. For any given pair, the crew will be based in the city, which results in shorter layover time. For each such pair, X would also have to mention the town where the crew would be based. Can you help X find the solution to the problem?

Solution To help X, one can proceed as follows:

First prepare the tables for the layover time between the flights. When the crew is based in Delhi, for flight No. 101, the time to stay in Mumbai will be the layover time away from home.

Layover time between flights

Flight 101 and 201 from 9.00 a.m. to 8.00 a.m. the next day; layover time is 23 h.
Flight 101 and 202 from 9.00 a.m. to 1.00 p.m. the next day; layover time is 28 h.
Flight 101 and 203 from 9.00 a.m. to 8.00 p.m; layover time is 11 h.
Flight 102 and 201 from 3.30 p.m. to 8.00 a.m. the next day; layover time is 16.5 h.
Flight 102 and 202 from 3.30 p.m. to 1.00 p.m. the next day; layover time is 21.5 h.
Flight 102 and 203 from 3.30 p.m. to 8.00 p.m. the next day; layover time is 28.5 h.
Flight 103 and 201 from 9.00 p.m. to 8.00 a.m. the next day; layover time is 11 h.
Flight 103 and 202 from 9.00 p.m. to 1.00 p.m. the next day; layover time is 16 h.
Flight 103 and 203 from 9.00 p.m. to 8.00 p.m. the next day; layover time is 23 h.

Thus the layover time between flights for

Crew based in Delhi

	201	202	203
101	23	28	11
102	16.5	21.5	28.5
103	11	16	23

Similarly, other layover times can be calculated.

Crew based in Mumbai

	101	102	103
201	21.5	28.5	9.5
202	16	23	28
203	9	16	21

(*Contd*)

(*Contd*)

Here, we assume either the layover times in terms of quarter hour (0.5 h or 30 minutes) as one unit of time or the layover times for weeks, to avoid fractions. Thus, multiplying each table by 2, we get

Crew based in Delhi

	201	202	203
101	46	56	22
102	33	43	57
103	22	32	46

Crew based in Mumbai

	101	102	103
201	43	57	19
202	32	46	56
203	18	32	42

For combining the aforegiven tables, we choose that base which gives a lesser layover time for each pairing. The layover times marked with '*' denote that the crew is based in Mumbai, and will have the layover time in Delhi; otherwise the crew is based in Delhi.

	201	202	203
101	43^*	32^*	18^*
102	33	43	32^*
103	19^*	32	42^*

1. Locate the minimum entry in each row of the cost matrix and subtract it from each entry of that row. The matrix thus obtained is called the reduced matrix having at least one zero entry in each row.

	201	202	203
101	25^*	14^*	0^*
102	1	11	0^*
103	0^*	13	23^*

2. Now subtract the smallest element of each column from every element of that column. We get the second reduced matrix containing at least one zero in every column. Thus steps 1 and 2 will bring at least one zero in every row and column of the reduced matrix.

	201	202	203
101	25^*	3^*	0^*
102	1	0	0^*
103	0^*	2	23^*

3. Now, examine the first row of the matrix obtained in step 2. If there is only one zero, then surround it by a square and cross all the other zeros in the column. Repeat it for all the rows. If there is more than one zero in any row, skip it and pass on to the next row. Repeat this procedure for the columns as well. By doing so, we get the following table:

	201	202	203
101	25^*	3^*	$\boxed{0^*}$
102	1	$\boxed{0}$	$\times 0^*$
103	$\boxed{0^*}$	2	23^*

(*Contd*)

(*Contd*)

In the aforegiven table, we see that all the zeros are either marked by □ or cancelled, and there is an assignment □ in every row and column.

The optimal assignment from the aforegiven table is as follows:

Flight pair	Crew	Units
101–203	Mumbai	18
102–202	Delhi	43
103–201	Mumbai	19
		80

The optimum solution is 101 → 203, 102 → 202, 103 → 201, that is, the crew will be based in Mumbai-Delhi-Mumbai. Therefore, the minimum layover time is 80 units or 40 h.

10 Sequencing

After studying this chapter, the reader will be able to

- understand this sequencing problem of processing n jobs through two machines
- solve the sequencing problem of processing n jobs through k machines
- obtain the solution to the sequencing problem of processing two jobs through k machines
- determine the total elapsed time
- calculate the idle time for each machine

10.1 INTRODUCTION

There are situations when a number of jobs have to be processed through a specific number of machines. There can be many sequences that can be followed in such situations to process a job. The sequencing problem determines the optimal order in which jobs should be arranged for processing through a certain number of machines, so as to minimize the total cost or time. In this chapter, the following three types of sequencing problems are discussed:

1. n jobs to be processed through two machines, say M_1 and M_2, in the order M_1M_2
2. n jobs to be processed through k machines, say M_1, M_2, M_3, ..., M_k, in the order $M_1M_2M_3 \dots M_k$ (k is a finite number)
3. Two jobs to be processed through k machines

Let us consider a sequencing problem to determine the order in which n jobs should be performed on each of the k different machines. The time required to perform these jobs on each machine and the order of the machines on which each job should be performed are given. Here, $(n!)^k$ is the number of possible sequences or orders in which n jobs can be performed through k machines. For example, if $n = 3$ and $k = 3$, then the total number of possible sequences would be $(3!)^3 = 216$. While it may be possible in theory to compare all the sequences to determine the optimum sequence, it would be rather difficult practically, because of the large number of calculations to be done.

The main objective of solving a sequencing problem is to determine the optimum sequence out of $(n!)^k$ possible sequences, which will minimize the total elapsed time between the first job on the first machine and the completion of the final job on the last machine.

Let us first discuss certain assumptions, notations, and terminologies that are used in defining the sequencing problems.

10.2 ASSUMPTIONS, NOTATIONS, AND TERMINOLOGIES

The following assumptions, notations, and terminologies are used to solve sequencing problems using Johnson's algorithm.

10.2.1 Assumptions

The following assumptions are made while solving a sequencing problem:

1. The exact processing time on each machine is known and is independent of the order of the jobs in which they are processed.
2. Each machine is allowed to process only one job at a time.
3. A job started on a machine must be completed before initiating another job on that machine.
4. The time taken by a job in going from one machine to another machine is negligible.
5. All jobs are ready to process before the processing starts.
6. Every machine is of a different type.
7. A job is processed without delay, subject to ordering requirements.

10.2.2 Notations

The main notations used in a sequencing problem are as follows:

M_{ij} = Processing time for job j on machine M_i ($i = 1, 2, 3, \ldots, k$)
T = Total elapsed time (including the idle time) for processing all the jobs
I_{ij} = Idle time of machine i between the completion of $(j - 1)$th job and the start of the jth job.

10.2.3 Terminologies

The following terminologies are used in sequencing problems:

1. *Number of machines*: The service facilities that a job passes through before its completion.
2. *Processing order*: The order in which the machines are placed for processing the jobs.
3. *Total elapsed time*: The time between the initiation of the first job and the completion of the last job including idle time (if any).
4. *Idle time on a machine*: The time for which a machine does not work (i.e., the time during which there is no job to process on it).
5. *No passing rule*: The order of machines is fixed, that is, there is no change in the given order of machines for performing different jobs.

10.3 JOHNSON'S ALGORITHM FOR PROCESSING *n* JOBS THROUGH TWO MACHINES

In sequencing problems for *n* jobs on two (or more) machines, the processing time on the machines is usually known with certainty. However, it is difficult to get the desired sequence with the minimum total elapsed time. The algorithm designed by Johnson for determining the total elapsed time and the optimum sequence makes this task simpler. These types of sequencing problems for processing *n* jobs through two machines can be described as follows:

1. Given jobs to be performed = *n*.
2. Two machines M_1 and M_2 are involved in performing these *n* jobs.
3. Each job is performed on machines in the order M_1M_2.
4. The time required for performing each job on each machine is known, as shown in the following table:

Jobs → Machines ↓	1	2	3	...	*j*	...	*n*
M_1	M_{11}	M_{12}	M_{13}	...	M_{1j}	...	M_{1n}
M_2	M_{21}	M_{22}	M_{23}	...	M_{2j}	...	M_{2n}

The main objective is to determine the optimum sequence or order of jobs such that the total elapsed time *T* is minimum. Another purpose is to determine the respective idle times for machines M_1 and M_2.

Solution Procedure

The solution procedure for solving the aforementioned sequencing problems is as follows:

1. Go through the processing times of the given jobs in the aforegiven table and select the minimum of M_{11}, M_{12}, M_{13}, …, M_{1n} and M_{21}, M_{22}, M_{23}, …, M_{2n}. If the minimum processing time is M_{1p}, then perform the *p*th job first and if it is for M_{2q}, then perform the *q*th job at the end (as the given order is M_1M_2).
2. If there is a tie in selecting the minimum processing time, then it can be resolved using one of the following solutions:
   (a) If the minimum values are for M_{1i} and M_{1j}, that is, $M_{1i} = M_{1j}$ for the *i*th and *j*th jobs on machine M_1, then select the job that requires longer processing time on machine M_2 and place it at the beginning of the job sequence.
   (b) If the minimum values are for M_{2i} and M_{2j}, that is, $M_{2i} = M_{2j}$ for the *i*th and *j*th jobs on machine M_2, then select the job that requires longer processing time on machine M_1 and place it at the end of the job sequence.
   (c) If the minimum values are for M_{1i} and M_{2j}, that is, $M_{1i} = M_{2j}$, then process the *i*th job at the beginning and the *j*th job at the end of the job sequence.
3. Delete the jobs processed and repeat steps 1 and 2 till all the jobs are processed and an optimum sequence is obtained.

4. Tabulate the data to calculate the total elapsed time and the respective idle time for machines M_1 and M_2.

Machines → Optimum job sequence ↓	M_1		M_2		Idle time for M_1	Idle time for M_2
	In	Out	In	Out		
–	0	…	…	…	…	…
–	…	…	…	…	…	…

5. Calculate the idle time for machines M_1 and M_2 using the following formulations:
   (a) Idle time for machine M_1 = Total elapsed time – Time when the last job in the sequence is completed on machine M_1
   (b) Idle time for machine M_2 = Time duration in which the first job in the sequence is completed on machine $M_1 + \sum_{j=2}^{n}$ {(Time when the jth job in a sequence starts on machine M_2) – (Time when the $(j-1)$th job in the sequence is completed on machine M_2)}
6. The total elapsed time = Total time from the beginning of the first job on machine M_1 to the completion of the last job on Machine $M_2 = \sum_{j=1}^{n} M_{2j} + \sum_{j=1}^{n} I_{2j}$

where M_{2j} = Time required for processing the jth job on machine M_2

and I_{2j} = Time duration in which machine M_2 remains idle after processing the $(j-1)$th job and before initiating work on the jth job.

Example 10.1 Solve the following sequencing problem to minimize the total elapsed time:

Job→ Department ↓	I	II	III
A	8	6	5
B	8	3	4

Solution The minimum processing time in the given problem is 3 h for Department B. Therefore, place job II at the end in the job sequence as shown Table 10.1.

Deleting job II, the processing times for the remaining jobs are given in Table 10.2.

The minimum processing time in the reduced table is 4 for Department B on job III. Therefore, place job III at the end (before II) and job I at the beginning of the empty cells to obtain the optimal job sequence as given in Table 10.3.

For the total elapsed time, we have Table 10.4.

Table 10.1 Iteration 1

		II

Table 10.2 Iteration 2

Jobs	I	II
Department A	8	5
Department B	8	4

Table 10.3 Iteration 3

I	III	II

Table 10.4 Iteration 4

Jobs	Department A		Department B		Idle time of A	Idle time of B
	In	Out	In	Out		
I	0	8	8	16	0	8
III	8	13	16	20	0	–
II	13	19	20	23	4	–

We arrive at the following conclusion:

The total minimum elapsed time (T) is 23 h and the idle times of Departments A and B are (23 – 19) = 4 h and 8 h respectively.

Example 10.2 A company has two machines, M_1 and M_2, on which six jobs are to be processed in the order M_1M_2. The processing time (in h) for each job on these machines is given in the following table:

Jobs → Machines ↓	J_1	J_2	J_3	J_4	J_5	J_6
M_1	1	3	8	5	6	3
M_2	5	6	3	2	2	10

Find (a) the optimal job sequence, (b) the total elapsed time, and (c) the idle time for each machine M_1 and M_2.

Solution The minimum processing time in the given problem is 1 h on machine M_1 for J_1. Therefore, place J_1 at the beginning of the job sequence as shown in Table 10.5.

Deleting J_1, the processing times for the remaining jobs are as shown in Table 10.6.

The minimum processing time in the reduced table is 2 h on machine M_2 for J_4 and J_5. Since the corresponding processing time of J_5 on machine M_1 is more than the corresponding time of J_4 on machine M_1, J_5 will be scheduled at the end and J_4 will be scheduled just before J_5. The resulting job sequence is as shown in Table 10.7.

The processing times for the remaining jobs are given in Table 10.8.

Now, the minimum processing time in the reduced table is 3 h, which corresponds to J_3 on machine M_2 and J_2 and J_6 on machine M_1. Thus, place J_3 at the end and J_6 and J_2 at the beginning of the empty cells of the job sequence. Place J_2 after J_6 as the corresponding processing time is more for J_6 than for J_2 on M_2. The resulting optimal job sequencing is given in Table 10.9.

For the total elapsed time, we arrive at the following schedule for optimum sequencing as shown in Table 10.10.

Table 10.6 Iteration 2

Jobs → Machines ↓	J_2	J_3	J_4	J_5	J_6
M_1	3	8	5	6	3
M_2	6	3	2	2	10

Table 10.5 Iteration 1

J_1					

Table 10.7 Iteration 3

J_1				J_4	J_5

Table 10.8 Iteration 4

Jobs → Machines ↓	J_2	J_3	J_6
M_1	3	8	3
M_2	6	3	10

Table 10.9 Iteration 5

J_1	J_6	J_2	J_3	J_4	J_5

Table 10.10 Iteration 6

Job	Machine M_1		Machine M_2		Idle time of M_1	Idle time of M_2
	In	Out	In	Out		
J_1	0	1	1	6	0	1
J_6	1	4	6	16	0	–
J_2	4	7	16	22	0	–
J_3	7	15	22	25	0	–
J_4	15	20	25	27	0	–
J_5	20	26	27	29	3	–

We arrive at the following conclusions:

(a) The total minimum elapsed time (T) is 29 h.
(b) The idle time of machine M_1 is $(29 - 26) = 3$ h.
(c) The idle time of machine M_2 is 1 h.

Example 10.3 A company has two machines, M_1 and M_2, on which five jobs are to be processed in the order M_1M_2. The processing time, in hours (h), for each job on these machines is given in the following table:

Jobs → Machines ↓	J_1	J_2	J_3	J_4	J_5
M_1	5	1	9	3	10
M_2	2	6	7	8	4

Find (a) the optimum job sequence, (b) the elapsed time, and (c) the idle times for each machine M_1 and M_2.

Solution The minimum processing time in the given problem is 1 h on machine M_1 of job 2 (J_2). Therefore, place job 2 (J_2) at the beginning of the job sequence as shown in Table 10.11.

Deleting J_2, the processing times for the remaining jobs are as shown in Table 10.12.

The minimum processing time in the reduced table is 2 h on machine M_2 of J_1. Therefore, place J_1 at the end of the job sequence as shown in Table 10.13.

Deleting J_1, the processing times for the remaining jobs are as shown in Table 10.14.

The minimum processing time in the reduced table is 3 h on machine M_1 of J_4. Therefore, place J_4 next to J_2 as shown in Table 10.15.

Deleting J_4, the processing times for the remaining jobs are as shown in Table 10.16.

Now, the minimum processing time in the reduced table is 4 h on machine M_2 of J_5. Therefore, place J_5 just before J_1 and J_3 next to J_4 to get the final optimal sequence as given in Table 10.17.

For the total elapsed time, we have Table 10.18.

Table 10.11 Iteration 1

J_2				

Table 10.12 Iteration 2

Jobs → Machines ↓	J_1	J_3	J_4	J_5
M_1	5	9	3	10
M_2	2	7	8	4

Table 10.13 Iteration 3

J_2				J_1

Table 10.14 Iteration 4

Jobs → Machines ↓	J_3	J_4	J_5
M_1	9	3	10
M_2	7	8	4

Table 10.15 Iteration 5

J_2	J_4			J_1

Table 10.16 Iteration 6

Jobs → Machines ↓	J_3	J_5
M_1	9	10
M_2	7	4

Table 10.17 Iteration 7

J_2	J_4	J_3	J_5	J_1

Table 10.18 Iteration 8

Job	Machine M_1		Machine M_2		Idle time of M_1	Idle time of M_2
	In	Out	In	Out		
J_2	0	1	1	7	0	1
J_4	1	4	7	15	0	–
J_3	4	13	15	22	0	–
J_5	13	23	23	27	0	1
J_1	23	28	28	30	2	1

We arrive at the following conclusions:

(a) The total minimum elapsed time (T) is 30 h.
(b) The idle time of machine M_1 is $(30 - 28) = 2$ h.
(c) The idle time of machine M_2 is 3 h.

10.4 JOHNSON'S ALGORITHM FOR PROCESSING *n* JOBS THROUGH *k* MACHINES

For finding an optimal sequence for n (more than two) jobs through k (more than two) machines, the rule used in processing n jobs through two machines discussed in Section 10.3 can be used under certain conditions:

1. Minimum $M_{1j} \geq$ Maximum M_{ij} $(i = 2, 3, \ldots, k - 1; j = 1, 2, \ldots, n)$ or
2. Minimum $M_{kj} \geq$ Maximum M_{ij} $(i = 2, 3, \ldots, k - 1; j = 1, 2, \ldots, n)$

where M_{ij} is the processing time on machine i of job j. The problem can be solved only if either of the aforementioned two or both the conditions are satisfied. In that case, the procedure explained in the subsequent subsection can be applied.

Solution procedure

Suppose there are n jobs, each of which is to be processed on k machines, say M_1, M_2, M_3, …, M_k in the order $M_1M_2M_3 \ldots M_k$. The processing time for each job on each machine is shown in the following table:

Jobs → Machines ↓	1	2	3	…	j	…	n
M_1	M_{11}	M_{12}	M_{13}	…	M_{1j}	…	M_{1n}
M_2	M_{21}	M_{22}	M_{23}	…	M_{2j}	…	M_{2n}
M_3	M_{31}	M_{32}	M_{33}	…	M_{3j}	…	M_{3n}
…	…	…	…	…	…	…	…
M_i	M_{i1}	M_{i2}	M_{i3}	…	M_{ij}	…	M_{in}
…	…	…	…	…	…	…	…
M_k	M_{k1}	M_{k2}	M_{k3}	…	M_{kj}	…	M_{kn}

The solution procedure for solving this sequencing problem is as follows:

1. Find the minimum M_{1j}, minimum M_{kj} and maximum processing time on all machines except M_1 and M_k, $j = 1, 2, 3, \ldots, n$ (i.e., minimum of M_1 and M_k rows and maximum of intermediate rows).
2. Check for the following conditions:
   (a) Minimum $M_{1j} \geq$ Maximum M_{ij} $(i = 2, 3, 4, \ldots, k-1;\ j = 1, 2, 3, \ldots, n)$
   (b) Minimum $M_{kj} \geq$ Maximum M_{ij} $(i = 2, 3, 4, \ldots, k-1;\ j = 1, 2, 3, \ldots, n)$
3. If either or both conditions (a) and (b) of step 2 are satisfied, then move to the next step, else this method fails to solve the given problem.
4. Convert the n jobs and k machines problem in to n jobs and two machines. The two fictitious machines, denoted by G and H, and corresponding processing times G_{gj} and H_{hj} $(j = 1, 2, 3, \ldots, n)$ are assumed, such that

   $$G_{gj} = M_{1j} + M_{2j} + M_{3j} + \ldots + M_{(k-1)j} \text{ (for } g = 1, 2, 3, \ldots, k-1;\ j = 1, 2, 3, \ldots, n)$$

   and $H_{hj} = M_{2j} + M_{3j} + M_{4j} + \ldots + M_{kj}$ (for $h = 2, 3, 4, \ldots, k;\ j = 1, 2, 3, \ldots, n$)

Thus, the sequencing problem of n jobs on k machines is converted into the sequencing problem of n jobs on two k machines, G and H, as shown in the following table:

Jobs → Machines ↓	1	2	3	…	j	…	n
G	M_{11} $+M_{21}$ $+M_{31}$ $+\ldots$ $+M_{(k-1)1}$	M_{12} $+M_{22}$ $+M_{32}$ $+\ldots$ $+M_{(k-1)2}$	M_{13} $+M_{23}$ $+M_{33}$ $+\ldots$ $+M_{(k-1)3}$	… … … … …	M_{1j} $+M_{2j}$ $+M_{3j}$ $+\ldots$ $+M_{(k-1)j}$	… … … …	M_{1n} $+M_{2n}$ $+M_{3n}$ $+\ldots$ $+M_{(k-1)n}$

(Contd)

(*Contd*)

Jobs → Machines ↓	1	2	3	...	*j*	...	*n*
H	M_{21} + M_{31} + M_{41} + ... + M_{k1}	M_{22} + M_{32} + M_{42} + ... + M_{k2}	M_{23} + M_{33} + M_{43} + ... + M_{k3}	...	M_{2j} + M_{3j} + M_{4j} ... + M_{kj}	...	M_{2n} + M_{3n} + M_{4n} + ... + M_{kn}

Now, one can solve this sequencing problem as *n* jobs on two machines with the prescribed order GH as usual. The resulting sequence will be optimum for the given problem.

Example 10.4 Six jobs are to be performed on three machines, M_1, M_2, and M_3, in the order $M_1M_3M_2$. The processing times are given as per the following table:

Jobs → Machines ↓	J_1	J_2	J_3	J_4	J_5	J_6
M_1	12	10	9	14	7	9
M_2	7	6	6	5	4	4
M_3	6	5	6	4	2	4

The suggested sequence: Jobs J_5, J_3, J_6, J_2, J_1, J_4

(a) Determine the total elapsed time for the suggested sequence.
(b) Is the given sequence optimal?
(c) If your answer to (b) is 'No', then determine the optimal sequence and the total elapsed time associated with it.

Solution Here, minimum $M_{1j} = 7$, minimum $M_{2j} = 4$, and maximum $M_{3j} = 6$. Hence, the condition

$$\text{Minimum } M_{1j} \geq \text{Maximum } M_{3j} \quad (j = 1, 2, \ldots, 6)$$

is satisfied. Thus, the given sequencing problem can be converted into a problem involving six jobs and two machines. The two fictitious machines denoted by G and H and the corresponding processing times are shown in the following table:

Jobs → Machines ↓	J_1	J_2	J_3	J_4	J_5	J_6
G	18	15	15	18	9	13
H	13	11	12	9	6	8

Here, $G_j = M_{1j} + M_{3j}$ and $H_j = M_{3j} + M_{2j}$ for $j = 1, 2, \ldots, 6$.

Now, solving this new problem using the *n* jobs and two machines method, we get the optimal sequence given in Table 10.19.

Table 10.19 Iteration 1

J_1	J_3	J_2	J_4	J_6	J_5

For the total elapsed time, we have Table 10.20.

Table 10.20 Iteration 2

Machines → Job ↓	M_1	M_3	M_2
	In–Out	In–Out	In–Out
J_1	0–12	12–18	18–25
J_3	12–21	21–27	27–33
J_2	21–31	31–36	36–42
J_4	31–45	45–49	49–54
J_6	45–54	54–58	58–62
J_5	54–61	61–63	63–67

The total minimum elapsed time (T) is 67 h.

Given sequence of jobs—J_5, J_3, J_6, J_2, J_1, J_4.

Now, for the total elapsed time according to the given sequence J_5, J_3, J_6, J_2, J_1, J_4, we have Table 10.21.

Table 10.21 Iteration 3

Machines → Job ↓	M_1	M_3	M_2
	In–Out	In–Out	In–Out
J_5	0–7	7–9	9–13
J_3	7–16	16–22	22–28
J_6	16–25	25–29	29–31
J_2	25–35	35–40	40–46
J_1	35–47	47–53	53–60
J_4	47–61	61–65	65–70

We arrive at the following conclusions:

(a) The total minimum elapsed time (T) is 70 h.
(b) The given sequence is not optimal.
(c) As the given sequence is not optimal, the associated total minimum elapsed time is 67 h.

Example 10.5 A company has six machines M_i, $i = 1, 2, 3, \ldots, 6$ in the order $M_1M_2M_3M_4M_5M_6$ on which five jobs are to be processed. The processing timing (in h) for each job on these machines is as given in the following table:

Job → Machine ↓	A	B	C	D
M_1	18	17	11	20
M_2	8	6	5	4
M_3	7	9	8	3
M_4	2	6	5	4
M_5	10	8	7	8
M_6	25	19	15	12

Determine a sequence of these four jobs that minimizes the total elapsed time.

Solution Here, minimum $M_{1j} = 11$, minimum $M_{6j} = 12$, and the maximum of M_{2j}, M_{3j}, M_{4j}, and M_{5j} are 8, 9, 6, and 10 respectively. Since the conditions

$$\text{minimum } M_{1j} \geq \text{maximum } M_{ij}\ (i = 2, 3, 4, 5)$$
$$\text{and minimum } M_{6j} \geq \text{maximum } M_{ij}\ (i = 2, 3, 4, 5)$$

are satisfied, the given sequencing problem can be converted into a four jobs and two-machines problem. The two fictitious machines, denoted by G and H, and the corresponding processing times are shown in the following table:

Jobs → Machines ↓	A	B	C	D
G	45	46	36	39
H	52	48	40	31

Here, $G_j = \sum_{i=1}^{5} M_{ij}$ and $H_j = \sum_{i=2}^{6} M_{ij}$ for $j = 1, 2, 3, 4$.

Now solving this new problem using the n jobs and two machines method, we obtain the following optimal job sequence:

C	A	B	D

For total elapsed time, we have

Job	Machine					
	M_1	M_2	M_3	M_4	M_5	M_6
	In–Out	In–Out	In–Out	In–Out	In–Out	In–Out
C	0–11	11–16	16–24	24–29	29–36	36–51
A	11–29	29–37	37–44	44–46	46–56	56–81
B	29–46	46–52	52–61	61–67	67–75	81–100
D	46–66	66–70	70–73	73–77	77–85	100–112

The total minimum elapsed time (T) is 112 h.

10.5 PROCESSING TWO JOBS THROUGH *k* MACHINES

A sequencing problem with two jobs to be processed through k machines with different orders, that is, a single order not being followed, can be put in the following form:

1. Consider jobs J_1 and J_2 as well as k machines say $M_1, M_2, M_3, \ldots, M_k$.
2. The order of each job on k machines is known in advance, though the order may not be the same for jobs J_1 and J_2.
3. The exact or expected processing time M_{ij} ($i = 1, 2, 3, \ldots, k$; $j = 1, 2$) is known as shown in Table 10.22.

Here, all α_i's and β_i's are any of the numbers $i = 1, 2, 3, \ldots, k$. We obtain the Table 10.23.

Table 10.22 Table depicting exact or expected processing time

Jobs → Machines ↓	Jobs J_1	J_2
Machines	$M_{\alpha 1}$	$M_{\beta 1}$
	$M_{\alpha 2}$	$M_{\beta 2}$
	$M_{\alpha 3}$	$M_{\beta 3}$
	…	…
	$M_{\alpha k}$	$M_{\beta k}$

Table 10.23 Iteration 1

Jobs → Machines ↓	Jobs J_1	J_2
M_1	M_{11}	M_{12}
M_2	M_{21}	M_{22}
M_3	M_{31}	M_{32}
…	…	…
M_k	M_{k1}	M_{k2}

The optimal sequence in this type of sequencing problem can be obtained by the graphical method, that is, by drawing up a graph.

Solution procedure

The solution procedure of this problem is as follows:

1. First, draw two perpendicular lines. One of the horizontal lines, that is, the x-axis, represents the processing time for the job J_1 on different k machines and the vertical line, the y-axis, represents the processing time for job J_2 on different k machines when job J_1 remains idle.
2. Mark the given processing times for job J_1 on x-axis and job J_2 on y-axis according to the given order of k machines.
3. Construct the various block diagrams on the graph by pairing the same machine from the origin to the end.
4. After constructing the blocks, draw an optimal path moving horizontally, vertically, or at an angle of 45° with the x-axis. Ensure that the line with an inclination of 45° should not cross any of the constructed blocks. It is observed that when we move horizontally, job 1 is at work (busy), while job 2 is idle. On the other hand, if we move vertically, job 1 is idle and job 2 is at work (busy). When we move at an inclination of 45° with the horizontal, both jobs are at work (busy). With the help of this concept, we can find the total elapsed time to complete both jobs along with finding the idle times for job 1 and job 2.
5. The path having the maximum diagonal movement should be chosen. An optimum path is that which minimizes the idle time for jobs J_1 and J_2, that is, a path in which diagonal movement is maximum.
6. The total elapsed time is obtained by adding the idle time for either of the jobs to the processing time for that job, that is,

   Total elapsed time = Processing time of job J_1 + Idle time for job J_1
   or Total elapsed time = Processing time of job J_2 + Idle time for job J_2

Example 10.6 Use the graphical method to minimize the total time elapsed to complete jobs 1 and 2 based on the following details:

Job 1 Machines					
Sequence	A	B	C	D	E
Time (h)	3	4	2	6	2

Job 2 Machines					
Sequence	B	C	A	D	E
Time (h)	5	4	3	2	6

Solution The following steps are involved in solving this problem:

(a) First, we mark the processing times for job 1 and job 2 on the x axis and y axis respectively (Fig. 10.1) according to the given sequential order of the five machines.

(b) Now, construct the rectangular blocks by pairing the same machines.

(c) Mark a path from the origin to the end point by moving along the 45° line as much as possible.

Hence, the total elapsed time = Processing time of job J_1 + Idle time for job J_1

$$= 17 + (2 + 3) = 22\,\text{h}$$

or the total elapsed time = Processing time of job J_2 + Idle time for job J_2

$= 20 + 2 = 22\,h$

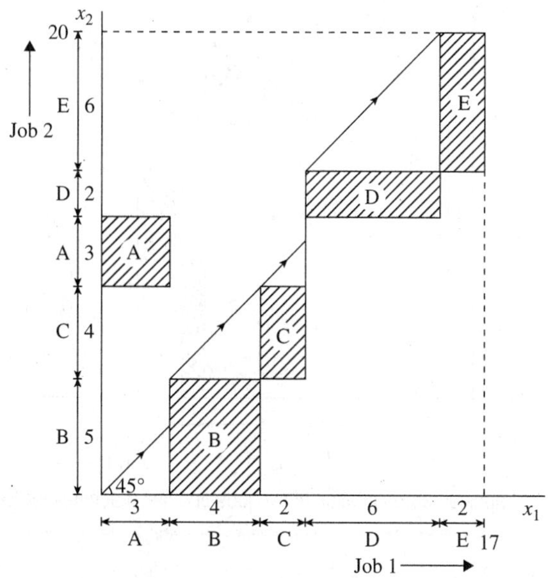

Figure 10.1 Graphical representation of two jobs and five machines for Example 10.6

Example 10.7 A company has five machines A, B, C, D, and E. Two jobs J_1 and J_2 must be processed through each of these machines. The processing times (in h) for each job on different machines are as follows:

Job 1 Machines					
Sequence	A	B	C	D	E
Time (h)	2	4	5	1	2

Job 2 Machines					
Sequence	D	E	A	C	B
Time (h)	6	4	2	3	6

Use the graphical method to determine the total minimum elapsed time.

Solution The following steps are involved in solving this problem:

(a) First, we mark the processing times for job 1 and job 2 on the *x*- and *y*-axis (Fig. 10.2) respectively according to the given sequential order of five machines.
(b) Now, construct the rectangular blocks by pairing the same machines.
(c) Mark a path from the origin to the end point by moving along the 45° line as much as possible.

Hence, the total elapsed time = Processing time of job J_1 + Idle time for job J_1

$= 14 + 7 = 21\,h$

or the total elapsed time = Processing time of job J_2 + Idle time for job J_2 = 21 h

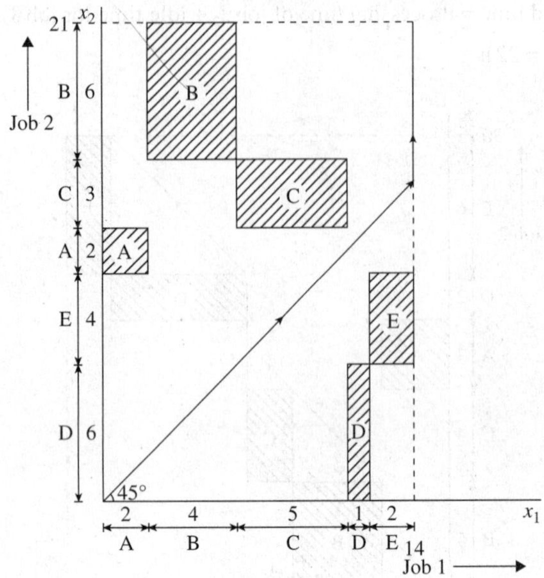

Figure 10.2 Graphical representation of two jobs and five machines for Example 10.7

RECAPITULATION

1. This chapter discusses the sequencing problem for processing n jobs on two machines, n jobs on k machines, and two jobs on k machines. It also illustrates the determination of the total elapsed time and the idle time on each machine.
2. In a sequencing problem, we first fix the order or sequence of the machines on which jobs are to be processed.
3. Once initiated, the process of each job continues till its completion.
4. The time taken by each job in moving from one machine to another is negligible.

EXERCISES

Multiple-choice Questions

10.1 The time for which a machine does not work is known as

(a) the total elapsed time

(b) the idle time

(c) the difference between the total elapsed time and idle time

(d) none of these

10.2 In a sequencing problem, the order is fixed for

(a) jobs

(b) machines

(c) both jobs and machines

(d) none of these

10.3 The total elapsed time to process n jobs through two machines is

(a) $\sum_{j=1}^{n} M_{2j} + \sum_{j=1}^{n} I_{2j}$

(b) $\sum_{j=1}^{n} M_{2j} + \sum_{j=1}^{n} M_{1j}$

(c) $\sum_{j=1}^{n} I_{1j} + \sum_{j=1}^{n} I_{2j}$

(d) none of these

State True or False

10.1 In a sequencing problem, each machine is of the same type.

10.2 In a sequencing problem, each machine is allowed to process only one job at a time.

10.3 If there are n jobs to be performed one at a time on each of the m machines, then the total number of possible sequences would be $(n!)^m$.

Fill in the Blanks

10.1 The time between starting the first job and completing the last job including the idle time is called ________.

10.2 The time during which there is no job for processing on the machine is called ________.

10.3 The total elapsed time to process n jobs through two machines is ________.

Review Questions

10.1 Explain the optimal sequencing procedure for solving the sequencing problem of processing n jobs through two machines.

10.2 Explain the optimal sequencing procedure for solving the sequencing problem of processing n jobs through k machines.

10.3 Explain the optimal sequencing procedure for solving the sequencing problem of processing two jobs through k machines.

Numerical Problems

10.1 A company has two machines, M_1 and M_2, on which six jobs are to be processed in the order M_1M_2. The processing time (in h) for each job on these machines is given in the following table:

Jobs → Machines ↓	J_1	J_2	J_3	J_4	J_5	J_6
M_1	4	8	3	6	7	5
M_2	6	3	7	2	8	4

Find the (a) optimal job sequence and (b) total elapsed time.

10.2 Solve the following sequencing problem to minimize the total elapsed ti ne:

Jobs → Machines ↓	J_1	J_2	J_3	J_4	J_5	J_6
M_1	5	9	4	7	8	6
M_2	7	4	8	3	9	5

10.3 A company has two machines, M_1 and M_2, on which six jobs are to be processed in the order M_1M_2. The processing time (in h) for each job on these machines is given in the following table:

Jobs → Machines ↓	J_1	J_2	J_3	J_4	J_5	J_6
M_1	30	120	50	20	90	100
M_2	80	100	90	60	30	10

Find the (a) optimal job sequence, (b) total elapsed time, and (c) idle time for each machine M_1 and M_2.

10.4 A company has two machines, M_1 and M_2, on which five jobs are to be processed in the order M_1M_2. The processing time (in h) for each job on these machines is given in the following table:

Jobs → Machines ↓	J_1	J_2	J_3	J_4	J_5
M_1	12	4	20	14	22
M_2	6	14	16	18	10

Find the (a) optimal job sequence, (b) total elapsed time, and (c) idle time for each machine, M_1 and M_2.

10.5 Solve the following sequencing problem to minimize the total elapsed time:

Jobs → Machines ↓	J_1	J_2	J_3	J_4	J_5	J_6
M_1	3	12	18	9	15	6
M_2	9	18	24	24	3	15

10.6 A company has two machines, M_1 and M_2, on which nine jobs are to be processed in the order M_1M_2. The processing time (in h) for each job on these machines is given in the following table:

Jobs → Machines ↓	J_1	J_2	J_3	J_4	J_5	J_6	J_7	J_8	J_9
M_1	2	5	4	9	6	8	7	5	4
M_2	6	8	7	4	3	9	3	8	11

Find the (a) optimal job sequence and (b) total elapsed time.

10.7 A company has two machines, M_1 and M_2, on which seven jobs are to be processed in the order M_1M_2. The processing time (in h) for each job on these machines is given in the following table:

Jobs → Machines ↓	J_1	J_2	J_3	J_4	J_5	J_6	J_7
M_1	10	12	13	7	14	5	16
M_2	15	11	8	9	6	7	16

Find the (a) optimal job sequence and (b) total elapsed time.

10.8 Solve the following sequencing problem to minimize the total elapsed time:

Jobs → Machines ↓	J_1	J_2	J_3	J_4	J_5	J_6	J_7
M_1	3	12	15	6	10	11	9
M_2	8	10	10	6	12	1	3

10.9 In a company, eight different products are being produced, each requiring time on two machines, M_1 and M_2, in the order M_1M_2 as given in the following table:

Products → Machines ↓	P_1	P_2	P_3	P_4	P_5	P_6	P_7	P_8
M_1	30	45	15	20	80	120	65	10
M_2	20	30	50	35	36	40	50	20

Find the (a) optimal product sequence and (b) total elapsed time.

10.10 Solve the following sequencing problem to minimize the total elapsed time:

Jobs → Machines ↓	J_1	J_2	J_3	J_4	J_5
M_1	10	2	18	6	20
M_2	4	12	14	16	18

10.11 A company has two machines, M_1 and M_2, on which five jobs are to be processed in the order M_1M_2. The processing time (in h) for each job on these machines is given in the following table:

Jobs → Machines ↓	J_1	J_2	J_3	J_4	J_5
M_1	6	2	10	4	11
M_2	3	7	8	9	5

(a) Find the optimal job sequence for five jobs that minimizes the total elapsed time and (b) calculate the total idle time for both the machines M_1 and M_2.

10.12 A company has two machines, M_1 and M_2, on which five jobs are to be processed in the order M_1M_2. The processing time (in h) for each job on these machines is given in the following table:

Jobs → Machines ↓	J_1	J_2	J_3	J_4	J_5
M_1	3	7	4	5	7
M_2	6	2	7	3	4

(a) Find the optimal job sequence in which the jobs should be processed (b) Determine the minimum elapsed time for an optimal sequence, (c) Calculate the idle time for machines M_1 and M_2 respectively.

10.13 Solve the following sequencing problem to minimize the total elapsed time:

Jobs → Machines ↓	I	II	III	IV
M_1	15	12	16	17
M_2	5	2	3	3
M_3	4	10	5	4
M_4	15	12	16	17

10.14 There are five jobs, each of which must go through three machines, M_1, M_2, and M_3, in the order $M_1M_2M_3$. The processing times are given in the following table:

Jobs → Machines ↓	J_1	J_2	J_3	J_4	J_5
M_1	7	9	5	6	10
M_2	4	5	1	2	3
M_3	3	8	7	5	4

Determine a sequence for these five jobs that will minimize the total elapsed time.

10.15 We have five jobs, each of which must go through machines M_1, M_2, and M_3 in the order $M_1M_2M_3$. The processing times are given in the following table:

Jobs → Machines ↓	J_1	J_2	J_3	J_4	J_5
M_1	8	10	6	7	11
M_2	5	6	2	3	4
M_3	4	9	8	6	5

Determine a sequence for these five jobs that will minimize the total elapsed time.

10.16 A company has six jobs to be processed on three machines, M_1, M_2, and M_3, in the order $M_1M_2M_3$. The processing time (in h) for each job on each machine is given as follows:

Jobs → Machines ↓	J_1	J_2	J_3	J_4	J_5	J_6
M_1	18	12	29	36	43	37
M_2	7	12	11	2	6	12
M_3	19	12	23	47	28	36

(a) What is the sequence that minimizes the total elapsed time? (b) Calculate idle times for machines M_1, M_2, and M_3.

10.17 Solve the following sequencing problem to minimize the total elapsed time:

Jobs → Machines ↓	J_1	J_2	J_3	J_4	J_5
M_1	10	12	8	15	16
M_2	3	2	4	1	5
M_3	5	6	4	7	3
M_4	14	7	12	8	10

10.18 We have five jobs, each of which must be processed on machines M_1, M_2, and M_3 in the order $M_1M_2M_3$. The processing time (in h) for each job on each machine is given in the following table.

Jobs → Machines ↓	J_1	J_2	J_3	J_4	J_5
M_1	8	5	4	6	5

(Contd)

(*Contd*)

Jobs → Machines ↓	J_1	J_2	J_3	J_4	J_5
M_2	6	2	9	7	4
M_3	10	13	11	10	12

(a) Determine a sequence that minimizes the total elapsed time. (b) Find the idle time on each machine.

10.19 Five jobs are to be processed on three machines M_1, M_2, and M_3 in the order $M_1M_2M_3$. The processing times are given in the following table:

Jobs → Machines ↓	J_1	J_2	J_3	J_4	J_5
M_1	5	7	6	9	5
M_2	2	1	4	5	3
M_3	3	7	5	6	7

(a) Determine a sequence for the jobs that minimizes the total elapsed time and (b) idle time on each machine.

10.20 Solve the following sequencing problem to minimize the total elapsed time:

Jobs → Machines ↓	J_1	J_2	J_3	J_4
M_1	3	7	8	10
M_2	4	5	7	15
M_3	10	8	5	6

10.21 Find the sequence that minimizes the total time required in performing the following jobs on three machines in the order $M_1M_2M_3$:

Jobs → Machines ↓	J_1	J_2	J_3	J_4	J_5	J_6
M_1	8	3	7	2	5	1
M_2	3	4	5	2	1	6
M_3	8	7	6	9	10	9

10.22 Solve the following sequencing problem and provide an optimal solution when passing is not allowed:

Jobs → Machines ↓	J_1	J_2	J_3	J_4
M_1	20	12	18	17
M_2	3	5	4	2
M_3	3	1	2	4
M_4	25	11	10	28

10.23 Determine the optimal sequence of jobs that minimizes the total elapsed time based on the following information; processing time (in h) and when passing is not allowed:

Jobs → Machines ↓	J_1	J_2	J_3	J_4	J_5
M_1	3	8	7	5	2
M_2	3	4	2	1	5
M_3	5	8	10	7	6

10.24 Solve the sequencing problem to determine an optimal sequence when passing is not allowed:

Jobs → Machines ↓	J_1	J_2	J_3	J_4
M_1	11	18	12	13
M_2	4	1	2	2
M_3	2	8	3	2
M_4	11	8	12	13

10.25 A company has six machines A, B, C, D, E, and F. Two jobs, J_1 and J_2, must be processed through each of the machines. The processing time and the necessary sequence of the jobs through the machines are given in the following table:

Job 1 Machines						
Sequence	A	C	D	B	E	F
Time (h)	20	10	10	30	25	16

Job 2 Machines						
Sequence	A	C	B	D	F	E
Time (h)	10	30	15	10	15	20

Determine the optimum sequence for the job in order to minimize the total time necessary to finish the jobs.

10.26 Two jobs are to be processed on four machines A, B, C, and D. The order of machines for each job is given in the following table:

Job 1	A	B	C	D
Job 2	D	B	A	C

The processing times (in h) are given in the following table:

Jobs → Machines ↓	A	B	C	D
1	4	6	7	3
2	4	7	5	8

Find the total minimum elapsed time to perform the jobs.

10.27 Use the graphical method to find the minimum total elapsed time sequence of two jobs and five machines, when we are given the following information:

Job 1 Machines					
Sequence	A	B	C	D	E
Time (h)	2	3	4	6	2

Job 2 Machines					
Sequence	C	A	D	E	B
Time (h)	4	5	3	2	6

10.28 Use the graphical method to find the minimum total elapsed time sequence of two jobs (1 and 2) and five machines (A, B, C, D, E), when we have the following information:

Job 1 Machines					
Sequence	A	B	C	D	E
Time (h)	6	8	4	12	4

Job 2 Machines					
Sequence	B	C	A	D	E
Time (h)	10	8	6	4	12

10.29 Use the graphical method to minimize the time needed to process jobs 1 and 2 on the machines as given in the following table:

Job 1 Machines						
Sequence	A	B	C	D	E	F
Time (h)	8	10	2	6	12	10

Job 2 Machines						
Sequence	B	A	C	F	D	E
Time (h)	12	6	4	8	6	10

CONCEPTUAL PROBLEM

A publishing house has to get five books published in the shortest possible time. It contacts a printing firm for the printing and binding of the books. The firm has two separate sections for these two activities. The printing and binding schedule is to be planned in such a way that the labour of neither units remains idle, and the best utilization of time is made. As soon as the printing of a book is complete, its binding has to be taken up by the binding section. By the time its binding is complete, the printing team has to deliver another book for binding. This is to ensure optimum use of time and eliminate wastage of time or labour. The printing firm has two machines, A (printing) and B (binding), on which five books are to be processed in the order AB. The processing time (in h) for each book on these machines is given in the following table:

Books → Machines ↓	**B_1**	**B_2**	**B_3**	**B_4**	**B_5**
A	3	5	4	1	6
B	6	7	10	5	2

Find (a) the optimal book sequence, (b) the total elapsed time, and (c) the idle time for each machine A and B.

Solution The minimum processing time in the given problem is 1 on machine A of B_4. Therefore, place B_4 at the beginning of the book sequence as shown here:

B_4				

Deleting B_4, the remaining processing times are as follows:

B_1	**B_2**	**B_3**	**B_5**
3	5	4	6
6	7	10	2

(Contd)

(Contd)

The minimum processing time in the reduced table is 2 on machine B for B_5. Therefore, B_5 will be scheduled at the end. The resulting book sequence is as follows:

B_4				B_5

The processing times for the remaining books are given in the following table:

B_1	B_2	B_3
3	5	4
6	7	10

Now, the minimum processing time in the reduced table is 3 on machine A of B_1. Therefore, B_1 will be scheduled next to B_4. The resulting book sequence is given in the following table:

B_4	B_1			B_5

The processing times for the remaining books are given in the following table:

B_2	B_3
5	4
7	10

Now, the minimum processing time in the reduced table is 4 on machine A of B_3. Therefore, B_3 will be scheduled next to B_1 and B_2 next to B_3. The resulting book sequence is as follows:

B_4	B_1	B_3	B_2	B_5

For total elapsed time, we have (scheduling for optimum sequencing):

Job	Machine A		Machine B		Idle time of A	Idle time of B
	In	Out	In	Out		
B_4	0	1	1	6	0	1
B_1	1	4	6	12	0	–
B_3	4	8	12	22	0	–
B_2	8	13	22	29	0	–
B_5	13	19	29	31	12	–

We arrive at the following conclusions:

(a) The total minimum elapsed time (T) is 31 h.
(b) The idle time of Machine A is (31 – 19) = 12 h.
(c) The idle time of Machine B is 1 h.

11 Project Scheduling

> After studying this chapter, the reader will be able to
> - understand the concept of project scheduling
> - draw the network diagram
> - understand the critical path method and program evaluation and review technique
> - find the optimum scheduling using the critical path method
> - learn the time–cost optimization algorithm

11.1 INTRODUCTION

Every organization, big or small, undertakes new projects for the expansion of its business. To reduce the expenditure burden, planning is done such that the project is completed in the minimum time duration, and with the minimum cost involved. In this chapter, we would discuss how to achieve this, both for the deterministic model (where the exact duration of each activity is known), and the probabilistic model (where the exact duration of the activity is not known, only certain time estimates of the activity are taken into consideration).

The concept of project scheduling is of great importance for the advancement of any organization, with respect to the completion of its projects. In this chapter, we shall discuss project management, construction of network diagrams, critical path method (CPM), program evaluation and review technique (PERT), and optimum scheduling using CPM.

It is the earnest endeavor of every government and industrial establishment to plan new projects for the future development and expansion of their industrial organizations. A project is normally defined as a collection of well-defined tasks, called activities, and efforts are made to carry out all these activities successfully. All these activities are non-repetitive and may be hundreds and thousands in numbers. A project may be of great size and complexity, for example, introduction of new products in the market, construction projects like building a bridge, and sending a spacecraft to Mars. All these projects require huge expenditure and time. Furthermore, the management cannot be guided just by past experience in their planning and control. Thus, there is a great need for careful study and analysis of the entire project, before getting the financial approval, and going ahead with the project. Therefore, it is essential that the planning and control on the project are precise, and provide ample warning of the difficulties that may arise later on.

11.2 PROJECT DEVELOPMENT

The three main steps involved in the development of a project are (a) planning, (b) scheduling, and (c) controlling.

11.2.1 Planning

In the first phase, the objectives of the project are set and all assumptions are made. A list of various activities that constitute the project, their completion time, that is, the duration of the activities, the requirement of resources, the total cost, etc., is made. In addition, the sequence in which the activities need to follow is penned down.

11.2.2 Scheduling

The second phase, that is, the scheduling involves the preparation of a time schedule for the project. Particularly, we determine the respective start and finish times of each activity, the relationship of each activity with other activities, and critical activities that require special attention for completion of the project within the scheduled time.

11.2.3 Controlling

This phase deals with the controlling and monitoring of the project, so that it is completed within the scheduled time. Periodical progress reports are made, and review and analysis of the progress of the project is conducted from time to time. Since there is a difference between the actual performance of an activity and the time schedule that is made, one tries to control and minimize this difference. Many techniques exist for managing a project and to determine the time schedule for a project. Some of these techniques are as follows:

1. Critical path method (CPM)
2. Programming evaluation and review technique (PERT)
3. Review analysis of multiple project (RAMP)

Both PERT and CPM network techniques are very useful for planning, scheduling, and executing large time-bound projects, which require a very careful coordination of a variety of complex as well as interrelated activities and resources. Both these techniques were developed in USA during the late 1950s. In fact, PERT was developed by the US navy engineers for the planning and control of the huge Polaris submarine programmes. CPM was conceived and developed by the DuPont and the Remington Rand companies to help scheduling maintenance of the chemical plants. For a long time, PERT and CPM techniques have been applied successfully to improve the efficiency of the working of large projects within a pre-determined time and cost limits. Both PERT and CPM are very useful, as these techniques

1. help in deciding how soon the project will be completed,
2. give an idea of scheduling for the start and finish of each individual phase of the project, and

3. highlight the critical phases of the project to be finished as per the scheduled time and indicate when managerial attention is required to avoid delay.

11.3 NETWORK

A network is a graphical representation of various tasks, which are to be put in a definite sequential order to complete the project in time. The network diagram plays an important role in managing a project. It graphically depicts the details of a project. The two major components of a network are (a) activities and (b) events.

Activity First, the activities and then their interrelationships are identified. Which activity will follow the completion of a certain activity, and what will be the next activity that starts after the completion of the present activity, and so on is identified. The time duration of an activity is also determined. An activity is denoted by an arrow (Fig. 11.1) with the time duration (indicated here as '2') over it.

The start and end points of an activity are denoted by a circle. Each one is called a node. For example, we graphically represent an activity of a time duration of 5 units, starting from node 1 and ending at node 2, as shown in Fig. 11.2.

A network diagram is a graphical representation of a project, representing the activities and nodes. It should fulfill the following conditions:

1. Each activity is represented by one and only one arrow.
2. Each activity must start from and end at a node.
3. More than one activity can start from a node.
4. More than one activity may end at a node.
5. No two distinct activities should have the same starting and ending nodes, that is, part of the network as given in Fig. 11.3 is not permissible.

In the case of Fig. 11.3, we split event 2 into events 2 and 3, and introduce a dummy (redundant) activity of '0' duration, connecting them as shown in Fig. 11.4:

A dummy (redundant) activity is denoted by a broken line.

6. Arrows should not form a closed loop. The network diagram in Fig. 11.5 is not permissible, as (1, 2), (2, 3), and (3, 1) is a closed loop.
7. Start and end nodes should be unique.
8. A dangling event is not permissible in a network diagram. In Fig. 11.6, '3' is the dangling event.

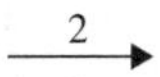

Figure 11.1 Arrow denoting activity

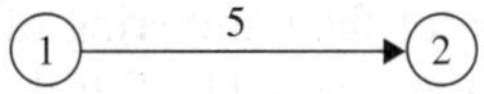

Figure 11.2 Graphical representation of activity with given duration and connecting nodes

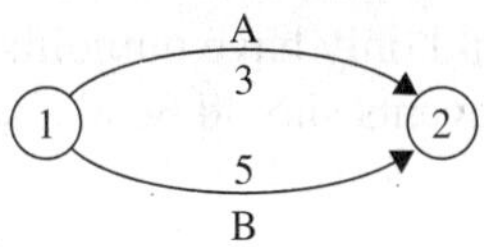

Figure 11.3 Events A and B (Starting and ending at same nodes)

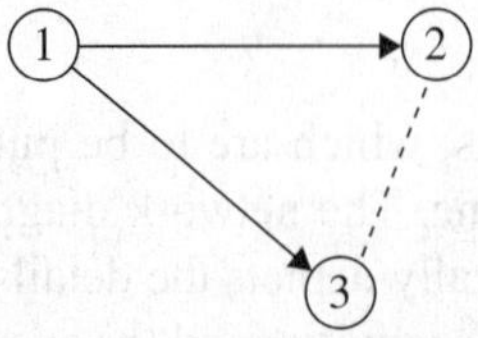

Figure 11.4 Correct network diagram

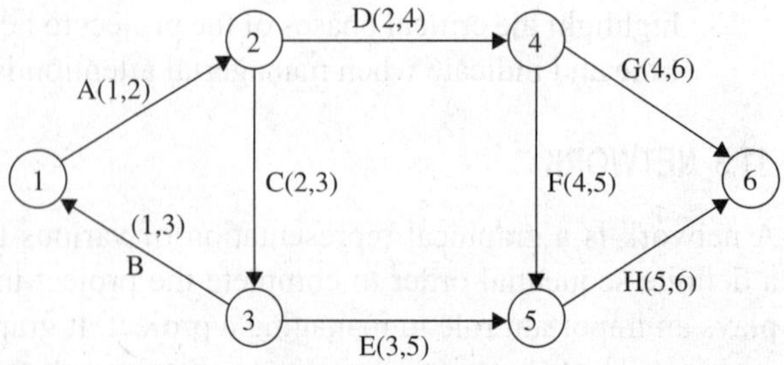

Figure 11.5 Looping

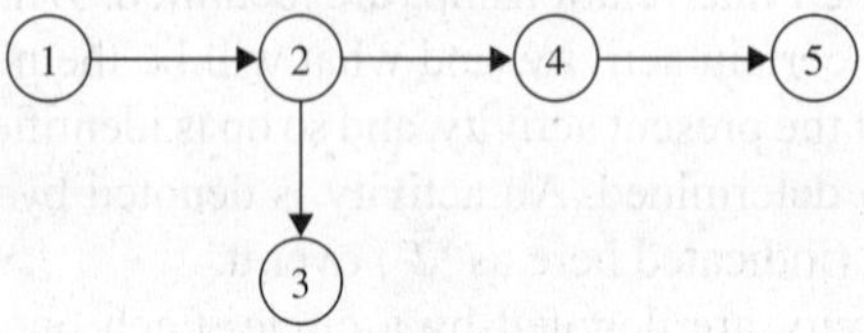

Figure 11.6 Dangling event

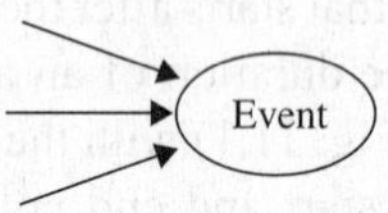

Figure 11.7 Merge event

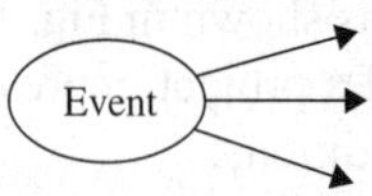

Figure 11.8 Burst event

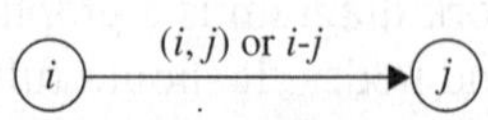

Figure 11.9 Activity (*i*, *j*) or *i-j* joining two nodes *i* and *j*

Event An event (node) in a network diagram is a specific instant of time at which an activity either starts or is completed. Events are usually represented by a circle and are called nodes. There are two types of events—merge and burst.

Merge event An event where the completion of more than one activity takes place is called a merge event (see Fig. 11.7).

Burst event An event which is the starting point of more than one activity is called a burst event (see Fig. 11.8).

11.3.1 Notations

The following notations are used to construct a network diagram:

1. The numerical value written inside the circle denotes the node number.
2. Capital alphabets such as A, B, and C are used for representing the activities.
3. Ordered pair (i, j) is used to represent an activity from the node 'i' to the node 'j' as shown in Fig. 11.9.

11.3.2 Fulkerson's Rule for Numbering Events

According to Fulkerson's rule for the numbering of events involved in a network diagram, the following conditions should be fulfilled:

1. Every event should have a unique number.
2. The starting event should only have outgoing arrows.
3. The numbering of the events should be in a sequential manner, from left to right.

These conditions are collectively known as the Fulkerson's rule.

Example 11.1 A project consists of activities A, B, C, D, E, F, G, and H. These activities along with their precedence relationships are given in the following table:

Activity	Predecessor activity	Activity	Predecessor activity
A	–	E	B, C
B	A	F	E
C	A	G	D, F
D	B	H	G

Draw the network diagram.

Solution The network diagram for the project is shown in Fig. 11.10.

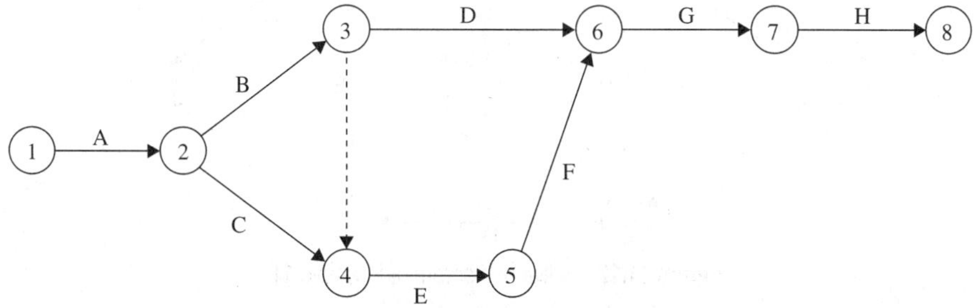

Figure 11.10 Network diagram of Example 11.1

Example 11.2 A project consists of activities A, B, C, D, E, F, G, H, I, and J with conditions given in the following table:

Activity	Predecessor activity	Activity	Predecessor activity
A	–	F	C
B	A	G	C
C	B	H	F, G
D	B	I	D, E, H
E	B	J	I

Draw the network diagram.

Solution The network diagram for the project is shown in Fig. 11.11.

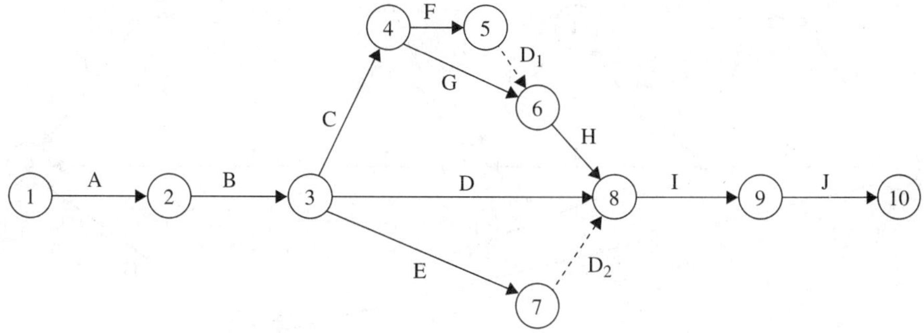

Figure 11.11 Network diagram of Example 11.2

Example 11.3 A project consists of activities A, B, C, D, E, F, and G, with precedence relationships as given in the following table:

Activity	Immediate predecessor (s)	Activity	Immediate predecessor (s)
A	–	E	D
B	–	F	C, E
C	A, B	G	D
D	B	–	–

where F and G are the terminal activities of the project. Draw the network diagram.

Solution The network diagram for the project is shown in Fig. 11.12.

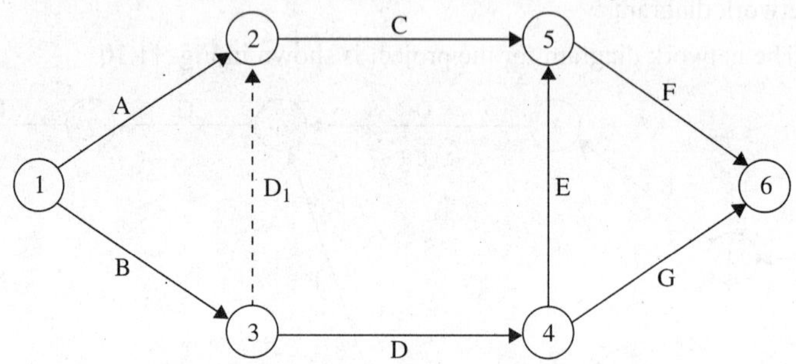

Figure 11.12 Network diagram of Example 11.3

Example 11.4 A project consists of activities A, B, C, D, E, F, G, H, I, J, K, and L that satisfy the following conditions:

(a) A, B, and C are the first activities of the project and can start simultaneously.
(b) A and B precede D.
(c) B precedes E, F, and H.
(d) F and C precede G.
(e) E and H precede I and J.
(f) D and J precede K.
(g) K precedes L.
(h) I, G, and L are the terminal activities of the project.

Draw the network diagram.

Solution The network diagram for the project is shown in Fig. 11.13.

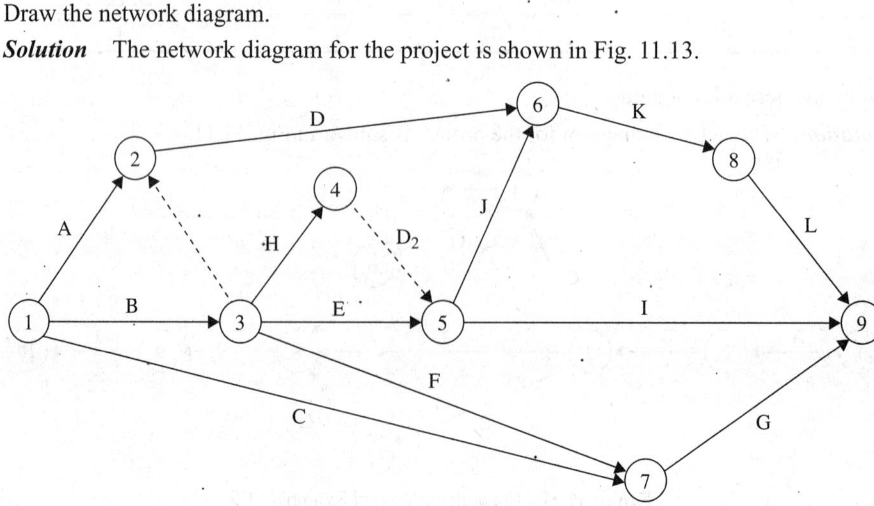

Figure 11.13 Network diagram of Example 11.4

Example 11.5 A project consists of activities B, C, ..., Q, and N with the following conditions: B < E, F; C < G, L; E, G < H; L, H < I; L < M; H < N; H < J; I, J < P; P < Q where the notation X < Y means that the activity X must be finished before Y can begin. Draw the network diagram.

Solution The network diagram for the project is shown in Fig. 11.14.

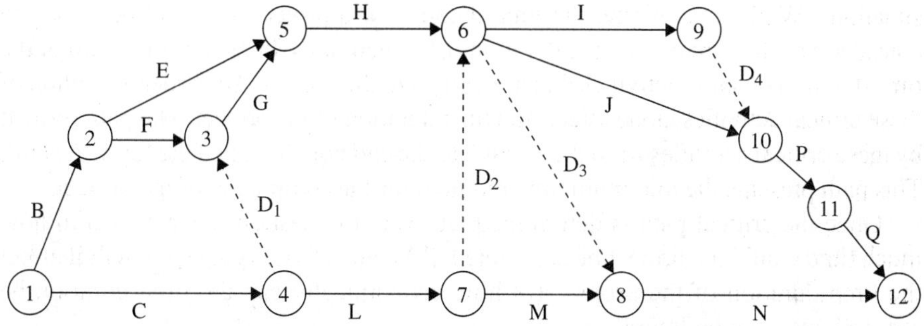

Figure 11.14 Network diagram of Example 11.5

Example 11.6 A project consists of activities A, B, C, D, E, F, G, and H, with their respective durations 3, 2, 4, 5, 7, 5, 8, and 6, which satisfy the following conditions:

(a) A, B, and C start immediately.
(b) D, E, and F start when A is completed.
(c) G starts after B and D are completed.
(d) H starts after C and E are completed.
(e) F, G, and H are terminal activities.

Draw the network diagram.

Solution The network diagram for the project is shown in Fig. 11.15.

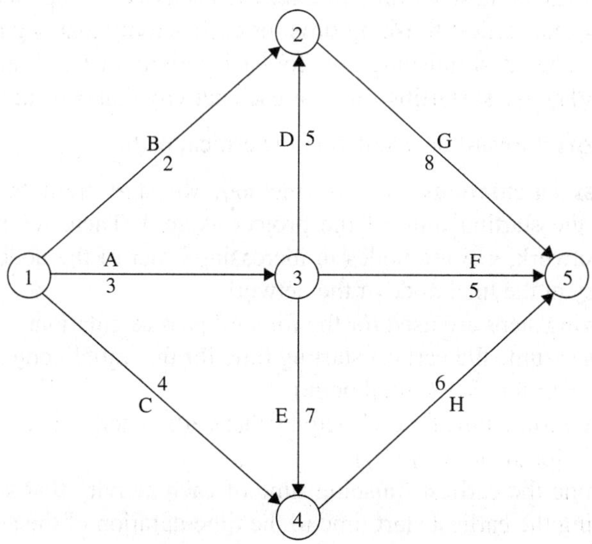

Figure 11.15 Network diagram of Example 11.6

11.4 CRITICAL PATH METHOD

In a project, the estimate of the time durations of activities may either be deterministic or probabilistic. For example, the time duration for erecting a wall is deterministic, whereas the time duration for conducting an experiment in a research lab is probabilistic. The critical path method (CPM) is normally used in deterministic situations. While determining the time duration of a project, it would be wrong to consider it as the sum of the time durations of all activities. In fact, it is the sum of the time duration of some activities that are critical to the project. An earlier execution of these critical activities alone affects the time duration of the project. The path formed by these critical activities from the starting to the end point is called the 'critical path'. This path presents the minimum time required for the completion of the project.

Once the critical path is determined, answers to questions such as, (a) in how much time would the project be completed, (b) if an activity is delayed, will it affect the time duration of the project, (c) how economically can the time duration be reduced, etc., can be found.

Special computations are carried out to obtain the following information regarding the project:

1. The total time duration that is needed for the completion of project.
2. The division of the activities of the project into critical and non-critical activities.

The following terms are used in CPM:

t_{ij} = The time duration of the activity (i, j)
E_i = The earliest happening time of any event i
L_j = The latest happening time of any event j
ES_{ij} = The earliest starting time for each activity that begins at node i
EF_{ij} = The earliest finishing time for each activity that begins at node i
LF_{ij} = The latest finishing time for each activity that ends at node j
LS_{ij} = The latest starting time for each activity that ends at node j

Let us now move forward in calculating the critical path.

Forward pass calculations In this method, we start from the initial node (event), with the starting time of the project as zero. Then, we move forward through the network, visiting nodes in increasing order of the node number, and end the journey at the final node of the network.

The following steps are used for the forward pass calculations:

1. Always assume the earliest starting time for the initial node of a project as zero. $E_i = 0$; for $i = 1$ (initial node)
2. Take the earliest time for each activity that starts at node i and ends at node j as $ES_{ij} = E_i$; for all activity (i, j)
3. Determine the earliest finishing time of each activity that starts at node i by adding the earliest start time to the time duration of the activity, that is, $EF_{ij} = ES_{ij} + t_{ij} = E_i + t_{ij}$

4. Move on to the next node, say j ($j > i$), and determine the earliest happening time for the node j. $E_j = \max_i\{EF_{ij}\} = \max_i\{E_i + t_{ij}\}$ for all immediate preceding activities.
5. Repeat steps 1–4 till you reach the end node.

Backward pass calculations In the backward pass calculations, we start from the final node (event) of the project, and move backwards through the network, visiting nodes in decreasing order of the node number, and stopping at the initial node of the network.

The following steps are used for the backward pass calculations:

1. The latest finishing time for the last node (n) of a project, is always equal to the earliest finishing time. $L_n = E_n$; for $j = n$ (last node)
2. One may take the latest finish time for each activity that ends at node j as $LF_{ij} = L_j$.
3. After that, determine the latest happening time of all the activities ending at j, by subtracting the time duration from the latest finished time of the activity, that is, $LS_{ij} = LF_{ij} - t_{ij} = L_j - t_{ij}$.
4. Move backwards to the node in sequence (i.e., decreasing j by 1), and determine the latest happening time for the node i ($i < j$). Thus, $L_i = \min_j\{LS_{ij}\} = \min_j\{L_j - t_{ij}\}$
5. Repeat steps 1–4 till you reach the starting node.

Critical activity and critical path An activity (i, j) is said to be critical, if the following conditions are satisfied:

1. $E_i = L_i$ 2. $E_j = L_j$ 3. $E_j - E_i = L_j - L_i = t_{ij}$

An activity which is not critical is called a non-critical activity.

The path along the critical activities is called a critical path. There may be more than one critical path in a network diagram.

Note In the network diagram, we enter the earliest time inside the rectangle and the latest time inside the triangle surmounted on the rectangle.

Total duration of project The total duration of a project is the sum total of the durations of all the activities along the critical path, and is denoted by T.

Event float The float (slack) of an event is the difference between its latest time and its earliest time, that is,

$$\text{Event float} = \text{Latest completion time} - \text{Earliest starting time} = L_i - E_i$$

where L_i is the latest completion time and E_i is the earliest completion time.

Total float The total float of any activity is given by the total time available for the completion of the activity minus the duration of the activity, that is, the difference between the latest and earliest finish times of an activity.

$$\text{Total float} = L_j - (E_i + t_{ij})$$

Free float The free float of any activity is the excess of available time over the time required, to complete the activity, that is, Free float $= E_j - (E_i + t_{ij})$.

Independent float The time in which any activity can be rescheduled, without affecting the preceding or succeeding activities is known as an independent float.

Independent float $= E_j - (L_i + t_{ij})$

Notes 1. The difference between the total float and free float is known as the interference float.
2. The basic difference between a slack and a float is that slack is used for events only, whereas float is applied for activities.

Example 11.7 In the following network diagram (Fig. 11.16), all the durations are in days. Find the critical path.

Solution First, we determine the forward and backward calculations.

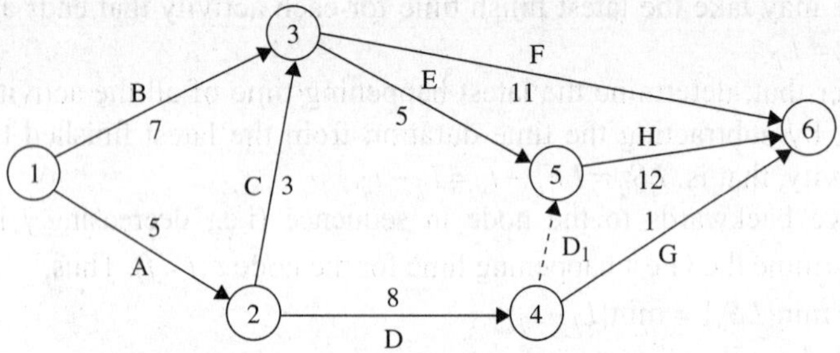

Figure 11.16 Network diagram of Example 11.7

Forward calculations

Node 1 $E_1 = 0$

Node 2 $E_2 = E_1 + t_{12} = 0 + 5 = 5$

Node 3 $E_3 = \max_{i=1,2}(E_i + t_{i3}) = \max(E_1 + t_{13}, E_2 + t_{23}) = \max(0 + 7, 5 + 3) = 8$

Node 4 $E_4 = E_2 + t_{24} = 5 + 8 = 13$

Node 5 $E_5 = \max_{i=3,4}(E_i + t_{i5}) = \max(E_3 + t_{35}, E_4 + t_{45}) = \max(8 + 5, 13 + 0) = 13$

Node 6 $E_6 = \max_{i=3,4,5}(E_i + t_{i6}) = \max(E_3 + t_{36}, E_4 + t_{46}, E_5 + t_{56}) = \max(8 + 11, 13 + 1, 13 + 12) = 25$

Backward calculations

Node 6 $L_6 = E_6 = 25$

Node 5 $L_5 = L_6 - t_{56} = 25 - 12 = 13$

Node 4 $L_4 = \min_{j=5,6}(L_j - t_{4j}) = \min(L_5 - t_{45}, L_6 - t_{46}) = \min(13 - 0, 25 - 1) = 13$

Node 3 $L_3 = \min_{j=5,6}(L_j - t_{3j}) = \min(L_5 - t_{35}, L_6 - t_{36}) = \min(13 - 5, 25 - 11) = 8$

Node 2 $L_2 = \min_{j=3,4}(L_j - t_{2j}) = \min(L_3 - t_{23}, L_4 - t_{24}) = \min(8 - 3, 13 - 8) = 5$

Node 1 $L_1 = \min_{j=2,3}(L_j - t_{1j}) = \min(L_2 - t_{12}, L_3 - t_{13}) = \min(5 - 5, 8 - 7) = 0$

Using the aforementioned calculations, we have Fig. 11.17, which depicts two critical paths.

The critical paths are $1 \rightarrow 2 \rightarrow 4 \rightarrow 5 \rightarrow 6$ and $1 \rightarrow 2 \rightarrow 3 \rightarrow 5 \rightarrow 6$, and the project duration is 25 days.

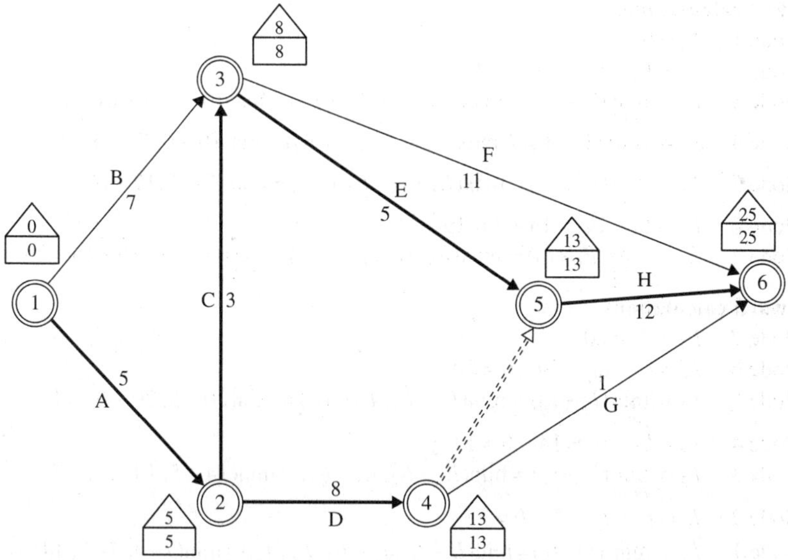

Figure 11.17 Network diagram depicting critical path

Example 11.8 For the following project draw the network diagram and find the project completion time. Find the total float for each of the activities as well.

Activity	Preceding activities	Activity duration (days)	Activity	Preceding activities	Activity duration (days)
A	–	4	E	A, B	7
B	–	7	F	C, D, E	6
C	–	6	G	C, D, E	5
D	A, B	5	–	–	–

Solution The network diagram for the project is shown in Fig. 11.18.

Now, to determine the project completion time, we find the earliest start time (E_i) and latest finish time (L_j) for each activity of the project.

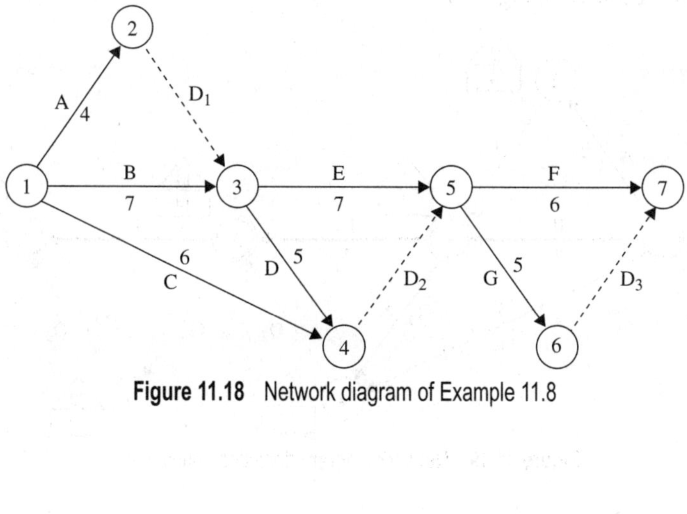

Figure 11.18 Network diagram of Example 11.8

Forward calculations

Node 1 $E_1 = 0$

Node 2 $E_2 = E_1 + t_{12} = 0 + 4 = 4$

Node 3 $E_3 = \max_{i=1,2}(E_i + t_{i3}) = \max(E_1 + t_{13}, E_2 + t_{23}) = \max(0 + 7, 4 + 0) = 7$

Node 4 $E_4 = \max_{i=1,3}(E_i + t_{i4}) = \max(E_1 + t_{14}, E_3 + t_{34}) = \max(0 + 6, 7 + 5) = 12$

Node 5 $E_5 = \max_{i=3,4}(E_i + t_{i5}) = \max(E_3 + t_{35}, E_4 + t_{45}) = \max(7 + 7, 12 + 0) = 14$

Node 6 $E_6 = E_5 + T_{56} = 14 + 5 = 19$

Node 7 $E_7 = \max_{i=5,6}(E_i + t_{i7}) = \max(E_5 + t_{57}, E_6 + t_{67}) = \max(14 + 6, 19 + 0) = 20$

Backward calculations

Node 7 $L_7 = E_7 = 20$

Node 6 $L_6 = L_7 - t_{67} = 20 - 0 = 20$

Node 5 $L_5 = \min_{j=6,7}(L_j - t_{5j}) = \min(L_6 - t_{56}, L_7 - t_{57}) = \min(20 - 5, 20 - 6) = 14$

Node 4 $L_4 = L_5 - t_{45} = 14 - 0 = 14$

Node 3 $L_3 = \min_{j=4,5}(L_j - t_{3j}) = \min(L_4 - t_{34}, L_5 - t_{35}) = \min(14 - 5, 14 - 7) = 7$

Node 2 $L_2 = L_3 - t_{23} = 7 - 0 = 7$

Node 1 $L_1 = \min_{j=2,3,4}(L_j - t_{1j}) = \min(L_2 - t_{12}, L_3 - t_{13}, L_4 - t_{14}) = \min(7 - 4, 7 - 7, 14 - 6) = 0$

To find the critical nodes and total float for each of the activities, all the calculations are shown in the following table:

Activity (i, j)	Normal time(days) t_{ij}	Earliest time		Latest finish time L_j	Total float $L_j - (E_i + t_{ij})$
		Start E_i	Finish $E_i + t_{ij}$		
(1, 2)	4	0	4	7	3
(1, 3)	7	0	7	7	0
(1, 4)	6	0	6	14	8
(3, 4)	5	7	12	14	2
(3, 5)	7	7	14	14	0
(5, 7)	6	14	20	20	0
(5, 6)	5	14	19	20	1

From the aforegiven table, we find that the critical nodes are (1, 3), (3, 5), and (5, 7). The critical path is depicted in Fig. 11.19.

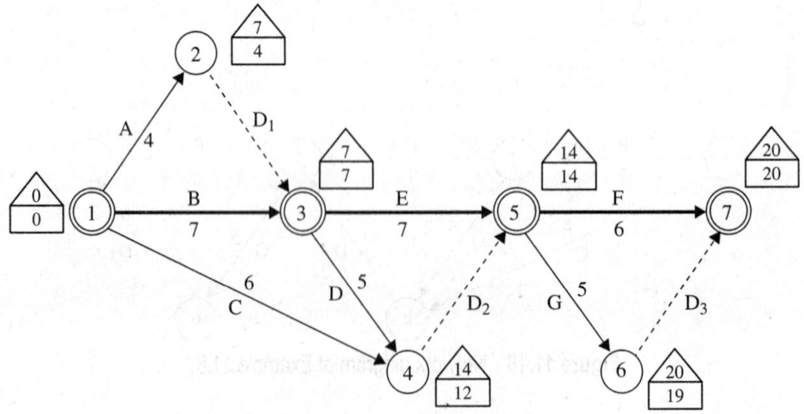

Figure 11.19 Network diagram depicting critical path

The critical path is $1 \rightarrow 3 \rightarrow 5 \rightarrow 7$ and the project completion time is 20 days.

Example 11.9 For the following project:

(a) Draw the network diagram.
(b) Find the earliest and latest event times.
(c) Find the critical path, total project duration, and total float for each activity.

Activity	Duration (days)	Proceeding activity
A	5	–
B	6	A
C	5	B
D	4	A
E	3	D
F	4	C, E

Solution (a) The network diagram for the project is shown in Fig. 11.20.

(b) Now, we determine the earliest event time and the latest event time for each activity of the project.

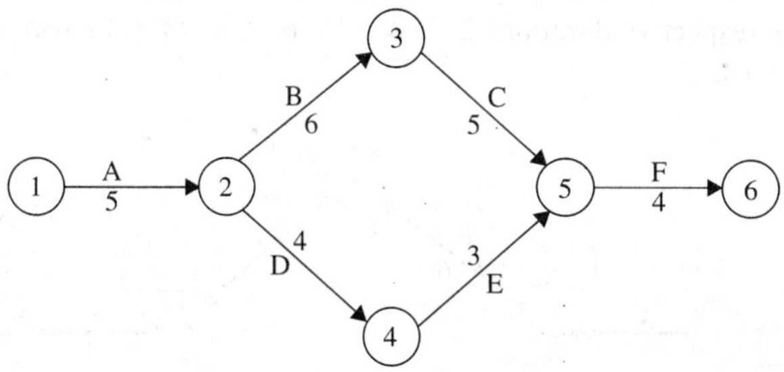

Figure 11.20 Network diagram of Example 11.9

Forward calculations

Node 1 $E_1 = 0$

Node 2 $E_2 = E_1 + t_{12} = 0 + 5$

Node 3 $E_3 = E_2 + t_{23} = 5 + 6 = 11$

Node 4 $E_4 = E_2 + t_{24} = 5 + 4 = 9$

Node 5 $E_5 = \max_{i=3,4}(E_i + t_{i5}) = \max(E_3 + t_{35}, E_4 + t_{45}) = \max(11 + 5, 9 + 3) = 16$

Node 6 $E_6 = E_5 + t_{56} = 16 + 4 = 20$

Backward calculations

Node 6 $L_6 = E_6 = 20$

Node 5 $L_5 = L_6 - t_{56} = 20 - 4 = 16$

Node 4 $L_4 = L_5 - t_{45} = 16 - 3 = 13$

Node 3 $L_3 = L_5 - t_{35} = 16 - 5 = 11$

Node 2 $L_2 = \min_{j=3,4}(L_j - t_{2j}) = \min(L_3 - t_{23}, L_4 - t_{24}) = \min(11 - 6, 13 - 4) = 5$

Node 1 $L_1 = L_2 - t_{12} = 5 - 5 = 0$

(c) To find the critical nodes and total float for each of the activities, all the calculations are shown in the following table:

Activity	Duration (days) t_{ij}	Earliest time		Latest finish time L_j	Total float $L_j - (E_i + t_{ij})$
		Start E_i	Finish $E_i + t_{ij}$		
A	5	0	5	5	0
B	6	5	11	11	0
C	5	11	16	16	0

(*Contd*)

(*Contd*)

Activity	Duration (days) t_{ij}	Earliest time		Latest finish time L_j	Total float $L_j-(E_i+t_{ij})$
		Start E_i	Finish E_i+t_{ij}		
D	4	5	9	13	4
E	3	9	12	16	4
F	4	16	20	20	0

From the aforegiven table, we find that the critical nodes are (1, 2), (2, 3), (3, 5), and (5, 6). The critical path is depicted in Fig. 11.21.

The critical path is $1 \to 2 \to 3 \to 5 \to 6$ and the total project duration time is 20 days.

Example 11.10 Consider a project which consists of the activities A, B, C, D, E, F, G, H, and I with the respective durations 2, 3, 4, 5, 5, 6, 7, and 8 (in days) and satisfies the following conditions:

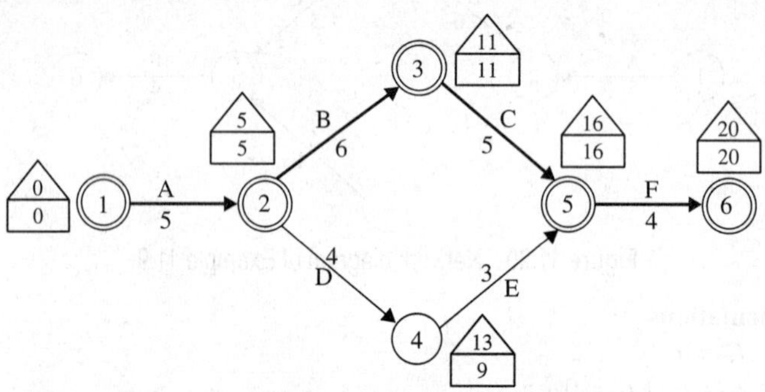

Figure 11.21 Network diagram depicting critical path

(a) A, B, and C start simultaneously.
(b) D and E cannot start unless A is completed.
(c) F cannot start unless B is completed.
(d) G cannot start unless C is completed.
(e) H cannot start unless B and G are completed.
(f) I cannot start unless F and G are completed.
(g) D, E, H, and I are terminal activities.

Draw the network diagram. Find the critical activities and the critical path for the project. Determine the project duration as well.

Solution The network diagram for the project is shown in Fig. 11.22.

Forward calculations

Node 1 $E_1 = 0$

Node 2 $E_2 = E_1 + t_{12} = 0 + 4 = 4$

Node 3 $E_3 = E_1 + t_{13} = 0 + 3 = 3$

Node 4 $E_4 = E_1 + t_{14} = 0 + 2 = 2$

Node 5 $E_5 = \max_{i=2,3}(E_i + t_{i5}) = \max(E_2 + t_{25}, E_3 + t_{35}) = \max(4 + 7, 3 + 0) = 11$

Node 6 $E_6 = \max_{i=3,5}(E_i + t_{i6}) = \max(E_3 + t_{36}, E_5 + t_{56}) = \max(3 + 6, 11 + 0) = 11$

Node 7 $E_7 = E_4 + t_{47}$ (we know t_{47} is a dummy activity)
$= 2 + 0 = 2$

Node 8 $E_8 = \max_{i=4,5,6,7}(E_i + t_{i8}) = \max(E_4 + t_{48}, E_5 + t_{58}, E_6 + t_{68}, E_7 + t_{78})$

$= \max(2 + 5, 11 + 8, 11 + 9, 2 + 5) = 20$

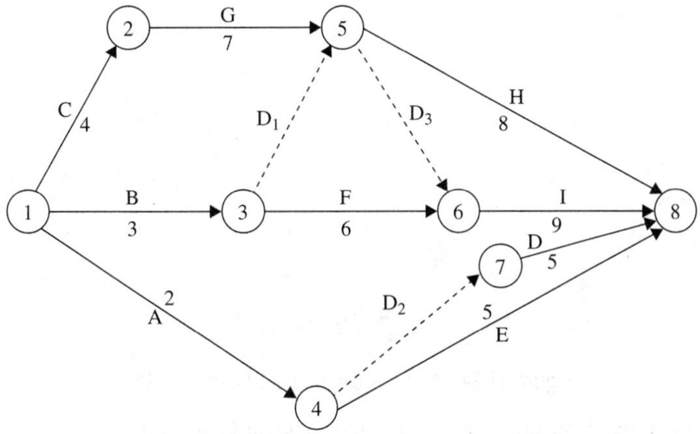

Figure 11.22 Network diagram of Example 11.10

Backward calculations

Node 8 $L_8 = E_8 = 20$

Node 7 $L_7 = L_8 - t_{78} = 20 - 5 = 15$

Node 6 $L_6 = L_8 - t_{68} = 20 - 9 = 11$

Node 5 $L_5 = \min(L_8 - t_{58}, L_6 - t_{56}) = \min(20 - 8, 11 - 10) = 11$

Node 4 $L_4 = \min_{j=7,8}(L_j - t_{4j}) = \min(L_7 - t_{47}, L_8 - t_{48}) = \min(15 - 0, 20 - 5) = 15$

Node 3 $L_3 = \min_{j=5,6}(L_j - t_{3j}) = \min(L_5 - t_{35}, L_6 - t_{36}) = \min(11 - 0, 11 - 6) = 5$

Node 2 $L_2 = L_5 - t_{25} = 11 - 7 = 4$

Node 1 $L_1 = \min_{j=2,3,4}(L_j - t_{1j}) = \min(L_2 - t_{12}, L_3 - t_{13}, L_4 - t_{14})$

$= \min(4 - 4, 5 - 3, 15 - 2) = 0$

Using the aforementioned calculations, we have Fig. 11.23, which depicts the critical path.

Here (1, 2), (2, 5), (5, 6), and (6, 8) are the critical activities. The critical path is $1 \rightarrow 2 \rightarrow 5 \rightarrow 6 \rightarrow 8$. The project duration is 20 days.

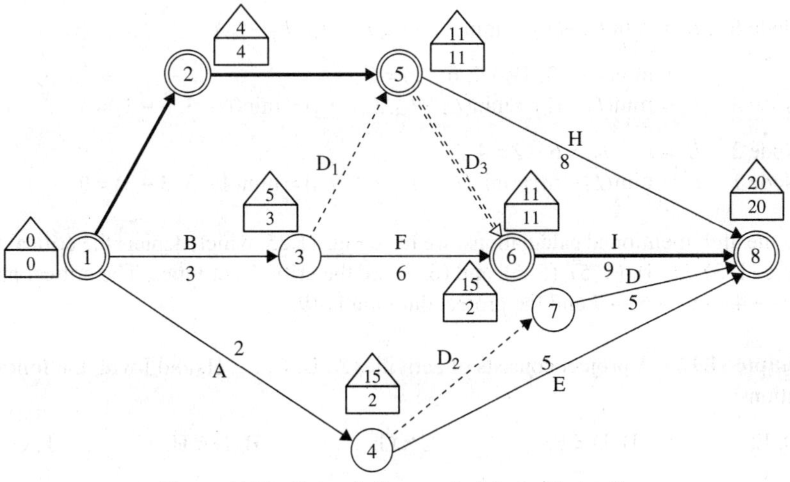

Figure 11.23 Network diagram depicting critical path

Example 11.11 Given the following network diagram (Fig. 11.24), find the critical path and project duration.

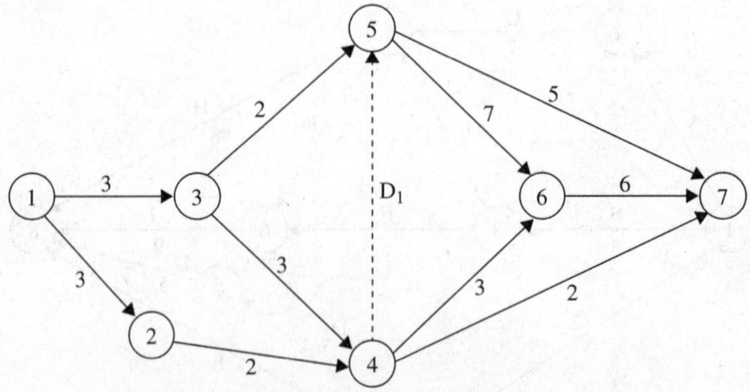

Figure 11.24 Network diagram of Example 11.11

Solution First, we determine the forward and backward calculations.

Forward calculations

Node 1 $E_1 = 0$

Node 2 $E_2 = E_1 + t_{12} = 0 + 3 = 3$

Node 3 $E_3 = E_1 + t_{13} = 0 + 3 = 3$

Node 4 $E_4 = \max_{i=2,3}(E_i + t_{i4}) = \max(E_2 + t_{24}, E_3 + t_{34}) = \max(3 + 2, 3 + 3) = 6$

Node 5 $E_5 = \max_{i=3,4}(E_i + t_{i5}) = \max(E_3 + t_{35}, E_4 + t_{45}) = \max(3 + 2, 6 + 0) = 6$

Node 6 $E_6 = \max_{i=4,5}(E_i + t_{i6}) = \max(E_4 + t_{46}, E_5 + t_{56}) = \max(6 + 3, 6 + 7) = 13$

Node 7 $E_7 = \max_{i=4,5,6}(E_i + t_{i7}) = \max(E_4 + t_{47}, E_5 + t_{57}, E_6 + t_{67})$

$= \max(6 + 2, 6 + 5, 13 + 5) = 19$

Backward calculations

Node 7 $L_7 = E_7 = 19$

Node 6 $L_6 = L_7 - t_{67} = 19 - 6 = 13$

Node 5 $L_5 = \min_{j=6,7}(L_j - t_{5j}) = \min(L_6 - t_{56}, L_7 - t_{57}) = \min(13 - 7, 19 - 5) = 6$

Node 6 $L_4 = \min_{j=5,6,7}(L_j - t_{4j}) = \min(L_5 - t_{45}, L_6 - t_{46}, L_7 - t_{47})$

$= \min(13 - 3, 19 - 2, 6 - 0) = 6$

Node 3 $L_3 = \min_{j=4,5}(L_j - t_{3j}) = \min(L_4 - t_{34}, L_5 - t_{35}) = \min(6 - 3, 6 - 2) = 3$

Node 2 $L_2 = L_3 - t_{23} = 6 - 2 = 4$

Node 1 $L_1 = \min_{j=2,3}(L_j - t_{1j}) = \min(L_2 - t_{12}, L_3 - t_{13}) = \min(4 - 3, 3 - 3) = 0$

Using the aforementioned calculations, we have Fig. 11.25, which depicts the critical path.

Here (1, 3), (3, 4), (4, 5), (5, 6), and (6, 7) are the critical activities. The critical path is $1 \to 3 \to 4 \to 5 \to 6 \to 7$ and the project duration is 19.

Example 11.12 A project consists of activities A, B, C, …, H, and I with the following conditions:

A < D, E; B, D < F; C < G; B, G < H; F, G < I

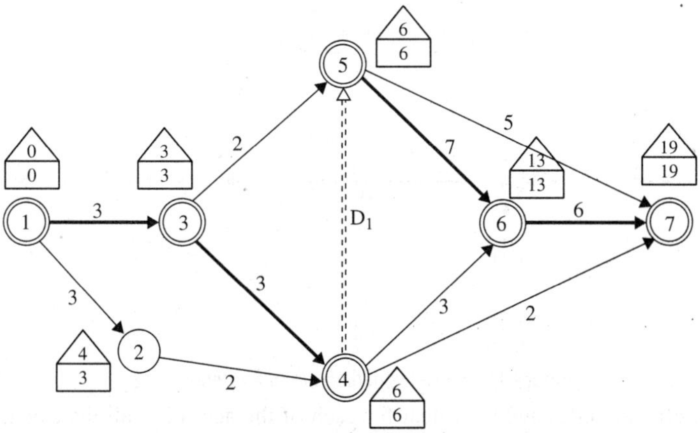

Figure 11.25 Network diagram depicting critical path

where, the notation W < X, Y means that X and Y cannot start until W is completed: X, Y < W means W cannot start until X and Y are completed. Draw the network diagram. Find the minimum time required for the completion of the project, when the time (in days) for the completion of each task is as follows:

Task	A	B	C	D	E	F	G	H	I
Time	23	8	20	16	24	18	19	4	10

Solution The network diagram for the project is shown in Fig. 11.26.

Now, to calculate the project completion time, we find the earliest time (E_i) and the latest finish time (L_j) for each event (node) of the project.

Forward calculations

Node 1 $E_1 = 0$

Node 2 $E_2 = E_1 + t_{12} = 0 + 20 = 20$

Node 3 $E_3 = E_1 + t_{13} = 0 + 23 = 23$

Node 4 $E_4 = \max\limits_{i=1,3}(E_i + t_{i4}) = \max(E_1 + t_{14}, E_3 + t_{34}) = \max(0 + 8, 23 + 16) = 39$

Node 5 $E_5 = \max\limits_{i=2,4}(E_i + t_{i5}) = \max(E_2 + t_{25}, E_4 + t_{45}) = \max(20 + 19, 39 + 0) = 39$

Node 6 $E_6 = \max\limits_{i=4,5}(E_i + t_{i6}) = \max(E_4 + t_{46}, E_5 + t_{56}) = \max(39 + 18, 39 + 0) = 57$

Node 7 $E_7 = \max\limits_{i=3,5,6}(E_i + t_{i7}) = \max(E_3 + t_{37}, E_5 + t_{57}, E_6 + t_{67})$

$= \max(23 + 24, 39 + 4, 57 + 10) = 67$

Backward calculations

Node 7 $L_7 = E_7 = 67$

Node 6 $L_6 = L_7 - t_{67} = 57$

Node 5 $L_5 = \min\limits_{j=6,7}(L_j - t_{5j}) = \min(L_6 - t_{56}, L_7 - t_{57}) = \min(57 - 0, 67 - 4) = 57$

Node 4 $L_4 = \min\limits_{j=5,6}(L_j - t_{4j}) = \min(L_5 - t_{45}, L_6 - t_{46}) = \min(57 - 0, 57 - 18) = 39$

Node 3 $L_3 = \min\limits_{j=4,7}(L_j - t_{3j}) = \min(L_4 - t_{34}, L_7 - t_{37}) = \min(39 - 16, 67 - 24) = 23$

Node 2 $L_2 = L_5 - t_{25} = 57 - 19 = 38$

Node 1 $L_1 = \min\limits_{j=2,3,4}(L_j - t_{1j}) = \min(L_2 - t_{12}, L_3 - t_{13}, L_4 - t_{14})$

$= \min(38 - 20, 23 - 23, 39 - 8) = 0$

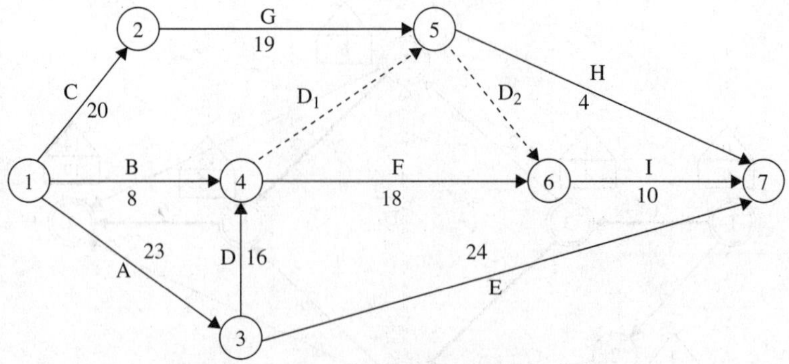

Figure 11.26 Network diagram of Example 11.12

To find the critical nodes and total float for each of the activities, all the calculations are shown in the following table:

Activity (task) (i, j)	Normal time (days) t_{ij}	Earliest time		Latest finish time L_j	Float		
		Start E_i	Finish $E_i + t_{ij}$		Total float $L_j - (E_i + t_{ij})$	Free $E_j - (E_i + t_{ij})$	Independent $E_j - (L_i + t_{ij})$
(1, 2)	20	0	20	38	18	0	0
(1, 3)	23	0	23	23	0	0	0
(1, 4)	8	0	8	39	31	31	31
(2, 5)	19	20	39	57	18	0	−18
(3, 4)	16	23	39	39	0	0	0
(3, 7)	24	23	47	67	20	20	20
(4, 5)	0	39	39	57	18	0	0
(4, 6)	18	39	57	57	0	0	0
(5, 6)	0	39	39	57	18	18	0
(5, 7)	4	39	43	67	24	24	6
(6, 7)	10	57	67	67	0	0	0

From the aforegiven table, we find that the critical nodes are (1,3), (3,4), (4,6), and (6,7).

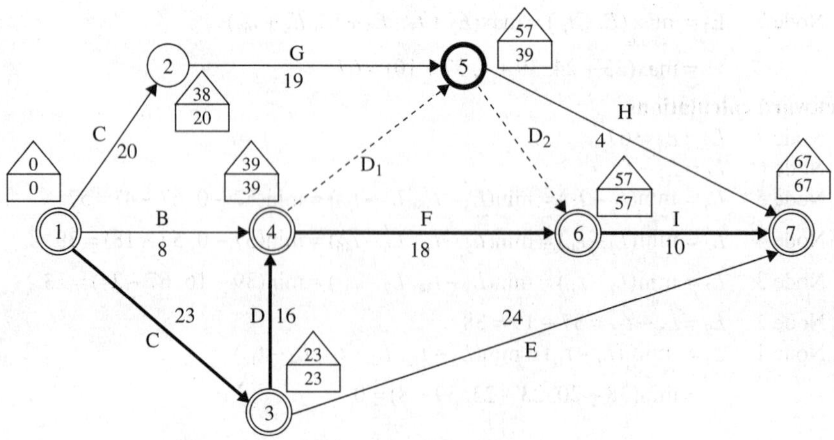

Figure 11.27 Network diagram depicting critical path

The critical path is depicted in Fig. 11.27.
The critical path is $1 \rightarrow 3 \rightarrow 4 \rightarrow 6 \rightarrow 7$, and the project completion time is 67 days.

11.5 PROGRAM EVALUATION AND REVIEW TECHNIQUE

Till now, we have only discussed projects that are deterministic. However, there are many projects for which we cannot determine the time duration of some activities. In such cases, we need a method with a probabilistic approach for scheduling. PERT was devised specifically for such projects. In PERT, a probability distribution is used for determining the schedules. There are three different time estimates made for each activity. The probability distribution of the activity time is based on these estimates. The three different time estimates are as follows:

1. *Optimistic time* (t_o): The shortest possible time to complete the activity, if the execution goes well.
2. *Pessimistic time* (t_p): The longest time that an activity could take, if the execution goes wrong.
3. *Most likely time* (t_m): The estimate of the normal time any activity would take. This estimate of time lies between the optimistic and pessimistic time.

The most likely time (t_m) estimate may not coincide with the mid-point of the optimistic time (t_o) and pessimistic time (t_p), that is, $(t_o + t_p)/2$ and may occur to its left or right as shown in Fig. 11.28.

In general, the normal distribution function is used to calculate the approximate probability density function; but here, since the end points are non-negative and the shapes are not symmetrical, we cannot use the normal distribution. Therefore, the beta distribution is used, as it fulfils the requirements. Thus, we arrive at the following:

$$\text{The expected time } T_e = \frac{t_o + 4t_m + t_p}{6}$$

$$\text{Standard deviation } \sigma_{ij} = \frac{t_p - t_o}{6}$$

$$\text{and variance } V_{ij} = \sigma_{ij}^2 = \left(\frac{t_p - t_o}{6}\right)^2$$

Now, our objective is to find the probability that the project will be completed in specific period of time. If T is the total duration of the project, then the probability of completing the project within the scheduled time, T_s is given as

$$P(T \le T_s) = P\left(\frac{T - T_e}{\sigma_t} \le \frac{T_s - T_e}{\sigma_t}\right) = P\left(Z \le \frac{T_s - T_e}{\sigma_t}\right)$$ where σ_t is the square root of the sum of variances along the critical path.

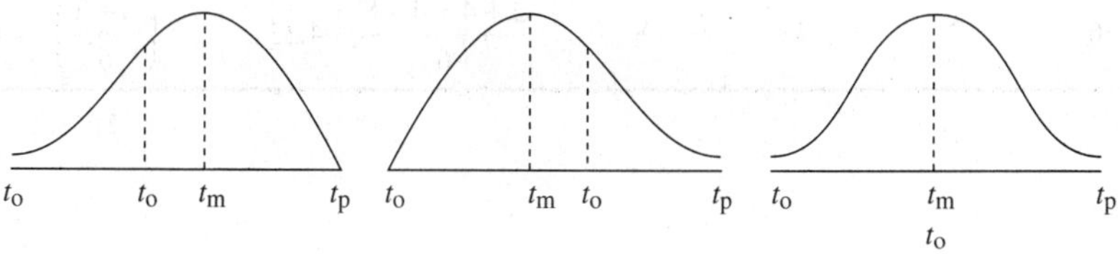

Figure 11.28 Relationship between t_o, t_m, and t_p

Where Z is the standard normal variable and T_e is the expected duration calculated along the critical path.

Example 11.13 Consider the following project.

Activity	t_o	t_m	t_p	Activity	t_o	t_m	t_p
1–2	3	5	8	3–5	3	5	9
1–3	2	4	8	4–6	3	6	10
1–4	6	8	12	5–6	2	4	8
2–5	5	9	12	–	–	–	–

(a) Draw the network diagram.
(b) Find the expected time (duration) and variance for each activity.
(c) Find the critical path and the project variance.
(d) What is the probability that the project will be completed in 22 days?

Solution We know that the expected time $T_e = \dfrac{t_o + 4t_m + t_p}{6}$ and variance $V_{ij} = \sigma_{ij}^2 = \left(\dfrac{t_p - t_o}{6}\right)^2$.

Activity	t_o	t_m	t_p	$T_e = \dfrac{t_o + 4t_m + t_p}{6}$	$V_{ij} = \sigma_{ij}^2 = \left(\dfrac{t_p - t_o}{6}\right)^2$
1–2	3	5	8	$\dfrac{3+4\times5+8}{6} = 5.167$	$\left(\dfrac{8-3}{6}\right)^2 = \dfrac{25}{36}$
1–3	2	4	8	$\dfrac{2+4\times4+8}{6} = 4.33$	$\left(\dfrac{8-2}{6}\right)^2 = 1$
1–4	6	8	12	$\dfrac{6+4\times8+12}{6} = 8.33$	$\left(\dfrac{12-6}{6}\right)^2 = 1$
2–5	5	9	12	$\dfrac{5+4\times9+12}{6} = 8.83$	$\left(\dfrac{12-5}{6}\right)^2 = \dfrac{49}{36}$
3–5	3	5	9	$\dfrac{3+4\times5+9}{6} = 5.33$	$\left(\dfrac{9-3}{6}\right)^2 = 1$
4–6	3	6	10	$\dfrac{3+4\times6+10}{6} = 6.16$	$\left(\dfrac{10-3}{6}\right)^2 = \dfrac{49}{36}$
5–6	2	4	8	$\dfrac{2+4\times4+8}{6} = 4.33$	$\left(\dfrac{8-2}{6}\right)^2 = 1$

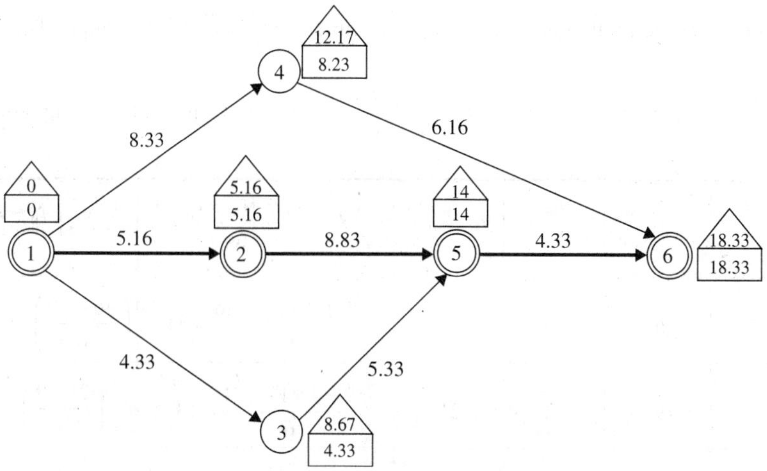

Figure 11.29 Network diagram of Example 11.13

(a) The network diagram for the project is shown in Fig. 11.29.
(b) The expected time T_e for the completion of the project is 18.33 days.
(c) The critical path is $1 \rightarrow 2 \rightarrow 5 \rightarrow 6$.

Variance along the project length critical path is

$$\sigma_t^2 = \frac{25}{36} + \frac{49}{36} + 1 = \frac{75}{36} = 2.08333$$

$$\text{or, } \sigma_t = 1.4433$$

$$\text{Now we have } Z = \frac{T_s - T_e}{\sigma_t}$$

$$Z = \frac{22 - 18.33}{1.4433} = \frac{3.67}{1.4433} = 2.540$$

Using the normal distribution table (for $Z = 2.54$), we have probability = 98.38%.

Example 11.14 Consider the following project.

Activity	t_o	t_m	t_p	Activity	t_o	t_m	t_p
1–2	20	30	46	2–4	2	3	4
1–3	9	12	21	3–5	1	2	3
2–3	3	5	7	4–5	12	18	24

(a) Draw the network diagram and find the total project duration.
(b) Find the critical path.
(c) Find the probability that the project will be completed in 50 days.

Solution The computation of the expected time $T_e = \dfrac{t_o + 4t_m + t_p}{6}$, standard deviation $\sigma_{ij} = \dfrac{t_p - t_o}{6}$, and variance $V_{ij} = \sigma_{ij}^2 = \left(\dfrac{t_p - t_o}{6}\right)^2$ are shown in the following table:

Activity	t_o	t_m	t_p	$T_e = \dfrac{t_o + 4t_m + t_p}{6}$	$V_{ij} = \left(\dfrac{t_p - t_o}{6}\right)^2$
1–2	20	30	46	$\dfrac{20+4\times30+46}{6} = 31$	$\left(\dfrac{46-20}{6}\right)^2 = 18.77$
1–3	9	12	21	$\dfrac{9+4\times12+21}{6} = 13$	$\left(\dfrac{21-9}{6}\right)^2 = 4$
2–3	3	5	7	$\dfrac{3+4\times5+7}{6} = 5$	$\left(\dfrac{7-3}{6}\right)^2 = 0.44$
2–4	2	3	4	$\dfrac{2+4\times3+4}{6} = 3$	$\left(\dfrac{4-2}{6}\right)^2 = 0.11$
3–4	1	2	3	$\dfrac{1+4\times2+3}{6} = 2$	$\left(\dfrac{3-1}{6}\right)^2 = 0.11$
4–5	12	18	24	$\dfrac{12+4\times18+24}{6} = 18$	$\left(\dfrac{24-12}{6}\right)^2 = 4$

The network diagram for the project is shown in Fig. 11.30.

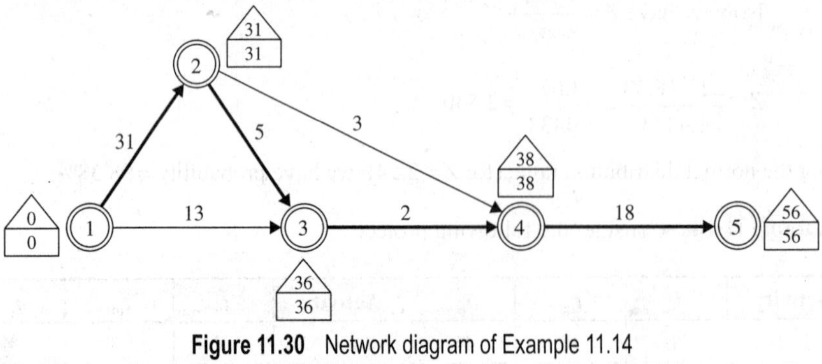

Figure 11.30 Network diagram of Example 11.14

The critical path is $1 \to 2 \to 3 \to 4 \to 5$.
Variance along the critical path

$$\sum V_{ij} = 18.77 + 0.44 + 0.11 + 4 = 23.32$$

The project duration time is 56 days.

Hence $$Z = \frac{T_s - T_e}{\sigma_t}$$

$$Z = \frac{50 - 56}{\sqrt{23.32}} = -1.24247$$

Using the normal distribution table (for $Z = -1.24247$), we have probability $(P) = 10.75\%$.

Example 11.15 A project consists of activities A, B, C, D, E, F, G, and H with the following details:

Activity	Immediate predecessor	t_o	t_m	t_p
A	–	1	1	7
B	–	1	4	7
C	–	2	2	8
D	A	1	1	1
E	B	2	5	14
F	C	2	5	8
G	D, E	3	6	15
H	F, G	1	2	3

(a) Draw the PERT network diagram.
(b) Find out the expected project duration.
(c) What duration will have 95% confidence for project completion?
(d) If the average duration for activity F increases to 14 days, what will be its effect on the expected project completion time, which has 95% confidence?

Solution The computations of the expected time $T_e = \frac{t_o + 4t_m + t_p}{6}$, standard deviation $\sigma_{ij} = \frac{t_p - t_o}{6}$, and variance $V_{ij} = \sigma_{ij}^2 = \left(\frac{t_p - t_o}{6}\right)^2$ are shown in following table:

Activity	t_o	t_m	t_p	$T_e = \frac{t_o + 4t_m + t_p}{6}$	$V_{ij} = \sigma_{ij}^2 = \left(\frac{t_p - t_o}{6}\right)^2$
1–2	1	1	7	$\frac{1+4\times1+7}{6} = 2$	$\left(\frac{7-1}{6}\right)^2 = 1$
1–3	1	4	7	$\frac{1+4\times4+7}{6} = 4$	$\left(\frac{7-1}{6}\right)^2 = 1$
1–4	2	2	8	$\frac{2+4\times2+8}{6} = 3$	$\left(\frac{8-2}{6}\right)^2 = 1$
2–4	1	1	1	$\frac{1+4\times1+1}{6} = 1$	$\left(\frac{1-1}{6}\right)^2 = 0$
3–5	2	5	14	$\frac{2+4\times5+14}{6} = 6$	$\left(\frac{14-2}{6}\right)^2 = 4$
4–6	2	5	8	$\frac{2+4\times5+14}{6} = 5$	$\left(\frac{8-2}{6}\right)^2 = 1$
5–6	3	6	15	$\frac{3+4\times6+15}{6} = 7$	$\left(\frac{15-3}{6}\right)^2 = 4$
6–7	1	2	3	$\frac{1+4\times2+3}{6} = 2$	$\left(\frac{3-1}{6}\right)^2 = 1/9$

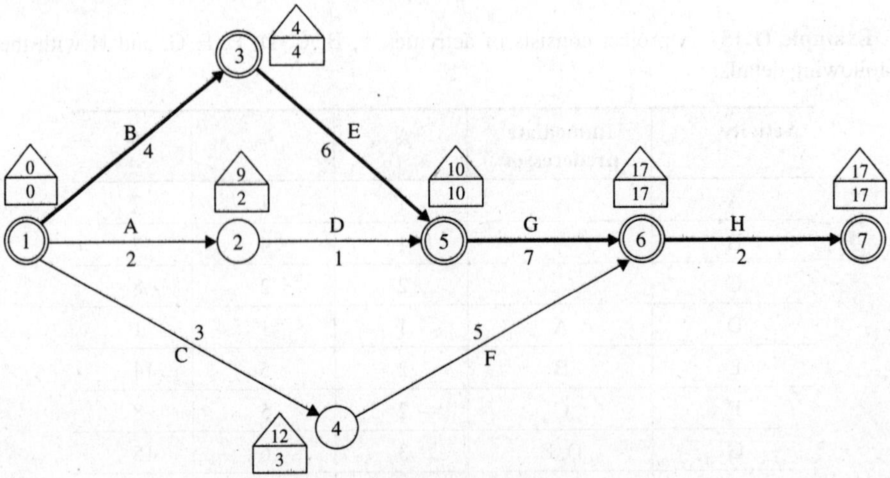

Figure 11.31 Network diagram of Example 11.15

(a) The network diagram or the project is shown in Fig. 11.31.
The critical path is $1 \rightarrow 3 \rightarrow 5 \rightarrow 6 \rightarrow 7$.

(b) The expected duration of the project is 19 days.

(c) Variance along the project length $\sigma_t^2 = 1 + 4 + 4 + 1/9 = 9.108$ or $\sigma_t = 3.02$
Now we have, $Z = \dfrac{T_s - T_e}{\sigma_t}$
For 95% probability, we have $Z = 1.65$ (using the normal distribution table)
$T_s = T_e + Z\sigma_t = 19 + (1.65 \times 3.02) = 24$ days

(d) If the average duration of activity F (4, 6) increases to 14, then the path $C \rightarrow F \rightarrow H$ also becomes a critical path.
The standard deviation of the $C \rightarrow F \rightarrow H$ path is $\sigma_t = \sqrt{1 + 9.108 + 1} = 3.36$.
For 95% probability, we have $Z = 1.65$
$T_s = 19 + (1.65 \times 3.36) = 24.52$ days.

Example 11.16 An industrial project is expected to take 12 months for its completion, having a standard deviation of 4 months. What is the probability of completing the project within (a) 12 months? (b) 16 months? (c) 8 months?

Solution (a) The probability factor is given by $Z = \dfrac{T_s - T_e}{\sigma_t} = \dfrac{12 - 12}{4} = 0$

From the normal distribution table (for $Z = 0$), we have probability (P) = 50%

(b) For $T_s = 16$ months, we have, $Z = \dfrac{16 - 12}{4} = 1$.

For $Z = 1$, we have probability (P) = 84.13%

(c) For $T_s = 8$ months, we have, $Z = \dfrac{8 - 12}{4} = -1$.

For $Z = -1$, we have probability (P) = 15.87%.

Example 11.17 Consider the following table:

(a) Find the values of the expected time (t_e), standard deviation (σ_t), and variance (V_{ij}) for each activity.
(b) Draw the network diagram.
(c) Find the earliest start, latest finish time, and total float for each activity.
(d) Find the critical path and project duration.

Activity	t_o	t_m	t_p	Activity	t_o	t_m	t_p
1–2	2	5	8	5–6	7	7	7
2–3	1	1	1	1–4	3	3	3
3–5	0	6	18	4–5	2	8	14

Solution (a) We know that the expected time $T_e = \dfrac{t_o + 4t_m + t_p}{6}$

Standard deviation $\sigma_{ij} = \dfrac{t_p - t_o}{6}$

Variance $V_{ij} = \sigma_{ij}^2 = \left(\dfrac{t_p - t_o}{6}\right)^2$

The computations for T_e, σ_{ij}, and σ_{ij}^2 are shown in the following table:

Activity	t_o	t_m	t_p	$T_e = \dfrac{t_o + 4t_m + t_p}{6}$	$\sigma_{ij} = \dfrac{t_p - t_o}{6}$	$V_{ij} = \sigma_{ij}^2 = \left(\dfrac{t_p - t_o}{6}\right)^2$
1–2	2	5	8	$\dfrac{2+4\times5+8}{6} = 5$	1	1
2–3	1	1	1	$\dfrac{1+4\times1+1}{6} = 1$	0	0
3–5	0	6	18	$\dfrac{0+4\times6+18}{6} = 7$	3	9
5–6	7	7	7	$\dfrac{7+4\times7+7}{6} = 7$	0	0
1–4	3	3	3	$\dfrac{3+4\times3+3}{6} = 3$	0	0
4–5	2	8	14	$\dfrac{2+4\times8+14}{6} = 8$	2	4

(b) The network diagram for the project is shown in Fig. 11.32.

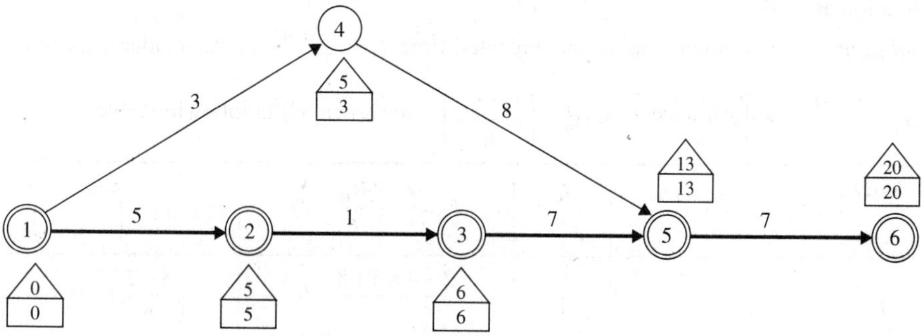

Figure 11.32 Network diagram of Example 11.17

(c)

Activity	Normal time t_{ij}	Earliest time		Latest finish time L_j	Total float (Slack) $L_j-(E_i+t_{ij})$
		Start E_i	Finish E_i+t_{ij}		
1–2	5	0	5	5	0
2–3	1	5	6	6	0
3–5	7	6	13	13	0
5–6	7	13	20	20	0
1–4	3	0	3	5	2
4–5	8	3	11	13	2

(d) The critical path is $1 \to 2 \to 3 \to 5 \to 6$. The project duration is 20 days.

Example 11.18 Consider the following project (Fig. 11.33).

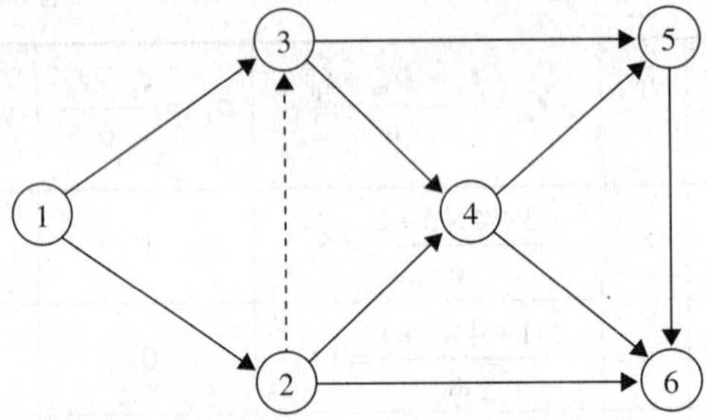

Figure 11.33 Network diagram of Example 11.18

Activity	t_o	t_m	t_p	Activity	t_o	t_m	t_p
(1, 2)	2	5	8	(2, 4)	2	4	6
(1, 3)	1	4	7	(2, 6)	5	7	12
(2, 3)	0	0	0	(4, 5)	4	6	10
(3, 4)	3	5	10	(4, 6)	2	5	8
(3, 5)	3	6	9	(5, 6)	2	4	6

Find the critical path and the expected time duration. Determine the variance of the time duration as well.

Solution The computations of the expected time $T_e = \dfrac{t_o + 4t_m + t_p}{6}$, standard deviation $\sigma_{ij} = \dfrac{t_p - t_o}{6}$, and variance $V_{ij} = \sigma_{ij}^2 = \left(\dfrac{t_p - t_o}{6}\right)^2$ are shown in the following table:

Activity	t_o	t_m	t_p	$T_e = \dfrac{t_o + 4t_m + t_p}{6}$	$V_{ij} = \sigma_{ij}^2 = \left(\dfrac{t_p - t_o}{6}\right)^2$
1–2	2	5	8	$\dfrac{2+4\times5+8}{6} = 5$	$\left(\dfrac{8-2}{6}\right)^2 = 1$
1–3	1	4	7	$\dfrac{1+4\times4+7}{6} = 4$	$\left(\dfrac{7-1}{6}\right)^2 = 1$

(*Contd*)

(*Contd*)

Activity	t_o	t_m	t_p	$T_e = \frac{t_o + 4t_m + t_p}{6}$	$V_{ij} = \sigma_{ij}^2 = \left(\frac{t_p - t_o}{6}\right)^2$
2–3	0	0	0	0	0
3–4	3	5	10	$\frac{3+4\times5+10}{6} = 5.5$	$\left(\frac{10-3}{6}\right)^2 = \frac{49}{36}$
3–5	3	6	9	$\frac{3+4\times6+9}{6} = 6$	$\left(\frac{9-3}{6}\right)^2 = 1$
2–4	2	4	6	$\frac{2+4\times4+6}{6} = 4$	$\left(\frac{6-2}{6}\right)^2 = \frac{16}{36}$
2–6	5	7	12	$\frac{5+4\times7+12}{6} = 7.5$	$\left(\frac{12-5}{6}\right)^2 = \frac{49}{36}$
4–5	4	6	10	$\frac{4+4\times6+10}{6} = 6.33$	$\left(\frac{10-4}{6}\right)^2 = 1$
4–6	2	5	8	$\frac{2+4\times5+8}{6} = 5$	$\left(\frac{8-2}{6}\right)^2 = 1$
5–6	2	4	6	$\frac{2+4\times4+6}{6} = 4$	$\left(\frac{6-2}{6}\right)^2 = \frac{16}{36}$

The network diagram for the project is shown in Fig. 11.34.

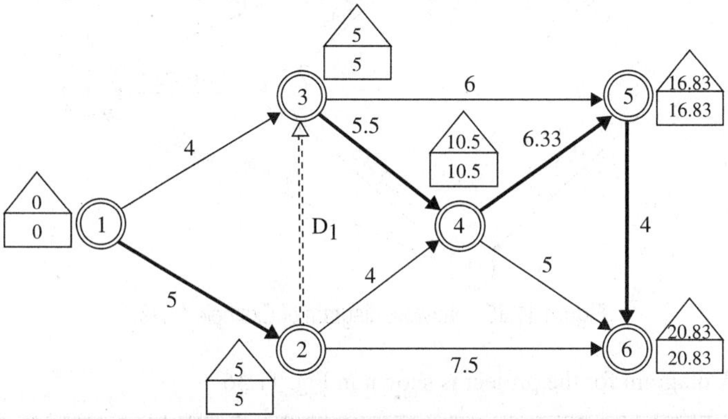

Figure 11.34 Network diagram depicting critical path

The critical path is $1 \rightarrow 2 \rightarrow 3 \rightarrow 4 \rightarrow 5 \rightarrow 6$, the expected time is 20.83, and the variance along the project length $\sigma_t^2 = 1 + 0 + \frac{49}{36} + 1 + \frac{16}{36} = \frac{16}{36} = 3.8055$.

Example 11.19 For the network diagram in Fig. 11.35, the time estimates $t_o - t_m - t_p$ (in days) for each activity are indicated. Find the expected time, standard deviation, and variance for each activity. In addition, find the project duration, the critical path, and the probability of completing the project in 32 days.

Solution We know that the expected time $T_e = \frac{t_o + 4t_m + t_p}{6}$, standard deviation $\sigma_{ij} = \frac{t_p - t_o}{6}$, and variance $V_{ij} = \sigma_{ij}^2 = \left(\frac{t_p - t_o}{6}\right)^2$.

Activity	t_o	t_m	t_p	$T_e = \frac{t_o + 4t_m + t_p}{6}$	$\sigma_{ij} = \frac{t_p - t_o}{6}$	$V_{ij} = \sigma_{ij}^2 = \left(\frac{t_p - t_o}{6}\right)^2$
1–2	6	9	18	$\frac{6+4\times9+16}{6} = 10$	2	4
1–3	5	8	17	$\frac{5+4\times8+17}{6} = 9$	2	4
2–4	4	7	22	$\frac{4+4\times7+22}{6} = 9$	3	9
3–4	4	7	16	$\frac{4+4\times7+16}{6} = 8$	2	4
4–5	4	10	22	$\frac{4+4\times10+22}{6} = 11$	3	9
2–5	4	7	10	$\frac{4+4\times7+10}{6} = 7$	1	1
3–5	2	5	8	$\frac{2+4\times5+8}{6} = 5$	1	1

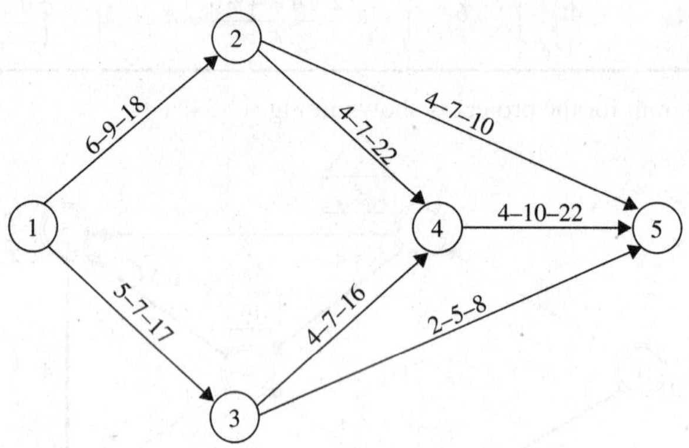

Figure 11.35 Network diagram of Example 11.19

The network diagram for the project is shown in Fig. 11.36

Activity	Normal time t_{ij}	Earliest time		Latest finish time L_j	Float $L_j - (E_i + t_{ij})$
		Start E_i	Finish $E_i + t_{ij}$		
1–2	10	0	10	10	0
1–3	9	0	9	11	2
2–4	9	10	19	19	0
3–4	8	9	17	19	2
4–5	11	19	30	30	0
2–5	7	10	17	30	13
3–5	5	9	14	30	16

The critical path is $1 \rightarrow 2 \rightarrow 4 \rightarrow 5$. The project duration is 30 days.

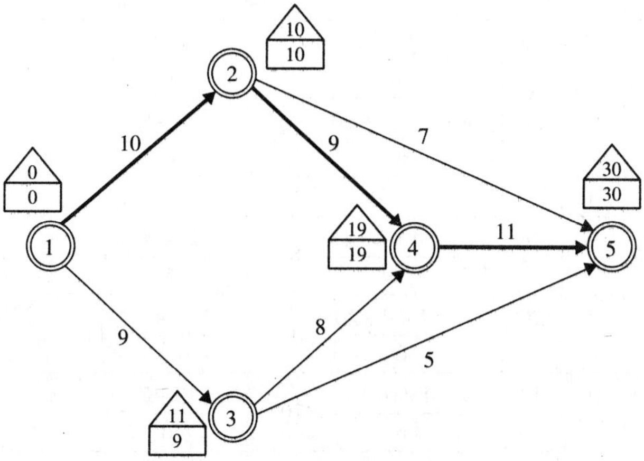

Figure 11.36 Network diagram depicting critical path

The variance along the critical path is determined as follows:

$$\sum V_{ij} = \sigma^2_{(1-2)} + \sigma^2_{(2-4)} + \sigma^2_{(4-5)} = 4 + 9 + 9 = 22$$

standard deviation along the critical path $\sigma_t = \sqrt{22} = 4.69$

$$\text{Hence, } Z = \frac{T_s - T_e}{\sigma_t} = \frac{32 - 30}{4.69} = 0.42$$

Thus, from the normal distribution table, we have $P = 66\% = 0.66$.

Example 11.20 For the project network shown in Fig. 11.37, find the following:

(a) The critical path and the variance.
(b) The probability of the completion of the project in 40 days.
(c) The time duration that will provide 95% probability of its completion time.

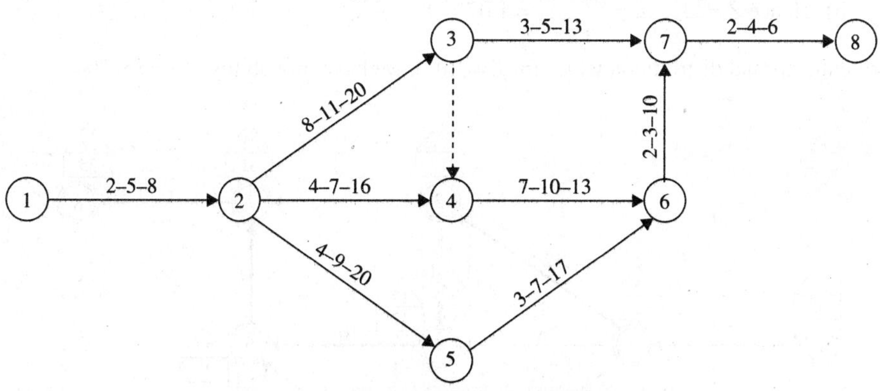

Figure 11.37 Network diagram of Example 11.20

Solution (a) The computations of the expected time $T_e = \dfrac{t_o + 4t_m + t_p}{6}$, standard deviation $\sigma_{ij} = \dfrac{t_p - t_o}{6}$, and variance $V_{ij} = \sigma_{ij}^2 = \left(\dfrac{t_p - t_o}{6}\right)^2$ are shown in the following table:

Activity	t_o	t_m	t_p	$T_e = \frac{t_o + 4t_m + t_p}{6}$	$\sigma_{ij} = \frac{t_p - t_o}{6}$	$V_{ij} = \sigma_{ij}^2$
1–2	2	5	8	$\frac{2+4\times5+8}{6}=5$	$\frac{8-2}{6}=1$	1
2–3	8	11	20	$\frac{8+4\times11+20}{6}=12$	$\frac{20-8}{6}=4$	4
3–4	0	0	0	0	0	0
2–4	4	7	16	$\frac{4+4\times7+16}{6}=8$	$\frac{16-4}{6}=4$	4
2–5	4	9	20	$\frac{4+4\times9+20}{6}=10$	$\frac{20-4}{6}=7.11$	7.11
4–6	7	10	13	$\frac{7+4\times10+13}{6}=10$	$\frac{13-7}{6}=1$	1
5–6	3	7	17	$\frac{3+4\times7+17}{6}=8$	$\frac{17-3}{6}=5.44$	5.44
3–7	3	5	13	$\frac{3+4\times5+13}{6}=6$	$\frac{13-3}{6}=2.78$	2.78
6–7	2	3	10	$\frac{2+4\times3+10}{6}=4$	$\frac{10-2}{6}=1.77$	1.77
7–8	2	4	6	$\frac{2+4\times4+6}{6}=4$	$\frac{6-2}{6}=0.44$	0.44

The network diagram for the project is shown in Fig. 11.38.

The critical path is $1 \to 2 \to 3 \to 4 \to 6 \to 7 \to 8$. The project duration is 35 days.

The variance along the critical path $\sum V_{ij} = 1 + 4 + 0 + 7.11 + 1.77 + 0.44 = 14.32$.

The standard derivation along the critical path $\sigma_t = \sqrt{14.32} = 3.78$.

(b) Hence $Z = \frac{T_s - T_e}{\sigma_t} = \frac{40-35}{3.78} = 1.07$

From the normal distribution table (for $Z = 1.07$), we have probability $(P) = 85.7\%$

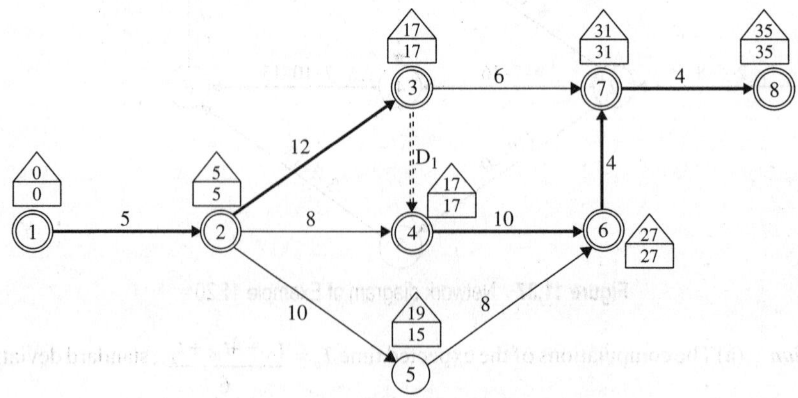

Figure 11.38 Network diagram depicting critical path

For 95% probability, we have $Z = 1.65$.

(c) Now, we have $Z = \dfrac{T_s - T_e}{\sigma_t}$.

Therefore, $T_s = T_e + Z\sigma_t = 35 + (3.78 \times 1.65) = 41.2$ days.

11.6 OPTIMUM SCHEDULING BY CRITICAL PATH METHOD

Every organization aims to reduce the total time of a project and accomplish the desired objective at the minimum cost. Organizers always remain conscious of time as well as cost, as a considerable amount is spent for completing a project with various activities. The optimum scheduling by CPM encompasses the study of the time–cost relationship. Every project involves two types of cost—direct cost and indirect cost. Direct costs include costs for materials, manpower loading, and equipment utilized and indirect costs include insurance, bonuses, administrative costs, and supervision costs. Direct costs are associated with individual activities and indirect costs are expenditures that cannot be allocated for individual activities of a project. The total cost of any project is the sum total of the direct and indirect costs.

An optimum project schedule implies the lowest possible cost along with the associated time for the project. If the direct cost of a project is increased and the indirect cost is decreased, then the time of the project is reduced. If the time is reduced, then the cost would be increased. However, there is a limit for this reduction. The time which cannot be further reduced is called the 'crash time'; otherwise it is called the normal time. The cost of completing an activity in normal time is called the 'normal cost' and that in crash time is called the 'crash cost'.

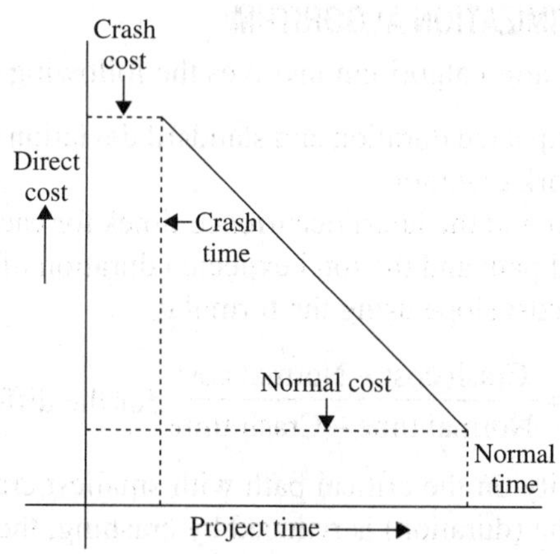

Figure 11.39 Relationship between project time and direct cost

Figure 11.39 depicts the relationship between the project time and direct cost, Fig. 11.40 depicts the relationship between the project time and the indirect cost, and Fig. 11.41 depicts the relationship between the project time and the total cost.

$$\text{The crash cost slope is } = \frac{\text{Crash cost} - \text{Normal cost}}{\text{Normal time} - \text{Crash time}}$$

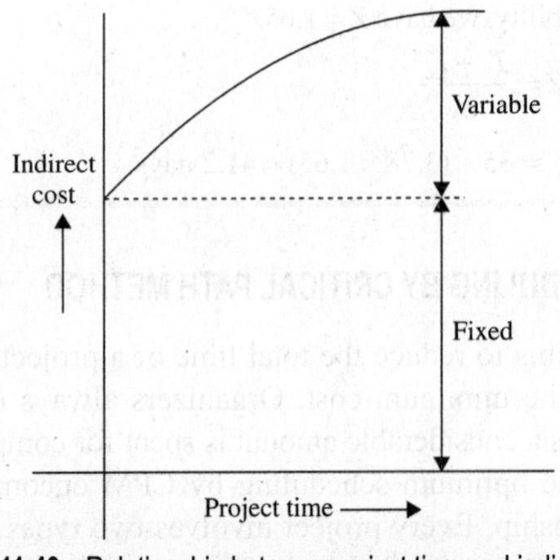

Figure 11.40 Relationship between project time and indirect cost

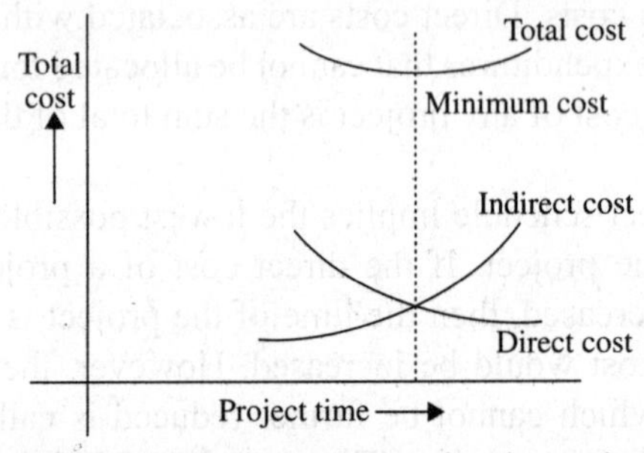

Figure 11.41 Relationship between project time and total cost

11.7 TIME–COST OPTIMIZATION ALGORITHM

The time–cost optimization algorithm involves the following steps:

1. Calculate the expected duration and standard deviation for all the activities.
2. Draw the network diagram.
3. Find the earliest and the latest occurrence times for each node.
4. Find the critical path and the total expected duration of the project.
5. Determine the cost slope using the formula:

$$\text{Crash slope} = \frac{\text{Crash cost} - \text{Normal cost}}{\text{Normal time} - \text{Crash time}} \text{ for the different activities.}$$

6. Crash the activity on the critical path with smallest crash slope. When the critical path time (duration) is reduced by crashing, then additional critical paths may be developed. Now, crash the activities on the critical paths.
7. Repeat step 6 till further crashing is not possible.
8. Now, calculate the crash duration of the project and the total project cost by adding the direct and indirect costs.

Example 11.21 Consider the following project with the crash and normal durations as well as the cost involved:

Activity	Normal		Crash	
	Duration (days)	Cost	Duration (days)	Cost
1–2	4	140	1	230
1–3	3	140	1	160
1–4	3	200	1	240
2–5	5	100	2	200
3–6	2	50	1	80
4–6	10	150	9	180
5–6	7	200	5	250

Find the different minimum cost schedule by compressing the project duration by one unit at a time.

Solution Using the normal time duration, we have Fig. 11.42.

Here, the critical path is $1 \rightarrow 2 \rightarrow 5 \rightarrow 6$ and the total cost = 980.

$$\text{The crash cost slope is } = \frac{\text{Crash cost} - \text{Normal cost}}{\text{Normal time} - \text{Crash time}}$$

To obtain the optimum cost schedule, we first determine the cost slope of activities, as follows:

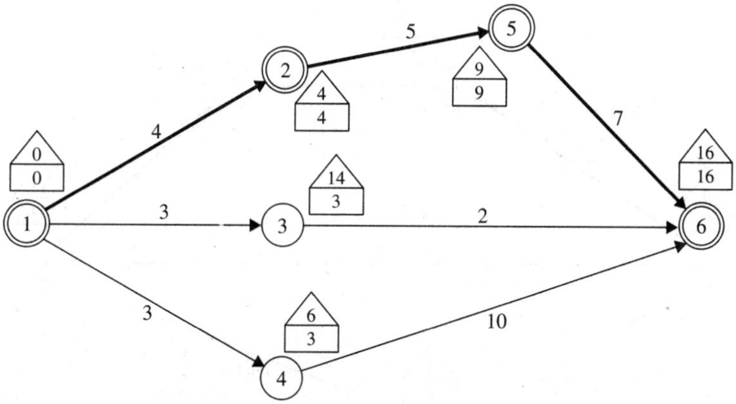

Figure 11.42 Network diagram of Example 11.21

Activities	Crash cost slope	Activities	Crash cost slope
1–2	$\frac{230-140}{4-1} = 30$	3–6	$\frac{80-50}{2-1} = 30$
1–3	$\frac{160-140}{3-1} = 10$	4–6	$\frac{180-150}{10-9} = 30$
1–4	$\frac{240-200}{3-1} = 20$	5–6	$\frac{250-200}{7-5} = 25$
2–5	$\frac{200-100}{5-2} = 33.33$	–	–

To reduce the duration of the activities (1, 2), (2, 5), and (5, 6), we first choose an activity from the critical activities with the minimum cost slope, that is, (5, 6), as it has a minimum cost slope of 25. So, we compress the (5, 6) activity by one unit and obtain Fig. 11.43:

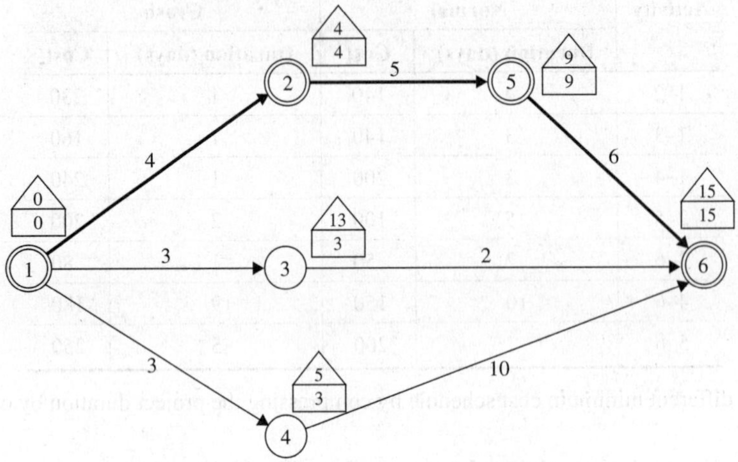

Figure 11.43 Network diagram after compression-I

Here the critical path is $1 \rightarrow 2 \rightarrow 5 \rightarrow 6$, the normal project duration is 15 days, and the total cost is $980 + 25 = 1005$. There are no additional critical paths. So, we compress the activity (5, 6) by one unit and obtain Fig. 11.44:

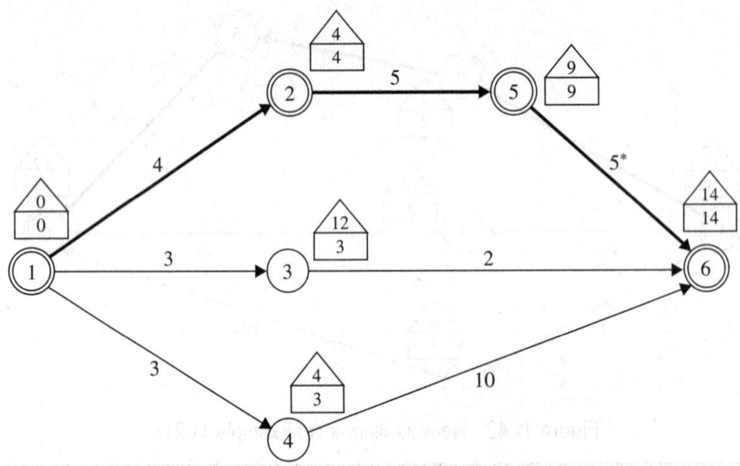

Figure 11.44 Network diagram after compression-II

Here the critical path is $1 \rightarrow 2 \rightarrow 5 \rightarrow 6$, the normal project duration is 14 days, and the total cost is $1005 + 25 = 1030$. Again, there are no more critical paths and the activity (5, 6) reaches its crash value. Hence, it cannot be reduced any further; so this duration is denoted with an * because we cannot compress an activity to less than its crash duration. Now, we scan the remaining critical activities for the next highest minimum cost slope, that is, the activity (1, 2), whose slope is 30. So we compress this activity by one unit and obtain Fig. 11.45.

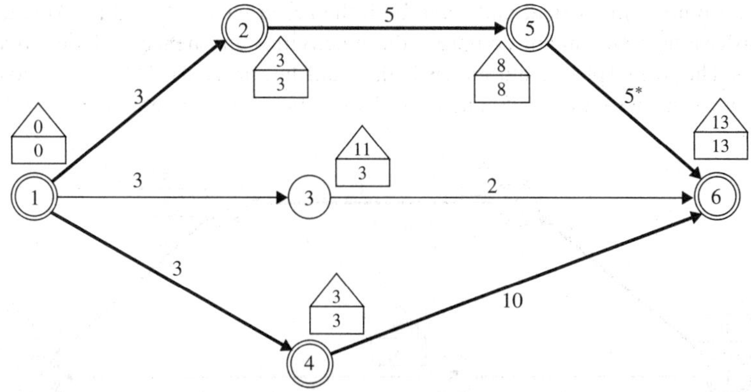

Figure 11.45 Network diagram after compression-III

Here, one more critical path has developed. The critical paths are $1 \rightarrow 2 \rightarrow 5 \rightarrow 6$ and $1 \rightarrow 4 \rightarrow 6$, the normal project duration is 13 days, and the total cost is $1030 + 30 = 1060$. Now, we compress the activities on both the paths by the same amount. So compressing the activity (1, 2) of the old path ($1 \rightarrow 2 \rightarrow 5 \rightarrow 6$) and (1, 4) of the new path ($1 \rightarrow 4 \rightarrow 6$) by one unit each, we obtain Fig. 11.46.

Here, there are no more critical paths, the normal project duration is 12 days and the total cost is $1060 + 30 + 20 = 1110$. So, we compress the (1, 2) and (1, 4) activities by one unit each and obtain Fig. 11.47.

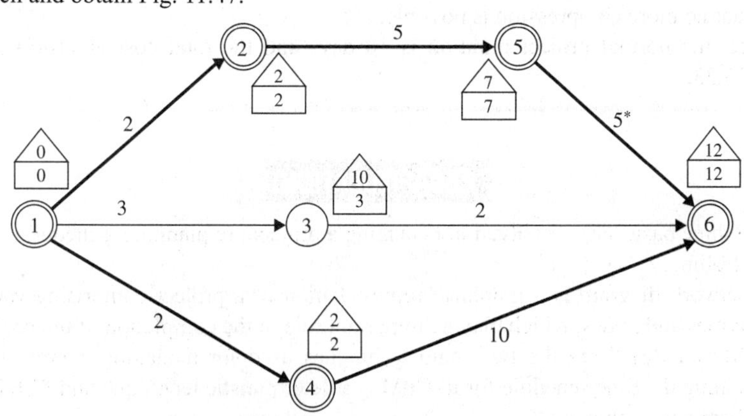

Figure 11.46 Network diagram after compression-IV

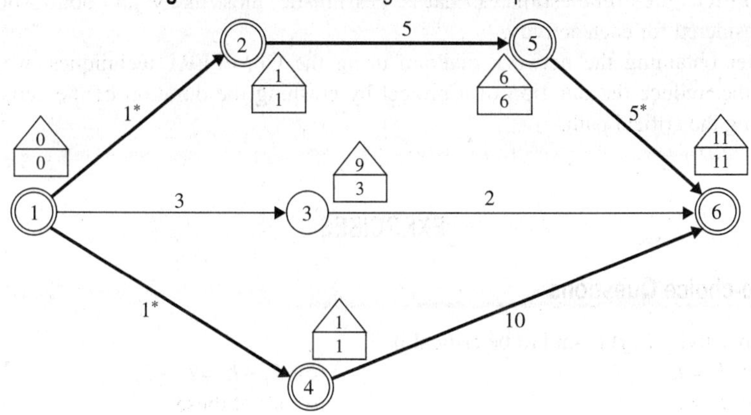

Figure 11.47 Network diagram after compression-V

There are no more critical paths and both the activities (1, 2) and (1, 4) reach their own crash values; we cannot compress these activities and hence include * over their durations. The normal project duration is 11 days and the total cost is 1110 + 30 + 20 = 1160. We compress the activities (2, 5) and (4, 6) by one unit each and obtain Fig. 11.48.

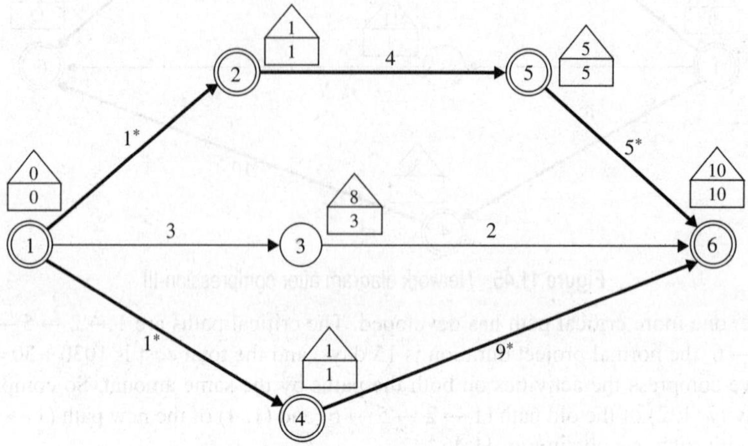

Figure 11.48 Network diagram after final compression

Again, there are no more critical paths and the (4, 6) activity reaches its crash value. So, all the activities of the critical path (1 → 4 → 6) have reached their crash values, which means that no more compression is possible.

Hence, the normal project duration is 10 days and the total cost is 1160 + 33.33 + 30 = 1223.33.

RECAPITULATION

1. The three basic steps involved in managing a project are planning, scheduling, and controlling.
2. A network diagram is a graphical representation of a project comprising various activities and nodes, which play an important role in the completion of the project.
3. CPM and PERT are the two main techniques used for managing a project and obtaining the time schedule for it. CPM is a deterministic technique and PERT is a probabilistic technique.
4. In PERT, three time estimates, that is, pessimistic, most likely, and optimistic are considered for each activity.
5. After obtaining the network diagram using the CPM/PERT techniques, we can further reduce the duration of a project by crashing the duration of the activities along the critical path.

EXERCISES

Multiple-choice Questions

11.1 An activity (i, j) is said to be critical if

(a) $E_i = L_i$
(b) $E_j = L_j$
(c) $E_j - E_i = L_j - L_i = t_{ij}$
(d) all of these

11.2 The basic steps in a project are
(a) planning and scheduling
(b) scheduling and controlling
(c) planning and controlling
(d) planning, scheduling, and controlling

11.3 CPM/PERT techniques were developed first in
(a) USA (b) UK (c) France (d) Japan

State True or False

11.1 In a network diagram, two activities can have the same starting and ending node.
11.2 In a network diagram, more than one activity can start from a node.
11.3 In a network diagram, not more than one activity can end at a node.
11.4 In a network diagram, the starting and ending nodes are unique.

Fill in the Blanks

11.1 CPM is a __________ technique for managing a project.
11.2 PERT is a __________ technique for managing a project.
11.3 In a network diagram, the starting and ending point of an activity is called __________.
11.4 The float of an event is the difference between it's __________ time and __________ time.

Review Questions

11.1 What is a network diagram?
11.2 Define the following terms:
(a) Activity
(b) Event
(c) Critical activity
(d) Critical path
(e) Crash duration
(f) Crash cost

11.3 Explain the critical path method.
11.4 Write a short note on floats associated with the project activity.
11.5 Explain the forward and backward pass calculations for the network diagram.
11.6 What is PERT? Define the optimistic time, pessimistic time, and most likely time and explain how you will estimate the expected time to complete the activity in the PERT technique.

Numerical Problems

11.1 A project consists of activities A, B, C, D, E, and F, satisfying the following conditions.
(a) B and C depend only on A.
(b) D depends on B but not on C.
(c) E depends on C and B.
(d) The project is completed when D and E are done.
Draw the network diagram.

11.2 A project consists of activities A, B, C, D, E, F, G, H, I, and J with conditions given in the following table. Draw the network diagram.

Activity	Predecessor activity	Duration (weeks)	Activity	Predecessor activity	Duration (weeks)
A	–	5	F	E	5
B	–	3	G	F	3
C	–	5	H	D	2
D	–	4	I	G, H	3
E	A, B	3	J	C, I	7

11.3 A project consists of activities 1, 2, 3, 4, 5, 6, 7, 8, 9, 10, and 11 with conditions given in the following table. Draw the network diagram.

Activity	Preceding activities	Activity	Preceding activities
1	–	7	5
2	1	8	7
3	1	9	4, 6
4	3	10	9, 8
5	2, 3	11	10
6	4, 5	–	–

11.4 Consider a project that consists of activities A, B, C, D, and E with the respective durations 15, 15, 10, 8, and 11 (in days), satisfying the following conditions:
(a) A and B can start simultaneously.
(b) C can start after 7 days and D after 5 days of starting A.
(c) D can start after 4 days of starting C and 7 days of starting B.
(d) E can start after A is 1/3 finished and B is completely finished.
(e) A, C, D, and E are terminal activities.
Draw the network diagram.

11.5 Consider the following project and draw the network diagram.

Activity	Immediate predecessor	Activity	Immediate predecessor
A	–	H	D, E, F
B	–	I	D
C	–	J	G
D	A	K	G
E	B, C	L	H, J
F	A	M	K
G	C	N	I, L

11.6 A project consists of activities A, B, C, D, E, F, G, H, and I with conditions given in the following table:

Activity	Predecessor	Estimated time (weeps)	Activity	Predecessor	Estimated time (weeps)
A	–	3	F	C	9
B	–	5	G	D, E	8
C	–	4	H	B	7
D	A	2	I	H, F	9
E	B	3	–	–	–

(a) Draw the network diagram.
(b) Find the critical path.
(c) Find the early start and late start times for each activity.
(d) Find the total float (slack) for each activity.

11.7 A project consists of activities A, B, C, …, G, H, and I, satisfying the following conditions:

$A < D$, $A < E$, $B < F$, $D < F$, $C < G$, $C < H$, $F < I$, $G < I$

(Where the notation $X < Y$ means that the activity X must be finished before Y can begin). Draw the network diagram for the project, and find the minimum time for the completion of the project, when the time (in days) for the completion of each task is as follows:

Task	A	B	C	D	E	F	G	H	I
Time (in days)	8	10	8	10	16	17	18	14	9

11.8 Consider the following project:
(a) Draw the network diagram of the project.
(b) Determine the total float and free-float for each activity.
(c) Find the critical path and the project duration.

Activity	Preceding activity	Time (in days)
(1, 2)	–	20
(1, 3)	–	25
(2, 3)	(1, 2)	10
(2, 4)	(1, 2)	12
(3, 4)	(1, 3), (2, 3)	5
(4, 5)	(2, 4), (3, 4)	10

11.9 Consider the following project:

Activity	Immediate predecessor	Time (days)	Activity	Immediate predecessor	Time (days)
1–2	–	5	4–6	2–4, 3–4	7
1–3	–	4	5–7	4–5	8
2–4	1–2	6	6–7	4–6	4
3–4	1–3	2	7–8	6–7, 5–7	3
4–5	2–4	1	–	–	–

Draw the network diagram. Find the earliest and latest times for each event. Find the critical path as well.

11.10 A project consists of activities a, b, c, d, e, f, g, h, i, and j with precedence relationships as given in the following table:

Activity	Node numbers	Estimated duration (days)	Activity	Node numbers	Estimated duration (days)
a	(1, 2)	2	f	(4, 6)	6
b	(2, 3)	3	g	(4, 7)	2
c	(2, 4)	5	h	(5, 8)	8
d	(3, 5)	4	i	(6, 8)	7
e	(3, 6)	1	j	(7, 8)	4

(a) Draw the network diagram.
(b) Find the earliest and latest starting and finishing times for each activity.
(c) Which activities are critical?
(d) How much float (slack) does activity (3, 5) have?

11.11 Consider the following project:

Activity	1–2	1–3	1–4	2–5	3–6	3–7	4–6	5–8	6–9	7–8	8–9
Time (in weeks)	2	2	1	4	8	5	3	1	5	4	3

(a) Draw the network diagram.
(b) Determine the total float for each activity.
(c) Find the critical path and project duration.

11.12 Consider the following project:

Activity	0–1	1–2	1–3	2–4	2–5	3–4	3–6	4–7	5–7	6–7
Duration (in days)	2	8	10	6	3	3	7	5	2	8

(a) Draw the network diagram.
(b) Find the critical path and the project duration.
(c) Determine the total, free, and independent floats.

11.13 Consider the following project:
(a) Find the critical path and the project duration.
(b) Find the earliest start time and the earliest finish time for each activity.
(c) Find the total float (slack) for each activity.

Activity	Immediate predecessors	Estimated time (weeks)
A	–	2
B	A	3
C	A	4
D	B, C	6
E	–	2
F	E	8

11.14 Consider the following project:

Activity	1	2	3	4	5	6
t_o	2	6	1	2	6	6
t_m	4	8	5	5	8	7
t_p	6	10	15	9	10	9

Find the expected duration, standard derivation, and variance for each activity.

11.15 Consider the following project:

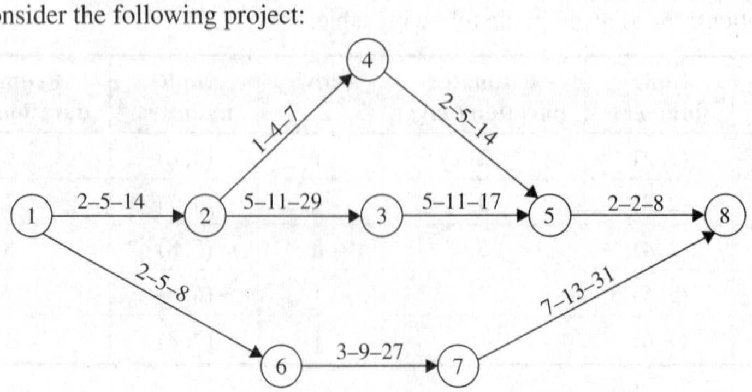

Find the critical path and the project duration.

11.16 Draw the critical path and find the project duration for the following network:

Activity	Immediate predecessor	t_o	t_m	t_p	Activity	Immediate predecessor	t_o	t_m	t_p
A	–	1	3	7	E	B	3	7	15
B	A	2	6	14	F	D, E	2	5	14
C	A	3	3	3	G	D	4	4	4
D	B, C	4	10	22	–	–	–	–	–

(a) Draw the network diagram.
(b) Find the critical path, the expected project duration completion time, and the next most critical path.
(c) What project duration will have 95% confidence of completion?

11.17 A project consists of activities A, B, C, D, E, F, G, H, and I, satisfying the following conditions:

$A < D$, $A < E$, $B < F$, $C < G$, $D < H$, $E, F < I$

(Where the notation $X<Y$ means that the activity X must be completed before Y can start, and $X, Y<W$ means that W can start only after the completion of the activities X and Y). The project has the following time schedules for the aforementioned activities:

Activity	t_o	t_m	t_p	Activity	t_o	t_m	t_p
A	5	8	10	F	6	9	12
B	18	20	22	G	7	10	12
C	26	33	40	H	7	8	9
D	16	18	20	I	3	4	5
E	15	20	25	–	–	–	–

(a) Draw the network diagram.
(b) Find the expected task times and their variance, total float, earliest expected time, latest expected time, and critical path.
(c) Find the probability of an event occurring at the proposed completion date, if the original contract time of completing the project is 41.5 weeks. What project duration will have 95% chance of being completed.

11.18 Consider the following project:

Activity	Optimistic	Most likely time	Pessimistic time
1–2	3	3	3
2–3	3	6	9
2–4	2	4	6
3–5	4	6	8
4–6	4	6	8
5–6	0	0	0
5–7	3	4	5
6–7	2	5	8

(a) Draw the network diagram.
(b) Find the variance of each activity, the critical path, and the expected project length.
(c) Find the probability that the project will be completed in 23 weeks.
It is given that

Z-value	1.90	1.91	1.92	1.93	1.44
Probability	0.9713	0.9719	0.9726	0.9732	0.9738

11.19 Consider the following project:

Activity	t_o	t_m	t_p	Activity	t_o	t_m	t_p
(1, 2)	6	8	11	(3, 6)	7	9	13
(1, 3)	19	20	23	(4, 7)	8	10	13
(1, 4)	27	33	41	(5, 7)	8	8	10
(2, 5)	17	18	21	(6, 7)	4	4	6
(2, 6)	16	20	26	–	–	–	–

Draw the network diagram. Find the critical path and the variance of the expected duration of the project.

11.20 Consider the following project whose details are as follows:

Activity	t_o	t_m	t_p	Activity	t_o	t_m	t_p
(1, 2)	3	6	15	(4, 5)	3	6	15
(1, 6)	2	5	14	(6, 7)	3	9	27
(2, 3)	6	12	30	(5, 8)	1	4	7
(2, 4)	2	5	8	(7, 8)	4	19	28
(3, 5)	5	11	17	–	–	–	–

(a) Draw the network diagram
(b) Find the critical path and the variance of the expected duration of the project

11.21 Consider the following project with crash and normal duration and cost involved:

Activity	Normal		Crash	
	Time	Cost	Time	Cost
1–2	6	1400	4	1900
1–3	8	2000	5	2800
2–3	4	1100	2	1500
2–4	3	800	2	1400
3–4	0	–	–	–
3–5	6	900	3	1600
4–6	10	2500	6	1500
5–6	3	500	2	800

Find the optimum duration and the minimum project cost, when the indirect cost for the project is ₹300 per day.

11.22 Consider the following project with crash and normal duration, and cost involved:

Activity	Normal		Crash	
	Duration (days)	Cost	Duration (days)	Cost
1–2	6	60	4	100
1–3	4	60	2	200
2–4	5	50	3	150
2–5	3	45	1	65
3–4	6	90	4	200
4–6	8	80	4	300
5–6	4	40	2	100
6–7	3	45	2	80

The indirect cost per day is ₹10.

(a) Draw the network diagram.

(b) Find the critical path.

(c) Determine the minimum total time and the corresponding cost.

11.23 Consider the following project with crash and normal duration and cost involved:

Activity	Normal		Crash	
	Time (days)	Cost (₹)	Time(days)	Cost (₹)
A (1–2)	8	1000	6	2000
B (1–3)	4	1500	2	3500
C (2–4)	2	500	1	900
D (2–5)	10	1000	5	4000
E (3–4)	5	1000	1	2000
F (4–5)	3	800	1	1000

It is given that the indirect cost involved in the project is ₹700 per day.

CONCEPTUAL PROBLEM

A construction company takes the advice of a team of engineers to plan, schedule, and control the construction of a building. The success of the construction venture depends on these three phases of the project. The first phase is planning by the engineers of the company and the second phase is the scheduling. A schedule is made for the completion of each phase for smooth operations. The schedule is to be strictly followed and is to be taken care of by the controlling authority. On the basis of the reports obtained from the engineers, the chief engineer of the construction company calls a meeting and concludes that mainly 7 activities are involved in completing the project, whose optimistic, most likely, and pessimistic durations are given in the adjacent table:

Activity	t_o	t_m	t_p
1–2	1	1	7
1–3	1	4	7
1–4	2	2	8
2–5	1	1	1
3–5	2	5	14
4–6	2	5	8
5–6	3	6	15

(Contd)

(Contd)

(a) Draw the network diagram.
(b) Find the expected duration and the variance of each activity. In addition, find the variance and the standard deviation of the project length.
(c) What is the probability that the project will be completed
  (i) at least 4 weeks later than expected?
  (ii) no more than 4 weeks later than expected?
(d) If the project due date is 19 weeks, what is the probability of meeting the due date?

Z	0.50	0.67	1.00	1.33	2.00
P	0.3085	0.2513	0.1587	0.0418	0.0228

Solution The computations of the expected time $T_e = \frac{t_o + 4t_m + t_p}{6}$, standard deviation $\sigma_{ij} = \frac{t_p - t_o}{6}$, and variance $V_{ij} = \sigma_{ij}^2 = \left(\frac{t_p - t_o}{6}\right)^2$.

Activity	t_o	t_m	t_p	$T_e = \frac{t_o + 4t_m + t_p}{6}$	$\sigma_{ij} = \left(\frac{t_p - t_o}{6}\right)$	$V_{ij} = \sigma_{ij}^2$
1–2	1	1	7	$\frac{1+4\times1+7}{6}=2$	1	1
1–3	1	4	7	$\frac{1+4\times4+7}{6}=4$	1	1
1–4	2	2	8	$\frac{2+4\times2+8}{6}=3$	1	1
2–5	1	1	1	$\frac{1+4\times1+1}{6}=1$	0	0
3–5	2	5	14	$\frac{2+4\times5+14}{6}=6$	2	4
4–6	2	5	8	$\frac{2+4\times5+8}{6}=5$	1	1
5–6	3	6	15	$\frac{3+4\times6+15}{6}=7$	2	4

(a) The network diagram (Fig. 11.49) for the project is as follows:
The critical path is $1 \rightarrow 3 \rightarrow 5 \rightarrow 6$.

(b) The expected duration of the project is 17 days.
The variance of the project length is $\sigma_t^2 = 1 + 4 + 4 = 9$. The standard deviation along the project length is $\sigma_t = 3$.

(c) We know that $Z = \frac{T_s - T_e}{\sigma_t}$

  (i) At least 4 weeks later, we have $Z = \frac{13-17}{3} = -\frac{4}{3} = -1.33$.

(Contd)

(*Contd*)

The probability for $Z = -1.33$ is 9.18% (from the normal table).

(ii) No more than 4 weeks taken: $Z = \dfrac{21-17}{3} = \dfrac{4}{3} = 1.33.$

(d) When the due date is 19 weeks: $Z = \dfrac{19-17}{3} = \dfrac{2}{3} = 0.67.$

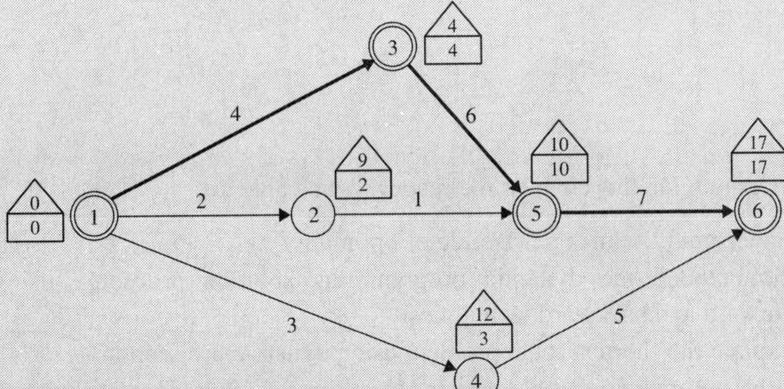

Figure 11.49 Network diagram

12 Dynamic Programming

After studying this chapter, the reader will be able to

- understand Bellman's principle of optimality
- comprehend the dynamic programming solution procedure using forward and backward techniques
- explain the shortest route problem using dynamic programming
- solve optimization problems using calculus or computational techniques based on whether the decision variables are continuous or discrete
- determine solutions for cargo loading, capital budgeting, linear programming, and reliability problems.

12.1 INTRODUCTION

In dynamic programming, the problem is broken into smaller subproblems for obtaining the optimal solution. Initially, the optimal solution of the first subproblem is obtained. To obtain the optimum solution to the second subproblem, the optimum solution of the first subproblem is used and so on. In other words, to find the optimum solution at any stage, we use the optimum solution of the immediately preceding stage. This is why it is referred to as dynamic programming.

In this chapter, we shall be considering problems that can be solved by breaking them into different parts called stages. While solving such problems, it is assumed that it is possible to split the problem into different stages and the problem at each stage is easily solvable. For this, we consider problems in the form of various models. Bellman's principle of optimality is used for obtaining the solutions to these models. In this chapter, we shall discuss the application of dynamic programming in solving problems of the shortest route, linear programming, cargo loading, capital budgeting, and reliability.

12.2 TERMINOLOGY USED IN DYNAMIC PROGRAMMING

When we solve a problem using dynamic programming, the following terms are used frequently:

Stage In dynamic programming, the problem is divided into various subproblems.

Each subproblem is called a stage. At each stage, various decision alternatives are available and out of these, the optimum decision is selected.

State At each stage, again, a number of choices are available. These choices are called states. For each such choice, there are alternatives called state variables. By analysing the effect of these state variables, the optimum decision is taken for its use in the next stage.

Optimal decision rule This rule specifies the decision to be made as a function of state variable and the stage number. This is also a policy that transforms the state at a given stage into a state associated with the next stage.

Optimal policy The rule that optimizes the value of the objective function at a particular stage is called the optimal policy.

Optimum subdivision problem Every problem cannot be solved using dynamic programming. A problem that can be solved using dynamic programming is called optimum subdivision problem.

12.3 MULTI-DECISION PROCESS

Business and industry often face situations where many decisions have to be taken to solve a problem. However, it is not possible to take all these decisions simultaneously. Decisions are taken one by one at different stages. The problems where decisions have to be taken at different stages are called multistage decision problems. A multistage decision problem can be solved using a mathematical technique called 'dynamic programming' on the basis of the principle developed by Richard Bellman in the early 1950s. This technique is called Bellman's principle of optimality.

12.4 BELLMAN'S PRINCIPLE OF OPTIMALITY

'*Whatever the initial state and the initial decisions are, the remaining decisions must constitute an optimal policy with regard to the state resulting from the first decision.*'

In the context of dynamic programming, a multistage process is one in which a number of single-stage processes, considered as a function of a single variable, are connected so that the optimum solution of one stage is used in the succeeding stage.

Thus, according to Bellman's principle of optimality, the problem of decision-making must be split into different stages and optimal decisions have to be taken sequentially. The various iterations of a decision-making process are shown in Fig. 12.1.

In Fig. 12.1, we observe that the entire problem is divided into smaller subproblems called stages. The optimum solution of the first subproblem is obtained. Then this optimum solution along with the second subproblem is considered to determine an optimum solution of the resulting problem. Proceeding this way, we sequentially obtain the solution to the given problem by breaking it into smaller subproblems (which are easy to solve).

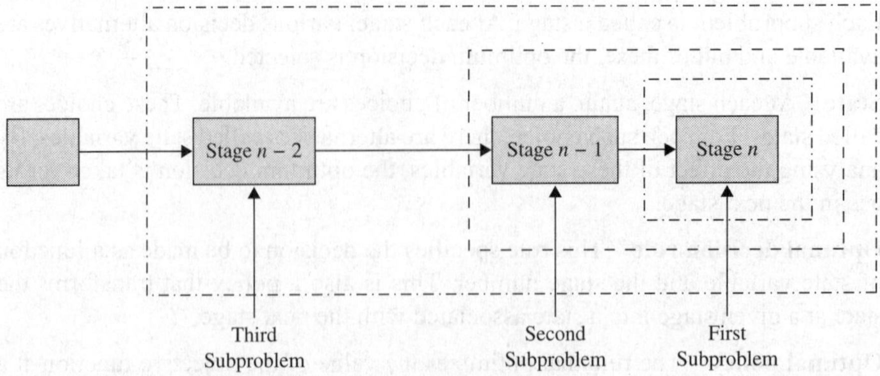

Figure 12.1 Division of optimization problem in stages

12.5 CHARACTERISTICS OF DYNAMIC PROGRAMMING PROBLEMS

The following are the characteristics of a dynamic programming problem:

1. In a dynamic programming problem, the given problem is divided into a number of smaller subproblems called stages, which are solved one by one for optimum returns at each stage.
2. The optimum return arrived at, at each stage, is used in obtaining the optimum returns in the succeeding stage.
3. In a dynamic programming problem, recursive relations are established, which, along with state transformations, are used to correlate the optimal policy at stage n to stage $(n - 1)$.
4. The solution procedure in dynamic programming moves from stage to stage, each time finding an optimal policy for each state at that stage until the optimal policy for the last stage is found.
5. In dynamic programming, once the n stage optimal policy is completed, the n component decision vector can be recovered by tracing back through transition functions.

12.6 DYNAMIC PROGRAMMING ALGORITHM

For dynamic programming, the iterative procedure is as follows:

1. Identify the decision variables for the problem.
2. Specify the objective function and the constraints present in the problem.
3. Divide the problem into smaller subproblems or stages.
4. Identify the state variables for each subproblem or stage.
5. Write the transformation function as a function of the state and decision variables in the next stage.
6. For computation of an optimal policy, write a general recursive relationship.
7. Decide on whether to follow the backward or the forward method for solving a problem.
8. Construct proper tables to indicate the return function values required at each stage.

9. Determine the overall optimal policy or decision and its value at each stage. There may be more than one such optimal policy.

12.7 DETERMINISTIC AND PROBABILISTIC DYNAMIC PROGRAMMING

Deterministic dynamic problems are those in which the state and policy decision of the current state decide the state at the next stage. In other words, the state of next stage is deterministic in nature. Examples include the shortest route problem, linear programming problem (LPP), cargo-loading problem, and capital budgeting problem.

In contrast, the next stage is unpredictable in case of probabilistic dynamic problems, that is, the state and policy variables of the current state do not decide the state of the next stage. Examples of such problems are reliability problems and inventory problems.

12.8 MODELS OF DYNAMIC PROGRAMMING

A few specific models of dynamic programming are discussed in the following subsections.

12.8.1 Model I—Shortest Route Problem

A variety of problems can be solved by dynamic programming using the *principle of optimality*. We know that dynamic programming can be used to solve any problem that can be broken into stages, wherein decisions can be taken stage-wise using the principle of optimality.

The computations in shortest route problem using dynamic programming can be mathematically expressed as follows:

Let $f_i^{**}(x_i)$ be the shortest distance (optimum return) to node x_i at the stage i. Define $d(x_{i-1}, x_i)$=distance from node x_{i-1} to node x_i. Compute f_i^{**} from f_{i-1}^{**} using the following recursive equation:

$$f_i^{**}(x_i) = \min_{\text{all feasible routes } (x_{i-1},\, x_i)} \left[d(x_{i-1}, x_i) + f_{i-1}^{**}(x_{i-1}) \right]$$

$$i = 1, 2, 3, \ldots, n \ (n = \text{number of stages})$$

This is forward dynamic programming. The backward recursive equation is

$$f_i^{**}(x_i) = \min_{\text{all feasible routes } (x_i,\, x_{i+1})} \left[d(x_i, x_{i+1}) + f_{i+1}^{**}(x_{i+1}) \right]$$

$$i = 1, 2, 3, \ldots, n \ (n = \text{number of stages})$$

To explain the application of dynamic programming to solve shortest route problems, let us consider the following examples.

Example 12.1 A travelling salesperson has to travel from city 1 to city 7. The distances (km) between the various cities are given in Fig. 12.2. Using dynamic programming, (a) find the minimum distance covered by the salesperson and (b) determine the optimal path.

Solution To solve this problem by dynamic programming, first decompose it into stages. Then, perform the computation for each stage separately. The optimum result at each stage is used in finding the optimum result at the succeeding stage.

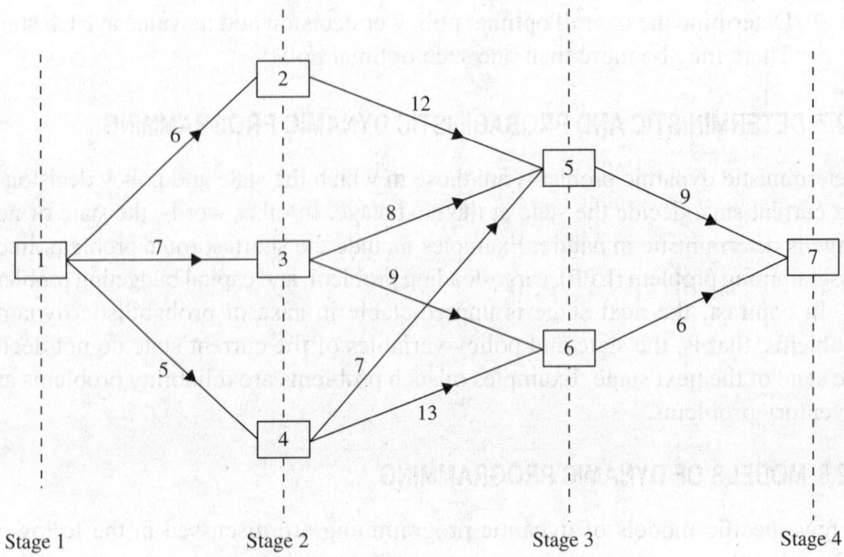

Figure 12.2 Shortest route problem for Example 12.1

Here, we solve this problem using the following backward recursive equation:

$$f_i^{**}(x_i) = \underset{\text{all feasible routes } (x_i,\, x_{i+1})}{\text{Min}} \left[d\left(x_i, x_{i+1}\right) + f_{i+1}^{**}\left(x_{i+1}\right) \right]$$

$$i = 1, 2, 3, \ldots, n \; (n = \text{number of stages})$$

Stage 4 $f_4(x_4) = 0$, for $x_4 = 7$. The associated order of computation is $f_4 \rightarrow f_3 \rightarrow f_2 \rightarrow f_1$.

Stage 3 Node 7 ($x_4 = 7$) is connected to nodes 5 and 6 ($x_3 = 5$ or 6) with exactly one route each. The results can be summarized as follows:

	$d(x_3, x_4)$	Optimum solution	
x_3	$x_4 = 7$	$f_3^{**}(x_3)$	x_4^{*}
5	9	9	7
6	6	6	7

*Optimal decision policy, **Optimal return

Stage 2 Route (2, 6) is not a feasible alternative, as it does not exist. The optimum solution at Stage 2 reads as follows.

	$d(x_2, x_3) + f_3^{**}(x_3)$		Optimum solution	
x_2	$x_3 = 5$	$x_3 = 6$	$f_2^{**}(x_2)$	x_3^{*}
2	$12 + 9 = 21$	–	21	5
3	$8 + 9 = 17$	$9 + 6 = 15$	15	6
4	$7 + 9 = 16$	$13 + 6 = 19$	16	5

*Optimal decision policy, **Optimal return

If a salesperson is in city 2, then only one route is available to city 5, which is 12 km away. Thus, to arrive at city 7 from city 2, he/she has to travel a minimum of $12 + 9 = 21$ km. When he/she is in city 3, then two routes are available to reach city 7—one via city 5, travelling a total of $8+9=17$ km and another via city 6, travelling a total $9+6=15$ km. Thus, if the salesperson is in city 3, then he/she has to travel a minimum of 15 km [min(17, 15)] to reach city 7 via city 6. Similarly, if a salesperson is in city 4 he/she has to travel $\min(7+9, 13+6)=16$ km to reach city 7 and for that he/she has to travel via city 5.

Thus, if the travelling salesperson is in cities 2 or 4, the shortest route is through city 5 and if he/she is in city 3, the shortest route is through city 6 to city 7, which yields the optimum solution (minimum distance $f_2^{**}(x_2)$ and optimal policy x_2^*).

Stage 1 From node 1, we have three alternative routes (1, 2), (1, 3), and (1, 4). Using $f_2(x_2)$ from Stage 2, we have the following table:

	$d(x_1, x_2) + f_2(x_2)$			Optimum solution	
x_1	$x_2=2$	$x_2=3$	$x_2=4$	$f_2^{**}(x_2)$	x_2^*
1	$6+21=27$	$7+15=22$	$5+16=21$	21	4

*Optimal decision policy, **Optimal return (minimum distance travelled)

Hence, (a) optimum distance is 21 km and (b) Tracing back, optimum route is $1 \rightarrow 4 \rightarrow 5 \rightarrow 7$.

Note We can solve this problem by exhaustively enumerating all the routes between nodes 1 and 7 (there are five such routes). However, in a large network, the exhaustive enumeration method is not computationally efficient.

Example 12.2 A travelling salesperson has to go from city 1 to city 10. The costs (in ₹) between various cities are given in Fig. 12.3.

Using dynamic programming, (a) find the minimum cost incurred by the salesperson and (b) determine the optimal path.

Solution We can solve this problem using either the forward or backward dynamic programming methods. Here, we solve the problem using backward dynamic programming.

Let us divide the problem into five stages as depicted in Fig. 12.4.

Stage 5 $f_5^{**}(x_5)=0$ for $x_5^*=10$, where * means optimal decision policy and ** means optimal return (minimum cost incurred).

Stage 4 In stage 4, there are two states 8 and 9, and decision variables 18 and 14.

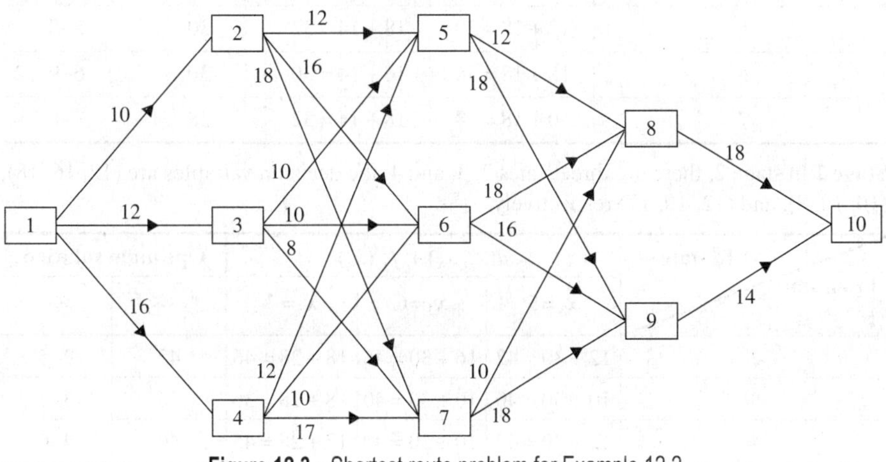

Figure 12.3 Shortest route problem for Example 12.2

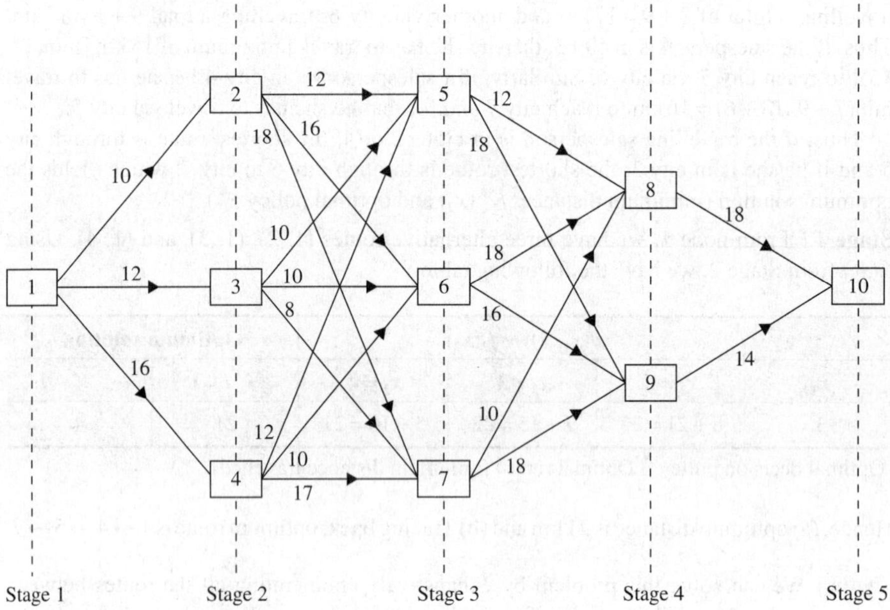

Figure 12.4 Minimum cost route problem for Example 12.2

From state ↓ x_4 / To state →	$d(x_4\ x_5)$	Optimum decision	Optimum solution	
	$x_5=10$	$d(x_4, x_5) + f_5^{**}(x_5)$	$f_4^{**}(x_4)$	x_4^*
8	18	18 + 0	18	8–10
9	14	14 + 0	14	9–10

Stage 3 In stage 3, there are three states 5, 6, and 7; the decision variables are (12, 18), (18, 16), and (10, 18) respectively.

From state ↓ x_3 / To state →	$d(x_3, x_4) + f_4^{**}(x_4)$		Optimum solution	
	$x_4=8$	$x_4=9$	$f_3^{**}(x_3)$	x_3^*
5	12 + 18 = 30	18 + 14 = 32	30	5–8
6	18 + 18 = 36	16 + 14 = 30	30	6–9
7	10 + 18 = 28	18 + 14 = 32	28	7–8

Stage 2 In stage 2, there are three states 2, 3, and 4; the decision variables are (12, 16, 18), (10, 10, 8), and (12, 10, 17) respectively.

From state ↓ x_2 / To state →	$d(x_2, x_3) + f_3^{**}(x_3)$			Optimum solution	
	$x_3=5$	$x_3=6$	$x_3=7$	$f_2^{**}(x_2)$	x_2
2	12 + 30 = 42	16 + 30 = 46	18 + 28 = 46	42	2–5
3	10 + 30 = 40	10 + 30 = 40	8 + 28 = 36	36	3–7
4	12 + 30 = 42	10 + 30 = 40	17 + 28 = 45	40	4–6

Stage 1 In stage 1, there is one state 1 and the decision variable is (8, 12, 14).

To state → / From state ↓	$d(x_1, x_2) + f_2^{**}(x_2)$			Optimum solution	
	$x_2=2$	$x_2=3$	$x_2=4$	$f_1^{**}(x_1)$	x_1^*
1	$10+42=52$	$12+36=48$	$16+40=56$	48	1–3

Hence, (a) optimum (minimum) cost is ₹48 and (b) tracing back optimum (shortest) route is $1 \to 3 \to 7 \to 8 \to 10$.

12.8.2 Model II—Solving Dynamic Programming using Calculus Method

Let us now discuss problems that can be solved using the dynamic programming techniques through the calculus method. We illustrate Model II with the help of the following cases:

Case 1 When both the objective function and the constraint are in additive form

Let us consider the following problem:

$$\text{Max (or Min) } Z = f_1(x_1) + f_2(x_2) + \ldots + f_n(x_n) = \sum_{i=1}^{n} f_i x_i$$

Subject to $a_1x_1 + a_2x_2 + a_3x_3 + \ldots + a_nx_n \geq \text{ or } = \text{ or } \leq b$

or $\sum_{i=1}^{n} a_i x_i \geq \text{ or } = \text{ or } \leq b$

and $(a_1, a_2, a_3, \ldots, a_n) \geq 0$, $(x_1, x_2, x_3, \ldots, x_n) \geq 0$ and $b > 0$

or $a_i \geq 0$, $x_i \geq 0$ and $b > 0$

To solve this type of problem, we first define the state variables s_i's as

$$s_n = a_1x_1 + a_2x_2 + a_3x_3 + \ldots + a_nx_n \geq \text{ or } = \text{ or } \leq b$$
$$s_{n-1} = a_1x_1 + a_2x_2 + a_3x_3 + \ldots + a_{n-1}x_{n-1} = s_n - a_nx_n = T_{n-1}(s_n, x_n)$$
$$s_{n-2} = a_1x_1 + a_2x_2 + a_3x_3 + \ldots + a_{n-2}x_{n-2} = s_{n-1} - a_{n-1}x_{n-1} = T_{n-2}(s_{n-1}, x_{n-1})$$
$$\ldots \quad \ldots \quad \ldots \quad \ldots \quad \ldots \quad \ldots \quad \ldots$$
$$s_i = a_1x_1 + a_2x_2 + a_3x_3 + \ldots + a_ix_i = s_{i+1} - a_{i+1}x_{i+1} = T_i(s_{i+1}, x_{i+1})$$
$$\ldots \quad \ldots \quad \ldots \quad \ldots \quad \ldots \quad \ldots \quad \ldots$$
$$s_2 = a_1x_1 + a_2x_2 = s_3 - a_3x_3 = T_2(s_3, x_3)$$
$$s_1 = a_1x_1 = s_2 - a_2x_2 = T_1(s_2, x_2)$$

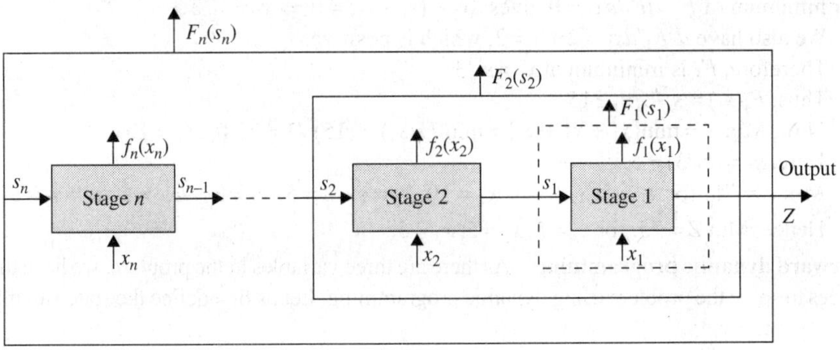

Figure 12.5 Optimization problem in stages with optimum returns

where $s_i = T_i(s_{i+1}, x_{i+1})$, $I = 1, 2, 3, \ldots, n-1$ are called state transformation functions connecting different states. The recursive relations are as follows:

$$F_i(s_i) = \text{Max (or Min)}[f_i(x_i) + F_{i-1}(s_{i-1})];\ i = 1, 2, 3, \ldots, n$$
$$\text{and } F_1(s_1) = f_1(x_1)$$

Now, we can start with $F_1(s_1)$ and recursively optimize to get $F_2(s_2)$, $F_3(s_3)$, …, $F_n(s_n)$. This is forward dynamic programming. We can also solve such problems using the backward dynamic programming technique. The aforementioned explanation can be shown diagrammatically (see Fig. 12.5).

Example 12.3 Solve the following problem using dynamic programming:

$$\text{Min } Z = x_1^2 + x_2^2 + x_3^2$$
$$\text{Subject to } x_1 + x_2 + x_3 \geq 15$$
$$\text{and } x_1, x_2, x_3 \geq 0$$

Solution We can solve this problem using either forward dynamic programming or backward dynamic programming.

Let us first solve it using backward dynamic programming.

Backward dynamic programming As there are three variables in the problem, we will have three stages to solve the problem using dynamic programming. We define the state variables as follows:

$$s_1 = x_1 + x_2 + x_3 \geq 15$$
$$s_2 = x_2 + x_3 = s_1 - x_1$$
$$s_3 = x_3 = s_2 - x_2$$

Recursive relations are

$$F_3(s_3) = x_3^2 = (s_2 - x_2)^2$$
$$F_2(s_2) = \min[x_2^2 + F_3(s_3)] = \min(x_2^2 + x_3^2) = \min[x_2^2 + (s_2 - x_2)^2] \text{ with respect to } x_2$$
$$F_1(s_1) = \min[x_1^2 + F_2(s_2)] \text{ with respect to } x_1$$

For minimum of F_2, $dF_2/dx_2 = 0$ gives $2x_2 - 2(s_2 - x_2) = 0 \Rightarrow x_2 = s_2/2$.
Further, $d^2F_2/dx_2^2 = 2 + 2 = 4$, which is positive.

Therefore, F_2 is minimum at $x_2 = s_2/2$ and $F_2(s_2) = \left[\left(\frac{s_2}{2}\right)^2 + \left(s_2 - \frac{s_2}{2}\right)^2\right] = \frac{s_2^2}{2}$.

Again, $F_1(s_1) = \min\left[x_1^2 + \frac{s_2^2}{2}\right] = \min\left[x_1^2 + \frac{(s_1 - x_1)^2}{2}\right]$.

For minimum of F_1, $dF_1/dx_1 = 0$ gives $2x_1 - (s_1 - x_1) = 0 \Rightarrow x_1 = s_1/3$.

We also have $d^2F_1/dx_1^2 = 2 + 1 = 3$, which is positive.

Therefore, F_1 is minimum at $x_1 = s_1/3$.

Thus, $F_1(s_1) = s_1^2/3$, $s_1 \geq 15$.

Now, Min $Z = \min[x_1^2 + x_2^2 + x_3^2] = \min F_1(s_1) = (15)^2/3 = 75$ for $s_1 = 15$

Here, $x_1 = (s_1/3) = 5$.

Min $Z = 75$, for $x_1 = 5$, $s_2 = s_1 - x_1 = 10$, $x_2 = s_2/2 = 5$, $s_3 = s_2 - x_2 = 5$, $x_3 = s_3 = 5$

Hence, Min $Z = 75$, for $x_1 = 5$, $x_2 = 5$, $x_3 = 5$.

Forward dynamic programming As there are three variables in the problem, we have three stages to solve the problem using dynamic programming. Let us first define the state variables:

$$s_1 = x_1$$
$$s_2 = x_1 + x_2$$
$$s_3 = x_1 + x_2 + x_3 \geq 15$$

This gives the following equations:

$$s_1 = x_1 = s_2 - x_2;\ s_2 = s_3 - x_3;\ s_3 = x_1 + x_2 + x_3 \geq 15$$

The recursive relations are as follows:

$$F_1(s_1) = x_1^2 = (s_2 - x_2)^2$$
$$F_2(s_2) = \min[x_2^2 + F_1(s_1)] = \min(x_2^2 + x_1^2) = \min[x_2^2 + (s_2 - x_2)^2] \text{ with respect to } x_2$$
$$F_3(s_3) = \min[x_3^2 + F_2(s_2)] \text{ with respect to } x_3$$

For minimum of F_2, $dF_2/dx_2 = 0$ gives $2x_2 - 2(s_2 - x_2) = 0 \Rightarrow x_2 = s_2/2$.
Further, $d^2F_2/dx_2^2 = 2 + 2 = 4$, which is positive.

Therefore, F_2 is minimum at $x_2 = s_2/2$ and $F_2(s_2) = \left[\left(\dfrac{s_2}{2}\right)^2 + \left(s_2 - \dfrac{s_2}{2}\right)^2\right] = \dfrac{s_2^2}{2}$.

Now, $F_3(s_3) = \min\left[x_3^2 + \dfrac{s_2^2}{2}\right] = \min\left[x_3^2 + \dfrac{(s_3 - x_3)^2}{2}\right]$.

For minimum of F_3, $dF_3/dx_3 = 0$ gives $2x_3 - (s_3 - x_3) = 0 \Rightarrow x_3 = s_3/3$.

Further, $d^2F_3/dx_3^2 = 2 + 1 = 3$, which is positive. Therefore, F_3 is minimum at $x_3 = s_3/3$ and $F_3(s_3) = s_3^2/3$.

Now, Min $Z = \min[x_1^2 + x_2^2 + x_3^2] = \min F_3(s_3) = (15)^2/3 = 75$ for $s_3 = 15$

Here, $x_3 = (s_3/3) = 5$.

Min $Z = 75$, for $x_3 = 5, s_3 = 15, s_2 = s_3 - x_3 = 10, x_2 = s_2/2 = 5,\ s_1 = s_2 - x_2 = 5, x_1 = s_1 = 5$

Hence, Min $Z = 75$, for $x_1 = 5, x_2 = 5, x_3 = 5$.

Case 2 When objective function is in additive form and constraint is in multiplicative form, we illustrate this case with the help of the following example:

Example 12.4 Solve the following problem using dynamic programming:

Min $Z = x_1^2 + x_2^2 + x_3^2 + x_4^2$

Subject to $x_1 x_2 x_3 x_4 = 16$

and $x_1; x_2, x_3, x_4 \geq 0$

Solution We define the state variables as follows:

$$s_1 = x_1$$
$$s_2 = x_1 x_2$$
$$s_3 = x_1 x_2 x_3$$
$$s_4 = x_1 x_2 x_3 x_4$$

This gives $s_1 = x_1 = s_2/x_2;\ s_2 = s_3/x_3;\ s_3 = s_4/x_4$

As there are four decision variables in the problem, we shall have four stages in the form of the following four recursive relations:

$$F_1(s_1) = x_1^2 = s_1^2$$
$$F_2(s_2) = \min[x_2^2 + F_1(s_1)] = \min\left[x_2^2 + \frac{s_2^2}{x_2^2}\right] \text{ with respect to } x_2$$
$$F_3(s_3) = \min[x_3^2 + F_2(s_2)] \text{ with respect to } x_3$$
$$F_4(s_4) = \min[s_4^2 + F_3(s_3)] \text{ with respect to } x_4$$

For minimum of F_2, $dF_2/dx_2 = 0$ gives $2x_2 - 2\, s_2^2/x_2^3 = 0 \Rightarrow x_2 = \sqrt{s_2}$.

Further, $dF_2/dx_2 = 2 + 6\, s_2^2/x_2^4$ which is positive.

Therefore, F_2 is minimum at $x_2 = \sqrt{s_2}$ and $F_2(s_2) = \left[\left(\sqrt{s_2}\right)^2 + \frac{s_2^2}{\left(\sqrt{s_2}\right)^2}\right] = 2s_2.$

Now, $F_3(s_3) = \min[x_3^2 + F_2(s_2)] = \min[x_3^2 + 2s_2] = \min[x_3^2 + 2s_3/x_3]$

For minimum of F_3, $dF_3/dx_3 = 0$ gives $2x_3 - 2s_3/x_3^2 = 0 \Rightarrow x_3 = s_3^{1/3}, \left(s_2 = \frac{s_3}{x_3}\right)$

Further, $\frac{d^2F_3}{dx_3^2} = 2 + 4\frac{s_3}{x_3^3}$, which is positive.

Therefore, F_3 is minimum at $x_3 = s_3^{1/3}$ and thus $F_3(s_3) = 3s_3^{2/3}$.

Now, $F_4(s_4) = \min[x_4^2 + F_3(s_3)] = \min\left[x_4^2 + 3s_3^{2/3}\right] = \min\left[x_4^2 + 3\frac{s_4^{2/3}}{x_4^{2/3}}\right]$

For minimum of F_4, $dF_4/dx_4 = 0$ gives $2x_4 - 2\frac{s_4^{2/3}}{x_4^{5/3}} = 0 \Rightarrow x_4 = s_4^{1/4}$.

Further, $\frac{d^2F_4}{dx_4^2} = 2 + \frac{10}{3}\frac{s_4^{2/3}}{x_4^{8/3}}$, which is positive. Therefore, F_4 is minimum at $x_4 = s_4^{1/4}$ and

thus, $F_4(s_4) = 4s_4^{1/2}$.

We have $x_4 = s_4^{1/4} = (16)^{1/4} = 2$. Now, $s_3 = s_4/x_4 = 16/2 = 8$, $x_3 = s_3^{1/3} = (8)^{1/3} = 2$, $s_2 = s_3/x_3 = 8/2 = 4$, and $x_2 = \sqrt{s_2} = \sqrt{4} = 2$, $s_1 = s_2/x_2 = 4/2 = 2 = x_1$.

Hence, Min $Z = 16$, for $x_1 = x_2 = x_3 = x_4 = 2$.

Case 3 When objective function is in multiplicative form and constraint is in additive form, we illustrate this case with the help of the following example:

Example 12.5 Solve the following problem using dynamic programming:

$$\text{Max } Z = x_1x_2x_3$$
$$\text{Subject to } \quad x_1 + x_2 + x_3 = 5$$
$$x_1, x_2, x_3 \geq 0$$

Solution Let us first define the state variables as

$$s_1 = x_1$$
$$s_2 = x_1 + x_2$$
$$s_3 = x_1 + x_2 + x_3 = 5$$

This gives $s_1 = x_1 = s_2 - x_2$; $s_2 = x_1 + x_2 = s_3 - x_3$; $s_3 = x_1 + x_2 + x_3 = 5$

The recursive relations are

$$F_1(s_1) = x_1 = s_1$$
$$F_2(s_2) = \max[x_2\, F_1(s_1)] = \max(x_2 \cdot s_1) = \max[x_2 \cdot (s_2 - x_2)], \text{ with respect to } x_2$$
$$F_3(s_3) = \max[x_3 \cdot F_2(s_2)] \text{ with respect to } x_3$$

For maximum of F_2, $dF_2/dx_2 = 0$ gives $s_2 - 2x_2 = 0 \Rightarrow x_2 = s_2/2$

Further, $d^2F_2/dx_2^2 = -2$, which is negative.

Therefore, F_2 is maximum at $x_2 = s_2/2$ and thus, $F_2(s_2) = s_2^2/4$.

Now, $F_3(s_3) = \max\left[x_3 \cdot \frac{s_2^2}{4}\right] = \max\left[x_3 \cdot \frac{(s_3 - x_3)^2}{4}\right]$.

For maximum of F_3, $dF_3/dx_3 = 0$ gives $(s_3 - x_3)^2 - 2x_3(s_3 - x_3) = 0$

$\Rightarrow (s_3 - x_3)\,(s_3 - 3x_3) = 0$ or $(s_3 - 3x_3) = 0$

$[s_3 - x_3 \neq 0$ because $x_2 = s_3 - x_3 > 0$, $x_2 \neq 0$ because if $x_2 = 0$, then max $Z = 0]$, that is, $x_3 = s_3/3$.

Further, $d^2F_3/dx_3^2 = 6x_3 - 4s_3$. At $x_3 = s_3/3$, $d^2F_3/dx_3^2 = -2s_3 < 0$, which is negative. Therefore, F_3 is maximum at $x_3 = s_3/3$ and thus, $F_3(s_3) = \frac{s_3^3}{27} = \frac{(5)^3}{27} = \frac{125}{27}$ for $s_3 = 5$

Now, $x_3 = \frac{s_3}{3} = \frac{5}{3}$, $s_2 = s_3 - x_3 = 5 - \frac{5}{3} = \frac{10}{3}$, $x_2 = \frac{s_2}{2} = \frac{10}{3} \cdot \frac{1}{2} = \frac{5}{3}$, $s_1 = s_2 - x_2 = \frac{10}{3} - \frac{5}{3} = \frac{5}{3}$,

$x_1 = s_1 = 5/3$. Hence, Max $Z = 125/27$, for $x_1 = x_2 = x_3 = 5/3$.

In all the three cases of Model II discussed in this section, the decision variables are continuous variables. However, we sometimes come across practical problems of importance where decision variables are not continuous. Therefore, in the next few models, we shall discuss problems where decision variables are discrete in nature. We cannot use the calculus method to obtain solutions to such problems. Therefore, we solve them using the tabular form or the enumeration method.

12.8.3 Model III

Cargo loading, capital budgeting, and reliability problems are discussed in this section.

Cargo Loading Problem

Dynamic programming plays an important role in business. For example, suppose a container is loaded with different items of different weights and values. This container can carry only a limited load. Thus, a problem arises as to how many items of different categories can be loaded in the container so that the value of the loaded items is maximum and the load constraint is also satisfied. Such a problem is called the cargo loading problem. A large variety of problems can be solved using this method. Let us consider an example for better understanding.

Example 12.6 A vessel that can carry a maximum load of seven units is loaded with three items. The weight per unit of different items and their values are given in the adjacent table. How many units of each item should be loaded in the vessel to maximize the value of the loaded items?

Item	Weight/Unit	Value/Unit
1	1	20
2	3	90
3	2	70

Solution Suppose x_1, x_2, and x_3 are, respectively, the number of units of items 1, 2, and 3 that have been loaded. The given problem can be formulated as follows:

$$\text{Max } Z = 20x_1 + 90x_2 + 70x_3$$
$$\text{Subject to} \quad x_1 + 3x_2 + 2x_3 \le 7$$
$$x_1, x_2, x_3 \ge 0 \text{ and integers}$$

As the decision variables are discrete here, we cannot use the calculus method; hence, solve the problem in tabular form. Now, we define the state variables as follows:

$$s_1 = x_1 \tag{12.1}$$
$$s_2 = x_1 + 3x_2 \tag{12.2}$$
$$s_3 = x_1 + 3x_2 + 2x_3 \le 7 \tag{12.3}$$

This gives $s_1 = x_1 = s_2 - 3x_2$; $s_2 = s_3 - 2x_3$; $s_3 = x_1 + x_2 + x_3 \le 7$

The recursive relations are

$$F_1(s_1) = f_1(x_1) = 20x_1 = 28s_1$$
$$F_2(s_2) = \text{Max}[f_2(x_2) + F_1(s_1)]$$
$$F_3(s_3) = \text{Max}[f_3(x_3) + F_2(s_2)]$$

From Eq. (12.3), we have $x_1 = 0, 1, 2, 3, 4, 5, 6, 7$; $x_2 = 0, 1, 2$; $x_3 = 0, 1, 2, 3$

The state transformations are shown in Tables 12.1 and 12.2.

Table 12.1 gives the possible values of s_2 and the entries are $s_2 = s_3 - 2x_3$. Next, we prepare Table 12.2, in which the entries are $s_1 = s_2 - 3x_2$, that is, s_1 can assume values for various combinations of s_2 and x_2. Now, we prepare Tables 12.3, 12.4, and 12.5 using the recursive relations.

The following points may be noted in Table 12.4:

1. We enter values in the possible positions. These are based on Table 12.1.
2. To get the $F_1(s_1)$ matrix, we read x_1 from Table 12.2 and then read $F_1(s_1)$ from Table 12.3.
3. The last column gives the maximum value with respect to x_2 for a fixed s_2.

Table 12.1 State transformation table with values for s_2

s_3 \ x_3	0	1	2	3	s_3 \ x_3	0	1	2	3
0	0	–	–	–	4	4	2	0	–
1	1	–	–	–	5	5	3	1	–
2	2	0	–	–	6	6	4	2	0
3	3	1	–	–	7	7	5	3	1

Table 12.2 State transformation table with values for s_1

s_2 \ x_2	0	1	2	s_2 \ x_2	0	1	2
0	0	–	–	4	4	1	–
1	1	–	–	5	5	2	–
2	2	–	–	6	6	3	0
3	3	0	–	7	7	4	1

Table 12.3 Recursive relation transformation table with values for $F_1(s_1)$

$x_1 = s_1$	0	1	2	3	4	5	6	7
$F_1(s_1) = f_1(x_1) = 20x_1$	0^{**}	20^{*}	40	60	80	100	120	140

*Optimal entries, **Alternate optimal solution.

Table 12.4 Recursive relation transformation table with values for $F_2(s_2)$

s_2 \ x_2	$f_2(x_2) = 90x_2$			$F_1(s_1)$			$f_2(x_2) + F_1(s_1)$			$F_2(s_2)$
	0	1	2	0	1	2	0	1	2	Maximum value of rows
0	0	–	–	0	–		0	–	–	0
1	0	–	–	20^*	–	–	20^*	–	–	20^*
2	0	–	–	40	–	–	40	–	–	40
3	0	90	–	60	0^{**}	–	60	90^*	–	90^{**}
4	0	90	–	80	20	–	80	110	–	110
5	0	90	–	100	40	–	100	130	–	130
6	0	90	180	120	0	0	120	150	180	180
7	0	90	180	140	80	20	140	170	200	200

*Optimal entries, **Alternate optimal solution

Table 12.5 Recursive relation transformation table with values for $F_3(s_3)$

s_3 \ x_3	$f_3(x_3) = 70x_3$				$F_2(s_2)$				$f_3(x_3) + F_2(s_2)$				$F_3(s_3)$
	0	1	2	3	0	1	2	3	0	1	2	3	Maximum value of rows
0	0	–	–	–	0	–	–	–	0	–	–	–	0
1	0	–	–	–	20	–	–	–	20	–	–	–	20
2	0	70	–	–	40	0	–	–	40	70	–	–	70
3	0	70	–	–	90	20	–	–	90	90	–	–	90
4	0	70	140	–	110	40	–	–	110	110	140	–	140
5	0	70	140	–	130	90	20	–	130	180	180	–	160
6	0	70	140	210	180	110	40	0	180	180	180	210	210
7	0	70	140	210	200	130	90^{**}	20^*	200	200	230^{**}	230^*	230^{***}

*Optimal entries, **Alternate optimal solution

Table 12.3 is the solution to the second subproblem. Now, let us proceed to calculate $F_3(s_3)$, which will give the optimal value for this problem. To get $F_3(s_3)$, we construct Table 12.5.

In Table 12.5 to get matrix $F_2(s_2)$, we read s_2 from Table 12.1 and then read the corresponding $F_2(s_2)$ from Table 12.4. From the last column of Table 12.5, we see that the largest value of $F_3(s_3)=230$ is for $s_3=7$ and $x_3=3$ and 2. Now, we trace the entries that gave this value, as these would provide the optimal solution. This is shown in Table 12.5 by marking the optimal entries with (*). From Tables 12.3 and 12.4, we have $s_2=1$, $x_2=0$, and $s_1=x_1=1$, respectively. The optimal values are $x_1=1$, $x_2=0$, $x_3=3$; the maximum value is 230.

In this problem, Table 12.5 shows that there is an alternative optimal solution. This solution is shown by marking entries with (**). The alternate optimal solution is $x_3=2$, $x_2=1$, and $x_1=0$ and the maximum value is 230.

Capital Budgeting Problem

Sometimes, we need to improve the working efficiency of a plant. While devising a plan for improving the state of the plants in use, the budgetary restrictions are taken into consideration. Here, the capital return depends on the cost of the plan. Such problems are categorized as capital budgeting problems.

Example 12.7 A company has to improve the working of three plants P, Q, and R, with the available capital of 5 units of money. The anticipated plans with costs and corresponding returns are given in the following table (in lakhs of ₹):

Plan no.	Plant P		Plant Q		Plant R	
	Cost	Return	Cost	Return	Cost	Return
1	0	0	0	0	0	0
2	1	5	2	8	1	3
3	2	6	3	9	–	–
4	–	–	4	12	–	–

Find the optimal policy for budgeting the available capital.

Solution Assume x_1, x_2, and x_3 are the amount allocated to plants P, Q, and R respectively. The given problem can be formulated as follows:

$$\text{Max } Z = f_1(x_1) + f_2(x_2) + f_3(x_3)$$

$$\text{Subject to} \quad x_1 + x_2 + x_3 \leq 5$$

$$x_1, x_2, x_3 \geq 0 \text{ and integers}$$

The values for $f_1(x_1)$, $f_2(x_2)$, and $f_3(x_3)$ are given in Tables 12.6, 12.7, and 12.8 respectively.

The calculation of values for s_2 and s_1 are shown in Tables 12.9 and 12.10 respectively.

Tables 12.11, 12.12, and 12.13 display the values of the recursive relations for the given problem.

Hence, one solution is $x_3 = 1$, $x_2 = 3$, $x_1 = 1$, and maximum is 17.

Table 12.6 Values for $f_1(x_1)$

x_1	$f_1(x_1)$
0	0
1	5
2	6

Table 12.7 Values for $f_2(x_2)$

x_2	$f_2(x_2)$
0	0
2	8
3	9
4	12

Table 12.8 Values for $f_3(x_3)$

x_3	$f_3(x_3)$
0	0
1	3

Table 12.9 Values for s_2

s_3 \ x_3	0	1
0	0	–
1	1	0
2	2	1
3	3	2
4	4	3
5	5	4

Table 12.10 Values for s_1

s_2 \ x_2	0	2	3	4
0	0	–	–	–
1	1	–	–	–
2	2	0	–	–
3	–	1	0	–
4	–	2	1	0
5	–	–	2	1

Table 12.11 Values for $F_1(s_1)$

$x_1 = s_1$	0	1^*	2
$F_1(s_1) = f_1(x_1)$	0	5^*	6

*Optimal entry

Table 12.12 Values for $F_2(s_2)$

s_2 \ x_2	$f_2(x_2)$				$F_1(s_1)$				$F_2(s_2)$
	0	**2**	**3**	**4**	**0**	**2**	**3***	**4**	**Maximum value of rows**
0	0	–	–	–	0	–	–	–	0
1	0	–	–	–	5	–	–	–	5
2	0	8	–	–	6	0	–	–	8
3	–	8	9	–	–	5	0	–	12
4	–	8	9	12	–	6	5*	0	14*
5	–	–	9	12	–		6	5	17

*Optimal entry

Table 12.13 Values for $F_3(s_3)$

s_3 \ x_3	$f_3(x_3)$		$F_2(s_2)$		$F_3(s_3)$
	0	**1**	**0**	**1***	**Maximum value of rows**
0	0	–	0	–	0
1	0	3	5	0	5
2	0	3	8	5	8
3	0	3	12	8	12
4	0	3	14	12	15
5	0	3	17	14*	17*

*Optimal entry

The optimal solution is $x_1 = 1$, $x_2 = 3$, $x_3 = 1$ and maximum return = ₹17lakhs.

Reliability Problem

Until now we have discussed deterministic problems, which can be solved using the dynamic programming problem. However, there are certain non-deterministic problems that can also be solved using dynamic programming.

Example 12.8 An instrument has three components 1, 2, and 3, which are connected in series (i.e., if one component fails then the instrument fails to work). To improve the reliability of the instrument each component has to be supplemented by either one, two, or three parallel units (in case the component fails then the supplementary unit can start working and ensure that the system remains functional). The total money available for improvement of the device is ₹10,000. The cost of parallel units and reliability of each component is given in the following table:

No. of parallel units	Component 1		Component 2		Component 3	
	Cost (in thousands of ₹)	*R*	Cost (in thousands of ₹)	*R*	Cost (in thousands of ₹)	*R*
1	1	0.4	2	0.5	3	0.7
2	3	0.6	4	0.6	4	0.8
3	4	0.8	5	0.7	5	0.9

In the aforegiven table, R denotes the reliability of the equipment. Let x_i denote the money spent on attaching supplementary units to i components. Find x_i, such that the reliability of the equipment is maximized.

Solution Suppose $f_i(x_i)$, $i = 1, 2, 3$ be the reliability of the ith component when x_i amount is spent on parallel units attached to the component. Here, the components 1, 2, and 3 denote the stage 1, 2, and 3, respectively. The function $f_i(x_i)$ is the return function of stage i.

Using the principle of reliability, the reliability of the equipment is the product of reliabilities of its components. Thus, the problem is as follows:

$$\text{Max } Z = f_1(x_1)\, f_2(x_2)\, f_3(x_3)$$

$$\text{Subject to } \quad x_1 + x_2 + x_3 \le 10$$

where x_1, x_2, and x_3 are specified in the aforegiven table.

Using the given data, we have the return functions as shown in Tables 12.14, 12.15, and 12.16.

Now, we define the state variables as follows:

$$s_1 = x_1$$
$$s_2 = x_1 + x_2$$
$$s_3 = x_1 + x_2 + x_3 \le 10$$

This gives $s_1 = x_1 = s_2 - x_2$; $s_2 = s_3 - x_3$; $s_3 = x_1 + x_2 + x_3 \le 10$

From table of the problem, we see that the least value of $x_1 + x_2 + x_3$ is 6; therefore from the aforementioned first relation, we have $6 \le s_3 \le 10$. We observe that the least value of $x_1 + x_2$ is 3. The largest value of $s_3 - x_3$ is $10 - 3 = 7$. Thus, $3 \le s_2 \le 7$. Finally, we know that $s_1 = x_1$.

In view of the aforementioned point, we subtract x_3 from s_3 to arrive at values for s_2 (Table 12.17) and x_2 from s_2 to get $s_1 (= x_1)$ (Table 12.18).

Table 12.14 Values for $f_1(x_1)$

x_1	$f_1(x_1)$
1	0.4
3	0.6
4	0.8

Table 12.15 Values for $f_2(x_2)$

x_2	$f_2(x_2)$
2	0.5
4	0.6
5	0.7

Table 12.16 Values for $f_3(x_3)$

x_3	$f_3(x_3)$
3	0.7
4	0.8
5	0.9

Table 12.17 Values for s_2

s_3 \ x_3	3	4	5
6	3	–	–
7	4	3	–
8	5	4	3
9	6	5	4
10	7	6	5

Table 12.18 Values for s_1

s_2 \ x_2	0	2	3
3	1	–	–
4	–	–	–
5	3	1	–
6	4-	–	1
7	–	3	–

Table 12.19 Values for $F_1(s_1)$

$x_1 = s_1$	1	3	4*
$F_1(s_1) = f_1(x_1)$	0.4	0.6	0.8*

Table 12.20 Values for $F_2(s_2)$

s_2 \ x_2	$f_2(x_2)$			$F_1(s_1)$			$f_2(x_2)\,F_1(s_1)$			$F_2(s_2)$
	2	4	5	2*	4	5	2	4	5	Max
3	0.5	–	–	0.4	–	–	0.2	–	–	0.2
4	0.5	–	–	–	–	–	–	–	–	–
5	0.5	0.6	–	0.6	0.4	–	0.3	0.24	–	0.3
6	0.5	0.6	0.7	0.8*	–	0.4	0.4*	–	0.28	0.4*
7	0.5	0.6	0.7	–	0.6	–	–	0.36	–	0.36

Table 12.21 Values for $F_3(s_3)$

s_3 \ x_3	$f_3(x_3)$			$F_2(s_2)$			$f_3(x_3)\,F_2(s_2)$			$F_3(s_3)$
	3	4	5	3	4*	5	3	4	5	Max
6	0.7	–	–	0.2	–	–	0.14	–	–	0.14
7	0.7	0.8	–	–	0.2	–	–	0.16	–	0.16
8	0.7	0.8	0.9	0.3	–	0.2	0.21	–	0.18	0.21
9	0.7	0.8	0.9	0.4	0.3	–	0.28	0.24	–	0.2
10	0.7	0.8	0.9	0.36	0.4*	0.3	0.252	0.32	0.27	0.32*

The optimal solution is marked with star. Tracing back from Table 12.21, we see that maximum reliability is 0.32 when $x_3 = 4$, $x_2 = 2$, and $x_1 = 4$.

12.9 SOLVING LINEAR PROGRAMMING PROBLEMS USING DYNAMIC PROGRAMMING

Let us consider an LPP with n decision variables and m constraints. This problem can be considered as an n stage dynamic programming problem with m state variables as there are n decision variables in the problem. However, the problem becomes complicated if we take $n \geq 3$. Therefore, we consider an LPP with two decision variables and m constraints as follows:

$$\text{Max } Z = c_1 x_1 + c_2 x_2$$

Subject to

$$\begin{array}{ccccc}
a_{11} x_1 & + & a_{12} x_2 & \leq & b_1 \\
a_{21} x_1 & + & a_{22} x_2 & \leq & b_2 \\
\cdots & \cdots & \cdots & & \cdots \\
a_{m1} x_1 & + & a_{m2} x_2 & \leq & b_n \\
\text{Stage 1} & & \text{Stage 2} & & \text{Resources}
\end{array}$$

We develop an algorithm for solving this problem using the forward dynamic programming method as follows:

Stage 1

$$F_1(x_1) = \text{Max } c_1 x_1,\ 0 \leq x_1 \leq \left(\frac{b_1 - a_{12}x_2}{a_{11}}, \frac{b_2 - a_{22}x_2}{a_{21}}, \ldots, \frac{b_m - a_{m2}x_2}{a_{m1}}\right)$$

$$= \text{Max } c_1 x_1, 0 \le x_1 \le \text{Min}\left(\frac{b_1 - a_{12}x_2}{a_{11}}, \frac{b_2 - a_{22}x_2}{a_{21}}, \ldots, \frac{b_m - a_{m2}x_2}{a_{m1}}\right)$$

Here,

$$x_1^* = \text{Min}\left(\frac{b_1 - a_{12}x_2}{a_{11}}, \frac{b_2 - a_{22}x_2}{a_{21}}, \ldots, \frac{b_m - a_{m2}x_2}{a_{m1}}\right) \tag{12.4}$$

$$F_1(x_2) = \text{Max } c_1\left(\frac{b_1 - a_{12}x_2}{a_{11}}, \frac{b_2 - a_{22}x_2}{a_{21}}, \ldots, \frac{b_m - a_{m2}x_2}{a_{m1}}\right)$$

Stage 2
We have

$$F_2(x_2) = \text{Max}\left[c_1 \text{ Min}\left(\frac{b_1 - a_{12}x_2}{a_{11}}, \frac{b_2 - a_{22}x_2}{a_{21}}, \ldots, \frac{b_m - a_{m2}x_2}{a_{m1}}\right) + c_2 x_2\right] \tag{12.5}$$

$$\text{Subject to} \quad x_2 \le \text{Min}\left\{\frac{b_1}{a_{12}}, \frac{b_2}{a_{22}}, \ldots, \frac{b_m}{a_{m2}}\right\} \tag{12.6}$$

Equation (12.6) gives the upper bound on x_2.

Now to calculate $\text{Min}\left(\frac{b_1 - a_{12}x_2}{a_{11}}, \frac{b_2 - a_{22}x_2}{a_{21}}, \ldots, \frac{b_m - a_{m2}x_2}{a_{m1}}\right)$, we proceed as follows:

(a) Let us consider $\frac{b_1 - a_{12}x_2}{a_{11}}$ to be less than or equal to the remaining $\frac{b_2 - a_{22}x_2}{a_{21}}, \frac{b_3 - a_{32}x_2}{a_{31}}, \ldots, \frac{b_m - a_{m2}x_2}{a_{m1}}$ one by one and find a condition on x_2. We obtain $(m - 1)$ conditions on x_2 and take the intersection of all these to get the maximum x_2 provided $a_{ij} \ge 0$ for i and j. Next, we find x_1 from Eq. (12.4) and Max Z from Eq. (12.5). This will give one solution set.

(b) Now let us consider $\frac{b_2 - a_{22}x_2}{a_{21}}$ to be less than or equal to (i.e., minimum of the remaining) $\frac{b_1 - a_{12}x_2}{a_{11}}, \frac{b_3 - a_{32}x_2}{a_{31}}, \ldots, \frac{b_m - a_{m2}x_2}{a_{m1}}$ and proceeding as in (a), we get another solution set.

Repeating this process, we get a maximum of m such solutions. Among these solutions, consider the one that provides the optimum solution to the problem.

Example 12.9 Solve the following problem using dynamic programming:

$$\text{Max } Z = 500x_1 + 800x_2$$

$$\begin{aligned} \text{Subject to} \quad & 5x_1 + 6x_2 \le 60 \\ & x_1 + 2x_2 \le 16 \\ & x_1 \le 8 \\ & x_2 \le 6 \\ \text{and} \quad & x_1, x_2 \ge 0 \end{aligned}$$

Solution The given problem can be written as follows:

$$\text{Max } Z = 500x_1 + 800x_2$$

$$\begin{aligned} \text{Subject to} \quad & 5x_1 + 6x_2 \le 60 \\ & x_1 + 2x_2 \le 16 \\ & x_1 + 0{\cdot}x_2 \le 8 \\ & 0{\cdot}x_1 + x_2 \le 6 \\ \text{and} \quad & x_1, x_2 \ge 0 \end{aligned} \tag{12.7}$$

As there are two variables in the problem, we shall have two stages to solve the problem using dynamic programming. We can solve it using either the forward or backward dynamic programming method. Let us first solve it using the forward dynamic programming method.

Stage 1

$$F_1(x_1) = \text{Max}\, 500x_1,\ 0 \le x_1 \le \left(\frac{60 - 6x_2}{5}, 16 - 2x_2, 8\right)$$

$$= \text{Max}\, 500x_1,\ 0 \le x_1 \le \text{Min}\left(\frac{60 - 6x_2}{5}, 16 - 2x_2, 8\right)$$

Thus, the optimum value of x_1 is $x_1^* = \text{Min}\left(\frac{60 - 6x_2}{5}, 16 - 2x_2, 8\right)$

$$F_1(x_2) = \text{Max } 500\left(\frac{60 - 6x_2}{5}, 16 - 2x_2, 8\right)$$

Stage 2 We have

$$F_2(x_2) = \text{Max}\left\{500 \text{ Min}\left(\frac{60 - 6x_2}{5}, 16 - 2x_2, 8\right) + 800x_2\right\} \tag{12.8}$$

$$\text{Subject to } x_2 \le \text{Min}\left\{\frac{60}{6}, \frac{16}{2}, 6\right\} = 6,\ 0 \le x_2 \le 6 \tag{12.9}$$

Now to calculate $\text{Min}\left(\frac{60 - 6x_2}{5}, 16 - 2x_2, 8\right)$, we proceed as follows:

(a) Assuming 8 to be minimum, we have $8 \le \frac{60 - 6x_2}{5} \Rightarrow x_2 \le 10/3$

In addition, $8 \le 16 - 2x_2 \Rightarrow x_2 \le 4$, but $x_2 \ge 0$

Thus, we have $0 \le x_2 \le 10/3$.

Using Eq. (12.8), we get

$$\begin{aligned} \text{Max } Z &= \text{Max}[500 \times 8 + 800x_2],\ \left(0 \le x_2 \le \frac{10}{3}\right) \\ &= \text{Max}[4000 + 800 \times 10/3] \\ &= 20{,}000/3 \text{ when } x_2 = 10/3 \text{ and } x_1^* = 8 \end{aligned} \tag{12.10}$$

(b) Let $\text{Min}\left(\frac{60 - 6x_2}{5}, 16 - 2x_2, 8\right) = \frac{60 - 6x_2}{5}$.

Then, we have $\frac{60 - 6x_2}{5} \le 8 \Rightarrow x_2 \ge 10/3$.

Further, $\frac{60 - 6x_2}{5} \le 16 - 2x_2 \Rightarrow x_2 \le 5,\ x_2 \ge 0$.

Thus, we have $10/3 \le x_2 \le 5$.

Using Eq. (12.8), we get

$$\text{Max } Z = \text{Max}\left[500\left(\frac{60-6x_2}{5}\right) + 800x_2\right] = \text{Max } [6000 + 200x_2], \left(\frac{10}{3} \le x_2 \le 5\right)$$
$$= 7000$$

when $x_2 = 5$ and $x_1^* = 6$ (12.11)

(c) Let $\text{Min}\left(\frac{60 - 6x_2}{5}, 16 - 2x_2, 8\right) = 16 - 2x_2.$

Then, we have $16 - 2x_2 \le 8 \Rightarrow x_2 \ge 4$

Further, $16 - 2x_2 \le \frac{60 - 6x_2}{5} \Rightarrow x_2 \ge 5, x_2 \ge 0$

Thus, we have $x_2 \ge 5$.

From Eq. (12.9), we have $x_2 \le 6$. Therefore, $5 \le x_2 \le 6$.

Now, using Eq. (12.8), we have

$$\text{Max } Z = \text{Max}[500(16 - 2x_2) + 800x_2], (5 \le x_2 \le 6)$$
$$= \text{Max}[8000 - 200x_2]$$
$$= 7000 \text{ when } x_2 = 5 \text{ and } x_1^* = 6 \quad (12.12)$$

From Eqs (12.10), (12.11), and (12.12), the optimal solution is

$$\text{Max } Z = \text{Max}[20000/3, 7000, 7000] = 7000 \text{ which is for } x_1^* = 6 \text{ and } x_2 = 5$$

Example 12.10 Solve the following problem using dynamic programming:

$$\text{Max } Z = 10x_1 + 8x_2$$

Subject to
$$2x_1 + x_2 \le 25$$
$$3x_1 + 2x_2 \le 45$$
$$x_2 \le 10$$
and
$$x_1, x_2 \ge 0$$

Solution The given problem can be written as follows:

$$\text{Max } Z = 10x_1 + 8x_2$$

Subject to
$$2x_1 + x_2 \le 25$$
$$3x_1 + 2x_2 \le 45$$
$$0 \cdot x_1 + x_2 \le 10$$
and
$$x_1, x_2 \ge 0 \quad (12.13)$$

As there are two variables in the problem, we shall have two stages to solve the problem using the dynamic programming method. We can solve it using either the forward or backward dynamic programming method. Let us first solve the problem using the backward dynamic programming method.

Stage 2

$$F_2(x_2) = \text{Max } 8x_2, 0 \le x_2 \le \left(25 - 2x_1, \frac{45 - 3x_1}{2}, 10\right)$$

$$= \text{Max } 8x_2, 0 \le x_2 \le \text{Min}\left(25 - 2x_1, \frac{45 - 3x_1}{2}, 10\right)$$

Here

Thus the optimum value of x_2 is $x_2^* = \text{Min}\left(25 - 2x_1, \dfrac{45 - 3x_1}{2}, 10\right)$

$$F_2(x_2) = F_2(x_1) = \text{Max } 8\left(25 - 2x_1, \frac{45 - 3x_1}{2}, 10\right)$$

Stage 1

We have

$$F_1(x_1) = \text{Max}\left[10x_1 + 8\left\{\text{Min}\left(25 - 2x_1, \frac{45 - 3x_1}{2}, 10\right)\right\}\right] \tag{12.14}$$

$$\text{Subject to} \quad x_1 \le \text{Min}\left\{\frac{25}{2}, \frac{45}{3}\right\} = \frac{25}{2}. \text{ Thus } 0 \le x_1 \le \frac{25}{2}. \tag{12.15}$$

Now to calculate $\text{Min}\left(25 - 2x_1, \dfrac{45 - 3x_1}{2}, 10\right)$, we proceed as follows:

(a) Assuming 10 to be minimum, we have

$10 \le 25 - 2x_1 \Rightarrow x_1 \le 15/2$

Further, $10 \le \dfrac{45 - 3x_1}{2} \Rightarrow x_1 \le 25/3, x_1 \ge 0.$

Thus, we have $0 \le x_1 \le 15/2$.

Using Eq. (12.14), we get

$$\text{Max } Z = \text{Max}[10x_1 + 8 \times 10], \left(0 \le x_1 \le \frac{15}{2}\right) = \text{Max}[10 \times 15/2 + 80]$$

$$= 155, \text{ when } x_1 = 15/2 \text{ and } x_2^* = 10. \tag{12.16}$$

(b) Let $\text{Min}\left(25 - 2x_1, \dfrac{45 - 3x_1}{2}, 10\right) = 25 - 2x_1.$

Then, we have $25 - 2x_1 \le \dfrac{45 - 3x_1}{2} \Rightarrow x_1 \ge 5.$

Further, $25 - 2x_1 \le 10 \Rightarrow x_1 \ge 15/2, x_1 \ge 0.$

Thus, we have $x_1 \ge 15/2$.

From Eq. (12.15) we have $x_1 \le 25/2$.

Therefore, we have $15/2 \le x_1 \le 25/2$.

Using Eq. (12.14), we get

$$\text{Max } Z = \text{Max}[10x_1 + 8(25 - 2x_1)] = \text{Max}[200 - 6x_1], (15/2 \le x_1 \le 25/2)$$

$$= 155, \text{ when } x_1 = 15/2 \text{ and } x_2^* = 10 \tag{12.17}$$

(c) Let $\text{Min}\left(25 - 2x_1, \dfrac{45 - 3x_1}{2}, 10\right) = \dfrac{45 - 3x_1}{2}.$

Then, we have $\dfrac{45 - 3x_1}{2} \le 25 - 2x_1 \Rightarrow x_1 \le 5$

Further, $\dfrac{45 - 3x_1}{2} \le 10 \Rightarrow x_1 \ge 25/3, x_1 \ge 0.$

Thus, there is no common region where $\text{Min}\left(25 - 2x_1, \dfrac{45 - 3x_1}{2}, 10\right) = \dfrac{45 - 3x_1}{2}.$

From Eqs (12.16) and (12.17), the optimal solution is obtained as follows:

$\text{Max } Z = \text{Max } [155, 155] = 155$, which is for $x_1 = 15/2$ and $x_2 = 10$.

12.10 DYNAMIC PROGRAMMING PROBLEM VS LINEAR PROGRAMMING PROBLEM

These problems can be distinguished on the basis of parameters such as objective function, constraints, technological coefficients, and solution procedure. Table 12.22 lists out the differences in detail.

Table 12.22 Differences between dynamic and linear programming problems

Characteristics	Dynamic programming problem	Linear programming problem
Objective function	May be linear or non-linear	Must be linear
Constraints	May be linear or non-linear	Must be linear
Technological coefficients	Must be positive	May be positive or negative
Solution procedure	By breaking it into different stages	Treated as a single problem

12.11 APPLICATIONS OF DYNAMIC PROGRAMMING

The following problems are the main applications of the dynamic programming method:

1. Shortest route problem
2. Capital budgeting problem
3. Cargo loading problem
4. Linear programming problem
5. Reliability problem
6. Inventory problem

RECAPITULATION

1. To solve an optimization problem using dynamic programming, first break it into different stages with each stage consisting of a single variable.
2. The optimal solution obtained at each stage is used as the input to obtain the optimum solution at the next stage.
3. A multistage decision problem can be solved using a mathematical technique known as dynamic programming based on the principle of Bellman's principle of optimality.
4. Dynamic programming can be used to solve an optimization problem with continuous or discrete decision variables. In the continuous case use the calculus method, whereas in the discrete case use the computational method.
5. Dynamic programming has a number of applications in solving problems of immense importance such as travelling salesmen problem, cargo loading, capital budgeting, and reliability problems.

EXERCISES

Multiple-choice Questions

12.1 Bellman's principle of optimality is used to solve
(a) the transportation problem
(b) an assignment problem
(c) the dynamic programming problem
(d) none of these

12.2 Dynamic programming can be solved using
(a) Hungarian algorithm
(b) Bellman's principle of optimality
(c) transportation problem algorithm
(d) none of these

12.3 The stages involved in solving an n variable dynamic programming problem are
(a) n
(b) $n - 1$
(c) $n + 1$
(d) none of these

State True or False

12.1 An LPP can be solved using dynamic programming.
12.2 Bellman's principle of optimality can be used to solve the travelling salesman problem.
12.3 A dynamic programming problem with n variables can be solved in less than n stages.
12.4 Every optimization problem can be solved using dynamic programming.

Fill in the Blanks

12.1 Bellman's principle of optimality is the main contribution of ________.
12.2 Bellman's principle of optimality is a framework for decomposing the problem into ________.
12.3 Dynamic programming determines the solution of an n variable problem by decomposing it into ________ stages.
12.4 In dynamic programming, the optimum solution of one subproblem is used as an ________ to the next subproblem.

Review Questions

12.1 State Bellman's principle of optimality.
12.2 Define following terms in dynamic programming:
(a) Stage
(b) State
(c) Principle of optimality
(d) Optimal decision rule
(e) Optimal policy
(f) Multistage decision problem
12.3 Outline the solution procedure of an LPP with two decision variables and m constraints using the dynamic programming problem method.

Numerical Problems

12.1 Determine the minimum path for the following travelling salesman problem represented graphically.

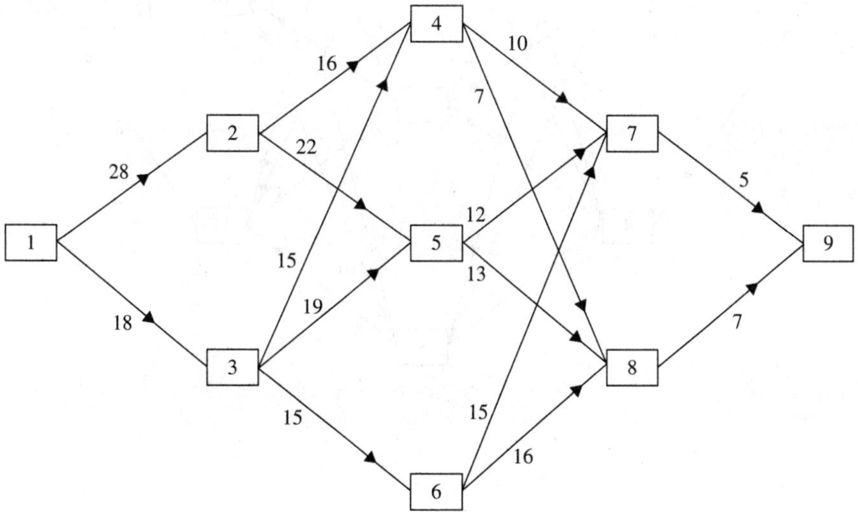

12.2 Solve the shortest route problem represented by the following network.

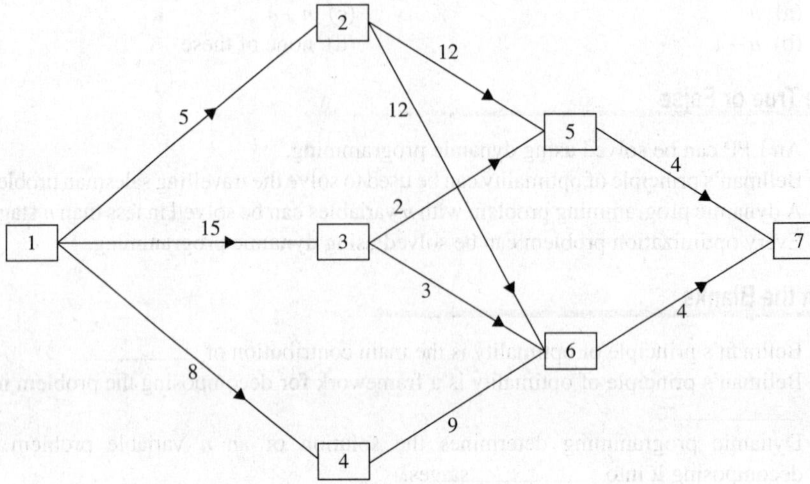

12.3 A salesperson has to travel from city 1 to city 12. The cost of travel and available routes are shown in the following diagram. Determine the minimum cost and the corresponding route.

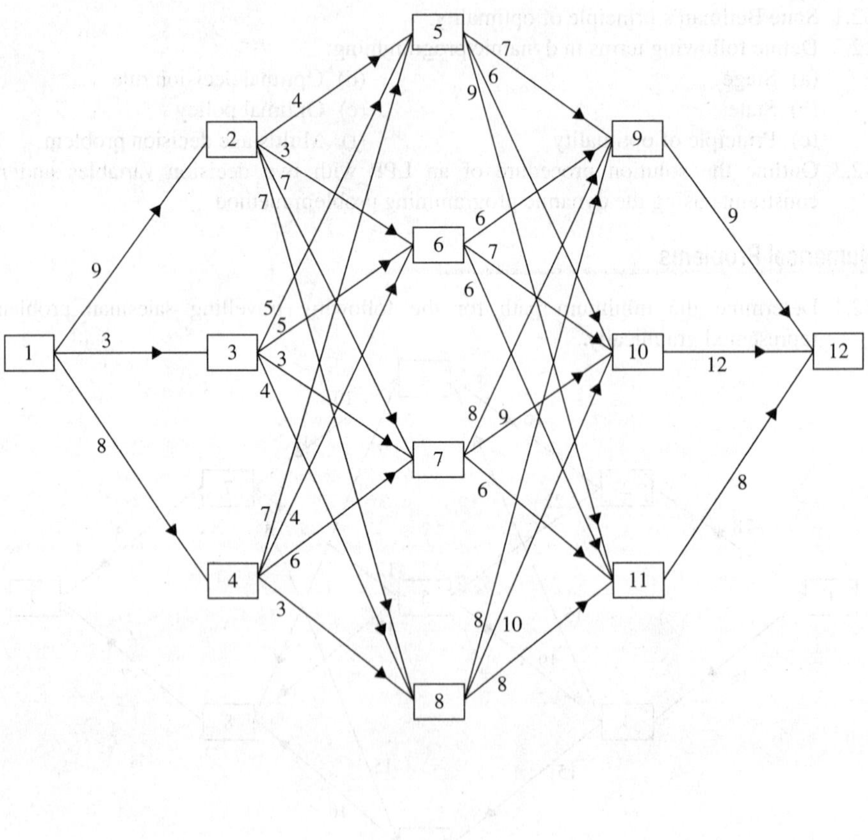

12.4 Solve the following travelling salesman problem using the dynamic programming technique.

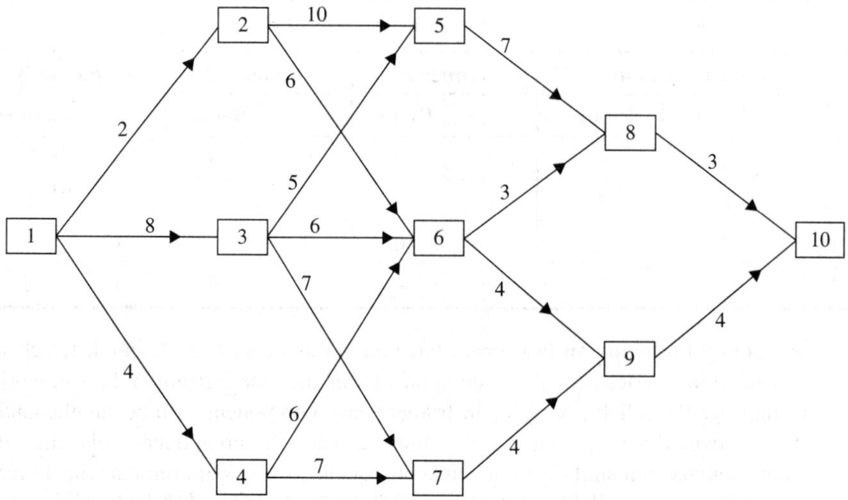

12.5 Solve the following problems using dynamic programming:

(a) Min $Z = x_1^2 + x_2^2 + x_3^2$
Subject to $x_1 + x_2 + x_3 \geq 10$
and $x_1, x_2, x_3 \geq 0$

(b) Min $Z = x_1 \log x_1 + x_2 \log x_2 + x_3 \log x_3$
Subject to $x_1 + x_2 + x_3 = 1$
and $x_1, x_2, x_3 \geq 0$

(c) Min $Z = x_1^2 + x_2^2 + 2x_3^2$
Subject to $x_1 + x_2 + 2x_3 = 12$
and $x_1, x_2, x_3 \geq 0$

(d) Max $Z = x_1^2 + 2x_2^2 + 4x_3$
Subject to $x_1 + x_2 + x_3 = 8$
and $x_1, x_2, x_3 \geq 0$

(e) Min $Z = x_1^2 + 2x_2^2 + 4x_3$
Subject to $x_1 + x_2 + x_3 = 6$
and $x_1, x_2, x_3 \geq 0$

(f) Min $Z = x_1^2 + x_2^2 + x_3^2$
Subject to $x_1 x_2 x_3 = 27$
and $x_1, x_2, x_3 \geq 0$

(g) Max $Z = x_1 x_2 x_3$
Subject to $x_1 + x_2 + x_3 = 12$
and $x_1, x_2, x_3 \geq 0$

(h) Max $Z = x_1 x_2 x_3 x_4$
Subject to $x_1 + x_2 + x_3 + x_4 = 12$
and $x_1, x_2, x_3, x_4 \geq 0$

(i) Min $Z = x_1^2 + x_2^2 + x_3^2$
Subject to $x_1 x_2 x_3 = 6$
$x_1, x_2, x_3 \geq 0$ and integers

12.6 How can a number c be divided into n parts such that the product of all the parts is maximum?
Max $Z = x_1, x_2, \ldots, x_n$
Subject to $x_1 + x_2 + \ldots + x_n = c$
and $x_1, x_2, \ldots, x_n \geq 0$

12.7 (Cargo loading problem) A vessel, which can carry a maximum load of 4 tons, is loaded with three items. The following table gives the unit weight, w_i, in tons and the unit revenue, r_i, in thousand ₹ for i items. How many units of each item should be loaded in the vessel to maximize the value of the loaded items.

Item i	w_i	r_i
1	2	31
2	3	47
3	1	14

12.8 (Capital budgeting problem) An industrialist has four companies in a hill station. He intends to spend ₹8 lakh in converting these companies to educational institutes.

The money x_i required to convert the ith company ($i = 1, 2, 3, 4$) into an E_3, E_2, or E_1 type educational institute and the corresponding returns are given in the following table. Maximize the total return. Note that company 2 cannot be converted into an E_1 class educational institute. All the data are in lakhs of ₹.

Edu. inst.	Industry Company 1		Company 2		Company 3		Company 4	
	x_1	Return	x_2	Return	x_3	Return	x_4	Return
	0	0	0	0	0	0	0	0
E_3	2	3	1	2	1	2	3	6
E_2	3	5	4	6	2	3	4	7
E_1	5	7	–	–	4	6	5	8

12.9 (Reliability Problem) An instrument has four components 1, 2, 3, and 4, which are connected in a series (i.e., if one component fails then the instrument fails to work). To improve the reliability of the instrument each component is to be supplemented by one, two, three, or four parallel units (in case the component fails, then the supplementary unit starts working and consequently the system remains functional). The total money available for improvement of the device is ₹15 lakh. The cost of parallel units and the reliability of each component are given in the following table:

No. of parallel units	Components—1		Components—2		Components—3		Components—4	
	Cost (in thousands of ₹)	R	Cost (in thousands of ₹)	R	Cost (in thousands of ₹)	R	Cost (in thousands of ₹)	R
1	2	0.6	1	0.4	1	0.3	2	0.4
2	4	0.7	3	0.5	2	0.4	3	0.5
3	6	0.8	5	0.7	3	0.5	4	0.6
4	8	0.9	7	0.8	4	0.6	6	0.8

In the aforegiven table, R denotes the reliability of the equipment. Let x_i denote the money spent on attaching supplementary units to components i. Find x_i, such that the reliability of the equipment is maximized.

CONCEPTUAL PROBLEM

The government is planning to launch a space satellite. For this project, three research teams have been constituted to conduct research on a technical problem to ensure successful launch of the satellite. The three teams, say T_1, T_2, and T_3, are trying three different approaches to solve the problem. The probability that these teams will not succeed in solving the problem is 0.4, 0.6, and 0.8 respectively. Thus, the probability of not succeeding in solving a problem by the three teams is 0.192. In order to reduce this probability of failure, the government has decided to have two additional experts for these teams. The following table gives the estimated probabilities that the respective teams will not succeed when the additional experts are attached to these teams:

(Contd)

(*Contd*)

Teams → / No. of experts ↓	T_1	T_2	T_3
0	0.4	0.6	0.8
1	0.2	0.4	0.5
2	0.15	0.2	0.3

Ascertain how many experts should be attached with each team so that the probability of not succeeding in solving the technical problem is minimized.

Solution Let x_1, x_2, and x_3 be the number of additional experts attached to teams T_1, T_2, and T_3 and $p_1(x_1)$, $p_2(x_2)$, and $p_3(x_3)$ be the reduced probabilities of failure after the addition of experts respectively. Then, the probability of failure to solve the problem collectively by the three teams is $Z = p_1(x_1)\,p_2(x_2)\,p_3(x_3)$.

Our objective is to minimize this subject to

$$x_1 + x_2 + x_3 = 2$$

and $x_1, x_2, x_3 \geq 0$ and integers.

This is a discrete case. Therefore, we cannot use the calculus method. We have to use the computational procedure of dynamic programming to solve this problem.

Let us define the state variables s_i's as

$$s_3 = x_1 + x_2 + x_3 = 2$$
$$s_2 = x_1 + x_2 = s_3 - x_3$$
$$s_1 = x_1 = s_2 - x_2$$

$x_1 = 0, 1, 2; s_1 = 0, 1, 2; x_2 = 0, 1, 2; x_3 = 0, 1, 2$

State transformations are

$$s_2 = s_3 - x_3 \text{ and } s_1 = s_2 - x_2; \text{ in addition, } s_2 = 2 - x_3, s_3 = 2.$$

Recursive relations are

$$F_1(s_1) = p_1(x_1) = p_1(s_1) \quad (12.19)$$
$$F_2(s_2) = \min[p_2(x_2) \cdot F_1(s_1)] \text{ with respect to } x_2 \quad (12.20)$$
$$F_3(s_3) = \min[p_3(x_3) \cdot F_2(s_2)] \text{ with respect to } x_3 \quad (12.21)$$

Using state transformations, we obtain the values of s_1, s_2, $F_1(s_1)$, $F_2(s_2)$, and $F_3(s_3)$ as shown in Tables 12.23 to 12.25.

Table 12.23 Values of s_1

s_2 \ x_2	0	1	2
0	0	–	–
1	1	0	–
2	2	1	0

Table 12.24 Values of s_2

s_3 \ x_3	0	1	2
2	2	1	0

(*Contd*)

(Contd)

Table 12.25 Values of $F_1(s_1)$

$x_1 = s_1$	0	1*	2
$F_1(s_1)$	0.4	0.2*	0.15

Table 12.26 Values of $F_2(s_2)$

s_2 \ x_2	$p_2(x_2)$			$F_1(s_1)$			$p_2(x_2)\ F_1(s_1)$			$F_2(s_2)$
	0	1	2	0	1	2	0*	1	2	Min over rows
0	0.6	–	–	0.4	–	–	0.24	–	–	0.24
1	0.6	0.4	–	0.2*	0.4	–	0.12*	0.16	–	0.12*
2	0.6	0.4	0.2	0.15	0.2	0.4	0.09	0.08	0.08	0.08

Table 12.27 Values of $F_3(s_3)$

s_3 \ x_3	$p_3(x_3)$			$F_2(s_2)$			$p_3(x_3)\ F_2(s_2)$			$F_3(s_3)$
	0	1	2	0	1	2	0	1*	2	Min over rows
2	0.8	0.5	0.3	0.08	0.12*	0.24	0.064	0.06	0.072	0.06*

Thus, retracing backwards we get Min $Z = F_3(s_3) = 0.06$ when $x_1 = 1, x_2 = 0$, and $x_3 = 1$. Hence, for the minimum probability of the launch of the satellite not succeeding, one expert each should be attached to teams T_1 and T_3 respectively.

13 Integer Programming

After studying this chapter, the reader will be able to understand the

- mathematical formulation of integer programming (IP)
- types of IP
- solution of IP using Gomory's cutting plane method
- solution of IP using branch and bound techniques
- Balas zero–one additive algorithm

13.1 INTRODUCTION

There are many problems wherein decision variables are required to be integers. In a linear programming problem (LPP), the decision variables can assume any non-negative value (integer or fractional). However, real life problems, such as capital budgeting, capacity expansion, shipping schedule, construction schedule, and location selection are a few examples where the fractional values of the decision variables are not acceptable. An LP in which some or all the decision variables in the optimal solution are restricted to non-negative integral values is called an integer programming problem (IPP).

In this chapter, we shall discuss (a) Gomory's cutting plane method for all integer programming problems (IPP), (b) Gomory's cutting plane method for mixed integer programming (MIP), (c) branch and bound technique, and (d) Balas additive algorithm to solve an IPP.

13.2 MATHEMATICAL FORMULATION OF INTEGER PROGRAMMING PROBLEMS

The mathematical formulation of an IP is as follows:

$$\text{Max or Min} \quad Z = C^{T}X$$

$$AX = b$$

$$\text{and} \quad X \geq 0$$

where $X = (x_1, x_2, \ldots, x_n)^T \geq 0$; x_j's are integers, $j = 1, 2, \ldots, n$; $C = (c_1, c_2, \ldots, c_n)^T$, $b = (b_1, b_2, \ldots, b_m)^T$

Example 13.1 A company has to manufacture the circular tops of cylindrical cans. The tops are to be made in two sizes, one of diameter 10 cm, and the other of diameter 20 cm. They are to be cut from metal sheets, which have the dimensions 20 cm by 70 cm. The requirement of the smaller size tops is 30,000 and of the larger size is 10,000. How should the tops from the metal sheets be cut so that the number of sheets used is minimum? Formulate the IPP.

Solution Let the sheets of size 20 cm by 70 cm be cut in four patterns. In pattern 1, the entire sheet is used to cut it in to 14 tops of 10 cm diameter. In pattern 2, one 20 cm diameter top and ten 10 cm diameter tops are cut. In pattern 3, two 20 cm diameter and six 10 cm diameter tops are cut. In pattern 4, three 20 cm diameter and two 10 cm diameter tops are cut from the given sheet.

Let x_1, x_2, x_3, and x_4 be the number of sheets cut according to the first, second, third, and fourth patterns respectively. Then the IPP is

$$\begin{aligned} &\text{Max} \quad Z = x_1 + x_2 + x_3 + x_4 \\ &\text{Subject to} \quad 14x_1 + 10x_2 + 6x_3 + 2x_4 \geq 30{,}000 \\ &\qquad\qquad x_2 + 2x_3 + 3x_4 \geq 10{,}000 \\ &\qquad\qquad x_1, x_2, x_3, x_4 \geq 0 \text{ are integers} \end{aligned}$$

13.3 TYPES OF INTEGER PROGRAMMING PROBLEMS

IPPs can be classified into the following three types:

All integer programming problem An LPP in which all the decision variables are restricted to the integer values is referred to as an all integer programming problem (AIPP).

Mixed integer programming problem An LPP in which some of the decision variables (not all) are restricted to the integer values is called a mixed integer programming problem (MIPP).

Zero–one integer programming problem An LPP in which all the decision variables are restricted to take the values 0 or 1 is called a zero–one IPP.

13.4 GOMORY'S CUTTING PLANE METHOD FOR AIPP

In 1956, R.E. Gomory developed a method to find the all integer solution of an IPP using the dual simplex method. The method is based on generating a sequence of linear inequalities (linear constraints) called cuts. The boundary of the cut is called the cutting plane. The cuts (additional linear constraints) thus generated are such that they never cut off the portion of the original feasible solution space that contains a feasible integer solution of the given problem.

Gomory's cutting plane method generates additional linear constraints (cuts) in a systematic way, to ensure an integer solution to a given IPP in a finite number of steps. In this method, first, the simplex method is used to solve the given IPP. If in the optimal simplex table, all the decision variables are integers, then it is the solution of the given IPP. However, if all the decision variables are not integers, then the basic variable, say x_r, is identified in the optimal table, having the largest fractional value among all the basic variables restricted to be integers. Then the row (chosen) corresponding to x_r in the optimal simplex table can be written as

$$x_{B_r} = 1 \cdot x_r + \sum_{j \neq r} a_{rj} x_j \text{ or } x_{B_r} = x_r + \sum_{j \neq r} a_{rj} x_j \quad (13.1)$$

where $x_j, j \neq r$ are all non-basic variables of the chosen row, x_{B_r} is the value of the rth basic variable, and a_{rj} is the coefficient of the variable x_r in the coefficient matrix.

Let us now decompose the coefficients of x_r, x_j, and x_{B_r} into integer and non-negative fractional parts in Eq. (13.1) to get

$$[x_{B_r}] + f_r = x_r + \sum_{j \neq r} \{[a_{rj}] + f_{rj}\} x_j \quad (13.2)$$

where $[x_{B_r}]$ and $[a_{rj}]$ are the greatest integers in x_{B_r} and $a_{rj}, j \neq r$ respectively.

Rearranging Eq. (13.2), so that all the integer coefficients appear on the LHS, we get

$$f_r + \left\{ [x_{B_r}] - x_r - \sum_{j \neq r} [a_{rj}] x_j \right\} = \sum_{j \neq r} f_{rj} x_j \quad (13.3)$$

where f_r is a strictly positive fraction $(0 < f_r < 1)$ and $f_{rj,\, r \neq j}$ is a non-negative fraction $(0 \leq f_{rj} < 1)$.

In Eq. (13.3), the RHS is positive, and the term f_r on the LHS is a strictly positive fraction. Thus, for Eq. (13.3) to hold good, the total of the terms in the bracket on the LHS must be a non-negative integer (as all the variables, including slack, assume integer values). Thus, Eq. (13.3) can be converted into an inequality as follows:

$$f_r \leq \sum_{j \neq r} f_{rj} x_j \text{ or } \sum_{j \neq r} f_{rj} x_j = f_r + s_g \quad (13.4)$$

$$\text{or} \quad -f_r = s_g - \sum_{j \neq r} f_{rj} x_j \quad (13.5)$$

where s_g is a non-negative slack variable. Equation (13.5) represents Gomory's cutting plane constraint. When this new constraint is added to the bottom of simplex table, it would create a new additional row in the table along with a column of the new basic variable s_g.

13.4.1 Algorithm for Gomory's Cutting Plane Method

The following steps can be used to solve an IPP by Gomory's cutting plane method.

1. Solve the given IPP to get its continuous solution using the simplex method. If all the decision variables in the optimal table are integers, then we have the solution of the problem. If not, go to step 2.
2. Identify the row of the basic variable, say x_r, with the largest fractional part in the optimal simplex table of step 1, and write it down as an equation. Decompose the coefficients of all the terms into the integer and non-negative fractional part, using the idea of the greatest integer function.

Note The greatest integer in x is the largest integer $\leq x$ and is denoted as $[x]$. For example, the greatest integer in 2.1 is denoted as $[2.1] = 2$ and the greatest integer in $-1.1 = [-1.1] = -2$.

3. Construct an additional constraint (Gomory's cut)

$$-f_r = s_g - \sum_{j \neq r} f_{rj} x_j$$

where f_r $(0 < f_r < 1)$ is a strictly positive fractional part of the basic variable x_r, s_g is the non-negative slack variable, and f_{rj}, $r \neq j$ are the non-negative fractions of the coefficients of the non-basic variables in the rth row equation.

4. Add the constraint of step 3 as a row at the bottom of the simplex table of step 1, and use the dual simplex method to get the optimal table. If the solution is an all integer solution, then the problem is solved. If not, repeat the process from step 2 to step 4 till the final optimal solution of the AIPP is obtained.

The aforementioned steps of Gomory's algorithm can be summarized in the form of a flow chart as shown in Fig. 13.1.

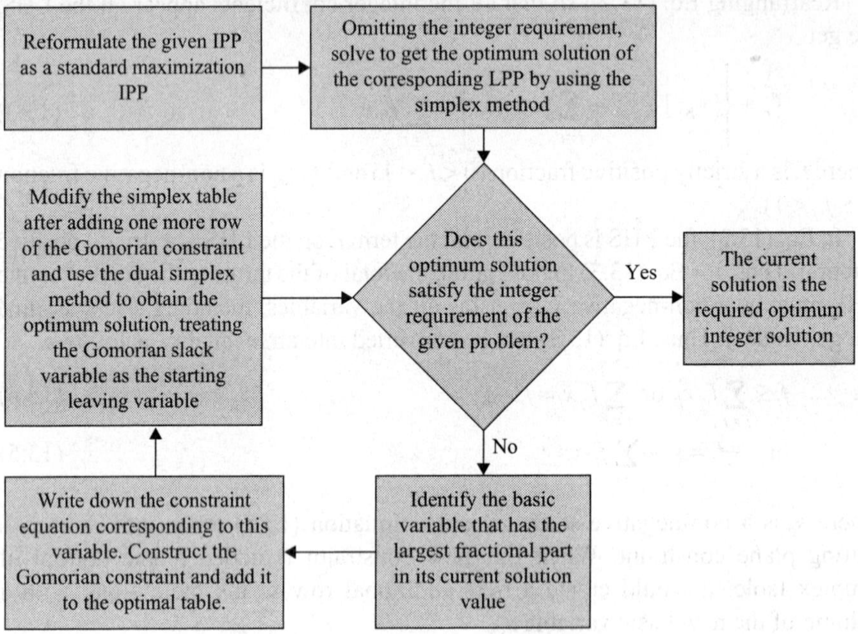

Figure 13.1 Gomory's algorithm

Let us understand Gomory's algorithm with the help of a few examples.

Example 13.2 Find the all integer solution to the following IPP:

Max $Z = 3x_1 + 2x_2$

Subject to $x_1 + x_2 \leq 4$

$x_1 - x_2 \leq 2$

$x_1, x_2 \geq 0$ and integers

Solution Reformulated IPP as a standard maximization IPP is

Max $Z = 3x_1 + 2x_2 + 0{\cdot}x_3 + 0{\cdot}x_4$

Subject to $x_1 + x_2 + x_3 = 4$

$x_1 - x_2 + x_4 = 2$

$x_1, x_2, x_3, x_4 \geq 0$ and integers; x_3 and x_4 are slack variables.

Now omitting the integer requirement, we solve the resulting LPP using the simplex method:

		Cost →		3	2	0	0	Minimum ratio
		Variable →		x_1	x_2	x_3	x_4	$\frac{X_{B_i}}{\alpha_i^j}$
Table no.	C_B	**Basic variable**	X_B	α^1	α^2	α^3	α^4	
1	0	x_3	4	1	1	1	0	4/1 = 4
	0	$\leftarrow x_4$	2	[1]	–1	0	1	2/1 = 2→
		$c_j - Z_j$		3↑	2	0	0	–
2	0	$\leftarrow x_3$	2	0	[2]	1	–1	2/2 = 1→
	3	x_1	2	1	–1	0	1	
		$c_j - Z_j$		0	5↑	0	–3	–
3	2	x_2	1	0	1	1/2	–1/2	–
	3	x_1	3	1	0	1/2	1/2	
		$c_j - Z_j$		0	0	–5/2	–1/2	–

The optimum solution is $x_1 = 3$, $x_2 = 1$, and Max $Z = 11$. This optimum solution satisfies the integer requirement of the given IPP and hence, is its optimum solution, and no further application of Gomory's cutting plane method is needed.

Example 13.3 Find the optimum all integer solution to the following IPP:

Max $Z = x_1 + 2x_2$

Subject to $2x_2 \leq 7$

$x_1 + x_2 \leq 7$

$2x_1 \leq 11$

$x_1, x_2 \geq 0$ and integers

Solution Omitting the integer condition, the LPP in the standard form is

Max $Z = x_1 + 2x_2 + 0 \cdot x_3 + 0 \cdot x_4 + 0 \cdot x_5$

Subject to $2x_2 + x_3 = 7$

$x_1 + x_2 + x_4 = 7$

$2x_1 + x_5 = 11$

$x_1, x_2, x_3, x_4, x_5 \geq 0$; x_3, x_4, and x_5 are slack variables

Using the simplex method, we have

		Cost →		1	2	0	0	0	Minimum ratio
		Variable →		x_1	x_2	x_3	x_4	x_5	$\frac{X_{B_i}}{\alpha_i^j}$
Table no.	C_B	**Basic variable**	X_B	α^1	α^2	α^3	α^4	α^5	
1	0	$\leftarrow x_3$	7	0	[2]	1	0	0	7/2 = 3.5→
	0	x_4	7	1	1	0	1	0	7/1 = 7
	0	x_5	11	2	0	0	0	1	…
		$c_j - Z_j$		1	2↑	0	0	0	

(*Contd*)

(*Contd*)

Cost →				1	2	0	0	0	Minimum ratio
Variable →				x_1	x_2	x_3	x_4	x_5	$\frac{X_{B_i}}{\alpha_i^j}$
Table no.	C_B	Basic variable	X_B	α^1	α^2	α^3	α^4	α^5	
2	2	x_2	7/2	0	1	1/2	0	0	…
	0	$\leftarrow x_4$	7/2	[1]	0	−1/2	1	0	7/2 = 3.5→
	0	x_5	11	2	0	0	0	1	11/2 = 5.5
	$c_j - Z_j$			1↑	0	−1	0	0	
3	2	x_2	7/2	0	1	1/2	0	0	
	1	x_1	7/2	1	0	−1/2	1	0	
	0	x_5	4	0	0	1	−2	1	
	$c_j - Z_j$			0	0	−1/2	−1	0	

The optimal solution of the LPP is $x_1 = 7/2$, $x_2 = 7/2$, and Max $Z = 21/2$, which is not an all integer solution. To get an all integer solution, we introduce the cut (additional constraint)

$$-\frac{1}{2} = -\frac{1}{2}x_3 + s_{g_1} \text{ (using } x_2 \text{ basic row)}$$

Adding this cut in the aforegiven simplex table, we get

Cost →				1	2	0	0	0	0
Variable →				x_1	x_2	x_3	x_4	x_5	s_{g_1}
Table no.	C_B	Basic variable	X_B	α^1	α^2	α^3	α^4	α^5	α^{g_1}
4	2	x_2	7/2	0	1	1/2	0	0	0
	1	x_1	7/2	1	0	−1/2	1	0	0
	0	x_5	4	0	0	1	−2	1	0
	0	$\leftarrow s_{g_1}$	−1/2	0	0	−1/2	0	0	1
	$c_j - Z_j$			0	0	−1/2↑	−1	0	0

The solution is infeasible. Restore feasibility using the dual simplex method. s_{g_1} leaves and x_3 enters to get the next simplex table.

Cost →				1	2	0	0	0	0
Variable →				x_1	x_2	x_3	x_4	x_5	s_{g_1}
Table no.	C_B	Basic variable	X_B	α^1	α^2	α^3	α^4	α^5	α^{g_1}
5	2	x_2	3	0	1	0	0	0	1
	1	x_1	4	1	0	0	1	0	−1
	0	x_5	3	0	0	0	−2	1	2
	0	x_3	1	0	0	1	0	0	−2
	$c_j - Z_j$			0	0	0	−1	0	−1

This is the all integer optimal simplex table. The all integer optimal solution of the given IPP is $x_1 = 4$, $x_2 = 3$, and Max $Z = 10$.

Example 13.4 Find the all integer solution to the following IPP:

Max $Z = 2x_1 - x_2$

Subject to $x_1 + x_2 \leq 2$

$x_1 - x_2 \leq 1$

$x_1, x_2 \geq 0$ and integers.

Solution The reformulated IPP as a standard maximization IPP is

Max $Z = 2x_1 - x_2 + 0{\cdot}x_3 + 0{\cdot}x_4$

Subject to $x_1 + x_2 + x_3 = 2$

$x_1 - x_2 + x_4 = 1$

$x_1, x_2, x_3, x_4 \geq 0$ and integers; x_3 and x_4 are slack variables.

Now omitting the integer requirement in the aforementioned IPP, we solve the resulting LPP as given here.

	Cost →			2	−1	0	0	Minimum ratio $\frac{X_{B_i}}{\alpha_i^j}$
	Variable →			x_1	x_2	x_3	x_4	
Table no.	C_B	**Basic variable**	X_B	α^1	α^2	α^3	α^4	
1	0	x_3	2	1	1	1	0	2/1 = 2
	0	←x_4	1	[1]	−1	0	1	1/1 = 1→
		$c_j - Z_j$		2↑	−1	0	0	
2	0	←x_3	1	0	[2]	1	−1	1/2→
	2	x_1	1	1	−1	0	1	...
		$c_j - Z_j$		0	1↑	0	−2	
3	−1	x_2	1/2	0	1	1/2	−1/2	
	2	x_1	3/2	1	0	1/2	1/2	
		$c_j - Z_j$		0	0	−1/2	−3/2	

This is the optimal simplex table. The optimum solution is $x_1 = 3/2$, $x_2 = 1/2$, and Max $Z = 5/2$, which is not an all integer solution. To get the all integer solution, we select the basic variable corresponding to the maximum fractional value f_{B_i}.

Here $x_{B_1} = I_{B_1} + f_{B_1} = 0 + 1/2$ and $x_{B_2} = I_{B_2} + f_{B_2} = 1 + 1/2$

Hence, $f_{B_1} = f_{B_2} = 1/2$

Therefore, $\max(f_{B_1}, f_{B_2}) = \max(1/2, 1/2) = 1/2$, which is the same for $x_2 (= x_{B_1})$ and $x_1 (= x_{B_2})$ basic variable rows, so, either of the two basic row equations can be chosen to construct a Gomorian cut.

Let us choose the basic variable x_1 row equation (second row of the optimal table). The cut is

$$-f_{B_2} = -f_{23}x_3 - f_{24}x_4 + s_{g_2}$$

or $$-\frac{1}{2} = -\frac{1}{2}x_3 - \frac{1}{2}x_4 + s_{g_2}$$

or $$-\frac{1}{2} = 0 \cdot x_1 + 0 \cdot x_2 - \frac{1}{2}x_3 - \frac{1}{2}x_4 + s_{g_2}$$

Adding this Gomorian cut in the aforegiven simplex table 3, we get

	Cost →			2	−1	0	0	0
	Variable →			x_1	x_2	x_3	x_4	s_{g_2}
Table no.	C_B	Basic variable	X_B	α^1	α^2	α^3	α^4	α^{g_2}
4	−1	x_2	1/2	0	1	1/2	−1/2	0
	2	x_1	3/2	1	0	1/2	1/2	0
	0	$\leftarrow s_{g_2}$	−1/2	0	0	[−1/2]	−1/2	1
		$c_j - Z_j$		0	0	−1/2↑	−2	0

The solution is infeasible. Restore feasibility using the dual simplex method. When s_{g_2} leaves, then x_3 enters $\left\{ \text{as } \min\left(\left|\frac{c_3 - z_3}{\alpha_3^3}\right|, \left|\frac{c_4 - z_4}{\alpha_3^4}\right|\right) = \min(1, 4) = 1 \text{ for } x_3 \text{ column} \right\}$ to get the next simplex table.

	Cost →			2	−1	0	0	0
	Variable →			x_1	x_2	x_3	x_4	s_{g_2}
Table no.	C_B	Basic variable	X_B	α^1	α^2	α^3	α^4	α^{g_2}
5	−1	x_2	0	0	1	0	−1	1
	2	x_1	1	1	0	0	0	1
	0	x_3	1	0	0	1	1	−2
		$c_j - Z_j$		0	0	0	−3/2	−1

This is the all integer optimal simplex table. The all integer optimal solution of the given IPP is $x_1 = 1$, $x_2 = 0$, and Max $Z = 2$.

Example 13.5 Solve the following IPP:

$$\text{Max } Z = 2x_1 + 20x_2 - 10x_3$$
$$\text{Subject to } \quad 2x_1 + 20x_2 + 4x_3 \leq 15$$
$$6x_1 + 20x_2 + 4x_3 = 20$$
$$x_1, x_2, x_3 \geq 0 \text{ and integers}$$

Solution Deleting the integer requirement, introducing the artificial variable x_5, and modifying the objective function, the standard form is given as

$$\text{Max } Z = 2x_1 + 20x_2 - 10x_3 + 0 \cdot x_4 - Mx_5$$
$$\text{Subject to } \quad 2x_1 + 20x_2 + 4x_3 + x_4 = 15$$
$$6x_1 + 20x_2 + 4x_3 + x_5 = 20$$

$x_1, x_2, x_3, x_4, x_5 \geq 0$; x_4 is the slack variable and x_5 is the artificial variable.

Now, solving the LPP using the simplex method, the optimum simplex table (deleting the x_5 column) is

Cost →				2	20	−10	0
Variable →				x_1	x_2	x_3	x_4
Table no.	C_B	Basic variable	X_B	α^1	α^2	α^3	α^4
1	20	x_2	5/8	0	1	1/5	3/40
	2	x_1	5/4	1	0	0	−1/4
	$c_j - Z_j$			0	0	−14	−1

The non-integer optimum solution is $x_1 = 5/4$, $x_2 = 5/8$, and Max Z =15. To obtain the integer valued solution, we consider

$$\max(f_{B_1}, f_{B_2}) = \max\left(\frac{5}{8}, \frac{1}{4}\right) = \frac{5}{8}, \text{ which is for } x_2 (\text{first row}).$$

Gomory's cut (additional constraint), using the first row of optimal simplex table 1 is

$$-\frac{5}{8} = 0{\cdot}x_1 + 0{\cdot}x_2 - \frac{1}{5}x_3 - \frac{3}{40}x_4 + s_{g_1}$$

Inserting the additional constraint in the aforegiven table and using dual simplex iterations, we get

Cost →				2	20	−10	0	0	Minimum ratio
Variable →				x_1	x_2	x_3	x_4	s_{g_1}	$\min_j \left\lvert \frac{c_j - Z_j}{\alpha_r^j} \right\rvert$
Table no.	C_B	Basic variable	X_B	α^1	α^2	α^3	α^4	α^{g_1}	
2	20	x_2	5/8	0	1	1/5	3/40	0	$\min\left(\left\lvert\frac{-14}{-1/5}\right\rvert, \left\lvert\frac{-1}{-3/40}\right\rvert\right) = 40/3$
	2	x_1	5/4	1	0	0	−1/4	0	
	0	$\leftarrow s_{g_1}$	−5/8	0	0	−1/5	[−3/40]	1	for x_3 column
	$c_j - Z_j$			0	0	−14	−1↑	0	–
3	20	x_2	0	0	1	0	0	1	
	2	x_1	10/3	1	0	2/3	0	−10/3	–
	0	$\leftarrow x_4$	25/3	0	0	8/3	1	−40/3	
	$c_j - Z_j$			0	0	−34/3	0	−40/3	–

This is not an all integer solution. To obtain the integer valued solution, we consider

$\max(f_{B_2}, f_{B_3}) = \max(1/3, 1/3) = 1/3$, which is the same for the second and third rows.

Choose a second row arbitrarily (x_1).

From the x_1 row, we have $10/3 = 1{\cdot}x_1 + 0{\cdot}x_2 + \frac{2}{3}x_3 + 0{\cdot}x_4 - \frac{10}{3}s_{g_1}$

$$\text{or} \quad 3 + \frac{1}{3} = (1+0)x_1 + (0+0)x_2 + \left(0 + \frac{2}{3}\right)x_3 + 0x_4 + \left(-4 + \frac{2}{3}\right)s_{g_1}$$

The Gomorian cut is $-\frac{1}{3} = -\frac{2}{3}x_3 - \frac{2}{3}s_{g_1} + s_{g_2}$

Adding this cut in the aforegiven simplex table, we get

		Cost →		2	20	−10	0	0	0
		Variable →		x_1	x_2	x_3	x_4	s_{g_1}	s_{g_2}
Table no.	C_B	**Basic variable**	X_B	α^1	α^2	α^3	α^4	α^{g_1}	α^{g_2}
4	20	x_2	0	0	1	0	0	1	0
	2	x_1	10/3	1	0	2/3	0	−10/3	0
	0	x_4	25/3	0	0	8/3	1	−40/3	0
	0	$\leftarrow s_{g_2}$	−1/3	0	0	−2/3	0	−2/3	1
		$c_j - Z_j$		0	0	−34/3↑	0	−40/3	0

The solution is infeasible. Restore feasibility using the dual simplex method. When s_{g_2} leaves, then x_3 enters $\left\{\text{as } \min\left(\left|\frac{c_3 - z_3}{\alpha_4^4}\right|, \left|\frac{c_5 - z_5}{\alpha_5^4}\right|\right) = \min(17, 20) = 17 \text{ for } x_3 \text{ column}\right\}$ to get the next simplex table.

		Cost →		2	20	−10	0	0	0
		Variable →		x_1	x_2	x_3	x_4	s_{g_1}	s_{g_2}
Table no.	C_B	**Basic variable**	X_B	α^1	α^2	α^3	α^4	α^{g_1}	α^{g_2}
5	20	x_2	0	0	1	0	0	1	0
	2	x_1	3	1	0	0	0	−4	1
	0	x_4	7	0	0	0	1	−16	4
	−10	x_3	1/2	0	0	1	0	1	−3/2
		$c_j - Z_j$		0	0	0	0	−2	−17

Since the solution is still a non-integer, a third fractional cut is required. From x_3 row (last row of the aforegiven table 5), we construct a Gomorian cut $-\frac{1}{2} = -\frac{1}{2}s_{g_2} + s_{g_3}$. Adding this cut in the aforegiven simplex table, we get

		Cost →		2	20	−10	0	0	0	0
		Variable →		x_1	x_2	x_3	x_4	s_{g_1}	s_{g_2}	s_{g_3}
Table no.	C_B	**Basic variable**	X_B	α^1	α^2	α^3	α^4	α^{g_1}	α^{g_2}	α^{g_3}
6	20	x_2	0	0	1	0	0	1	0	0
	2	x_1	3	1	0	0	0	−4	1	0
	0	x_4	7	0	0	0	1	−16	4	0
	−10	x_3	1/2	0	0	1	0	1	−3/2	0
	0	$\leftarrow s_{g_3}$	−1/2	0	0	0	0	0	[−1/2]	1
		$c_j - Z_j$		0	0	0	0	−2	−17↑	0

(Contd)

(Contd)

Cost →				2	20	−10	0	0	0	0
Variable →				x_1	x_2	x_3	x_4	s_{g_1}	s_{g_2}	s_{g_3}
Table no.	C_B	Basic variable	X_B	α^1	α^2	α^3	α^4	α^{g_1}	α^{g_2}	α^{g_3}
7	20	x_2	0	0	1	0	0	1	0	0
	2	x_1	2	1	0	0	0	−4	0	2
	0	x_4	3	0	0	0	1	−16	0	8
	−10	x_3	2	0	0	1	0	1	0	−3
	0	$\leftarrow s_{g_2}$	1	0	0	0	0	−1/2	1	−2
	$c_j - Z_j$			0	0	0	0	−2	0	−34

This is all integer solution. The optimal integer solution of the given IPP is $x_1 = 2$, $x_2 = 0$, $x_3 = 2$, and Max $Z = -16$.

13.5 GOMORY'S CUTTING PLANE METHOD FOR MIPP

If all the variables in an integer linear programming problem are not restricted as integers, it is called an MIPP. The construction of a cut in an MIPP is different from that in an AIPP. To construct a cut in case of an MIPP, we start with the optimal solution of the LPP obtained using the simplex method after ignoring the integer requirement(s). Now, from the optimal table thus obtained, choose the ith row corresponding to a basic variable x_{B_i}, which has the largest fractional value amongst those required to be integers. The x_{B_i} row in the optimal table has the following forms:

$$x_i + \sum_{j=1}^{p} f_{ij}^{+} x_j + \sum_{j=p+1}^{q} f_{ij}^{-} x_j = b_i$$

or $$x_i + \sum_{j=1}^{p} f_{ij}^{+} x_j + \sum_{j=p+1}^{q} f_{ij}^{-} x_j = [b_i] + f_{b_i} \tag{13.6}$$

where x_j, ($j = 1$ to p and $j = p + 1$ to q) are the non-basic variables with positive coefficients f_{ij}^{+} and negative coefficients f_{ij}^{-} respectively. f_{b_i}, $0 < f_{b_i} < 1$ is the fractional part of the right-hand side and $[b_i]$ is greater integer in b_i, which is less than or equal to b_i.

Equation (13.6) can be written as

$$\sum_{j=1}^{p} f_{ij}^{+} x_j + \sum_{j=p+1}^{q} f_{ij}^{-} x_j = f_{b_i} + \{[b_i] - x_i\} \tag{13.7}$$

The right-hand side in Eq. (13.7) is a number which may be ≤ 0 or ≥ 0.

Case 1 Consider the case when $f_{b_i} + \{[b_i] - x_i\} \geq 0$. Then $[b_i] - x_i = 0, 1, 2, \ldots$, as $[b_i]$ and x_i are integers. Thus Eq. (13.7) becomes,

$$\sum f_{ij}^{+} x_j + \sum f_{ij}^{-} x_j \geq f_{b_i} \tag{13.8}$$

and clearly $\sum f_{ij}^{+} x_j \geq \sum f_{ij}^{-} x_j + \sum f_{ij}^{-} x_j$ (13.9)

as x_j's ≥ 0 and $f_{ij}^{-} < 0$.

From Eqs (13.8) and (13.9), we have $\sum f_{ij}^{+} x_j \geq f_{b_i}$ (13.10)

Case 2 In this case, $f_{b_i} + \{[b_i] - x_i\} < 0$ will imply $[b_i] - x_i = -1, -2, \ldots,$ as the difference is an integer.

Using this fact from Eq. (13.7), we have

$$\sum f_{ij}^{+} x_j + \sum f_{ij}^{-} x_j \leq f_{b_i} - 1 \tag{13.11}$$

Hence, clearly, $\sum f_{ij}^{-} x_j \leq \sum f_{ij}^{+} x_j + \sum f_{ij}^{-} x_j$ (13.12)

From Eqs (13.11) and (13.12), we have

$$\sum f_{ij}^{-} x_j \leq f_{b_i} - 1 \text{ or } \frac{f_{b_i}}{f_{b_i} - 1} \sum f_{ij}^{-} x_j > f_{b_i} \text{ as } 0 < f_{b_i} < 1 \tag{13.13}$$

From Eqs (13.10) and (13.13), we have

$$\sum f_{ij}^{+} x_{ij} + \frac{f_{b_i}}{f_{b_i} - 1} \sum f_{ij}^{-} x_j \geq f_{b_i}$$

$$\text{or } -f_{b_i} = -\sum f_{ij}^{+} x_j - \frac{f_{b_i}}{f_{b_i} - 1} \sum f_{ij}^{-} x_j + s_{g_i} \tag{13.14}$$

This is Gomory's cut for the MIPP.

Note For an MIPP in Gomory's cut, f_{ij}^{+}, f_{ij}^{-} are just the positive and negative coefficients of non-basic variables, without being converted into positive fractional values, but f_{b_i} is a non-negative fractional part of a basic variable x_i.

Example 13.6 Solve the following MIPP:

Max $Z = 2x_1 + x_2$

Subject to $3x_1 + 2x_2 \leq 5$

$x_2 \leq 2$

$x_1, x_2 \geq 0$ and x_1 is an integer.

Solution The standard form of the given MIPP is

Max $Z = x_1 + x_2 + 0{\cdot}x_3 + 0{\cdot}x_4$

Subject to $3x_1 + 2x_2 + x_3 = 5$

$x_2 + x_4 = 2$

$x_1, x_2, x_3, x_4 \geq 0$ and x_1 is an integer; x_3 and x_4 are slack variables.

Ignoring the integer requirement, the optimal solution using the simplex method is obtained as given in the following simplex table:

Cost →				**3**	**2**	**0**	**0**
Variable →				x_1	x_2	x_3	x_4
Table no.	C_B	**Basic variable**	X_B	α^1	α^2	α^3	α^4
1	1	x_1	1/3	1	0	1/3	−2/3
	1	x_2	2	0	1	0	1
		$c_j - Z_j$		0	0	−1/3	−1/3

Here, x_1 is not an integer which is required to be an integer. From x_1 row (first row), we have

$$\frac{1}{3} = x_1 + \frac{1}{3}x_3 - \frac{2}{3}x_4$$

Thus, the Gomory's cut is

$$-\frac{1}{3} = -\frac{1}{3}x_3 - \frac{1/3}{\frac{1}{3}-1}\left(-\frac{2}{3}\right)x_4 + s_{g_1} \quad \left[\text{using } -f_{b_i} = -\sum f_{ij}^{+}x_j - \frac{f_{b_i}}{f_{b_i}-1}\sum f_{ij}^{-}x_j + s_{g_i}\right]$$

or $\quad -\frac{1}{3} = -\frac{1}{3}x_3 - \frac{1}{3}x_4 + s_{g_1}$

Adding this additional constraint in simplex table 1, we get

		Cost →		3	2	0	0	0	**Minimum ratio**
		Variable →		x_1	x_2	x_3	x_4	s_{g_1}	$\min_j \left\lvert \frac{c_j - Z_j}{\alpha_r^j} \right\rvert$
Table no.	C_B	**Basic variable**	X_B	α^1	α^2	α^3	α^4	α^{g_1}	
	1	x_1	1/3	1	0	1/3	−2/3	0	$\min\left(\left\lvert\frac{-1/3}{-1/3}\right\rvert, \left\lvert\frac{-1/3}{-1/3}\right\rvert\right) = (1, 1) = 1$
	1	x_2	2	0	1	0	1	0	(tie between x_3 and x_4)
	0	$\leftarrow s_{g_1}$	−1/3	0	0	[−1/3]	−1/3	1	when s_{g_1} leaves, choose x_3 arbitrarily to enter.
		$c_j - Z_j$		0	0	−1/3↑	−1/3	0	–
	1	x_1	0	1	0	0	−1	1	
	1	x_2	2	0	1	0	1	0	–
	0	x_3	1	0	0	1	1	−3	
		$c_j - Z_j$		0	0	0	0	−1	–

In this solution, x_1 is an integer. Thus the optimum solution is $x_1 = 0$, $x_2 = 2$, and Max $Z = 2$.

Example 13.7 Solve the following MIPP:

Max $Z = 4x_1 + 6x_2 + 2x_3$

Subject to $4x_1 - 4x_2 \le 5$

$-x_1 + 6x_2 \le 5$

$-x_1 + x_2 + x_3 \le 5$

$x_1, x_2, x_3 \ge 0$ and x_2 is an integer.

Solution The given problem can be written in the standard form as

Max $Z = 4x_1 + 6x_2 + 2x_3 + 0{\cdot}x_4 + 0{\cdot}x_5 + 0{\cdot}x_6$

Subject to $4x_1 - 4x_2 + x_4 = 5$

$-x_1 + 6x_2 + x_5 = 5$

$-x_1 + x_2 + x_3 + x_6 = 5$

$x_1, x_2, x_3, x_4, x_5, x_6 \ge 0$; x_2 is an integer and x_4, x_5, and x_6 are the slack variables.

Ignoring the integer requirement, the optimal solution using the simplex method is obtained as given in the following simplex table 1:

		Cost →		4	6	2	0	0	0
		Variable →		x_1	x_2	x_3	x_4	x_5	x_6
Table no.	C_B	Basic variable	X_B	α^1	α^2	α^3	α^4	α^5	α^6
1	4	x_1	5/2	1	0	0	3/10	1/5	0
	6	x_2	5/4	0	1	0	1/20	1/5	0
	2	x_3	25/4	0	0	1	1/4	0	1
	$c_j - Z_j$			0	0	0	−2	−2	−2

The non-integer optimal solution is $x_1 = 5/2$, $x_2 = 5/4$, $x_3 = 25/4$, and Max $Z = 30$. Here $x_2 = 5/4$, which is a non-integer and is required to be an integer. Thus, the x_2-row (second row in the aforegiven simplex table) can be written as

$$\frac{5}{4} = 0 \cdot x_1 + 1 \cdot x_2 + 0 \cdot x_3 + \frac{1}{20}x_4 + \frac{1}{5}x_5 + 0 \cdot x_6$$

The Gomory's cut for the MIPP is $-f_{b_i} = -\sum f_{ij}^{+} x_j - \dfrac{f_{b_i}}{f_{b_i} - 1} \sum f_{ij}^{-} x_j + s_{g_i}$

$$\text{Here } i = 2,\ f_{b_i} = \frac{1}{4},\ f_{24}^{+} = \frac{1}{20},\ f_{25}^{+} = \frac{1}{5},\ f_{26}^{-} = 0$$

Thus, the Gomory's cut is $-\dfrac{1}{4} = -\dfrac{1}{20}x_4 - \dfrac{1}{5}x_5 + s_{g_2}$

Adding this constraint in the aforegiven table, the new simplex table 2 is

		Cost →		4	6	2	0	0	0	0
		Variable →		x_1	x_2	x_3	x_4	x_5	x_6	s_{g_2}
Table no.	C_B	Basic variable	X_B	α^1	α^2	α^3	α^4	α^5	α^6	α^{g_2}
2	4	x_1	5/2	1	0	0	3/10	1/5	0	0
	6	x_2	5/4	0	1	0	1/20	1/5	0	0
	2	x_3	25/4	0	0	1	1/4	0	1	0
	0	$\leftarrow s_{g_2}$	−1/4	0	0	0	−1/20	−1/5	0	1
	$c_j - Z_j$			0	0	0	−2	−2↑	−2	0

The solution is infeasible. Restore feasibility using the dual simplex method. When s_{g_2} leaves, then x_5 enters

$$\left\{ \text{as min} \left(\left| \frac{c_4 - z_4}{\alpha_4^4} \right|, \left| \frac{c_5 - z_5}{\alpha_5^4} \right| \right) = \text{min}\,(40, 10) = 10 \text{ for } x_5 \text{ column} \right\}.$$

The next simplex table 3 is

		Cost →		4	6	2	0	0	0	0
		Variable →		x_1	x_2	x_3	x_4	x_5	x_6	s_{g_1}
Table no.	C_B	Basic variable	X_B	α^1	α^2	α^3	α^4	α^5	α^6	α^{g_1}
3	4	x_1	9/4	1	0	0	1/4	0	0	1
	6	x_2	1	0	1	0	0	0	0	1
	2	x_3	25/4	0	0	1	1/4	0	1	0
	0	x_5	5/4	0	0	0	1/4	1	0	−5
		$c_j - Z_j$		0	0	0	−3/2	0	−2	−10

This is the optimal table with x_2 as an integer. Thus, the optimum solution is $x_1 = 9/4, x_2 = 1$, $x_3 = 25/4$, and Max $Z = 55/2$.

13.6 DIFFERENCE BETWEEN GOMORY'S CUTTING PLANE METHOD FOR AIPP AND MIPP

In an AIPP, for Gomory's cut, we use the fractional parts of all non-basic variables of the optimum simplex table, including the basic variable.

In an MIPP, we only use the coefficients and not the fractional parts of all non-basic variables. However, we use the fractional part of basic variables having the most positive fractional part in an optimal simplex table.

On the basis of the aforementioned differences, the Gomory's cut is different for AIPP and MIPP.

13.7 BRANCH AND BOUND TECHNIQUE TO FIND SOLUTION OF IPP

This method was first introduced by A.H. Land and A.G. Doig in 1960. The branch and bound technique is applicable if the problem only has a finite number of feasible solutions, which are examined to find the optimum solution. The number of feasible solutions becomes larger with the increase in the variables involved in the problem. It is not possible to examine all these solutions. In this technique, we go on filtering large subsets of feasible solutions in a systematic way to get the solution of the problem in a finite number of steps. The following steps are used to find the optimal solution of an IPP using branch and bound techniques:

Step 1 Let the given IPP be a maximization problem. If it is a minimization problem, convert it into a maximization problem by multiplying the objective function with (−1).

Step 2 Solve the given IPP using the simplex method ignoring the integer requirement. If the solution obtained is an integer, it is the optimal solution. If it is not an integer solution, then go to Step 3.

Step 3 Consider the objective function value Z in a maximization problem as upper bound, along with the solution obtained in Step 2 and call it node 1. Now, if x_k^* is the non-integer value of x_k in node 1, which is required to be an integer, then introduce two branches (sub problems): (a) $x_k \le [x_k^*]$ and (b) $x_k \ge [x_k^*] + 1$, where $[x_k^*]$ is the greatest integer $\le x_k^*$.

Step 4 Solve sub problems (a) and (b) using sensitivity analysis and dual simplex method, getting two nodes say, node 2 and node 3. Associate the value of Z as upper bound with these nodes for comparison.

Step 5 From the terminal nodes, branch off from the node for which the value of Z (upper bound) is the largest. Finally, the process will end if the terminal nodes are (a) with integer solution, (b) non-feasible solution, or (c) fathomed nodes. The best out of the nodes of type (a) gives the optimal solution of IPP.

Note Let α be upper bound with the best node, i.e., with the largest Z and β be the Z value of some other node such that $\beta \le \alpha$. Then this node is not branched off further, as its branching will not give a better Z-value. Such nodes are called fathomed nodes.

Example 13.8 Solve the following IPP using the branch and bound techniques:

Max $Z = 2x_1 + 3x_2$

Subject to $-x_1 + x_2 \le 1$

$2x_1 + x_2 \le 2$

$x_1, x_2 \ge 0$ and integers.

Solution The given problem in standard form can be written as

Max $Z = 2x_1 + 3x_2 + 0 \cdot x_3 + 0 \cdot x_4$

Subject to $-x_1 + x_2 + x_3 = 1$

$2x_1 + x_2 + x_4 = 2$

$x_1, x_2, x_3, x_4 \ge 0$ and integers; x_3, x_4 are the slack variables.

Now, omitting the integer requirement in the aforementioned IPP, we solve the resulting LPP as follows:

		Cost →		2	3	0	0	Minimum ratio
		Variable →		x_1	x_2	x_3	x_4	$\frac{X_{B_i}}{\alpha_i^j}$
Table no.	C_B	Basic variable	X_B	α^1	α^2	α^3	α^4	
1	0	$\leftarrow x_3$	1	−1	[1]	1	0	1/1 = 1→
	0	x_4	2	2	1	0	1	2/1 = 2
		$c_j - Z_j$		2	3↑	0	0	–
2	3	x_2	1	−1	1	1	0	…
	0	$\leftarrow x_4$	1	[3]	0	−1	1	1/3→
		$c_j - Z_j$		5↑	0	−3	0	–
3	3	x_2	4/3	0	1	2/3	1/3	–
	2	x_1	1/3	1	0	−1/3	1/3	
		$c_j - Z_j$		0	0	−4/3	−5/3	–

The optimum solution is $x_1 = 1/3$, $x_2 = 4/3$, and Max $Z = 14/3$ (node 1), which is not an all integer solution. Here, both x_1 and x_2 are non-integers, we choose x_1 arbitrarily and introduce two branches (sub problems) (a) $x_1 \geq 1$ and (b) $x_1 \leq 0$, where $[x_1] = 0$.

Now solve the aforementioned two sub problems using the sensitivity analysis and dual simplex method, getting two nodes say, node 2 and node 3 (Fig. 13.2).

(a) For node 2, the sub problem is

Max $Z = 2x_1 + 3x_2 + 0 \cdot x_3 + 0 \cdot x_4 + 0 \cdot x_5$

Subject to $-x_1 + x_2 + x_3 = 1$

$2x_1 + x_2 + x_4 = 2$

$-x_1 + x_5 = -1$

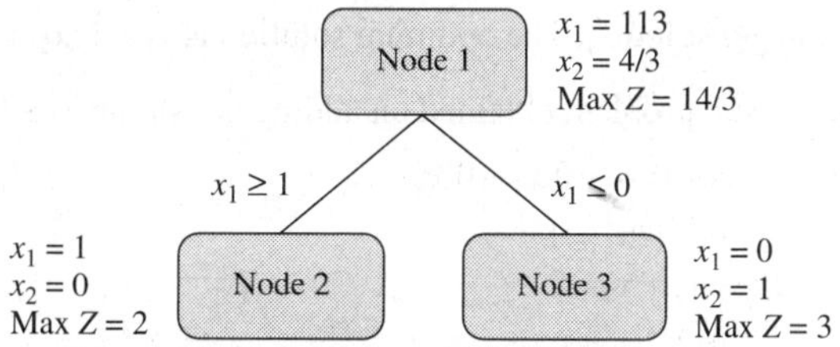

Figure 13.2 Branching at Node 1

$x_1, x_2, x_3, x_4, x_5 \geq 0$ and integers; x_3, x_4, and x_5 are the slack variables.

Adding this additional constraint in simplex table 3, the new simplex table is

Cost →				**2**	**3**	**0**	**0**	**0**	**Minimum ratio**
Variable →				x_1	x_2	x_3	x_4	x_5	$\frac{X_{B_i}}{\alpha_i^j}$
Table no.	C_B	**Basic variable**	X_B	α^1	α^2	α^3	α^4	α^5	
4	3	x_2	4/3	0	1	2/3	1/3	0	
	2	x_1	1/3	1	0	−1/3	1/3	0	–
	0	x_5	−1	−1	0	0	0	1	
	$c_j - Z_j$			0	0	−4/3	−5/3	0	–

The aforegiven simplex table is not in the standard format (as column α^1 is not an identity column). To reduce it to the standard format, add x_1-row entries of table 4 in the corresponding x_5-row entries to get the following simplex table 5:

Cost →				**2**	**3**	**0**	**0**	**0**
Variable →				x_1	x_2	x_3	x_4	x_5
Table no.	C_B	**Basic variable**	X_B	α^1	α^2	α^3	α^4	α^5
5	3	x_2	4/3	0	1	2/3	1/3	0
	2	x_1	1/3	1	0	−1/3	1/3	0
	0	$\leftarrow x_5$	−2/3	0	0	−1/3	1/3	1
	$c_j - Z_j$			0	0	−4/3↑	−5/3	0

The solution is infeasible. Restore feasibility using the dual simplex method. When x_5 leaves, x_3 enters, to get the next simplex table 6:

	Cost →			2	3	0	0	0
	Variable →			x_1	x_2	x_3	x_4	x_5
Table no.	C_B	**Basic variable**	X_B	α^1	α^2	α^3	α^4	α^5
6	3	x_2	0	0	1	0	1	2
	2	x_1	1	1	0	0	0	−1
	0	x_3	2	0	0	1	−1	−3
	$c_j - Z_j$			0	0	0	−3	−4

This is an optimum integer solution. The optimum solution is $x_1 = 1, x_2 = 0$, and Max $Z = 2$.

(b) For node 3, the sub problem obtained on adding constraint $x_1 \le 0$ is

$$\text{Max } Z = 2x_1 + 3x_2 + 0{\cdot}x_3 + 0{\cdot}x_4 + 0{\cdot}x_5$$

Subject to $-x_1 + x_2 + x_3 = 1$

$2x_1 + x_2 + x_4 = 2$

$x_1 + x_6 = 0$

$x_1, x_2, x_3, x_4, x_5 \ge 0$ and integers; x_3, x_4, x_5 are the slack variables.

Adding this additional constraint again in simplex table 3, we get

	Cost →			2	3	0	0	0	Minimum ratio $\frac{X_{B_i}}{\alpha_i^j}$
	Variable →			x_1	x_2	x_3	x_4	x_6	
Table no.	C_B	**Basic variable**	X_B	α^1	α^2	α^3	α^4	α^6	
7	3	x_2	4/3	0	1	2/3	1/3	0	
	2	x_1	1/3	1	0	−1/3	1/3	0	–
	0	x_6	0	1	0	0	0	1	
	$c_j - Z_j$			0	0	−4/3	−5/3	0	–

The aforegiven simplex table is not in the standard format (as column α^1 is not an identity column). To reduce it to the standard format, subtracting x_1-row entries from the corresponding x_6-row entries, we get the following simplex table:

	Cost →			2	3	0	0	0
	Variable →			x_1	x_2	x_3	x_4	x_6
Table no.	C_B	**Basic variable**	X_B	α^1	α^2	α^3	α^4	α^6
8	3	x_2	4/3	0	1	2/3	1/3	0
	2	x_1	1/3	1	0	−1/3	1/3	0
	0	$\leftarrow x_6$	−1/3	0	0	1/3	−1/3	1
	$c_j - Z_j$			0	0	−4/3	−5/3↑	0

The solution is infeasible. Restore feasibility using the dual simplex method. When x_6 leaves, x_4 enters. The next simplex table is as given here.

Cost →				2	3	0	0	0
Variable →				x_1	x_2	x_3	x_4	x_6
Table no.	C_B	Basic variable	X_B	α^1	α^2	α^3	α^4	α^6
9	3	x_2	1	0	1	1	0	1
	2	x_1	0	1	0	0	0	1
	0	x_4	1	0	0	–1	1	–3
		$c_j - Z_j$		0	0	–3	0	–5

The optimum solution is $x_1 = 0$, $x_2 = 1$, and Max $Z = 3$, which is an all integer solution.

Now with nodes 2 and 3, associate the value of Z as the upper bounds 2 and 3 respectively for the purpose of comparison. Now with nodes 2 and 3, associate upper bounds $Z=2$ and $Z=3$ respectively. The branching process ends as both the terminal nodes 2 and 3 have integer solution. The optimum solution is $x_1=0$, $x_2=1$ with Max $Z=3$ at node 3.

Example 13.9 Solve the following IPP using the branch and bound techniques:

Max $Z = 5x_1 - 3x_2$

Subject to $-x_1 + x_2 \leq 1$

$2x_1 + x_2 \leq 2$

$2x_1 \leq 1$

$x_1, x_2 \geq 0$ and integers.

Solution The given problem can be written in standard form as

Max $Z = 5x_1 - 3x_2 + 0{\cdot}x_3 + 0{\cdot}x_4 + 0{\cdot}x_5$

Subject to $-x_1 + x_2 + x_3 = 1$

$2x_1 + x_2 + x_4 = 2$

$2x_1 + x_5 = 1$

$x_1, x_2, x_3, x_4, x_5 \geq 0$ and integers; x_3, x_4, and x_5 are the slack variables.

Now omitting the integer requirement in aforementioned IPP, we solve the resulting LPP as follows:

Cost →				5	–3	0	0	0	Minimum ratio
Variable →				x_1	x_2	x_3	x_4	x_5	$\frac{X_{B_i}}{\alpha_i^j}$
Table no.	C_B	Basic variable	X_B	α^1	α^2	α^3	α^4	α^5	
1	0	x_3	1	–1	1	1	0	0	2/2 = 1
	0	x_4	2	2	1	0	1	0	1/2→
	0	$\leftarrow x_5$	1	[2]	0	0	0	1	
		$c_j - Z_j$		5↑	–3	0	0	0	–
2	0	x_3	3/2	0	1	1	0	1/2	
	0	x_4	1	0	1	0	1	–1	–
	5	x_1	1/2	1	0	0	0	1/2	
		$c_j - Z_j$		0	–3	0	0	–5/2	–

The optimum solution is $x_1 = 1/2$, $x_2 = 0$, and Max $Z = 5/2$ (node 1), which is not an all integer solution. Here, 1/2 is the non-integer value of x_1 at node 1, which is required to be an integer. Since $[x_1] = 0$, introduce two branches (sub problems) (a) $x_1 \geq 1$ and (b) $x_1 \leq 0$, where 0 is the greatest integer $\leq x_1$.

Now, solve the aforementioned two subproblems, using sensitivity analysis and dual simplex method getting two nodes say, node 2 and node 3 (Fig. 13.3).

(a) For node 2, the sub problem becomes

$$\text{Max } Z = 5x_1 - 3x_2 + 0{\cdot}x_3 + 0{\cdot}x_4 + 0{\cdot}x_5$$

$$\text{Subject to} \quad -x_1 + x_2 + x_3 = 1$$

$$2x_1 + x_2 + x_4 = 2$$

$$2x_1 + x_5 = 1$$

$$-x_1 + x_6 = -1$$

$x_1, x_2, x_3, x_4, x_5, x_6 \geq 0$ and integers; x_3, x_4, x_5, and x_6 are the slack variables.

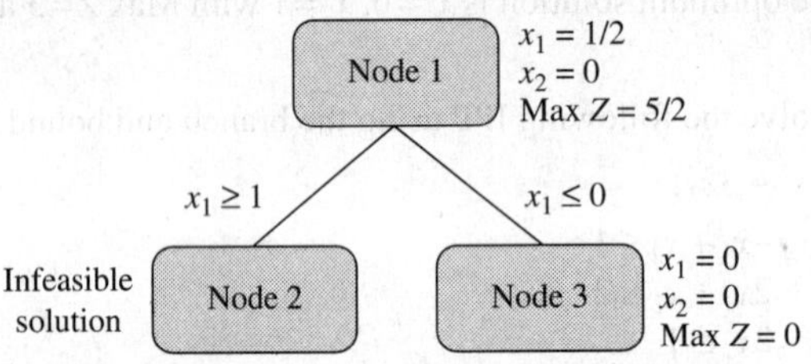

Figure 13.3 Branching at Node 1

Adding this additional constraint, the new simplex table is as follows:

		Cost →		5	−3	0	0	0	0
		Variable →		x_1	x_2	x_3	x_4	x_5	x_6
Table no.	C_B	**Basic variable**	X_B	α^1	α^2	α^3	α^4	α^5	α^6
3	0	x_3	3/2	0	1	1	0	1/2	0
	0	x_4	1	0	1	0	1	−1	0
	5	x_1	1/2	1	0	0	0	1/2	0
	0	x_6	−1	−1	0	0	0	0	1
		$c_j - Z_j$		0	−3	0	0	−5/2	0

The aforegiven simplex table is not in the standard format (as column α^1 is not an identity column). To reduce it to the standard format, add x_1-row entries in the corresponding x_6-row entries. Then, we get the following simplex table:

		Cost →		2	3	0	0	0	0
		Variable →		x_1	x_2	x_3	x_4	x_5	x_6
Table no.	C_B	**Basic variable**	X_B	α^1	α^2	α^3	α^4	α^5	α^6
4	0	x_3	3/2	0	1	1	0	1/2	0
	0	x_4	1	0	1	0	1	−1	0
	5	x_1	1/2	1	0	0	0	1/2	0
	0	x_6	−1/2	0	0	0	0	1/2	1
		$c_j - Z_j$		0	−3	0	0	−5/2	0

The solution is infeasible. Restore feasibility using the dual simplex method; variable x_6 leaves but there is no variable to enter (as there is no negative entry in the x_6 row). Thus, the solution is infeasible. Let us call this node 2.

(b) For node 3, the sub problem becomes

Max $Z = 5x_1 - 3x_2 + 0 \cdot x_3 + 0 \cdot x_4 + 0 \cdot x_5$

Subject to $-x_1 + x_2 + x_3 = 1$

$2x_1 + x_2 + x_4 = 2$

$2x_1 + x_5 = 1$

$x_1 + x_7 = 0$

$x_1, x_2, x_3, x_4, x_5, x_7 \geq 0$ and integers; x_3, x_4, x_5, and x_7 are the slack variables.

Adding this additional constraint in simplex table 2, the new simplex table obtained is:

		Cost →		5	−3	0	0	0	0
		Variable →		x_1	x_2	x_3	x_4	x_5	x_6
Table no.	C_B	Basic variable	X_B	α^1	α^2	α^3	α^4	α^5	α^6
5	0	x_3	3/2	0	1	1	0	1/2	0
	0	x_4	1	0	1	0	1	−1	0
	5	x_1	1/2	1	0	0	0	1/2	0
	0	x_7	0	1	0	0	0	0	1
		$c_j - Z_j$		0	−3	0	0	−5/2	0

The aforegiven simplex table is not in the standard format (as the column α^1 is not an identity column). To reduce it to the standard format, subtract x_1-row entries from the corresponding x_7-row entries. Then, we get the following simplex table:

		Cost →		5	−3	0	0	0	0
		Variable →		x_1	x_2	x_3	x_4	x_5	x_6
Table no.	C_B	Basic variable	X_B	α^1	α^2	α^3	α^4	α^5	α^6
6	0	x_3	3/2	0	1	1	0	1/2	0
	0	x_4	1	0	1	0	1	−1	0
	5	x_1	1/2	1	0	0	0	1/2	0
	0	x_7	−1/2	0	0	0	0	−1/2	1
		$c_j - Z_j$		0	−3	0	0	−5/2	0

The solution is infeasible. Restore feasibility using the dual simplex method. When x_7 leaves, then x_5 enters, to get the next simplex table 7:

		Cost →		5	−3	0	0	0	0
		Variable →		x_1	x_2	x_3	x_4	x_5	x_6
Table no.	C_B	Basic variable	X_B	α^1	α^2	α^3	α^4	α^5	α^6
7	0	x_3	3/2	0	1	1	0	0	1
	0	x_4	1	0	1	0	1	0	−2
	5	x_1	0	1	0	0	0	0	1
	0	x_5	1	0	0	0	0	1	−2
		$c_j - Z_j$		0	−3	0	0	0	−5

The optimum solution is $x_1=0, x_2=0$, and Max $Z=0$, which is the all integer solution at node 3.

Finally, the process is terminated because node 2 has an infeasible solution and node 3 has an integer solution. Hence the optimum solution is $x_1 = 0, x_2 = 0$, and Max $Z = 0$ of the given ILPP at node 3.

13.8 ZERO–ONE INTEGER PROGRAMMING PROBLEM

In an LPP, when all the decision variables are restricted to assume the values 0 or 1, then such an LPP is called zero–one integer programming. The first zero–one algorithm, called the additive algorithm, was proposed in 1965 by E. Balas, nearly seven years after the development of the branch and bound technique. In the beginning, the zero–one IPP appeared to be unrelated to the branch and bound technique, in the sense that it did not require solving LP problems, and the main computational work included simple additions and subtractions. However, soon it became evident that the zero–one algorithm is itself a special case of the branch and bound algorithm.

13.8.1 Format of Balas Zero–One Additive Algorithm

$$\left.\begin{aligned} &\text{Min } Z = \sum_{j=1}^{n} C_j x_j \\ &\sum_{j=1}^{n} a_{ij} x_j \le b_i,\ i = 1, 2, \ldots, m \\ &x_j = 0 \text{ or } 1, C_j > 0, \forall j \end{aligned}\right\} \quad (13.15)$$

$$\left.\begin{aligned} &\text{or Min } Z = \sum_{j=1}^{n} C_j x_j \\ &s_i = b_i - \sum_{j=1}^{n} a_{ij} x_j,\ i = 1, 2, \ldots, m \\ &x_j = 0 \text{ or } 1,\ C_j > 0, \forall j \\ &\text{and } s_i \text{ is the slack variable added to the } i\text{th constraint, } i = 1, 2, \ldots, m \end{aligned}\right\} \quad (13.16)$$

Thus, for Balas additive algorithm, the following conditions must be satisfied:

1. The problem must be of minimization type.
2. All the objective function coefficients C_j's must be non-negative.
3. All the constraints must be of '≤' type.
4. Any zero–one LP (where variables assume the value 0 or 1) can be put in the form of additive algorithm.

13.8.2 Some Important Terms used in Balas Additive Algorithm

The following terms are used in Balas additive algorithm to solve a zero–one IPP:

Branching nodes The partial solution (solution of the subproblem) from where branching is done is called a branching node.

Branching variable The variable selected to raise it from level 0 to level 1 is called the branching variable.

Slack feasibility The sum of non-positive slack values, given that a zero variable x_j is elevated to level 1, is called the slack infeasibility of slack variable s_i, denoted by I_j and given as

$$I_j = \sum_i \min[s_i - a_{ij}, 0], \tag{13.17}$$

where s_i = current value of slack variable i and a_{ij} = constraint coefficient of the jth variable x_j in constraint i.

Criterion of selection of branching variables The variable x_j for which the amount added to bring a non-positive I_j at zero level is minimum, is selected for branching. If the minimum amount is the same for more than one variable, then the variable whose objective function coefficient is smaller should be preferred for branching. In case objective function coefficients are also equal, break the tie arbitrarily.

Note Branching is done in both branch and bound and zero–one IPP methods, but the method and the selection criteria of the branching variable are different for these.

Fathomed nodes A node (subproblem solution) is said to be a fathomed node (a node from where further branching does not improve the solution) if one of following situations arise:

1. Branching from this node cannot give a feasible solution.
2. Branching from this node cannot yield a better upper bound.
3. Branching from this node cannot yield a feasible integer solution.

Free variables A variable that is free to assume a value of 0 or 1 is called a free variable.

13.8.3 Solution Procedure of Zero–One IPP

The following steps are used to solve a zero–one IPP:

1. Bring the problem in the initial form required by the additive algorithm, that is, the problem must be a minimization problem with all the objective function coefficients non-negative, and all the constraints of '≤' type.
2. Express the slack variable s_i added to the ith constraint in terms of the remaining variables by transposing every term except s_i on the RHS of the equality.
3. Find an initial all zero binary solution (slack solution) and call it node 1, attaching with it the Z-value (objective function value) as upper bound. If all the slacks are non-negative, we conclude that all the zero binary solutions are optimum. If not, go to step 4.
4. If some of the slack variable(s) are infeasible (negative), then using the slack infeasibility test, identify the branching variable. Let it be x_j. The two branches with x_j as the branching variables are $x_j = 1$ and $x_j = 0$, creating nodes 2 and 3 respectively. Proceeding this way, we arrive at a situation where all the terminal nodes are fathomed. Out of these, we pick up the best one, giving the optimal solution of the given minimization problem.

Example 13.10 Convert the following 0–1 problem to satisfy the starting requirements of the additive algorithm:

$$\text{Max } Z = 2x_1 - 7x_2$$
$$\text{Subject to } \quad x_1 + x_2 = 5$$
$$4x_1 + 6x_2 \geq 4$$
$$x_1 \geq 4$$
$$x_2 \leq 5$$
$$x_i\text{'s} = 0 \text{ or } 1, i = 1, 2.$$

Solution Convert the given problem to minimization with all '≤' constraints as follows:

(a) Multiplying the objective function row (Z-row) by (–1) to get min $Z' = -2x_1 + 7x_2$.
(b) Convert the first constraint equation into two constraints of '≤' type to obtain $x_1 + x_2 \leq 5$ and $-x_1 - x_2 \leq -5$.
(c) Multiply the second and third constraints by (–1) to obtain constraints $-4x_1 - 6x_2 \leq -4$ and $-x_1 \leq -4$ respectively.

Now using the aforementioned computations, the problem is written after adding slacks as

$$\text{Min } Z' = -Z = -2x_1 + 7x_2$$
$$\text{Subject to } \quad x_1 + x_2 + s_1 = 5$$
$$-x_1 - x_2 + s_2 = -5$$
$$-4x_1 - 6x_2 + s_3 = -4$$
$$-x_1 + s_4 = -4$$
$$x_2 + s_5 = 5$$
$$s_i\text{'s} \geq 0, i = 1, 2, \ldots, 5 \text{ are slack variables and } x_1, x_2 = 0 \text{ or } 1$$

To ensure that the coefficients in an objective function are non-negative, substitute $x_1 = 1 - y_1$ (as x_1 is with a negative coefficient) and $x_2 = y_2$ (as x_2 is with a positive coefficient). Change the LHS of the constraints accordingly, to get the given 0–1 problem, fulfilling the requirements of the additive algorithm. The required form for additive algorithm is

$$\text{Min } Z' = -2(1 - y_1) + 7y_2$$
$$\text{Subject to } \quad (1 - y_1) + y_2 + s_1 = 5$$
$$-(1 - y_1) - y_2 + s_2 = -5$$
$$-4(1 - y_1) - 6y_2 + s_3 = -4$$
$$-(1 - y_1) + s_4 = -4$$
$$y_2 + s_5 = 5$$
$$s_i\text{'s} \geq 0, i = 1, 2, \ldots, 5 \text{ are slack variables, and } y_1, y_2 = 0 \text{ or } 1.$$

or

$$\text{Min } Z' = 2y_1 + 7y_2 - 2$$
$$\text{Subject to } \quad -y_1 + y_2 + s_1 = 4$$
$$y_1 - y_2 + s_2 = -4$$
$$4y_1 - 6y_2 + s_3 = 0$$
$$y_1 + s_4 = -3$$
$$y_2 + s_5 = 5$$
$$s_i\text{'s} \geq 0, i = 1, 2, \ldots, 5 \text{ are slack variables, and } y_1, y_2 = 0 \text{ or } 1.$$

Example 13.11 Solve the following 0–1 problem:

$$\text{Min } Z = 2x_1 + 3x_2$$

$$\begin{aligned} \text{Subject to} \quad & x_1 + x_2 \le 2 \\ & -x_1 + x_2 \le 1 \\ & x_1, x_2 = 0 \text{ or } 1. \end{aligned}$$

Solution The problem can be set according to the initial form required by the additive algorithm using the following operations:

Adding the two slack variables s_1 and s_2 to convert the two constraints into equations, we get.

$$x_1 + x_2 + s_1 = 2 \text{ and } -x_1 + x_2 + s_2 = 1.$$

$$\text{Min } Z = 2x_1 + 3x_2$$

$$\begin{aligned} \text{Subject to} \quad & s_1 = 2 - x_1 - x_2 \\ & s_2 = 1 + x_1 - x_2 \end{aligned}$$

As we seek the minimum of the objective function, a logical starting solution is when all binary variables are at the zero level. In this case, slacks will act as basic variables and the initial all zero binary solution is $s_1 = 2$, $s_2 = 1$, $Z = 0$. Here, all the slacks are non-negative, that is, the all zero binary solution is the optimum solution.

Hence, the optimum solution is $x_1 = x_2 = 0$ and $Z = 0$.

Example 13.12 Solve the following 0–1 problem:

$$\text{Max } W = 3x_1 + 2x_2 - 5x_3 - 2x_4 + 3x_5$$

$$\begin{aligned} \text{Subject to} \quad & x_1 + x_2 + x_3 + 2x_4 + x_5 \le 4 \\ & 7x_1 + 3x_3 - 4x_4 + 3x_5 \le 8 \\ & 11x_1 - 6x_2 + 3x_4 - 3x_5 \ge 3 \\ & x_i\text{'s} = 0 \text{ or } 1,\ i = 1, 2, \ldots, 5. \end{aligned}$$

Solution The problem can be set according to the initial form required by the additive algorithm using the following operations:

(a) Multiplying the objective function by (–1)
(b) Multiplying the third constraints by (–1)
(c) Adding the three slack variables s_1, s_2, and s_3 to convert the three constraints into equations
(d) After using operation in (a), (b), and (c), substituting $x_1 = 1 - y_1$, $x_2 = 1 - y_2$, $x_3 = y_3$, $x_4 = y_4$, and $x_5 = 1 - y_5$ to produce all positive objective coefficients.

$$\begin{aligned} \text{Min } W' = -W &= -3x_1 - 2x_2 + 5x_3 + 2x_4 - 3x_5 \\ &= -3(1 - y_1) - 2(1 - y_2) + 5y_3 + 2y_4 - 3(1 - y_5) \\ &= 3y_1 + 2y_2 + 5y_3 + 2y_4 + 3y_5 - 8 \end{aligned}$$

For convenience, we replace $W' + 8$ with Z, so that the resulting converted problem reads as

$$\text{Min } Z = 3y_1 + 2y_2 + 5y_3 + 2y_4 + 3y_5$$

$$\begin{aligned} \text{Subject to} \quad & s_1 = 1 + y_1 + y_2 - y_3 - 2y_4 + y_5 \\ & s_2 = -2 + 7y_1 + 0y_2 - 3y_3 - 4y_4 + 3y_5 \\ & s_3 = -1 - 11y_1 + 6y_2 + 0y_3 + 3y_4 + 3y_5 \end{aligned}$$

As we seek the minimum of the objective function, a logical starting solution is when all the binary variables are at the zero level. In this case, slacks will act as basic variables and the initial all zero binary solution is $s_1 = 1$, $s_2 = -2$, $s_3 = -1$, $Z = 0$.

Let us denote this solution by node 1 with $Z=0$ as upper bound. Here s_2 and s_3 are infeasible, so elevate one of the binary variables to level 1 from level 0. For this we use the following test:

(a) Infeasibility test for node 1[using Eq. (13.17)]:

$$I_1 = \min(1 + 1, 0) + \min(-2 + 7, 0) + \min(-1 - 11, 0) = 0 + 0 - 12 = -12$$
$$I_2 = \min(1 + 1, 0) + \min(-2 + 0, 0) + \min(-1 + 6, 0) = 0 - 2 + 0 = -2$$
$$I_3 = \min(1 - 1, 0) + \min(-2 - 3, 0) + \min(-1 + 0, 0) = -1 - 50 + 0 = -6$$
$$I_4 = \min(1 - 2, 0) + \min(-2 + 4, 0) + \min(-1 + 3, 0) = -1 + 0 + 0 = -1$$
$$I_5 = \min(1 + 1, 0) + \min(-2 + 3, 0) + \min(-1 + 5, 0) = 0 + 0 + 0 = 0$$

Here, we see that y_5 yields the smallest measure of infeasibility (as I_5 comes to zero level when 0 is added to 0. In case of I_4, I_3, I_2, and I_1, we have to add more than 0 to bring them to zero level, so 0 is the smallest number). Hence, we select y_5 as a branching variable at node 1. The two branches each associated with $y_5=1$ and $y_5=0$ enter into nodes 2 and 3 (Fig. 13.4).

For node 2, $y_1 = y_2 = y_3 = y_4 = 0$, $y_5 = 1$, and $Z = 3$ with slack values $(s_1, s_2, s_3) = (2, 1, 2)$, all feasible.

For node 3, $y_1 = y_2 = y_3 = y_4 = 0$, $y_5 = 0$, and $Z = 0$ with slack values $(s_1, s_2, s_3) = (1, -2, -1)$, which is infeasible. Hence, we branch off from node 3 where y_1, y_2, y_3, y_4 (free variables) are candidates for becoming branching variables. It is worth observing here that although the solutions at node 1 and node 3 are identical, node 3 differs in that y_5 is no longer a branching candidate at node 3.

(b) Infeasibility test for node 3:

For variables y_1, y_2, y_3, y_4, infeasibility measures are $I_1 = -12$, $I_2 = -2$, $I_3 = -6$, and $I_4 = -1$(calculated in a similar way as in infeasibility test (a)).

From these calculations, we see that y_4 yields the smallest measure (one) of infeasibility. Hence we select y_4 as the branching variable at node 3. The two branches at node 3 result in node 4 and node 5, with slack solutions $(s_1, s_2, s_3) = (-1, 2, 2)$, $Z = 2$, and $(s_1, s_2, s_3) = (1, -2, -1)$, $Z = 0$ respectively (Fig. 13.5).

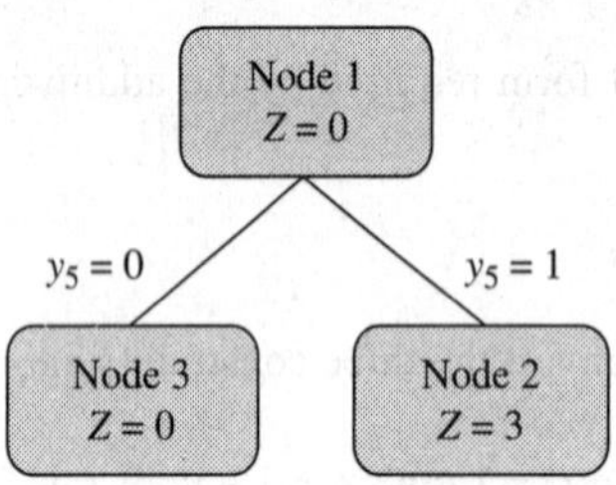

Figure 13.4 Branching at Node 1

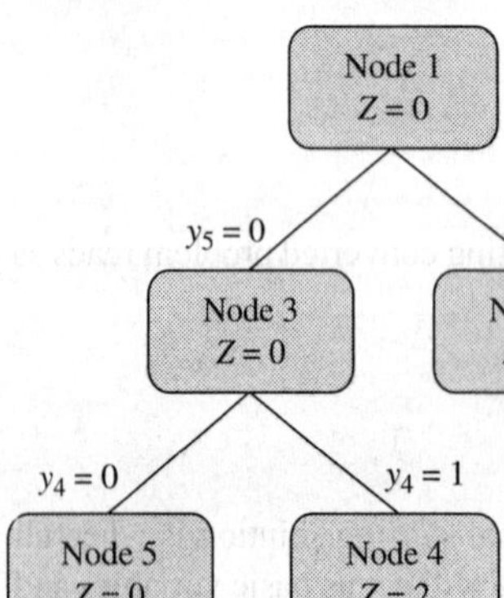

Figure 13.5 Branching at Node 3

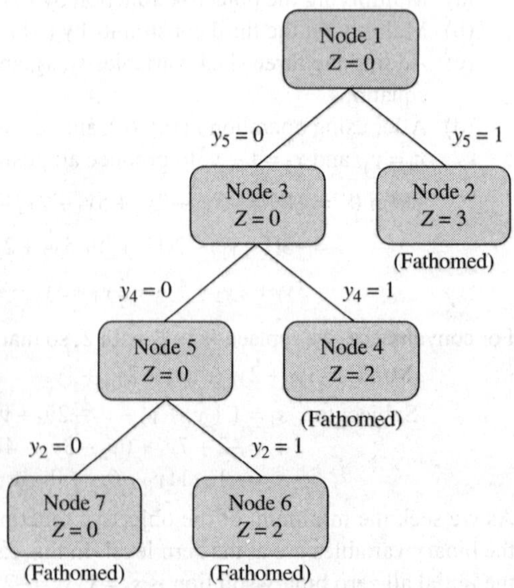

Figure 13.6 Branching at Node 5

The candidates for branching at node 4 are y_1, y_2, and y_3. However, elevating any of these variables to level 1 will worsen the value of Z relatively. The current upper bound $Z = 3$. Thus, all the candidate variables are excluded and the node 4 is fathomed.

At node 5, defined by $y_4 = y_5 = 0$, slack solutions are $(s_1, s_2, s_3) = (1, -2, -1)$, $Z = 0$. The variables y_1 and y_3 are excluded by the upper bound test; y_3 can also be excluded because it cannot reduce the slack infeasibility. The remaining variable y_2, thus, is the branching variable at node 5. Figure 13.6 shows the branches $y_2 = 1$ and $y_2 = 0$ leading to nodes 6 and 7.

All the dangling nodes in Fig. 13.6 are fathomed and branch and bound ends. The optimum solution is $y_1 = y_2 = y_3 = y_4 = 0$, $y_5 = 1$, and $Z = 3$ (node 2).

In terms of original variables, $x_1 = 1$, $x_2 = 1$, $x_3 = x_4 = x_5 = 0$, and $W = 5$.

RECAPITULATION

1. An LPP in which some or all decision variables in the optimal solution are restricted to non-negative integral values is called an integer programming problem (ILPP).
2. An LPP in which all the decision variables are restricted to integer values is called an AILPP.
3. An LPP in which some of the decision variables (not all) are restricted to integer values is called an MIPP.
4. An LPP in which all the decision variables are restricted to take the values 0 or 1 is called a zero–one IPP.
5. To solve an IPP using Gomory's cutting plane method, we first solve it using the simplex method, neglecting the integer requirement of the decision variables.
6. The cut for AIPP in Gomory's cutting plane method is

$$-f_r = s_g - \sum_{j \neq r} f_{rj} x_r$$

where f_r $(0 < f_r < 1)$ is the strictly positive fractional part of the basic variable x_r, s_g is the non-negative slack variable, and f_{rj}, $r \neq j$ are the non-negative fractions of the coefficients of non-basic variables in the rth row equation.

7. The cut for MIPP in Gomory's cutting plane method is

$$-f_{b_i} = -\sum f_{ij}^{+} x_j - \frac{f_{b_i}}{f_{b_i} - 1} \sum f_{ij}^{-} x_j + s_{g_i}$$

where f_{ij}^{+}, f_{ij}^{-} are the positive and negative coefficients of the non-basic variables, $0 < f_{b_i} < 1$ is the fractional part of the basic variable x_i required for an integer and s_{g_i} is the Gomory's slack variable.

8. Branch and bound technique is used only for AIPP.
9. To solve a zero–one IPP using Balas additive algorithm, the problem must be (i) a minimization problem (ii) all objective function coefficients should be non-negative, and (iii) all of the constraints should be of '≤' type.

EXERCISES

Multiple-choice Questions

13.1 In an AIPP, the decision variables are required to be

(a) all integers
(b) only 0 or 1
(c) between 0 and 1
(d) none of these

13.2 To solve an AIPP, the method used is
(a) branch and bound method (c) both (a) and (b)
(b) cutting plane method (d) none of these

13.3 The requirement of the objective function coefficients to be non-negative should be fulfilled in
(a) Gomory's cutting plane method for an AIPP
(b) Gomory's cutting plane method for an MIPP
(c) branch and bound method
(d) Balas additive algorithm

13.4 In the cutting plane method, dual simplex method is used to maintain
(a) feasibility (c) both feasibility and optimality
(b) optimality (d) none of these

13.5 Additive algorithm is used in
(a) transportation problem (c) cutting plane method
(b) assignment problem (d) none of these

13.6 Additive algorithm is a part of
(a) branch and bound technique (c) transportation algorithm
(b) Hungarian algorithm (d) none of these

13.7 The situation of multiple solutions arises in
(a) branch and bound technique (c) zero–one additive method
(b) cutting plane method (d) none of these

State True or False

13.1 To solve an IPP using Gomory's cutting plane method, first solve it using the simplex method, neglecting the integer requirements of the decision variables.
13.2 The AILPP was first suggested by R.E. Gomory.
13.3 The cut used in Gomory's cut for AIPP and MIPP is exactly the same.
13.4 In MIPP, all decision variables are not required to be integers.
13.5 In the zero–one IPP, all the decision variables are integers.
13.6 Balas additive algorithm is applicable only if the objective function coefficients are non-negative.
13.7 In the zero–one IPP, the decision variables can take values 0 or 1.
13.8 In a zero–one IPP, if some of the objectives function coefficients are negatives, then we cannot solve it using the additive algorithm.

Fill in the Blanks

13.1 The Gomory's cut for ________ is $-f_r = s_g - \sum_{j \neq r} f_{rj} x_j$.
13.2 The Gomory's cut for ________ is $-f_{b_i} = -\sum f_{ij}^{+} x_j - \frac{f_{b_i}}{f_{b_i} - 1} \sum f_{ij}^{-} x_j + s_{g_i}$.
13.3 A variable that is free to assume the values 0 or 1 is called ________ variable.
13.4 A node is ________ if branching from it cannot yield a feasible integer solution.
13.5 In Balas additive algorithm, all the constraints are of the ________ type.
13.6 In ________ a few decision variables are required to be integers.

Review Questions

13.1 Write a short note on IPP.

13.2 Write the mathematical formulation of an IPP. Discuss the types of IPP in brief.
13.3 Explain Gomory's cutting plane method for an AIPP.
13.4 Discuss Gomory's cutting plane method for an MIPP.
13.5 Construct the flow chart of Gomory's cutting plane method for an IPP.
13.6 What is the zero–one IPP?

Numerical Problems

13.1. Solve the following IPPs:

(a) Max $Z = 3x_2$
Subject to $3x_1 + 2x_2 \leq 7$
$-x_1 + x_2 \leq 2$
$x_1, x_2 \geq 0$ and integers

(b) Max $Z = x_1 + 4x_2$
Subject to $2x_1 + 4x_2 \leq 7$
$5x_1 + 3x_2 \leq 15$
$x_1, x_2 \geq 0$ and integers

(c) Max $Z = 5x_1 + 7x_2$
Subject to $-2x_1 + 3x_2 \leq 6$
$6x_1 + x_2 \leq 30$
$x_1, x_2 \geq 0$ and integers

(d) Max $Z = x_1 + x_2$
Subject to $3x_1 + 2x_2 \leq 5$
$x_2 \leq 2$
$x_1, x_2 \geq 0$ and integers

13.2 Use the branch and bound method to solve the following IPPs:

(a) Min $Z = 4x_1 + 3x_2$
Subject to $5x_1 + 3x_2 \geq 30$
$x_1 \leq 4,$
$x_2 \leq 6$
$x_1, x_2 \geq 0$ and integers

(b) Max $Z = 4x_1 + 3x_2$
Subject to $5x_1 + 3x_2 \geq 30$
$x_1 \leq 4,$
$x_2 \leq 6$
$x_1, x_2 \geq 0$ and integers

(c) Max $Z = 6x_1 + 8x_2$
Subject to $4x_1 + 16x_2 \leq 32$
$14x_1 + 4x_2 \leq 28$
$x_1, x_2 \geq 0$ and integers

(d) Max $Z = 2x_1 + 3x_2$
Subject to $5x_1 + 7x_2 \leq 35$
$4x_1 + 9x_2 \leq 36$
$x_1, x_2 \geq 0$ and integers

13.3 Solve the following MIPP:

Max $Z = x_1 + 2x_2 - 3x_3$
Subject to $20x_1 + 15x_2 - x_3 \leq 10$
$12x_1 - 3x_2 - 4x_3 \leq 14$
$3x_1 + 5x_2 + x_3 \leq 6$
$x_j = 0$ or 1 for all j

13.4 Suppose five items are to be loaded on the vessel. The weight W, volume V, and price P are given in the following table:

Item →	1	2	3	4	5
W	5	8	3	2	7
V	1	8	6	5	7
P	4	7	6	5	4

The maximum cargo weight and cargo volume are $W = 112$ and $V = 109$ respectively. Determine the most valuable cargo load in the discrete unit of each item. Formulate the problem as an IPP.

CONCEPTUAL PROBLEM

Travelling and tourism companies face some problems during the summer vacation when there is a greater demand for reservations. The companies want to plan their activities in such a way that there are no vacant seats in the buses and optimum use of the capacity is made. A tourism company based in Delhi is planning to run air-conditioned luxury buses to three hill stations—Nainital, Srinagar, and Simla. The company has two types of luxury buses—T_1 and T_2. The data regarding bus capacity, the number of trips each bus makes to a hill station if it is put on that route, and the number of passengers for each hill station is given in the following table:

Bus type	Capacity	Number of buses	Number of daily trips to		
			Nainital	Srinagar	Simla
T_1	80	6	2	1	3
T_2	50	9	4	2	5
Number of daily passengers			200	100	250

After reaching a hill station, the bus returns without taking passengers. The expenditure of each trip and the loss due to a vacant seat in the bus is given as per the following table:

Bus type	Expenditure (in thousands) per trip to		
	Nainital	Srinagar	Simla
T_1	10	10	15
T_2	6	7	9
Loss per vacant seat (in hundreds)	3	2	4

Formulate the aforementioned problem as an IPP to minimize the total cost.

Solution Let x_{i1}, x_{i2}, and x_{i3} be the number of buses of type T_i, $i = 1, 2, 3$ allocated to Nainital, Srinagar, and Simla respectively. The total cost of the system is $2000x_{11} + 1000x_{12} + 4500x_{13} + 2400x_{21} + 1400x_{22} + 4500x_{23} + 300s_3 + 200s_4 + 400s_5$.

Our objective is to minimize this cost. Thus the objective function is

$$\text{Min } Z = 2000x_{11} + 1000x_{12} + 4500x_{13} + 2400x_{21} + 1400x_{22} + 4500x_{23} + 300s_3 + 200s_4 + 400s_5$$

Subject to

$$x_{11} + x_{12} + x_{13} + s_1 = 6$$
$$x_{21} + x_{22} + x_{23} + s_2 = 9$$
$$160x_{11} + 200x_{12} - s_3 = 200$$
$$80x_{12} + 100x_{22} - s_4 = 100$$
$$240x_{13} + 250x_{23} - s_5 = 250$$

All variables ≥ 0 and integers; s_1, s_2 are slack variables and s_3, s_4, and s_5 are surplus variables.

This is the formulation of the given IPP to get the allocation of buses to hill stations for the total cost of the system to be minimum.

14 Queuing Theory

After studying this chapter, the reader will be able to

- understand the basic elements of the queuing system
- get acquainted with the basic terms and symbols used in the queuing system
- explain the probability distribution in a queuing system
- classify the queuing models

14.1 INTRODUCTION

The waiting line or queue formation is a common occurrence in our day-to-day life. People purchasing grocery items in stores stand in a line waiting for their turn; ships wait in the harbour to be unloaded; machines that are out of order are repaired in workshops in a particular sequence. All these are a few examples of queue formation. Customers constitute that part of the population group who need some service. They arrive for a service following a certain pattern. On arrival, a customer is delivered the required service provided the service facility is idle at that particular time. The customer joins the queue if the service facility is busy. Eventually when the facility becomes free, the customer in the queue is given the required service by one of its servers. Once the service is complete, the customer leaves the service facility or system. Figure 14.1 illustrates the queuing process.

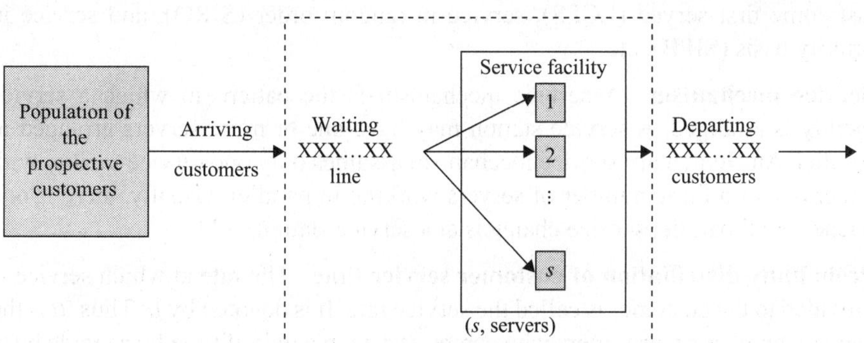

Figure 14.1 Queuing system

14.2 BASIC ELEMENTS OF QUEUING SYSTEMS

There are a variety of queuing systems and each of these has its own unique set of characteristics. However each queuing system has some basic elements, as described in the following subsections.

Population of prospective customers This is the input source for customers requiring a service. For example, in a factory, when a machine breaks down, then the machine (customer) is sent (for service) to the company's workshop (service facility) for repair.

The most important characteristic of a population is its size, which may be finite or infinite. In actual practice, a finite population that is so large that the arrival rate (the rate at which the customers are arriving) is unaffected by the number of customers in the queuing system is assumed to be infinite.

Arrival pattern or probability distribution of inter-arrival times The arrival of customers is usually not regular and is thus described by a probability distribution. A common assumption is that arrivals occur in a Poisson pattern with a mean arrival rate λ. It means that the number of customers that arrive in any interval of length t has a Poisson distribution with parameters λt. Thus, the random number of arrivals (X) in interval $(0, t)$ has a Poisson distribution with parameter λt as follows:

$$P(X = n) = e^{-\lambda t} \frac{(\lambda t)^n}{n!}, n = 0, 1, 2, \ldots$$

Queue A queue is characterized by the maximum permissible number of customers that it can contain. It is called finite or infinite based on whether the number of customers in the system is finite or infinite respectively. A finite queue is quantified by the maximum number of customers it can hold. This number is denoted as N.

Queue discipline or service discipline The method of selection of customers from a queue for providing service is called queue discipline. In general, customers are served on a first come first served (FCFS) basis. While solving a queuing problem, this basis is assumed, unless otherwise stated. The other disciplines are last come first served (LCFS), service in random order (SIRO), and service in priority basis (SIPB), etc.

Service mechanism A service mechanism is the pattern in which a service facility is arranged. A service station may have one or more servers arranged in parallel. An elementary service mechanism assumes only one service facility with either one or a finite number of servers working in parallel. Usually, there is one queue for all parallel service channels at a service station.

Probability distribution of customer service time The rate at which service is provided to the customer is called the service rate. It is denoted by μ. Thus, μ is the average number of customers who can be served in a unit of time by a single busy server. Consequently, the mean service time of a customer is $1/\mu$. If the service facility has several servers, say s, working in parallel, we assume that each server has the same mean service rate μ. The most common assumption about the service time distribution is that it is exponential with parameter μ.

State of systems The state of the system at a particular instant of time refers to the total number of customers in the system at that instant. During transition, the state of the system has a probability distribution, which is time-dependent; however, after the infinite time has elapsed, the state of the system may have a probability distribution, which is independent of time t. This distribution is called the steady state probability distribution of the system. It gives the probability that the system has n customers at any instant of time and is denoted by P_n. These probabilities (for $n = 0, 1, 2, \ldots, m$) are used to find various characteristics of the queuing system as discussed in Section 14.10.

In queuing theory, a large number of mathematical models are used to describe different waiting line situations. In this chapter, we will briefly discuss the Markovaian queue models.

14.3 MARKOVIAN QUEUES

The queuing system is a random process. Customers who want to avail a particular service arrive at a service station at random intervals of time. The service time of customers is also random. This random nature of arrivals and service time causes congestion, and customers form a queue. We make the assumption that customers arrive at the service station one by one and not in batches.

Consider moment of time (time at which service starts) $t = t_o$ in the queue system. The randomness of the queue system for $t > t_o$ in general depends on the history of the queuing process at $t = t_o$ and on the history of the Markovian queue system process for $t > t_o$. A random process is called a Markovian process if its probabilistic character in the future depends only on its present state at $t = t_o$ and not on how this present state was acquired. The queuing systems in which the aforementioned assumption is valid are called Markovian queues. Queues that are not based on this assumption are called non-Markovian queues. Thus, in Markovian queues, we assume that the past history is completely summarized in the specification of the current state and therefore, it is not necessary to specify how long the queue system has been in its current state. This implies that the random distribution of time must be a 'memory-less' distribution. The exponential distribution is the only distribution among continuous distributions that possesses this property. Therefore, it is an integral part of Markovian queues and it is assumed that the inter-arrival time and service time are exponentially distributed.

14.4 TERMINOLOGY AND NOTATIONS USED IN QUEUING SYSTEMS

The following standard terminology and notations are used in a queuing system:

State of the system = Number of customers in the system say n

= Number of customers in the queue + the number of customers under the service

$N(t)$ = Number of customers in the system at time t (in transient state)

N = Number of customers in the system in steady state

n = Observed value of N or $N(t)$

Q = Number of customers in the queue in steady state

q = Observed value of Q

$P_n(t)$ = Probability that there are exactly n customers in the system at time t
= $P[N(t) = n]$, $n = 0, 1, 2, \ldots$

λ_n = Mean arrival rate (expected number of arrival of customers per unit of time) when the state of the system is n.

μ_n = Mean service rate of the whole service facility (expected number of customers having their service completed per unit of time) when the state of the system is n

λ = Mean rate of arrivals per unit of time in steady state

μ = Mean service rate of a customer per unit of time

s = Number of servers working in parallel

$\rho = \lambda/\mu$ = Traffic intensity (service utilization factor)

L_s = Line length
= Expected or average number of customers present in the system (in queue + under service)

L_q = Queue length
= Average or expected number of customers waiting in the queue

W_s = Expected waiting time per customer present in the system

W_q = Expected waiting time for those who are waiting in the queue
= Expected waiting time in the system-service time of one customer (if $s = 1$)

$E(W/W > 0)$ = Expected waiting time of a customer

$E(L/L > 0)$ = Expected length of a non-empty queue
= Average length of queue formed from time to time.

14.5 SYMBOLIC REPRESENTATION OF QUEUING MODELS

The following abbreviated notation, called Kendall's notation, is used to present information about the elements of a queuing system. The queuing system is expressed in the symbolic form as $(a/b/c):(d/e/f)$

where

a = Distribution of inter-arrival time
b = Service time distribution
c = Number of servers working in parallel
d = Queue discipline
e = Maximum number of customers allowed in the system
f = Population size of the prospective customers (finite or infinite)

While discussing specific queuing models here, we shall ignore f because the size of the population requiring service is assumed to be very large. Hence, the model that we shall consider throughout our discussion will be $(a/b/c):(d/e)$, assuming that size of the population is infinite.

If the inter-arrival time or the service time distribution is exponential, then it is represented by M for *a* (or *b*). M indicates the Markovian (memory-less) property of exponential distribution. For example when we write (M/M/2):(FCFS/50), it indicates the following—that the arrival rate has Poisson distribution and both inter-arrival time and service time have exponential distributions; there are two service channels (servers); the service discipline is FCFS; the maximum number of customers that can be allowed in the system is 50; and the population size is ∞.

Now, we shall use the aforementioned notations to drive the probability distribution of some specific queue models such as pure birth, pure death, and the birth and death processes.

The *pure birth* process refers to only the arrival of customers; *pure death* process means only departures are taking place; and *birth and death* process means both arrivals and departures are taking place.

14.6 PROBABILITY DISTRIBUTION OF *n* ARRIVALS IN TIME INTERVAL $(t, t+\Delta t)$ IN PURE BIRTH PROCESSES

We shall assume that

1. for any interval $(t, t+\Delta t)$, there can be either no arrival or exactly one arrival (as we are considering a pure birth process).
2. For any number (n) of customers present in the system the arrival rate λ per unit time is fixed (constant).

The probability that there are n customers in the system at time t is denoted by $P_n(t)$. Let us now consider $P_n(t+\Delta t)$, that is, the probability that there are n customers in the system at time $t+\Delta t$. This situation is possible if

1. there are n customers in the system at time t and there is no arrival in time interval Δt, or
2. there are $(n-1)$ customers in the system at time t and there is exactly one arrival in time interval Δt. Situations 1 and 2 are mutually exclusive.

Thus,

$$P_n(t+\Delta t) = \text{Probability of 1} + \text{Probability of 2}$$

or $$P_n(t+\Delta t) = P_n(t)(1-\lambda\Delta t) + P_{n-1}(t)\,\lambda\Delta t,\ n > 0 \tag{14.1}$$

When $n = 0$, that is, there is no customer present at time $(t+\Delta t)$, then there is no customer present at time t also and $P_{n-1}(t) = 0$. Thus, from Eq. (14.1), we have

$$P_0(t+\Delta t) = P_0(t)(1-\lambda\Delta t) \tag{14.2}$$

From Eq. (14.1), we also have $\dfrac{P_n(t+\Delta t)-P_n(t)}{\Delta t} = -\lambda P_n(t) + \lambda P_{n-1}(t)$.

Now, taking the limit as $\Delta t \to 0$, on simplification we have

$$\frac{dP_n(t)}{dt} + \lambda P_n(t) = \lambda P_{n-1}(t) \tag{14.3}$$

From Eq. (14.2), we have

$$\lim_{\Delta t\to 0}\frac{P_0(t+\Delta t)-P_0(t)}{\Delta t} = \lim_{\Delta t\to 0}\left(-\lambda P_0(t)\right)$$

$$\frac{dP_0(t)}{dt} + \lambda P_0(t) = 0 \tag{14.4}$$

Solving Eq. (14.4), we get $P_0(t) = ce^{-\lambda t}$

Taking $t = 0$, $P_0(0) = ce^0$

$\Rightarrow c = 0$ as $P_0(0) = 1$

$$\therefore P_0(t) = e^{-\lambda t} \tag{14.5}$$

Again, taking $n = 1$ in Eq. (14.3), we have $\dfrac{dP_1(t)}{dt} + \lambda P_1(t) = \lambda P_0(t)$.

$$\text{or } \frac{dP_1(t)}{dt} + \lambda P_1(t) = \lambda e^{-\lambda t} \text{ [from Eq. (14.5)]} \tag{14.6}$$

Eq. (14.6) is a first order linear differential equation. Its solution is

$$P_1(t)\, e^{\lambda t} = \int \lambda e^{-\lambda t} e^{\lambda t} dt + c$$

or $P_1(t) = \lambda t e^{-\lambda t} + c e^{-\lambda t}$

When $t = 0$, $P_1(0) = 0 \quad \Rightarrow \quad c = 0$

Thus, $P_1(t) = \dfrac{e^{-\lambda t}(\lambda t)^1}{1!}$

Proceeding this way, we get $P_n(t) = \dfrac{e^{-\lambda t}(\lambda t)^n}{n!}, \quad n = 0, 1, 2, 3, \ldots$

Thus, the number of pure arrivals in the system in the interval $(0, t)$ follows Poisson distribution.

14.7 DISTRIBUTION OF INTER-ARRIVALS TIME

Consider time interval $(0, t)$. Let the two consecutive arrivals in Poisson pattern take place at time $t = 0$ and $t = T$. Then T is the inter-arrival time, which is a random variable. Let $f(t)$ be the probability density function (PDF) of T.

$P(T > t) = P(\text{no arrival by time } t)$

$= P_0$

$= \dfrac{e^{-\lambda t}(\lambda t)^0}{o!} = e^{-\lambda t}$ as the number of arrivals have Poisson distribution and λ is the arrival rate of customers in the system per unit time.

Further, $P(T \le t) + P(T > t) = 1$

Distribution function $F(t)$ of T is given as

$$F(t) = P(T \le t) = 1 - P(T > t) = 1 - P_0 = 1 - e^{-\lambda t}$$

Thus, $f(t) = F'(t) = \lambda e^{-\lambda t}$, $t > 0$ which is the probability distribution of an exponentially distributed random variable. Thus, the inter-arrival time T has an exponential distribution with parameter λ, if the customer arrivals follow Poisson distribution with λ per unit time.

14.8 DISTRIBUTION OF DEPARTURES IN PURE DEATH PROCESSES

Let us consider the case when only departures can take place according to Poisson distribution and there are no arrivals. Again, let there be N customers in the system at time $t = 0$.

Now consider $P_n(t + \Delta t)$, (the probability of n customers in the system at time $(t + \Delta t)$) which is possible in two mutually exclusive ways:

1. There are n customers in the system at time t and no departure takes place during small time Δt.
2. There are $(n + 1)$ customers present in the system at time t and one departure takes place during Δt.

Thus, $P_n(t + \Delta t) = P_n(t)\,(1 - \mu\Delta t) + P_{n+1}(t)\mu\Delta t$

Transposing $P_n(t)$ on the left, dividing by Δt throughout, and taking the limit as $\Delta t \to 0$, we get

$$\frac{d}{dt}P_n(t) = -\mu P_n(t) + \mu P_{n+1}(t) \tag{14.7}$$

When $n = N$, $\dfrac{d}{dt}P_N(t) = -\mu P_N(t)$ as $P_{N+1}(t) = 0$

The aforementioned equation is a linear differential equation of the first order. On solving, we get $P_N(t) = c_1 e^{-\mu t}$

When $t = 0$, $P_N(0) = 1$, we get $c_1 = 1$. Thus, $P_N(t) = e^{-\mu t}$ (14.8)

Taking $n = N - 1$ in Eq. (14.7), we get

$$\frac{d}{dt}P_{N-1}(t) = -\mu P_{N-1}(t) + \mu P_N(t)$$

$$\text{or } \frac{d}{dt}P_{N-1}(t) = -\mu P_{N-1}(t) + \mu e^{-\mu t} \text{ [using Eq. (14.8)]}$$

or $\dfrac{d}{dt}P_{N-1}(t) + \mu P_{N-1}(t) = \mu e^{-\mu t}$, which is a linear differential equation.

On solving, we get

$$P_{N-1}(t)\,e^{\mu t} = \int \mu e^{-\mu t}.e^{\mu t}\,dt + c_2$$

$$\text{or } P_{N-1}(t)\,e^{\mu t} = \int \mu dt + c_2 \text{ or } P_{N-1}(t)\,e^{\mu t} = \mu t + c_2$$

$$\text{or } P_{N-1}(t) = \mu t e^{-\mu t} + c_2 e^{-\mu t}$$

When $t = 0$, $P_{N-1}(0) = 0$, we get $c_2 = 0$

$$\text{Thus, } P_{N-1}(t) = \mu t\, e^{-\mu t} \tag{14.9}$$

When $n = N - 2$ from Eq. (14.7), we have

$$\frac{d}{dt}P_{N-2}(t) = -\mu P_{N-2}(t) + \mu P_{N-1}(t)$$

$$\text{or } \frac{d}{dt}P_{N-2}(t) = -\mu P_{N-2}(t) + \mu(\mu t)e^{-\mu t} \text{ [using (14.9)]}$$

or $\frac{d}{dt}P_{N-2}(t)+\mu P_{N-2}(t)=\mu^2 t\,e^{-\mu t}$, which is a linear differential equation of order one and on solving it, we get

$$P_{N-2}(t)e^{\mu t}=\int \mu^2 te^{-\mu t}.e^{\mu t}dt+c_3$$

$$\text{or } P_{N-2}(t)e^{\mu t}=\int \mu^2 tdt+c_3 \text{ or } P_{N-2}(t)e^{\mu t}=\mu^2\frac{t^2}{2}+c_3$$

$$\text{or } P_{N-2}(t)=\mu^2\frac{t^2}{2}e^{-\mu t}+c_3e^{-\mu t}$$

When $t=0$, $P_{N-2}(0)=0$, and we get $c_3=0$

$$\text{Thus, } P_{N-2}(t)=\frac{e^{-\mu t}\mu^2 t^2}{2!} \tag{14.10}$$

Proceeding this way we get $P_{N-k}(t)=\dfrac{e^{-\mu t}\mu^k t^k}{k!}$

Replacing k by $N-n$, we get

$$P_n(t)=\frac{e^{-\mu t}\mu^{N-n}t^{N-n}}{(N-n)!},\ n=1,2,3,\ldots,N \tag{14.11}$$

$$\text{Further, } \sum_{n=0}^{N}P_n=1 \Rightarrow P_0=1-\sum_{n=1}^{N}P_n(t) \tag{14.12}$$

Thus, probability P_n can be obtained from Eq. (14.11) for $n=1, 2, 3, \ldots, N$ and P_0 can be obtained from Eq. (14.12).

14.9 BIRTH AND DEATH PROCESS

In birth and death process, birth means an 'entry into the system' and death means 'departure from the system after completion of the service'. The birth and death process will be used in Section 14.10 to define the various parameters of some specific queue systems. All Poisson queues satisfy the following axioms of the birth and death model mentioned herewith.

Axiom 1 Given that there are n customers in the system at time t, that is, $N(t)=n$, the conditional probability distribution of the time (reckoned from the instant t) until the next birth has exponential distribution with parameter λ_n, $n=0, 1, 2, 3, \ldots$

Axiom 2 Given $N(t)=n$, the conditional distribution of the time (reckoned from the instant t) until the next death is exponentially distributed with parameters μ_n, $n=1, 2, 3, \ldots, \mu_0=0$

Axiom 3 Either birth or death (not both) can occur at a particular instance of time, that is, in an interval of small length Δt. In other words, probability of more than one event occuring in an interval of small length Δt is $O(\Delta t)$.

Axiom 4 Birth and death occur independently for a given population of size n. Let λ_n and μ_n be respectively the mean arrival rate and the mean service rate of customers in the system at the instant t, when the state of the system is n.

Let $P_n(t)$ denote the probability that there are n units (customers) in the system at time t, that is, $P_n(t) = P(N(t) = n)$, where $N(t)$ = number of units in the system at time t, which is a random variable. Now, let us consider $P_n(t + \Delta t)$. There are three mutually exclusive possibilities for this situation of n customers in the system at time $(t + \Delta t)$:

1. When there are n customers in the system at time t and neither an arrival nor a departure takes place in duration Δt of time after t, then $P_n(t+\Delta t)$
   $= P_n(t)\,(1 - \lambda_n \Delta t)\,(1 - \mu_n \Delta t)$
   $= P_n(t)\,\{1 - (\lambda_n + \mu_n)\Delta t\}$ as $\lambda_n \mu_n (\Delta t)^2 = o(\Delta t)$ and is neglected.
2. When there are $(n - 1)$ customers in the system at time t and exactly one arrival and no departure takes place during Δt length of time after t, then $P_n(t+\Delta t)$
   $= P_{n-1}(t)\lambda_{n-1}\Delta t\,(1 - \mu_{n-1}\Delta t)$
   $= P_{n-1}(t)\,(\lambda_{n-1}\Delta t)$ as $\lambda_{n-1}\mu_{n-1}(\Delta t)^2 = o(\Delta t)$ and is neglected.
3. When there are $(n + 1)$ customers in the system at time t with no arrivals and only one departure takes place from t to $t + \Delta t$, then $P_n(t+\Delta t)$
   $= P_{n+1}(t)\,(1 - \lambda_{n+1}\Delta t)\,\mu_{n+1}\Delta t$
   $= P_{n+1}(t)\,\mu_{n+1}\Delta t$ (neglecting $\lambda_{n+1}\mu_{n+1}(\Delta t)^2$)

Since all the aforementioned possibilities are mutually exclusive, we get

$$P_n(t + \Delta t) = P_n(t)\,\{1 - (\lambda_n + \mu_n)\Delta t\} + P_{n-1}(t)\,(\lambda_{n-1}\Delta t) + P_{n+1}(t)\,\mu_{n+1}\Delta t$$

Transposing $P_n(t)$ on the left, dividing by Δt and taking the limit as $\Delta t \to 0$, we get the following on simplification:

$$\frac{dP_n(t)}{dt} = -(\lambda_n + \mu_n)\,P_n(t) + \lambda_{n-1}\,P_{n-1}(t) + \mu_{n+1}\,P_{n+1}(t) \tag{14.13}$$

Taking $n = 0$ (for $n = 0$, $P_{n-1}(t) = 0$ and $\mu_0 = 0$), we get

$$\frac{dP_0(t)}{dt} = -\lambda_0 P_0(t) + \mu_1 P_1(t) \tag{14.14}$$

When the steady state is reached $(t \to \infty)$, then $\lim_{t\to\infty} P_n(t) = P_n$ that is, $P_n(t)$ becomes independent of time and $\frac{dP_n(t)}{dt} = 0$, $\frac{dP_0(t)}{dt} = 0$ and from Eqs (14.13) and (14.14), respectively, we get

$$\mu_{n+1}P_{n+1} = (\lambda_n + \mu_n)\,P_n - \lambda_{n-1}P_{n-1},\ n \geq 1 \tag{14.15}$$

$$\mu_1 P_1 - \lambda_0 P_0 = 0 \tag{14.16}$$

From Eqs (14.15) and (14.16) and the fact that sum of all probabilities is one, that is,

$$\sum_{n=0}^{\infty} P_n = 1 \tag{14.17}$$

we can determine P_n, $n = 0, 1, 2, 3, \ldots$. As a result, the probability distribution of the birth and death process is known. Thus, from Eq. (14.16), we have

$$P_1 = \frac{\lambda_0}{\mu_1} P_0 \tag{14.18}$$

Now in Eq. (14.15), taking $n = 1$ and using Eq. (14.18), on simplification we get

$$P_2 = \frac{\lambda_0 \lambda_1}{\mu_1 \mu_2} P_0$$

Proceeding in a similar way by taking $n = 2, 3, 4, \ldots$ in Eq. (14.15) and simplifying each time, we get

$$P_n = \frac{\lambda_0 \lambda_1 \ldots \lambda_{n-1}}{\mu_1 \mu_2 \ldots \mu_n} P_0, \; n = 1, 2, 3, \ldots \tag{14.19}$$

To determine P_0 from Eq. (14.17), we have

$$P_0 + \sum_{n=0}^{\infty} P_n = 1 \Rightarrow P_0 + P_0 \sum_{n=1}^{\infty} \frac{\lambda_0 \lambda_1 \ldots \lambda_{n-1}}{\mu_1 \mu_2 \ldots \mu_n} = 1 \text{ using Eq. (14.19)}$$

Thus, $$P_0 = \frac{1}{1 + \sum_{n=1}^{\infty} \frac{\lambda_0 \lambda_1 \ldots \lambda_{n-1}}{\mu_1 \mu_2 \ldots \mu_n}} \tag{14.20}$$

From Eq. (14.19), we can see that for P_n to be meaningful P_0 must be greater than zero ($P_0 > 0$). This in turn implies that there is a positive probability of the system being empty sometimes. For this, the infinite series in Eq. (14.20) must be a convergent series. This series is convergent if there exists a positive number n_0 such that $\lambda_n/\mu_n < 1$ for $n \geq n_0$. This means that the average arrival rate per unit time when the state of the queue system is n, is less than the average service rate (departure rate) from the system. This condition should be satisfied in most of the queue systems, otherwise the situation would become explosive.

14.10 VARIOUS QUEUING MODELS WITH THEIR CHARACTERISTIC PROPERTIES

Now, we shall discuss some special queuing models, which will symbolically be represented as follows:

1. *Model 1—(M/M/1):(FCFS/∞)*: This denotes Poisson arrival, exponential service time, single server, first come first served discipline, infinite capacity.
2. *Model 2—(M/M/1):(FCFS/N)*: In this model, the maximum number of customers allowed in the system is *N*.
3. *Model 3—(M/M/S):(FCFS/∞)*: In this model, *S* is the number of servers and the system capacity is unlimited.
4. *Model 4—(M/M/S):(FCFS/N)*: Here, there are *S* servers and limited capacity is *N* of the system.

14.10.1 Model-I—(M/M/1):(FCFS/∞)

In this system, there is one service channel (one server) with infinite capacity and service discipline is FCFS (which is same as FIFO, first in first out).

Here, $\lambda_n = \lambda$, $n = 0, 1, 2, \ldots$ and $\mu_n = \mu$, $n = 1, 2, \ldots$

From Eq. (14.20), we get

$$P_0 = \frac{1}{\sum_{n=0}^{\infty} \rho^n} = 1 - \rho, \text{ where } \rho = \frac{\lambda}{\mu} \tag{14.21}$$

and from Eq. (14.19), we get

$$P_n = P_0 \rho^n = (1-\rho)\rho^n, \; n = 1, 2, 3, \ldots \tag{14.22}$$

Now, we derive the different characteristics of queuing model I.

1. *Expected number of customers in the system*: The expected number of customers in the system (or line length) is denoted by L_s and is given as

   $$L_s = E(N) = \sum_{n=0}^{\infty} n\, P_n$$

   Here N is the number of customers in the system, which is a random variable.
   Using Eqs (14.21) and (14.22), we get

   $$L_s = \sum_{n=0}^{\infty} n\,(1-\rho)\rho^n$$

   $$= (1-\rho)(\rho + 2\rho^2 + 3\rho^3 + \ldots) = (1-\rho)\frac{\rho}{(1-\rho)^2} = \frac{\rho}{1-\rho} = \frac{\lambda}{\mu - \lambda}$$

   $$\text{Thus, } L_s = \frac{\lambda}{\mu - \lambda} \tag{14.23}$$

   where λ and μ are the mean arrival and service rates, respectively.

2. *Expected number of customers in the queue*: The expected number of customers in the queue is denoted by L_q and is given as

   $$L_q = \sum_{n=0}^{\infty} (n-1)P_n$$

   $$= \sum_{n=0}^{\infty} nP_n - \sum_{n=0}^{\infty} P_n$$

   $$= \sum_{n=0}^{\infty} nP_n - [1 - P_0] \quad \text{as} \quad \sum_{n=0}^{\infty} P_n = 1$$

   $$= \frac{\rho}{1-\rho} - \rho$$

   $$= \rho\left(\frac{1}{1-\rho} - 1\right) = \frac{\rho^2}{1-\rho} = \rho L_S$$

   $$= \frac{\lambda^2}{\mu(\mu - \lambda)}$$

$$\text{Thus, } L_q = \rho L_s = \frac{\rho^2}{1-\rho} = \frac{\lambda^2}{\mu(\mu-\lambda)} \tag{14.24}$$

3. *Expected waiting time in the system*: The expected waiting time in the system denoted by W_s is $W_s = \dfrac{L_s}{\overline{\lambda}}$

(where $\overline{\lambda}$ = effective arrival rate = $\sum_{n=0}^{\infty} \lambda P_n = \lambda \sum_{n=0}^{\infty} P_n = \lambda$)

$$\text{Thus, } W_s = \frac{L_s}{\overline{\lambda}} = \frac{\rho}{1-\rho} \cdot \frac{1}{\lambda} = \frac{\lambda}{\mu-\lambda} \cdot \frac{1}{\lambda} = \frac{1}{\mu-\lambda} \tag{14.25}$$

4. *Expected waiting time in the queue*: The expected waiting time in the queue denoted as W_q is

$$W_q = \frac{L_q}{\overline{\lambda}}$$

$$= \frac{\rho^2}{1-\rho} \cdot \frac{1}{\lambda} = \frac{\lambda}{\mu(\mu-\lambda)}, \ (\overline{\lambda} = \lambda)$$

$$\text{Thus, } W_q = \frac{L_q}{\overline{\lambda}} = \frac{\lambda}{\mu(\mu-\lambda)} \tag{14.26}$$

$$\text{or } W_q = \frac{\lambda}{\mu} W_s = \rho W_s \tag{14.27}$$

5. *Expected length of a non-empty queue:* Let Q be the number of customers in a queue where $Q > 0$, then Q is a random variable and the expected value of Q, given that $Q > 0$ is denoted by $E(Q/Q > 0)$ and is given as

$$E(Q/Q > 0) = \frac{E(Q)}{P(Q > 0)} = \frac{L_q}{\sum_{k=2}^{\infty} P_k}$$

where k varies from 2 to ∞ as one person is availing service and at least one person is in the queue.

$$\text{Thus,} \quad E(Q/Q > 0) = \frac{L_q}{1 - P_O - P_1} = \frac{\rho^2}{(1-\rho)\rho^2}$$

$$= \frac{1}{1-\rho}, \ \rho = \frac{\lambda}{\mu}$$

6. *Probability that at least k customers are in the queue system*: The probability that at least k customers are in the queue system is as follows:

$$= \sum_{i=k}^{\infty} P_i = \sum_{i=k}^{\infty} (1-\rho)\rho^i = (1-\rho)\rho^k \sum_{j=0}^{\infty} \rho^j = \frac{(1-\rho)\rho^k}{(1-\rho)}$$

$$= \rho^k, \ \rho = \frac{\lambda}{\mu}$$

7. *Probability density function (PDF) of waiting time in the system*: The PDF of waiting time in the system W_s, that is, the time a customer spends in the system, can be derived as follows:

   Let W be the time spent by a customer in the system, which is a random variable. Let F be the distribution function of W, then $F(w) = P(W \le w)$.

   Let n be the number of customers already present in the system at the time of entry of a new customer for availing the service. Let $W = T + T_1 + T_2 + \ldots + T_n$, where T is the service time for the newly arrived customer and $T_1, T_2, T_3, \ldots, T_n$ are the service times for the n customers ahead of this new customer. $T, T_1, T_2, T_3, \ldots, T_n$ are the independent exponentially distributed random variables each with parameter μ. Hence the conditional distribution of W, given that there are n customers in the system when this customer arrives, is a gamma distribution with parameters $(n + 1)$ and μ. Thus, we have

$$F(w) = 0, \; w < 0$$
$$= P(W \le w), \; w \ge 0$$
$$= \sum_{n=0}^{\infty} P(W \le w, N = n), \; w \ge 0$$
$$= \sum_{n=0}^{\infty} P(N = n)\, P(W \le w / N = n), \; w \ge 0$$
$$= \sum_{n=0}^{\infty} (1-\rho)\, \rho^n \int_0^w \frac{\mu(\mu t)^n}{n!} e^{-\mu t} dt, \; w \ge 0$$

   Now differentiating this with respect to w, the PDF $f(w)$ of waiting time in the system is

$$f(w) = \begin{cases} \sum_{n=0}^{\infty} (1-\rho)\, \rho^n \dfrac{\mu(\mu w)^n}{n!} e^{-\mu w}, & w \ge 0 \\ = 0, & w < 0 \end{cases}$$

   or

$$f(w) = \mu(1-\rho)\, e^{-\mu w} \sum_{n=0}^{\infty} \frac{(\rho\mu w)^n}{n!}, \; w \ge 0$$
$$= \mu(1-\rho)\, e^{-\mu w} e^{\rho\mu w}$$
$$= \mu(1-\rho)\, e^{-\mu(1-\rho)w}, \; w \ge 0$$
$$= (\mu - \lambda)\, e^{-(\mu-\lambda)w}, \; w \ge 0$$

   Thus W, the waiting time in the system, has an exponential distribution with parameter $(\mu - \lambda)$ and the mean waiting time, W_s, of the customer in the system is $W_s = \dfrac{1}{\mu - \lambda}$, which is same as derived in Eq. (14.25).

8. *Distribution of waiting time and expected time in a queue with discipline FCFS*: Following the aforementioned approach in deriving probability density for waiting in the system, the PDF $g(w)$ of the waiting time in a queue in a queuing system is given as

$$g(w) = \lambda(1-\rho)e^{-(\mu-\lambda)w},\ w \ge 0$$
$$= 0,\ w < 0$$

and W_q = Expected waiting time in queue

$$= \int_0^\infty w\, g(w)\,dw$$

$$= \int_0^\infty \lambda(1-\rho)w\, e^{-(\mu-\lambda)w}\, dw$$

$$= \frac{\lambda}{\mu}(\mu-\lambda)\int_0^\infty w\, e^{-(\mu-\lambda)w}\, dw = \frac{\lambda(\mu-\lambda)}{\mu(\mu-\lambda)^2} = \frac{\lambda}{\mu(\mu-\lambda)}$$

Thus, the mean waiting time W_q of the customer in the queue under the queuing system is $W_q = \dfrac{\lambda}{\mu(\mu-\lambda)}$.

Note We are discussing Markovian queues where arrivals take place in Poisson process and inter-arrival time (and service time) has an exponential distribution. For such queues following an FCFS discipline, the waiting time in the system also has an exponential distribution with a PDF as given here.

$$f(t) = \mu(1-\rho)e^{-\mu(1-\rho)t},\ t \ge 0$$

Example 14.1 A tollgate is operated on a highway where vehicles arrive according to Poisson distribution with a mean frequency of 1.2 vehicles per minute (min). The time of completing toll tax payment is an exponential distribution with mean 20 seconds. Find (a) P_0, (b) L_s, (c) L_q, (d) W_s, (e) W_q, (f) the probability that a vehicle spends more than 30 seconds in the system, and (g) the average waiting time in the queue for those who wait.

Solution Here $\lambda = 1.2/60 = 0.02$ per second and $\mu = 1/20$ per second, therefore, $\rho = 2/5$.

(a) $P_0 = 1-\rho = 1-\dfrac{2}{5} = \dfrac{3}{5}$

(b) $L_s = \dfrac{\rho}{1-\rho} = \dfrac{2}{3}$

(c) $L_q = \rho L_s = \dfrac{2}{5}\times\dfrac{2}{3} = \dfrac{4}{15}$

(d) $W_s = \dfrac{L_s}{\lambda} = \dfrac{2}{3}\times\dfrac{1}{0.02} = \dfrac{100}{3}$ seconds

(e) $W_q = \dfrac{L_q}{\lambda} = \dfrac{4}{15}\times\dfrac{1}{0.02} = \dfrac{40}{3}$ seconds

(f) P(time spent in the system by a vehicle is > 30 seconds)

$= P(T_s > 30)$

$= \displaystyle\int_{30}^\infty \mu(1-\rho)e^{-\mu(1-\rho)t}dt,\quad \mu(1-\rho) = 0.03 = e^{-0.9} = 0.4065$

(g) $E(T_q/T_q > 0) = \dfrac{E(T_q)}{P(T_q > 0)}$

$$\left[\therefore \text{ as } P(T_q > 0) = \int_0^\infty (\mu - \lambda)e^{-(\mu-\lambda)t}dt = \rho\right]$$

$$= \frac{\lambda}{\mu(\mu-\lambda)\rho} = \frac{1}{\mu-\lambda} = \frac{100}{3}$$

Example 14.2 A firm has several machines and wants to install its own service facility for the repair of its machines. The average breakdown rate of the machines is three per day. Assume that the inter-arrival times are independent exponential variables and the repair time has exponential distribution. The loss incurred due to the lost time caused by the breakdown of an inoperative machine is ₹40 per day. The firm has two repair facilities—A and B. While facility A requires an installation cost of ₹20,000, facility B costs ₹40,000. The total labour cost per year is ₹5000 and ₹8000 per year for A and B respectively. While A can repair 4.5 machines per day, B can repair five machines per day. Both these facilities have a life of four years. Which facility should be installed?

Solution We shall compare the total annual cost of the two facilities:

Facility A: Annual cost of capital $= \dfrac{20,000}{4} =$ ₹5000

Annual labour cost = ₹5000

Expected number of customers $L_s = \dfrac{\lambda}{\mu-\lambda} = \dfrac{3}{4.5-3} = 2$

The daily cost of lost time $= 40 \times 2 =$ ₹80 per day.

Annual cost of lost machine time $= 365 \times 80 =$ ₹29,200

Thus, the total annual cost incurred for facility A = 5000 + 5000 + 29,200 = ₹39,200.

Facility B: Annual cost of capital $= \dfrac{40,000}{4} =$ ₹10,000

Annual labour cost = ₹8000

Expected number of customers in the system $= L_s = \dfrac{\lambda}{\mu-\lambda} = \dfrac{3}{5-3} = \dfrac{3}{2}$

The daily cost of lost time $= 40 \times \dfrac{3}{2} =$ ₹60.

Annual cost of lost machine time $= 365 \times 60 =$ ₹21,900

Thus, the total annual cost for facility B = 10,000 + 8000 + 21,900 = ₹39,900.

Hence, we conclude that facility B is more expensive and hence, facility A should be preferred.

Example 14.3 At a telephone booth, arrivals of customers follow the Poisson process with an average inter-arrival time of 10 min. The length of a phone call is exponentially distributed with mean three min. Find the

(a) average numbers of customers in the (i) system (ii) queue
(b) expected waiting time of the customer in the (i) system (ii) queue for getting service
(c) probability that a customer spends more than 10 min in the booth
(d) probability that an arriving customer at the booth (i) has to wait (ii) does not have to wait
(e) fraction of the time the booth is busy

Solution Here, we use (M/M/1):(FCFS/∞). One customer arrives in every 10 min at the booth, so the mean arrival rate $\lambda = 60/10 = 6$ customers/hour (h)

Mean service rate $\mu = \frac{60}{3} = 20$ customers/h

Then, $\rho = \frac{\lambda}{\mu} = \frac{6}{20} = \frac{3}{10}$.

(a) (i) Average number of customers in the system $L_s = \frac{\lambda}{\mu - \lambda} = \frac{6}{20 - 6} = \frac{3}{7}$

(ii) Average number of customers in the queue $L_q = \rho L_s = \frac{3}{10} \times \frac{3}{7} = \frac{9}{70}$

Alternatively, $L_q = \frac{\lambda^2}{\mu(\mu - \lambda)} = \frac{36}{20(20 - 6)} = \frac{9}{5 \times 14} = \frac{9}{70}$.

(b) (i) Expected waiting time of the customer in the system for receiving service

$W_s = \frac{L_s}{\lambda} = \frac{3/7}{6} = \frac{1}{14}$ h

$= \frac{1}{14} \times 60 = \frac{30}{7}$ min

(ii) Expected waiting time of the customer in the queue for receiving service

$W_q = \frac{L_q}{\lambda} = \frac{9}{70 \times 6} = \frac{3}{140}$ h $= \frac{3}{140} \times 60 = \frac{9}{7}$ min

Alternatively, $W_q = \frac{\lambda}{\mu} W_S = \rho W_S = \frac{3}{10} \times \frac{30}{7} = \frac{9}{7}$ min

(c) Probability density function $f(t)$ of waiting time in the system is

$f(t) = (\mu - \lambda)e^{-(\mu - \lambda)t}, t \geq 0.$

Thus, the probability of spending more than 10 min $\left(\frac{10}{60} = \frac{1}{6} \text{ h}\right)$ in the booth is

$= \int_{1/6}^{\infty} f(t)dt$

$= (\mu - \lambda) \int_{1/6}^{\infty} e^{-(\mu - \lambda)t} dt, t \geq 0, \mu = 20, \lambda = 6$

$= \left[-e^{-(\mu - \lambda)t}\right]_{1/6}^{\infty} = e^{-7/3} = 0.0969$

(d) (i) An arriving customer at the booth has to wait if there is at least one customer in the system. The required probability $= P_1 + P_2 + \ldots = 1 - P_0 = \rho = \frac{\lambda}{\mu} = \frac{6}{20} = 0.3$

(ii) Probability of receiving service immediately on arrival

$= P$(there is no customer in the system) $= P_0 = 1 - \rho = 1 - 0.3 = 0.7$

(e) The fraction of the time the booth (server) is busy $= \rho = 0.3$. Thus, if a booth is opened for 10 h in a day for service, then it is busy for $0.3 \times 10 = 3$ h out of 10 h of the day.

Example 14.4 At a cycle shop with one person repairing the cycles, on an average a customer arrives every 5 min and on an average the service time is 4 min per customer. Suppose the inter-arrival time and the service time are exponentially distributed, discuss the different characteristics of the queuing system.

Solution (a) Average number of arrivals per min $\lambda = \frac{1}{5} = 0.20$

Average number of customers served per min $\mu = \frac{1}{4} = 0.25$

(b) The fraction of the time the repairman is busy $= P_1 + P_2 + \ldots = 1 - P_0 = \rho$ $= \frac{\lambda}{\mu} = \frac{0.20}{0.25} = 0.80$, that is, the repairman is busy 80% of the time.

Thus, if the cycle shop opens for 10 h in a day, then the repairman is busy for 8 h.

(c) $P_0 = 1 - \rho = 1 - \frac{\lambda}{\mu} = 0.20$

Thus, the probability that a customer will be able to avail some service immediately is 0.20.

$$P_1 = \rho P_0 = 0.80 \times 0.20 = 0.16$$

Thus, the probability that the customer will have to stand alone in the queue is 0.16.

$$P_2 = \rho P_1 = 0.80 \times 0.16 = 0.128$$

Thus, the probability that the customer will have to stand second in the queue and so on is 0.128.

$P_3 = \rho P_2 = 0.80 \times 0.128 = 0.1024$ and so on.

(d) L_s = Expected number of customers in the system $= \frac{\lambda}{\mu - \lambda} = \frac{0.20}{0.25 - 0.20} = 4$

(e) The average length of the queue $= L_q = \rho L_s = 0.80 \times 4 = 3.2$ customers.

(f) The mean time a customer spends in the system $= W_s = \frac{L_s}{\lambda} = \frac{4}{0.20} = 20$ min.

(g) The mean time a customer spends in the queue

$= W_q = \rho W_s = 0.80 \times 20 = 16$ min.

Note that as the utilization factor ρ is quite high, the waiting time W_s is also quite long. Hence to reduce the average waiting time in the system, the shop should think of employing another repairman.

(h) Expected length of the non-empty queue:

Let N_Q be the number of customers in the queue. Then, the expected length of the non-empty queue is $= E[N_Q/N_Q > 0]$

$$= \frac{E(N_Q)}{P(N_Q > 0)} = \frac{L_q}{\sum_{k=2}^{\infty} P_k}$$

$$= \frac{\lambda^2}{\mu(\mu - \lambda)} \cdot \frac{1}{1 - P_0 - P_1} = \frac{\rho^2}{1 - \rho} \cdot \frac{1}{\rho^2} = \frac{1}{1 - \rho} = \frac{1}{1 - 0.8} = 5.$$

(i) Probability that at least k customers are there in the system:

Required probability $= \sum_{i=k}^{\infty} P_i$

$= \sum_{i=k}^{\infty} (1 - \rho)\rho^i$ as $P_n = P_0 \rho^n$

$$= (1-\rho)\rho^k \sum_{j=0}^{\infty} \rho^j$$

$$= (1-\rho)\rho^k (1-\rho)^{-1}$$

$$= \rho^k, \ \rho = \frac{\lambda}{\mu}$$

$$= (0.8)^k.$$

14.10.2 Finite Storage Queue System with One Server (M/M/1):(FCFS/N)

Let us now consider the queue system with one server. The system can hold at most N customers including the customer in the service. Any customer arriving thereafter is refused entry into the system. Thus, we have

$$\lambda_n = \lambda, \ n < N$$
$$= 0, \ n \geq N$$

and

$$\mu_n = \mu, \ n = 1, 2, \ldots, N$$

where we have assumed that $\lambda_n = \lambda$ and $\mu_n = \mu$ for all n (i.e., arrival and service rates in the system are constant).

Thus, from Eqs (14.19) and (14.20), we have

$$P_n = \begin{cases} \rho^n P_0, & n = 1, 2, \ldots, N \\ 0, & n > N \end{cases} \tag{14.28}$$

and $P_0 = \dfrac{1}{\left(1 + \sum\limits_{n=1}^{N} \rho^n\right)}, \ \rho = \dfrac{\lambda}{\mu}$

$$= \begin{cases} \dfrac{1-\rho}{1-\rho^{N+1}}, & \text{if } \rho \neq 1 \\ \dfrac{1}{N+1}, & \text{if } \rho = 1 \end{cases} \tag{14.29}$$

Some characteristics of this model are discussed in the following text:

1. *Expected number of customers* (L_s) *in the system*: The expected number of customers in the system (or line length) is denoted by L_s and is given as follows:

$$L_s = \sum_{n=0}^{N} nP_n$$

$$= P_0 \sum_{n=0}^{N} n\rho^n = P_0 \sum_{n=0}^{N} \rho \frac{d}{d\rho}\rho^n$$

$$= P_0 \rho \frac{d}{d\rho}(1 + \rho + \rho^2 + \ldots + \rho^N)$$

$$= P_0 \rho \frac{d}{d\rho} \frac{1-\rho^{N+1}}{1-\rho}$$

$$= P_0 \rho \left(\frac{-(N+1)\rho^N (1-\rho) + (1-\rho^{N+1})}{(1-\rho)^2} \right)$$

Now, substituting the value of P_0, on simplification we get

$$L_s = \frac{\rho}{1-\rho} - \frac{(N+1)\rho^{N+1}}{1-\rho^{N+1}}, \rho \neq 1$$

When $\rho = 1$, $L_s = \sum_{n=0}^{N} nP_n = P_0 \sum_{n=0}^{N} n\rho^n = \frac{1}{N+1} \frac{N(N+1)}{2} = \frac{N}{2}$ as $P_0 = \frac{1}{N+1}$ (where $\rho = 1$)

$$\text{Thus, } L_s = \begin{cases} \frac{\rho}{1-\rho} - \frac{(N+1)\rho^{N+1}}{1-\rho^{N+1}}, & \rho \neq 1 \\ \frac{N}{2} & \rho = 1 \end{cases} \tag{14.30}$$

2. *Expected length of the queue (L_q)*:

$$L_q = \sum_{n=1}^{N} (n-1)P_n = \sum_{n=1}^{N} nP_n - \sum_{n=1}^{N} P_n$$

$$= L_s - [1 - P_0] \text{ as } \sum_{n=0}^{N} P_n = 1 \tag{14.31}$$

The values of L_q for cases $\rho \neq 1$ and $\rho = 1$ are obtained by substituting the corresponding values of L_s and P_0 in Eq. (14.31).

3. *Expected waiting time in the system*: The effective mean arrival rate is denoted by $\bar{\lambda}$ and is given by

$$\bar{\lambda} = \sum_{n=0}^{N-1} \lambda_n P_n \quad \text{as } \lambda_n = 0 \quad \text{for} \quad n \geq N$$

$$= \sum_{n=0}^{N-1} \lambda P_n \quad \text{as } \lambda_n = \lambda \quad \text{for} \quad n < N$$

$$= \lambda(1 - P_N) \quad \text{as} \quad \sum_{n=0}^{N} P_n = 1$$

Thus, the waiting time in the system W_s is given by $W_s = \frac{L_s}{\bar{\lambda}}$

$$= \frac{L_s}{\lambda(1-P_N)} \tag{14.32}$$

Hence, W_s can be obtained for the cases $\rho \neq 1$ and $\rho = 1$ by substituting the corresponding values of L_s and P_N in Eq. (14.32).

4. *Expected waiting time in the queue*: Using Eq. (14.31), the waiting time in the queue W_q is given as follows:

$$W_q = \frac{L_q}{\bar{\lambda}} = \frac{L_s - (1-P_0)}{\bar{\lambda}} = \frac{L_s - (1-P_0)}{\lambda(1-P_N)} \tag{14.33}$$

which can be obtained for the cases $\rho \neq 1$ and $\rho = 1$ by substituting the corresponding values of L_s, P_0, and P_N.

From Eqs (14.32) and (14.33), we get

$$W_q = W_s - \frac{1}{\mu} \text{ or } W_s = W_q + \frac{1}{\mu}$$

For both the cases (a) $\rho \neq 1$ and (b) $\rho = 1$ ($\lambda = \mu$), the mean time spent by a customer in the queue is obtained by subtracting the mean service time of the customer from the mean time spent by him/her in the system.

Further, $L_s = L_q + \rho$

This shows that the expected number of customers in the service $= \rho$ or the expected number of busy servers $= \rho$

Thus, the proportion of the expected time for which a server is busy $= \rho/s$ and the proportion of expected time for which a server isidle $= 1 - \rho/s$.

Example 14.5 In a doctor's clinic, patients arrive according to Poisson distribution at an average rate of 15 patients per h. The waiting room can accommodate 14 patients at the maximum. The examining time per patient is exponentially distributed with the mean rate 10 per h. Find the

(a) effective arrival rate of patients at the clinic.
(b) probability that an arriving patient does not have to wait.
(c) probability that an arriving patient will find a vacant seat in the room.
(d) Find the expected waiting time of a patient until he/she is discharged from the clinic?

Solution Here $\lambda = 15$, $\mu = 10$, and $N = 15$.

$$\text{Thus, } \rho = \frac{\lambda}{\mu} = \frac{15}{10} = 1.5$$

$$P_0 = \frac{1-\rho}{1-\rho^{N+1}} = \frac{1-1.5}{1-(1.5)^{16}} = 0.00076$$

$$P_N = \rho^N P_0 = (1.5)^{15} \times (0.00076) = 0.333$$

(a) Effective arrival rate at the clinic $\bar{\lambda} = \lambda(1 - P_N) = 15 \times 0.667 = 10.005$ per h

(b) Probability that an arriving patient does not have to wait
$=$ probability that there is no patient at the clinic
$= P_0 = 0.00076$

(c) Probability that an arriving customer (patient) finds a vacant seat
$= P$ (that the number of the patient in the clinic is ≤ 14)
$= P_0 + P_1 + P_2 + \ldots + P_{14} = 1 - P_{15} = 1 - 0.333 = 0.667$

(d) Expected number of patients in the clinic

$$N_s = \frac{\rho}{1-\rho} - \frac{(N+1)\rho^{N+1}}{1-\rho^{N+1}}, \rho \neq 1$$

$= -3 + 16 \times 1.001 = 13.016$

(e) Expected time spent by a patient in the clinic $=$ Expected waiting time of the patient in the clinic

$$= W_s = \frac{N_s}{\bar{\lambda}} = \frac{13.016}{10.005} = 1.301\,\text{h}$$

where $\bar{\lambda} = \lambda(1 - P_N) = 15(1 - P_{15}) = 15 \times 0.667 = 10.005$

Example 14.6 At a washing service facility, cars arrive for servicing in a Poisson process at a mean arrival rate of 5 per h. The total servicing time of cars is exponentially distributed with a mean of 10 min per car. The service facility handles one car at a time and has 5 parking spaces for waiting if more than one car arrives at the service station. Find the

(a) effective arrival rate of cars at the service station
(b) probability that an arriving car does not have to wait for the service
(c) expected number of parking spaces occupied
(d) proportion of time the service station is busy
(e) total time of the day the service station is idle if it operates 10 h a day

Solution Here $\lambda = 5$ per h, $\mu = 1/10$ per min $= 6$ per h, $N = 1 + 5 = 6$.

Thus, $\rho = \dfrac{\lambda}{\mu} = \dfrac{5}{6}$

$$P_0 = \frac{1-\rho}{1-\rho^{N+1}} = \frac{1-\dfrac{5}{6}}{1-(5)^7} = \frac{6^6}{6^7-5^7} = \frac{46}{201} = 0.23$$

$$P_N = \rho^N P_0 \Rightarrow P_N = P_6 = \rho^6 P_0$$

$$= \left(\frac{5}{6}\right)^6 \times 0.23 = (0.3364) \times (0.23) = 0.076.$$

Thus, $P_6 = 0.076$

(a) Effective arrival rate $\bar{\lambda} = \lambda(1 - P_6) = 15 \times 0.924 = 4.620$ per h
(b) Probability that an arrival car does not have to wait
$= P$ (that there is no car at the service station) $= P_0 = 0.23$
(c) Expected number of parking spaces occupied $= L_q = \sum\limits_{n=1}^{5} (n-1)P_n$

$$= 0P_1 + 1P_2 + 2P_3 + 3P_4 + 4P_5 = 0P_1 + \rho^2 P_2 + 2\rho^3 P_3 + 3\rho^4 P_4 + 4\rho^5 P_5$$

$$= \rho^2 P_0 (1 + 2\rho + 3\rho^2 + 4\rho^3)$$

$$= \left(\frac{5}{6}\right)^2 (0.23)\left[1 + 2\left(\frac{5}{6}\right) + 3\left(\frac{5}{6}\right)^2 + 4\rho^3\right] = 1.219 \approx 1.$$

Thus, on an average one parking space is occupied.

(d) Proportion of the time the service station is busy $= \dfrac{\bar{\lambda}}{\mu} = \dfrac{4.62}{6} = 0.77$ h/unit h

(e) Proportion of the time the service station is idle $= 1 - \dfrac{\bar{\lambda}}{\mu} = 1 - 0.77 = 0.33$ h

Thus during 10 h of the day, the total idle time $= 0.33 \times 10 = 3.3$ h.

14.10.3 S-server Case (M/M/S):(FCFS/∞)

Here, we consider the case when there are S number of servers in the system with unlimited accommodation space and constant arrival ratio λ. Using Eqs 14.19 and 14.20, we have

$$\lambda_n = \lambda,\ n = 0, 1, 2, \ldots$$

$$\mu_n = \begin{cases} n\mu, 0 \le n \le s \\ s\mu, n \ge s \end{cases}$$

where μ is the average number of services per unit time per server

$$P_n = \frac{\lambda_0 \lambda_1 \ldots \lambda_{n-1}}{\mu_1 \mu_2 \ldots \mu_n} P_0$$

$$= \begin{cases} \dfrac{\lambda^n P_0}{\mu(2\mu)(3\mu)\ldots(n\mu)}, \quad n = 1, 2, \ldots, s \\ \dfrac{\lambda^n P_0}{\underbrace{\mu(2\mu)(3\mu)\ldots(s\mu)}_{s \text{ products}}\underbrace{(s\mu)\ldots(s\mu)}_{(n-s) \text{ products}}}, \quad n \ge s \end{cases}$$

On simplification, we get

$$P_n = \begin{cases} \rho^n P_0, 1 \le n \le s \\ \dfrac{\rho^n P_0}{s!\, s^{n-s}}, n \ge s \end{cases} \qquad \text{where } \rho = \lambda/\mu \qquad (14.34)$$

and $P_0 = \dfrac{1}{\left(1 + \sum\limits_{n=1}^{N} \dfrac{\lambda_0 \lambda_1 \ldots \lambda_{n-1}}{\mu_1 \mu_2 \ldots \mu_n} P_0\right)},$

On simplification using the aforementioned data, we get

$$P_0 = \left[\sum_{n=0}^{s-1} \frac{\rho^n}{n!} + \frac{\rho^s}{\left(1 - \dfrac{\rho}{s}\right) s!}\right]^{-1} \qquad (14.35)$$

For steady state solution to exist, we must have $\rho < s$, that is, $\lambda < \mu s$.

1. *Expected number of customer in the queue (L_q) is given as follows:*

$$L_q = \sum_{n=s}^{\infty} (n - s) P_n = \sum_{n=s}^{\infty} \frac{(n-s)\rho^n}{s^{n-s}\, s!} P_0$$

$$= \sum_{x=0}^{\infty} \frac{x \rho^{x+s}}{s^x\, s!} P_0 \text{ where } x = n - s$$

$$= \frac{\rho^s}{s!} \sum_{x=0}^{\infty} x \left(\frac{\rho}{s}\right)^x P_0 = \frac{\rho^s}{s!} P_0 y \frac{d}{dy} \sum_{x=0}^{\infty} y^x \text{ where } \quad y = \frac{\rho}{s}$$

$$= \frac{\rho^s}{s!} P_0 y \frac{d}{dy}\left(\frac{1}{1-y}\right) = \frac{\rho^s}{s!} \frac{y}{(1-y)^2} P_0 = \frac{\lambda \mu}{(s-1)!} \frac{\rho^s P_0}{(\mu s - \lambda)^2}.$$

2. *Expected number of customers (L_s) in the system*:

The expected number of customers in the system (or line length) is denoted by L_s and can be calculated as follows:

The effective rate $\bar{\lambda}$ is given as follows:

$$\bar{\lambda} = \sum_{n=0}^{\infty} \lambda_n P_n = \lambda \sum_{n=0}^{\infty} P_n = \lambda$$

We also know that $W_s = W_q + 1/\mu$ and hence

$$\lambda W_s = \lambda W_q + \frac{\lambda}{\mu} \qquad \left[\text{Here } \bar{\lambda} = \lambda\right]$$

$$\Rightarrow L_s = L_q + \rho, \qquad \left[\because \lambda W_s = L_s,\ \lambda W_q = L_q, \text{ and } \rho = \frac{\lambda}{\mu}\right]$$

Knowing the value of L_q and ρ, we can find L_s, W_s (the expected waiting time in the system), and W_q (the waiting time in the queue).

Here ρ = expected number of customers in the service = expected number of busy servers.

Hence, the expected time for which a server is busy = ρ/s and the expected time for which a server is idle = $1 - \rho/s$.

3. *P(all servers are busy)*

$= P(\text{number of customers } n \text{ in the system} \geq s)$

$= P_s + P_{s+1} + \ldots$

$$= \sum_{n=s}^{\infty} P_n = \sum_{n=s}^{\infty} \frac{\rho^n P_0}{s!s^{n-s}}$$

$$= \frac{P_0}{s!}\left(\rho^s + \frac{\rho^{s+1}}{s} + \frac{\rho^{s+2}}{s^2} + \ldots\right)$$

$$= \frac{\rho^s P_0}{s!}\left(1 + \frac{\rho}{s} + \frac{\rho^2}{s^2} + \ldots\right) = \frac{\rho^s}{s!} P_0 \left(1 - \frac{\rho}{s}\right)^{-1} = P(\text{that an arrival has to wait})$$

Thus, in this model, the probability that an arrival has to wait equals the probability that all servers are busy.

Example 14.8 In a bank, clerk A deals with the withdrawals and clerk B deals with the payments. The service time of both is exponentially distributed with average 3 min per customer. The arrival rate of both the withdrawers and the payment takers is in the Poisson pattern with average rate of 14 and 16 per h respectively.

(a) Find the expected waiting time in the system for (i) withdrawers (ii) payment takers.

(b) If both the clerks are pooled together to handle both withdrawals and payments, what will be the effect on waiting time in the system for a customer in comparison to (a).

Solution (a) (i) For the clerk A dealing with withdrawals:

λ = 14 per h = 14/60 per min and μ = 1/3 per min

The system is (M/M/1)/FCFS/∞), for withdrawals. Thus, the withdrawers expected time in the system is $W_s = \dfrac{1}{\mu - \lambda} = 10$ min.

(ii) For the clerk B dealing with payments: $\lambda = 16/60$ per min and $\mu = 1/3$ per min Again, the system being (M/M/1)/FCFS/∞), the expected waiting time is

$$W_s = \frac{1}{\mu - \lambda} = 15 \text{ min}$$

(b) If the clerks are pooled, the system becomes (M/M/2) with

$$\lambda = \frac{14}{60} + \frac{16}{60} = \frac{1}{2} \text{ per min and } \mu = 1/3 \text{ per min, } \rho = \frac{\lambda}{\mu} = \frac{3}{2}$$

It is to be noted that μ remains the same.

We know that

$$P_0 = \left[\sum_{n=0}^{s-1} \frac{\rho^n}{n!} + \frac{\rho^s}{\left(1 - \dfrac{\rho}{s}\right)s!}\right]^{-1} = \left[1 + \rho + \frac{\rho^2}{2\left(1 - \dfrac{\rho}{2}\right)}\right]^{-1} = \frac{1}{7} \left[\because \rho = \frac{3}{2}\right]$$

Now, the waiting time $W_s = W_q + \dfrac{1}{\mu}$

$$= \frac{L_q}{\lambda} + \frac{1}{\mu} = \frac{\mu \rho^2 P_0}{(2\mu - \lambda)^2} + \frac{1}{\mu} = \frac{27}{7} + 3 = 6.86 \text{ min.}$$

Comparing this with the results of part (a), we find that the efficiency improves by pooling.

Example 14.9 A supermarket has two salesgirls at the counter. Answer the following questions, if the service time for each customer is exponentially distributed with mean 4 min and the customers arrive in Poisson pattern at the rate of 10 per h.

(a) What is the probability that a customer has to wait in the queue for service?
(b) What is the idle time for each girl?
(c) What is the expected number of customers in the queue?
(d) What is the expected time spent in the queue?

Solution Here $s = 2$, $\lambda = 1/6$ per min, $\mu = 1/4$ per min for each girl.
(It is to be noted that μ is the rate of service per min per server)

$$\rho = \frac{\lambda}{\mu} = \frac{2}{3}$$

We know that

$$P_0 = \left[\sum_{n=0}^{s-1} \frac{\rho^n}{n!} + \frac{\rho^s}{\left(1 - \dfrac{\rho}{s}\right)s!}\right]^{-1} = \left[1 + \rho + \frac{\rho^2}{2\left(1 - \dfrac{\rho}{2}\right)}\right]^{-1} = \frac{1}{2} \left[\because \rho = \frac{2}{3}\right]$$

and $\quad P_1 = \rho P_0 = \dfrac{2}{3} \cdot \dfrac{1}{2} = \dfrac{1}{3}$

(a) Probability of waiting in queue $= P(n \geq 2)$

$$= P_2 + P_3 + \ldots = 1 - P_0 - P_1 = 1 - \frac{1}{2} - \frac{1}{3} = \frac{1}{6}$$

(b) Proportion of idle time for each girl

$$= P_0 + \frac{1}{2} P_1 = \frac{1}{2} + \frac{1}{6} = \frac{2}{3}$$

Alternately, proportion of the busy time $= \frac{\rho}{s} = \frac{1}{3}$

Thus, the proportion of idle time $= 1 - \frac{1}{3} = \frac{2}{3}$.

(c) Expected number of customers L_q in the queue is

$$L_q = \frac{\lambda\mu}{(s-1)!}\frac{\rho^s P_0}{(\mu s - \lambda)^2} = \frac{1}{24}\frac{\rho^2 P_0}{(2\mu - \lambda)^2} = \frac{1}{24}\cdot\frac{4}{9}\cdot\frac{1}{2}\cdot 9 = \frac{1}{12}$$

(d) Expected time in queue is $\frac{L_q}{\lambda} = \frac{1}{2}$ min.

14.10.4 S-server Case with Finite Accommodation Capacity (M/M/S):(FCFS/N)

There are S servers in the system. The system can accommodate not more than N customers. Thus, for this system

$$\lambda_n = \lambda,\ n = 0, 1, 2, \ldots, N-1$$
$$= 0,\ n \geq N$$

$$\mu_n = \begin{cases} n\mu, & 0 \leq n \leq s \\ s\mu, & s \leq n \leq N \end{cases}$$

Thus, as derived in the (M/M/S):(FCFS/∞)model, we have

$$P_n = \frac{\rho^n P_0}{n!}, \quad 0 \leq n \leq s$$

$$= \frac{\rho^n P_0}{s!\,s^{n-s}}, \quad s \leq n \leq N, \rho = \frac{\lambda}{\mu} \tag{14.36}$$

$$\text{and } P_0 = \left[\sum_{n=0}^{s-1} \frac{\rho^n}{n!} + \sum_{n=s}^{N} \frac{\rho^s}{s!\,s^{n-s}}\right]^{-1}$$

$$= \left[\sum_{n=0}^{s-1} \frac{\rho^n}{n!} + \frac{\rho^s(1-x)^{N-s+1}}{s!(1-x)}\right]^{-1}, x = \frac{\rho}{s} \tag{14.37}$$

1. *Expected number of customers in the queue*:

$$L_q = \sum_{n=s}^{N} (n-s)\,P_n = \sum_{n=s}^{N} (n-s)\frac{\rho^{x+s}}{s^{n-s}\,s!} P_0 = \frac{P_0}{s!}\sum_{n=s}^{N} \frac{(n-s)\rho^s}{s^{n-s}}$$

Replacing n by $n + s$, we get

$$= \frac{P_0\rho^s}{s!}\sum_{n=0}^{N-s} n\left(\frac{\rho}{s}\right)^n = \frac{\rho^s}{s!}P_0 \sum_{n=0}^{N-s} nx^x \text{ where } x = \frac{\rho}{s}$$

On simplification, we get

$$L_q = \frac{P_0}{s!} \frac{\rho^{s+1}}{(s-\rho)^2} \left\{1 - x^{N-s} - (N-s)x^{N-s}(1-x)\right\}$$

2. *Expected number of customers in the system*: From (1), we have

$$L_q = \sum_{n=s}^{N} (n-s) P_n$$

$$= \sum_{n=s+1}^{N} (n-s) P_n \text{ as the term corresponding to } n = s \text{ is zero.}$$

$$= \sum_{n=0}^{N} nP_n - \sum_{n=0}^{s} nP_n - s \sum_{n=s+1}^{N} P_n$$

$$= L_s - \sum_{n=0}^{s} nP_n - s\left(\sum_{n=0}^{N} P_n - \sum_{n=0}^{s} P_n\right)$$

$$= L_s - \sum_{n=0}^{s} nP_n - s\left(1 - \sum_{n=0}^{s} P_n\right)$$

$$= L_s - s + \sum_{n=0}^{s} (s-n) P_n = L_s - s + \overline{s}$$

where $\overline{s} = \sum_{n=0}^{s} (s-n) P_n$ = expected number of idle servers.

Thus, the expected number of idle servers $\overline{s} = \sum_{n=0}^{s} (s-n) P_n$

and $L_s = L_q + (s - \overline{s})$

where $s - \overline{s}$ = expected numbers of customers under service

or $s - \overline{s}$ = expected numbers of busy servers.

Further, the proportion of busy server $= \dfrac{s - \overline{s}}{s}$.

3. *Expected waiting time in the queue and system*:

W_q = Expected waiting time in the queue $= L_q/\lambda$,

where $\overline{\lambda}$ = mean effective arrival rate

$$= \sum_{n=0}^{N} \lambda_n P_n = \lambda \sum_{n=0}^{N} P_n \quad (\lambda_n = 0 \text{ when } n = N)$$

$$= \lambda(1 - P_N), \text{ as } \sum_{n=0}^{N} P_n = 1$$

$$W_q = \frac{L_q}{\overline{\lambda}} = \frac{L_q}{\lambda(1-P_N)}$$

Substituting the value of L_q, we can get W_q.

Now, we have

$$W_s = \text{Expected waiting time of a customer in the system}$$
$$= \text{Expected waiting time of a customer in the queue} + \text{Mean service time}$$
$$= W_q + \frac{1}{\mu}.$$

Multiply by $\bar{\lambda}$, we get $\bar{\lambda} W_s = \bar{\lambda} W_q + \dfrac{\bar{\lambda}}{\mu}$

or $L_s = L_q + \dfrac{\bar{\lambda}}{\mu}$.

Example 14.10 Customers arrive at a barber's shop in a Poisson pattern at a mean rate of 6 per h. The barber's shop has two servers whose service time is exponentially distributed at a mean service rate of 4 per h. There are only two seats for waiting. If the seats are filled, the customers are lost. Find the

(a) expected number of customers in the queue
(b) expected number of idle servers
(c) expected number of customers in the system
(d) proportion of customers lost
(e) expected number of customers lost per h
(f) expected number of customers in the service
(g) effective service rate
(h) proportion of idle time for a server

Solution Here, we have to apply queue model (M/M/S):(FCFS/*N*).

In this case $s = 2$, $N = 4$, $\lambda = 6$ per h, $\mu = 4$ per h, $\rho = \dfrac{\lambda}{\mu} = \dfrac{3}{2}, x = \dfrac{\rho}{s} = \dfrac{3}{4}$.

(a) We have $P_1 = \rho P_0 = \dfrac{3}{2} P_0$

$$P_2 = \frac{\rho^2}{2} P_0 = \frac{9}{8} P_0$$

$$P_3 = \frac{\rho^3}{4} P_0 = \frac{27}{32} P_0$$

$$P_4 = \frac{\rho^4}{8} P_0 = \frac{81}{128} P_0$$

Using $P_0 + P_1 + P_2 + P_3 + P_4 = 1$

We get $P_0 = \dfrac{128}{653}$

Now, $L_q = \sum_{n=2}^{4} (n-2) P_n = P_3 + 2P_4 = \left(\dfrac{27}{32} + \dfrac{81}{64}\right) P_0 = \dfrac{270}{653}$.

(b) We know that $\bar{s}$ = expected number of idle servers

$$= \sum_{n=0}^{s} (s-n) P_n$$

$$= \sum_{n=0}^{2} (2-n) P_n$$

$$= 2P_0 + P_1 = \frac{3}{2}P_0 + 2P_0 = \frac{7}{2}P_0 = \frac{7}{2} \times \frac{270}{653} = \frac{448}{653}$$

(c) L_s = Expected number of customers in the system

$$= L_q + (s - \bar{s})$$

$$= \frac{270}{653} + \left(2 - \frac{448}{653}\right) = \frac{1128}{653}$$

(d) Proportion of customers lost $= P_4$

$$= \frac{81}{128} P_0 = \frac{81}{653}$$

(e) Expected number of customers lost per h $= \lambda P_4$

$$= 6 \times \frac{81}{653} = \frac{486}{653}$$

(f) Expected number of customers in the service $= 0 \cdot P_0 + 1 \cdot P_1 + 2P_2 + 2P_3 + 2P_4$

$$= \left(\frac{3}{2} + \frac{9}{4} + \frac{27}{16} + \frac{81}{64}\right) P_0 = \frac{858}{653}$$

(g) The effective service rate $= \bar{\mu}$

$$= \mu P_1 + 2\mu P_2 + 2\mu P_3 + 2\mu P_4 = \frac{3432}{653}$$

which can also be calculated using $\bar{\mu} = \bar{\lambda} = \lambda(1 - P_4) = 6(1 - P_4) = 3432/653$.
or we can use $\bar{\lambda} = \mu(s - \bar{s})$

$$= 4\left(2 - \frac{448}{653}\right) = \frac{3432}{653}.$$

(h) Proportion of idle time for a server $= 1 - \dfrac{(s - \bar{s})}{s}$

$$= 1 - \frac{\bar{\lambda}}{\mu s} = \frac{224}{653}.$$

Example 14.11 In a shop there are two computers for carrying out the job work. The average time per job on each computer is 20 min per job and the average arrivals rate is 2 jobs per h. Assuming that the job times are exponentially distributed and the maximum number of jobs accepted on a day are 6, find the

(a) probability of a new job not having to wait
(b) expected number of idle computers
(c) expected queue length
(d) expected waiting time in the queue

Solution The model to be used here is (M/M/2):(FCFS/6).

Here $\lambda = 2$ jobs per h, service rate $\mu = 3$ jobs per h, $\rho = \dfrac{\lambda}{\mu} = \dfrac{2}{3}$, $x = \dfrac{\rho}{s} = \dfrac{1}{3}$,
the number of servers = 2, and N (the capacity of the system) = 6

$$P_0 = \left[\sum_{n=0}^{s-1} \frac{\rho^n}{n!} + \frac{\rho^s (1-x)^{N-s+1}}{s!(1-x)}\right]^{-1}, \quad x = \frac{\rho}{s}$$

$$P_n = \begin{cases} \dfrac{\rho^n P_0}{n!}, & 0 \le n \le s \\ \dfrac{\rho^n P_0}{s!\, s^{n-s}}, & s \le n \le N, \rho = \dfrac{\lambda}{\mu} \end{cases}$$

$$P_0 = \left[1 + \rho + \frac{\rho^2}{2}\frac{1 - x^5}{1 - x}\right]^{-1} = \left[1 + \frac{2}{3} + \frac{1}{3}\left(1 - \frac{1}{243}\right)\right]^{-1} = \frac{729}{143} = 0.5$$

Taking $n = 1, 2, 3, 4, 5$, and 6, we have the following:

$$P_1 = \rho P_0;\ P_2 = \frac{\rho^2}{2}P_0;\ P_3 = \frac{\rho^3}{4}P_0;\ P_4 = \frac{\rho^4}{8}P_0;\ P_5 = \frac{\rho^5}{16}P_0;\text{ and } P_6 = \frac{\rho^6}{32}P_0.$$

(a) A new job entering the system does not have to wait if the number of jobs in the system is less than or equal to one.

Thus, $P(\text{not to wait}) = P_0 + P_1 = P_0 + \rho P_1 = \frac{5}{3} \times \frac{1}{2} = \frac{5}{6}$.

(b) Expected number of idle computers

$$= \sum_{n=0}^{s} (s - n)\, P_n$$

$$= \sum_{n=0}^{2} (2 - n)\, P_n = 2P_0 + P_1 = 2P_0 + \rho P_0 = (2 + \rho)P_0 = \left(2 + \frac{2}{3}\right)P_0 = \frac{8}{3} \times \frac{1}{2} = \frac{4}{3}$$

(c) Expected queue length $= L_q = \sum_{n=2}^{6} (n - 2)\, P_n$

$$= P_3 + 2P_4 + 3P_5 + 4P_6$$

$$= \left(\frac{\rho^3}{4} + 2\frac{\rho^4}{8} + 3\frac{\rho^5}{16} + 4\frac{\rho^6}{32}\right)P_0 = \frac{\rho^3}{4}\cdot\frac{1}{2}\left(1 + \rho + \frac{3}{4}\rho^2 + \frac{\rho^3}{2}\right) = \frac{1}{27}\left(1 + \frac{2}{3} + \frac{1}{3} + \frac{4}{27}\right)$$

$$= \frac{58}{729} = 0.08$$

(d) Effective arrival rate of a job in the system $= \bar{\lambda}$

$$= \lambda(1 - P_6) = 2\left(1 - \frac{1}{3^6}\right)$$

Thus, the expected waiting time for a job $= L_q / \bar{\lambda}$

$$= \frac{0.08}{2\left(1 - \frac{1}{3^6}\right)} = 0.04 \text{ h.}$$

14.11 ADVANTAGES OF QUEUING THEORY

The following are some of the main advantages of the queuing theory:

1. Preparing a plan for transporting fleets
2. Managing and controlling inventory
3. Minimizing congestion caused by traffic delays at red light
4. Preparing a plan for aircraft landing and take-off

RECAPITULATION

1. The objective of the queuing theory is to bring about a balance between the waiting time of a customer and the idle time of the server.
2. In the queuing system, whenever the arrivals take place in Poisson pattern, the service as well as the inter-arrival times have an exponential distribution.
3. The probability distributions of arrivals in time interval $(t, t+\Delta t)$ in a pure birth process refers to only the probability of the arrival of customers in time Δt. The probability distribution of a pure death process means only the probability of the departure of customers.
4. In the birth and death process, birth means an 'entry into the system' and death means 'departure from the system after delivery of the service'.
5. Queuing model (M/M/1):(FCFS/∞) implies Poisson arrival, exponential service time, single server, and first come first served discipline with infinite capacity.
6. Queuing model (M/M/1):(FCFS/*N*) implies Poisson arrival, exponential service time, single server, and first come first served discipline with capacity *N*.
7. Queuing model (M/M/*S*):(FCFS/*N*) implies Poisson arrival, exponential service time, *S* servers, and first come first served discipline with capacity *N*.
8. Queuing model (M/M/*S*):(FCFS/∞) implies Poisson arrival, exponential service time, *S* servers, and first come first served discipline with infinite capacity.

EXERCISES

Multiple-choice Questions

14.1 If the arrival rate of customers is λ per unit time, then the mean service rate is
(a) $1/\lambda$ (b) λ (c) λ^2 (d) none of these

14.2 In Poisson queues, the number of arrivals has
(a) an exponential distribution (c) a normal distribution
(b) a Poisson distribution (d) none of these

14.3 In steady state, the probability that there are n customers in the system at any time t
(a) depends upon t (c) may or may not depend on t
(b) is independent of t (d) none of these

14.4 Which of the following characteristics apply to the queuing system?
(a) Arrival process (c) Arrival and customer process
(b) Customer process (d) None of these

14.5 The service mechanism in a queuing system is characterized by the
(a) customer's behaviour (c) customers in the system
(b) server's behaviour (d) none of these

14.6 Which of the following relationships is not true?
(a) $L_s = \lambda W_s$ (c) $W_s = W_q + \frac{1}{\mu}$
(b) $L_q = \lambda W_q$ (d) $L_s = L_q + \frac{1}{\lambda}$

State True or False

14.1 If the number of arrivals of customers has Poisson distribution, then inter-arrival times also have Poisson distribution.

14.2 If the number of arrivals of customers has Poisson distribution, then the inter-arrival time also has exponential distribution.
14.3 In the queuing system, both arrival and departure can take place simultaneously in a small interval of length Δt.
14.4 In a pure death process, there can be an arrival of a customer.
14.4 In a pure birth process, no departure can take place.

Fill in the Blanks

14.1 If the arrival rate of customers in the system is λ per unit time, than the mean service rate is ____________.
14.2 In birth and death process, ____________ arrival and departure can take place but not simultaneously.
14.3 In a queuing system in a small time of length Δt either an ____________ or a ____________ can take place.

Review Questions

14.1 What is a queue?
14.2 Explain the basic elements of queues.
14.3 Explain the Markovian queues.
14.4 What do you mean by the pure death process?
14.5 Explain the probability distribution of n arrivals in time interval $(t, t + \Delta t)$ in a pure birth process.
14.6 Write a short note on the birth and death process.
14.7 Write a short note on M/M/1 queue.
14.8 Discuss the various queuing models with their characteristic properties.
14.9 Write the benefits of the queuing theory.

Numerical Problems

14.1 In a transportation company, trucks arrive at a rate of 30 trucks per day. Determine the following assuming that the inter-arrival time follows an exponential distribution and the service time distribution is also exponential with an average of 36 min:
(a) Mean line length
(b) Probability that the queue size exceeds 10
(c) Expected waiting time in the queue
(d) Average number of trucks in the queue

14.2 A mobile phone repairman determines that the time spent on his job has an exponential distribution with a mean of 20 min. Assume, he repairs sets in the order in which they come in, and the arrival of sets is approximately Poisson with an average rate 10 in an 8 h day. What is the repairman's expected idle time each day?

14.3 At the Pilani reservation counter there is one clerk on duty. The clerk handles information regarding passenger reservations and train timings. Determine the following assuming that the number of customers arriving during any given period is Poisson distributed with an arrival rate of five per h and that the reservation clerk attends to a customer in 6 min on an average, with an exponentially distributed service time:
(a) Probability that the system is busy
(b) Average time a customer spends in the system
(c) Average length of the queue and the number of customers in the system

14.4 A supermarket has a single cashier. During peak hours, customers arrive at a rate of 25 per h. The average number of customers who can proceed to the cashier is 30 per h.

(a) Probability that the cashier is idle
(b) Average number of customers in the queuing system
(c) Average time a customer spends in the system
(d) Average time a customer spends in the queue waiting for the service

14.5 A workshop attended by a single mechanic has an average of five customers an hour who bring small appliances for repair. The mechanic inspects these for defects and often repairs them right away or renders a diagnosis. Arrivals follow a Poisson pattern and service time has exponential distribution.

(a) Calculate the proportion of time during which the shop is empty.
(b) Find the probability of determining at least one customer in the shop.
(c) Find the average number of customers in the system.
(d) Determine the average time spent, including service.

14.6 The arrival of customers at a telephone booth is considered to be following a Poisson pattern, with an average time of 10 min between one arrival and the next. The duration of phone calls is assumed to be distributed exponentially, with a mean of 3 min. Find the probability that a person arriving at the booth will have to wait.

14.7 In a supermarket, customers arrive in a Poisson process at an average time of 5 min between arrivals. The time interval between servers at the supermarket follows an exponential distribution and the mean time taken to service a unit is 2 min. Find the expected average queue length and average time a customer has to wait in the queue.

14.8 On an average, 72 patients per 24 h day require the service of an emergency clinic and on an average a patient requires 10 min of active attention. The facility can handle only one emergency at a time. Assuming that it costs the clinic ₹100 per patient treated to obtain an average servicing time of 10 patients and that each min of decrease in this average time would cost ₹10 per patient treated, how much would have to be budgeted by the clinic to decrease the average size of the queue from one and one third patients to half a patient.

14.9 Persons with problems arrive at a computer centre in Poisson pattern at an average rate of six per day. The rules of the computing centre are that any person waiting to get his/her problem solved must aid the person whose problem is being solved. If the time to solve a problem with one person has an exponential distribution with mean time of 1/3 day and if the average solving time is inversely proportional to the number of people working on the problem, determine the approximate expected time in the centre for a person entering the line.

14.10 A Maruti car servicing station has three stalls where service can be offered simultaneously. The cars wait in such a way that when a stall becomes vacant, the car at the head of the line pulls up to it. The station can accommodate a maximum of four cars waiting (seven in the station) at one time. The arrival pattern is Poisson with a mean of one car per min during the peak h. The service time is exponential with mean 6 min. Find the

(a) average number of cars in the queue
(b) average number of cars in the service station during peak hour
(c) average waiting time
(d) average number of cars per hour that cannot enter the station because of full capacity

CONCEPTUAL PROBLEM

Suppose there is a supermarket in a small town, which functions from 10.00 a.m. to 1.00 p.m. and from 5.00 p.m. to 9.00 p.m. In general, 5.00 to 9.00 p.m. are the rush hours (h). It has only one cashier. Naturally, the cashier at the counter is busy during rush hour and idle during the morning hours. During the busy hours, customers arrive at a rate of 20 customers per h. The average number of customers that can be attended to by the cashier is 24 per h. Find the

1. probability that the cashier is idle
2. proportion of time the cashier is busy and hence the total number of hours the super market is busy if it is opened for 12 h in a day
3. average number of customers in the queuing system
4. average time a customer spends in the system
5. average number of customers in the queue
6. average service time for a customer

Solution Here $\lambda = 20$ per h and service rate $\mu = 24$ customers per h. Then, $\rho = \dfrac{\lambda}{\mu} = \dfrac{20}{24} = \dfrac{5}{6}$

1. Probability that the cashier is idle $= P_0$

   $$= 1 - \rho = 1 - \frac{5}{6} = \frac{1}{6}$$

2. Proportion of busy time of the cashier $= 1 - P_0$

   $$= 1 - \frac{1}{6} = \frac{5}{6}$$

   Thus, the period in a day of 12 h. when the cashier is busy $= \dfrac{5}{6} \times 12 = 10\,\text{h}$.

3. Average (expected) number of customers in the system L_s is given as

   $$L_s = \frac{\lambda}{\mu - \lambda} = \frac{\rho}{1 - \rho} = \frac{5/6}{1 - (5/6)} = 5 \text{ customers.}$$

4. The average (expected) time W_s a customers spends in the system is given as

   $$W_s = \frac{1}{\mu - \lambda} = \frac{1}{24 - 20} = \frac{1}{4}\text{ h} = \frac{1}{4} \times 60 = 15 \text{ min.}$$

5. The average number of customers L_q in the queue is given as

   $$L_q = \frac{\lambda^2}{\mu(\mu - \lambda)}$$

   $$= \frac{\rho^2}{1 - \rho} = \frac{25}{36[1 - (5/6)]} = \frac{25}{36} \times 6 = \frac{25}{6} \approx 4 \text{ customers.}$$

6. The average service time for a customer $= \dfrac{1}{\mu}\text{ h} = \dfrac{1}{24} \times 60 = 2.5 \text{ min.}$

15 Goal Programming

After studying this chapter, the reader will be able to

- discuss the formulation of goal programming (GP)
- understand single-goal objective programming
- explain the multiple-goal model with equal or no priorities
- get acquainted with the GP algorithm or modified simplex method
- understand and explain the multiple-goal model with priorities and weights
- provide a graphical solution for a GP problem

15.1 INTRODUCTION

Goal programming (GP) is an extension of linear programming (LP). It is a multi-objective programming that intends to minimize the deviations between the actual results and the desired goals and works according to the priorities assigned to different goals. This model was first presented by Charnes and Copper as an extension to the LP model without naming it as the GP model.

15.2 FORMULATION OF GOAL PROGRAMMING

If there are m goals and p resource constraints in a problem, then the most general GP model can be written in the following form:

$$\text{Min } Z = \sum_{i=1}^{m} p_i \,(w_i^+ \, d_i^+ + w_i^- \, d_i^-)$$

$$\text{Subject to} \quad \sum_{j=1}^{n} a_{ij} x_j + d_i^- - d_i^+ = b_i, \; i = 1, \ldots, m$$

$$\sum_{j=1}^{n} a_{ij} x_j \leq = \geq b_i, \; i = m+1, \ldots, m+p$$

where x_j, $j = 1, \ldots, n$ are n decision variables, p_i, $i = 1, \ldots, m$ are the priorities associated with m goals, d_i^- = deviational variable representing underachievement in the ith goal, d_i^+ = deviational variable representing overachievement in the ith goal, w_i^+ is the relative weight of d_i^+ in the ith priority level, and w_i^- is the relative weight of d_i^- in the ith priority level.

Here priorities p_i's are not assigned any values, but this is simply a convenient way of providing more importance to one goal over another. Thus, if p_i is given more importance than p_j, it means $p_i > np_j$ however large n may be. We also indicate this by writing $p_i >>> p_j$. At the same priority level p_i, the deviational variables d_i^- and d_i^+ may be provided different weights in the objective function. The person formulating the GP model has to analyse each of the m goals carefully. If overachievements are acceptable, d_i^+ (called surplus variable in LP) is removed from the objective function and if underachievement is acceptable, then d_i^- (called slack variable in LP formulation) is removed from the objective function. If the exact achievement of the ith goal is desired, then both d_i^- and d_i^+ must be included in the objective function and ranked according to their order of priority. In this manner, the higher priority goals are considered before the lower priority goals.

15.3 BASIC TERMINOLOGIES

Before providing a detailed description of GP, we shall discuss certain necessary preliminaries:

Goal equation A goal expressed in the form of an equation by using variables d_i^- (underachievement) and d_i^+ (overachievement) is called the goal equation. The variables d_i^- and d_i^+ are called the deviational variables.

Priorities in GP model The coefficients of deviational variables in the objective function of the GP are called priorities. The magnitude of priorities reflects the order of preference of a goal.

Multiple goals with priorities and weights A multiple-goal model in which different weights are employed in one or more priority levels, to distinguish the goal preference, are called multiple goals with priorities and weights.

GP simplex method The modified version of simplex method used in solving a GP problem is called the GP simplex method.

Resource or structure constraints The constraints that are not directly related to the goals of the problem are called structural or resource constraints. Deviational variables are not incorporated into these constraints. These constraints have to be satisfied by the solution.

Trade-off function The objective function of a GP is called the trade-off function. This may be linear or non-linear.

Example 15.1 A company manufactures two products A and B. These products are manufactured in two different departments D_1 and D_2 of the company. While product A requires 2h in department D_1 and 3h in department D_2, product B requires 2h in department D_1 and 5h in department D_2. The production time is limited to 60h in department D_1 and 80h in department D_2. The profit per unit of products A and B is ₹3 and ₹6 respectively. To maximize profit, the company has set a high profit goal of ₹1500. The management of the company desires to produce at least 30 units of each product.

The management considers this second goal to be as equally important as the first goal of maximizing the profit. Formulate the given problem as a GP problem.

Solution

Profit constraint: Let x_1 be the units of product A to be produced and x_2 be the units of product B to be produced. Therefore, the profit goal is $3x_1 + 6x_2 \approx 1500$ and the profit goal equation is $3x_1 + 6x_2 + d_1^- - d_1^+ = 1500$, where d_1^- = amount of underachievement of profit and d_1^+ = amount of overachievement of profit.

Resource or structure constraints:

$$2x_1 + 3x_2 \le 60 \text{ or } 2x_1 + 3x_2 + x_3 = 60$$

$$2x_1 + 5x_2 \le 80 \text{ or } 2x_1 + 5x_2 + x_4 = 80$$

Production constraint: Since at least 30 units of products A and B have to be produced, the production constraints can be written as follows:

$$x_1 + d_2^- - d_2^+ = 30 \text{ and } x_2 + d_3^- - d_3^+ = 30$$

Here $x_1 \ge 30$ and $x_2 \ge 30$. Thus d_2^- and d_3^- are the deviational variables and have to be included in the objective function.

Objective function: With priority-ranked goals, one objective function has to be formulated for each goal in the GP. The management has given equal priority to all the goals. Thus, the complete GP model is formulated as follows:

$$\text{Min } Z = d_1^- + d_2^- + d_3^- \quad \text{(Objective function)}$$

$$\text{Subject to} \quad 3x_1 + 6x_2 + d_1^- - d_1^+ = 1500 \quad \text{(Goal 1)}$$

$$x_1 + d_2^- - d_2^+ = 30 \quad \text{(Goal 2)}$$

$$x_2 + d_3^- - d_3^+ = 30 \quad \text{(Goal 3)}$$

$$\left.\begin{aligned} 2x_1 + 3x_2 + x_3 &= 60 \\ 2x_1 + 5x_2 + x_4 &= 80 \end{aligned}\right\} \text{Resource constraints}$$

$x_1, x_2, x_3, x_4, d_1^-, d_1^+, d_2^-, d_2^+, d_3^-, d_3^+ \ge 0; x_3, x_4$ are slack variables and d_1^+, d_2^+, and d_3^+ are overachievements for goals 1, 2, and 3 respectively. Here overachievements d_1^+, d_2^+, and d_3^+ in goals 1, 2, and 3 are desired and hence have not been included in the objective function.

15.4 SINGLE-GOAL MODELS

To solve single-goal programming problems, we use the standard simplex method. To have a clear understanding of the GP–LP relationship, we consider the following example.

Example 15.2 A manufacturer produces two models P_1 and P_2 which have to go through two machines M_1 and M_2 before the final output is obtained. The machine time available, time required by each product on two machines, and the profit on each product is given in the adjacent table:

Product→	Product		Available time (h)
Machine↓	P_1	P_2	
M_1	2	3	60
M_2	2	5	80
Profit	3	5	–

Determine how the manufacturer can earn maximum profit. Further, if the manufacturer sets the target of achieving a maximum profit of ₹200, then formulate the problem in GP model.

Solution The LP formulation of the problem is as follows

$$\left.\begin{aligned} &\text{Max } Z = 3x_1 + 5x_2 \\ &\text{Subject to} \quad 2x_1 + 3x_2 \le 60 \\ &\qquad 2x_1 + 5x_2 \le 80 \\ &\qquad x_1, x_2 \ge 0 \end{aligned}\right\} \tag{15.1}$$

where x_1 units of product P_1 are produced and x_2 units of product P_2 are produced.

The manufacturer's objective is to determine x_1 and x_2 so that the profit is maximum. If we solve the problem using the simplex method, the maximum profit $Z = 95$, for $x_1 = 15$ and $x_2 = 10$.

Now in Eq. (15.1), if the manufacturer fixes a target of achieving a maximum profit of ₹200, then the GP formulation of the problem can be given as follows:

$$\left.\begin{aligned} &\text{Min } Z = d_1^- \text{ or Max } -Z = -d_1^- \\ &\text{Subject to} \quad 2x_1 + 3x_2 \le 60 \\ &\qquad 2x_1 + 5x_2 \le 80 \\ &\qquad 3x_1 + 5x_2 + d_1^- - d_1^+ = 200 \\ &\qquad x_1, x_2, d_1^-, d_1^+ \ge 0. \end{aligned}\right\} \tag{15.2}$$

Here x_i's are the decision variables and d_i^- (underachievement) and d_i^+ (overachievement) are the deviational variables. Taking x_3 and x_4 as slack variables added to constraints of Eq. (15.2), we solve the GP models as follows:

	Costs →			0	0	0	0	0	−1	
Table no.	C_B	Basic variables	X_B	x_1	x_2	d_1^+	x_3	x_4	d_1^-	Min ratio
1	0	x_3	60	2	3	0	1	0	0	60/3 = 20
	0	x_4	80	2	[5]	0	0	1	0	80/5 = 16→
	−1	d_1^-	200	3	5	−1	0	0	1	200/5 = 40
		$c_j - Z_j$		3	5↑	−1	0	0	0	–
2	0	x_3	12	[4/5]	0	0	1	−3/5	0	60/4 = 15→
	0	x_2	16	2/5	1	0	0	1/5	0	16/(2/5) = 40
	−1	d_1^-	120	1	0	−1	0	−1	1	120
		$c_j - Z_j$		1↑	0	−1	0	−1	0	–
3	0	x_1	15	1	0	0	5/4	−3/4	0	
	0	x_2	10	0	1	0	−1/2	1/2	0	–
	0	d_1^-	105	0	0	−1	−5/4	−1/4	0	
		$c_j - Z_j$		0	0	−1	−5/4	−1/4	0	–

Therefore, the optimum solution is $x_1 = 15, x_2 = 10, d_1^{-1} = 105$. In this problem, 105 is the underachievement of the maximum profit goal of ₹200. Thus the actual profit is 200 − 105 = ₹95, which is the same as the maximum profit obtained using the LP model. Therefore in a single-goal programming problem, the solution derived is same as that obtained using the LP model of the problem.

The main difference between the LP model shown in Eq. (15.1) and the GP model in Eq. (15.2) is the objective function. In GP model, we include deviational variable d_1^- in the objective function. Further, while the problem in the LP model may be either maximization or minimization, it is always a minimization problem in GP model where we minimize the sum (weighted) of deviations. In addition, the profit objective function in Eq. (15.1) has been considered as the goal equation in Eq. (15.2).

15.5 GP ALGORITHM OR MODIFIED SIMPLEX METHOD

The standard simplex method can easily be used to solve GP problems. This is accomplished by assigning values to the priority coefficients in the objective function of GP formulation, so that the values reflect the same order of relationship as the priorities.

In Chapter 4, we have already described in detail the steps and procedure of the simplex method. Here we shall describe how the simplex method algorithm can be modified to solve a GP problem. The following steps are used in a GP algorithm:

Step 1 Construct the initial modified simplex table similar to that of a simplex table with the only difference in the $(c_j - Z_j)$ row, where it is split into as many different goal rows as the number of priorities assigned.

Step 2 Check for optimality. If there is no positive entry in the $(c_j - Z_j)$ row for the highest priority row, say p_k, then the priority p_k goal has been met; go to Step 6, otherwise go to Step 3.

Step 3 Determine the new entering variable by identifying the largest positive entry in the p_k row. This fixes the column of the entering variable.

Step 4 Determine the departing variable by considering the minimum of the ratios of the X_B column entries with the corresponding non-negative entries of the column fixed in Step 3, as is done in standard simplex method.

Step 5 Develop the new table to update the entries in the body of the table by using elementary row operations. The new $(c_j - Z_j)$ rows are computed in the same manner as in the simplex method. The only difference is the tabular representation. As an example, compute Z_j by multiplying the values in the jth column with the corresponding entries of the C_B column. Then subtract it from c_j to obtain $(c_j - Z_j)$. Break $(c_j - Z_j)$ into parts, where the parts are associated with the priority levels.

To be specific, if there are three priorities p_1, p_2, and p_3 arranged in the order of their importance and if $c_j - Z_j = 0 + 3p_3 - 4p_1 + 0$, then in the jth column we split it as

$$\begin{array}{lll} (c_j - Z_j)\text{ row}: p_1 & -4 & (p_1 >> p_2 >> p_3) \\ \qquad\qquad\qquad\;\; p_2 & 0 & \\ \qquad\qquad\qquad\;\; p_3 & 3 & \end{array}$$

In this way continue Steps 3, 4, and 5 until optimality conditions are satisfied for the highest priority p_k row and then go to Step 6.

Step 6 Evaluate the next lower priority level row p_{k+1} ($p_{k+1} << p_k$) by identifying the largest positive coefficient for which there are no negative coefficients at a higher priority in the same column. If a tie exists in the values of the coefficients that determine the entering variable, break it arbitrarily.

Note The lower priority goals must not be satisfied at the expense of higher priority goals.

15.6 MULTIPLE-GOAL MODELS

In this section, we shall discuss the following three types of the multiple-goal model:

1. Multiple-goal models with equal or no priority
2. Multiple-goal models with priority
3. Multiple-goal models with priority and weights

In general, multiple-goal models with priority and weights are considered most useful in day-to-day life.

15.6.1 Multiple-goal Models with Equal or No Priorities

The multiple-goal with equal-priority model of GP is not of much practical value when compared to GP models with different priorities attached to different goals. Nevertheless, we have illustrated it in Example 15.3 to have a better understanding of the concept of priorities and weights.

Example 15.3 A firm is manufacturing B_1 and B_2 types of bags; while B_1 bags are ordinary, B_2 are luxury bags. These bags are processed through two machines M_1 and M_2 to give them a final form. B_1 requires 20 h of machine M_1 and 10 h of machine M_2. B_2 requires 10 h each of machines M_1 and M_2. The time available on machines M_1 and M_2 is 60 h and 40 h respectively. The profit earned per bag is ₹40 and ₹80 on bags B_1 and B_2 respectively. The firm wants to maximize profit to as much as ₹1000. In addition to the profit goal, the firm wants to produce at least two bags of each type and considers this second goal as equally important as the first profit goal. Formulate and solve the given problem as a GP problem. Interpret the solution.

Solution

Formulation: Let x_1 be the number of type B_1 bags manufactured and x_2 be the number of type B_2 bags manufactured.

Resource constraints: Time taken on machine M_1 to produce x_1 and $x_2 = 20x_1 + 10x_2$. Thus $20x_1 + 10x_2 \leq 60$.

Similarly for machine M_2, we have $10x_1 + 10x_2 \leq 40$.

Profit constraints: $40x_1 + 80x_2 + d_1^- - d_1^+ = 1000$

Production constraints: As $x_1 \geq 2$, $x_2 \geq 2$, the production constraints can be written as $x_1 + d_2^- - d_2^+ = 2$, $x_2 + d_3^- - d_3^+ = 2$.

As equal priority (no priority) has been given to all the goals, the objective function is to minimize the sum of deviational: Min $Z = d_1^- + d_2^- + d_3^-$.

Thus, the complete GP model can be described as follows:

$$\text{Min } Z = d_1^- + d_2^- + d_3^-$$

$$\text{Subject to } \quad 20x_1 + 10x_2 \leq 60$$

$$10x_1 + 10x_2 \leq 40$$

(*Contd on page 411*)

Table 15.1 Solution to problem in Example 15.3

		$c_j \rightarrow$		0	0	−1	0	−1	0	−1	0	0	0	
Table no.	C_B	**Basic variable**	X_B	x_1	x_2	d_1^-	d_1^+	d_2^-	d_2^+	d_3^-	d_3^+	x_3	x_4	**Min ratio**
1	0	x_3	60	20	10	0	0	0	0	0	0	1	0	60/10 = 6
	0	x_4	40	10	10	0	0	0	0	0	0	0	1	40/10 = 4
	−1	d_1^-	1000	40	80	1	−1	0	0	0	0	0	0	1000/80
	−1	d_2^-	2	1	0	0	0	1	−1	0	0	0	0	…
	−1	d_3^-	2	0	[1]	0	0	0	0	1	−1	0	0	2/1 = 2 min
		$c_j - Z_j$		41	81↑	0	−1	0	−1	0	−1	0	0	–
2	0	x_3	40	20	0	0	0	0	0	−10	10	1	0	40/10 = 4
	0	x_4	20	10	0	0	0	0	0	−10	[10]	0	1	20/10 = 2 min
	−1	d_1^-	840	40	0	1	−1	0	0	−80	80	0	0	840/80
	−1	d_2^-	2	1	0	0	0	1	−1	0	0	0	0	…
	0	x_2	2	0	1	0	0	0	0	1	−1	0	0	…
		$c_j - Z_j$		41	0	0	−1	0	−1	−81	80↑	0	0	–
3	0	x_3	20	10	0	0	0	0	0	0	0	1	−1	
	0	d_3^+	2	1	0	0	0	0	0	−1	1	0	1/10	
	−1	d_1^-	680	−40	0	1	1	0	0	0	0	0	−8	–
	−1	d_2^-	2	1	0	0	0	1	−1	0	0	0	0	
	0	x_2	4	1	1	0	0	0	0	0	0	0	1/10	
		$c_j - Z_j$		−39	0	0	−1	0	−1	−1	0	0	−8	–

$$40x_1 + 80x_2 + d_1^- - d_1^+ = 1000 \quad \text{(Goal 1)}$$
$$x_1 + d_2^- - d_2^+ = 2 \quad \text{(Goal 2)}$$
$$x_2 + d_3^- - d_3^+ = 2 \quad \text{(Goal 3)}$$

$x_1, x_2, d_1^-, d_1^+, d_2^-, d_2^+, d_3^-, d_3^+ \geq 0$. d_1^-, d_2^-, d_3^- are the underachievements d_1^+, d_2^+, d_3^+ are the overachievements of goals 1, 2, and 3 respectively. The canonical form is as follows:

$$\text{Min } Z = d_1^- + d_2^- + d_3^- \text{ or Max } -Z = -d_1^- - d_2^- - d_3^-$$

Subject to

$$20x_1 + 10x_2 + x_3 = 60$$
$$10x_1 + 10x_2 + x_4 = 40$$
$$40x_1 + 80x_2 + d_1^- - d_1^+ = 1000 \quad \text{(Goal 1)}$$
$$x_1 + 0{\cdot}x_2 + d_2^- - d_2^+ = 2 \quad \text{(Goal 2)}$$
$$0{\cdot}x_1 + x_2 + d_3^- - d_3^+ = 2 \quad \text{(Goal 3)}$$

$x_1, x_2, x_3, d_1^-, d_1^+, d_2^-, d_2^+, d_3^-, d_3^+ \geq 0$. x_3, x_4 are the slack variables, d_1^-, d_2^-, d_3^- are the underachievements, d_1^+, d_2^+, d_3^+ are the overachievements of goals 1, 2, and 3 respectively.

To solve the problem we use the modified simplex method (Table 15.1). The solution is $x_1 = 0$, $x_2 = 4$, $x_3 = 20$, $x_4 = 0$, $d_1^- = 680$, $d_1^+ = 0$, $d_2^- = 2$, $d_2^+ = 0$, $d_3^- = 0$, $d_3^+ = 2$ and $Z=682$.

Interpretation of solution obtained Here $d_1^- = 680$ implies that the profit goal of ₹1000 is underachieved by ₹680. Thus the actual profit is ₹1000 – ₹680 = ₹320. Further $d_2^- = 2$ indicates that the production goal of type B_1 bags was missed by 2. Further $d_3^+ = 2$ indicates that the production goal of type B_2 bags was overachieved by 2.

15.6.2 Multiple-goal Models with Priorities

In Section 15.6.1, we have considered a GP problem with no priorities (having equal priorities). To solve those problems, we used the standard simplex method. In this section, we shall consider multiple-goal programming problems with priorities. To solve these problems, we shall use the modified simplex method.

Example 15.4 A manufacturer produces two models 1 and 2 of a product that first requires machining and then assembling. The time available for machining and assembling is respectively 60 h and 80 h. The time required (h) and profit per piece of a model are given as per the adjacent table:

Activity	Model 1	Model 2	Available time (h)
Machining	2	3	60
Assembling	2	5	80
Profit	3	6	–

Now, the manufacturer, who is planning to increase the manufacturer production to earn more profit, decides to use overtime to maximize profit. Assume that he sets the following priorities for the attainment of the goal: (a) first priority p_1 is to limit the total overtime in assembly to 30 h and (b) second priority p_2 is set to earn maximum profit with an arbitrary high goal of ₹200. Formulate and solve the problem using GP.

Table 15.2 Solution to problem in Example 15.4

		$c_j \rightarrow$		0	0	0	0	0	$-p_2$	0	0	$-p_1$	
Table no.	C_B	**Basic variable**	X_B	x_1	x_2	x_3	d_1^-	d_1^+	d_2^-	d_2^+	d_3^-	d_2^+	**Min ratio**
1	0	x_3	60	2	3	1	0	0	0	0	0	0	60/3 = 20
	0	d_1^-	80	2	[5]	0	1	−1	0	0	0	0	80/5 = 16 min
	$-P_2$	d_2^-	200	3	6	0	0	0	1	−1	0	0	200/6 = 33.3
	0	d_3^-	30	0	0	0	0	1	0	0	1	−1	…
$c_j - Z_j$			p_1	0	0	0	0	0	0	0	0	−1	–
			p_2	3	6↑	0	0	0	0	−1	0	0	
2	0	x_3	12	4/5	0	1	−3/5	[3/5]	0	0	0	0	20 min
	0	x_2	16	2/5	1	0	1/5	−1/5	0	0	0	0	…
	$-P_2$	d_2^-	104	3/5	0	0	−6/5	6/5	1	−1	−6	6	520/6
	0	d_3^-	30	0	0	0	0	1	0	0	1	−1	30
$c_j - Z_j$			p_1	0	0	0	0	0	0	0	0	−1	–
			p_2	3/5	0	0	−6/5	6/5↑	0	−1	0	0	
3	0	d_1^+	20	4/3	0	1	−1	1	0	0	0	0	–
	0	x_2	20	2/3	1	1/5	0	0	0	0	0	0	
	$-P_2$	d_2^-	80	−1	0	−6/5	0	0	1	−1	−6	6	
	0	d_3^-	10	−4/3	0	−1	1	0	0	0	1	−1	
$c_j - Z_j$			p_1	0	0	0	0	0	0	0	0	−1	–
			p_2	−1	0	−6/5	0	0	0	−1	0	0	

Solution The GP model of the problem is formulated as follows:

$$\text{Min } Z = p_1 d_3^+ + p_2 d_2^- \text{ or Max } (-Z) = -p_1 d_3^+ - p_2 d_2^-$$

$$\text{Subject to } \quad 2x_1 + 3x_2 + x_3 = 60$$

$$2x_1 + 5x_2 + d_1^- - d_1^+ = 80$$

$$3x_1 + 6x_2 + d_2^- - d_2^+ = 200 \quad \text{(Goal 2)}$$

$$d_1^+ + d_3^- - d_3^+ = 30 \quad \text{(Goal 1)}$$

$$x_1, x_2, x_3, d_1^+, d_1^-, d_2^+, d_2^-, d_3^+, d_3^- \geq 0.$$

The procedure for constructing the initial table is identical to that of the simplex table. The only difference between the initial table for GP and that of the LP is the $(c_j - Z_j)$ row. For example, in Table 15.2 for x_1 column,

$$Z_1 = (0)(2) + (0)(2) + (-p_2)(3) + (0)(0)$$

$$\text{and } c_1 - z_1 = 0 - (0 + 0 - 3p_2 + 0) = 3p_2 = 0p_1 + 3p_2$$

Therefore in $(c_j - Z_j)$ row under the x_1 column, we write 0 in front of p_1 and 3 in front of p_2, instead of writing a single entry $3p_2$.

Thus, the solution is $x_1 = 0$, $x_2 = 20$, $x_3 = 0$, $d_1^- = 0$, $d_1^+ = 20$, $d_2^- = 80$, $d_2^+ = 0$, $d_3^- = 10$, $d_3^+ = 0$.

Here goal 1, with priority p_1, is fully achieved as the overtime in assembly is exactly 30 h. Goal 2, with priority p_2, is also completely achieved. The profit is exactly ₹200. The results indicate that the profit goal is achieved by extending 20 h overtime in assembly.

15.6.3 Multiple-goal Models with Priorities and Weights

This is the GP model where a differential weight is used to reflect the difference in importance within the same priority level. To illustrate let us consider Example 15.5.

Example 15.5 A company produces two types of products—A (ordinary) and B (fancy). These products are processed through two machines M_1 and M_2 to get them in their final forms. Type A requires 20 h of machine M_1 and 10 h of machine M_2. Type B requires 10 h each of machine M_1 and M_2. The time available on machine M_1 and M_2 is 60 h and 40 h respectively. The profit earned per product is ₹40 and ₹80 on bags B_1 and B_2 respectively. The firm wants to maximize profit to as much as ₹800. The management of the company has set the following priorities as goals to achieve.

p_1 (Priority 1): Maintain the minimum production goals of 4 units of product A and 6 units of product B. Use the differential weights of 1 and 2 respectively, since these weights reflect the 'weighted contribution to profit' of ₹40 and ₹80.

p_2 (Priority 2): Limit the total overtime on two machines to 50 h.

p_3 (Priority 3): Maximize the profit.

Formulate and solve the given problem as a GP problem.

Solution The GP model of the problem is formulated as follows:

$$\text{Min } Z = 1p_1 d_2^- + 2p_1 d_3^- + p_2 d_6^+ + p_3 d_1^-$$

$$\text{or Max } (-Z) = -1p_1 d_2^- - 2p_1 d_3^- - p_2 d_6^+ - p_3 d_1^-$$

(Contd on page 417)

Table 15.3 Solution to problem in Example 15.5

$c_j \rightarrow$			0	1	$-p_3$	0	$-p_1$	0	$-2p_1$	0	0	0	0	0	0	$-p_2$	
C_B	Basic variables	X_B	x_1	x_2	d_1^-	d_1^+	d_2^-	d_2^+	d_3^-	d_3^+	d_4^-	d_4^+	d_5^-	d_5^+	d_6^-	d_6^+	Min ratio
0	d_4^-	60	20	10	0	0	0	0	0	0	1	–1	0	0	0	0	6
0	d_5^-	40	10	[10]	0	0	0	0	0	0	0	0	1	–1	0	0	4 min →
$-p_3$	d_1^-	800	40	80	1	–1	0	0	0	0	0	0	0	0	0	0	10
$-p_1$	d_2^-	4	1	0	0	0	1	–1	0	0	0	0	0	0	0	0	…
$-2p_1$	d_3^-	6	0	1	0	0	0	0	1	–1	0	0	0	0	0	0	6
0	d_6^-	50	0	0	0	0	0	0	0	0	0	1	0	1	1	–1	…
$c_j - Z_j$		p_1	1	2↑	0	0	0	–1	0	–2	0	0	0	0	0	0	–
		p_2	0	0	0	0	0	0	0	0	0	0	0	0	0	–1	
		p_3	40	80	0	–1	0	0	0	0	0	0	0	0	0	0	
0	d_4^-	20	10	0	0	0	0	0	0	0	1	–1	–1	1	0	0	20
0	x_2	4	1	1	0	0	0	0	0	0	0	0	1/10	–1/10	0	0	…
$-p_3$	d_1^-	480	–40	0	1	–1	0	0	0	0	0	0	–8	8	0	0	60
$-p_1$	d_2^-	4	1	0	0	0	1	–1	0	0	0	0	0	0	0	0	…
$-2p_1$	d_3^-	2	–1	0	0	0	0	0	1	–1	0	0	–1/10	[1/10]	0	0	20 →
0	d_6^-	50	0	0	0	0	0	0	0	0	0	1	0	1	1	–1	50
$c_j - Z_j$		p_1	–1	0	0	0	0	–1	0	–2	0	0	–1/5	1/5↑	0	0	–
		p_2	0	0	0	0	0	0	0	0	0	0	0	0	0	–1	
		p_3	–40	0	0	–1	0	0	0	0	0	0	–8	8	0	0	

(*Contd*)

(*Contd*)

	$c_j \rightarrow$		**0**	**1**	**$-p_3$**	**0**	**$-p_1$**	**0**	**$-2p_1$**	**0**	**0**	**0**	**0**	**0**	**0**	**$-p_2$**	
C_B	**Basic variables**	**X_B**	**x_1**	**x_2**	**d_1^-**	**d_1^+**	**d_2^-**	**d_2^+**	**d_3^-**	**d_3^+**	**d_4^-**	**d_4^+**	**d_5^-**	**d_5^+**	**d_6^-**	**d_6^+**	**Min ratio**
0	d_4^-	0	[20]	0	0	0	0	0	–10	10	1	–1	0	0	0	0	0 (min) →
0	x_2	6	0	1	0	0	0	0	1	–1	0	0	0	0	0	0	…
$-p_3$	d_1^-	320	40	0	1	–1	0	0	–80	80	0	0	0	0	0	0	8
$-p_1$	d_2^-	4	1	0	0	0	1	–1	0	0	0	0	0	0	0	0	4
0	d_5^+	20	–10	0	0	0	0	0	10	–10	0	0	–1	1	0	0	…
0	d_6^-	30	10	0	0	0	0	0	–10	10	0	1	1	0	1	–1	3
$c_j - Z_j$		p_1	1↑	0	0	0	0	–1	–2	0	0	0	0	0	0	0	–
		p_2	0	0	0	0	0	0	0	0	0	0	0	0	0	–1	
		p_3	40	0	0	–1	0	0	–80	80	0	0	0	0	0	0	
0	x_1	0	1	0	0	0	0	0	–1/2	1/2	1/20	–1/20	0	0	0	0	–
0	x_2	6	0	1	0	0	0	0	1	–1	0	0	0	0	0	0	–
$-p_3$	d_1^-	320	0	0	1	–1	0	0	–120	120	0	0	0	0	0	0	–
$-p_1$	d_2^-	4	0	0	0	0	1	–1	1/2	–1/2	–1/20	1/20	0	0	0	0	80
0	d_5^+	20	0	0	0	0	0	0	5	–5	1/2	–1/2	–1	1	0	0	–
0	d_6^-	30	0	0	0	0	0	0	–5	5	–1/2	[3/2]	1	0	1	–1	20 min →
$c_j - Z_j$		p_1	0	0	0	0	0	–1	–3/2	–1/2	–1/20	1/20↑	0	0	0	0	–
		p_2	0	0	0	0	0	0	0	0	0	0	0	0	0	–1	
		p_3	0	0	0	–1	0	0	–60	60	–2	2	0	0	0	0	

(*Contd*)

(*Contd*)

	$c_j \rightarrow$		0	1	$-p_3$	0	$-p_1$	0	$-2p_1$	0	0	0	0	0	0	$-p_2$	
C_B	Basic variables	X_B	x_1	x_2	d_1^-	d_1^+	d_2^-	d_2^+	d_3^-	d_3^+	d_4^-	d_4^+	d_5^-	d_5^+	d_6^-	d_6^+	Min ratio
0	x_1	10	1	0	0	0	0	0	–2/3	2/3	1/30	0	1/30	0	1/30	–1/30	–
0	x_2	6	0	1	0	0	0	0	1	–1	0	0	0	0	0	0	–
$-p_3$	d_1^-	320	0	0	1	–1	0	0	–120	120	0	0	0	0	0	0	–
$-p_1$	d_2^-	3	0	0	0	0	1	–1	2/3	–2/3	–1/30	0	–1/30	0	–1/30	[1/30]	→
0	d_5^+	30	0	0	0	0	0	0	10/3	–10/3	1/3	0	–2/3	0	1/3	–1/3	–
0	d_4^-	20	0	0	0	0	0	0	–10/3	10/3	–1/3	1	2/3	0	2/3	–2/3	–
	$c_j - Z_j$	p_1	0	0	0	0	0	–1	–11/4	–3/4	–2/30	0	–1/30	0	–1/30	1/30↑	
		p_2	0	0	0	0	0	0	0	0	0	0	0	0	0	–1	–
		p_3	0	0	0	–1	0	0	–160/3	160/3	–11/60	–	–4/3	0	–4/3	4/3	
0	x_1	4	1	0	0	0	1	–1	0	0	0	0	0	0	0	0	–
0	x_2	6	0	1	0	0	0	0	1	–1	0	0	0	0	0	0	–
$-p_3$	d_1^-	320	0	0	1	–1	0	0	–120	120	0	0	0	0	0	0	–
0	d_6^-	90	0	0	0	0	30	–30	20	–20	–1	0	–1	0	–1	1	–
0	d_5^+	60	0	0	0	0	10	–10	10	–10	0	0	–1	0	0	0	–
0	d_4^-	80	0	0	0	0	20	–20	10	–10	–1	1	0	0	0	0	–
	$c_j - Z_j$	p_1	0	0	0	0	–1	0	–41/12	–1/12	–1/30	0	0	0	0	0	
		p_2	0	0	0	0	30	–30	20	–20	–1	0	–1	0	–1	0	–
		p_3	0	0	0	–1	–10	10	–60	60	–3/2	0	–1	0	–1	0	

$$\text{Subject to} \quad 20x_1 + 10x_2 + d_4^- - d_4^+ = 60$$
$$10x_1 + 10x_2 + d_5^- - d_5^+ = 40$$
$$40x_1 + 80x_2 + d_1^- - d_1^+ = 800$$
$$x_1 + d_2^- - d_2^+ = 4$$
$$x_2 + d_3^- - d_3^+ = 6$$
$$d_4^+ + d_5^+ + d_6^- - d_6^+ = 50$$
$$\text{and} \quad x_1, x_2, d_1^+, d_1^-, d_2^+, d_2^-, d_3^+, d_3^-, d_4^-, d_4^+, d_5^-, d_5^+, d_6^-, d_6^+ \geq 0.$$

In Table 15.3 priority p_1 is satisfied. In priority p_2 row, 30 is the most positive entry in column d_2^-. However, it cannot be entered because of the negative entry above it in the higher priority row (p_1 row). The next most positive entry is 20 and is in column d_3^-. However, d_3^- cannot be entered as well because of a negative entry above it in the p_1 row. In the next lower priority row p_3, the most positive entry, 60, is in d_3^+ and 10 is in the d_2^+ column. However, these deviational variables as well cannot enter because of the same reasons as stated for the entries in the p_2 row. Thus, the optimal solution is $x_1 = 3$, $x_2 = 6$, $d_1^- = 320$, $d_6^- = 90$, $d_5^+ = 60$, $d_4^+ = 80$, and the remaining variables are zero, being non-basic.

Example 15.6 A manufacturer produces two types of products A and B. The time required (in h) per unit of these products along with profit (in ₹) is given in the adjacent table. Time (in h) available in departments D_1 and D_2 is respectively 60 h and 80 h. The manufacturer wants to earn profit as much as ₹600 and sets the following priorities, in order to achieve this.

Department	Product		Available time
	A	B	
D_1	2	3	60
D_2	2	5	80
Profit	3	6	–

P_1 (Priority 1): Meet the total over time limit of two departments to 44 h.

P_2 (Priority 2): Meet the production goals of 5 and 8 units of type A and B products respectively. On the basis of profit from these products, attach weights 1 and 2 with the priority P_2.

P_3 (Priority 3): Meet the profit goal.

Formulate and solve the given problem as a GP problem.

Solution The GP model of the problem is formulated as follows:

$$\text{Min } Z = p_1 d_6^+ + 1p_2 d_2^- + 2p_2 d_3^- + p_3 d_1^-$$
$$\text{or} \quad \text{Max } (-Z) = -p_1 d_6^+ - 1p_2 d_2^- - 2p_2 d_3^- - p_3 d_1^-$$
$$\text{Subject to} \quad 2x_1 + 3x_2 + d_4^- - d_4^+ = 60$$
$$2x_1 + 5x_2 + d_5^- - d_5^+ = 80$$
$$3x_1 + 6x_2 + d_1^- - d_1^+ = 600$$
$$x_1 + d_2^- - d_2^+ = 5$$
$$x_2 + d_3^- - d_3^+ = 8$$
$$d_4^+ + d_5^+ + d_6^- - d_6^+ = 44$$
$$\text{and} \quad x_1, x_2, d_1^+, d_1^-, d_2^+, d_2^-, d_3^+, d_3^-, d_4^-, d_4^+, d_5^-, d_5^+, d_6^-, d_6^+ \geq 0.$$

Thus, the optimal solution is $x_1 = 20, x_2 = 8, d_1^- = 492, d_6^- = 40, d_2^+ = 15, d_4^+ = 4$, and the remaining variables are zero, being non-basic (Table 15.4).

Table 15.4 Solution to problem in Example 15.6

c_j			0	1	$-p_3$	0	$-p_2$	0	$-2p_1$	0	0	0	0	0	0	$-p_1$	
C_B	Basic variables	X_B	x_1	x_2	d_1^-	d_1^+	d_2^-	d_2^+	d_3^-	d_3^+	d_4^-	d_4^+	d_5^-	d_5^+	d_6^-	d_6^+	Min ratio
0	d_4^-	60	2	3	0	0	0	0	0	0	1	–1	0	0	0	0	20
0	d_5^-	80	2	5	0	0	0	0	0	0	0	0	1	–1	0	0	16
$-p_3$	d_1^-	600	3	6	1	–1	0	0	0	0	0	0	0	0	0	0	100
$-p_2$	d_2^-	5	1	0	0	0	1	–1	0	0	0	0	0	0	0	0	…
$-2p_2$	d_3^-	8	0	[1]	0	0	0	0	1	–1	0	0	0	0	0	0	8 →
0	d_6^-	44	0	0	0	0	0	0	0	0	0	1	0	1	1	–1	…
$c_j - Z_j$		p_1	0	0	0	0	0	0	0	0	0	0	0	0	0	–1	
		p_2	1	2↑	0	0	0	–1	0	–2	0	0	0	0	0	0	–
		p_3	3	6	0	–1	0	0	0	0	0	0	0	0	0	0	
0	d_4^-	36	2	0	0	0	0	0	–3	3	1	–1	0	0	0	0	18
0	d_5^-	40	2	0	0	0	0	0	–5	5	0	0	0	0	0	0	20
$-p_3$	d_1^-	552	3	0	1	–1	0	0	–6	6	0	0	0	0	0	0	552/3
$-p_2$	d_2^-	5	[1]	0	0	0	1	–1	0	0	0	0	0	0	0	0	5 min →
0	x_2	8	0	1	0	0	0	0	1	–1	0	0	0	0	0	0	…
0	d_6^-	44	0	0	0	0	0	0	0	0	0	1	0	1	1	–1	…
$c_j - Z_j$		p_1	0	0	0	0	0	0	0	0	0	0	0	0	0	–1	
		p_2	1↑	0	0	0	0	–1	–2	0	0	0	0	0	0	0	–
		p_3	3	0	0	–1	0	0	–6	6	0	0	0	0	0	0	

(*Contd*)

(*Contd*)

	c_j		0	1	$-p_3$	0	$-p_2$	0	$-2p_1$	0	0	0	0	0	0	$-p_1$	
C_B	**Basic variables**	X_B	x_1	x_2	d_1^-	d_1^+	d_2^-	d_2^+	d_3^-	d_3^+	d_4^-	d_4^+	d_5^-	d_5^+	d_6^-	d_6^+	**Min ratio**
0	d_4^-	26	0	0	0	0	−2	2	−3	3	1	−1	0	0	0	0	26/3
0	d_5^-	30	0	0	0	0	−2	2	−5	[5]	0	0	0	0	0	0	6 min →
$-p_3$	d_1^-	537	0	0	1	−1	−3	3	−6	6	0	0	0	0	0	0	537/6
0	x_1	5	1	0	0	0	1	−1	0	0	0	0	0	0	0	0	…
0	x_2	8	0	1	0	0	0	0	1	−1	0	0	0	0	0	0	…
0	d_6^-	44	0	0	0	0	0	0	0	0	0	1	0	1	1	−1	…
$c_j - Z_j$		p_1	0	0	0	0	0	0	0	0	0	0	0	0	0	−1	–
		p_2	0	0	0	0	−1	0	−2	0	0	0	0	0	0	0	
		p_3	0	0	0	−1	−3	3	−6	6↑	0	0	0	0	0	0	
0	d_4^-	8	0	0	0	0	−4/5	[4/5]	0	0	1	−1	0	0	0	0	10 →
0	d_3^+	6	0	0	0	0	−2/5	2/5	−1	1	0	0	0	0	0	0	15
$-p_3$	d_1^-	501	0	0	1	−1	−3/5	3/5	0	0	0	0	0	0	0	0	…
0	x_1	5	1	0	0	0	1	−1	0	0	0	0	0	0	0	0	35
0	x_2	14	0	1	0	0	−2/5	2/5	0	0	0	0	0	0	0	0	…
0	d_6^-	44	0	0	0	0	0	0	0	0	0	1	0	1	1	−1	…
$c_j - Z_j$		p_1	0	0	0	0	0	0	0	0	0	0	0	0	0	−1	–
		p_2	0	0	0	0	−1	0	−2	0	0	0	0	0	0	0	
		p_3	0	0	0	−1	−3/5	3/5↑	0	0	0	0	0	0	0	0	

(*Contd*)

(*Contd*)

c_j			**0**	**1**	**$-p_3$**	**0**	**$-p_2$**	**0**	**$-2p_1$**	**0**	**0**	**0**	**0**	**0**	**0**	**$-p_1$**	
C_B	**Basic variables**	**X_B**	**x_1**	**x_2**	**d_1^-**	**d_1^+**	**d_2^-**	**d_2^+**	**d_3^-**	**d_3^+**	**d_4^-**	**d_4^+**	**d_5^-**	**d_5^+**	**d_6^-**	**d_6^+**	**Min ratio**
0	d_2^+	10	0	0	0	0	–1	1	0	0	5/4	–5/4	0	0	0	0	…
0	d_3^+	2	0	0	0	0	0	0	–1	1	–1/2	[1/2]	0	0	0	0	4 min →
$-p_3$	d_1^-	495	0	0	1	–1	0	0	0	0	–3/4	3/4	0	0	0	0	large
0	x_1	15	1	0	0	0	0	0	0	0	5/4	–5/4	0	0	0	0	…
0	x_2	10	0	0	0	0	0	0	0	0	–1/2	1/2	0	0	0	0	20
0	d_6^-	44	0	1	0	0	0	0	0	0	0	1	0	1	1	–1	44
$c_j - Z_j$		p_1	0	0	0	0	0	0	0	0	0	0	0	0	0	–1	
		p_2	0	0	0	0	–1	0	–2	0	0	0	0	0	0	0	–
		p_3	0	0	0	–1	0	0	0	0	–3/4	3/4↑	0	0	0	0	
0	d_2^+	15	0	0	0	0	–1	1	–5/2	5/2	0	0	0	0	0	0	–
0	d_4^+	4	0	0	0	0	0	0	–2	2	–1	1	0	0	0	0	–
$-p_3$	d_1^-	492	0	0	1	–1	0	0	–3/2	3/2	0	0	0	0	0	0	–
0	x_1	20	1	0	0	0	0	0	–5/2	5/2	0	0	0	0	0	0	–
0	x_2	8	1	0	0	0	0	0	1	–1	0	0	0	0	0	0	–
0	d_6^-	40	0	0	0	0	0	0	2	–2	1	0	0	1	1	–1	–
$c_j - Z_j$		p_1	0	0	0	0	0	0	0	0	0	0	0	0	0	–1	
		p_2	0	0	0	0	–1	0	–2	0	0	0	0	0	0	0	–
		p_3	0	0	0	–1	0	0	3/2	–3/2	0	0	0	0	0	0	

15.7 GRAPHICAL SOLUTION OF GOAL PROGRAMMING PROBLEMS

We can use the graphical approach to solve GP problems with two variables as used for LP problems. After formulating the problem, the following procedural steps are used for obtaining the solution:

1. Plot all resource constraints and identify the feasible region. If no resource constraints are present in the problem, the feasible region is the area where both x_1 and x_2 are ≥ 0 (the non-negative quadrants).
2. Graph the lines corresponding to the goal constraints. This can be done by equating the deviational variables to zero in the goal constraints and plotting the resulting equation.
3. Determine the points within the feasible region that satisfy the highest priority goal.
4. Move to the next priority goal and determine the best solution which does not degrade the solutions already achieved for higher priority goals.
5. Repeat step 4 until all priority levels have been investigated.

Example 15.7 A company is producing two types of items—I_1 and I_2. As per past experience, the production of either item requires an average of 1h in the plant. The plant has a normal production capacity of 300h fortnightly. The company's marketing department reports that the maximum number of items I_1 and I_2 that can be sold fortnightly are 200 and 250 respectively. The net profit from the sales of items I_1 and I_2 are respectively ₹300 and ₹600. The management of the company has set the following goals arranged in the order of their importance. Priority p_1 is to avoid any underutilization of idle time of the production capacity, priority p_2 is to minimize the overtime operation of the plant as much as possible, and priority p_3 is to sell the maximum possible units of items I_1 and I_2. The management has twice as much desire to achieve the same sales as I_1 for I_2 since the profit from the sales of item I_2 is the double of item I_1 ($p_1 >> p_2 >> p_3$). Formulate and solve the problem using the graphical method of GP.

Solution Let x_1 and x_2 be the number of units of items I_1 and I_2 respectively to be produced.

Plant capacity constraint equation can be written as $x_1 + x_2 + d_1^- - d_1^+ = 300$ where d_1^- = underutilization of the production capacity and d_1^+ = overutilization of the production capacity.

Production goal constraint equation can be written as $x_1 + d_2^- = 200$ and $x_2 + d_3^- = 250$ where d_2^- = underachievement of the sales goal for item I_1 and d_3^- = underachievement of the sales goal for item I_2.

Here we have not considered d_2^+ and d_3^+ in goal constraints as the maximum units that can be sold are 200 and 250 units for items I_1 and I_2 respectively.

Combining the aforementioned constraints, the GP model of the problem is formulated as follows:

$$\text{Min } Z = p_1 d_1^- + p_2 d_1^+ + p_3(2d_3^- + d_2^-) \quad (15.3)$$

$$\text{Subject to} \quad x_1 + x_2 + d_1^- - d_1^+ = 300 \quad (15.4)$$

$$x_1 + d_2^- = 200 \quad (15.5)$$

$$\text{and} \quad x_2 + d_3^- = 250 \quad (15.6)$$

$$x_1, x_2, d_1^-, d_2^-, d_3^-, d_1^+ \geq 0$$

The first step to solve a GP problem by graphical method is to plot all the resource (structured) constraints. Since there are no resource constraints in this problem, the feasible region is the area where $x_1 \geq 0$ and $x_2 \geq 0$, that is, the entire first quadrant. Next, sketching the goal constraints (15.4), (15.5), and (15.6) by setting the deviational variables to zero in each constraint, the feasible region is represented by the shaded region OCNM (Fig. 15.1). Since the underutilization and overtime operations of production capacity are permissible, both the deviational variables d_1^- and d_1^+ are indicated by arrows and are shown in Fig. 15.1. Similarly, d_2^- and d_3^- indicated by arrows are also shown in Fig. 15.1.

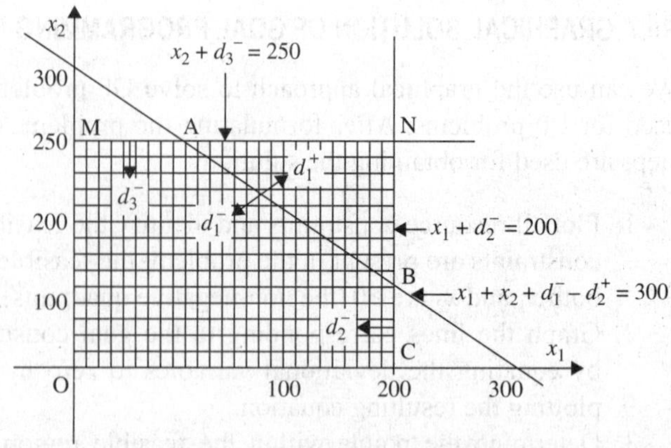

Figure 15.1 Graphical solution

For achivement of priority p_1 goal, set $d_1^- = 0$ in Eq. (15.4) to get $x_1 + x_2 - d_1^+ = 300$, reducing the shaded feasible region OCNM to region ABN. For achieving the next priority p_2 goal set $d_1^+ = 0$ to get $x_1 + x_2 = 300$. The feasible region ABN is further reduced to the points on line segment AB where the underutilization of idle time and overtime operations of production capacity are fully achieved. In the next lower priority p_3 goal, the sale constraint of item I_2 is given more weightage. Therefore to demonstrate the importance of item I_2 over item I_1, set $d_3^- = 0$ in Eq. (15.6) to get $x_2 = 250$, further reducing the feasible region of points on line segment AB to a single point A on AB. At point A, $x_2 = 250$ along with $x_1 + x_2 = 300$ gives $x_1 = 50$, which when put in constraint Eq. (15.5) gives $d_2^- = 150$. Thus the solution to the GP problem is $x_1 = 50$, $x_2 = 250$, $d_2^- = 150$, and $d_1^- = d_1^+ = d_3^- = 0$.

RECAPITULATION

1. Goal programming model is an extension of the LP model. A single objective GP solution is similar to the solution obtained by the LP method.
2. In GP, assigning priorities to different goals helps achieve the most important goal first. In other words, important or higher priority goals can be achieved at the cost of lower priority goals.
3. Goal programming is always a minimization problem where deviations from the established goals within the given set of constraints are minimized.
4. Deviational variables in GP are the overachievement or the underachievement of the goals.

EXERCISES

Multiple-choice Questions

15.1 In GP model, deviational variables added to the ith constraints must satisfy the following conditions:

(a) $d_i^+ \times d_i^+ = 0$
(b) $d_i^- \times d_i^- = 0$
(c) $d_i^- \times d_i^+ = 0$
(d) None of these

15.2 In the optimal simplex table of GP model, two or more $c_j - z_j$ rows indicate
(a) unequal priority goals
(b) equal priority goals
(c) both (a) and (b)
(d) none of these

15.3 Goal programming model is preferred when
(a) more than one objective is set to be achieved
(b) goals are multiple and incommensurable
(c) goals are satisfied in an ordinal sequence
(d) all of these

15.4 In GP model, goals are assigned priorities such that
(a) goals of greatest importance are given less priority
(b) goals may not have equal priority
(c) higher priority goals must be achieved before lower priority goals
(d) none of these

State True or False

15.1 In GP, if there are two or more z_j or $c_j - z_j$ rows, then the problem has prioritized goals.

15.2 Goal programming is used for solving a multi-objective optimization problem that trade-offs conflicting objectives.

15.3 A single-objective GP problem ignores the concept of a goal.

15.4 The deviational variable in the basis of the initial simplex table of a GP problem is a negative deviational variable.

Fill in the Blanks

15.1 In a GP problem, a goal constraint having ________ and overachievement variables is expressed as equality constraints.

15.2 In a GP formulation, each goal generates new ________ and adds at least one new variable to the objective function.

15.3 Goal programming is an extension of the ________.

15.4 In simplex method of GP, the variable to enter the solution mix is selected from the highest priority row and ________ $c_j - Z_j$ value in it.

Review Questions

15.1 What is goal programming?

15.2 Write a short note on single-goal programming.

15.3 What do you mean by multiple-goal programming with priority?

15.4 Write the procedure of modified simplex method for a GP problem.

15.5 Explain the graphical solution to a GP problem.

Numerical Problems

15.1 A manufacturer produces two types of products A and B. The production of each A and B requires 1h of production capacity in the workshop. The workshop of the company has a maximum production capacity of 80h per week. On account of restriction on the sales capacity of these products, the maximum number of A and B that can be sold is 8 and 10 per week respectively. The profit on the sale of A is ₹100, whereas that on B is ₹60. The manager wants to determine the number of units of each product that should be produced per week and sets the following goals to achieve with equal priority:

Goal 1: The production capacity should not exceed 50 h per week.

Goal 2: The sales of the two products A and B must be as much as possible.

Formulate the problem as a GP problem.

15.2 A company produces two products A_1 and A_2 which yield a per unit profit of ₹400 and ₹300 respectively. These products require a certain raw material whose supply is limited to 300 kg per month. The raw material requirement is 1 kg per unit of each product. Further while one unit of A_1 requires 2h of processing, each unit of A_2 requires 1h of processing. The company has a normal processing capacity of 400h per month. The marketing department has stated that under the prevailing conditions, which are likely to continue in the foreseeable future, the maximum number of units of A_1 and A_2 that can be sold each month are 100 and 250 respectively. The manager of the company has set the following goals in the order of their importance, the priorities being pre-emptive.

Goal 1: Avoid the underutilization of normal production capacity.

Goal 2: Sell the maximum number of units of products A_1 and A_2. Since the two products yield profit in the ratio 400:300, the company manager would give the weightage to the sale of the products in the same ratio.

Goal 3: Minimize overtime of processing capacity.

Formulate this problem as a GP problem.

15.3 A company produces two different types of products E and F. A product mix is to be obtained by utilizing two limited resources—labour and raw material. Each unit of E required 2h of labour and 3 units of raw material, whereas the respective figures for F are 4h of labour and 4 units of raw material. Every day 20 h of labour and 24 units of raw material are available. The goals before the management according to priorities are as follows:

(a) The profit per day should be at least ₹36 assuming that the profit per unit of the products E and F is ₹8 and ₹6 respectively.

(b) Due to market conditions as well as product substitutability, the number of units of E should be twice the number of units of F.

(c) The labour should be fully utilized.

Formulate this problem as a GP problem.

15.4 Solve the following GP problems:

(a) Min $Z = p_1d_1^- + p_2d_2^- + p_3d_3^-$

Subject to $200x_1 + 300x_2 + d_1^- - d_1^+ = 4800$

$x_2 + d_2^- - d_2^+ = 15$

$x_1 + d_3^- - d_3^+ = 5$

$4x_1 + 2x_2 + x_3 = 60$

$4x_1 + 4x_2 + x_4 = 70$

and $x_1, x_2, x_3, x_4, d_1^-, d_1^+, d_2^-, d_2^+, d_3^-, d_3^+ \geq 0.$

(b) Min $Z = p_1d_1^+ + p_2d_2^- + p_3d_3^-$

Subject to $4x_1 + 5x_2 + d_1^- - d_1^+ = 80$

$100x_1 + 100x_2 + d_2^- - d_2^+ = 2000$

$x_2 + d_3^- - d_3^+ = 10$

and $x_1, x_2, d_1^-, d_1^+, d_2^-, d_2^+, d_3^-, d_3^+ \geq 0.$

15.5 Solve the following GP problems using graphical method:

(a) Min $Z = p_1d_1^- + p_2(4d_2^- + 3d_3^-) + p_3d_1^+$

Subject to $x_1 + x_2 \leq 300$

$2x_1 + x_2 + d_1^- - d_1^+ = 400$

$x_1 + d_2^- = 150$

$$x_2 + d_3^- = 350$$

$$x_1, x_2, d_1^-, d_1^+, d_2^-, d_3^- \geq 0.$$

(b) Min $Z = p_1 d_1^- + p_2 d_3^- + p_3 d_2^-$

Subject to $2x_1 + 3x_2 \leq 30$

$$6x_1 + 4x_2 \leq 60$$

$$x_1 + x_2 + d_1^- - d_1^+ = 10$$

$$x_1 + d_2^- - d_2^+ = 8$$

$$x_2 + d_3^- - d_3^+ = 7$$

$$x_1, x_2, d_1^-, d_1^+, d_2^-, d_2^+, d_3^-, d_3^+ \geq 0.$$

CONCEPTUAL PROBLEM

A manufacturing company produces two types of products A and B. The production of A and B requires an average of 1 h in the plant. The plant has a normal production capacity of 400 h per week. A survey revealed that the maximum number of products A and B that can be sold in a week are 240 and 300 respectively. The net profit from the sales of A and B are ₹800 and ₹400 respectively. The manager of the company has set the following goals in the order of importance ($p_1 >> p_2 >> p_3$), that is, $p_i > np_{i+1}$, however large n may be. Priority p_1 is the avoidance of any underutilization of the normal production capacity. Priority p_2 is the sale of maximum possible units of A and B. As the net profit from the sale of product A is double than that of product B, the manager has twice as much desire to achieve the same sales as product B for Product A. Priority p_3 is the manager's aim to minimize the overtime operation of the plant as much as possible. Formulate as a GP problem and solve using the modified simplex method.

Solution The formulation of the GP problem is as follows:

Let x_1 and x_2 be the number of units to be produced of products A and B respectively.

The plant capacity constraint equation is $x_1 + x_2 + d_1^- - d_1^+ = 400$ where d_1^- = underutilization or idle time of the normal production capacity and d_1^+ = overutilization of the normal production capacity of the plant.

The production goal or sales goal constraint equations are $x_1 + d_2^- = 240$ and $x_2 + d_3^- = 300$ where d_2^- and d_3^- are the underachievement of sales goals for products A and B respectively. Here d_2^+ and d_3^+ have not been incorporated because the given sales goals are the maximum possible volumes (sales not to exceed 240 and 300 units for products A and B respectively).

Since profit is to be maximized, the objective as per the order of priorities given to different goals is as follows:

$$\text{Min } Z = p_1 d_1^- + p_2(2d_2^- + d_3^-) + p_3 d_1^+$$

where in sales priority p_2, double weightage has been given to the sale of product A over the sale of product B because of double profit obtained in the sales of A.

Combining all the aforementioned goals, the GP model of the problem becomes

$$\text{Min } Z = p_1 d_1^- + p_2(2d_2^- + d_3^-) + p_3 d_1^+$$

Subject to $x_1 + x_2 + d_1^- - d_1^+ = 400$

$$x_1 + d_2^- = 240$$

$$x_2 + d_3^- = 300$$

$$x_1, x_2, d_1^-, d_2^-, d_3^-, d_1^+ \geq 0$$

(Contd)

(*Contd*)

		c_j		0	0	$-p_3$	$-p_1$	$-2p_2$	$-p_2$	
Table no.	C_B	**Basic variables**	X_B	x_1	x_2	d_1^+	d_1^-	d_2^-	d_3^-	**Min ratio**
1	$-p_1$	d_1^-	400	1	1	−1	1	0	0	400
	$-2p_2$	d_2^-	240	1	0	0	0	1	0	...
	$-p_2$	d_3^-	300	0	[1]	0	0	0	1	300 min→
	$c_j - Z_j$		p_1	1	1↑	−1	1	0	0	
			p_2	2	1	0	0	0	0	
			p_3	0	0	−1	0	0	0	
2	$-p_1$	d_1^-	100	[1]	0	−1	1	0	−1	100 min→
	$-2p_2$	d_2^-	240	1	0	0	0	1	0	240
	0	x_2	300	0	1	0	0	0	1	...
	$c_j - Z_j$		p_1	1↑	0	−1	0	0	−1	
			p_2	2	0	0	0	0	−1	
			p_3	0	0	−1	0	0	0	
3	0	x_1	100	1	0	−1	1	0	−1	...
	$-2p_2$	d_2^-	140	0	0	[1]	−1	1	1	140 min→
	0	x_2	300	0	1	0	0	0	1	...
	$c_j - Z_j$		p_1	0	0	0	−1	0	0	
			p_2	0	0	2↑	−2	0	1	
			p_3	0	0	−1	0	0	1	
4	0	x_1	240	1	0	0	0	1	0	
	$-p_3$	d_2^+	140	0	0	1	−1	1	1	
	0	x_2	300	0	1	0	0	0	1	
	$c_j - Z_j$		p_1	0	0	0	−1	0	0	
			p_2	0	0	0	0	−2	−1	
			p_3	0	0	0	−1	1	2	

The optimal solution is $x_1 = 240$, $x_2 = 300$, $d_1^+ = 140$ and all the remaining variables are zero, being non-basic. In the final (modified) simplex table, the third goal is not completely attained as there are positive values 1 and 2 in $(c_j - Z_j)$ row at p_3 level in d_2^- and d_3^- columns respectively. As evident, we can attain the third goal to a greater extent if we introduce d_3^- in the solution. However, there is a negative value (−1) at the higher priority level p_2 in the same column (this implies that if we introduce d_3^- in the solution, we can improve the achievement of the third goal at the expense of achieving the second goal which is not desirable). Similarly, we cannot enter d_2^- as there is a negative value (−2) at the higher priority level p_2 in the same column.

16 Game Theory

After studying this chapter, the reader will be able to

- understand the game theory
- grasp the characteristics of a game
- get acquainted with some basic terminology used in the game theory
- comprehend the lower and upper value of the game
- explain the minimax principle with pure strategies
- be aware of the procedure to find the saddle point
- explain the dominance principle of the game theory
- solve games without saddle point, 2×2 games without saddle point (using algebraic and arithmetic method), and (3×3) games with no saddle point
- Use the graphical method to solve $(2\times n)$ and $(m\times2)$ games, and the method of submatrices to solve $(2\times n)$ or $(m\times2)$ games with no saddle point
- explain the two person zero-sum game with mixed strategies (linear programming method)
- deduce the limitations of the game theory

16.1 INTRODUCTION

There are many situations when one has to analyse problems where more than one party is pursuing conflicting objectives and the outcome of every action of one party depends on the action taken by the other. Such situations are called conflicting situations. For example, such situations can be seen in a number of economic situations, particularly those involving free competition where the contending parties can be firms, industrial enterprises, etc.

Although in real life, the conflicting situations are quite complicated, our objective in this chapter is to consider simple situations and construct a simplified formalized model of such situations. Such a formalized model of conflicting situations is called a *game*. A set of definite rules are followed while constructing this model or game. Thus, a game is quite different from a real conflict situation as it is played according to definite rules. A game can also be defined as a set of rules for playing.

A game may be considered as a clash of interests between two or more opponents (parties). In the first case, it is called a two-person game and in the latter, a multiperson game. In practice, two-person games are the most important. Conventionally, the parties taking part in a game are called the players and the outcome of an encounter is known as a party's pay-off (or gain).

One of the basic concepts of the game theory is the concept of strategy. A strategy is considered as a set of unambiguous rules determining the choice of every personal move of the player, depending on the situation that arises in the course of the game.

A game is said to be finite, if either player has only a finite number of strategies. A finite game in which player A has m strategies, say $A_1, A_2, \ldots, A_m$ and player B has n strategies say $B_1, B_2, \ldots, B_n$ is called an $m \times n$ game. The choice of strategies A_i by player A and B_j by player B determines the pay-off (gain) to player A and is denoted as a_{ij}. Values of a_{ij} can be arranged in the matrix form as shown in Table 16.1.

This matrix is called the pay-off (gain) matrix of player A. Entry a_{ij} can either be positive (gain to player A), or negative (loss to player A), or zero (no loss, no gain to A).

The matrix of an $m \times n$ game of the form shown in Table 16.1 is denoted as $(a_{ij})_{m \times n}$. The pair strategy (A_i, B_j), when used rendering to a player the maximum possible gain (or minimum possible loss), is called the optimal strategy and the corresponding value a_{ij} is called the game value, denoted by α.

16.2 CHARACTERISTICS OF GAMES

The following are the main characteristics of a game:

1. *Number of activities or courses of actions*: In a game, the number of activities may be finite or infinite. Based on this, a game may be called a finite or an infinite game.
2. *Number of persons*: If n persons are participating in a game, then it is called an n-persons game. Here, person refers to an individual or a group of participants.
3. *Pay-off*: The payment to a person as a result of the game is called pay-off.
4. *Courses of action available to a player*: In a game, the courses of actions available to each player are known.

Table 16.1 Pay-off matrix for A

B / A	B_1	B_2	…	B_1	…	B_m
A_1	a_{11}	a_{12}	…	a_{1j}	…	a_{1m}
A_2	a_{21}	a_{22}	…	a_{2j}	…	a_{2m}
…	…	…	…	…	…	…
A_i	a_{i1}	a_{i2}	…	a_{ij}	…	a_{2m}
…	…	…	…	…	–	…
A_n	a_{n1}	a_{n2}	…	a_{nj}	–	a_{nm}

16.3 BASIC TERMINOLOGY USED IN GAME THEORY

Before having a detailed discussion on the game theory, we shall discuss some basic terms used in this theory.

Two-Person Zero-Sum or Rectangular Games

A game with only two players (say, player A and player B) is called a two-person zero-sum game, if the gain of one player, say A, is equal to the loss of the other player, say B, such that the sum of their net gain is zero. It is also known as a rectangular game as the pay-off matrix is rectangular.

Pay-off Matrix

Suppose the player A has m strategies $A_1, A_2, \ldots, A_m$ and player B has n activities $B_1, B_2, \ldots, B_n$, then a pay-off matrix can be formed by adopting the following rules:

1. The row designations of each matrix are strategies $A_1, A_2, \ldots, A_i, \ldots, A_m$ that are available to player A.
2. The column designations of each matrix are strategies $B_1, B_2, \ldots, B_i, \ldots, B_n$ that are available to player B.
3. The cell entry 'a_{ij}' is the payment to player A in A's pay-off matrix (Table 16.2) when A chooses the activity A_i and B chooses the activity B_j.
4. In a zero-sum two-person game, the cell entry in B's pay-off matrix (Table 16.3) will be the negative of the corresponding cell entry 'a_{ij}' in the player A's pay-off matrix, such that the sum of the pay-off matrices of both players is ultimately zero.

Table 16.2 Representation of A's pay-off matrix

		B_1	B_2	$\cdots$	B_j	$\cdots$	B_n
	A_1	a_{11}	a_{12}	$\cdots$	a_{1j}	$\cdots$	a_{1n}
	A_2	a_{21}	a_{22}	$\cdots$	a_{2j}	$\cdots$	a_{2n}
	$\vdots$	$\vdots$	$\vdots$		$\vdots$		$\vdots$
Player A	A_i	a_{i1}	a_{i2}	$\cdots$	a_{ij}	$\cdots$	a_{in}
	$\vdots$	$\vdots$	$\vdots$		$\vdots$		$\vdots$
	A_m	a_{m1}	a_{m2}	$\cdots$	a_{mj}	$\cdots$	a_{mn}

Table 16.3 Representation of B's pay-off matrix

		B_1	B_2	$\cdots$	B_j	$\cdots$	B_n
	A_1	$-a_{11}$	$-a_{12}$	$\cdots$	$-a_{1j}$	$\cdots$	$-a_{1n}$
	A_2	$-a_{21}$	$-a_{22}$	$\cdots$	$-a_{2j}$	$\cdots$	$-a_{2n}$
	$\vdots$	$\vdots$	$\vdots$		$\vdots$		$\vdots$
Player B	A_i	$-a_{i1}$	$-a_{i2}$	$\cdots$	$-a_{ij}$	$\cdots$	$-a_{in}$
	$\vdots$	$\vdots$	$\vdots$	$\cdots$	$\vdots$	$\cdots$	$\vdots$
	A_m	$-a_{m1}$	$-a_{m2}$	$\cdots$	$-a_{mj}$	$\cdots$	$-a_{mn}$

Note In practical problems, there is no need of writing B's pay-off matrix as it is just the negative of A's pay-off matrix. Thus, if 'a_{ij}' is the gain to player A, then $-a_{ij}$ is the gain to player B, such that the net gain is zero.

Strategies

When two players play a game, then they have different alternatives (course of actions) at their disposal to go ahead with the game. These alternatives are called strategies. In Table 16.3, $A_1, A_2, \ldots, A_m$ are the strategies available to player A and $B_1, B_2, \ldots, B_i, \ldots, B_n$ are the strategies available to player B.

Saddle Point

A saddle point (A_r, B_s) of a pay-off matrix is the position of an element with minimum value in its row and maximum in its column. Thus in the game of Table 16.4, position (1, 1) is the saddle point with optimum pay-off 6 to player A.

Or mathematically, if the pay-off matrix (a_{ij}) is such that $\max_i \left(\min_j (a_{ij}) \right) = \min_j \left(\max_i (a_{ij}) \right) = a_{rs}$, then the matrix is said to have a saddle point at (A_r, B_s).

Optimal Strategy

If the pay-off matrix (a_{ij}) has a saddle point (r, s), then A_r and B_s are called the optimal strategies of players A and B respectively. In Table 16.4, $r = s = 1$ and A_1 and B_1 are the optimal strategies for players A and B respectively.

Pure Strategy

When only a single alternative is used in the game, it is called pure strategy. As an example, the game problem of Table 16.4 shows pure strategies. For optimum gain, both players A and B use single courses of actions A_1 and B_1 respectively.

Table 16.4 Representation of the saddle point

B / A	B_1	B_2	B_3	α_i
A_1	6	8	10	6
A_2	4	12	2	2
β_j	6	12	10	–

Mixed Strategy

If several alternatives with different values are used to play the game, then it is called a game with mixed strategy. For example, the game problem with pay-off matrix $\begin{bmatrix} 1 & -1 \\ -1 & 1 \end{bmatrix}$ is a game with mixed strategies, as this game does not have a saddle point. When a saddle point does not exist, mixed strategies are used to find the value of the game.

Value of Game

The optimum pay-off to player A is called the value of the game. In Table 16.4, 6 is the game value with optimum strategies A_1 and B_1. If any of the players deviates from his/her optimal strategies, he/she will be in loss. Alternatively, the pay-off a_{rs} at the saddle point (A_r, B_s) is called the value of the game.

16.4 LOWER AND UPPER VALUE OF GAME—'MINIMAX' PRINCIPLE WITH PURE STRATEGIES

Consider the pay-off matrix $(a_{ij})_{m \times n}$ of the player A (see Table 16.1). If player A chooses strategy A_i, then he/she is sure to get $\min_j a_{ij}$; j varies from 1 to n. Thus, A would like to choose that strategy A_i ($i = 1, 2, \ldots, m$) for which $\min_j a_{ij}$ is maximized to get $\max_i\left(\min_j a_{ij}\right)$. It is denoted as $\underline{a}$, and is called the lower value of the game.

$$\text{Thus, } \max_i\left(\min_j a_{ij}\right) = \underline{a} \tag{16.1}$$

On the other hand, if player B chooses strategy B_j, then he/she is sure that player A will not get more than $\max_i a_{ij}$. Thus, B would like to choose the strategy B_j, which minimizes the maximum gain to player A, reducing it to $\min_j\left(\max_i a_{ij}\right)$. It is denoted by $\overline{a}$, and is called the upper value of the game.

$$\text{Thus, } \min_j\left(\max_i a_{ij}\right) = \overline{a} \tag{16.2}$$

Equations (16.1) and (16.2) are called the maximin and minimax criteria, respectively. If $\underline{a} = \overline{a} = \alpha$ (say), then the game is said to have a saddle point (i.e., a solution to the game exists) and α is the game value (pay-off to player A). If $\alpha = a_{ij}$ (see Table 16.1), then we conclude that the optimal strategy of player A is A_i and that of B is B_j.

Note A game is said to be fair if both the lower and upper values of the game are equal to zero.

Theorem 16.1 If $\underline{a} = \max_i\left(\min_j a_{ij}\right)$ and $\overline{a} = \min_j\left(\max_i a_{ij}\right)$ are the lower and upper values of the game respectively, then the lower value is always less than or equal to the upper value of the game, that is, $\underline{a} = \max_i\left(\min_j a_{ij}\right) \le \min_j\left(\max_i a_{ij}\right) = \overline{a}$.

Proof It is evident that

$$\min_j a_{ij} \le a_{ij} \text{ [for any } j \text{ and } i \text{ fixed]} \tag{16.3}$$

$$\text{Further, } \max_i a_{ij} \ge a_{ij} \text{ for any } i \text{ and } j \text{ fixed} \tag{16.4}$$

$$\text{Let } \min_j a_{ij} = a_{ir} \text{ [for } j = r] \tag{16.5}$$

$$\text{and } \max_i a_{ij} = a_{sj} \text{ [for } i = s] \tag{16.6}$$

Thus from Eqs (16.5) and (16.6), we have $a_{ir} \le a_{ij} \le a_{sj}$ for all i and j.

$$\text{Hence, } \max_i a_{ir} \le a_{ij} \le \min_j a_{sj} \tag{16.7}$$

Using Eqs (16.5), (16.6), and (16.7), we have $\max_i \min_j a_{ij} \le a_{ij} \le \min_j \max_i a_{ij}$. Thus, $\underline{a} = \max_i \min_j a_{ij} \le \min_j \max_i a_{ij} = \overline{a}$.

16.5 PROCEDURE TO DETERMINE SADDLE POINT

The following steps are used to determine the saddle point of a given game problem:

1. Choose the minimum element of each row i (α_i's) of the pay-off matrix and write it on the extreme right of that row i (refer Fig. 16.4).
2. Choose the greatest element of each column j (β_j's) of the pay-off matrix and write it against (below) that column j (refer Fig. 16.4).

3. If maximum of α_i's are equal to minimum of β_j's, then the common value is the game value and thus, we can conclude that the saddle point exists. Otherwise, the saddle point does not exist in pure strategies.
4. When the saddle point does not exist, use mixed strategies to determine the value of the game (see Section 16.7).

Example 16.1 Consider a problem of military operations between two countries A and B. While country A has three kinds of aircraft A_1, A_2, and A_3, country B has three kinds of aircraft B_1, B_2, and B_3. Country A's goal is to hit its enemy's aircraft, whereas B's goal is to avoid being hit. When aircraft A_1 is used, aircraft B_1, B_2, and B_3 are hit with probabilities 0.9, 0.4, and 0.2 respectively. When aircraft A_2 is used, they are hit with probabilities 0.3, 0.6, and 0.8. Further, when aircraft A_3 is used, they are hit with probabilities 0.5, 0.7, and 0.2. Formulate the problem in terms of the game theory and determine the lower and upper values of the game.

Solution The game theory is formulated in the adjoining table:

B→ A↓	B_1	B_2	B_3	α_i
A_1	0.9	0.4	0.2	0.2
A_2	0.3	0.6	0.8	0.3
A_3	0.5	0.7	0.2	0.2
β_j	0.9	0.7	0.8	

Here $\alpha_1 = 0.2, \alpha_2 = 0.3, \alpha_3 = 0.2, \beta_1 = 0.9, \beta_2 = 0.7$, and $\beta_3 = 0.8$.

Hence, the lower value of the game $\underline{a} = \max(0.2, 0.3, 0.2) = 0.3$ and the upper value of the game $\overline{a} = \min(0.9, 0.7, 0.8) = 0.7$.

Example 16.2 Two players A and B, each write down simultaneously and independently of each other, one of the four numbers 1, 2, 3, or 4 on a chit. If the sum of the numbers they have written down is even, then B pays that sum to A; if the sum is odd, then A pays that sum to B.

(a) Construct the game matrix.
(b) Find the lower and upper value of the game.
(c) Find the maximin and minimax strategies for players A and B.

Solution (a) There are four strategies for player A: writing down 1, A_1; writing down 2, A_2; writing down 3, A_3; and writing down 4, A_4. Opponent B also has the same four strategies: B_1, B_2, B_3, and B_4. This is a 4×4 game with the game matrix (pay-off matrix of player A) as shown in the following table:

A \ B	B_1	B_2	B_3	B_4	α_i (row minimum)
A_1	2	−3	4	−5	−5
A_2	−3	4	−5	6	−5
A_3	4	−5	6	−7	−7
A_4	−5	6	−7	8	−7
β_j(column maximum)	4	6	6	8	

where α_i = minimum of ith row and β_j = maximum of jth column. Evidently, B can respond to any strategy that would be the worst for A. Indeed if A chooses strategy A_1, for instance, B will always counter it with strategy B_4. Strategy A_2 will always be countered by B_3. Strategy A_3 will be countered by B_4 and A_4 will be countered by B_3.

(b) The lower value of the game $\underline{a} = \max(-5, -5, -7, -7) = -5$
The upper value of the game $\overline{a} = \min(4, 6, 6, 8) = 4$

(c) Player A's maximin strategy is either A_1 or A_2; employing them systematically he/she can in any case guarantee that his/her gain is not less than −5 (a loss not greater than 5). B's minimax strategy is B_1; employing it systematically he/she can guarantee that his/her loss is not greater than 4.

Notes
1. If any of the players A or B deviates from these maximin or minimax strategies respectively, their gain is adversely affected.
2. In Examples 16.1 and 16.2, we had come across a situation wherein when both the players employ the minimax strategies then $\underline{a} \neq \overline{a}$. Such games are called unstable games. However there are some games, for which minimax strategies are stable ($\underline{a} = \overline{a}$). These are the games for which the lower value of the game is equal to the upper one, that is, $\underline{a} = \overline{a} = \alpha$ (i.e., saddle point and hence game value exists).

Example 16.3 Find the lower and upper values of a game for the following game matrix (pay-off matrix of player A). Determine whether the saddle point exists.

Player B

$$\text{Player A} \begin{bmatrix} 4 & 0 \\ 0 & 2 \end{bmatrix}$$

Solution The game matrix (pay-off matrix of player A) is shown in the adjacent table.
where α_i = minimum of ith row and β_j = maximum of jth column.

$\text{Max}\{\alpha_i\} = 0$, $\min\{\beta_j\} = 2$. While player A employs a maximin strategy, player B uses the minimax strategy. The lower and upper values of the game are 0 and 2 respectively. As the lower and upper game values are not equal, we can conclude that the saddle point does not exist.

B→ A↓	**B_1**	**B_2**	α_i
A_1	4	0	0
A_2	0	2	0
β_j	4	2	

Example 16.4 Find the lower and upper values of a game for the following game matrix (pay-off matrix of player A). Determine whether the saddle point exists.

Player B

$$\text{Player A} \begin{bmatrix} 1 & -1 \\ 2 & 3 \end{bmatrix}$$

Solution The game matrix (pay-off matrix of player A) is shown in the adjacent table.
where α_i = minimun of ith row and β_j = maximum of jth column.

$\text{Max}\{\alpha_i\} = 2$, $\min\{\beta_j\} = 2$. While player A employs a maximin strategy, player B uses the minimax strategy. Both the lower and upper game value is 2. As the lower and upper game values are equal, we can conclude that the saddle point exists.

B→ A↓	**B_1**	**B_2**	α_i
A_1	1	−1	−1
A_2	2	3	2
β_j	2	3	

Example 16.5 Find the lower and upper values of a game for the following game matrix (pay-off matrix of player A). Determine whether the saddle point exists.

$$\text{Player A}\overset{\text{Player B}}{\begin{bmatrix} 0 & 2 \\ -1 & 4 \end{bmatrix}}$$

Solution The game matrix (pay-off matrix of player A) is shown in the adjacent table.

B→ A↓	B_1	B_2	α_i
A_1	0	2	0
A_2	–1	4	–1
β_j	0	4	

where α_i = minimun of the *i*th row and β_j = maximum of *j*th column.

Max$\{\alpha_i\} = 0$, min$\{\beta_j\} = 0$. While player A employs a maximin strategy, player B uses the minimax strategy. Both the lower value and upper game value is 0. As the lower and upper game values are equal, we can conclude that the saddle point exists and the game is fair.

Example 16.6 Two players A and B, without looking at each other, place a coin face up on a table, showing either heads or tails, as they wish. If both choose the same side of the coin, player A gains ₹1, else (if they choose the opposite side) player B gain ₹1. Construct the pay-off matrix of the player A and find the lower and upper values of the game. Does the saddle point exist?

Solution There are only two strategies (moves) for player A in the game—choosing heads A_1 and choosing tails A_2 for player A; similar is the case for player B, that is, choosing heads B_1 and choosing tails B_2. Hence, this is a 2×2 game. The game matrix (pay-off matrix of player A) is as follows:

B / A	B_1 (heads)	B_2 (tails)	α_i (row maximum)
A_1 (heads)	1	–1	–1
A_2 (tails)	–1	1	–1
β_j (column maximum)	1	1	

Where α_i = minimun of the *i*th row and β_j = maximum of the *j*th column.

Max$\{\alpha_i\} = -1$ for both A_1 and A_2, min$\{\beta_j\} = 1$ for both B_1 and B_2. Here, either of the strategies of player A is maximin and either of the strategies of player B is minimax. While, the lower value of the game is –1, its upper value is 1. Thus, player A is sure that his/her loss is not greater than 1 and the same is the case for player B. As the lower and upper game values are not equal, we can conclude that the saddle point does not exist.

Example 16.7 Find the range of *p* and *q*, which will render the entry (2, 1) as a saddle point for the game given in Table 16.5.

Solution Ignoring the A_3 row and B_3 column (the respective row and columns where *p* and *q* lie), the reduced pay-off matrix is given by Table 16.6.

Table 16.5 Problem

	B_1	B_2	B_3
A_1	5	4	3
A_2	8	9	q
A_3	p	6	10

Table 16.6 Solution

5	4	4
8	9	8
8	9	

Thus, minimax = maximin = 8, which is the entry in the position (2, 1). For (2, 1) to remain the saddle point of the given game, p cannot be greater than 8. Thus, $p \leq 8$. Further q cannot be less than 8 and hence, $q \geq 8$.

Example 16.8 Find the range of u and v, which will render the entry (2, 2) as a saddle point for the game given in Table 16.7.

Solution Ignoring the A_3 row and B_3 column (the respective row and columns where u and v lie), the reduced pay-off matrix is given in Table 16.8.

Table 16.7 Problem

	B_1	B_2	B_3
A_1	2	4	5
A_2	8	7	u
A_3	5	v	9

Table 16.8 Solution

2	4	2
8	7	7
8	7	

Thus, minimax = maximin = 7, which is the entry in the position (2, 2). For (2, 2) to remain the saddle point of the given game, v cannot be greater than 7. Thus, $v \leq 7$. Further, u cannot be less than 7 and hence $u \geq 7$.

16.6 MATRIX REDUCTION BY DOMINANCE PRINCIPLE

Sometimes, the size of a games's pay-off matrix can be reduced by eliminating an inferior course of action among those available, such that the inferior one is never used. Such a course of action is said to be dominated by the other course of action. The other course of action is called the dominating course of action. This principle of reducing a pay-off matrix is called the *principle of dominance*. The concept of dominance is particularly useful for the evaluation of two-person zero-sum games, where a saddle point does not exist. It reduces the size of the pay-off matrix and thus helps in swiftly obtaining a solution to a given problem.

The *dominance rule for column* implies that every entry in the dominating column(s) must be less than or equal to the corresponding entry of the dominated column.

The *dominance rule for rows* implies that every entry in the dominating row(s) must be greater than or equal to the corresponding entry of the dominated row.

Dominance rule using convex linear combination When each entry in the convex linear combination (CLC) of a certain number of pure strategies of player A (row in pay-off matrix) is greater than or equal to the corresponding entries of A's *i*th strategies (*i*th row in pay-off matrix), then the *i*th row is said to be the inferior or dominated row. If the *i*th row is deleted, then the value of the game and the optimal strategies of the player remain unaffected. When each entry in the convex linear combination of a certain number of pure strategies of player B (column in pay-off matrix) is less than or equal to the corresponding entries of B's *j*th strategies (*j*th column in pay-off matrix), then the *j*th column is said to be the inferior or dominated column. If the *j*th column is deleted, then the value of the game and optimal strategies of the player remain unaffected.

Example 16.9 Reduce the following pay-off matrix using the principle of dominance

	B_1	B_2	B_3	B_4
A_1	2	3	11	8
A_2	7	5	2	7
A_3	6	4	–4	–9

Solution Here, we notice that every entry of row 2 is greater than the corresponding entry of row 3 and so the latter (third row) can be removed using the dominance principle to get the truncated matrix (Table 16.9).

Table 16.9 Iteration 1

	B_1	B_2	B_3	B_4
A_1	2	3	11	8
A_2	7	5	2	7

Table 16.10 Iteration 2

	B_1	B_2	B_3
A_1	2	3	11
A_2	7	5	2

Further, in the truncated matrix column 4 is dominated by columns 1 and 2. Thus, column 4 can be removed as per the dominance principle to get the final truncated matrix given by Table 16.10.

Example 16.10 Two players P_1 and P_2 play a game where each of them has to choose one of the three coins—white (W), black (B), and red (R)—independently of the other. Then, the coins are compared. If both P_1 and P_2 have chosen white (W, W), neither gets anything. If P_1 selects white and P_2 black (W, B), then player P_1 loses ₹2 or player P_2 wins ₹2 and so on to get the following matrix. Reduce the pay-off matrix using the principle of dominance and hence find the saddle point if possible.

		Colour chosen by P_2		
		W	B	R
	W	0	–2	7
Colour chosen by P_1	B	2	5	6
	R	3	–3	8

Solution The game matrix (pay-off matrix of player P_1) is as follows:

	W	B	R	Min of rows
W	0	–2	7	–2
B	2	5	6	2
R	3	–3	8	–3
Max of columns	3	5	8	

It is a 3×3 game and the matrix has no saddle point. Looking carefully at the pay-off matrix, we see that every entry in column 3 is greater than those in columns 1 and 2. Hence, player P_2 will never play the third column strategy, as it will cause the heaviest losses to him/her and higher gains to player P_1. Thus column R has to be deleted. Strategy R is called the dominated strategy. Deleting column 3, we obtain the truncated matrix given in Table 16.11.

Looking again at the truncated matrix, we see that every entry of the first row is smaller than the corresponding entry of the second row, so player P_1 will not play the strategy of first row (W row) as it will give him/her greater losses. Thus, deleting the first row, the resulting truncated pay-off matrix is given by Table 16.12.

Table 16.11 Truncated matrix

P_1 \ P_2	W	B
W	0	–2
B	2	5
R	3	–3

Table 16.12 Reduced truncated matrix

P_1 \ P_2	W	B	
B	2	5	2
R	3	–3	–3
	3	5	

Thus, a 3×3 game has reduced to a 2×2 game with $\underline{a} = 2$ and $\overline{a} = 5$, we see that $\underline{a} \neq \overline{a}$ and hence saddle point does not exist.

Example 16.11 Solve the following game using the dominance principle:

Player A \ Player B	1	2	3	4	5
1	3	5	4	9	6
2	5	6	3	7	8
3	8	7	9	8	7
4	4	2	8	5	3

Solution In the given pay-off matrix, column 4 is dominated by column 1 and column 5 is dominated by column 2. Thus deleting columns 4 and 5, the reduced matrix is given by Table 16.13.

In the Table 16.13, again rows 1, 2, and 4 are dominated by row 3. Thus, deleting rows 1, 2, and 4, the new reduced matrix is given by Table 16.14.

Table 16.13 Reduced matrix

		Player B		
		1	2	3
Player A	1	3	5	4
	2	5	6	3
	3	8	7	9
	4	4	2	8

Table 16.14 New reduced matrix

		Player B		
		1	2	3
Player A	3	8	7	9

In the aforegiven pay-off matrix, columns 1 and 3 are dominated by column 2 and can be deleted. We arrive at the adjoining matrix. Thus, the solution to the game is as follows:

		Player B
		2
Player A	3	7

1. The best strategy for player A is 3
2. The best strategy for player B is 2
3. The value of game (pay-off to A) is 7

Note The aforementioned game has a saddle point. In such a case when the dominance principle has to be applied, the matrix is ultimately reduced to a single element.

16.7 GAMES WITHOUT SADDLE POINT

While solving a game problem, it is not necessary that the saddle point always exists. In such cases, we use mixed strategies to determine the value of the game. Here, we discuss some methods for solving game problems in which the saddle point does not exist.

16.7.1 2 × 2 Game Without Saddle Point

In many cases, the matrix games (rectangular games) can be reduced to a 2×2 game (to be discussed later in Section 16.11). It is therefore worthwhile to determine the solution of a 2×2 game. We describe here two methods to solve 2×2 games without saddle points.

1. Algebraic method
2. Arithmetic method (Oddment method).

Algebraic Method

In this method, for any two-person zero-sum game where optimal strategies are not pure (i.e., there are no saddle points), player A's pay-off matrix, the optimal strategies (x_1, x_2) of A and (y_1, y_2) of B, and the value of the game V are as given in the adjacent table.

		Player B	
		y_1	y_2
Player A	x_1	a_{11}	a_{12}
	x_2	a_{21}	a_{22}

Therefore, $\dfrac{x_1}{x_2} = \dfrac{a_{22} - a_{21}}{a_{11} - a_{21}}$, $\dfrac{y_1}{y_2} = \dfrac{a_{22} - a_{12}}{a_{11} - a_{21}}$ and $V = \dfrac{a_{11}a_{22} - a_{12}a_{21}}{a_{11} + a_{22} - a_{12} - a_{21}}$.

Proof Let the mixed strategies for player A be (x_1, x_2) with $x_1 + x_2 = 1$ $x_1, x_2 \geq 0$. If player B uses the first strategy y_1, then the expected gain to A will be $E_1(x) = a_{11}x_1 + a_{21}x_2$. Further if B uses his/her second strategy y_2, then the expected gain to A will be $a_{12}x_1 + a_{22}x_2 = E_2(x)$, say. Now, player A would like to maximize his/her minimum expected gain. Therefore, the value of the game V must be maximum of $(E_1(x), E_2(x))$, that is,

$$E_1(x) = a_{11}x_1 + a_{21}x_2 \leq V \tag{16.8}$$

$$E_2(x) = a_{12}x_1 + a_{22}x_2 \leq V \tag{16.9}$$

$$x_1 + x_2 = 1 \tag{16.10}$$

The optimum strategies in Eqs (16.8) and (16.9) have become strict equations. Thus, we get the following equations:

$$a_{11}x_1 + a_{21}x_2 = V \tag{16.11}$$

$$a_{12}x_1 + a_{22}x_2 = V \tag{16.12}$$

Solving Eqs (16.11) and (16.12), we get

$$\frac{x_1}{x_2} = \frac{a_{22} - a_{21}}{a_{11} - a_{12}} \tag{16.13}$$

Using Eqs (16.10) and (16.13), we get

$$x_1 = \frac{a_{22} - a_{21}}{a_{11} + a_{22} - a_{12} - a_{21}} \tag{16.14}$$

and $x_2 = 1 - x_1 = \dfrac{a_{11} - a_{12}}{(a_{11} + a_{22}) - (a_{12} + a_{21})}$ (16.15)

Substituting the values of x_1 and x_2 in either Eq. (16.11) or Eq. (16.12), the value of the game is as follows:

$$V = \frac{a_{11}a_{22} - a_{21}a_{12}}{a_{11} + a_{22} - (a_{12} + a_{21})} \tag{16.16}$$

In a similar way, the optimal strategies for player B are as follows:

$$y_1 = \frac{a_{22} - a_{12}}{a_{11} + a_{22} - a_{12} - a_{21}} \tag{16.17}$$

$$y_2 = \frac{a_{11} - a_{21}}{a_{11} + a_{22} - a_{12} - a_{21}} \tag{16.18}$$

Arithmetic Method

Arithmetic method provides us an easy way for obtaining the optimum strategies for each player A and B in 2×2 games without saddle points. The following steps are to be followed to get the solution of a 2×2 game:

1. Find the difference between two numbers in the first row, neglecting the negative sign if it occurs. Call it R_1^- and write it against the second row.

2. Find the difference between two numbers in row 2, call it R_2^-, and write it against row 1 (neglecting the negative sign).
3. Repeat steps 1 and 2 for columns also, denoting the differences c_1^- and c_2^- (see the following figure). The differences thus obtained are called oddments. These are the frequencies with which the players use their courses of action in their optimum strategies.

	Player B		
Player A	a_{11}	a_{12}	$R_2^- = \|a_{21} - a_{22}\|$
	a_{21}	a_{22}	$R_1^- = \|a_{11} - a_{12}\|$
	$c_2^- = \|a_{12} - a_{22}\|$	$c_1^- = \|a_{11} - a_{21}\|$	

where R_i^-'s and c_i^-'s are notations used for the row and column entry differences.

The optimum strategies for A are $\left(\frac{R_2^-}{R_1^- + R_2^-}, \frac{R_1^-}{R_1^- + R_2^-}\right)$. The optimum strategies for B are $\left(\frac{c_2^-}{c_1^- + c_2^-}, \frac{c_1^-}{c_1^- + c_2^-}\right)$.

Example 16.12 Two players A and B play a game. For the following pay-off matrix, find the optimal strategies and the value of the game using the (a) algebraic method and (b) arithmetic method.

		Player B 1	2
Player A	1	2	4
	2	5	2

Solution (a) Algebraic method: Comparing the given pay-off matrix in Table 16.15 with a standard 2 × 2 pay-off matrix given by Table 16.16,

Table 16.15 Iteration 1

		Player B 1	2
Player A	1	2	4
	2	5	2

Table 16.16 General 2×2 matrix

		Player B y_1	y_2
Player A	x_1	a_{11}	a_{12}
	x_2	a_{21}	a_{22}

we have $a_{11} = 2, a_{12} = 4, a_{21} = 5, a_{12} = 2$.

The optimal strategies (x_1, x_2) of player A are as follows:

$$x_1 = \frac{a_{22} - a_{21}}{a_{11} + a_{22} - a_{12} - a_{21}} = \frac{2 - 5}{2 + 2 - 4 - 5} = \frac{-3}{-5} = \frac{3}{5}$$

$$x_2 = 1 - x_1 = 1 - \frac{3}{5} = \frac{2}{5}$$

Game value $V = \frac{a_{11}a_{22} - a_{12}a_{21}}{a_{11} + a_{22} - a_{12} - a_{21}} = \frac{4 - 20}{2 + 2 - 4 - 5} = \frac{-16}{-5} = \frac{16}{5}.$

The optimal strategies (y_1, y_2) of player B are as follows:

$$y_1 = \frac{a_{22} - a_{12}}{a_{11} + a_{22} - a_{12} - a_{21}} = \frac{2 - 4}{2 + 2 - 4 - 5} = \frac{2}{5}$$

and $y_2 = \frac{a_{11} - a_{21}}{a_{11} + a_{22} - a_{12} - a_{21}} = \frac{2 - 5}{2 + 2 - 4 - 5} = \frac{3}{5}.$

(b) Arithmetic method: The given pay-off matrix is as follows:

		Player B 1	2	
Player A	1	2	4	2
	2	5	2	2
		5	4	

Since no saddle point is found, the optimal strategies will be mixed strategies.

	1	2		
1	2	4	3	$\frac{3}{3+2} = \frac{3}{5}$
2	5	2	2	$\frac{2}{3+2} = \frac{2}{5}$
	2	3		

1. Take the difference between two numbers in row 1. We find 4 – 2 = 2. Place it against row 2.
2. Taking the difference between two numbers in row 2, we find 2 – 5 = –3. Place 3 (neglecting the negative sign) against row 1.
3. Repeat the aforementioned two steps for column 1 and column 2 as well. Thus for optimum gain, player A must use the first strategy with probability 3/5 and the second strategy with probability 2/5.

The optimum strategies (y_1, y_2) for player B are (2/5, 3/5). Using any one oddment of A or B the game value $V = (2 \times 3/5) + (5 \times 2/5) = 16/5$.

Example 16.13 Define the oddment for player A and B in a 2×2 game without saddle points. Find oddments of player A and B in the following game:

		Player B 1	2
Player A	1	8	–3
	2	–3	1

Solution Oddments are the frequencies (or the probabilities) with which a player should use his/her strategies to obtain the optimum game value. First, we find the difference between two numbers in row 1 and place it against the second row, neglecting the negative sign if it occurs. Here, it is $8 - (-3) = 11$.

The difference between two numbers in row 2 is $-3-1=-4$; write 4 (neglecting negative sign) against row 1.

Thus, oddments for player A are 4/15 and 11/15. As a result, player A should use strategies 1 and 2 with probabilities 4/15 and 11/15 respectively.

The difference between two entries in column $1=8-(-3)=11$ and the difference between two entries in column $2=-3-1=4$. Thus, write 11 below columns 2 and 4 (neglecting negative sign) below column 1. Further oddments of player B are 4/15 and 11/15, which are the probabilities with which player B must use strategies 1 and 2 for obtaining the optimum game value.

8	−3	4
−3	1	11
4	11	

Example 16.14 Solve the following game using the principle of dominance:

		Player B			
		1	2	3	4
Player A	1	2	2	4	0
	2	3	4	2	4
	3	4	2	4	0
	4	0	6	0	8

Table 16.17 Iteration 1

		Player B				
		1	2	3	4	
Player A	1	2	2	4	0	0
	2	3	4	2	4	2
	3	4	2	4	0	0
	4	0	6	0	8	0
		4	8	4	8	

Solution First we calculate the minimum of rows and maximum of columns and obtain Table 16.17.

Maximin = 2 and minimax = 4. This game does not have a saddle point. Try to reduce the size of the given pay-off matrix. Every entry of row 1 is less than or equal to the corresponding entry of row 3, is hence dominated by the latter, and can be removed from the matrix to get the reduced matrix as shown in Table 16.18.

Table 16.18 Iteration 2

		B			
		1	2	3	4
A	2	3	4	2	4
	3	4	2	4	0
	4	0	6	0	8

In the aforegiven reduced matrix, every entry in column 1 is greater than or equal to the corresponding entry of column 3 and hence can be deleted from the table, without affecting the optimal strategies and game value to obtain a further reduced matrix given by Table 16.19.

Table 16.19 Iteration 3

		B		
		2	3	4
A	2	4	2	4
	3	2	4	0
	4	6	0	8

To check whether further reduction of this matrix is possible, we find the average of the player A's third and fourth pure strategies as $[(2+6)/2,\ (4+0)/2,\ (0+8)/2]=(4,2,4)$. The average so derived dominates player A's second strategy as every entry in a CLC of the 3rd and 4th rows and is greater than or equal to the corresponding entries of the 2nd row. Thus, the 2nd row can be deleted to get a further reduced matrix given by Table 16.20.

For player B, the average of the third and fourth pure strategies is $[(4+0)/2, (0+8)/2]=(2, 4)$, which dominates the first strategy as every entry in the CLC of the third and fourth pure strategies is less than or equal to the corresponding entries of the second pure strategy of player B. Thus, the column corresponding to the second strategy is deleted to get the reduced matrix given by Table 16.21.

Table 16.20 Iteration 4

		B 2	3	4
A	3	2	4	0
	4	6	0	8

Table 16.21 Iteration 5

		B 3	4
A	3	4	0
	4	0	8

Thus, using the dominance principle, the given game is reduced to a 2×2 game with no saddle points and can be solved using the algebraic method, by solving the following equations:

$$\left.\begin{aligned} 4x_3 + 0x_4 &= V \\ 0x_3 + 8x_4 &= V \\ x_1 + x_2 + x_3 + x_4 &= 1 \end{aligned}\right\} \qquad (16.19)$$

and
$$\left.\begin{aligned} 4y_3 + 0y_4 &= V \\ 0y_3 + 8y_4 &= V \\ y_1 + y_2 + y_3 + y_4 &= 1 \end{aligned}\right\} \qquad (16.20)$$

where (x_1, x_2, x_3, x_4) and (y_1, y_2, y_3, y_4) are the strategies of players A and B respectively and V is the game value. Solving the Eqs (16.19) and (16.20), we get

$$x_3 = 2/3, \quad x_4 = 1/3, \quad x_1 = x_2 = 0, \quad y_3 = 2/3, \quad y_4 = 1/3, \quad y_1 = y_2 = 0, \quad \text{and} \quad V = 8/3$$

Thus, the optimal strategies (x_1, x_2, x_3, x_4) of player A are = (0, 0, 2/3, 1/3) and the optimal strategies (y_1, y_2, y_3, y_4) of player B are = (0, 0, 2/3, 1/3). The value of the game $V = 8/3$.

16.8 (3 × 3) GAMES WITH NO SADDLE POINT

A general and simple way of solving a (3×3) game with no saddle points is illustrated through Example 16.15.

Example 16.15 Solve the following (3×3) game:

		B 1	2	3
	1	1	−1	−1
A	2	−1	−1	3
	3	−1	2	−1

Solution The given problem has no saddle points. Let (x_1, x_2, x_3), $x_1 + x_2 + x_3 = 1$ and (y_1, y_2, y_3), $y_1 + y_2 + y_3 + y_4 = 1$ be the optimal strategies of players A and B respectively.

For player A:

$(1)x_1+(-1)x_2+(-1)x_3 \geq V$
$(-1)x_1+(-1)x_2+(2)x_3 \geq V$
$(-1)x_1+3x_2+(-1)x_3 \geq V$
$x_1+x_2+x_3=1,\ x_1, x_2, x_3 \geq 0$

For player B:

$(1)y_1+(-1)y_2+(-1)y_3 \leq V$
$(-1)y_1+(-1)y_2+3(y_3) \leq V$
$(-1)y_1+2y_2+(-1)y_3 \leq V$
$y_1+y_2+y_3=1,\ y_1, y_2, y_3 \geq 0$

Now to find the values of $x_1, x_2, x_3, y_1, y_2, y_3$, and V, let all the aforementioned inequalities hold with equality. Thus, we get

$$x_1 - x_2 - x_3 = V \tag{16.21}$$

$$-x_1 - x_2 + 2x_3 = V \tag{16.22}$$

$$-x_1 + 3x_2 - x_3 = V \tag{16.23}$$

$$x_1 + x_2 + x_3 = 1 \tag{16.24}$$

$$y_1 - y_2 - y_3 = V \tag{16.25}$$

$$-y_1 - y_2 + 3y_3 = V \tag{16.26}$$

$$-y_1 + 2y_2 - y_3 = V \tag{16.27}$$

$$y_1 + y_2 + y_3 = 1 \tag{16.28}$$

From Eq. (16.24), we get $x_3 = 1 - x_1 - x_2$. Substituting the value of x_3 in Eqs (16.21), (16.22), and (16.23) and solving them, we get $x_1 = 6/13, x_2 = 3/13, x_3 = 4/13$, and $V = -1/13$.

Using the same way to solve Eqs (16.25), (16.26), (16.27), and (16.28), we get $y_1 = 6/13$, $y_2 = 4/13, y_3 = 3/13$, and $V = -1/13$.

Hence, the solution to the game is as follows:

(a) Optimal strategies of player A are $(x_1, x_2, x_3) = (6/13, 3/13, 4/13)$
(b) Optimal strategies of player B are $(y_1, y_2, y_3) = (6/13, 4/13, 3/13)$

The game value (gain to player A) $V = -1/13$. The negative sign of the game value indicates a loss to player A.

16.9 GRAPHICAL METHOD FOR $(2 \times n)$ AND $(m \times 2)$ GAMES

In this section, we shall discuss the solution of $(2 \times n)$ and $(m \times 2)$ rectangular games, using the graphical method in cases where the saddle point does not exist. It should be clear that if one player has only two strategies, then the other player can also use only two strategies. Hence, the graphical method can be applied to determine the two optimal strategies of a player.

16.9.1 Graphical Method for $2 \times n$ Games

To illustrate the graphical method, let us consider a $(2 \times n)$ game that has the following pay-off matrix and does not have saddle points.

			Player B			
			y_1	y_2	…	y_n
			B_1	B_2	…	B_n
Player A	x_1	A_1	a_{11}	a_{12}	…	a_{1n}
	$x_2 = 1 - x_1$	A_2	a_{21}	a_{22}	…	a_{2n}

Player A has only two strategies (x_1, x_2), where $x_1+x_2=1$, x_1, $x_2 \geq 0$. Now according to the maximin criterion for mixed strategies, player A should select the value of x_1, $x_2=1-x_1$ so as to maximize his/her minimum expected pay-off.

B's pure strategies	A's expected pay-off ($E_j(x_1)$; $j = 1, 2, \ldots, n$)
B_1	$a_{11}x_1 + a_{21}x_2 = (a_{11} - a_{21})x_1 + a_{21} = E_1(x_1)$
B_2	$a_{12}x_1 + a_{22}x_2 = (a_{12} - a_{22})x_1 + a_{22} = E_2(x_1)$
...	...
B_n	$a_{1n}x_1 + a_{2n}x_2 = (a_{1n} - a_{2n})x_1 + a_{2n} = E_n(x_1)$

Thus, the expected pay-off lines will be as follows:

$$E_1(x_1) = (a_{11} - a_{21})x_1 + a_{21}$$

$$E_2(x_1) = (a_{12} - a_{22})x_1 + a_{22}$$

$$E_n(x_1) = (a_{1n} - a_{2n})x_1 + a_{2n} \text{ as a function of } x_1$$

The lowest boundary of these lines will be the minimum expected pay-off as a function of x_1. The highest point on these lowest boundaries would then give the maximum expected pay-off and the optimum value of x_1 say x_1^*.

Then, determine optimum strategy x_2, say $x_2^* = 1 - x_1^*$

Now, determine only two strategies for player B, corresponding to those two lines that pass through the maximum (highest) points.

Procedure to Find Game Value using Graphical Method

To determine the maximum value $\underline{v}$, we consider different values of x_1 on the horizontal line and the values of $E_j(x_1)$ on the vertical axis. As $0 \leq x_1 \leq 1$, the straight line $E_j(x_1)$ must pass through the points $(0, E_j(0))$ and $(1, E_j(1))$, where $E_j(0) = a_{2j}$ and $E_j(1) = a_{1j}$. Thus, the lines $E_j(x_1) = (a_{1j} - a_{2j})x_1 + a_{2j}$, $j = 1, 2, \ldots, n$ can be drawn using Fig. 16.1.

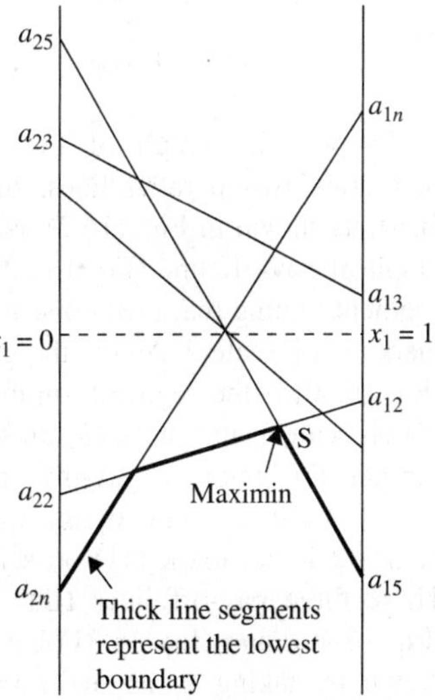

Figure 16.1 Graphical representation of $2 \times n$ games

1. Construct two vertical axes, namely axis 1 as the point $x_1=0$ and axis 2 as the point $x_1=1$.
2. Represent the pay-offs a_{2j}, $j = 1, 2, \ldots, n$ on axis 1 and pay-offs a_{1j}, $j = 1, 2, \ldots, n$ on axis 2.
3. Join the point representing a_{ij} on axis 2 to the points representing a_{2j} on axis 1. The representing straight line is the expected pay-off line $E_j(x_1)$, $j = 1, 2, \ldots, n$
4. Mark the lowest boundary of these lines $E_j(x_1)$, thus plotted, by thick line

segments. The highest point on this lowest boundary gives the maximin point S and specifies the two critical moves of player B. If more than two lines pass through the highest point S, it indicates the existence of an alternate points solution.

16.9.2 Graphical Method for $m \times 2$ Games

The $(m \times 2)$ games can be treated exactly in the same way as $(2 \times n)$ games, except in that the minimax point S in $m \times 2$ games will be the lowest point on the uppermost boundary, instead of the highest point on the lowest boundary in case of $(2 \times n)$ games.

Example 16.16 Solve the following game using the graphical method:

		Player B		
		1	2	3
Player A	1	1	3	11
	2	8	5	2

Solution In the given problem maximin = max(1, 2) = 2 and minimax = min(8, 5, 11) = 5. Thus, the saddle point does not exist. The adjoining table shows the expected pay-off corresponding to the player B's pure strategies:

B's strategies	A's expected pay-off $E(x_1)$
1	$E_1(x_1) = 1x_1 + 8(1 - x_1) = 8 - 7x_1$
2	$E_2(x_1) = 3x_1 + 5(1 - x_1) = 5 - 2x_1$
3	$E_3(x_1) = 11x_1 + 2(1 - x_1) = 2 + 9x_1$

The three expected pay-off lines are as follows:

$$E_1(x_1) = 8 - 7x_1 \quad (16.29)$$

$$E_2(x_1) = 5 - 2x_1 \quad (16.30)$$

$$E_3(x_1) = 2 + 9x_1 \quad (16.31)$$

To plot the graph of these lines, first draw two parallel lines, one unit apart, as shown in Fig. (16.2). Now, the graph of pay-off line (Eq. 16.29) is the segment joining mark (8) on scale 1 and mark (1) on scale 2. Again, the graph of (Eq. 16.30) is the segment joining mark (5) on scale 1 and mark (3) on scale 2. Further, the graph of pay-off line (Eq. 16.31) is the segment joining mark (2) on scale 1 and mark (11) on scale (2). These three pay-off lines (Eq. 16.29), (Eq. 16.30), and (Eq. 16.31) have been drawn by taking x_1 as the x-axis and $E(x_1)$ as the y-axis. Points P, A, B, and Q on the lowest boundary of the graph

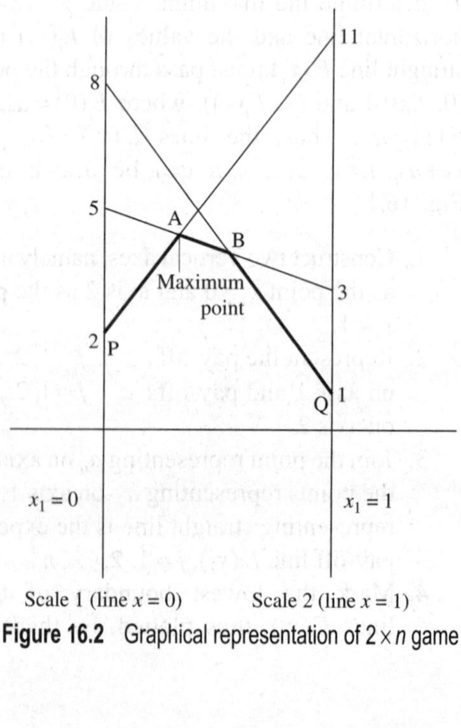

Figure 16.2 Graphical representation of $2 \times n$ game

in Fig (16.2) represent the smallest possible expected gain to player A for any value of x_1 between 0 and 1. According to the maximin criterion, player A chooses the best of these minimum expected gains. The highest point A on the lowest boundary will give the largest pay-off to A; thus the optimum strategies are those that pass through point A. Point A is the intersection of pay-off lines (16.30) and (16.31), which are corresponding to strategies 2 and 3 of player B. Thus, the game is reduced to 2×2.

The optimum strategies of player A and B can be obtained by solving the following pair of equations:

$$3x_1 + 5x_2 = V$$

$$11x_1 + 2x_2 = V, \; x_1 + x_2 = 1, \; x_1, x_2 \geq 0$$

$$\text{and } 3y_2 + 11y_3 = V$$

$$5y_2 + 2y_3 = V, \; y_2 + y_3 = 1, \; y_2, y_3 \geq 0 \quad \text{(Using algebraic method)}$$

where V is the value of the game.

Solving these equations, we get

$$x_1 = \frac{3}{11}, \; x_2 = \frac{8}{11}, \; V = \frac{49}{11}$$

$$\text{and } y_2 = \frac{9}{11}, \; y_3 = \frac{2}{11}$$

Thus, player A chooses optimal strategies $(x_1, x_2) = (3/11, 8/11)$ and player B chooses optimal strategies $(y_1, y_2, y_3) = (0, 2/11, 9/11)$. The value of the game is 49/11.

Example 16.17 Use the graphical method to solve the following game:

		Player B 1	2
	1	2	7
Player A	2	3	5
	3	11	2

Solution The game does not have a saddle point. Let y_1 and y_2 be the strategies of player B, $y_1 + y_2 = 1, \; y_1, y_2 \geq 0$.

The three strategy lines obtained are given here.

Strategies of player A	B's expectations
1	$E_1(y_1) = 2y_1 + 7y_2 = 2y_1 + 7(1 - y_1) = 7 - 5y_1$
2	$E_2(y_1) = 3y_1 + 5y_2 = 5 - 2y_1$
3	$E_3(y_1) = 11y_1 + 2y_2 = 2 + 9y_1$

$$E_1(y_1) = 7 - 5y_1 \tag{16.32}$$

$$E_2(y_1) = 5 - 2y_1 \tag{16.33}$$

$$E_3(y_1) = 2 + 9y_1 \tag{16.34}$$

These three lines can be drawn by taking y_1 as the x-axis and $E_j(y_1)$ as the y-axis, $j = 1, 2, 3$ between two scales one unit apart as shown in Fig. 16.3.

Point R is the lowest of the uppermost boundary of three lines, which is the intersection of lines represented by Eqs (16.32) and (16.34). Thus, the game is reduced to 2×2. The optimal strategies (x_1, x_2, x_3) of player A and (y_1, y_2) of player B are obtained by solving the following pairs:

$$2y_1 + 7y_2 = V; 11y_1 + 2y_2 = V;$$
$$y_1 + y_2 = 1, y_1, y_2 \geq 0$$

where V is the game value, and

$$2x_1 + 11x_3 = V$$

$$7x_1 + 2x_3 = V, x_1 + x_3 = 1, x_1, x_3 \geq 0$$

($x_2 = 0$ as the game is reduced to 2×2)

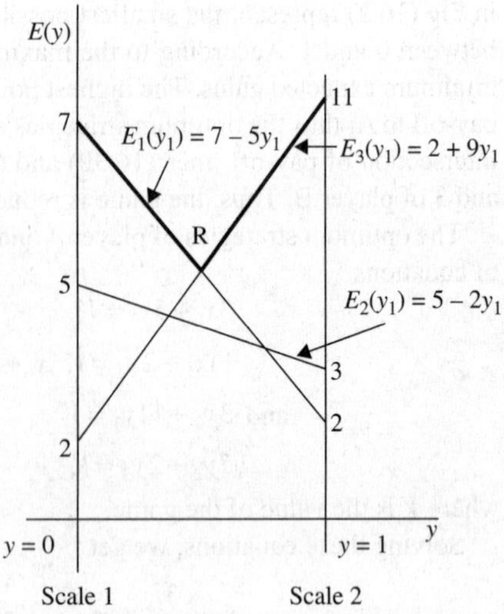

Figure 16.3 Graphical representation of $m \times 2$ game

Solving these pairs, we get

$$x_1 = 9/14,\ x_3 = 5/14,\ y_1 = 5/14,\ y_2 = 9/14 \Rightarrow \text{Game value } V = 73/14$$

Thus, the optimal strategies (x_1, x_2, x_3) of player A are 9/14, 0, and 5/14 and that of player B are $(y_1, y_2) = (5/14, 9/14)$

Example 16.18 Solve the following game:

		Player B			
		1	2	3	4
	1	2	3	11	8
Player A	2	7	5	2	7
	3	6	4	−4	−9

Table 16.22 Iteration 1

		Player B			
		1	2	3	4
Player A	1	2	3	11	8
	2	7	5	2	7

Solution Here, we see that row 2 dominates row 3. Therefore, using the dominance principle, row 3 can be deleted to get the reduced pay-off matrix shown in Table 16.22.

In the reduced matrix, column 4 is dominated by column 1 and column 2. Thus, column 4 can be removed without affecting the game value. The resulting reduced matrix is obtained after reassigning the strategies as shown in Table 16.23.

The game has no saddle point. The game value lies between 2 and 5. We use mixed strategies to obtain the game value. Consider A's problem. Let x be the probability of occurrence of strategies 1 and $(1-x)$ that of strategy 2. The expected pay-off corresponding to 1, 2, and 3 of player B are as follows:

Table 16.23 Iteration 2

		Player B			
		1	2	3	
Player A	1	2	3	11	[2] = maxmin
	2	7	5	2	[2] = maxmin
		7	[5] minimax	11	

$$E_1(x) = 2x + 7(1-x) = 7 - 5x \tag{16.35}$$

$$E_2(x) = 3x + 5(1-x) = 5 - 2x \tag{16.36}$$

$$E_3(x) = 11x + 2(1-x) = 2 + 9x \tag{16.37}$$

Now, plot the graph of these three lines, taking x as abscissa and the expected pay-off E as ordinate, as shown in Fig. (16.4). Draw these three lines by joining the extremities obtained by taking $x = 0$ and $x = 1$.

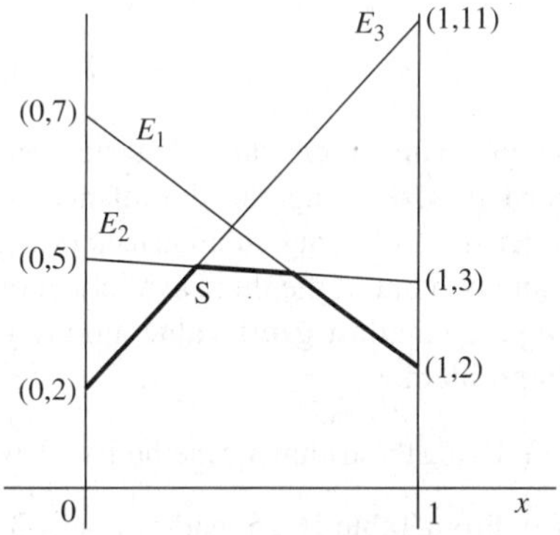

Figure 16.4 Graphical representation of 2×3 game

Since A's problem is maximin, we identify the lower boundary in the graph shown by thick lines. The maximin occurs at the highest extreme point S at the intersection of lines E_2 and E_3, that is, as $5 - 2x = 2 + 9x$, which gives $x = 3/1$. Therefore, the optimum strategies of A are (3/11, 8/11). Value of the game $V = 3x + 5(1-x) = 49/11$.

Now, consider B's game. In A's first strategy, one strategy of player B is ruled out as the line E_1 does not pass through the highest intersection point S. Now suppose the second and third strategies (i.e., B_2 and B_3) are y and $1 - y$ respectively, then

$$3y + 11(1-y) = V \tag{16.38}$$

$$5y + 2(1-y) = V \text{ (using algebraic method)} \tag{16.39}$$

Solving Eqs (16.38) and (16.39), we conclude the following:

(a) Optimal strategies of B are (0, 9/11, 2/11)
(b) Game value $V = 49/11$.

16.10 METHOD OF SUBMATRICES OR SUBGAMES FOR $(2 \times n)$ OR $(m \times 2)$ GAMES WITH NO SADDLE POINT

1. In $(2 \times n)$ games, $n > 2$, consider the set of all (2×2) submatrices of $(2 \times n)$ matrix for their solution. As $n > 2$, player B has more flexibility to play any two out of the n available courses of action. B would like to play those two courses of action for which the loss occurring to him/her (gain obtained by A) is minimum. Thus, the solution of that subgame (2×2) for which game value is minimum will be considered as the optimum solution of the given (original) game.
2. In case of $(m \times 2)$ games with $m > 2$, player A will have more flexibility of using any two of the available activities. Thus, in this case the solution of that (2×2) subgame out of all (2×2) subgames of $(m \times 2)$, for which the game value has the maximum value will be considered as the optimal solution of the given $(m \times 2)$ game.

Example 16.19 Solve the following game using the method of submatrices:

		B 1	2	3
A	1	3	−3	4
	2	−1	1	−3

Solution The given game has no saddle points. In addition, the game cannot be reduced in size using the dominance principle. To use the method of matrices, we consider the following submatrices (subgames). Here, player B has more flexibility to play any two out of the three available courses of action. Thus, the solution of that game having the smallest game value among the following will result in the solution to the given problem.

(a) Using the arithmetic method, we have Table 16.24. Game value $= \dfrac{3\times 1+(-1)\times 3}{4} = 0.$

(b) From Table 16.25, game value = 3.

(c) Using the arithmetic method, we have Table 16.26. Game value $= \dfrac{(-3)\times 4+(1)\times 7}{11} =$ −5/11.

Thus, the optimal strategies of players A and B are (4/11, 7/11) and (0, 7/11, 4/11) respectively for the game value $V = \min(0, 3, -5/11) = -5/11$.

Table 16.24 Iteration 1

	1	2		
1	3	−3	2	$\frac{2}{8}=\frac{1}{4}$
2	−1	1	6	$\frac{6}{8}=\frac{1}{4}$
	4	4		

Table 16.25 Iteration 2

	1	3	
1	3	4	3 = Maximum
2	−1	−3	−3
	3	4	
	Maximum		

Table 16.26 Iteration 3

	2	3		
1	−3	4	4	$\frac{4}{11}$
2	1	−3	7	$\frac{7}{11}$
	7	4		

Example 16.20 Solve the following game using the method of matrices:

	B	
A	2	1
	1	0
	0	3
	−2	3

Solution Consider the submatrices of order 2 and find their solution as described here.

(a) In Table 16.27, maximum = minimum = 1. This subgame has a saddle point. Game value = 1.

(b) The subgame in Table 16.28 has no saddle points. Using the algebraic method, the game value $V = \dfrac{2\times 3+(0)\times 1}{4} = \dfrac{3}{2} = 1.5.$

(c) The subgame in Table 16.29 has no saddle points. Using the algebraic method, the game value $V = \dfrac{2\times 5+(-2)\times 1}{6} = \dfrac{10-2}{6} = \dfrac{4}{3} = 1.33$

(d) The subgame in Table 16.30 has no saddle points. Using the algebraic method, the game value $V = \dfrac{3\times 1+0\times 1}{4} = \dfrac{3}{4} = 0.75$

(e) The subgame in Table 16.31 has no saddle points. Using the algebraic method, the game value $V = \dfrac{5\times 1+1\times(-2)}{6} = \dfrac{5-2}{6} = \dfrac{1}{2} = 0.5$

(f) The subgame in Table 16.32 has saddle points. Game value $V = 0$. Out of the aforementioned games, we select that one whose game value is maximum.

Table 16.27 Subgame (a)

		B 1	B 2	
A	1	2	1	[1] = maximum
	2	1	0	0
		2	[1] = maximum	

Table 16.28 Subgame (b)

		B			
A	1	2	1	3	3/4
	3	0	3	1	1/4
		2	2		
		2/4	2/4		

Table 16.29 Subgame (c)

		B 1	B 2		
A	1	2	1	5	5/6
	4	−2	3	1	1/6
		2	4		

Table 16.30 Subgame (d)

	B			
2	1	0	3	3/4
3	0	3	1	1/4
	3	1		

Table 16.31 Subgame (e)

		B 1	B 2		
A	2	1	0	5	5/6
	4	−2	3	1	1/6

Table 16.32 Subgame (f)

		B 1	B 2	
A	2	0	3	[0] = maximum
	4	−2	3	−2
		[0] Maximum	3	

The maximum game value 1.5 is for the subgame in (b). Thus, the optimum strategies of player A and B are (3/4, 0, 1/4, 0) and (1/2, 1/2) respectively.

Hence, the game value is 1.5.

16.11 TWO-PERSON ZERO-SUM GAME WITH MIXED STRATEGIES OR LINEAR PROGRAMMNING METHOD

Every two-person zero-sum game can be solved through mixed strategies (since a complex linear combination of all strategies is used in mixed strategies).

Let us assume that player A selects the strategy A_i with the probability p_i and player B selects the strategy B_j with probability q_j, $i = 1, 2, ..., m$; $j = 1, 2, ..., n$. A_i and B_j can be considered events with probabilities p_i and q_j, that is $P(A_c) = p_i$ and $p(B_j) = q_j$. Table 16.33 is the pay-off matrix for player A along with the corresponding probabilities with which both players, A and B, use their strategies:

To obtain the value of the game, we have to determine the value of p_i's and q_j's. We do this by using the minimax or maximin criteria. If B selects pure strategy B_j, then the expected pay-off to player A is to find,

$$a_{ij}p_1 + a_{2j}\cdot p_2 + \ldots + a_{mj}\, p_m = \sum_{i=1}^{m} a_{ij} p_i$$

Player B can select only from the pure strategies $B_j (j = 1, \ldots, n)$, and hence player A would like to select p_i's in such a way that it maximizes its smallest expected pay-off. Thus, A's problem is

$$\underset{p_i, p_2, \ldots, p_m}{\text{Max}} \left[\text{Min} \left\{ \sum_{i=1}^{m} a_{i1} p_i, \sum_{i=1}^{m} a_{i2} p_i, \ldots, \sum_{i=1}^{m} a_{in} p_i \right\} \right] \tag{16.40}$$

Table 16.33 Matrix representing pay-off to player A in mixed strategies

		q_1	q_2	$\cdots$	q_j	$\cdots$	q_n
		B_1	B_2	$\cdots$	B_j	$\cdots$	B_n
p_1	A_1	a_{11}	a_{12}	$\cdots$	a_{1j}	$\cdots$	a_{1n}
p_2	A_2	a_{21}	a_{22}	$\cdots$	a_{2j}	$\cdots$	a_{2n}
$\vdots$	$\vdots$						
p_i	A_i	a_{i1}	a_{i2}	$\cdots$	a_{ij}	$\cdots$	a_{in}
$\vdots$	$\vdots$	$\vdots$	$\vdots$		$\vdots$		$\vdots$
p_m	A_m	a_{m1}	a_{m2}	$\cdots$	a_{mj}	$\cdots$	a_{mn}

Subject to the condition that $\sum_{i=1}^{m} p_i = 1, p_i\text{'s} \geq 0$

Similarly, player B would like to select q_j's, which minimize the largest expected pay-off to A.

Thus, B's problem is

$$\underset{q_i, q_2, \ldots, q_n}{\text{Min}} \left[\text{Max} \left\{ \sum_{j=1}^{n} a_{1j} q_j, \sum_{j=1}^{n} a_{2j} q_j, \ldots, \sum_{j=1}^{n} a_{mj} q_j \right\} \right] \tag{16.41}$$

Subject to the condition $\sum_{j=1}^{n} q_j = 1, q_j\text{'s} \geq 0$ for $\forall_j$.

The value in Eqs (16.40) and (16.41) are maximin and minimax expected pay-offs to A respectively. Let these be denoted by $\underline{a}$ and $\overline{a}$ respectively.

$$\underline{a} = \text{Min} \left\{ \sum_{i=1}^{m} a_{i1} p_i, \sum_{i=1}^{m} a_{i2} p_i, \ldots, \sum_{i=1}^{m} a_{in} p_i \right\}$$

Then, A's problem given in Eq. (16.40) reduces to

$$\text{Max } Z_A = \underline{a} \tag{16.42}$$

$$\text{Subject to } \sum_{i=1}^{m} a_{ij} p_i \geq \underline{a}, j = 1, \ldots, n, \sum_{i=1}^{m} p_i = 1, p_i\text{'s} \geq 0.$$

In this case, we do not know the value of $\underline{a}$—it may be positive or non-positive. We shall assume that $\underline{a}$ is equal to some positive number. However, to have $\underline{a}$ as positive, it is evidently sufficient for all elements a_{ij} of the pay-off matrix to be non-negative. This can always be attained by adding a sufficiently large positive number C to elements a_{ij}. This way, the value of the game is increased by C, whereas the solution remains unchanged. The optimum value of the objective function is then obtained by subtracting the added constant C.

Thus, without any loss of generality, assume $\underline{a}$ to be positive. Divide constraints in (16.42) by $\underline{a}$ and let $x_i = p_i/\underline{a}, i = 1, \ldots, m;$ then the problem in (16.42) reduces to

$$\text{Max } Z_A = \underline{a} \tag{16.43}$$

$$\text{Subject to } \sum_{i=1}^{m} a_{ij} x_i \geq 1, \; j = 1, 2, \ldots, n$$

Further,

$$\sum_{i=1}^{m} x_i = \sum_{i=1}^{m} \frac{p_i}{\underline{a}} = \frac{1}{\underline{a}} \sum_{i=1}^{m} p_i = \frac{1}{\underline{a}}$$

Thus,

$$\sum_{i=1}^{m} x_i = x_1 + x_2 + \ldots + x_m = \frac{1}{\underline{a}} \tag{16.44}$$

$$\text{Max } Z_A = \underline{a} \text{ is equivalent to min } Z_A' = \frac{1}{\underline{a}} \tag{16.45}$$

Thus, A's problem can be written in the following form:

$$\left.\begin{aligned} &\text{Min } Z_A' = x_1 + x_2 + \ldots + x_m \\ &\text{Subject to } \sum_{i=1}^{m} a_{ij} x_i \geq 1 \\ &x_i\text{'s} \geq 0, \; i = 1, 2, \ldots, m \end{aligned}\right\} \tag{16.46}$$

Next, to solve B's problem, let $\overline{a} > 0$. This can always be achieved by adding a positive constant C to make all entries a_{ij} to make them positive in the pay-off matrix. Thus, proceeding exactly in the same way as in deriving A's problem to the form of an LPP, B's problem in Eq. (16.41) can be reduced to the following LPP form,

$$\left.\begin{aligned} &\text{Max } Z_B = y_1 + y_2 + \ldots + y_n \\ &\text{Subject to } \sum_{r=1}^{n} a_{ij} y_j \leq 1, \; i = 1, \ldots, m \\ &y_j\text{'s} \geq 0, \; j = 1, 2, \ldots, n \end{aligned}\right\} \tag{16.47}$$

It is important to observe that problems presented in (16.46) and (16.47) are the dual of each other. There are feasible p_i's and q_j's $(i = 1, \ldots, m; \; j = 1, \ldots, n)$ because each player can use pure strategies. Consequently, we have feasible x_i's and y_j's as $\underline{a}$ and $\overline{a}$ are positive. As a result, problems in (16.46) and (16.47) have optimal solutions. If one problem has a solution, then the other problem also has a solution. We prefer to solve those problems that have lesser number of constraints.

Example 16.21 Solve the following game using the method of linear programming:

Player B

$$\text{Player A} \begin{bmatrix} 4 & -1 \\ 3 & 5 \end{bmatrix}$$

Solution The game matrix (pay-off matrix of player A) is as shown in 16.34.

Table 16.34 Pay-off matrix

B→ A↓	B_1	B_2	α_i
A_1	4	−1	−1
A_2	3	5	3
β_j	4	5	

The lower value of the game $\underline{a} = \max(-1, 3) = 3$ and the upper value of the game $\overline{a} = \min(4, 5) = 4$. Therefore, the value of the game lies between 3 and 4.

Let us assume that player A uses strategies A_i's with probabilities p_i, $\sum p_i = 1$, $i = 1, 2$ and player B uses strategies B_j's with probabilities q_j's, $\sum q_j = 1, j = 1, 2$.

Player B's problem is given as Max $Z_B = y_1 + y_2 = 1/V$

$$\begin{aligned} \text{Subject to} \quad & 4y_1 - y_2 \le 1 \\ & 3y_1 + 5y_2 \le 1 \\ & y_j \ge 0,\ y_j = q_j/V,\ j = 1, 2 \end{aligned}$$

$$\begin{array}{cccrr} & & & y_1 & y_2 \\ & & & q_1 & q_2 \\ & & & B_1 & B_2 \\ x_1 & p_1 & A_1 & 4 & -1 \\ x_2 & p_2 & A_2 & 3 & 5 \end{array}$$

Introducing slack variables y_3 and y_4, the standard form obtained is as follows:

$$\begin{aligned} \text{Max } Z_B = 1/V\ & y_1 + y_2 + 0 \cdot y_3 + 0 \cdot y_4 \\ \text{Subject to} \quad & 4y_1 - y_2 + y_3 = 1 \\ & 3y_1 + 5y_2 + y_4 = 1 \\ & y_j \ge 0 \end{aligned}$$

Using the simplex method, we get

		Cost →		1	1	1	0	Minimum ratio
		Variable →		y_1	y_2	y_3	y_4	
Table no.	C_B	Basic variable	X_B	α^1	α^2	α^3	α^4	$\frac{X_{B_i}}{\alpha_i^j}$
1	0	y_3	1	4	−1	1	0	…
	0	y_4	1	3	[5]	0	1	1/5→
		$c_j - Z_j$		1	1↑	0	0	
2	0	y_3	6/5	[3/5]	0	1	1/5	6/23→
	1	y_2	1/5	3/5	1	0	1/5	1/3
		$c_j - Z_j$		2/5↑	0	0	−1/5	
3	1	y_1	6/23	1	0	5/23	1/23	
	1	y_2	1/23	0				
		$c_j - Z_j$		0	0	−2/23	−5/23	

This is an optimal table. Hence, $y_1 = 6/23$, $y_2 = 1/23$

Now, we have $1/V = y_1 + y_2 = 6/23 + 1/23 = 7/23 \Rightarrow V = 23/7$

The actual value is 23/7.

Using $y_j = q_j/V$, we arrive at the following values for q_1 and q_2:

$$q_1 = 23/7 \times 6/23 = 6/7,\ q_2 = 23/7 \times 1/23 = 1/7.$$

Thus, player B should use strategies B_1 and B_2 with probabilities $q_1 = 6/7, q_2 = 1/7$

A_i's best strategies appear in the $c_j - Z_j$ row under slack variables y_3 and y_4. Thus,

$$x_1 = 2/23, \quad x_2 = 5/23$$

Using $x_i = p_i/V$, we have $p_1 = 23/7 \times 2/23 = 2/7, p_2 = 23/7 \times 5/23 = 5/7$.

Example 16.22 Solve the following game using the linear programming (LP) method:

$$\begin{array}{cc} & \text{B} \\ & \begin{array}{ccc} B_1 & B_2 & B_3 \end{array} \\ \text{A} \begin{array}{c} A_1 \\ A_2 \\ A_3 \end{array} & \begin{bmatrix} 1 & 1 & 1 \\ 2 & -2 & 2 \\ 3 & 3 & -3 \end{bmatrix} \end{array}$$

Solution The game matrix (pay-off matrix of player A) is as follows:

A \ B	B_1	B_2	B_3	α_i
A_1	1	1	1	1
A_2	2	−2	2	−2
A_3	3	3	−3	−3
β_j	3	3	2	

The lower value of the game $\underline{a} = \max(1, -2, -3) = 1$ and the upper value of the game $\overline{a} = \min(3, 3, 2) = 2$. Therefore, the value of the game lies between 1 and 2.

Let player A use strategies A_i's with probabilities p_i, $\sum p_i = 1$, $i = 1, 2, 3$ and player B uses strategies B_j's with probabilities q_j's, $\sum q_j = 1, j = 1, 2, 3$.

$$\begin{array}{ccc} & & \begin{array}{ccc} y_1 & y_2 & y_3 \\ q_1 & q_2 & q_3 \\ B_1 & B_2 & B_3 \end{array} \\ \begin{array}{c} x_1 \\ x_2 \\ x_3 \end{array} & \begin{array}{c} p_1 \\ p_2 \\ p_3 \end{array} \begin{array}{c} A_1 \\ A_2 \\ A_3 \end{array} & \begin{bmatrix} 1 & 1 & 1 \\ 2 & -2 & 2 \\ 3 & 3 & -3 \end{bmatrix} \end{array}$$

Player B's problem is formulated as follows:

$$\text{Max } Z_B = y_1 + y_2 + y_3 = \frac{1}{V}$$

$$\begin{aligned} \text{Subject to } \quad & y_1 + y_2 + y_3 \le 1 \\ & 2y_1 - 2y_2 + 2y_3 \le 1 \\ & 3y_1 + 3y_2 - 3y_3 \le 1 \\ & y_j \ge 0, y_j = q_j/V, j = 1, 2, 3 \end{aligned}$$

Introducing slack variables y_4, y_5, and y_6, the standard form is as follows:

$$\text{Max } Z_B = \frac{1}{V} = y_1 + y_2 + y_3 + 0 \cdot y_4 + 0 \cdot y_5 + 0 \cdot y_6$$

$$\begin{aligned} \text{Subject to } \quad & y_1 + y_2 + y_3 + y_4 = 1 \\ & 2y_1 - 2y_2 + 2y_3 + y_5 = 1 \\ & 3y_1 + 3y_2 - 3y_3 + y_6 = 1 \\ & y_j \ge 0. \end{aligned}$$

Using the simplex method, we get the following table:

		Cost →		1	1	1	0	0	0	Minimum ratio
		Variable →		y_1	y_2	y_3	y_4	y_5	y_6	
Table no.	C_B	Basic variable	X_B	α^1	α^2	α^3	α^4	α^5	α^6	$\frac{X_{B_i}}{\alpha_i^j}$
1	0	y_4	1	1	1	1	1	0	0	1
	0	y_5	1	2	−2	[2]	0	0	1	1/2→
	0	y_6	1	3	3	−3	0	0	1	…
		$c_j - Z_j$		1	1	1↑	0	0	0	
2	0	y_4	1/2	0	[2]	0	1	−1/2	0	1/4→
	1	y_3	1/2	1	−1	1	0	1/2	0	…
	0	y_6	5/2	6	0	0	0	3/2	1	…
		$c_j - Z_j$		0	2↑	0	0	−1/2	0	
3	1	y_2	1/4	0	1	0	1/2	−1/4	0	
	1	y_3	3/4	1	0	1	1/2	1/4	0	
	0	y_6	5/2	6	0	0	0	3/2	1	
		$c_j - Z_j$		0	0	0	−1	0	0	

This is an optimal table. Hence, $y_1 = 0, y_2 = 1/4$, and $y_3 = 3/4$

Now, we have $1/V = y_1 + y_2 + y_3 = 0 + 1/4 + 3/4 = 1 \Rightarrow V = 1$

The actual value is 1. Using $y_j = q_j/V$, we have

$$q_1 = 0 \times 1 = 0, \quad q_2 = \frac{1}{4} \times 1 = \frac{1}{4}, \quad q_3 = \frac{3}{4} \times 1 = \frac{3}{4}$$

Thus, player B should use strategies B_1, B_2, and B_3 with probabilities $q_1 = 0,\ q_2 = 1/4,\ q_3 = 3/4$.

A_i's best strategies appear in the $c_j - Z_j$ row under slack variables y_4, y_5, and y_6. Thus, $x_1 = 1$, $x_2 = 0$, and $x_3 = 0$.

Using $x_i = p_i/V$, we have $p_1 = 1 \times 1 = 1$, $p_2 = 0 \times 1 = 0$, and

$$p_3 = 0 \times 1 = 0.$$

Example 16.23 Solve the following game using the LP method:

$$\begin{array}{cc} & \text{B} \\ & \begin{array}{ccc} B_1 & B_2 & B_3 \end{array} \\ \text{A} \begin{array}{c} A_1 \\ A_2 \\ A_3 \end{array} & \begin{bmatrix} 3 & -4 & 2 \\ 1 & -3 & -7 \\ -2 & 4 & 7 \end{bmatrix} \end{array}$$

Table 16.35 Pay-off matrix

A \ B	B_1	B_2	B_3	α_i
A_1	3	−4	2	−4
A_2	1	−3	−7	−7
A_3	−2	4	7	−2
β_j	3	4	7	

Solution The game matrix (pay-off matrix of player A) is as shown in Table 16.35.

The lower value of the game $\underline{a}$ = max(−4, −7, −2) = −2 and the upper value of the game $\overline{a}$ = min(3, 4, 7) = 3.

The value of the game lies between −2 and 3. It is possible that the value of the game V may be negative or zero. So, let a constant k=3 (k must be ≥ 2) be added to all the elements of the matrix to ensure that the game value of the revised matrix ≥ 0. The modified game matrix is

$$\begin{array}{cc} & \begin{array}{ccc} B_1 & B_2 & B_3 \end{array} \\ \begin{array}{c} A_1 \\ A_2 \\ A_3 \end{array} & \begin{bmatrix} 6 & -1 & 5 \\ 4 & 0 & -4 \\ 1 & 7 & 10 \end{bmatrix} \end{array}$$

Let player A use strategies A_i's with probabilities p_i, $\sum p_i = 1$, $i = 1, 2, 3$ and player B use strategies B_j's with probabilities q_j's, $\sum q_j = 1, j = 1, 2, 3$.

$$\begin{array}{cccc} & y_1 & y_2 & y_3 \\ & q_1 & q_2 & q_3 \\ & B_1 & B_2 & B_3 \\ x_1\ p_1\ A_1 & 6 & -1 & 5 \\ x_2\ p_2\ A_2 & 4 & 0 & -4 \\ x_3\ p_3\ A_3 & 1 & 7 & 10 \end{array}$$

Player B's problem is formulated as follows:

$$\text{Max } Z_B = y_1 + y_2 + y_3 = 1/V$$
$$\text{Subject to } \quad 6y_1 - y_2 + 5y_3 \leq 1$$
$$4y_1 + 0 \cdot y_2 - 4y_3 \leq 1$$
$$y_1 + 7y_2 + 10y_3 \leq 1$$
$$y_j \geq 0,\ y_j = q_j/V,\ j = 1, 2, 3$$

Introducing slack variables y_4, y_5, and y_6, we get the following:

$$\text{Max } Z_B = \frac{1}{V} = y_1 + y_2 + y_3 + 0 \cdot y_4 + 0 \cdot y_5 + 0 \cdot y_6$$
$$\text{Subject to } \quad 6y_1 - y_2 + 5y_3 + y_4 = 1$$
$$4y_1 + 0 \cdot y_2 - 4y_3 + y_5 = 1$$
$$y_1 + 7y_2 + 10y_3 + y_6 = 1$$
$$y_j \geq 0.$$

Using the simplex method, we get the following table:

Cost →				1	1	1	0	0	0	Minimum ratio
Variable →				y_1	y_2	y_3	y_4	y_5	y_6	$\frac{X_{B_i}}{\alpha_i^j}$
Table no.	C_B	Basic variable	X_B	α^1	α^2	α^3	α^4	α^5	α^6	
1	0	y_4	1	6	−1	5	1	0	0	...
	0	y_5	1	4	0	−4	0	1	0	...
	0	y_6	1	1	[7]	10	0	0	1	1/7→
		$c_j - Z_j$		1	1↑	1	0	0	0	
2	0	y_4	8/7	[43/7]	0	45/7	1	0	1/7	8/43→
	0	y_5	1	4	0	−4	0	1	0	1/4
	1	y_2	1/7	1/7	1	10/3	0	0	1/7	1
		$c_j - Z_j$		6/7↑	0	−3/7	0	0	−1/7	
3	1	y_1	8/43	1	0	45/43	7/43	0	1/43	
	0	y_5	11/43	0						
	1	y_2	5/43	0						
		$c_j - Z_j$		0	0	−57/43	−6/43	0	−7/43	

This is an optimal table. Hence, $y_1 = 8/43$, $y_2 = 5/43$, and $y_3 = 0$.

Now, we have $1/V = y_1 + y_2 + y_3 = 8/43 + 5/43 + 0 = 13/43 \Rightarrow V = 43/13$

The actual value is $V-3=43/13-3=4/13$.

Using $y_j=q_j/V$, we have $q_1=8/43\times 43/13=8/13$, $q_2=5/43\times 43/13=5/13$, $q_3=0\times 43/13=0$.

Thus, player B should use strategies B_1, B_2, and B_3 with probabilities $q_1=8/13, q_2=5/13$, and $q_3=0$.

A_i's best strategies appear in the $c_j - Z_j$ row under slack variables y_4, y_5, and y_6. Thus,

$$x_1 = 6/43, x_2 = 0, \text{ and } x_3 = 7/43.$$

Using $x_i = p_i/V$ we have

$$p_1 = \frac{6}{43}\times\frac{43}{13}=\frac{6}{13},\ p_2 = 0\times\frac{43}{13}=0, \text{ and } p_3 = \frac{7}{43}\times\frac{43}{13}=\frac{7}{13}.$$

16.12 LIMITATIONS OF GAME THEORY

In spite of the usefulness of the game theory, it has some limitations, which restrict its popularity. A few of these limitations are as follows:

1. The assumptions of maximin and minimax criterions show that the players always know each other's strategies and have complete awareness about the risks involved. The game theory therefore, does not seem to be practical.
2. As the number of players increase in a game, the analysis of game strategies becomes complicated and thus, the game theory is not so helpful in such situations.
3. The players have knowledge about each other's strategies. Thus, sharing the secrets of business and colluding with each other is not healthy.

RECAPITULATION

1. A game can be defined as a conflict of interests between two or more opponents.
2. A game can either be a two-person or a multiperson game.
3. Two-person zero-sum games are also known as rectangular games as they are represented by a pay-off matrix in rectangular form.
4. A saddle point of a pay-off matrix is the position of such an element in the pay-off matrix, which is minimum in its row and maximum in its column.
5. When only a single alternative is used in the game, it is called pure strategy.
6. If several alternatives with different strategies are used to play the game, then the game is known as a game with mixed strategies.
7. A game is said to be fair if both the lower and upper values of the game are equal to zero.
8. The principle of dominance is particularly useful for the evaluation of two-person zero-sum games where a saddle point does not exist. It reduces the size of the pay-off matrix and thus facilitates in obtaining the solution to the problem quickly.
9. While solving a game problem, it is not necessary that the saddle point will always exist. We use mixed strategies to determine the value of the game.

(Contd)

(*Contd*)

10. In the algebraic method for any two-person zero-sum game where optimal strategies are not pure (i.e., there are no saddle points), player A's pay-off matrix, as shown gives the optimal strategies (x_1, x_2) of A and (y_1, y_2) of B and the value of the game V are given by

$$\frac{x_1}{x_2} = \frac{a_{22} - a_{21}}{a_{11} - a_{21}}, \quad \frac{y_1}{y_2} = \frac{a_{22} - a_{12}}{a_{11} - a_{21}} \text{ and } V = \frac{a_{11}a_{22} - a_{12}a_{21}}{a_{11} + a_{22} - a_{12} - a_{21}}.$$

11. Arithmetic method provides us an easy way of obtaining the optimum strategies for each player A and B in 2×2 games without the saddle points.
12. In $(2 \times n)$ games, where $n > 2$, consider the set of all (2×2) submatrices of $(2 \times n)$ matrix for their solution. As $n > 2$, player B has more flexibility to play any two out of n available courses of action. B would like to play those two courses of action for which the loss (gain obtained by A) is minimum. Thus, the solution of that subgame (2×2) for which game value is minimum will be considered as the optimum solution of the given (original) game.
13. In case of $(m \times 2)$ games with $m > 2$, player A will have more flexibility of using any two of the available activities. Thus in this case, the solution of (2×2) subgame out of all (2×2) subgames of $(m \times 2)$ for which game value has maximum value will be considered as the optimal solution of the given $(m \times 2)$ game.
14. Every two-person zero-sum game has a solution in mixed strategies.

EXERCISES

Multiple-choice Questions

16.1 In a two-person pure strategy game, each player plays only
(a) one strategy at a time
(b) two strategies at a time
(c) more than two strategies at a time
(d) none of these

16.2 In a mixed strategy game, each player uses
(a) a fixed strategy
(b) a random strategy
(c) both (a) and (b)
(d) none of these

16.3 A saddle point exists if
(a) maximin value = maximax value
(b) minimax value = minimin value
(c) minimax value = maximin value
(d) none of these

16.4 A game is said to be fair if both the lower and upper values of the game are
(a) equal to zero
(b) greater than zero
(c) less than zero
(d) none of these

16.5 Dominance principle in game theory
(a) reduces the size of the pay-off matrix
(b) reduces the game value
(c) both (a) and (b)
(d) none of these

16.6 Dominance principle in game theory
(a) facilitates in obtaining the solution to the problem
(b) increases the game value
(c) decreases the game value
(d) none of these

State True or False

16.1 In a pure strategy game, each player can use a combination of the available strategies.
16.2 In a mixed strategy game, each player uses only one strategy at a time.
16.3 If the maximin value is equal to the minimax value, then the game is said to have a saddle point.
16.4 In a mixed strategy game, no saddle point exists.
16.5 If the upper and lower values of the game are the same but not equal to zero, then the game is said to be fair.
16.6 Game theory is the study of selecting optimal strategies.
16.7 If each element of a pay-off matrix is multiplied by a constant C, then the game value becomes C times the actual value.
16.8 If a constant C is added to each element of a pay-off matrix, then the optimal strategies of a player can change.
16.9 In the dominance principle, if a column dominates, it is removed to reduce the size of the pay-off matrix without affecting the game value.
16.10 In the dominance principle, if a row dominates, it is retained and the dominated row is removed to reduce the size of the pay-off matrix without affecting the game value.

Fill in the Blanks

16.1 In a pure strategy game, each player uses __________ strategy.
16.2 In a mixed strategy game, each player uses __________ strategies.
16.3 If the maximin value is __________ to the minimax value, then the game value exists.
16.4 A game is said to be __________ if both the lower and upper values of the game are equal to zero.
16.5 When maximin and minimax value of the game are __________, then the saddle point exists.
16.6 In game theory, if every entry of a row is greater than or equal to the corresponding entries of another row, then it is called __________ row.

Review Questions

16.1 Define the following terms:
(a) Pay-off matrix
(b) Pure and mixed strategies
(c) Saddle point
(d) Optimal strategies
(e) Rectangular game
16.2 Define saddle point and value of the game with examples. Is it necessary that a game always has a saddle point?
16.3 Describe a two-person zero-sum game.
16.4 Show that a game can be formulated in the form of an LPP.
16.5 Describe the maximin principle of the game theory. What do you understand by the terms pure strategy and saddle point?
16.6 What is the difference between the saddle point and value of the game?
16.7 Explain the theory of dominance in the solution of rectangular games.
16.8 How can a game problem be solved when (a) a saddle point exists and (b) a saddle point doesn't exist.
16.9 Explain the algebraic method for solving a rectangular game.
16.10 What are the limitations of the game theory?

Numerical Problems

16.1 Find the lower and upper values of the game and check for the existence of saddle points for the following game matrices (pay-off matrices of player A):

(a) Player A (rows) vs Player B (columns): $\begin{bmatrix} 10 & 6 \\ 8 & 2 \end{bmatrix}$

(b) Player A (rows) vs Player B (columns): $\begin{bmatrix} 3 & 0 \\ 0 & 2 \end{bmatrix}$

(c) Player A (rows) vs Player B (columns): $\begin{bmatrix} -5 & 2 \\ -7 & -4 \end{bmatrix}$

(d) Player A (rows) vs Player B (columns): $\begin{bmatrix} 5 & 4 \\ 3 & 7 \end{bmatrix}$

(e) Player A (rows) vs Player B (columns): $\begin{bmatrix} 7 & 10 \\ 5 & 6 \end{bmatrix}$

16.2 Find the optimum strategies and the value of the game using the algebraic method for the following pay-off matrix:

Player A (rows) vs Player B (columns): $\begin{bmatrix} 5 & 1 \\ 3 & 4 \end{bmatrix}$

16.3 Use the graphical method to solve the following games:

(a) Player A vs Player B

	B_1	B_2	B_3	B_4
A_1	2	1	0	−2
A_2	1	0	3	2

(b) Player A vs Player B

	B_1	B_2
A_1	1	−3
A_2	3	5
A_3	−1	6
A_4	4	1
A_5	2	2
A_6	−5	0

16.4 Solve the following game:

Player A vs Player B

	B_1	B_2	B_3	B_4
A_1	3	2	4	0
A_2	3	4	2	4
A_3	4	2	4	0
A_4	0	4	0	8

16.5 Solve the following 3×3 game:

Player A vs Player B

	B_1	B_2	B_3
A_1	3	−2	4
A_2	−1	4	2
A_3	2	2	6

16.6 Solve and determine the value of the following game:

$$\text{Player A}\overset{\text{Player B}}{\begin{bmatrix} 4 & -4 \\ -4 & 4 \end{bmatrix}}$$

16.7 Obtain the optimal strategies for player A and B for the following game using the LP method:

$$\text{Player A}\begin{matrix} A_1 \\ A_2 \\ A_3 \end{matrix}\overset{\begin{matrix}\text{Player B} \\ B_1 \quad B_2 \quad B_3\end{matrix}}{\begin{bmatrix} 1 & -1 & -1 \\ -1 & -1 & 3 \\ -1 & 2 & -1 \end{bmatrix}}$$

16.8 Solve the following games using the LP method:

(a)

$$\text{Player A}\begin{matrix} A_1 \\ A_2 \end{matrix}\overset{\begin{matrix}\text{Player B} \\ B_1 \quad B_2\end{matrix}}{\begin{bmatrix} 1 & -1 \\ -1 & 1 \end{bmatrix}}$$

(b)

$$\text{Player A}\begin{matrix} A_1 \\ A_2 \\ A_3 \end{matrix}\overset{\begin{matrix}\text{Player B} \\ B_1 \quad B_2 \quad B_3\end{matrix}}{\begin{bmatrix} 1 & -1 & 3 \\ 3 & 5 & -3 \\ 6 & 2 & -2 \end{bmatrix}}$$

CONCEPTUAL PROBLEM

Assume the pay-off matrix of two persons in a game is an important issue. The pay-off matrix can be solved (explained) with the following case study. Two computer manufacturers C_1 and C_2, are attempting to sell computer systems to two banks, 1 and 2. Company C_1 has four salesmen, whereas company C_2 has only three available. The computer companies must decide on the number of salesmen to be assigned to sell at each bank. Thus, company C_1 can assign four salesmen to bank 1 and none to bank 2 or three to bank 1 and one to bank 2, and so on. Each bank will buy one computer system. The probability that a bank will buy a computer from a particular company is directly related to the number of salesmen calling from that company relative to the total salesmen calling. Thus, if company C_1 assigned three salesmen to bank l and company C_2 assigned two salesman, the odds would be 3 out of 5 that bank 1 would purchase company C_1's computer system. If none calls from either of the companies, the odds are 50 per cent for buying either company's computer. Let the pay-off be the expected number of computer systems that company C_1 sells. What stategy would company C_1 use in allocating its salesmen? What is the value of the game to company C_1.

(Contd)

(*Contd*)

		C_2				α_i
		0	**1**	**2**	**3**	
	0	0	0	0	0	0
	1	1	1/2	3/4	1/4	1/4
C_1	2	1	2/3	1/2	2/5	2/5
	3	1	3/4	3/5	1/2	1/2
	4	1	4/5	2/3	4/7	4/7
β_j		1	4/5	2/3	4/7	

Here maximin = minimax = 4/7. Thus the saddle point exists, the game value is 4/7, and the game is said to have a pure strategy. Therefore, the optimum strategy for company C_1 and C_2 is deputing 4 and 3 salesmen respectively.

17 Decision Theory—Analysis

After studying this chapter, the reader will be able to understand

- the decision models
- decision-making under certainty
- decision-making under risk
- decision-making under uncertainty
- posterior probability and Bayesian analysis
- decision-making under conflict

17.1 INTRODUCTION

Decision analysis is a set of concepts and systematic procedures to help improve the decision-making process, particularly in problems involving uncertainty. It helps the decision-maker select the best course of action out of the available options across different types of environment. The concept of decision analysis can be best described by quoting R.H. Howard (1966)—'A decision is an irrevocable allocation of resources, irrevocable in the sense that it is impossible or extremely costly to change back to the situation that existed before making the decision'. Thus a decision is not a mental commitment to follow a course of action, but the actual pursuit of that course of action.

17.2 DECISION MODELS

Decision-making is one of the primary functions of the management of any organization, so as to determine the future course of action. Decision models can be classified into various categories. One such model is the allocation model, wherein the management has to allocate limited resources to various activities, so as to optimize the objective function under certain restrictions. Any decision model of any category has the following essential characteristics in common:

1. Decision alternatives
2. States of nature (events)
3. Pay-off

17.2.1 Decision Alternatives

The possible courses of action (strategies) available before taking a decision are called decision alternatives (or decision choices). For a given problem, all the courses of action, which maybe small or large in number, must be included. These courses of action are under the control of the decision-maker and are only known to him.

17.2.2 States of Nature or Events

The decision-maker should have an exhaustive list of all possible future events (happenings). However, he/she has no direct control over the occurrences of a particular event. Such future happenings are referred to as states of nature or events that are assumed to be mutually exclusive and collectively exhaustive.

The state of nature (event) can have a non-numeric description, such as drill or do not drill high, medium, or low, or may have a numerical description, such as the shortage or demand of some item.

17.2.3 Pay-off

It is the payment associated with a specified combination of events and strategies. Pay-offs are known as conditional values.

Let $N_1, N_2, \ldots, N_m$ be the m events and $S_1, S_2 \ldots, S_n$ be the n strategies associated with these events. Further, let a_{ij} be the pay-off corresponding to the combination of event N_i with strategy S_j; then the pay-offs can be tabulated, called the pay-off matrix, as follows:

States of nature (Event)	Courses of action (Strategies)						
	S_1	S_2	...	S_j	...	S_n
N_1	a_{11}	a_{21}	...	a_{ij}	...	a_{1n}
N_2	a_{21}	a_{22}	...	a_{2j}	...	a_{2n}
.	.	.	...	.	...	.
N_i	a_{i1}	a_{i2}	...	a_{ij}	...	a_{in}
.	.	.	...	.	...	.
N_m	a_{m1}	a_{m2}	...	a_{mj}	...	a_{mn}

Thus, we observe that the pay-off matrix for a given problem consists of the states of nature and the courses of action along with the pay-offs.

17.3 DECISION-MAKING SITUATIONS

Decisions taken by the managers fall under the following categories:

1. Decision-making under certainty
2. Decision-making under risk
3. Decision-making under uncertainty
4. Decision-making under conflict

17.3.1 Decision-making Under Certainty

In decision-making under certainty, the data is assumed to be known deterministically. So, decision-making under certainty is a deterministic problem. For example, the linear programming problem (LPP) falls in the category of decision-making under certainty, where the amount of resources required and the corresponding unit profit/cost is known with certainty. Other techniques used for solving problems under certainty are the transportation, assignment, goal programming, and inventory models.

17.3.2 Decision-making Under Risk

In this category, the information regarding the states of nature is probabilistic, and hence, the decision-maker is not sure about the pay-off that will occur as a consequence of the selection of a particular course of action. Further, every course of action has several outcomes; it is difficult to find the exact monetary pay-off for various combinations of courses of action and the states of nature. However, the past records or data enable the decision-maker to assign the probability p_i to the likely possible occurrence of each state of nature N_i. Thus after knowing the probability distribution of the states of nature, the best decision is taken by selecting that course of action that has the largest expected pay-off. In short, we can say that in decision-making under risk, there exists several states of nature with which the decision-maker can attach probabilistic values on the basis of the past records available to him. Having known the probability distribution of the states of nature, the selection of that course of action that has the largest expected pay-off value is the best decision.

The criteria for evaluating the courses of action to take a decision under risk are expected monetary value (EMV), expected opportunity loss (EOL), expected profit of perfect information (EPPI), expected value of perfect information (EVPI), and conditional opportunity loss (COL). Each of these has been discussed in the succeeding subsections.

Expected Monetary Value

The EMV for a given course of action is the sum of the products of pay-offs of that course of action with the corresponding probabilities attached with the states of nature. Mathematically, EMV for a particular course of action, say S_j, can be expressed as follows:

$$\text{EMV}(S_j) = \sum_{i=1}^{m} a_{ij} p_i$$

where m = number of possible states of nature

p_i = probability of occurrence of the state of nature N_i

a_{ij} = conditional pay-off associated with the state of nature N_i and course of action S_j

The following steps can be used to calculate the EMV for various courses of action:

1. First of all, identify the courses of action and the states of nature. Calculate the conditional pay-offs attached with each possible combination of states of nature and courses of action along with the probabilities of occurrence of each state of nature. Tabulate all of these.

2. After tabulating, evaluate the EMV for each course of action by multiplying its conditional pay-offs with the corresponding probability associated with the given states of nature and adding all the values thus obtained.
3. Among the courses of action, identify the one for which the EMV obtained is optimal. The course of action thus identified will yield the optimum decision, along with its associated EMV.

Note Optimum value of EMV is denoted as EMV*.

Similar to EMV, expected opportunity loss (EOL) is another decision-making criterion for decision-making under risk. The decision results obtained under the EOL criterion will be the same as those obtained using the EMV criterion. The only difference between EMV and EOL is that in the former, monetary value is maximized and in the latter, the opportunity loss is minimized.

Before going into the details of EOL, we will explain two other terms; EPPI and EVPI.

Expected Profit with Perfect Information

Expected profit with perfect information (EPPI) is defined as the maximum monetary value that can be obtained on the basis of having perfect information about the state of nature that will occur. To calculate EPPI, the following steps should be followed.

1. Choose the optimal course of action for each state of nature with its associated conditional pay-off from the pay-off table.
2. Multiply each conditional pay-off, chosen in step 1, with the corresponding probability of occurrence of that state of nature.
3. Add all these products to get the EPPI.

Expected Value of Perfect Information

Expected value of perfect information (EVPI) is defined as the maximum amount that the decision-maker would be willing to pay to obtain the perfect information about the state of nature that would occur.

EVPI can be obtained as $\text{EVPI} = \text{EPPI} - \text{EMV}^*$

Note EMV* represents the maximum EMV, under given prior outcome probabilities with no information as to which state of nature will exactly occur. Thus, the perfect information would lift the profit value EMV* to EPPI.

Expected Opportunity Loss

The expected opportunity loss for each state S_j is denoted as EOL (S_j) and is defined as follows:

$$\text{EOL}(S_j) = \sum_{i=1}^{n} \text{COL}(N_i, S_j)p_i$$

where $\text{COL}(N_i, S_j)$ = conditional opportunity loss due to the state of nature N_i and course of action S_j, and p_i = probability of occurrence of the state of nature N_i.

Conditional Opportunity Loss

Conditional opportunity loss (COL) is defined as the difference between the highest pay-off value for a particular state of nature and the actual pay-off obtained for the particular course of action. The COL with the state of nature N_i and the course of action S_j is denoted as COL (N_i, S_j). The COL can be obtained for each course of action. We illustrate all these criteria used in decision-making under risk through examples given here.

Example 17.1 A person who just retired from service has received a big provident fund (PF) amount. He wants to invest this amount in the share market. A broker offers him the following investment alternatives and percentage return rates:

Investment	Market conditions (Return rates)		
	Low	Medium	High
Regular shares	–5%	10%	15%
Risky shares	–8%	12%	20%
Property	–10%	15%	25%

Over the past 200 days, the market conditions have had medium return rates for 100 days and high return rates for 40 days.

On the basis of the given data, as a decision-making policy, answer the following.

(a) Identify the states of nature and find the probabilities of their occurrence.
(b) Identify the courses of action.
(c) Write the pay-off matrix and find the EMV for each course of action and state the optimum investment policy for the investment.

Solution (a) There are three states of nature—low, medium, and high market conditions—and are not under the control of the decision-maker. Further over 200 days, low, medium, and high market conditions prevailed for 60, 100, and 40 (200–100–60) days respectively. So, the associated probabilities associated with states of nature—low, medium, and high are 0.3, 0.5, and 0.2 respectively.

(b) Here the courses of action are the investment alternatives—regular shares, risky shares, or properties—which are in the hands of the investor.

(c) The pay-off matrix is as follows:

States of nature (Market conditions)	Probability	Courses of action (Strategies)		
		S_1 Regular shares	S_2 Risky shares	S_3 Property
Low	0.30	–0.05	–0.08	–0.10
Medium	0.50	0.10	0.12	0.15
High	0.20	0.15	0.20	0.25

$$\text{EMV}(S_1) = -0.05 \times 0.3 + 0.10 \times (0.50) + 0.15 \times (0.20)$$
$$= -0.015 + 0.050 + 0.030 = 0.080 - 0.015 = 0.065 = 6.5\%$$
$$\text{EMV}(S_2) = -0.08 \times 0.3 + 0.12 \times 0.5 + 0.20 \times 0.2 = -0.024 + 0.060 + 0.040$$
$$= 0.100 - 0.024 = 0.076 = 7.6\%$$

$$\text{EMV}(S_3) = -0.10 \times 0.3 + 0.15 \times 0.50 + 0.25 \times 0.20$$
$$= -0.030 + 0.0750 + 0.050 = 0.125 - 0.030 = 0.095 = 9.5\%$$

Since the expected return of 9.5% is the highest for property, the investor should invest in property as it would be the optimum investment policy.

Example 17.2 An oil exploration company is expanding its business and has to decide whether or not to drill for oil at a particular location. The results of drilling are classified into three categories—high yield worth 12 units, moderate yield with 6 units, and no oil. The drilling operation costs 5 units. At a similar location, 50, 30, and 20 per cent of previous drillings have given high, moderate, and no yield respectively. Using the EMV concept, identify the best course of action.

Solution There are three states of nature here–N_1 (high yield), N_2 (moderate yield), and N_3 (no yield). All of these are not under the control of the decision-maker.

There are two courses of action, S_1 (drill) and S_2 (do not drill). The probabilities of occurrence of N_1, N_2, and N_3 are 0.5, 0.3, and 0.2 respectively.

The conditional profit values (pay-offs) for each pair of courses of action and state of nature combinations are given in the following table.

States of nature N_i	Courses of action	
	Drill (S_1)	Do not drill (S_2)
N_1 (high yield)	$12 - 5 = 7$	0
N_2 (moderate yield)	$6 - 5 = 1$	0
N_3 (no yield)	$0 - 5 = -5$	0

The EMV for S_1 and S_2 is as follows:

States of nature	Probabilities	Courses of action		EMV	
		S_1	S_2	S_1	S_2
N_1	0.5	7	0	0.5×7	0.5×0
N_2	0.3	1	0	0.3×1	0.3×0
N_3	0.2	–5	0	0.2×-5	0.2×0
EMV				2.8	0

Thus, $\text{EMV}(S_1) = 2.8$ and $\text{EMV}(S_2) = 0$

$$\text{EMV}^* = \text{EMV}(S_1)$$

Hence, the best course of action is S_1. Thus, the company should opt for drilling to have a maximum pay-off worth 2.8 units.

Example 17.3 A bread seller purchases bread packets every day at ₹23 per packet and sells them at ₹25. The packets unsold at the end of the day are disposed of at ₹22. The sales in the past have ranged from 10 to 14 packets per day and the record of sales for the past 100 days is as follows:

Packets sold	10	11	12	13	14
Number of days	10	20	40	20	10

Using the decision theory, find out how many packets should the bread seller purchase every day to have the maximum profit.

Solution Let N_i ($i = 1, 2, 3, 4, 5$) be the daily demand (the states of nature) and S_j ($j = 1, \ldots, 5$) be the courses of action.

Marginal profit (MP) per bread packet sold = Selling price − Cost price = 25 − 23 = ₹2

Marginal loss (ML) per bread packet unsold = 23 − 22 = ₹1

Conditional profit = MP × Number of packets sold − ML × Number of packets unsold

$$= 2S - 1N = 2S - N$$

where S = number of bread packets sold

N = number of bread packets not sold

The conditional profit values are shown in the following pay-off matrix.

States of nature (Demand/ day)	Probability	Courses of action (Purchase/day) conditional profit value					Courses of action (Purchase/day) expected profit in ₹				
		10	**11**	**12**	**13**	**14**	**10**	**11**	**12**	**13**	**14**
	(1)	**(2)**	**(3)**	**(4)**	**(5)**	**(6)**	**(1) × (2)**	**(1) × (3)**	**(1) × (4)**	**(1) × (5)**	**(1) × (6)**
10	10/100 = 0.1	20	19	18	17	16	2	1.9	1.8	1.7	1.6
11	20/100 = 0.2	20	22	21	20	19	4	4.4	4.2	4.0	3.8
12	40/100 = 0.4	20	22	24	23	22	8	8.8	9.6	9.2	8.8
13	20/100 = 0.2	20	22	24	26	25	4	4.4	4.8	5.2	5.0
14	10/100 = 0.1	20	22	24	26	28	2	2.2	2.4	2.6	2.8
EMV							20.0	21.7	22.8	22.7	22.0

The highest EMV of ₹22.8 corresponds to course of action 12. Therefore, the bread seller must purchase 12 packets of bread every day.

17.3.3 Decision-making Under Uncertainty or Fuzzy Environment

When the decision-maker has no knowledge about the probabilities of occurrence of the states of nature, he/she must base his/her decisions on the actual conditional pay-off together with a policy. In such a situation, different decision-making criteria may come to the fore. These will be discussed in the succeeding subsections.

Maximax or Minimin Decision Criterion

This criterion is also called the criterion of optimism (or optimistic decision criterion), where the decision-maker ensures that the opportunity to achieve the greatest pay-off (or the lowest possible cost) is not missed. Thus, the selection of the courses of action is made so as to obtain the maximum of the maxima in case of profit and minimum of the minima in the case of cost or loss pay-offs.

The functioning of this criterion is given in the following steps:

1. Locate the maximum (or minimum) pay-off value for profit (or loss) corresponding to each course of action and write it down at the bottom of its column.
2. Locate the maximum for profit (or minimum for cost or loss) among the pay-offs obtained in step 1 and identify the corresponding course of action that will yield the optimum decision.

Minimax or Maximin Decision Criterion

This criterion is also called Wald's criterion or the pessimistic decision criterion, where the decision-maker ensures to earn no less (or pay no more) than a specific amount. Thus, the selection among the courses of action is made on the basis that yield the maximum of the minima in case of profit (or minimum of the maxima in case of cost or loss) pay-offs.

The working rules of the criterion are as follows:

1. Locate the minimum pay-off in case of profit corresponding to each course of action (or maximum in the case of cost or loss).
2. Locate the maximum for profit (or minimum for the cost) among those evaluated in step 1 and identify the corresponding optimum course of action.

Equally Likely Decisions or Laplace Criterion

When the probabilities of states of nature are not known, all the states of nature are assumed to occur with equal probability. This is also known as the criterion of insufficient reason.

The working rules for this criterion are summarized as follows:

1. Assign equal probabilities to each state of nature and find the expected pay-off corresponding to each course of action.
2. Locate the best expected pay-off (maximum for profit and minimum for cost or loss) and identify the corresponding course of action yielding the optimum decision.

Hurwitz Criterion or Criterion of Realism

Hurwitz, who suggested this criterion, introduced the idea of a coefficient of optimism, denoted by α to measure the decision-maker's degree of optimism. This α lies between 0 and 1, where 0 represents a completely pessimistic attitude and 1 a completely optimistic attitude of the decision-maker about the state of nature. If α is the coefficient of optimism, then $(1 - \alpha)$ is the coefficient of pessimism.

Hurwitz approach suggests that the decision-maker must select a course of action that maximizes

$$H(S_j) = \alpha(\text{maximum in column } S_j) + (1 - \alpha)(\text{minimum in columns } S_j)$$

of the pay-off matrix.

The working rule for this criterion is explained as follows:

1. Identify the coefficient of optimism α. It is either given or has to be determined by finding $(1 - \alpha)$, the coefficient of pessimism.
2. For each course of action S_j, select the largest and the lowest pay-off value and multiply these with α and $(1 - \alpha)$ values and add to get $H(S_j)$ given as

$$H(S_j) = \alpha(\text{largest pay-off}) + (1 - \alpha)(\text{lowest pay-off})$$

The course of action S_j for which $H(S_j)$ is maximum will yield the optimum decision.

Criterion of Regret or Savage Criterion

This criterion is also called the opportunity loss decision criterion or the minimax regret decision criterion. The decision-maker regrets after adopting a wrong course of action, resulting in an opportunity loss of pay-offs that he/she would have liked to minimize.

The working procedure for this criterion is as follows:

1. Write down the pay-off matrix (conditional profit or cost pay-off table) from the given description of the problem.
2. Prepare a conditional opportunity loss (or regret) table where regret pay-off is obtained by taking the difference between the maximum pay-off of a course of action and some other courses of action resulting from a particular state of nature.
3. Find the maximum regret for each course of action. Then choose that course of action with the minimum of these maximum values as the optimum decision.

Example 17.4 The conditional pay-offs (profit in ₹) of different courses of action S_1, S_2, and S_3 against the states of nature (events) N_1, N_2, N_3, and N_4 are given in the adjoining table.

Indicate the decision taken under the following approaches:

(a) Pessimistic (b) Optimistic (c) Equal probability (d) Regret and (e) Hurwitz criterion. The degree of optimism is given as $\alpha = 0.7$.

States of nature	Courses of action		
	S_1	S_2	S_3
N_1	4000	20,000	20,000
N_2	–100	5000	15,000
N_3	6,000	400	–2000
N_4	18,000	0	1000

Solution (a) *Pessimistic criterion*: Under this criterion, the minimum pay-off in each course of action is taken and then the maximum of these minima is taken as the optimum decision. The following table indicates the minimum (the worst) outcome associated with each course of action:

States of nature	Courses of action		
	S_1	S_2	S_3
N_1	4000	20,000	20,000
N_2	–100	5000	15,000
N_3	6000	400	–2000
N_4	18,000	0	1000
Minimum pay-off	–100	0	–2000

Now maximum(–100, 0, –2000) = 0 for S_2. Thus, the optimum course of action is S_2.

(b) *Optimistic criterion*: Under the optimistic decision criterion, the maximum of pay-offs from each course of action is taken and then the maximum of these maxima is taken as the optimum decision. The following table indicates the maximum (the best) outcome associated with each course of action:

States of nature	Courses of action		
	S_1	S_2	S_3
N_1	4000	20,000	20,000
N_2	−100	5000	15,000
N_3	6000	400	−2000
N_4	18,000	0	1000
Maximum pay-off	18,000	20,000	20,000

Now, maximum(18,000, 20,000, 20,000) = 20,000 for S_2 and S_3.
Thus, both S_2 and S_3 are the optimum courses of action.

(c) *Equal probability (Laplace) criterion*: Here, we find the expected pay-off of each course of action as given here, $p_1 = p_2 = p_3 = p_4 = 1/4$.

Courses of action	Expected return
S_1	$\text{Expected pay-off} = \frac{1}{4}(4000 - 100 + 6000 + 18{,}000) = \frac{27{,}900}{4} = 6975$
S_2	$\text{Expected pay-off} = \frac{1}{4}(20{,}000 + 5000 + 400 + 0) = \frac{25{,}400}{4} = 6350$
S_3	$\text{Expected pay-off} = \frac{1}{4}(20{,}000 + 15{,}000 - 2000 + 1000) = \frac{34{,}000}{4} = 85{,}000$

The expected return is maximum for S_3 and hence S_3 is the optimum decision.

(d) *Regret criterion*: From the given pay-offs, we calculate the regret pay-offs as shown in the following table.

States of nature	Conditional pay-offs—Courses of action			Regret (opportunities)—Courses of action		
	S_1	S_2	S_3	S_1	S_2	S_3
N_1	4000	20,000	20,000	16,000	0	0
N_2	−100	5000	15,000	15,100	10,000	0
N_3	6000	400	−2000	0	5600	8000
N_4	18,000	0	1000	0	18,000	17,000
	Maximum regret			16,000	18,000	17,000

The maximum regrets due to S_1, S_2, and S_3 are 16,000, 18,000, and 17,000 respectively. Now we take the minimum of these maxima. Thus, minimum(16,000, 18,000, 17,000) = 16,000 which is S_1. Therefore, S_1 is the optimum course of action.

(e) *Hurwitz criterion*: Given the coefficient of optimism $\alpha = 0.7$, the coefficient of pessimism is $(1 - \alpha) = 0.3$. Then according to Hurwitz, select the course of action that optimizes (maximize for profit and minimize for cost or loss) the pay-off value.

$H = \alpha$ (best pay-off) + $(1 - \alpha)$ (worst payoff)
$= \alpha$(maximum in column) + $(1 - \alpha)$ (minimum in column)

This is given in the adjoining table

S_1	S_2	S_3
4000	20,000	20,000
−100	5,000	15,000
6000	400	−2000
18,000	0	1,000
Max = 18,000	20,000	20,000
Min = −100	0	−2000

$H = (S_1) = 0.7 \times 18{,}000 + 0.3 \times (-100) =$ ₹12,570
$H = (S_2) = 0.7 \times 20{,}000 + 0.3 \times (0) =$ ₹14,000
$H = (S_3) = 0.7 \times 20{,}000 + 0.3 \times (-2000) =$ ₹13,400
$\text{Max}(H_i) = \text{Max}(12{,}570, 14{,}000, 13{,}400) =$ ₹14,000 $= H(S_2)$

Since course of action S_2 provides the maximum profit = ₹14,000, it is the optimum course of action (decision).

Example 17.5 The chemical division of GRASIM manufactures a product, of which a principal ingredient is chemical A. At present, the company spends ₹10,000 per month for the supply of A. However, because of its acute shortage, the price may increase fourfold. There is another chemical B that can be used along with chemical C, to impart the same effect as chemical A. Both B and C taken together cost ₹30,000 per month and their prices are unlikely to rise. Indicate the decision taken regarding the use of A or B and C under the following approaches: (a) pessimistic (b) optimistic (c) regret.

Solution There are two states of nature here: N_1 (increase in the price of A) and N_2 (no increase in the price of A). The two courses of action are S_1 (use chemicals B and C), S_2 (use chemical A).

States of nature	Courses of action	
	S_1 (use B and C)	S_2 (use A)
N_1	−30,000	−40,000
N_2	−30,000	−10,000

The given data is shown in the aforegiven table. Negative figures in the table represent the profit.

(a) Pessimistic (maximin) criterion: From the adjoining table, we see that the maximum of column minima = −30,000. Hence, the management of the company should follow action S_1, that is, use chemicals B and C.

States of nature	Courses of action	
	S_1	S_2
N_1	−30,000	−40,000
N_2	−30,000	−10,000
Column minimum	−30,000	−40,000

(b) Optimistic (maximax) criterion: In this case, consider the maximum of maxima. Details are given in the adjoining table: Maximum of maxima = max(−30,000, −10,000) = −10,000, which is for S_2 (i.e., use chemical A).

States of nature	Courses of action	
	S_1	S_2
N_1	−30,000	−40,000
N_2	−30,000	−10,000
Column maximum	−30,000	−10,000

(c) Regret (profit loss) criterion or minimax criterion: From the given data, we have the following conditional opportunity loss table:

States of nature	Courses of action	
	S_1	S_2
N_1	$-30{,}000 + 30{,}000 = 0$	$-30{,}000 - (-40{,}000) = 10{,}000$
N_2	$-10{,}000 - (-30{,}000) = 20{,}000$	$-10{,}000 - (-10{,}000) = 0$
Column maximum	20,000	10,000

Minimum of maxima = minimum (20,000, 10,000) = 10,000, which is for S_2. Thus, S_2 (i.e., use chemical A) should be adopted as the optimal decision under the regret criterion.

17.3.4 Posterior Probability and Bayesian Analysis

The initial probabilities to get the probability distribution of the states of nature that are supposed to be known are called prior probabilities. If some extra information is required on the states of nature with a new outcome, then the prior probabilities require revision. Such revised probabilities in decision theory are called posterior (conditional) probabilities. These probabilities can be calculated using Bayes' theorem.

Bayes' theorem Let N_i with prior probabilities $P(N_i)=p_i$, $i=1, 2, \ldots, m$ be the m mutually exclusive and collectively exhaustive states of nature of S. Let the conditional probabilities $P(X/N_i)$, $i = 1, 2, \ldots, m$ be known for an outcome X. Given that outcome X has occurred, the revised posterior (conditional) probabilities $P(N_i/X)$ with this additional information are given as follows:

$$P\left(\frac{N_i}{X}\right) = \frac{P(X/N_i)P(N_i)}{\sum_{i=1}^{m} P(X/N_i)P(N_i)}$$

Proof Here $S = N_1 \cup N_2 \cup, \ldots, \cup N_m$

$$X = X \cap S = X \cap (N_1 \cup N_2 \cup, \ldots, \cup N_m)$$

or $X = (X \cap N_1) \cup (X \cap N_2) \cup, \ldots, \cup (X \cap N_m)$

$$P(X) = \sum_{i=1}^{m} P(X \cap N_i) = \sum_{i=1}^{m} P(X/N_i)P(N_i)$$

Thus $P\left(\dfrac{N_i}{X}\right) = \dfrac{P(X/N_i)P(N_i)}{P(X)} = \dfrac{P(X/N_i)P(N_i)}{\sum_{i=1}^{m} P(X/N_i)P(N_i)}$

Example 17.6 An oil company that is engaged in oil explorations has to decide whether or not to drill for oil in the Kaveri basin. There are three categories into which the drilling results are classified—high yield with 12 units, moderate yield with 6 units, and no yield. The drilling operation costs 5 units. At similar places, 50, 30, and 20 percent of previous drillings have given high, moderate, or no yield respectively. A seismic test is available, which would indicate a good, fair, or bad prospect of drilling. Past experience indicates the probabilities shown in the adjoining table for the result of seismic tests for the three ultimate possible yields:

Prospect	Yield		
	High	Moderate	Nil
Good	0.7	0.5	0.1
Fair	0.2	0.2	0.3
Bad	0.1	0.3	0.6

Find the best decision rule on the assumption that a test is made and the maximum amount that it would be worth paying for such a test. If this is the actual cost of the test, what is the expected value of the potential well?

Solution There are three states of nature: N_1 (high), N_2 (moderate), and N_3 (no yield), and two courses of action: S_1 (drill) and S_2 (do not drill). Let us denote the good, fair, or bad prospectus of drilling by X, Y, and Z respectively. The initial pay-off table along with the conditional probabilities is given as follows:

States of nature N_i	Probability	Courses of action		Regrets		Yield prospects		
		Drill (x_1)	Do not drill (x_2)	x_1	x_2	$P(X/N_i)$	$P(Y/N_i)$	$P(Z/N_i)$
N_1 (High)	0.5	7	0	0	7	0.7	0.2	0.1
N_2 (moderate)	0.3	1	0	0	1	0.5	0.2	0.3
N_3 (no yield)	0.2	–5	0	5	0	0.1	0.3	0.6

Using Bayes' theorem for the aforementioned data, we have the following:
For outcome X: $P(X) = P(X/N_1)\,P(N_1) + P(X/N_2)\,P(N_2) + P(X/N_3)\,P(N_3)$

$$\text{or } P(X) = 0.7 \times 0.5 + 0.5 \times 0.3 + 0.1 \times 0.2 = 0.52$$

$$P(N_1/X) = \frac{P(X/N_1)\,P(N_1)}{P(X)} = \frac{0.7 \times 0.5}{0.52} = 0.67$$

$$P(N_2/X) = \frac{P(X/N_2)\,P(N_2)}{P(X)} = \frac{0.5 \times 0.3}{0.52} = 0.29$$

$$P(N_3/X) = \frac{P(X/N_3)\,P(N_3)}{P(X)} = \frac{0.1 \times 0.2}{0.52} = 0.04$$

Thus, $ER(x_1/X) = 0.67 \times 0 + 0.29 \times 0 + 0.04 \times 5 = 0.20$ units

and $ER(x_2/X)$ = expected regret due to not drilling when the prospects are good, and $= 0.67 \times 7 + 0.29 \times 1 + 0.04 \times 0 = 4.98$ units

From the aforementioned information, the expected regret is minimum for x_1 (drilling) with the value 0.20 units.

For outcome Y, $P(Y) = P(Y/N_1)\,P(N_1) + P(Y/N_2)\,P(N_2) + P(Y/N_3)\,P(N_3)$

$$\text{or } P(Y) = 0.5 \times .02 + 0.3 \times .02 + 0.2 \times 0.3 = 0.22$$

$$P(N_1/Y) = \frac{0.5 \times 0.2}{0.22} = 0.46,\ P(N_2/Y) = \frac{0.3 \times 0.2}{0.22} = 0.27 = P(N_3/Y)$$

Hence, $ER(x_1/Y) = 0.46 \times 0 + 0.27 \times 0 + 0.27 \times 5 = 1.35$ units

$ER(x_2/Y) = 0.46 \times 7 + 0.27 \times 1 + 0.27 \times 0 = 3.49$ units

The expected regret (ER) is minimum for x_1 with a value of 1.35 units.

For outcome Z, $P(Z) = P(Z/N_1)P(N_1) + P(Z/N_2)P(N_2) + P(Z/N_3)P(N_3)$

or $P(Z) = 0.1 \times 0.5 + 0.3 \times 0.3 + 0.6 \times 0.2 = 0.26$

$$P(N_1/Z) = \frac{0.1 \times 0.5}{0.26} = 0.19, \; P(N_2/Z) = \frac{0.3 \times 0.3}{0.26} = 0.35,$$

$$P(N_3/Z) = \frac{0.6 \times 0.2}{0.26} = 0.46$$

Thus, $ER(x_1/Z) = 0.19 \times 0 + 0.35 \times 0 + 0.46 \times 5 = 2.30$ units

$ER(x_2/Z) = 0.19 \times 7 + 0.35 \times 1 + 0.46 \times 0 = 1.68$ units

The expected regret is now minimum for x_2 with a value of 1.68 units.
The total expected regrets for X, Y, and Z together are as follows:

$ER(x_1) = 0.52 \times 0.20 + 0.22 \times 1.35 + 0.26 \times 2.30 = 0.999 = 1$ unit (approximate)

$ER(x_2) = 0.52 \times 4.98 + 0.22 \times 3.49 + 0.26 \times 1.68 = 3.794$ units

From the aforementioned data, we see that the expected regret is minimum for x_1 (drilling) with a value of 1 unit.

Hence, the best decision for the company would be to go for drilling while paying a maximum of 1 unit for the seismic test. If this is done, the pay-off for x_1 in the aforementioned table would be 7-1,1-1, and 5-1 for the states N_1, N_2, and N_3, and the expected value of the potential well will be $EV(x_1) = 0.5 \times 6 + 0.3 \times 0 - 0.2 \times 6 = 1.8$ units

17.3.5 Decision-making Under Conflict—Game Theory

The problem of a simple game of conflict between two players can be thought of as a decision-making problem under conflict. In this situation, the combined gain or loss for the two players is zero. For this reason, such problems are called two-person zero-sum games. We have discussed these games in detail in Chapter 15.

RECAPITULATION

1. Decision-making environments are certainty, risk, uncertainty, and conflict.
2. In decision theory, states of nature, pay-offs, and courses of action are the main components and the states of nature are probabilistic.
3. The techniques used in solving the problems of decision-making under certainty are LPPs, transportation, assignment, goal programming, and inventory models.
4. The techniques used for decision-making under risk are expected monetary value (EMV), expected value of perfect information (EVPI), expected profit with perfect information (EPPI), conditional opportunity loss (COL), and expected opportunity loss (EOL).
5. The techniques used for decision-making under uncertainty are maximin, maximax, minimax (regret), Laplace, and Hurwitz's methods.

EXERCISES

Multiple-choice Questions

17.1 The EMV criterion is used for decision-making under
(a) certainty (b) risk (c) uncertainty (d) none of these

17.2 The decision-making criterion that should be used to achieve the maximum long-term pay-off is
(a) EMV (b) EOL (c) Hurwitz (d) none of these

17.3 The minimum EOL is equal to
(a) EMV (b) EVPI (c) COL (d) none of these

17.4 The EVPI is equal to
(a) EPPI – EMV* (c) Maximum EOL
(b) EPPI + EMV* (d) none of these

17.5 The essential characteristic of a decision model are
(a) decision alternatives (c) pay-offs
(b) states of nature (d) all of these

State True or False

17.1 The decision-maker cannot choose a course of action of his/her choice.

17.2 The pay-off due to the equally likely criterion of decision-making is the same as the minimum EOL criterion.

17.3 The probability of an outcome is not known in decision-making environment under risk.

17.4 The difference between the highest EMV and EPPI is called EVPI.

17.5 In decision theory, the probabilities are associated with the states of nature and not with the courses of action.

Fill in the Blanks

17.1 The EMV criterion is used for decision-making under ________.

17.2 The alternative that may be chosen by a decision-maker is called ________.

17.3 States of nature are ________ and are not under the control of the decision-maker.

17.4 Certainty, risk, conflict, and ________ are the decision-making environments.

17.5 The EVPI is equal to minimum ________.

Review Questions

17.1 What do you mean by decision-making situations?

17.2 Explain the term EMV.

17.3 Discuss the various techniques used for decision-making under risk.

17.4 Write a short note on decision-making under uncertainty.

17.5 Explain the following decision-making criteria:
(a) Minimax (c) Maximax (e) Hurwitz
(b) Maximin (d) Laplace

17.6 Write a short note on EOL.

17.7 Explain EPPI and EVPI.

Numerical Problems

17.1 A tube and tyre manufacturing company is planning to increase its existing capacity. Among various available alternatives, its management ultimately approves two plans:

(a) modernization at a cost of ₹50 lakhs or (b) expansion at a cost of ₹80 lakhs. Both these plans would require the same time for implementation. The management believes that over the required payback period, the demand will either be high or moderate. Since high demand is considered less likely, the probability of its occurrence is set at 0.35. If the demand is high, modernization would gross an estimated additional ₹60 lakhs, whereas expansion would gross an additional ₹120 lakhs. On the other hand, if the demand is moderate, the corresponding figures would be ₹50 lakhs for modernization and ₹70 lakhs for expansion.

(a) Identify the states of nature and courses of action.
(b) Calculate the expected profit values for the courses of action.
(c) Calculate EVPI.

17.2 A flower dealer purchases flowers in the evening and delivers them to an organization by 8.00 a.m. the next day. The daily demand for roses is as follows:

Dozens of flowers	7	8	9	10
Probability	0.1	0.2	0.4	0.3

The management purchases flowers at ₹10 per dozen and sells them for ₹30. All the unsold roses are donated to a nearby hospital. How many dozens of flowers should the manager order each evening to maximize its profits? What is the optimum profit?

17.3 In a mining project at a remote area, a sophisticated part of equipment has to be purchased. This equipment part is expensive and subject to random failure. If the spares of this part are purchased at the time of purchase, then the cost would be ₹1500 per unit and they would have no scrap value. On the other hand, if the part fails on the job and no spare is available, the part will have to be manufactured on a special order basis and the cost would approximately be ₹9000 for each such occurrence. Based on previous experience, the following probability estimate of the number of failures over the duration of the project is given:

Failure	0	1	2
Probability	0.80	0.15	0.05

(a) Identify the states of nature and courses of action.
(b) Construct a table for the conditional costs for each pair of 'course of action' and 'state of nature'.
(c) Determine the optimal EMV* and optimal number of spares to purchase initially.
(d) Based on opportunity losses, determine the optimal course of action and optimal value of EOL.
(e) Determine the EPPI and EVPI.

Hint (a) Number of failures: N_1 (no failure), N_2 (one failure), and N_3 (two failure).

Courses of action: Number of spares purchased S_1 (no spare purchased), S_2 (one spare purchased), S_3 (two spare purchased).

(b) Pair-wise conditional cost table

States of nature (Spares required)	Courses of action (Spares purchased)	Purchase cost (₹)	Emergency cost (₹)	Total conditional cost
0	0	0	0	0
0	1	1500	0	1500
0	2	3000	0	3000
1	0	0	9000	9000

(*Contd*)

(Contd)

States of nature (Spares required)	Courses of action (Spares purchased)	Purchase cost (₹)	Emergency cost (₹)	Total conditional cost
1	1	1500	0	1500
1	2	3000	0	3000
2	0	0	1800	18,000
2	1	1500	9000	10,500
2	2	3000	0	3000

(c) $\text{EMV}^* = \text{EMV}(S_2) = 1950$, one spare part
(d) $\text{EOL}^* = \text{EOL}(S_2) =$ ₹1575; S_2 (one spare part)
(e) EPPI = ₹375, EVPI = ₹1575

17.4 An electric equipment manufacturer has estimated, on the basis of past data, the following weekly demand of distribution for the equipment:

Number demanded	0	1	2	3	4	5	6
Probability	0.14	0.27	0.27	0.18	0.09	0.04	0.01

Each equipment costs ₹7000 and is sold for ₹10,000 each. The equipment left unsold at the end of the week is disposed of for ₹6000 each. What should be the weekly equipment stock so as to maximize the expected profit? What is the optimum profit per week?

17.5 A newspaper distributor distributes a particular newspaper in a local market. The cost price of the newspaper is ₹1.05 and the selling price is ₹1.50. An unsold copy is disposed of in the reuse market at ₹0.05. The estimates for the number of copies in demand are as follows:

Demand	15	16	17	18	19	20
Probability	0.04	0.19	0.33	0.26	0.11	0.07

How many copies of the newspaper should the distributor order so that the expected profit is maximum? Find the maximum expected profit.

17.6 A person has ₹20,000 to invest. He was given the option to invest in three companies: A, B, and C. The return per year on investment depends on the performance measures that are classified into excellent, good, and normal. The possible returns, depending upon performance, are given in the following table:

Courses of action	States of nature		
	Excellent	Good	Normal
A	4000	2400	3000
B	6000	1600	2000
C	5000	2000	3600

What should the person decide with regard to investment on using the following criteria: (a) maximin, (b) maximax, (c) equally likely, and (d) regret?

17.7 A milk vending booth sells milk at ₹11 per litre, the cost price being ₹10.87 per litre. The owner desires to give away unsold milk at the end of the day towards charity. The milk owner collects data based on past sales for a 100 day period:

Litre per day	600	610	620	630	640	650	660	670	680	690	700
No. of days	1	1	2	6	10	20	25	15	10	7	3

What should be the optimal inventory, based on the aforementioned data and the net profit for a day?

17.8 A company wants to introduce a new product to its existing product range. The company has to decide whether to market the product or not and wants to take the decision on the basis of two levels of sales—high or low. The estimated changes that will occur in each market level, together with their costs and consequent profits or losses, are given in the following table:

States of nature of product	Probability	Courses of action	
		Market the product (₹1000)	Not market the product (₹1000)
High sales	0.3	150	0
Low sales	0.7	–40	0

The company's marketing manager suggests for a market research survey to gather more information. On the basis of past experience, the marketing manager in the light of actual sales achievement provides additional information given in the following table

Market research (Survey outcome)	Actual sales	
	High market	Low market
High sales	0.5	0.1
Indecisive	0.3	0.4
Low sales	0.2	0.1

The market research survey costs ₹20,000. State whether or not there is a case for employing the market research organization.

17.9 In numerical problem 17.2, find the optimal course of action using the concept of opportunity loss. In addition, find the EPPI and EVPI.

CONCEPTUAL PROBLEM

A toy manufacturing company producing dancing dolls is planning to expand its business beyond its existing production capacity. The management of the company has two plans to increase the maximum production capacity (a) modernization at the cost of 7 million and (b) expansion at the cost of 10 million. Both of these plans require the same amount of time to put them in action. The management is of the opinion that over the required payback period, the demand may be high or moderate. As moderate demand is considered to be more likely than high demand, the probability of moderate demand has been

(Contd)

(Contd)

set at 0.65. If the demand is moderate, expansion would gross an estimated additional amount of ₹9 million, but modernization would bring an additional amount of ₹7 million only, due to the lower production capacity. On the other hand if the demand is high, the respective figures would be 9 million for modernization and 14 million for expansion.

1. Construct the conditional profit table along with the given probability.
2. Based on the EMV, should the company opt for modernization or expansion?
3. Calculate EVPI.
4. Calculate conditional opportunity loss.

Solution 1. Here, the states of nature are N_1 = moderate demand, N_2 = high demand, and the courses of action S_1 = modernization, S_2 = expansion.
The probability of occurrence of moderate demand $p_1 = 0.65$. Thus, the probability of occurrence of N_1, say $p_2 = 1 - 0.65 = 0.35$. The calculations are shown in Table 17.1.

Table 17.1 Representing conditional profit (million ₹)

States of nature demand N_i	Courses of action	
	Modernization S_1	Expansion S_2
N_1 (Moderate)	$7 - 7 = 0$	$9 - 10 = -1$
N_2 (High)	$9 - 7 = 2$	$14 - 10 = 4$

2. Incorporating the probability of occurrence of states of nature, we get Table 17.2.

Table 17.2 Representation of conditional profit along with probability

States of nature demand N_i	Probability	Courses of action	
		Modernization S_1	Expansion S_2
N_1 (Moderate)	0.65	0	–1
N_1 (High)	0.35	2	4

The expected monetary value EMV $(S_1) = 0.65 \times (0) + 0.35 \times (2) =$ ₹0.70 million and EMV $(S_2) = 0.65 \times (-1) + 0.35 \times (4) =$ ₹0.75 million. The maximum EMV denoted as EMV* = EMV (S_1) = ₹0.75 million.

To have maximum expected monetary value (EMV), the company must choose expansion (i.e., course of action S_2).

3. To calculate EPPI, we identify the optimal course of action for each state of nature and multiply its conditional profit with the corresponding probability to get the weighted profit and then add these as shown in Table 17.3:

Table 17.3 Computations to find out EPPI

States of nature N_i	Probability	Courses of action		Optimal courses of action	Conditional profit	Weighted profit
		S_1	S_2			
N_1 (Moderate)	0.65	0	–1	S_1	0	0×0.65
N_1 (High)	0.35	2	4	S_2	4	4×0.35
Expected profit with perfect information (EPPI) = $\sum$(Weightned profit)						1.40

(Contd)

(*Contd*)

The max EMV = EMV* = ₹0.75 million is corresponding to the course of action S_2 (i.e., expansion).

Now, the expected value of perfect information (EVPI) is given as follows:

$$\text{EVPI} = \text{EPPI} - \text{EMV}^* = \text{EPPI} - \text{EMV}(S_2) = 1.40 - 0.75 = ₹0.65 \text{ million}$$

This means if the company could get the perfect information of demand, it should have paid ₹0.65 million for the receipt of information.

4. To construct the conditional opportunity loss table, subtract from the highest pay-off value for a particular state of nature, the actual pay-off obtained for the various courses of action. Referring to Table 17.2, the highest pay-off for state of nature N_1 is 0 by taking course of action S_1 (modernization). Thus, the conditional losses corresponding to N_1 are $0 - 0 = 0$ for S_1, and $0 - (-1) = 1$ for course of action S_2 (expansion). Similarly, the highest pay-off for state of nature N_2 is 4 corresponding to S_2 and thus the conditional losses for N_2 are $4 - 2 = 2$ corresponding to S_2, and $4 - 4 = 0$ for S_2. The calculated values are shown in the following Table 17.4:

Table 17.4 Representation of conditional profit/losses

States of nature	Probability	Conditional profit		Conditional profit/loss for courses of action	
				S_1	S_2
N_1 (Moderate)	0.65	0	−1	0 [0 − 0]	1 [= (0 − (−1))]
N_1 (High)	0.35	2	4	2 [4 − 2]	0 [= (4 − 4)]

Now, the expected opportunity loss $(S_1) = 0.65 \times (0) + 0.35 \times (2) = 0.70$ and the expected opportunity loss $(S_2) = 0.65 \times (1) + 0.35 \times (0) = 0.65$.

As the objective of decision-maker is to minimize the expected opportunity loss, which is for course of action S_2 (expansion), the manufacturer must opt for the expansion plan.

Notes

1. For an optimum decision, the expected opportunity loss should be minimum and the expected monetary value should be maximum.
2. To understand the concept of opportunity loss, refer to Table 17.4. If state of nature N_2 occurs, then the maximum profit of ₹4 million is achieved by selecting the course of action S_2 (expansion). As a result opportunity loss is 0 (=4−4), S_2 being the best course of action for occurring state of nature N_2. However if the course of action S_1 is chosen, then the conditional profit is ₹2 million and it will result in a opportunity loss of 4−2=₹2 million.

18 Networking

> After studying this chapter, the reader will be able to
>
> - comprehend the definitions and notations used in networking
> - discuss the shortest route problem
> - explain the minimum spanning tree problem
> - understand the maximum flow problem

18.1 INTRODUCTION

Network analysis finds great importance in electrical circuits, pipelines, mobiles, roads, railways, airlines, analysis of blood vessels, production, resource management, distribution, planning, scheduling, and controlling of research and development projects. Transportation and assignment problems (both linear programming problems) are network optimization models. In this chapter, we shall discuss the shortest route, minimum spanning tree, and maximum flow problems.

18.2 DEFINITIONS AND NOTATIONS USED IN NETWORKING

In this section, we shall discuss terms such as graphs, trees, and networks.

Graph

A graph G is a collection of vertices that may or may not be connected to each other by edges. In other words, a graph G is a set of points (called vertices denoted by v) connected by lines (called edges denoted by e). For example, the graph with six vertices, $v = (v_1, v_2, v_3, v_4, v_5, v_6)$ and eight edges, $e = (e_1, e_2, e_3, e_4, e_5, e_6, e_7, e_8)$ is given in Fig. 18.1.

A graph G is said to be a *weighted graph* when weights are assigned to each edge. For example, Fig. 18.2 shows a weighted graph.

A graph G is said to be a *labelled graph* when each vertex is assigned a unique name or label. Fig. 18.3 shows a labelled graph. A graph G is said to be *complete* when each vertex is connected to every other vertex, that is, at least one edge exists between every pair of vertices. It is denoted by K_n. Figure 18.4 shows a K_4 graph (complete graph).

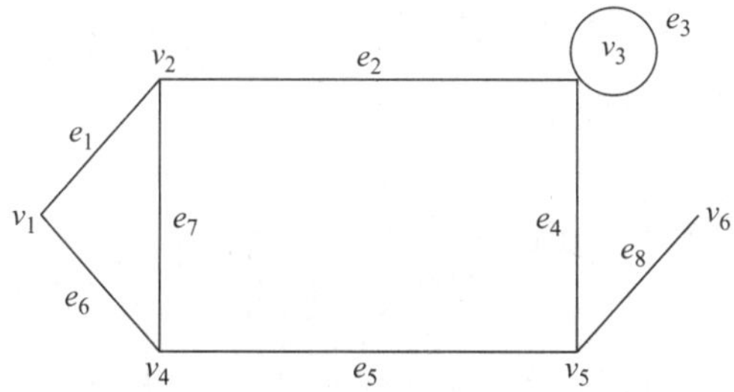

Figure 18.1 Graph with 6 vertices and 8 edges

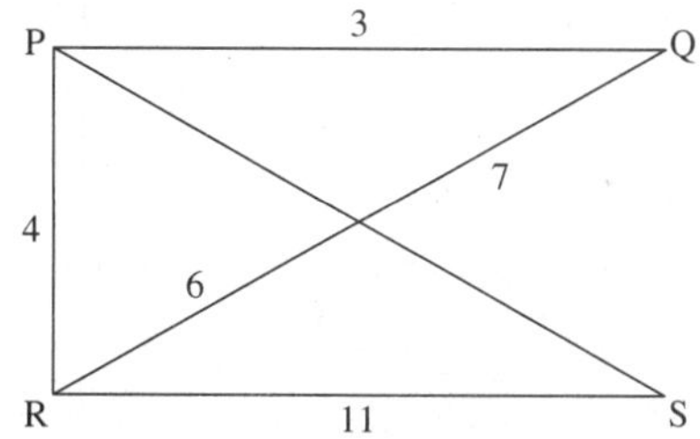

Figure 18.2 Weighted graph

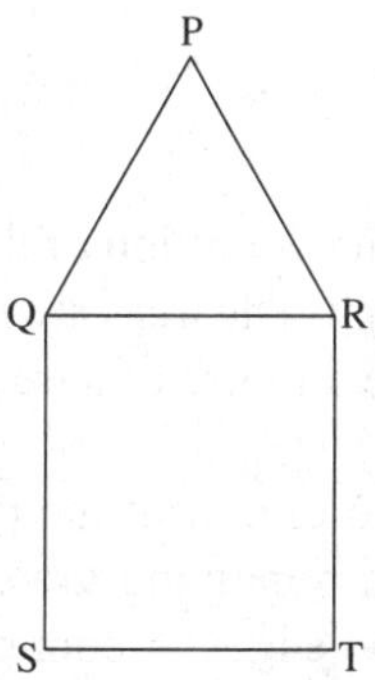

Figure 18.3 Labelled graph

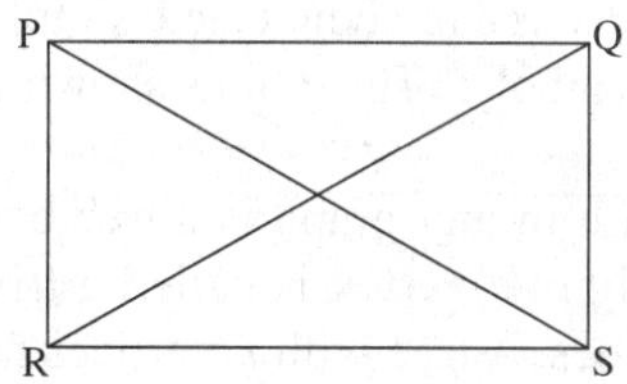

Figure 18.4 Complete graph K_4

A graph is said to be *finite* when it has a finite number of vertices as well as a finite number of edges; otherwise it is called an *infinite graph*. An edge is an *incident* with a vertex if the edge is joined to the vertex. If there is an edge joining a pair of vertices, those vertices are said to be *adjacent*, otherwise they are *non-adjacent*. The number of edges connected to a given vertex constitutes the *degree* of that vertex. It is denoted by $d(v)$. The degree of self-loops is counted twice. In Fig. 18.1, $d(v_1)=2$, $d(v_2)=3$, $d(v_3)=4$, $d(v_5)=3$, $d(v_4)=3$, $d(v_6)=1$. A vertex is said to be *isolated* if it has no incident edge. The degree of an isolated vertex is zero. A vertex of degree one is called a *pendant vertex*. A *walk* (defined of a graph G) is an alternating sequence of finite vertices and edges that begins and ends with vertices. No edge should be traversed more than once in a *walk* whenever a vertex appears more than once. A walk that begins and ends at different vertices is called an *open walk*. A walk that begins and ends at the same vertex is called a *closed walk*. An open walk in which one vertex appears more than once is called a *path* (also called a *simple* or an *elementary path*). In a path, the total number of edges is called the *length of a path*. A self-loop can be included in a walk but not in a path. In Fig. 18.1, $v_6e_8v_5e_4v_3e_2v_2e_1v_1e_6v_4$ is a path. A closed walk in which no vertex (except the beginning and ending vertex) appears more than once is called a *circuit* and is also called a cycle, elementary cycle, circular path, or polygon. Every circuit is not a self-loop but every self-loop is a circuit. In Fig. 18.1, $v_1e_1v_2e_2v_3e_4v_5e_5v_4e_6v_1$ is a circuit. A graph G is said to be connected if there is at least one edge between every pair of vertices in G, else we can infer that the graph is disconnected. A disconnected graph consists of two or more connected graphs, and each of these connected graphs is known as a component.

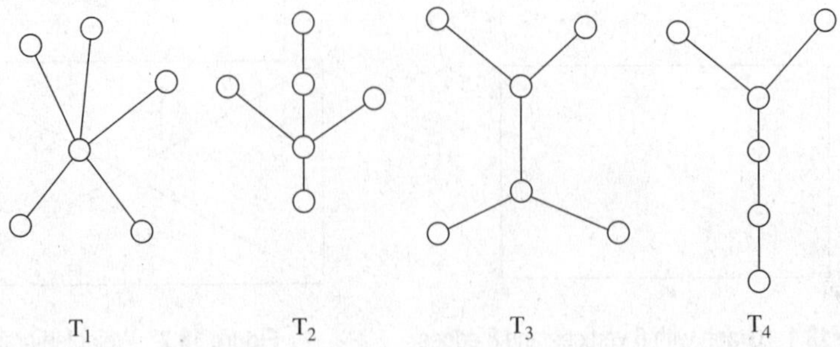

Figure 18.5 Trees of order 6

Tree

A *tree* is a connected graph without any circuit. By the definition of the tree, it is clear that a tree is a connected, simple, and cyclic graph. The trees are represented by the symbol T. The graphs shown in Fig. 18.5 are examples of a tree having an order 6.

A path in any graph is a tree because there is no circuit in the path. A tree with only one vertex is called a trivial tree and the remaining are called non-trivial trees. A tree with n vertices has exactly $(n - 1)$ edges. A connected graph G is said to be *minimally connected* if deletion of any edge from G renders the graph to be disconnected. The graph shown in Fig. 18.6 is a minimally connected graph.

The distance between the two vertices of a tree is easy to obtain as there is no circuit and therefore there is one and only one path between every pair of vertices. The distance between the vertices of tree T, as shown in Fig. 18.7, is as follows:

$d(a, b) = 1$, $d(a, d) = 2$, $d(a, c) = 1$, $d(a, e) = 2$, $d(a, f) = 2$, $d(c, b) = 2$, $d(c, d) = 3$, $d(e, d) = 4$, $d(f, d) = 4$, and so on. There are two special types of trees, called the rooted and the binary tree. A tree in which one vertex is distinct from all other vertices is called a rooted tree. A tree in which there is exactly one vertex of degree 2 and each of the remaining vertices are of degree 1 or 3 is called a *binary tree*. A binary tree is also a rooted tree because the vertex of degree 2 is distinct from all other vertices; hence, the vertex with degree 2 is the root of the binary tree. Binary trees are widely used in computer applications such as searching methods known as binary search, and sorting methods known as heap sort. The vertices with degree 1 in a binary tree are called external or terminal vertices and the remaining are called internal vertices. The number of vertices in a binary tree is always odd. While the number of external vertices in a binary tree is $(n + 1)/2$, the number of internal vertices in a binary tree is $(n - 1)/2$. The source is

Figure 18.6 Minimally connected graph

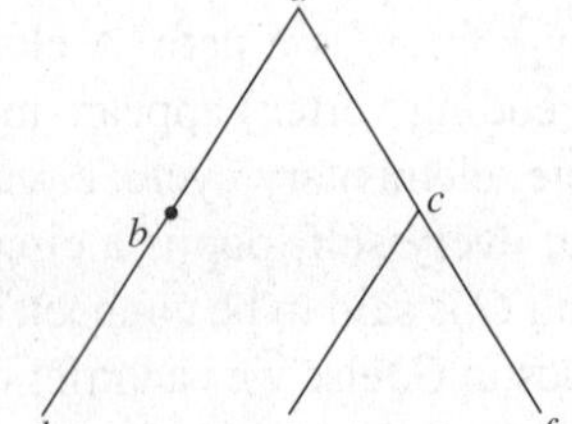

Figure 18.7 Tree

the starting event of a project with only succeeding but no preceding activity. The sink is the last event indicating the end of a project with only preceding but no succeeding activity.

Network

A network can be defined as a set of points or nodes that are connected by links or arrows. In other words, a network is a graph that facilitates a flow in the branches of the graph. In a network, the branches are connected only at nodes. The objective of networks is to determine a course of action, which minimizes certain measures of performance, such as time, distance, and cost.

18.3 SHORTEST ROUTE PROBLEM

The determination of the shortest route from the origin to the destination in a network is known as the shortest route problem. It is also known as the minimum path problem. For solving the problem of determining the minimum path, we discuss the Dijkstra method developed by F.W. Dijkstra in 1959. Let u_i be the shortest path from node 1 to node i and d_{ij} be an arc length (i, j). The label for node j is defined as follows:

$$[u_j, i] = [u_i + d_{ij}, i], \ d_{ij} \geq 0$$

In Dijkstra algorithm, there are two types of node labels: temporary and permanent. A temporary label can be replaced with another label, if the shortest route to the same node can be found. If no shortest route can be found, then it becomes a node with a permanent label.

The procedure to find the shortest route using Dijkstra algorithm consists of the following steps:

Step 1 First label the node 1, with permanent label [0, -], and set $j = 1$.

Step 2 Obtain the temporary labels $[u_i + d_{ij}, i]$ for each node j that can be reached from node i. If node j has already been labeled $[u_j, r]$, reached from another node r, such that $u_i + d_{ij} < u_j$, then replace $[u_j, r]$ with $[u_i + d_{ij}, i]$; else $[u_i + d_{ij}, i]$ is the permanent label of the node j.

Step 3 For $j < n$, set j = next j, to reach the permanent label. If all the nodes have permanent labels, then stop.

Example 18.1 Find the shortest route in the network shown in Fig. 18.8.

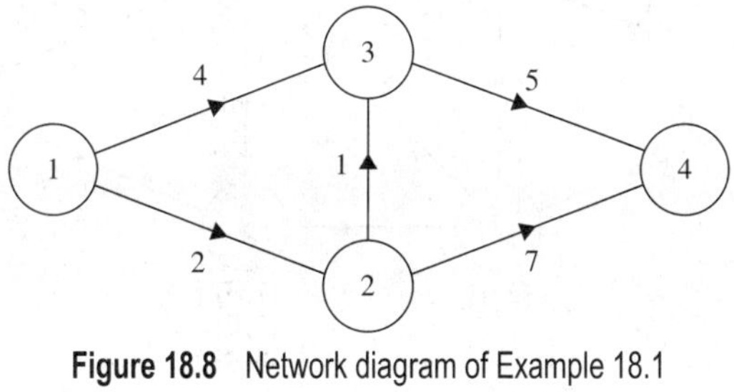

Figure 18.8 Network diagram of Example 18.1

Solution It is given that the network has four nodes. The calculations for labelling these nodes are as follows:

Node 1: [0, –]

Node 2: [0 + 2, 1] = [2, 1]

Node 3: [0 + 4,1] = [4,1] and [2 + 1, 2] = [3, 2]

Node 4: Reached from nodes 2 and 3—[2 + 7, 2] = [9, 2] and [3 + 5, 3] = [8, 3]

Thus, the shortest route covers a distance of eight units. Now, tracking back from node 4 (Fig. 18.9), by checking the second elements of the labels, we find the shortest route as 1→2→3→4.

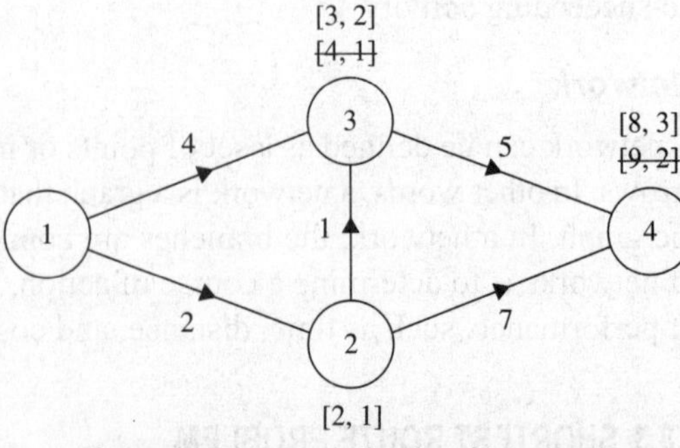

Figure 18.9 Network showing shortest route

Example 18.2 Find the shortest route in the network shown in Fig. 18.10.

Solution It is given that the network has six nodes. The calculations for labelling the nodes are as follows:

Node 1: [0, –]

Node 3: [0 + 2, 1] = [2, 1]

Node 2: [0 + 4, 1] = [4, 1] and [2 + 3, 3] = [5, 3]

Node 4: [4 + 6, 2] = [10, 2] and [2 + 3, 3] = [5, 3]

Node 5: [4 + 3, 2] = [7, 2], [2 + 5, 3] = [7, 3], and [5 + 3, 4] = [8, 4]

Node 6: [5 + 5, 4] = [10, 4] and [7 + 6, 5] = [13, 5]

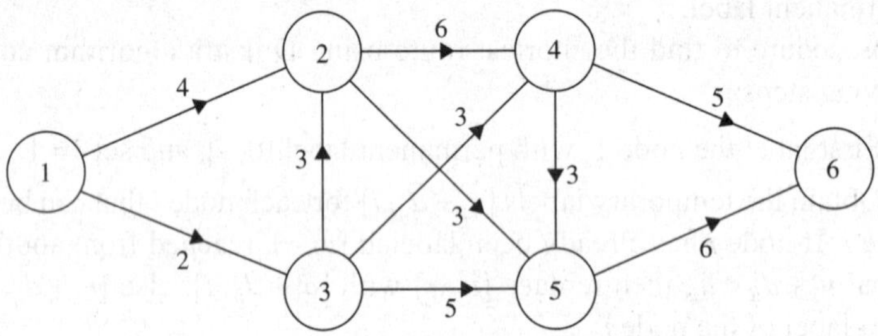

Figure 18.10 Network diagram of Example 18.2

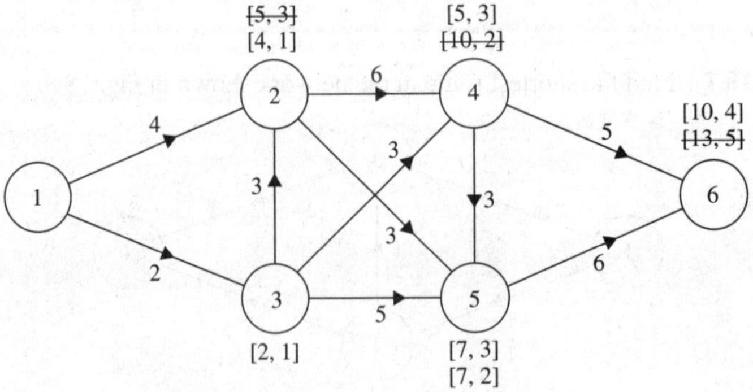

Figure 18.11 Network showing shortest route

Thus, the shortest route covers a distance of 10 units. Now tracking back from node 6 (Fig. 18.11) by checking the second elements of the labels, we find the possible route as 1→3→4→6.

Example 18.3 Find the shortest route in the network shown in Fig. 18.12.

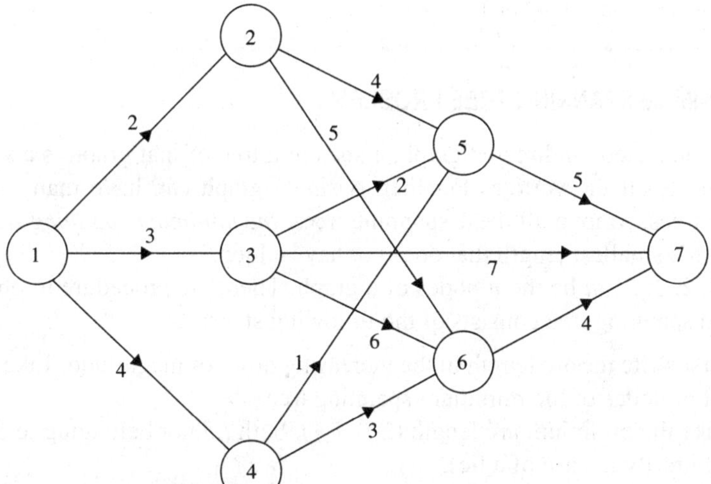

Figure 18.12 Network diagram of Example 18.3

Solution It is given that the network has seven nodes. The calculations for labelling the nodes are as follows:
Node 1: [0, –]
Node 2: [0 + 2, 1] = [2, 1]
Node 3: [0 + 3, 1] = [3, 1]
Node 4: [0 + 4, 1] = [4, 1]
Node 5: [2 + 4, 2] = [6, 2], [3 + 2, 3] = [5, 3], and [4 + 1, 4] = [5, 4]

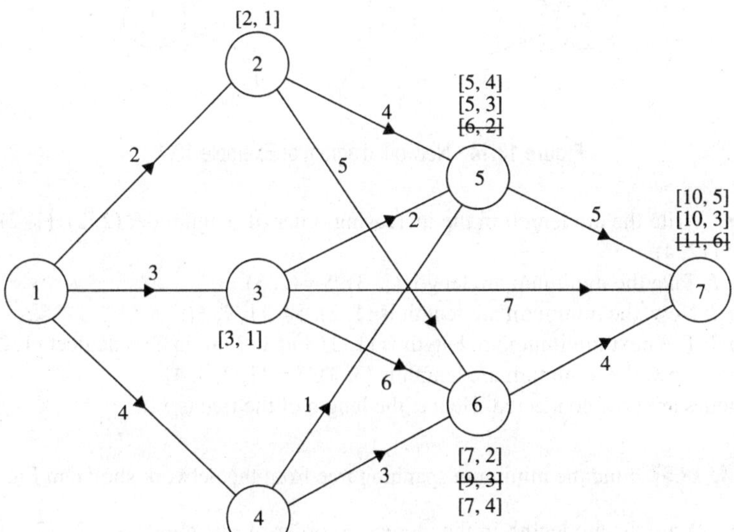

Figure 18.13 Network showing shortest route

Node 6: [2 + 5, 2] = [7, 2], [3 + 6, 3] = [9, 3], and [4 + 3, 4] = [7, 4]
Node 7: [3 + 7, 3] = [10, 3], [5 + 5, 5] = [10, 5], and [7 + 4, 6] = [11, 6]
Thus, the shortest route covers a distance of 10 units. Now tracking back from node 6 (Fig. 18.13), by checking the second elements of the labels, the possible routes are 1→3→5→7 or 1→4→5→7 or 1→3→7.

18.4 MINIMUM SPANNING TREE PROBLEM

Given a connected, undirected graph, a spanning tree of that graph is a subgraph that connects all the vertices together. A single graph can have many different spanning trees. Among all these spanning trees, *the minimum spanning tree* is the tree with the smallest length that does not have a loop.

Let 1, 2, 3, …, n be the n nodes of a graph. Then, the procedure to obtain the minimum spanning tree consists of the following steps:

1. First write the arc length in the increasing order of magnitude. Take S as the set of nodes of the minimum spanning tree.
2. Take the minimum arc length (i, j), $i \neq j$, both i, j not belonging to S (select arbitrarily in case of a tie).
3. Now set $S = S \cup \{i, j\}$. If S is connecting all nodes then stop, and check if S consists of unconnected branches. If yes, then go to step 4 otherwise go to step 2.
4. Connect the unconnected branches with minimum arcs and stop.

Example 18.4 Find the minimum spanning tree from the network shown in Fig. 18.14.

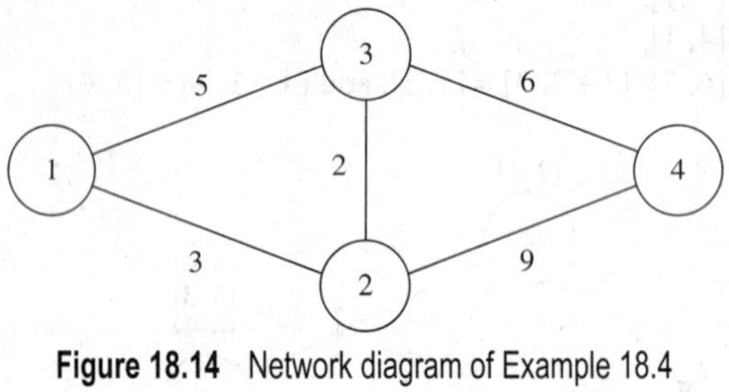

Figure 18.14 Network diagram of Example 18.4

Solution Write the arc length in the increasing order of magnitude: (2, 3), (1, 2), (1, 3), (3, 4), and (2, 4).
Iteration 1: Take the minimum arc length (2, 3), S = (2, 3).
Iteration 2: Now, the minimum arc length is (1, 2), $S = \{1, 2, 3\}$.
Iteration 3: The next minimum arc length is (1, 3), but 1, 3 are in S so neglect (1, 3).
Iteration 4: Now, the minimum arc length is (3, 4), $S = \{1, 2, 3, 4\}$.
All the nodes are now connected. Hence, the length of the tree is 11.

Example 18.5 Find the minimum spanning tree from the network shown in Fig. 18.15.

Solution Write the arc length in the increasing order of magnitude: (1, 2), (2, 3), (1, 6), (1, 5), (4, 5), (3, 4), (3, 6), (5, 6), and (2, 4).

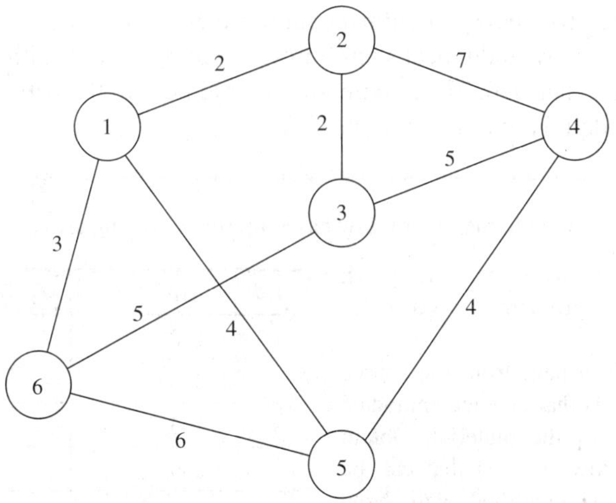

Figure 18.15 Network diagram of Example 18.5

Iteration 1: Take the minimum length arc (1, 2), $S = \{1, 2\}$.
Iteration 2: Now, the minimum arc length is (2, 3), $S = \{1, 2, 3\}$.
Iteration 3: The next minimum arc length is (1, 6), $S = \{6, 1, 2, 3\}$.
Iteration 4: The next minimum arc length is (4, 5) or (1, 5). Considering (4, 5), $S = \{6, 1, 2, 3, 4, 5\}$.
Iteration 5: The next minimum arc length is (1, 5), $S = \{6, 1, 2, 3, 4, 5\}$.
All the nodes are now connected. Hence, the length of the tree is 15.

18.5 MAXIMUM FLOW PROBLEMS

There are many real-life situations where we encounter flow, such as flow of oil in pipelines and flow of traffic. In such problems, the objective is to maximize the flow. A maximum flow problem can be formulated as a linear programming problem. In most flow problems, flows (through a directed and connected network) originate at one node, known as the source, and terminate at another node known as the sink. All the remaining nodes are transshipment nodes. The flow through an arc is allowed only in the direction indicated by the arrowhead, where the maximum amount of flow is given by the capacity of that arc. While all the arcs point away from the node at the source, they point towards the node at the sink. The main objective is to maximize the total amount of flow from the source to the sink. This amount is measured in two equivalent ways, namely the amount leaving the source or the amount entering the sink. The procedure to determine the maximum flow consists of the following steps:

1. First identify a path from the source to sink that can accommodate a positive flow of the material.
2. Obtain the maximum flow that can be shipped along the identified path and denote it by Z.
3. Now decrease the capacity of each branch in the direction of flow by Z units, and increase the reverse capacity by Z. Add Z units to the amount delivered to the sink.

4. Next, identify a new path from the source to the sink which can accommodate a positive flow, and repeat steps, through 3 until all such paths with positive flow are exhausted. The maximum possible flow in the network is the total amount delivered (added) to the sink.

Example 18.6 Find the maximum flow based on the following arcs and capacities:

Arc	Capacity
1–2	10
2–3	5
1–3	7
2–4	4
3–4	10

Solution The network for the given maximum flow problem is given in Fig. 18.16(a).

First, identify a path from the source to the sink (1-3-4) that can accommodate a positive flow of the material. Obtain the maximum flow (i.e., 7) that can be shipped along the identified path. Now decrease the capacity of each branch in the direction of the flow by 7 units along this path (1-3-4), increase the reverse capacity by 7, and add 7 units to the amount delivered to the sink as shown in Fig. 18.16(b).

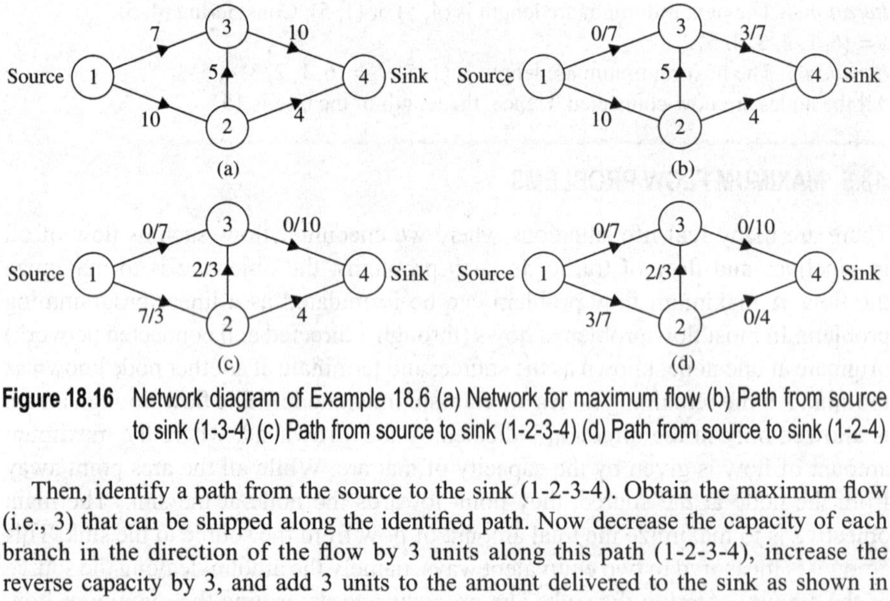

Figure 18.16 Network diagram of Example 18.6 (a) Network for maximum flow (b) Path from source to sink (1-3-4) (c) Path from source to sink (1-2-3-4) (d) Path from source to sink (1-2-4)

Then, identify a path from the source to the sink (1-2-3-4). Obtain the maximum flow (i.e., 3) that can be shipped along the identified path. Now decrease the capacity of each branch in the direction of the flow by 3 units along this path (1-2-3-4), increase the reverse capacity by 3, and add 3 units to the amount delivered to the sink as shown in Fig. 18.16(c).

Again, identify a path from the source to the sink (1-2-4). Obtain the maximum flow (i.e., 4) that can be shipped along the path. Now decrease the capacity of each branch in the direction of the flow by 4 units along this path (1-2-4), increase the reverse capacity by 4, and add 4 units to the amount delivered to the sink as shown in Fig. 18.16(d).

Now, no path exists from the source to sink that can accommodate a positive flow of the material. Calculate the maximum flow, by summing the maximum flow through the three identified paths, that is, via paths (1-3-4), (1-2-3-4), and (1-2-4).

Maximum flow = Max flow (1-3-4) + Max flow (1-2-3-4) + Max flow (1-2-4) = 7 + 3 + 4 = 14
Hence, the maximum material delivered to the sink is 14 units.

Example 18.7 Find the maximum flow of the network shown in Fig. 18.17.

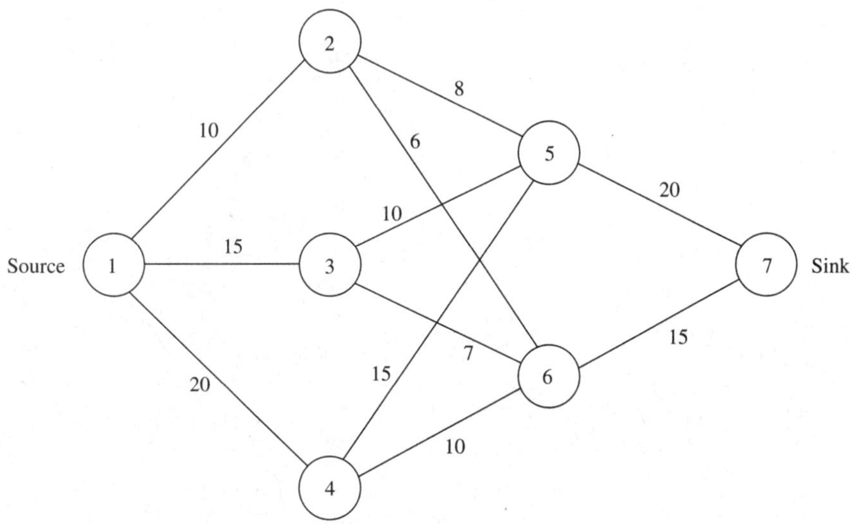

Figure 18.17 Network diagram of Example 18.7

Solution First identify a path from the source to the sink (1-2-5-7) that can accommodate a positive flow of the material. Obtain the maximum flow (i.e., 8) that can be shipped along this path. Now, decrease the capacity of each branch in the direction of flow by 8 units along this path (1-2-5-7), increase the reverse capacity by 8, and add 8 units to the amount delivered to the sink as shown in Fig. 18.18(a).

Next, identify a path from source to sink (1-2-6-7) that can accommodate a positive flow of the material. Obtain the maximum flow (i.e., 2) that can be shipped along this path.

Now decrease the capacity of each branch in the direction of flow by 2 units along this path (1-2-6-7) and increase the reverse capacity by 2. Add 2 units to the amount delivered to the sink as shown in Fig. 18.18(b).

Again, we identify a path from source to sink (1-3-5-7). Obtain the maximum flow (i.e., 10) that can be shipped along this path. Now decrease the capacity of each branch in the direction of flow by 10 units along this path (1-3-5-7) and increase the reverse capacity by 10. Add 10 units to the amount delivered to the sink as shown in Fig. 18.18(c).

Again, identify a path from the source to the sink (1-3-6-7). Obtain the maximum flow (i.e., 5) that can be shipped along the path. Now decrease the capacity of each branch in the direction of flow by 5 units along this path (1-3-6-7) and increase the reverse capacity by 5. Add 5 units to the amount delivered to the sink as shown in Fig. 18.18(d).

Again, identify a path from the source to the sink (1-4-5-7) and obtain the maximum flow (i.e., 2) that can be shipped along the path. Now, decrease the capacity of each branch in the direction of flow by 2 units along this path (1-4-5-7) and increase the reverse capacity by 2. Add 2 units to the amount delivered to the sink as shown in Fig. 18.18(e).

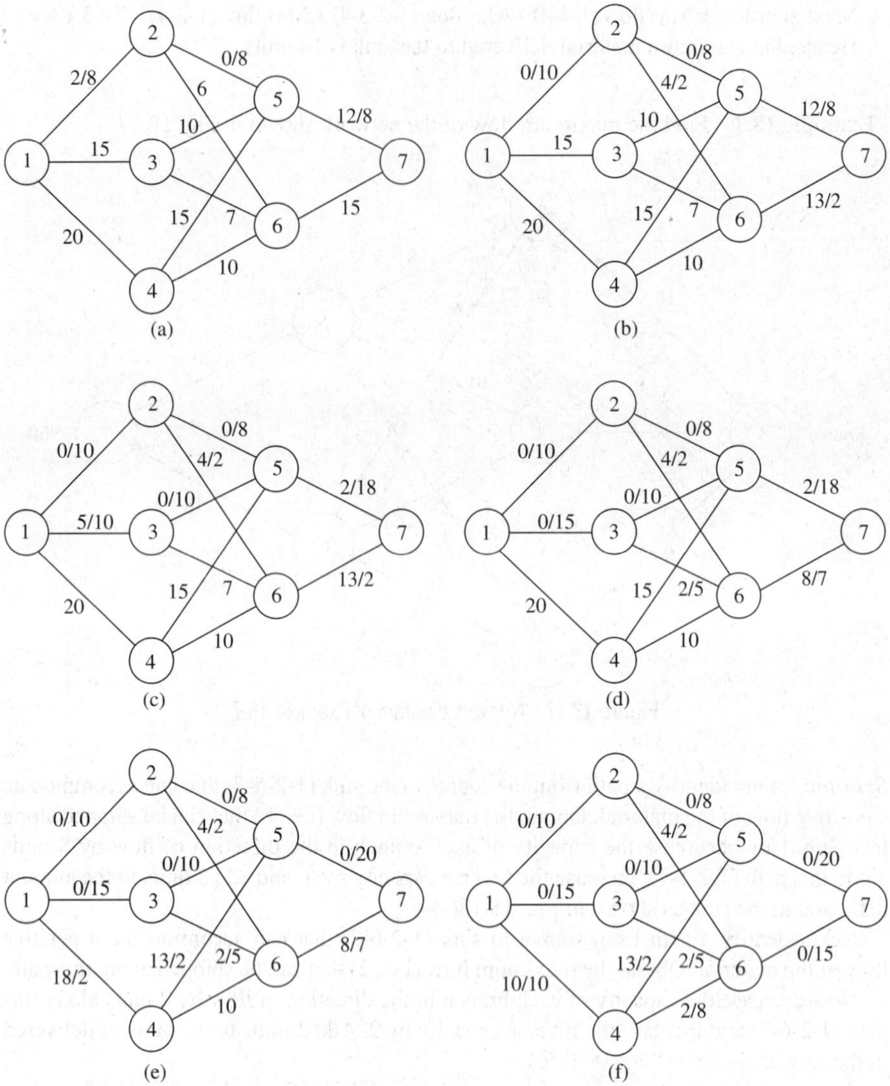

Figure 18.18 Network diagram of Example 18.7 to depict maximum flow through various paths (a) (1-2-5-7) (b) (1-2-6-7) (c) (1-3-5-7) (d) (1-3-6-7) (e) (1-4-5-7) (f) (1-4-6-7)

Again, identify a path from the source to the sink (1-4-6-7). Obtain the maximum flow (i.e., 8) that can be shipped along the path. Now decrease the capacity of each branch in the direction of flow by 8 units along this path (1-4-6-7) and increase the reverse capacity by 8. Add 8 units to the amount delivered to the sink as shown in Fig. 18.18(f).

Now, no path exists from the source to the sink that can accommodate a positive flow of the material. The maximum flow is 35 $(8 + 2 + 10 + 5 + 2 + 8)$, that is, through paths 1-2-5-7, 1-2-6-7, 1-3-5-7, 1-3-6-7, 1-4-5-7, and 1-4-6-7 of the problem and this is the amount of material delivered to the sink.

Example 18.8 Find the maximum flow of the network shown in Fig. 18.19.

Solution First, we identify a path from the source to the sink (1-2-5-7) that can accommodate a positive flow of the material. Obtain the maximum flow (i.e., 10) that can be shipped along the path. Now, decrease the capacity of each branch in the direction of flow by 10 units along this path (1-2-5-7) and increase the reverse capacity by 10. Add 10 units to the amount delivered to the sink as shown in Fig. 18.20(a).

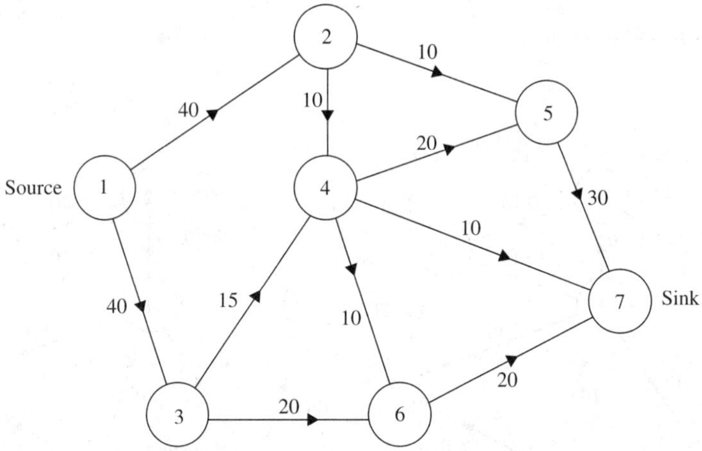

Figure 18.19 Network diagram of Example 18.8

Again, identify a path from source to sink (1-2-4-5-7) that can accommodate a positive flow of the material. Obtain the maximum flow (i.e., 10) that can be shipped along the path. Now, decrease the capacity of each branch in the direction of flow by 10 units along this path (1-2-4-5-7) and increase the reverse capacity by 10. Add 10 units to the amount delivered to the sink as shown in Fig. 18.20(b).

Now, we identify a path from source to sink (1-3-6-7) that can accommodate a positive flow of the material. Obtain the maximum flow (i.e., 20) that can be shipped along this path. Then, decrease the capacity of each branch in the direction of flow by 20 units along this path (1-3-6-7) and increase the reverse capacity by 20. Add 20 units to the amount delivered to the sink as shown in Fig. 18.20(c).

We can now identify a path from source to sink (1-3-4-7) that can accommodate a positive flow of the material. Obtain the maximum flow (i.e., 10) that can be shipped along this path. Now, decrease the capacity of each branch in the direction of the flow by 10 units along this path (1-3-4-7) and increase the reverse capacity by 10. Add 10 units to the amount delivered to the sink as shown in Fig. 18.20(d).

Next identify a path from the source to the sink (1-3-4-5-7) that can accommodate a positive flow of the material. Obtain the maximum flow (i.e., 5) that can be shipped along this path. Decrease the capacity of each branch in the direction of flow by 5 units along this path (1-3-4-5-7) and increase the reverse capacity by 5. Add 5 units to the amount delivered to the sink as shown in Fig. 18.20(e).

Now, no path exists from the source to the sink that can accommodate a positive flow of the material. Therefore, we can arrive at the maximum flow as $(10 + 10 + 20 + 10 + 5) = 55$,

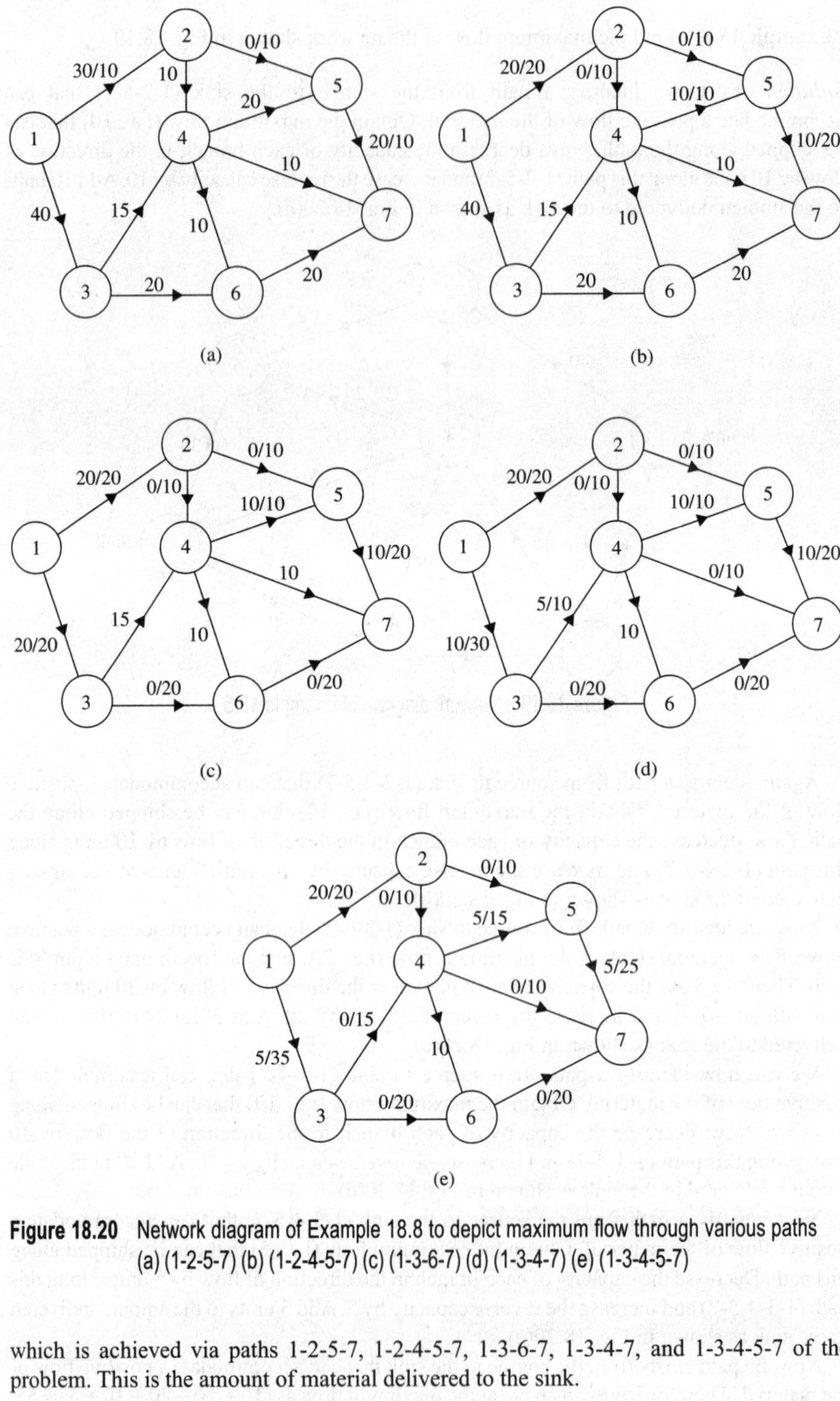

Figure 18.20 Network diagram of Example 18.8 to depict maximum flow through various paths (a) (1-2-5-7) (b) (1-2-4-5-7) (c) (1-3-6-7) (d) (1-3-4-7) (e) (1-3-4-5-7)

which is achieved via paths 1-2-5-7, 1-2-4-5-7, 1-3-6-7, 1-3-4-7, and 1-3-4-5-7 of the problem. This is the amount of material delivered to the sink.

RECAPITULATION

1. In Dijkstra algorithm, there are two types of node labels: temporary and permanent. A temporary label can be replaced with another label if the shortest route to the same node can be determined. If the shortest route cannot be identified, then it becomes a node with a permanent label.
2. To obtain the minimum spanning tree, we first write the arc length in the increasing order of magnitude. Take S, the set of nodes of the minimum spanning tree. Now choose the minimum arc length (i, j), $i \neq j$, both i, j not belonging to S (select arbitrarily in the event of a tie).
3. In maximum flow problems, the main objective is to maximize the total amount of flow from the source to the sink. This amount is measured in two equivalent ways—either the amount leaving the source or the amount entering the sink.

EXERCISES

Multiple-choice Questions

18.1 In Dijkstra algorithm, all initial node labels
(a) are temporary
(b) are permanent
(c) may or may not be temporary
(d) none of these

18.2 In a minimum spanning tree, the arc length is taken in the
(a) increasing order of magnitude
(b) decreasing order of magnitude
(c) any order of magnitude
(d) none of these

18.3 A tree is a connected graph without any
(a) circuit (b) walk (c) path (d) none of these

State True or False

18.1 In Dijkstra algorithm, a temporary label can be replaced with another label if the shortest route to the same node can be identified.
18.2 A vertex is said to be isolated if it has no incident edge.
18.3 Every path in a graph is a chain.

Fill in the Blanks

18.1 In Dijkstra algorithm, there are two types of node labels: __________ and __________.
18.2 In a minimum spanning tree, take the minimum arc length (i, j), $i \neq j$, both i, j __________ to S.
18.3 A graph G is said to be __________ in which each vertex is assigned a unique name or label.

Review Questions

18.1 What do you mean by networking?
18.2 Write a short note on the shortest route problem.
18.3 Explain the procedure to obtain a minimum spanning tree.
18.4 Discuss the maximum flow problem.

Numerical Problems

18.1 Find the shortest route for the following networks:

(a)

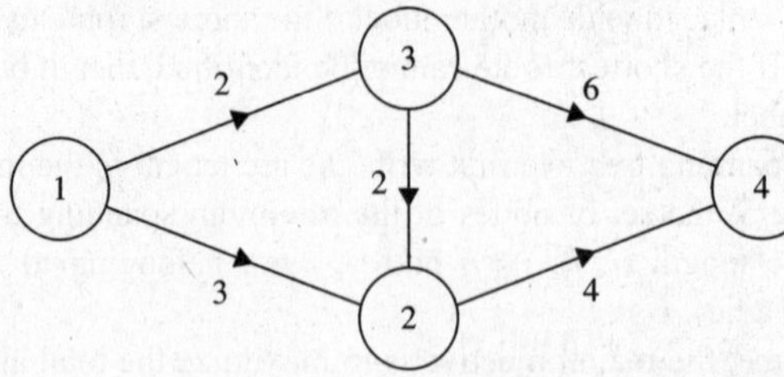

(b)

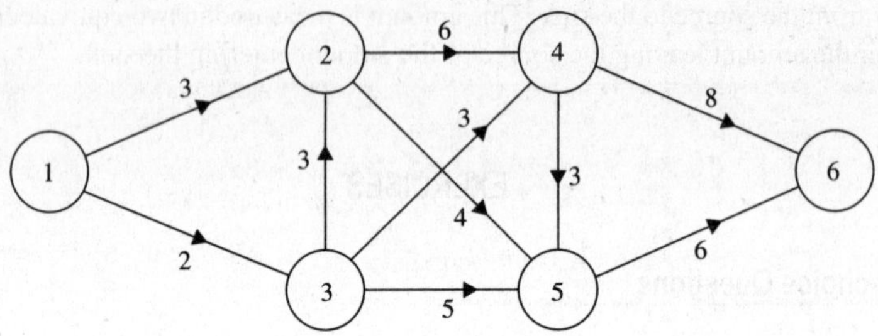

(c)

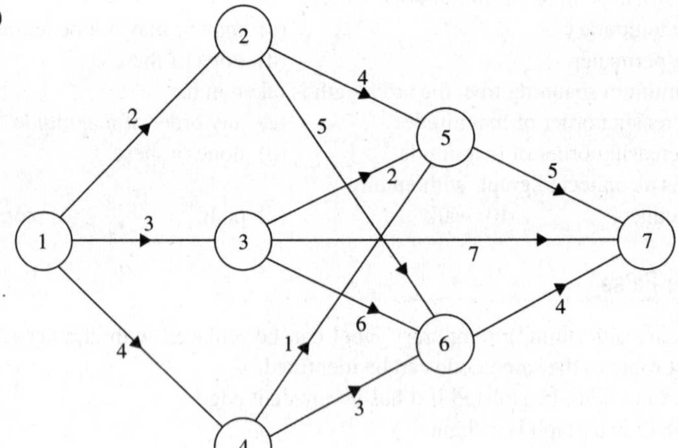

18.2 Find the minimum spanning tree of the following networks:

(a)

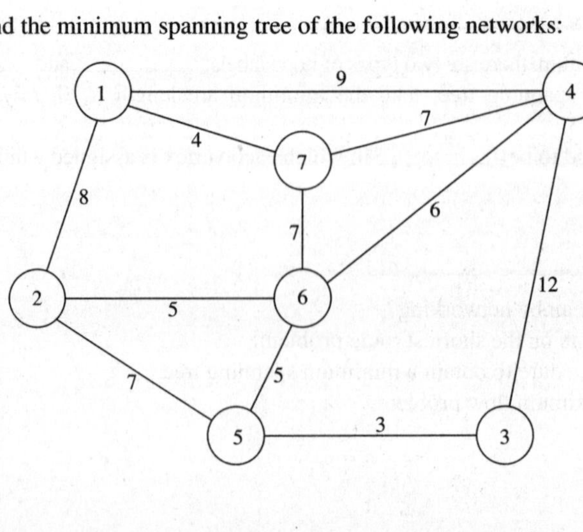

(b)

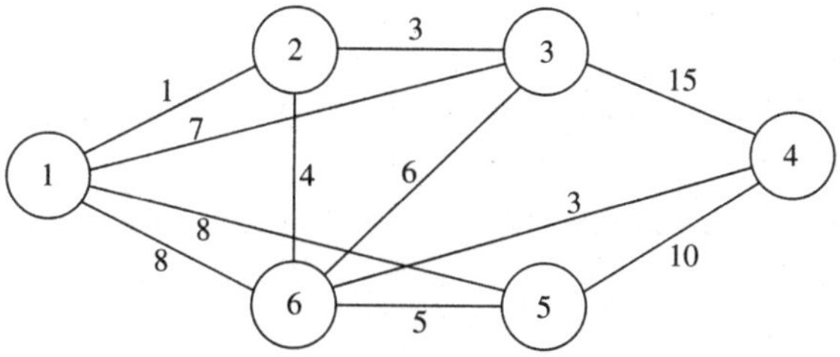

18.3 Find the maximum flow, for the following arcs and capacities:

Arc	Capacity	Arc	Capacity	Arc	Capacity
1–2	40	1–4	30	4–6	20
1–3	60	1–5	40	3–6	30
2–4	20	3–5	20	5–6	30

18.4 Find the maximum flow of the following network:

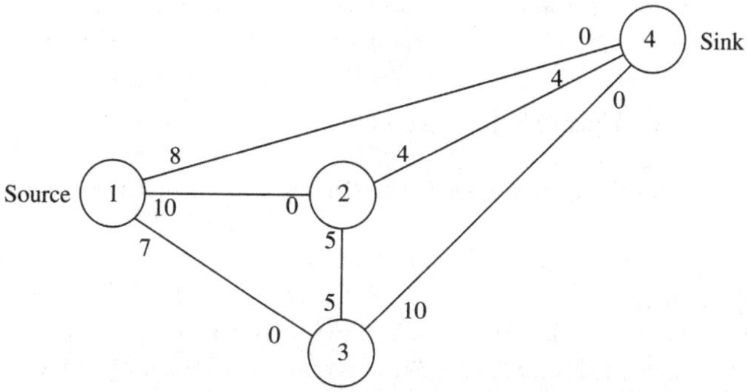

CONCEPTUAL PROBLEM

A manufacturing company has been awarded a contract to produce casings. The contract is for four years and is not expected to be renewed. The production process requires a specialized machine that the company does not have at present. It can buy the machine, maintain it for the four years of the contract, and then sell it at the scrap value, or it can replace the machine at the end of any given year by a new model. New models require less maintenance than the older ones. The estimated net operating cost (purchase price + maintenance – trade-in) for buying a machine at the beginning of year p and trading it at the beginning of year q is provided in the following table:

p \ q	1	2	3	4	5
1	–	13	16	34	50
2	–	–	15	24	39
3	–	–	–	17	27
4	–	–	–	–	14

(*Contd*)

(*Contd*)

Construct the network. Obtain a replacement policy that will minimize the total operating cost for the machine over the life of the contract.

Solution Using the aforegiven data, the network can be depicted as shown in Fig. 18.21.

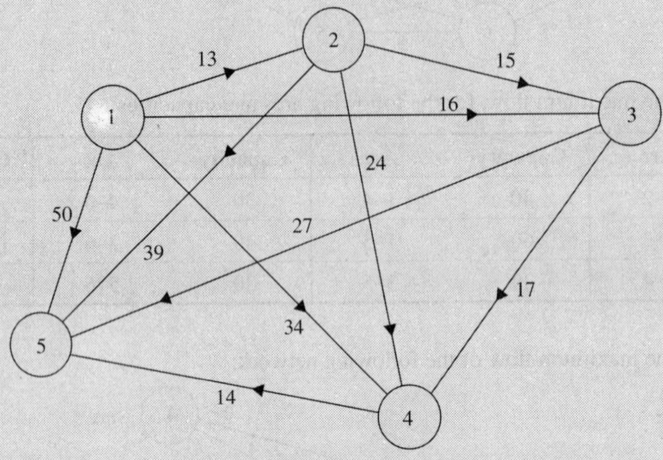

Figure 18.21 Network diagram

It is given that the network has five nodes. The calculations for these nodes are as follows:

Node 1: [0, -].
Node 2: [0 + 13, 1] = [13, 1].
Node 3: Reached from nodes 2 and 1: [13 + 15, 2] = [28, 2] and [0 + 16, 1] = [16, 1].
Node 4: Reached from nodes 1, 2, and 3: [0 + 34, 1] = [34, 1], [13 + 24, 2] = [37, 2] and [16 + 17, 3] = [33, 3].
Node 5: Reached from nodes 1, 2, 3, and 4: [0 + 50, 1] = [50, 1], [13 + 39, 2] = [52, 2], [16 + 27, 3] = [43, 3] and [33 + 14, 4] = [47, 4].

Thus, the shortest route covers a distance of 43 units. Now tracking back from node 5, by checking the second elements of the labels, we find the possible route as 1→3→5. Hence the path 1→3→5 represents the policy of buying a machine at the beginning of year 1, trading it for a new machine at the beginning of year 3, and finally scrapping the two-year-old machine at the beginning of year 5.

19 Replacement Models

After going through this chapter, the reader will be able to understand:

- the replacement policy
- the effect on the replacement policy when the value of money does not change with time
- the effect on the replacement policy when the value of money changes with time
- the variation in the replacement policy when the items/units fail completely all of a sudden

19.1 INTRODUCTION

In real-life situations, replacement of items such as machines and electric light bulbs, is needed due to the deteriorating efficiency, failure, or breakdown. Some of the conditions for the equipment to be changed are better design of equipment available in the market, deterioration of the old equipment that leads to expensive maintenance, or complete failure of the equipment. The deteriorating efficiency or complete breakdown may either be gradual or sudden. For example, vehicles such as cars, bikes, and trucks become increasingly expensive to maintain after a number of years. Similar instances can be a pipeline being blocked, an employee leaving his job, or an electric light bulb failing all of a sudden.

19.2 REPLACEMENT POLICY MODELS

Sometimes the cost of maintenance and repairing of certain items increases with time, and a stage might come when these costs become so high that it is more economical to opt for a replacement. Replacement policy models are classified into the following categories:

1. When the value of money does not change with time
2. When the value of money changes with time
3. When the items/units fail completely all of a sudden

We will now study in detail, the aforementioned categories of replacement policy models in this chapter.

19.3 REPLACEMENT POLICY WHEN THE VALUE OF MONEY DOES NOT CHANGE WITH TIME

In this section, we will find the optimum replacement age of an equipment/item whose maintenance cost increases with time. The value of money remains static during that time.

Let us consider C as the capital cost of the equipment, S as the scrap value of the equipment, N as the number of years the equipment would be in use, $f(t)$ as the maintenance cost function, and $A(n)$ as the average total annual cost. Two cases might arise:

Case 1 Let t be a continuous variable.
When the equipment is used for n years, the total cost (TC) incurred during this period is given as follows:

Total cost = Capital cost – Scrap value + Maintenance cost

$$= C - S + \int_0^n f(t)\,dt$$

And the average annual cost $A(n) = \dfrac{\text{TC}}{n} = \dfrac{C-S}{n} + \dfrac{1}{n}\int_0^n f(t)dt$

For the minimum cost, we have $\dfrac{dA(n)}{dn} = 0$

$$\Rightarrow -\frac{C-S}{n^2} - \frac{1}{n^2}\int_0^n f(t)\,dt + \frac{1}{n}f(n) = 0$$

$$\Rightarrow f(n) = \frac{C-S}{n} + \frac{1}{n}\int_0^n f(t)\,dt = A(n)$$

Now we calculate $d^2A(n)/dn^2$, which is greater than zero at $f(n) = A(n)$, that is, $d^2A(n)/dn^2 \geq 0$ at $f(n) = A(n)$.

Hence, if the maintenance cost is equal to the average total annual cost, the equipment can be replaced.

Case 2 Let t be a discrete variable.
The period of time is considered fixed at n, t takes the values 1, 2, 3, …, n

$$\text{Then } A(n) = \frac{C-S}{n} + \frac{1}{n}\sum_{t=1}^{n} f(t)$$

$A(n)$ will be minimum if

$$A(n+1) \geq A(n) \Rightarrow A(n+1) - A(n) \geq 0$$

$$\text{and } A(n-1) \geq A(n) \Rightarrow A(n) - A(n-1) \leq 0$$

Now, we have

$$A(n+1) = \frac{C-S}{n+1} + \frac{1}{n+1}\sum_{t=1}^{n+1} f(t)$$

$$= \frac{1}{n+1}\left[C - S + \sum_{t=1}^{n} f(t)\right] + \frac{1}{n+1} f(n+1)$$

$$= \frac{1}{n+1}\left[nA(n) + f(n+1)\right]$$

$$A(n+1) - A(n) = \frac{1}{n+1}\left[nA(n) + f(n+1)\right] - A(n)$$

$$\Rightarrow A(n+1) - A(n) = \frac{1}{n+1}\left[nA(n) + f(n+1) - (n+1)A(n)\right]$$

$$= \frac{1}{n+1}\left[f(n+1) - A(n)\right]$$

Thus, $A(n+1) - A(n) \geq 0$, if $f(n+1) \geq A(n)$

Similarly, $A(n) - A(n-1) \leq 0$ when $f(n) \leq A(n-1)$

We have

$$A(n) = \frac{1}{n}\left[C - S + \sum_{t=1}^{n} f(t)\right]$$

$$= \frac{1}{n}\left[C - S + \sum_{t=1}^{n-1} f(t)\right] + \frac{1}{n} f(n)$$

$$= \frac{1}{n}\left[(n-1)A(n-1)\right] + \frac{1}{n} f(n)$$

$$= \frac{1}{n}\left[(n-1)A(n-1) + f(n)\right]$$

$$A(n) - A(n-1) = \frac{1}{n}\left[f(n) - A(n-1)\right]$$

$$A(n) - A(n-1) \leq 0 \text{ when } f(n) \leq A(n-1)$$

Hence the replacement of the equipment is required at the end of n years, if the maintenance cost in the $(n+1)$th year is more than the average total cost in the nth year, and the maintenance cost of the nth year is less than the average total cost of the previous year.

Example 19.1 A company wants to replace an existing machine whose cost price is ₹17,500 and the scrap value is ₹500.

The maintenance costs (in ₹) are as follows:

Year	1	2	3	4	5	6	7	8
Maintenance cost (₹)	200	300	3500	1200	1800	2400	3300	4500

When should the machine be replaced?

Solution Given the running cost $f(n)$, the scrap value $S = ₹500$, and the cost of the machine $C = ₹17{,}500$. For considering the replacement, computations are given in the following table:

Year n	Running cost (₹) $f(n)$	Cumulative running cost (₹) $\sum f(n)$	Depreciation cost (₹) $C-S$	Total cost (₹) TC	Average cost (₹) $A(n)$
1	200	200	17,000	17,200	17,200
2	300	500	17,000	17,500	8750
3	3500	4000	17,000	21,000	7000
4	1200	5200	17,000	22,200	5550
5	1800	7000	17,000	24,000	4800
6	2400	9400	17,000	26,400	4400
7	3300	12,700	17,000	29,700	4242
8	4500	17,200	17,000	34,200	4275

On the basis of the computations made in the aforegiven table, we find that the best time for the replacement of the machine is at the end of the 7th year.

Example 19.2 The following table gives the running costs (per year) and resale price of certain equipment, whose purchase price is ₹5000.

Year	1	2	3	4	5	6	7	8
Running cost (₹)	1500	1600	1800	2100	2500	2900	3400	4000
Resale value (₹)	3500	2500	1700	1200	800	500	500	500

When should the equipment be replaced?

Solution Given, the purchase price is ₹5000.
Depreciation = Purchase price (cost of machine) – Resale price
To decide the best time for the replacement, we consider the following table:

Year n	Running cost (₹) $f(n)$	Cumulative running cost (₹) $\sum f(n)$	Depreciation cost (₹) $C-R$	Total cost (₹) TC	Average cost (₹) $A(n)$
1	1500	1500	1500	3000	3000
2	1600	3100	2500	5600	2800
3	1800	4900	3300	8200	2733.30
4	2100	7000	3800	10,800	2700
5	2500	9500	4200	13,700	2740
6	2900	12,400	4500	16,900	2816.6
7	3400	15,800	4500	20,300	2900
8	4000	19,800	4500	24,300	3037.30

From the aforegiven table, we observe that the best time to replace the equipment is at the end of the 4th year.

Example 19.3 A transport company finds from its past records that the maintenance costs per year of a truck, whose purchase price is ₹8000, are as follows:

Year	1	2	3	4	5	6	7	8
Maintenance cost (₹)	1000	1300	1700	2200	2900	3800	4800	6000
Resale price (₹)	4000	2000	1200	600	500	400	400	400

Determine the time at which it is profitable to replace the truck.

Solution Given the cost price (C) of the truck = ₹8000
Depreciation cost = $C - R$
To decide about the replacement of the truck, consider the following table:

Year n	Running cost (R) (₹) $f(n)$	Cumulative running cost (₹) $\sum f(n)$	Depreciation cost (₹) $C - R$	Total cost (₹) TC	Average cost (₹) $A(n)$
1	1000	1000	4000	5000	5000
2	1300	2300	6000	8300	4150
3	1700	4000	6800	10,800	3600
4	2200	6200	7400	13,600	3400
5	2900	9100	7500	16,600	3320
6	3800	12,900	7600	20,500	3416.60
7	4800	17,700	7600	25,300	3614.20
8	6000	23,700	7600	31,300	3916.20

On the basis of the aforementioned calculations, we find that the best time to replace the truck is at the end of the 5th year.

Example 19.4 A travelling company manager finds from his past records that the cost (per year) of running a car whose purchase price is ₹6000, is as follows:

Years	1	2	3	4	5	6	7	8
Running cost (₹)	1000	1200	1400	1800	2300	2800	3400	4000
Resale value (₹)	3000	1500	750	375	200	200	200	200

Find out at what age is replacement due.

Solution Given the purchase price of car is ₹6000.
Depreciation cost = $C - S$

To consider the replacement, computations are given in the following table:

Year n	Running cost (₹) $f(n)$	Cumulative running cost (₹) $\sum f(n)$	Depreciation cost (₹) $C-R$	Total cost (₹) TC	Average cost (₹) $A(n)$
1	1000	1000	3000	400	4000
2	1200	2200	4500	6700	3350
3	1400	3600	5250	8850	2950
4	1800	5400	5625	11,025	2756
5	2300	7700	5800	13,500	2700
6	2800	10,500	5800	16,300	2716
7	3400	13,900	5800	19,700	2814
8	4000	17,900	5800	23,700	2962

From the aforegiven table, we observe that the best time to replace the car is at the end of the 5th year.

19.4 REPLACEMENT POLICY WHEN THE VALUE OF MONEY CHANGES WITH TIME

Consider the scenario when the value of money changes with time. Let us assume that (a) the equipment has no salvage value (selling price), and (b) the maintenance costs are incurred at the beginning of different time periods.

Let the interest rate be r per year. Now if ₹1 is invested, then after one year it will be worth $(1 + r)$, since the maintenance cost increases with time, and each cost has to be incurred at the start of the time period. Similarly after 2 years, it would be worth $(1 + r)^2$ and so on. Hence, if a rupee is invested today, it will be worth $(1 + r)^n$ after n years. The expression $(1 + r)^n$ is called the payment compound amount factor of one rupee spent in n years' time. In other words, if a payment of one rupee is to be made over a period of n years, it is equivalent to paying $(1 + r)^{-n}$ rupees today. $(1 + r)^{-n}$ is called the present worth factor of one rupee spent in n years' time from now.

Suppose C is the initial cost of the equipment, R_n is the operating cost in year n, and v is the rate of interest such that the discount rate (present worth factor) is $v=(1+r)^{-1}$. Hence, the present value of all the future discounted costs V_n associated with a policy of replacing the equipment at the end of each n years is

$$V_n = \left[(C + R_0) + vR_1 + v^2R_2 + \dots + v^{n+1}R_{n-1} \right]$$

$$+ \left\{ (C + R_0)v^n + v^{n+1}R_1 + v^{n+2}R_2 + \dots + v^{2n-1}R_{n-1} \right\} + \dots$$

$$= \left\{ (C + R_0) + R_1v + R_2v^2 + \dots + R_{n-1}v^{n-1} \right\} \left\{ 1 + v^n + v^{2n} + \dots \right\}$$

$$= \left\{ C + \sum_{k=0}^{n-1} v^k R_k \right\} \times \sum_{k=0}^{\infty} (v^n)^k$$

$$V_n = \frac{\left\{C + \sum_{k=0}^{n-1} v^k R_k\right\}}{\left[1 - v^n\right]} \tag{19.1}$$

It is noteworthy that $\sum_{k=0}^{\infty} (v^n)^k = 1 + v^n + v^{2n} + \ldots = \frac{1}{1 - v^n}$ and

$$(1 - v^n)^{-1} = \frac{1}{1 + v^n + v^{2n} + \ldots}$$

Now, V_n will be minimum for that n for which the following conditions are satisfied:

$$V_{n+1} - V_n \geq 0 \text{ and } V_{n-1} - V_n > 0$$

Using Eq. (19.1), we have

$$V_{n+1} - V_n = \frac{\left\{C + \sum_{k=0}^{n} v^k R_k\right\}}{\left(1 - v^{n+1}\right)} - V_n$$

$$= \left\{C + \sum_{k=0}^{n-1} v^k R_k\right\}\left(1 - v^{n+1}\right)^{-1} + v^n R_n \left(1 - v^{n+1}\right)^{-1} - V_n$$

$$= \left(1 - v^n\right) V_n \left(1 - v^{n+1}\right)^{-1} + v^n R_n \left(1 - v^{n+1}\right)^{-1} - V_n$$

$$= \frac{1}{\left(1 - v^{n+1}\right)}\left[\left(1 - v^n\right) V_n + v^n R_n - \left(1 - v^{n+1}\right) V_n\right]$$

$$= \frac{1}{\left(1 - v^{n+1}\right)}\left[V_n - v^n V_n + v^n R_n - V_n + v^{n+1} V_n\right]$$

$$= \frac{v^n}{\left(1 - v^{n+1}\right)}\left[R_n - V_n + n V_n\right]$$

$$V_{n+1} - V_n = \frac{v^n\left[R_n - V_n (1 - v)\right]}{\left(1 - v^{n+1}\right)}$$

$$\Rightarrow V_{n+1} - V_n > 0 \text{ when } R_n - V_n (1 - v) > 0 \Rightarrow R_n > (1 - v) V_n$$

Similarly, we have $V_n - V_{n-1} = -\frac{\left[C + \sum_{k=0}^{n-2} v^k R_k\right]}{\left(1 - v^n\right)} + V_n$

$$= -\frac{\left[C + \sum_{k=0}^{n-1} v^k R_k\right] + v^{n-1} R_{n-1} + \left(1 - v^{n-1}\right) V_n}{\left(1 - v^{n-1}\right)}$$

$$= \frac{-(1-v^n)V_n + v^{n-1}R_{n-1} + V_n - v^{n-1}V_n}{(1-v^{n-1})}$$

$$= \frac{-V_n + v^n V_n + v^{n-1}R_{n-1} + V_n - v^{n-1}V_n}{(1-v^{n-1})}$$

$$= \frac{v^{n-1}[R_{n-1} + vV_n - V_n]}{(1-v^{n-1})}$$

$$V_n - V_{n-1} = \frac{v^{n-1}[R_{n-1} - (1-v)V_n]}{(1-v^{n-1})}$$

$$\Rightarrow V_n - V_{n-1} < 0 \text{ if } R_{n-1} - (1-v)V_n < 0 \Rightarrow R_{n-1} < (1-v)V_n$$

Hence, $R_{n-1} < (1-v)V_n < R_n$

$$\Rightarrow R_{n-1} < \frac{C + R_0 + vR_1 + v^2R_2 + \ldots + v^{n-1}R_{n-1}}{1 + v + v^2 + \ldots + v^{n-1}} < R_n$$

$$\left[\because \frac{1-v^n}{1-v} = \sum_{n=0}^{n-1} v^k = 1 + v + v^2 + \ldots + v^{n-1}\right]$$

The expression $\dfrac{C + R_0 + vR_1 + v^2R_2 + \ldots + v^{n-1}R_{n-1}}{1 + v + v^2 + \ldots + v^{n-1}}$ lying between R_{n-1} and R_n is called the *weighted average cost* of all previous n years, with weights $1, v, v^2, \ldots, v^{n-1}$ respectively. Thus, the optimal replacement policy of the equipment after n periods of time is as follows:

1. If the next period's operating cost is less than the weighted average of the previous costs, then do not replace the equipment.
2. If the next period's operating cost is greater than the weighted average of the previous costs, then replace the equipment.

Note When the time value of money is not taken into consideration, the rate of interest becomes zero, and hence, $v \to 1$; as a result we get,

$$R_{n-1} < \frac{C + R_0 + R_1 + R_2 + \ldots + R_{n-1}}{n} < R_n \text{ or } R_{n-1} < W(n) < R_n$$

19.5 PROCEDURE TO SELECT THE BETTER EQUIPMENT

To select the better of two given equipments, we take the following steps:

1. Consider the two types of equipment, say A and B. First, we find the best replacement age for both the equipment by using the following formula:

   $R_{n-1} < (1-v)V_n < R_n$

Suppose we get the optimum replacement age for A and B as n_1 and n_2 respectively.

2. Now, find the fixed annual payment (or weighted average cost) for each equipment, using the formula $W(n) = \dfrac{C + R_0 + vR_1 + v^2R_2 + \ldots + v^{n-1}R_{n-1}}{1 + v + v^2 + \ldots + v^{n-1}}$ and substitute $n = n_1$ for equipment A and $n = n_2$ for equipment B.
3. Now if
   (a) $W(n_1) < W(n_2)$, choose equipment A.
   (b) $W(n_1) > W(n_2)$, choose equipment B.
   (c) $W(n_1) = W(n_2)$, means both equipment are equally good.

19.6 REPLACEMENT OF EQUIPMENT THAT FAILS SUDDENLY

It is very difficult to predict the failure of any equipment in advance. As a result, our objective is to find a value of t that minimizes the total cost of replacement for the following policies:

Individual replacement policy Under this policy, the equipment is replaced immediately after it fails.

Group replacement policy Under this policy, equipment is replaced in a group, irrespective of whether they fail or not, with the condition that any item can be replaced individually if it fails before the optimum time.

19.7 GROUP REPLACEMENT THEOREM

Consider the replacement of all the items in a system after a time interval t, with the condition that an individual item can be replaced as and when they fail during this time period. Then

1. The group replacement must be made at the end of the tth period, if the cost of the individual replacement for the period is greater than the average cost per unit time period, through the end of the tth period.
2. The group replacement is not advisable at the end of period t, if the cost of the individual replacement at the end of the period $t - 1$ is less than the average cost per unit period through the end of the period t.

Proof We use the following notations:

N = Total number of items in the system
C_2 = Cost of replacing an individual item
C_1 = Cost of replacing an item in a group
$C(t)$ = Total cost of group replacement after time period t, $0 \le t < t$
$f(t)$ = Number of failures during the time period t

Then, we have $C(t) = NC_1 + C_2 \sum_{t=0}^{t-1} f(t)$.

The average cost of the group replacement per unit period of time during a period t, is thus given by

$$A(t)=\frac{C(t)}{t}=\frac{\left[NC_1+C_2\sum_{t=0}^{t-1}f(t)\right]}{t}$$

Now our objective is to find out the optimum t, which minimizes $C(t)/t$. In other words, we want to find out that value of t for which

$$\frac{C(t-1)}{t-1}>\frac{C(t)}{t} \text{ and } \frac{C(t+1)}{t+1}>\frac{C(t)}{t} \tag{19.2}$$

is better to replace all items the after the time period t.

Conditions given in Eq. (19.2) will constitute our policy for group replacement after time period t. On further simplification of inequalities in Eq. (19.2), we proceed as follows:

$$\frac{C(t+1)}{t+1}-\frac{C(t)}{t}>0$$

$$\Rightarrow \frac{NC_1+C_2\sum_{t=0}^{t}f(t)}{t+1}-\frac{C(t)}{t}>0$$

$$\Rightarrow \frac{NC_1+C_2\sum_{t=0}^{t-1}f(t)+C_2f(t)}{t+1}-\frac{C(t)}{t}>0$$

$$\Rightarrow \frac{t\left[NC_1+C_2\sum_{t=0}^{t-1}f(t)+C_2f(t)\right]-tC(t)-C(t)}{t(t+1)}>0$$

$$\Rightarrow \frac{tC(t)+tC_2f(t)-tC(t)-C(t)}{t(t+1)}>0$$

$$\Rightarrow tC_2f(t)-C(t)>0 \Rightarrow C_2f(t)>\frac{C(t)}{t} \tag{19.3}$$

$$\text{and } \frac{C(t)}{t}-\frac{C(t-1)}{t-1}<0$$

$$\Rightarrow \frac{C(t)}{t}-\frac{\left[NC_1+C_2\sum_{t=0}^{t-2}f(t)\right]}{t-1}<0$$

$$\frac{tC(t)-C(t)-t\left[NC_1+C_2\sum_{t=0}^{t-1}f(t)-C_2f(t-1)\right]}{t(t-1)}<0$$

$$\text{or } \frac{tC(t)-C(t)-tC(t)+tC_2f(t-1)}{t(t-1)}<0$$

$$\text{or } -C(t)+tC_2f(t-1)<0$$

$$\text{or } \frac{C(t)}{t}>C_2f(t-1) \tag{19.4}$$

From Eqs (19.3) and (19.4), we have

$$C_2f(t-1)<\frac{C(t)}{t}<C_2f(t)$$

or $tC_2 f(t-1) < C(t) < t - C_2 f(t)$

or $tC_2 f(t-1) < \left[NC_1 + C_2 \sum_{t=0}^{t-1} f(t) \right] < t - C_2 f(t)$

or $tf(t-1) - \sum_{t=0}^{t-1} f(t) < \frac{NC_1}{C_2} < t f(t) - \sum_{t=0}^{t-1} f(t).$

Example 19.5 Machine A is replaced every 3 years and machine B is replaced every 6 years. The appreciation of the value of money is assumed to be 10% per year. The yearly costs of both the machines are given as follows:

Year	1	2	3	4	5	6
Machine A	1000	200	400	1000	200	400
Machine B	1700	100	200	300	400	500

Find out which machine should be purchased.

Solution As the money carries the rate of interest, the present worth of the money to be spent over a period of one year is $v = \frac{100}{100+10} = \frac{10}{11} = 0.9091.$

The total discount cost (present worth) of A for 3 years is as follows:

$$1000 + 200 \times (0.9091) + 400 \times (0.9091)^2 = 1512$$

The total discount cost of B for 6 years is as follows:

$$1700 + 100 \times (0.9091) + 200 \times (0.9091)^2 + 300 \times (0.9091)^3 + 400 \times (0.9091)^4 + 500 \times (0.9091)^5 = ₹2765$$

The average yearly cost of machine A $= \frac{1512}{3} = ₹504.$

The average yearly cost of machine B $= \frac{2765}{6} = ₹461.$

This shows that the apparent advantage is with machine B. However, this comparison is unfair, as the periods for which the costs are considered are different. Therefore, if we consider a 6 year period for machine A as well, then the total discounted cost of A will be $1000 + 200 \times (0.9091) + 400 \times (0.9091)^2 + 1000 \times (0.9091)^3 + 200 \times (0.9091)^4 + 400 \times (0.9091)^5 = ₹26,470.$

This is ₹118 less then machine B over the same period. Thus, machine A should be purchased.

Example 19.6 For a particular type of transistors in a digital computer, the following failure rates have been noted:

End of the week	1	2	3	4	5	6	7	8
Probability of failure to date	0.05	0.13	0.25	0.43	0.68	0.88	0.96	1.00

The replacement cost of an individual failed transistor is ₹1.25. The decision is made to replace all these transistors simultaneously at fixed intervals, and to replace the individual transistors as they fail in service. If the cost of group replacement is 30 paise per transistor, what is the best interval between group replacements? At what group replacement price per transistor would a policy of strictly individual replacement become preferable to the adopted policy?

Solution Let's assume that there are 1000 transistors in use. Let p_i be the probability that a transistor, which was new when placed in position for use, fails during the ith week of its life. Thus, we have

$$p_1 = 0.05 \qquad p_5 = 0.68 - 0.43 = 0.25$$

$$p_2 = 0.13 - 0.05 = 0.08 \qquad p_6 = 0.88 - 0.68 = 0.20$$

$$p_3 = 0.25 - 0.13 = 0.12 \qquad p_7 = 0.96 - 0.88 = 0.08$$

$$p_4 = 0.43 - 0.25 = 0.18 \qquad p_8 = 1.00 - 0.96 = 0.04$$

Let N_i denote the number of replacements made at the end of the ith week. Then, we have

$$N_0 = \text{Number of transistors in the beginning} = 1000$$

$$N_1 = N_0p_1 = 1000 \times 0.05 = 50$$

$$N_2 = N_0p_2 + N_1p_1 = 1000 \times 0.08 + 50 \times 0.05 = 82.5$$

$$N_3 = N_0p_3 + N_1p_2 + N_2p_1 = 1.000 \times 0.12 + 50 \times 0.08 + 82 \times 0.05 = 128$$

$$N_4 = N_0p_4 + N_1p_3 + N_2p_2 + N_3p_1$$

$$= 1000 \times 0.18 + 50 \times 0.12 + 82 \times 0.08 + 128 \times 0.05$$

$$= 180 + 6 + 6.46 + 6.40 = 199$$

$$N_5 = N_0p_5 + N_1p_4 + N_2p_3 + N_3p_2 + N_4p_1$$

$$= 1000 \times 0.25 + 50 \times 0.18 + 82 \times 0.12 + 128 \times 0.08 + 199 \times 0.05$$

$$= 250 + 9 + 9.84 + 10.24 + 9.95 = 289$$

$$N_6 = N_0p_6 + N_1p_5 + N_2p_4 + N_3p_3 + N_4p_2 + N_5p_1$$

$$= 1000 \times 0.20 + 50 \times 0.25 + 82 \times 0.18 + 199 \times 0.08 + 289 \times 0.5$$

$$= 200 + 12.5 + 16.56 + 15.97 + 14.44 = 272$$

$$N_7 = N_0p_7 + N_1p_6 + N_2p_5 + N_3p_4 + N_4p_3 + N_5p_2 + N_6p_1$$

$$= 1000 \times 0.08 + 50 \times 0.20 + 82 \times 0.25 + 128 \times 0.18 + 199 \times 0.12 + 272 \times 0.05$$

$$= 80 + 10 + 16.4 + 20.50 + 29.22 + 24.88 + 13.6 = 194$$

$$N_8 = N_0p_8 + N_1p_7 + N_2p_6 + N_3p_5 + N_4p_4 + N_5p_3 + N_6p_2 + N_7p_1 = 195$$

From the aforementioned, it is observed that the expected number of transistors failing each week increases till the 5th week, and then starts decreasing, and later again increases from the 8th week. Therefore, N_i will oscillate till the system acquires a steady state. The expected life of each transistor is

$$1 \times p_1 + 2 \times p_2 + 3 \times p_3 + 4 \times p_4 + 5 \times p_5 + 6 \times p_6 + 7 \times p_7 + 8 \times p_8$$

$$= 1 \times 0.05 + 2 \times 0.08 + 3 \times 0.12 + 4 \times 0.18 + 5 \times 0.25 + 6 \times 0.2 + 7 \times 0.08 + 8 \times 0.04$$

$$= 0.05 + 0.16 + 0.48 + 0.54 + 1.25 + 1.2 + .056 + 0.32 = 4.62 \text{ weeks}$$

The average number of failures per week = 1000/4.62 = 216 (approximately)
Thus, the cost of individual replacement = 216 × 1.25 = ₹270 per week.

Now, as the replacement of all the 1000 transistors costs 30 paise per transistor, and the replacement of an individual transistor on failure costs ₹1.25, the average cost for different group replacement policies is given as follows:

End of week	Individual replacement	Total cost (₹) Individual group	Average cost (₹)
1	50	$50 \times 1.25 + 1000 \times 0.30 = 363$	363
2	132	$132 \times 1.25 + 1000 \times 0.30 = 465$	232.50
3	260	$260 \times 1.25 + 1000 \times 0.30 = 625$	208.30
4	459	$459 \times 1.25 + 1000 \times 0.30 = 874$	218.50

Here, the average cost is the lowest in week 3; the optimum interval between group replacements is 3 weeks. Further, since the average cost is less than ₹270 (for individual replacement), the policy of group replacement is better.

Example 19.7 At the starting time when $t = 0$, all items in a system are brand new. Each item has a probability of failing immediately before the end of the first month of life, and a probability $q = 1 - p$ of failing immediately before the end of the second month (i.e., all items fail by the end of the second month). If all items are replaced as they fail, show that the expected number of failures $f(x)$ at the end of month x is given by

$$f(x) = \frac{N}{1+q}\left[1-(-q)^{x+1}\right]$$

where N is the number of items in the system.

If the cost per item of individual replacement is C_1 and the cost per item of group replacement is C_2, find the condition under which

(a) Group replacement policy at the end of each month is the most profitable.
(b) No group replacement policy is better than a policy of pure individual replacement.

Solution We use the following notations:
N = Number of items in the system at the beginning

N_1 = Number of items expected to fail at the end of the first month

$$= N_0 p = N(1-q), \text{ as } p = 1-q$$

N_2 = Number of items expected to fail at the end of the second month

$$= N_0 q + N_1 p = Nq + N(1-q)^2 = N(1-q+q^2)$$

N_3 = Number of items expected to fail at the end of the third month

$$= N_1 q + N_2 p = N(1-q)q + N(1-q+q^2)(1-q)N(q-q^2+1-q+q^2-q+q^2-q^3)$$

$$= N(1-q+q^2-q^3) \text{ and so on}$$

In general, $N_r = N\left\{1-q+q^2-q^3+\ldots+(-q)^r\right\}$

$$N_{r+1} = N_{r-1}q + N_r p$$

$$= N\left[1-q+q^2-\ldots+(-q)^{r-1}\right]q + N\left[1-q+q^2-\ldots+(-q)^r\right](1-q)$$

$$= N\left[1-q+q^2+\ldots+(-q)^{r+1}\right]$$

Using mathematical induction, the expected number of failures at the end of month x will be given by

$$f(x) = N\left[1-q+q^2-q^3+\ldots+(-q)^x\right] = \frac{N\left(1-(-q)^{x+1}\right)}{1+q}.$$

The value of $f(x)$ at the end of month x will vary for different values of $(-q)^{x+1}$ and it will reach the steady state at $x \to \infty$.
Thus in the steady state case, the expected number of failures will be

$$\lim_{x\to\infty} f(x) = \lim_{x\to\infty} N\left(1-(-q)^{x+1}\right)/(1+q)$$

$$= \frac{N}{(1+q)} \text{ as } q<1 \text{ and } (-q)^{x+1} \to 0 \text{ as } x \to \infty$$

$$= \text{Total number of items in the system/Mean age}$$

Now, since C_1 is the cost of replacement per item individually and C_2 is the cost of an item in a group, therefore

(i) If we have a group replacement at the end of each month, then the cost of replacement is NC_2

(ii) If we have a group replacement policy at the end of every other month, then the cost is

$$NC_2 + NpC_1 \tag{19.5}$$

(iii) The average life of an item $= 1\times p + 2\times q = 1\times(1-q) + 2q = 1+q$

Thus, the average number of failures is $N/(1+q)$ and hence, the cost of individual replacement

$$= \frac{NC_1}{(1+q)} \tag{19.6}$$

(a) The group replacement policy at the end of first month will be better than individual replacement, if the total cost of group replacement is less than the average monthly cost of individual replacement.

Hence

$$N(1-q)C_1 + NC_2 < \frac{NC_1}{1+q}$$

$$\Rightarrow C_2 < \frac{C_1-(1-q)(1+q)C_1}{1+q}$$

$$\Rightarrow C_2 < \frac{C_1-\left(1-q^2\right)C_1}{1+q}$$

$$\Rightarrow C_2 < \frac{q^2C_1}{1+q} \tag{19.7}$$

End of month	Individual replacement	Total cost (₹)	Average cost (₹)
1	N_1	$N_1C_1 + NC_2$	$N_1C_1 + NC_2$
2	$N_1 + N_2$	$(N_1+N_2)C_1 + NC_2$	$\frac{(N_1+N_2)C_1 + NC_2}{2}$

Now for a group replacement at the end of every second month, the total cost of replacement will be $(N_1 + N_2)\, C_1 + NC_2 = N(2-2q-q^2)\, C_1 + NC_2$.

The average monthly cost of group replacement at the end of the second month is $[N(2-2q-q^2)\, C_1 + NC_2]/2$.

In this case, the group replacement policy will be better than the individual replacement policy, if average monthly cost of group replacement < Average monthly cost of individual replacement.

$$\text{or } N\left[1-q+\frac{q^2}{2}\right]C_1+\frac{1}{2}NC_2<\frac{NC_1}{1+q}$$

$$\text{or } \frac{1}{2}NC_2<\frac{NC_1}{1+q}-NC_1\left(1-q+\frac{q^2}{2}\right)$$

$$\text{or } \frac{1}{2}C_2<\frac{C_1}{1+q}-C_1\left(1-q+\frac{q^2}{2}\right)$$

$$\text{or } \frac{1}{2}C_2<C_1\left[\frac{2-(2-2q+q^2)(1+q)}{2(1+q)}\right]$$

$$\text{or } C_2<C_1\left[\frac{2-(2-2q+q^2+2q-2q^2+q^3)}{1+q}\right]$$

$$\text{or } C_2<C_1\left[\frac{(2-2+2q-q^2-2q+2q^2-q^3)}{1+q}\right]$$

$$\text{or } C_2<\frac{C_1q^2(1-q)}{1+q} \qquad (19.8)$$

(b) For the individual replacement policy to be better than any of the group replacement policies discussed so far, from Eqs (19.7) and (19.8) we have:

$$C_2>\frac{C_1q^2}{1+q} \text{ or } C_1<\frac{C_2(1+q)}{q^2}$$

$$\text{and } C_2>\frac{C_1q^2(1-q)}{1+q} \text{ or } \Rightarrow C_1<\frac{(1+q)C_2}{q^2(1-q)}$$

Since $q<1$, so we have $\frac{(1+q)}{q^2}<\frac{1+q}{q^2(1-q)}$

Hence, $C_1<\frac{(1+q)C_2}{q^2}$.

RECAPITULATION

1. The replacement of the equipment/items at the end of n years is required, if the maintenance cost in the $(n+1)$th year is more than the average total cost in the nth year and the maintenance cost of the nth year is less than the previous $(n-1)$th years average total cost.
2. For the optimal replacement policy of the equipment after n periods, do not replace the equipment if the next period's operating cost is less than the weighted average of the previous costs.

(Contd)

(Contd)

3. For the optimal replacement policy of the equipment after n periods, replace the equipment if the next period's operating cost is greater than the weighted average of the previous cost.
4. Group replacement must be made at the end of tth period, if the cost of individual replacement for the period is greater than the average cost per unit time period at the end of the tth period.
5. Group replacement is not advisable at the end of the period t, if the cost of individual replacement at the end of period $t - 1$ is less than the average cost per unit period at the end of period t.

EXERCISES

Multiple-choice Questions

19.1 Replacement of an item is required when
(a) the present year's running cost is less than the previous year's average cost
(b) the average annual cost for n years becomes equal to the current annual running cost
(c) the running cost for next year is more than the average cost of the nth year
(d) none of these

19.2 The group replacement policy is suitable for similar low cost items when
(a) the failure is sudden
(b) the failure is complete and sudden
(c) the failure occurs over a period of time
(d) none of these

19.3 The problem of replacement is faced when a working item fails
(a) gradually
(b) suddenly
(c) both gradually and suddenly
(d) none of these

19.4 The present worth factor of one rupee spent in n years' time from now onwards is
(a) $(1 + r)^n$
(b) $(1 + r)^{-n}$
(c) $(1 - r)^{-n}$
(d) none of these

State True or False

19.1 The group replacement policy is suitable for similar low cost items that fail suddenly.

19.2 Machines become less effective either due to gradual or sudden deterioration in their efficiency.

19.3 The replacement of an item is required when the average annual cost for n years becomes equal to the current annual running cost.

19.4 When the operational efficiency of an item deteriorates with time, it is economical to replace the same with a new one.

Fill in the Blanks

19.1 If the present year's running cost is __________ than the previous year's average cost, then do not replace.

19.2 Replacement of an item is needed when it fails ________.

19.3 If the effect of the time-value of money is to be considered, then the replacement decision analysis must be based upon an equivalent ________ cost.

19.4 The payment compound amount factor of one rupee spent in n years' time is ________.

Review Questions

19.1 What do you mean by replacement policy?

19.2 Explain the various types of replacement situations.

19.3 Write the replacement policy when the value of money does not change with time.

19.4 Write the replacement policy when the value of money changes with time.

19.5 Write a short note on group replacement.

Numerical Problems

19.1 A company is thinking to replace an existing machine, whose cost price is ₹12,200, and the scrap value is ₹200.

The running (maintenance and operating) costs (in ₹) are as follows:

Year	1	2	3	4	5	6	7	8
Running cost	200	500	800	1200	1800	2500	3200	4000

When should the machine be replaced?

19.2 The purchasing cost of machine A is ₹9000, its annual operating cost is ₹200 for the first year, and then it increases by ₹2000 every year. Determine the best age at which to replace the machine. If the optimum replacement policy is followed, what will be the average yearly cost of owning and operating the machine?

19.3 The purchasing cost of machine B is ₹10,000, its annual operating cost is ₹400 for the first year, and then it increases by ₹800 every year. Now, if you have a machine A (from the problem 19.2), which is one year old, should you replace it with B? If so, when is the ideal time to replace it?

19.4 The data collected in running a machine, whose cost is ₹60,000 is as follows:

Year	1	2	3	4	5
Resale value (₹)	42,000	30,000	20,400	14,400	9,650
Cost of spares (₹)	4000	4270	4880	5700	6800
Cost of labour (₹)	14,000	16,000	18,000	21,000	25,000

Determine the optimum period for the replacement of the machine.

19.5 The cost of a machine is ₹6100 and its scrap value is ₹100. The maintenance costs are as follows:

Year	1	2	3	4	5	6	7	8
Maintenance cost (₹)	100	250	400	600	900	1200	1600	2000

When should the machine be replaced?

19.6 A vehicle company manager finds from his past records that the costs (per year) of running a vehicle, whose purchase price is ₹50,000 are as follows:

Year	1	2	3	4	5	6	7
Running cost (₹)	5000	6000	7000	9000	11,500	16,000	18,000
Resale value (₹)	30,000	15,000	7500	9750	2000	2000	2000

Thereafter, the running cost increases by ₹2000, but the resale value remains constant at ₹2000. Find out at what age is the replacement due.

19.7 A gas pipeline is due for repairs. It will cost ₹10,000 and last for 3 years. Alternatively, a new gas pipeline can be laid at a cost of ₹30,000 and would last for 10 years. Assuming the cost of capital to be 10% and ignoring the salvage value, which alternative should be chosen?

19.8 For a particular type of light bulb, the following failure rates have been observed:

Week	1	2	3	4	5
% Failing by end of week	10	25	50	80	100

There are 1000 bulbs in use, and it costs ₹2 to replace an individual bulb that has burnt out. If all the bulbs were replaced simultaneously, it would cost 50 paise per bulb. It is proposed to replace all the bulbs at fixed intervals, whether or not they have burnt out or not and to continue replacing burnt out bulbs as they fail. When should all the bulbs be replaced?

CONCEPTUAL PROBLEM

Companies now focus on providing better quality of equipment and producing better designs, which have a longer life as compared to inferior products. A company is using a machine whose purchase price is ₹13,000. The installation charges amount to ₹3600 and the machine has a scrap value of only ₹1600, because the firm has a monopoly. The maintenance cost in various years is given in the following table:

Year	1	2	3	4	5	6	7
Cost (in ₹)	250	750	1000	1500	2100	2900	4000

After how many years should the machine be replaced on economic considerations, assuming that the machine replacement can only be done at the year end?

Solution Given, the purchase price of the machine is ₹13,000 and the installation charges are ₹3600.

Then, $C = 13,000 + 3600 = ₹16,600$
Scrap value $(S) = ₹1600$
and $C - S = 16,600 - 1600 = ₹15,000$

(Contd)

(*Contd*)

Computations to determine the time period after which the machine should be replaced are given in the following table:

Year n	Running cost (₹) $f(n)$	Cumulative running cost (₹) $\sum f(n)$	Depreciation cost (₹) $C-S$	Total cost (₹) TC	Average cost (₹) $A(n)$
1	250	250	15,000	15,250	15,250
2	750	1000	15,000	16,000	8000
3	1000	2000	15,000	17,000	5666
4	1500	3500	15,000	18,500	4625
5	2100	5600	15,000	20,600	4125
6	2900	8500	15,000	23,500	3916
7	4000	12,500	15,000	27,500	3928

On the basis of computations made in the aforegiven table, we find that the best replacement year for the machine is the end of the sixth year. Hence, the machine should be changed after the sixth year.

20 Simulation

After studying this chapter, the reader will be able to

- comprehend random numbers and pseudo-random numbers
- explain the mid-square method of generating pseudo-random numbers
- get acquainted with the multiplicative and mixed congruential method
- explain the generation of random variates
- understand the applications of simulation in queuing models
- grasp the advantages and disadvantages of simulation models
- discuss the simulation languages

20.1 INTRODUCTION

Simulation is probably the most important technique used for analysing a number of complex systems and estimating their characteristics. This technique is used when the analytical methods for analysing the characteristics of a complex system do not exist. Basically, simulation is a technique of manipulating the model of a system through the process of imitation. The characteristics of a system are estimated by artificially obtaining a large number of samples of the variables of the system. In general, simulation involves a large number of computations. With the advent of computers, these computations can now be carried out in much lesser time as compared to performing them manually.

In operations research, due to the development of superior models such as transportation, assignment, inventory, and queuing models, simulation is adopted as the last resort when everything else fails. Nevertheless, it has found extremely useful applications in complex queues, inventory control, financial planning, capital budgeting, etc. In this chapter, we shall discuss some of these applications.

20.2 BASIC TERMINOLOGIES

Before providing a detailed description of simulation, we shall discuss certain necessary preliminaries.

Sample A set of finite objects extracted from a population of certain objects, for a specific purpose is called a sample.

Random sample A sample is called a random sample if in every draw of units from the sample, each unit has an equal opportunity to be drawn, that is, the probability of drawing every unit from the sample is the same. A random sample is the true representation of a large population.

Random variable Random variable is a real valued function defined on a sample space associated with an experiment.

Random variate The numerical value of a random variable is called a random variate.

Random number A number in a sequence of numbers, whose probability of occurrence is the same as that of any other number in that sequence, is called a random number. A random number is usually denoted by u, and always lies between 0 and 1, that is, $0 \le u < 1$. Thus, a random number u is always greater than or equal to zero but is strictly less than one.

Pseudo-random numbers Random numbers are called pseudo-random numbers when they are generated using some deterministic process but qualify the predetermined statistical test for randomness.

Seed The initial number used in generating pseudo-random numbers is called seed.

20.3 RANDOM NUMBERS AND PSEUDO-RANDOM NUMBERS

Random numbers must satisfy the basic requirement that the order in which they occur should be independent of any deterministic order. It is not practical to obtain random numbers manually, because it requires much labour. The most common method to generate random numbers is by using computers.

Many tables of random numbers are available in related literature. To generate random numbers, many sophisticated mechanical or electronic devices are used. Tables of random numbers generated by these methods are available in statistical texts. These were provided by RAND Corporation (1955). While selecting random numbers from these tables, the starting point is immaterial. We may start with any number of any row or column and proceed in the same row or column for the next number. Thus, a consistent pattern should be used in drawing random numbers from these tables. For selecting random numbers for more than one variable, different lists of random numbers for each variable should be used.

We shall now discuss some of the methods of generating pseudo-random numbers.

20.3.1 Mid-square Method or Technique of Generating Pseudo-random Numbers

The mid-square technique of generating pseudo-random numbers is one of the oldest techniques. For generating a pseudo-random number using the mid-square method, consider an initial four digit number (seed) x_0 in [0, 1]. The seed is squared, and the resulting number is supposed to have eight digits (in case of less than eight, leading zeros are inserted). From this number, the middle four digits are extracted as the required number. Subsequently, this number is again squared. In this manner, a new pseudo-random number is generated each time by squaring the previously generated random number. This process is continued till we obtain the already generated random number.

Example 20.1 Using the mid-square method, generate pseudo-random numbers starting with seed 0300.

Solution Consider the seed $x_0 = 0300$. This number x_0 is multiplied with itself to obtain an eight-digit number. Else, insert the requisite number of zeros at the beginning to make it an eight-digit number. Now, retain the middle four digits cancelling the first two and last two digits. The number thus obtained is 0900. The procedure for generating the requisite pseudo-random numbers is as follows:

$u_0 =$	$x_0 =$ **0300**	$u_3 =$	65**6100**00	$u_5 =$	04**4100**00
	× 0300		× 6100		× 4100
$u_1 =$	00**0900**00	$u_4 =$	37**2100**00	$u_6 =$	16**8100**00 $= u_2$
	× 0900		× 2100		× 8100
$u_2 =$	00**8100**00				
	× 8100				

From the aforegiven data, we observe that the sequence of random numbers obtained is $u_0, u_1, u_2, u_3, u_4, u_5$.

20.3.2 Limitations of Mid-square Method

The following are the limitations of the mid-square method:

1. This method tends to degenerate rapidly. A loop may generate and as a result, the same sequence of random numbers may be repeated. For example, $u_6 = u_2$ in Example 20.1 and the sequence u_2, u_3, u_4, u_5 is repeated.
2. This method is time-consuming as numerous multiplications are required to access the middle digits.

20.3.3 Multiplicative Congruential or Power Residual Technique

Two integers a and b are called congruent modulo (m), written as $a \equiv b \pmod{m}$, m being a positive integer if and only if there exists an integer k such that $b - a = km$ and is read as 'a is congruent to b modulo m'. In $a \equiv b \pmod{m}$, a is the remainder on dividing b by m. Thus, a can take values $0, 1, 2, \ldots, (m - 1)$. For example, $1 \equiv 7 \pmod{3}$ means when 7 is divided by 3, 1 is the remainder and $k = 2$. Thus, $7 = 2 \times 3 + 1$ with $b = 7$, $k = 2$, $m = 3$, and $a = 1$.

In the multiplicative congruential method, we select the nth number x_n as the remainder of the division of the product of a constant number a and the $(n-1)$th number x_{n-1} by another constant m, denoted as follows:

$$x_n \equiv ax_{n-1} \pmod m), n = 1, 2, 3, \ldots \tag{20.1}$$

where '$\equiv ax_{n-1}(\text{mod } m)$' means remainder from the division of product (ax_{n-1}) by m. The seed x_0 should be selected as randomly as possible. The first number (for $n = 1$) is $x_1 \equiv ax_0 \,(\text{mod } m)$.

The second number $(n = 2)$ is $x_2 = ax_1(\text{mod } m) = a^2x_0 \,(\text{mod } m)$.

In general, $x_k = a^kx_0 \,(\text{mod } m)$.

Then, we can obtain the pseudo-numbers $u_k \in [0,1)$ after dividing x_k by m. That is,

$$u_k = x_k/m, \quad k = 1, 2, 3, \ldots \tag{20.2}$$

Thus, Eqs (20.1) and (20.2), taken together, give a scheme of generating pseudo-random numbers in the multiplicative congruential technique.

Example 20.2 Generate pseudo-random numbers by taking $a = 2$, $m = 5$, and $x_0 = 3$ using the multiplicative congruential method.

Solution The generated pseudo-random numbers are provided in the adjacent table. Thus, the random numbers generated are 0.2, 0.4, 0.8, and 0.6.

k	x_k	$u_k = X_k/m$
1	$(2 \times 3) \,(\text{mod } 5) = 1$	0.2
2	$(2 \times 1) \,(\text{mod } 5) = 2$	0.4
3	$(2 \times 2) \,(\text{mod } 5) = 4$	0.8
4	$(2 \times 4) \,(\text{mod } 5) = 3$	0.6
5	$(2 \times 3) \,(\text{mod } 5) = 1$	0.2 (cycling starts)

Note The numbers generated by algorithm Eqs (20.1) and (20.2) may begin to recycle from the mth iteration ($m = 5$ in Example 20.2). It has been proved that (for decimal computers) by using x_0 and a according to the following specifications, one can attain maximal cycle length of $5 \times 10^{d-3}$ (provided $d \geq 4$):

$x_0 = a$, a number not divisible by 2 or 5

$a = 200t + p$, where t is a positive integer and p is one of the following numbers:

$\pm(3, 11, 13, 19, 21, 27, 29, 37, 53, 59, 61, 67, 69, 77, 83, 91)$

20.3.4 Mixed Congruential Method

A lot of research has been conducted on the multiplicative congruential method, and as its generalization, the mixed congruential method has been devised. This method is considered to be one of the best methods to generate the pseudo-random numbers. The recursion formula for this method is given in Eq. (20.3).

$$x_n = (ax_{n-1} + c)(\text{mod } m), \; n = 1, 2, \ldots \tag{20.3}$$

Then, divide x_n by m to get pseudo-random numbers u_n, that is,

$$u_n = x_n / m \tag{20.4}$$

where additive constant c, multiplier a, and seed x_0 are all positive integers less than m.

Again, m is chosen to be 2^b or 10^d according to the computer-used binary or decimal. By a proper choice of c and a, all non-negative integers less than m can be generated, before any one of them is repeated. The selection of c and a is done according to the following scheme:

$$\left.\begin{array}{ll} c \text{ must be relatively prime to } m & \text{(a)} \\ a \equiv 1 \pmod{p} \text{ for each prime factor } p \text{ of } m & \text{(b)} \\ a = 1 \pmod{4} \text{ if 4 is a factor of } m & \text{(c)} \end{array}\right\} \quad (20.5)$$

Example 20.3 Generate single decimal pseudo-random numbers by choosing c and a according to the specification mentioned in Eq. (20.5).

Solution Let us take $m = 10$ (for one-digit pseudo-random numbers)

(a) Choose $c = 3$, which is relatively prime to 10, thus satisfying Eq. (20.5a).

(b) Choose $a = 1$.

Thus, selection of $c = 3$ and $a = 1$ satisfies both (a) and (b). Now let us select seed $x_0 = 3$, then the resulting pseudo-random numbers (u_i) are listed using Eqs (20.3) and (20.4):

k	x_k	$u_k = x_k/m$, $m = 10$
1	$(1\times3+3) \pmod{10} = 6$	0.6
2	$(1\times6+3) \pmod{10} = 9$	0.9
3	$(1\times9+3) \pmod{10} = 2$	0.2
4	$(1\times2+3) \pmod{10} = 5$	0.5
5	$(1\times5+3) \pmod{10} = 8$	0.8
6	$(1\times8+3) \pmod{10} = 1$	0.1
7	$(1\times1+3) \pmod{10} = 4$	0.4
8	$(1\times4+3) \pmod{10} = 7$	0.7
9	$(1\times7+3) \pmod{10} = 0$	0.0
10	$(1\times0+3) \pmod{10} = 3$	0.3

Thus, the ten pseudo-random numbers generated are 0.3, 0.6, 0.9, 0.2, 0.5, 0.8, 0.1, 0.4, 0.7, 0.0 as shown in tne adjacent table.

Note $x_0 = 0.3$ (the seed) is the first pseudo-random number.

Example 20.4 Generate two-digit pseudo-random numbers using the recursion formulae provided in Eqs (20.3) and (20.4) by selecting a and c as per the specifications prescribed in Eq. (20.5).

Solution For two-digit pseudo-random numbers, we must take $m = 100$. Let us take $c = 7$, so that the requirement in Eq. (20.5a) is satisfied. We have $p = 2$ or 5 as prime factors of 100. Further, 4 is a factor of 100 (= m). To satisfy Eqs (20.5b) and (20.5c), we must have $a = 1 \pmod 2$, $a = 1 \pmod 5$, and $a = 1 \pmod 4$. These give $a = 1 \pmod{20}$.

This means we can take $a = 1$, 21, 41, etc., as 1 – 1, 21 – 1, 41 – 1 are divisible by 20. Let us choose $a = 1$. Further let $x_0 = 1$. Then the pseudo-random numbers ($u_k = x_k/m$), generated using the scheme in Eqs (20.3) and (20.4) are listed as follows:

$$x_1 = (ax_0 + c) \pmod{100} = (1 \times 1 + 7) \pmod{100} = 8 \pmod{100} = 08, \text{ thus } u_1 = \frac{08}{100} = 0.08$$

$$x_2 = (ax_1 + c) \pmod{100} = (1 \times 8 + 7) \pmod{100} = 15 \pmod{100} = 15, \text{thus } u_2 = \frac{15}{100} = 0.15$$

$$x_3 = (ax_2 + c) \pmod{100} = (1 \times 15 + 7) \pmod{100} = 22 \pmod{100} = 0.22, \text{thus } u_3 = \frac{22}{100} = 0.22$$

Hence, the pseudo-random numbers generated are listed column-wise in the following table:

k	x_k	$u_k = x_k/100$	k	x_k	u_k	k	x_k	u_k	k	x_k	u_k	k	x_k	u_k
1	8	0.08	21	48	0.48	41	88	0.88	61	28	0.28	81	68	0.68
2	15	0.15	22	55	0.55	42	95	0.95	62	35	0.35	82	75	0.75
3	22	0.22	23	62	0.62	43	02	0.02	63	42	0.42	83	82	0.82
4	29	0.29	24	69	0.69	44	09	0.09	64	49	0.49	84	89	0.89
5	36	0.36	25	76	0.76	45	16	0.16	65	56	0.56	85	96	0.96
6	43	0.43	26	83	0.83	46	23	0.23	66	63	0.63	86	03	0.03
7	50	0.50	27	90	0.90	47	30	0.30	67	70	0.70	87	10	0.10
8	57	0.57	28	97	0.97	48	37	0.37	68	77	0.77	88	17	0.17
9	64	0.64	29	04	0.04	49	44	0.44	69	84	0.84	89	24	0.24
10	71	0.71	30	11	0.11	50	51	0.51	70	91	0.91	90	31	0.31
11	78	0.78	31	18	0.18	51	58	0.58	71	98	0.98	91	38	0.38
12	85	0.85	32	25	0.25	52	65	0.65	72	05	0.05	92	45	0.45
13	92	0.92	33	32	0.32	53	72	0.72	73	12	0.12	93	52	0.52
14	99	0.99	34	39	0.39	54	79	0.79	74	19	0.19	94	59	0.59
15	06	0.06	35	46	0.46	55	86	0.86	75	26	0.26	95	66	0.66
16	13	0.13	36	53	0.53	56	93	0.93	76	33	0.33	96	73	0.73
17	20	0.20	37	60	0.60	57	00	0.00	77	40	0.40	97	80	0.80
18	27	0.27	38	67	0.67	58	07	0.07	78	47	0.47	98	87	0.87
19	34	0.34	39	74	0.74	59	14	0.14	79	54	0.54	99	94	0.94
20	41	0.41	40	81	0.81	60	21	0.21	80	61	0.61	100	01	0.01

Thus in mixed congruential method, we can see that all non-negative integers less than m (= 100) have been generated through proper selection of c and a. If more than 100 pseudo-random numbers are needed, we can generate another sequence by changing the seed or parameter. These sequences are used one by one. Further if four-digit pseudo-random numbers are required, we can take $m = 1000$.

Example 20.5 Using mixed congruential method, generate two-digit pseudo-random numbers when $a = 21$, $m = 100$, $c = 7$, and $x_0 = 20$.

Solution Using the procedure applied in Example 20.4, the pseudo-random numbers generated are listed in the adjacent table arranged row-wise:

20	27	74	61	88	55	62	09	96	23
90	97	44	31	58	25	32	79	66	93
60	67	14	01	28	95	02	49	36	63
30	37	84	71	98	65	72	19	06	33
00	07	54	41	68	35	42	89	76	03
70	77	24	11	38	05	12	59	46	73
40	47	94	81	08	75	82	29	16	43
10	17	64	51	78	45	52	99	86	13
80	87	34	21	48	15	22	69	56	83
50	57	04	91	18	85	92	39	26	53

20.4 MONTE CARLO SIMULATION

The Monte Carlo simulation technique involves repetitively conducting experiments on the model of the system under study, with a known probability distribution. To arrive at the random samples, random numbers are used. If a system cannot be described using a standard probability distribution such as exponential, normal, or Poisson, an empirical probability distribution can be constructed. The Monte Carlo simulation technique consists of the following steps:

1. Setting-up probability distribution(s) for values to be analysed
2. Building a cumulative probability distribution for involved random variable(s)
3. Assigning a given set of random numbers, values, or a range of values, for involved random variables
4. Conducting the simulation a requisite number of times (The more number of times the simulation is conducted, the better will be the simulated result.)

20.5 GENERATION OF RANDOM VARIATES

In Section 20.3, we have discussed random numbers and pseudo-random numbers. In this section, our aim is to generate a random sample from any known probability distribution. It is an interesting fact of probability theory, that a random sample can be generated from any known probability distribution (continuous or discrete). This means in Monte Carlo method, the sequence of random numbers becomes the basic sequence.

20.5.1 Continuous Random Variable *X*

First, we consider the continuous case. Let X be a random variable with the probability density function (PDF) $f(x)$ and cumulative density function (CDF) $F(X)$, then

$$F(x) = P(X \leq x) = \int_{-\infty}^{x} f(t)dt$$

Now, consider a new random variable U defined by

$$U = F(X) = \int_{-\infty}^{X} f(t)dt \tag{20.6}$$

where F is CDF of X

$$\text{Then } u = F(x) = \int_{-\infty}^{x} f(t)dt \text{ and } \frac{du}{dx} = f(x) \tag{20.7}$$

From the probability theory, the PDF $g(u)$ of U is $g(u) = \dfrac{dx}{du} f(x)$ (20.8)

Substituting Eq. (20.7) in Eq. (20.8), we have $g(u) = \dfrac{f(x)}{f(x)} = 1, 0 \leq u < 1$, $\left[\dfrac{du}{dx} = F'(x) = f(x)\right]$ which is the PDF of uniform distribution on [0, 1].

This proves that random variable U defined in Eq. (20.6) is a uniform distribution over [0, 1] for any continuous random variables X.

Since $F(x)$ is monotonic, from Eq. (20.7) we have $x = F^{-1}(u)$.

Thus, given a random number $u \in [0, 1)$ a random variate x from X can be obtained by solving

$$u = F(x) \qquad \text{for } x \text{ (Fig. 20.1)}$$

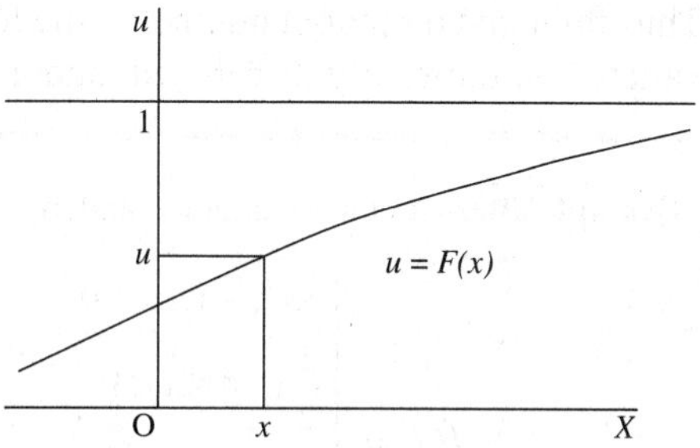

Figure 20.1 Representation of graph of $u = F(x)$

Thus, $x = F^{-1}(u)$ (20.9)

It is called the inverse transformation method for generating a random variate.

Generation of Exponential Random Variate

Exponential distribution is useful in queues for the distribution of inter-arrival and service times. If the mean inter-arrival time of customers in a queuing system is $1/\lambda$, then the probability density $f(t)$ of service time T is given as $f(t) = \lambda e^{-\lambda t}, t \geq 0, \lambda \geq 0$.

The distribution function $F(t) = \int_{-\infty}^{t} f(t)dt = \int_{0}^{t} \lambda e^{-\lambda t} dt = 1 - e^{-\lambda t}, t \geq 0.$

Thus, if u is a random number in [0, 1) then using Eq. (20.7) $u = F(t) = 1 - e^{-\lambda t}, t \geq 0$

or $t = -\dfrac{1}{\lambda} \ln(1 - u)$

or $t = -\dfrac{1}{\lambda} \ln u, \; [u \in [0,1) \Rightarrow (1 - u) \in [0,1)]$ (20.10)

Thus for a given random number u, the corresponding exponential random variate t of random variable T is obtained using Eq. (20.10).

Generation of Uniform Random Variate Over [a, b]

The probability density $f(x)$ of uniform distribution over $[a, b]$ is

$$f(x) = \begin{cases} \dfrac{1}{b-a}, & a \leq x \leq b \\ 0, & \text{elsewhere} \end{cases}$$

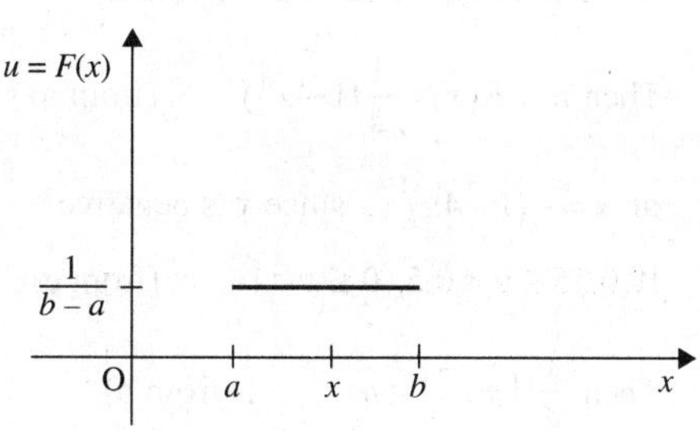

Figure 20.2 Representation of graph of $u = F(x)$

PDF $F(x)$ is

$$F(x) = \int_{a}^{x} f(x)dx = \frac{x-a}{b-a}, \; a \leq x \leq b$$

Let $u = F(x)$, then $u = \dfrac{x-a}{b-a}$

or $x = a + (b - a)u$ (20.11)

Thus for a given random number u, the formula in Eq. (20.11) generates a random variate x of uniformly distributed random variable X over $[a, b]$.

Example 20.6 Let a continuous random variable X have the following PDF:

$$f(x) = \begin{cases} -x^3, & -1 \le x < 0 \\ \frac{1}{2}x, & 0 \le x < 1 \\ \frac{1}{2}, & 1 \le x < 2 \\ 0, & \text{elsewhere} \end{cases}$$

Using the inverse transformation method [i.e., Eq. (20.9)], determine a scheme for generating random variates x of X.

Solution The distribution function (or cumulative distribution function) $F(x)$ of X is

$$F(x) = 0,\ x < -1$$

$$= \frac{1}{4}(1 - x^4),\ -1 \le x < 0;\ F(0) = 1/4 \qquad \text{(a)}$$

$$= \frac{1}{4}(1 + x^2),\ 0 \le x < 1,\ F(1) = 1/2 \qquad \text{(b)}$$

$$= \frac{1}{2}x,\ 1 < x < 2 \qquad \text{(c)}$$

Thus, we note that

$$\text{for } -1 \le x < 0,\ 0 \le F(x) < 1/4 \qquad \text{(d)}$$

$$\text{for } 0 \le x < 1,\ 1/4 \le F(x) < 1/2 \qquad \text{(e)}$$

$$\text{and for } 1 \le x < 2,\ 1/2 \le F(x) < 1 \qquad \text{(f)}$$

Now let $u = F(x)$, then from these equations we have,

If $0 \le u < 1/4,\ -1 \le x < 0$ (from d)

Then $u = F(x) = \frac{1}{4}(1 - x^4)$ (from a)

or $x = -(1 - 4u)^{1/4}$, since x is negative

If $0.25 \le u < 0.5,\ 0 \le x < 1$ (from e)

Then $\frac{1}{4}(1 + x^2) = u$ (from b)

or $x = (4u - 1)^{1/2}$, the positive root to be taken as x is positive

If $0.5 \le u < 1,\ 1 \le x < 2$ (from f)

Then $\frac{1}{2}x = u$ (from c)

or $x = 2u$

Thus, we have the following algorithm (scheme) to generate a value x from random variable X for a given random number u:

For $0 \le u < 0.25$, set $x = -(1-4u)^{1/4}$

For $0.25 \le u < 0.50$, set $x = (4u-1)^{1/2}$

For $0.5 \le u < 1$, set $x = 2u$

20.5.2 Discrete Case

Discrete random variables describe a variety of probabilistic situations. In general, the technique of inverse transformation can be used to generate a random sample from a discrete random variable X by using the basic sequence of random numbers. For this let X be a discrete random variable, assuming values $x_1, x_2, \ldots, x_n$ with $p(X = x_i) = p_i$. Arrange x_i in ascending order $(x_1 < x_2 < x_3 < \ldots)$. The distribution function (*cdf*) $F(x)$ is as follows:

$$F(x) = \begin{cases} 0, & x < x_1 \\ p_1, & x_1 \le x < x_2 \\ p_1 + p_2, & x_2 \le x < x_3 \\ p_1 + p_2 + p_3, & x_3 \le x < x_4 \end{cases}$$

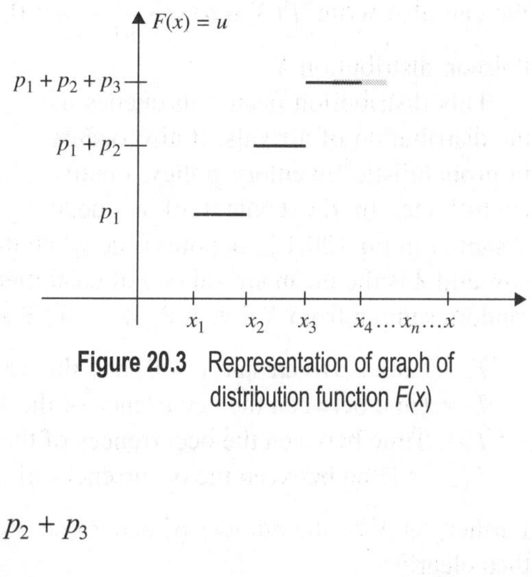

Figure 20.3 Representation of graph of distribution function $F(x)$

The CDF $F(x)$ has been plotted in Fig. 20.3; its graph is a step function. Now for a given sequence of random numbers (u_i), consider

$$x = x_1 \text{ if } 0 \le u < p_1$$

$$x = x_2 \text{ if } p_1 \le u < p_1 + p_2$$

$$x = x_3 \text{ if } p_1 + p_2 \le u < p_1 + p_2 + p_3$$

Example 20.7 Consider the random variable X given with the following probability distribution:

X	−1	0	1	3
$P(x) = P(X = x)$	0.2	0.3	0.4	0.1

Generate a random sample from X of size 5 using the sequence of random numbers 0.2, 0.0, 0.3, 0.5, 0.8.

Solution The CDF $F(x)$ of X is

$$F(x) = \begin{cases} 0, & x < -1 \\ 0.2, & -1 \le x < 0 \\ 0.5, & 0 \le x < 1 \\ 0.9, & 1 \le x < 3 \\ 1, & 3 \le x < \infty \end{cases}$$

Taking $u = F(x)$, we have $x = -1$ for $0 \le u < 0.2$, $x = 0$ for $0.2 \le u < 0.5$, $x = 1$ for $0.5 \le u < 0.9$, and $x = 3$ for $0.9 \le u < 1$.

The aforementioned scheme generates random variates x of X for a given random number u. Thus, using this scheme we have $x = 0$ for $u = 0.2$, $x = -1$ for $u = 0.0$, $x = 0$ for $u = 0.3$, $x = 1$ for $u = 0.5$, and $x = 1$ for $u = 0.8$.

Thus, 0, −1, 0, 1, 1 is the sequence of random variate x of random variable X corresponding to random numbers 0.2, 0.0, 0.3, 5, 0.8.

To Generate Random Variate from Poisson Distribution

Let X be the number of occurrences of events in an interval $[0, t)$, with mean occurrence rate λ, then X has Poisson distribution $p(\lambda t)$ with probability function of the form

$$f_x(t) = \frac{e^{-\lambda t}(\lambda t)^x}{x!}, x = 0, 1, 2, \ldots \tag{20.12}$$

We can also write $P(X = n) = \dfrac{e^{-\lambda}\lambda^n}{n!}$, $n = 0, 1, 2, \ldots$ where λ is the parameter of Poisson distribution X.

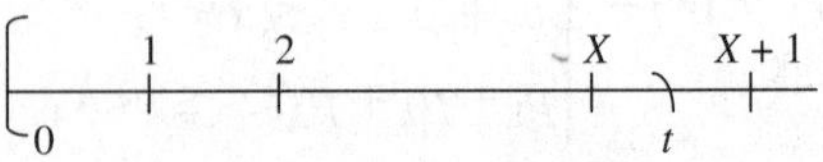

This distribution occurs in queues as the distribution of arrivals. It also occurs in probabilistic inventory policy, quality control etc. In the context of a queue system, t in Eq. (20.12), denotes time when the number of customers in the queue is x, and λ is the mean arrival rate of customers. Our problem here is to generate a random sample from X. Let 1, 2, 3, …, X, $X + 1$ be the events.

T_1 = Time between start ($t = 0$) and the occurrence of the 1st event
T_2 = Time between the occurrence of the 1st and 2nd events
T_X = Time between the occurrences of the $(X-1)$th and the Xth events
T_{X+1} = Time between the occurrences of the Xth and the $(X+1)$th events

Further, let X be the number of occurrences of events during time interval $[0, t)$, then clearly

$$T_1 + T_2 + \ldots + T_X \le t < T_1 + T_2 + \ldots + T_{X+1} \tag{20.13}$$

where LHS in Eq. (20.13) is the total time of X occurrences and RHS is the total time of $(X + 1)$ occurrences of the event.

In Eq. (20.13), let each T_i (inter-arrival time) be exponentially distributed random variables with parameter λ. Then X, the number of occurrences of events in $[0, t)$ is Poisson distribution $P(\lambda t)$ with parameter λt. From Eq. (20.10), each T_i being exponentially distributed can be generated from random numbers U_i's as $T_i = -\dfrac{1}{\lambda}\ln U_i$.

Substituting this $i = 1, 2, \ldots, X, X + 1$ in Eq. (20.13), we get

$$-\frac{1}{\lambda}\ln(U_1 U_2 \ldots U_X) \le t < -\frac{1}{\lambda}L_n(U_1 U_2 \ldots U_{X+1})$$

or $U_1 U_2 \dots U_{X+1} < e^{-\lambda t} \le U_1\, U_2 \dots U_X$

Assuming $t = 1$ we get $U_1 U_2 \dots U_{X+1} < e^{-\lambda} \le U_1\, U_2 \dots U_X$

Let $e^{-\lambda} = A$ then $U_1 U_2 \dots U_{X+1} < A \le U_1\, U_2 \dots U_X$ (20.14)

The inequality in Eq. (20.14) provides a scheme of generating random variate x from the Poisson random variable X with parameter λ. In this scheme, we cannot determine in advance as to how many random numbers (u_i's) will be needed to generate one Poisson random variate x.

Example 20.8 Generate a random sample of size five from Poisson distribution X with parameter $\lambda = 3$ by using random numbers (30, 37, 84, 71, 98, 65, 72, 19, 06, 33).

Solution Here $A = e^{-\lambda} = e^{-3} = 0.498$, $u_1 = 0.30$ is less than A, and the first random variate $x_1 = 0$.

The second number $u_1 = 0.37 < A$, so the second random variate $x_2 = 0$. For the third random variate $u_1 = 0.84$, which is not less than A.

$$u_1\, u_2 = 0.84 \times 0.71 = 0.5964 > A$$

$$u_1 u_2 u_3 = 0.5964 \times 0.98 = 0.5844 > A$$

$$u_1 u_2 u_3 u_4 = 0.5844 \times 0.65 = 0.03798 < A$$

Thus, $u_1\, u_2\, u_3\, u_4 < A < u_1\, u_2\, u_3$

Hence, the third random variate $x_3 = 3$. To generate fourth random variate, we have

$$u_1 = 0.72 > A$$

$$u_1\, u_2 = 0.72 \times 0.19 = 1368 < A = 0.498$$

Thus, $u_1\, u_2 < A < u_1$

Hence, the fourth random variate $x_4 = 1$. To generate the fifth random variate, we have $u_1 = 0.06$ less than A.

Hence, the fifth random variate $x_5 = 0$.

Thus, the sample of size 5 from Poisson random variable is 0, 0, 3, 1, 0.

20.6 APPLICATIONS OF SIMULATION IN QUEUING MODELS

Here we consider two types of queuing models:

1. Queuing models with fixed internal arrivals and fixed services
2. Queuing models where inter-arrival and service times are random variables

Example 20.9 In a queuing system with a single server, assume that the customers arrive such that the inter-arrival time for each customer is two minutes and the service time for each customer is 3.5 minutes. Simulate the system using the 'next event increment' method for 15 minutes.

Solution Let us assume that the first arrival takes place at $t = 0$. The simulated data is provided in the following table:

Time	Event	Total arrivals so far	Total no. of customers departed	Total no. of customers in the system	Total no. of customers in the queue	Time spent between the present and preceding instants in the system	queue
						system	queue
0.0	A	1	0.0	1	0	0.0	0.0
2.0	A	2	0	2	1	2.0	0.0
3.5	D	2	1	1	0	3.0	1.5
4.0	A	3	1	2	1	0.5	0.0
6.0	A	4	1	3	2	4.0	2.0
7.0	D	4	2	2	1	3.0	2.0
8.0	A	5	2	3	2	2.0	1.0
10.0	A	6	2	4	3	6.0	4.0
10.5	D	6	3	3	2	2.0	1.5
12.0	A	7	3	4	3	4.5	3.0
14.0	A, D	8	4	4	3	8.0	6.0
15.0	End	8	4	4	3	4.0	3.0
Total	–	–	–	–	–	39.0	24.0

Average waiting time in the system $(W_s) = 39.0/8 = 4.875$

Average waiting time in the queue $(W_q) = 24/8 = 3.0$

Average number of customers in the system $(N_s) = 39/15 = 2.60$

Average number of customers in the queue $(N_q) = 24/15 = 1.60$

Note In Example 20.9, inter-arrival and service time have been taken as fixed. However, such situations are rare in practical life.

Let us consider the case when both the arrival and departure patterns are taking place in a random fashion. In the next example, we will consider the case where both the inter-arrival and service time are exponential distributions. We consider a single-sever queuing system where customers arriving at the service facility are served on a first come, first served basis. Our interest is in determining the average number of customers in the system (or queue), mean waiting time of a customer (or queue), proportion of idle (or busy) time of the server, etc. The status of the queue changes when an arrival or departure takes place. The server is either idle or busy and accordingly service is either provided immediately or the customer has to wait. The point in time, when the status of the system changes is called an event and the simulation is carried on from event to event.

While simulating a single server queuing system, we shall use the following notations:

T = Time, that is, the elapsed simulated time

TBE = Time between the present and previous events

CNA = Cumulative (total) number of arrivals so far
CND = Cumulative (total) number of departures so far
NS = Number of customers in the system
NQ = Number of customers in the queue
CWTS = Cumulated (total) waiting time of all customers in the system

where CWTS = TBE × NS + CWTS until previous events
CWTQ = Cumulated (total) waiting time of all customers in the queue

where CWTQ = TBE × NQ + CWTQ until the previous event

STS = State of the system, where STS = 1 if NS = 0 and STS = 0 if NS ≥ 1
CIDT = Cumulated idle time of the server

where CIDT = STS in the previous event × TBE + CIDT in the previous event.

UA = The random number for simulating inter-arrival time (IAT)
US = The random number for simulating service time (ST)
NAT = Next arrival time, where NAT=T+IAT and IAT= $-\frac{1}{\lambda}\ln$ (UA) if inter-arrival time is exponentially distributed with parameter λ.
NDT = Next departure time
where NDT = T + ST (service time)

and ST = $-\frac{1}{\mu}\ln$ (US) if service time is an exponentially distributed random variable with parameter μ.

Example 20.10 In a queuing system comprising one server that follows first come first served discipline with unlimited queue length and unlimited input source, customers are arriving to avail the service. The inter-arrival time distribution X and service time distribution Y are exponential with parameters $\lambda = 1/2$ and $\mu = 3/4$ respectively. Use random numbers for inter-arrival time and service time respectively as 20, 23, 86, 09, 92, 35, 38, 01, 24, 07, 50 and 21, 44, 27, 70, 73, 36, 59, 42, 85, 88, 51. Simulate the system for 25 units of time starting with time $t = 0$ and estimate the following characteristics of the queuing system.

1. Average number of customers in the system
2. Average number of customers in the queue
3. Mean waiting time of a customer in the system
4. Mean waiting time of a customer in the queue
5. The proportion of idle time of the server

Solution To simulate the model, we need random samples from X and Y using $x = -\frac{1}{0.5}\ln u_1$ and $y = -\frac{1}{0.75}\ln u_2$ where u_1 = UA: 20, 23, 86, 09, 92, 35, 38, 01, 24, 07, 50,

u_2 = US: 21, 44, 27, 70, 73, 36, 59, 42, 85, 88, 51,

and X and Y are exponentially distributed random variables with parameters $\lambda = 0.5$ and 0.75 respectively.

The generated random samples from X and Y with event-wise time is shown in the following table:

u_1	x	Arrival at	u_2	y	Departure at	Event-wise time	
20	3.22	3.22	21	2.08	5.30	3.22	A
23	2.94	6.16	44	1.09	7.25	5.30	D
86	0.30	6.46	27	1.75	9.00	6.16	A
09	4.82	11.28	70	0.48	11.76	6.46	A
92	0.17	11.45	73	0.42	12.18	7.25	D
35	2.10	13.55	36	1.36	14.91	9.00	D
38	1.94	15.49	59	0.70	16.19	11.28	A
01	9.21	24.7	42	1.16	25.86	11.45	A
24	2.85	27.55	85	0.22	27.77	11.76	D
07	5.32	32.87	88	0.17	33.04	12.18	D
50	1.37	34.24	51	0.90	35.14	13.55	A
–	–	–	–	–	–	14.91	D
–	–	–	–	–	–	15.49	A
–	–	–	–	–	–	16.19	D
–	–	–	–	–	–	24.7	A
–	–	–	–	–	–	25.86	D

Now, event-wise time obtained in the aforegiven table is used to simulate the system. The details are given in the following table:

T	Event	TBE	CNA	CND	NS	NQ	WTS	WTQ	STS	CIDT
0.00	–	–	0	0	0	0	0.00	0	1	0.00
3.22	A	3.22	1	0	1	0	0.00	0	0	3.22
5.30	D	2.08	1	1	0	0	2.08	0	1	3.22
6.16	A	0.86	2	1	1	0	0.00	0	0	4.08
6.46	A	0.30	3	1	2	1	0.30	0	0	4.08
7.25	D	0.79	3	2	1	0	1.58	0.79	0	4.08
9.00	D	1.75	3	3	0	0	1.75	0	1	4.08
11.28	A	2.28	4	3	1	0	0.00	0	0	6.36
11.45	A	0.17	5	3	2	1	0.17	0	0	6.36
11.76	D	0.31	5	4	1	0	0.62	0.31	0	6.36
12.18	D	0.42	5	5	0	0	0.42	0	1	6.36
13.55	A	1.37	6	5	1	0	0.00	0	0	7.73
14.91	D	1.36	6	6	0	0	1.36	0	1	7.73

(Contd)

(Contd)

T	Event	TBE	CNA	CND	NS	NQ	WTS	WTQ	STS	CIDT
15.49	A	0.58	7	6	1	0	0.00	0	0	8.31
16.19	D	0.70	7	7	0	0	0.70	0	1	8.31
24.70	A	8.51	8	7	1	0	0.00	0	0	16.82
25.00	End	0.30	8	7	1	0	0.30	0	0	16.82
Total	–	–	–	–	–	–	9.28	1.10	–	–

From the aforegiven table, we conclude the following:

1. Average number of customers in the system $= \text{CWTS}/T = 9.28/25 = 0.37$
2. Average number of customers in the queue $= \text{CWTQ}/T = 1.10/25 = 0.044$
3. Mean waiting time in the system $= \text{CWTS}/\text{NS} = 9.28/8 = 1.16$ units of time
4. Mean average waiting time in the queue $= \text{CWTQ}/\text{NQ} = 1.10/8 = 0.138$ units of time
5. Proportion of idle time of the server $= \text{CIDT}/T = 16.82/25 = 0.6728$

20.7 ADVANTAGES AND DISADVANTAGES OF SIMULATION

In comparison to the mathematical programming and standard statistical analysis, the simulation technique has many advantages. A few of these have been listed here.

1. One advantage of simulation is that it does not involve use of complicated mathematics and thus can easily be understood by the operating staff and non-technical managers.
2. Simulation models offer flexibility and thus can be modified to adjust the environmental variations of real situations.
3. Simulation provides the solution using experimentation with a model of the system without interfering with the real system.
4. With the advent of computers, the time for performing large and complex calculations has been greatly reduced.
5. By the application of the simulation technique, the management can foresee the difficulties and bottlenecks that may arise due to the addition of new equipment or a process. It eliminates the need of costly trial-and-error methods before adding new equipment or introducing some new process.

However, there are certain disadvantages of the simulation technique; these are listed here.

1. Simulation provides a way of obtaining a solution but does not provide the solution technique. The user has to develop the solution approach they want to test.
2. Simulation is not precise. It is not an optimum process and does not yield an exact answer.
3. A good simulation model may be too expensive. In fact, it takes years to develop a usable corporate planning model.

4. All situations cannot be tackled using simulation. Only those involving uncertainties can be tackled using the simulation technique as without a random component all simulated experiments will render the same result.

20.8 SIMULATION LANGUAGES

The conventional computer languages such as FORTRAN (FORmula TRANslation), COBOL (Common Business-oriented Language), and ALGOL (ALGOrithmic Language) can be used to write a simulation computer program. However, these languages require extensive programming experience. At the same time, with the development of special languages for simulation with features specially designed for different types of models and systems, the program preparation time and cost has been reduced considerably. A few such special-purpose simulation languages include GPSS (general-purpose simulation system), SIMSCRIPT, GASP, SIMPAC, DYNAMO (DYNAmic MOdels), SIMULATE, and CSL. The most popular of these languages are GPSS, SIMSCRIPT, and DYNAMO. These three languages provide automatic instructions for timekeeping and other common simulation operations.

The GPSS finds use in a wider class of systems to maintain a fixed set of procedures for performing the simulation automatically. It has four entities—(a) transactions (b) facilities (c) storages, and (d) blocks—constituting its foundation.

The SIMSCRIPT is more flexible than the GPSS. However, it requires the knowledge of FORTRAN. Both GPSS and SIMSCRIPT are appropriate for the 'queuing type' problems.

The DYNAMO although similar to SIMSCRIPT is not suitable for queuing-type problems. It simulates continuous-type models, instead of those involving individual events like the production scheduling inventory.

RECAPITULATION

1. Simulation technique is useful when analytical methods fail to analyse the characteristics of a complex system.
2. Random numbers play an important role in simulation. A random number lies between 0 and 1. It can be zero but never equal to 1.
3. The random numbers are generated by computers. In case they are generated by some deterministic process but qualify the predetermined statistical test for randomness, we call them pseudo-random numbers. There are many methods to generate pseudo-random numbers. We have discussed here three of them, namely the mid-square method with its limitations, multiplicative congruential technique, and the mixed congruential method.
4. The Monte Carlo technique is an important part of simulation. This technique involves the selection of random observations within the simulated model. The main objective of this technique is to replace the actual statistical universe with another universe described by some probability distribution and then carrying out sampling for this assumed distribution using random numbers.
5. Simulation techniques have found some extremely important applications in queuing systems such as inventory control, financial planning, and capital budgeting. A few of these have been discussed in this chapter using the Monte Carlo technique.

EXERCISES

Multiple-choice Questions

20.1 If u is a random number then
- (a) $0 < u < 1$
- (b) $0 \leq u < 1$
- (c) $0 < u < 1$
- (d) $0 \leq u \leq 1$

20.2 In Monte Carlo method,
- (a) only random variables are used
- (b) only random numbers are used
- (c) both random numbers and variables are used
- (d) none of these

20.3 The result obtained by applying the simulation technique is
- (a) exact
- (b) approximate
- (c) unrealistic
- (d) none of these

State True or False

20.1 The random numbers generated by computer software are uniformly distributed fractions between 0 and 1.

20.2 Simulation is defined as a technique which does not involve use of computers.

20.3 The Monte Carlo method of simulation is developed using probability distribution and random numbers.

20.4 Pseudo-random numbers are the numbers that satisfy the predetermined statistical tests for randomness.

20.5 In the mixed congruential technique of generating pseudo-random numbers, the prescribed specifications of parameters should be satisfied to avoid early repetition of random numbers.

Fill in the Blanks

20.1 ________ is not an analytical method.

20.2 While assigning random numbers in Monte Carlo simulation, it is necessary to develop a ________ probability distribution.

20.3 The step required for simulation approach in solving a problem is to ________ and ________ the experiment.

20.4 The initial number used in generating pseudo-random numbers is known as ________.

20.5 ________ is the numeric value of a random variable.

Review Questions

20.1 What are pseudo-random numbers? Describe the mid-square method of generating pseudo-random numbers.

20.2 State the major reasons of using simulation. Explain the basic steps involved in Monte Carlo simulation.

20.3 Do you think that there will be a substantial increase in the application of simulation in the next ten years? State the reasons for your answer.

20.4 Describe the advantages and disadvantages of simulation models.

20.5 Write short notes on simulation and its applications.

20.6 Suggest any two probability distributions applied to discrete random variables. Give expressions for their probability density functions.

Numerical Problems

20.1 Generate four three-digit pseudo-random numbers using the mixed congruential method. Assume $c = 9$, $a = 21$, and $x_0 = 20$.

20.2 Generate pseudo-random numbers by assuming $m = 10^4$ and c, a, and x_0 of your choice, but according to the specifications of mixed congruential method.

20.3 Consi'ering, the seed x_0=0400, generate four-digit random numbers using mid-square method.

20.4 Generate a random sample of size 5 from a uniform distribution X on [−2, 3] by using random numbers 0.020, 0.429, 0.018, 0.387, 0.136.

20.5 The output of a firm's product over 50 typical days is provided in the following table. Vehicles are available to deliver goods. Each vehicle makes one trip a day and can deliver 105 units when fully loaded. The maximum number of available trucks are fully loaded each day. Units in excess of total vehicle capacity are stored overnight and are included in the output of the next day. The number of vehicles available for the same 50-day period is also given in the following table. Simulate the model for 10 days to determine the (a) average number of units produced daily, (b) average number of vehicles available, and (c) average number of units stored overnight.

Units produced	No. of days	Vehicles available	No. of days
500	3	5	8
550	7	6	18
600	10	7	10
650	20	8	8
700	10	9	6

To generate a random sample for units produced and number of vehicles available on different days, make use of the following random numbers respectively: (i) 26, 85, 06, 45, 57, 54, 27, 03, 20, 96 and (ii) 12, 68, 46, 55, 89, 60, 09, 71, 87, 89

20.6 The probability distributions of units produced (X) and the number of vehicles available (Y) are as follows:

Units produced (x_i) X	$P(X = x_i)$	Vehicles available (y_i) Y	$P(Y = y_i)$
500	0.06	5	0.16
550	0.14	6	0.36
600	0.20	7	0.20
650	0.40	8	0.16
700	0.20	9	0.12

Simulate the model for 10 days in respect of the aforementioned changes to determine the average daily output, average number of vehicles available, and average number of units stored overnight.

20.7 A firm has three vehicles and uses whichever vehicle that is available to transport stock to its five branches. The average petrol consumption by the vehicles and the distance (to and fro) in kms of the branches are provided in the following table:

Vehicles	Distance travelled (km) (per gallon)	Branch	Distance (km) (to and fro)
A	25	1	100
B	35	2	130
C	40	3	160
–	–	4	200
–	–	5	260

Assuming that a random choice of destinations and vehicles is made, simulate 10 trips and hence estimate the average petrol consumption per trip.

Hint Use a dice to generate a random sequence for selection of vehicles and use the seconds on a digital watch for the selection of branches, as per the following table:

Outcome	Vehicle	Time (seconds)	Branch
1 or 2	A	0–11	1
3 or 4	B	12–23	2
5 or 6	C	24–35	3
–	–	36–47	4
–	–	48–59	5

20.8 A shopkeeper maintains a stock of Britannia bread. On the basis of the shopkeeper's previous experience, the daily demand for the bread along with associated probabilities is given in the following table:

Daily demand (number)	0	10	20	30	40	50
Probability	0.01	0.20	0.15	0.50	0.12	0.02

Estimate the daily average demand for the bread by simulating the demand for 10 days. Use the following random numbers: 25, 39, 65, 76, 12, 05, 73, 89, 19, 49.

20.9 In a transportation company, the finished job arriving at the checkout area is recorded for 100 trucks. Checkout takes five minutes and the checker takes care of only one truck at a time. The data is as follows:

Truck inter-arrival time (min)	1	2	3	4	5	6	7	8	9	10	
Frequency	1	4	7	17	31	23	7	5	3	2	(Total = 100)

After checkout, the drivers take their trucks to the next department. Using Monte Carlo method, determine the following:

(a) Average waiting time before service

(b) Expected longest period of waiting

Use the following random numbers: 12, 81, 36, 82, 21, 74, 90, 55, 79, 70, 14, 59, 62, 57, 15, 18, 74, 11, 41, 29.

CONCEPTUAL PROBLEM

A truck company with its head office in Delhi is engaged in the booking and delivery of various items of daily use, in Delhi and nearby areas. The excess units of booked items are stored overnight and the company has to pay storage cost. For the optimum use of its trucks, the company collected the following data for 100 days.

No. of units booked (X)	No. of days	No. of trucks available (Y)	No. of days
300	8	3	16
350	18	4	36
400	24	5	20
450	38	6	16
500	12	7	12

Each truck makes one trip a day and can deliver 100 units when fully loaded. As many trucks as possible are fully loaded each day. If certain units, which cannot be completely loaded in a truck, are left, and another truck is available, then the available truck will be partially loaded and sent. The unused trucks are turned away and are not required to wait for the next day. Units in excess of the total trucks' capacity are stored overnight and added to the next day's booking for delivery.

Determine the (a) average daily booking, (b) average number of units stored overnight, (c) average number of trucks that are fully (or partially) filled, (d) average number of trucks not used. Simulate the system for 10 days. (For simulation, use random numbers (i) 80, 87, 34, 21, 48, 15, 22, 69, 56, 83 for units booked and (ii) 50, 57, 04, 91, 18, 85, 92, 31, 26, 53 for the number of available trucks.)

Solution It is reasonable to assume the number of units booked as a random variable, say X, and the number of trucks available as another random variable, say Y. The probability distribution of these random variables is described in the following tables:

No. of units booked (X)	300	350	400	450	500
Probability	0.08	0.18	0.24	0.38	0.12

No. of trucks available (Y)	3	4	5	6	7
Probability	0.16	0.36	0.20	0.16	0.12

(*Contd*)

(*Contd*)

The process of generating random variates is as follows:

Random no. (u)	X-random variate (x)	Random no. (u)	y-random variate (y)
$0.00 \le u < 0.08$	300	$0.00 \le u < 0.16$	3
$0.08 \le u < 0.26$	350	$0.16 \le u < 0.52$	4
$0.26 \le u < 0.50$	400	$0.52 \le u < 0.72$	5
$0.50 \le u < 0.88$	450	$0.72 \le u < 0.88$	6
$0.88 \le u < 1.00$	500	$0.88 \le u < 1.00$	7

Now, we use the following notations to simulate the system and the simulated results are shown in the following table:

RNB = Random number for units booked
NUB = Number of units booked
UDB = Undelivered units on the day before
TND = Total number of units booked for delivery
RNT = Random number of trucks available
NTA = Number of trucks available
NTU = Number of full trucks used
CTU = Capacity of the full trucks
LPT = Load of partially filled trucks
NUT = Number of unused trucks
NUS = Number of units stored overnight

The random numbers given to simulate the system are as follows:

1. 0.80, 0.87, 0.34, 21.0, 48.0, 15.0, 22.0, 69.0, 56.0, 83 for X
2. 0.50, 0.57, 0.04, 0.91, 0.18, 0.85, 0.92, 0.31, 0.26, 0.53 for Y

Now using the aforementioned random numbers for X and Y, and the scheme of generating random variates, we obtain the following table:

DAY	RNB	NUB	UDB	TND	RNT	NTA	NTU	CTU	LPT	NUT	NUS
1	0.80	450	–	450	0.50	4	4	400	–	–	50
2	0.87	450	50	500	0.57	5	5	500	–	–	–
3	0.34	400	–	400	0.04	3	3	300	–	–	100
4	0.21	350	100	450	0.91	7	4	400	50	2	–
5	0.48	400	–	400	0.18	4	4	400	–	–	–
6	0.15	350	–	350	0.85	6	3	300	50	2	–
7	0.22	350	–	350	0.92	7	3	300	50	3	–
8	0.69	450	–	450	0.31	4	4	400	–	–	50
9	0.56	450	50	500	0.26	4	4	400	–	–	100
10	0.83	450	100	550	0.53	5	5	500	–	–	50
Total	–	4100	300	4350	–	–	39	–	–	7	350

From the aforegiven table, we see that the total number of units booked for delivery = 4100. Thus, we have

(a) Average daily booking $= 4100/10 = 410$ units
(b) Average number of trucks not used and returned $= 7/10 = 0.7 \approx 1$
(c) Average number of trucks fully used $= 39/10 = 3.9 \approx 4$
(d) Average number of units stored overnight $= 350/10 = 35$

Inventory Models

After studying this chapter, the reader will be able to understand

- inventory theory and the types of inventory
- some basic terminologies used in inventory
- inventory control and inventory costs
- inventory management and its benefits
- economic order quantity (EOQ) and EOQ problems with price break
- deterministic and probabilistic inventory models
- some important inventory control techniques

21.1 INTRODUCTION

Holding and managing of inventory is essential for the efficient and smooth running of any business or organization, such as manufacturing industries, five-star hotels, printing press, and hospitals. Inventory models are very helpful in the development of inventory policies. The main objective of an inventory in any business is to decrease the cost of set-up, the carrying cost, and the shortage cost. Failure in fulfilling the demands of the customers may result in losing the goodwill of the customers. Cancellation of orders would mean loss of business. Thus, the only solution to these problems is to maintain an inventory.

21.2 INVENTORY

An inventory can be defined as a stock of goods, which is held for the purpose of future production or sales. The stock of goods may be kept in the form of raw materials, partly finished items, finished items, and spare parts. Inventory can be of materials, commodities, or products, which are usually stocked in order to be consumed or benefitted from when needed. Keeping an inventory in the store has its own expenses, which may, sometimes, be more than the value of the commodity kept there. Inventories may be held for a variety of purposes, but in general, there are five types of inventories based on their purpose. These are as follows:

Movement inventories These are also called transit or pipeline inventories. They are needed for the shipment of inventory items to the distribution centres and from the various production centres. The quantum of this inventory depends on the time involved in transportation and the nature of demand.

Buffer inventories When it comes to meeting the uncertainties of demand and supply, these come in handy. Such inventories are in excess when compared to the inventories that are necessary just to meet the average demand during the lead time (the time between placing an order and receiving it).

Anticipation inventories These are also known as seasonal inventories. They are held because a future demand for the product is anticipated.

Decoupling inventories These are in the form of raw materials, which are used to decouple the produce from the supplies and finished goods.

Lot-size inventories They are also known as cycle inventories and exact quantities of these are held for their use at an appropriate time.

21.3 SOME BASIC TERMINOLOGIES USED IN INVENTORY

The following terms will be frequently used in our discussion:

Back order It is an unfilled request for the issue of warehouse stock. The term requires the generation of a purchase order for stock replenishment, if it has not been ordered as yet.

Demand The number of units required per period is called demand. The demand pattern of a commodity may either be deterministic or probabilistic.

Lot size The quantity of an item to be ordered is called the lot size.

Deterioration It is defined as the decay, evaporation, pilferage, spoilage, damage, obsolescence, or loss of utility of the product, which results in a decrease in the usefulness of the product. The process of deterioration is observed in volatile liquids, beverages, medicines, blood components, foodstuff, and dairy items. The decrease or loss of utility due to decay is usually a function of the on-hand inventory. For items such as steel, hardware, glassware, and toys, the rate of deterioration is so low that there is little need for considering deterioration in the determination of the economic lot size. However, some items such as milk, fish, fruits, alcohol, gasoline, radioactive chemicals, medicines, and food grains deteriorate remarkably over time.

Profit The amount earned from an investment or operation after the deduction of expenses is the gross profit or the gross income. Net profit includes the deduction of interest, taxes, depreciation, and other expenses not associated with the cost of sales.

Reorder point The inventory level set to trigger the reorder of a specific item is called the reorder point. It is generally calculated as per the expected usage (demand) during the lead time plus the safety stock.

Reorder level The level between the maximum and minimum stocks at which the purchasing for replenishment should commence is known as the reorder level.

Stock replenishment It may be uniform or instantaneous. Uniform replenishment occurs when the company manufactures the product. Instantaneous replenishment occurs in case of the stock, which is purchased from external sources.

Time horizon The time period over which the inventory level is controlled is called the time horizon. It may be finite or infinite.

Partial backlogging In most developed inventory models, researchers assumed that the shortages are either completely backlogged or completely lost. The first case, known as the back-ordered or backlogging case, represents a situation where the unfulfilled demand is completely back-ordered. In the second case, also known as the lost sale case, we assume that the unfulfilled demand is completely lost. Furthermore, when shortages occur, some customers are willing to wait for a back order, whereas others turn to alternative sellers. In many cases, customers are conditioned to a shipping delay and may be willing to wait for a short time, in order to get their first choice.

Unit cost The costs of material, labour, and overheads, when clubbed together and allocated to a single unit, is known as the cost of that unit.

Warehouse It is a large building where goods are stored. In real life, enterprises usually purchase more goods than can be stored in their own warehouse (OW). The excess quantities are usually stored in an additional storage space known as the rented warehouse (RW).

Periodic review system It is also known as the cycle review system or fixed order interval system. In this system, the ordering interval is fixed but the size of the order quantity may vary with fluctuations in demand.

21.4 INVENTORY CONTROL

Inventory control is the process of deciding what and how much of an item is to be kept in stock. It also determines the time and quantity of various items to be procured. The basic objective of inventory control is to reduce the investment in inventories and ensure that the production process does not suffer as a result. To attain the various objectives in an inventory control situation, there are three basic questions to be answered.

1. How much to order?—that is, what is the optimum quantity of an item that should be ordered?
2. When to order?
3. Where should the order be placed?

The objective of an inventory problem is to minimize the total (actual or expected) cost or to maximize (actual or expected) profit. The following factors affect inventory control:

Demand It is the number of units required per period. It may either be known exactly or in terms of probabilities, or may also be completely unknown. Further, if the demand is known, it may either be fixed or variable per unit time.

The problems in which the demand is known and fixed are called deterministic problems, whereas those problems in which the demand is assumed to be random are called stochastic or probabilistic problems.

Lead time The time gap between the placing of an order and its actual arrival in the inventory is known as the lead time.

Order cycle The time period between the placements of two successive orders is called the order cycle.

Continuous review The record of the inventory level is continuously checked until a specified point (known as the reorder point) is reached, when a new order is placed.

Periodic review The inventory levels are viewed at equal time intervals and orders are placed at such intervals. The quantity ordered each time depends on the available inventory level at the time of review.

21.5 INVENTORY COSTS

Inventory costs play an important role while making decisions pertaining to the maintenance of inventory in any organization. These costs are as follows:

Ordering costs It includes order placing cost, cost due to follow-up measures, and receiving cost.

Set-up/Items cost It is the cost of purchasing or manufacturing items whose inventory is to be maintained.

Inventory carrying cost It is also known as the holding cost. It has already been indicated that the items stored for future consumption incur a certain amount of expenditure. This is commonly referred to as the inventory carrying cost. The carrying costs include the storage cost and the operational cost.

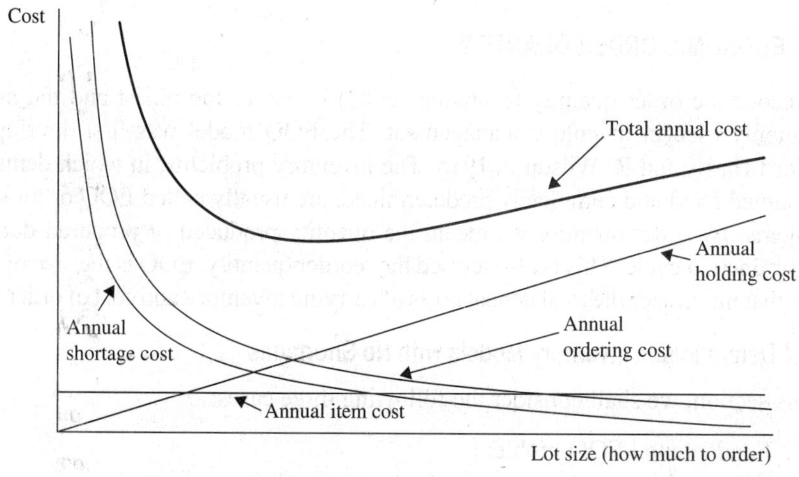

Figure 21.1 Total inventory cost

Salvage cost When the demand for a certain commodity is affected by the quantity stocked, its decision problem is based on a profit maximization criterion that includes the revenue from selling. Salvage value (selling price) may be combined with the cost of storage and is hence generally neglected.

Shortage costs Companies incur shortage costs whenever the demand for an item exceeds the available inventory. These costs can manifest themselves in the form of lost sales, loss of goodwill, customer irritation, back order, and expediting charges. Fig. 21.1 shows the total inventory cost incurred.

21.6 INVENTORY MANAGEMENT AND ITS BENEFITS

Inventory management can help remove the barriers between manufacturers and retailers, and establish a closer relationship between them. Ideally, inventory management is considered easy, as the main aim is to reduce inventories. If the items wanted are not in hand, sales will be lost to competitors. Precise control of the inventory is an essential ingredient for a successful company. The main objectives of inventory management are improved customer service, reduced inventory investment, and increased productivity. Successful inventory management involves balancing the costs of the inventory with the benefits of the inventory. Many small business owners fail to fully appreciate the true costs of carrying inventory, which not only includes the direct costs of storage, insurance, and taxes, but also the cost of money tied up in inventory. The benefits of inventory management are as follows:

1. Complete control of the inventory
2. Comprehensive information about the value of the inventory
3. Reduced response time to demand changes
4. Increased sales
5. Knowledge of the exact size of the merchandizing inventory
6. Frequent analysis of purchases, sales, and inventory records
7. Avoiding the stocking of unneeded items in warehouses

21.7 ECONOMIC ORDER QUANTITY

The economic order quantity technique (EOQ) is one of the oldest and the most commonly used in inventory management. The EOQ model was first developed by Ford Harris and R. Wilson in 1915. The inventory problems, in which demand is assumed fixed and completely predetermined, are usually called EOQ or lot-size problems. By order quantity we mean, the quantity produced or procured during one production cycle. This is also termed the reorder quantity. EOQ is the size of the order that minimizes the total annual costs of carrying inventory and cost of ordering.

21.7.1 Deterministic Inventory Models with No Shortages

In this section, we shall consider the following three cases:

1. Fundamental EOQ problem
2. EOQ problem with several production runs of unequal length
3. EOQ problem with finite replenishment (production)

Case 1—Fundamental EOQ Problem

The main objective of this problem is to determine an optimum order quantity, such that the total inventory cost is minimized. The following assumptions and notations are used:

1. Demand (D) is known and uniform.
2. Let Q denote the lot size in each production run.
3. Shortages are not allowed.
4. The production or supply of the commodity is instantaneous (abundant availability).
5. Lead time is taken to be zero.
6. The set-up cost per production run is denoted by C_1.
7. The holding cost is C_h per unit of inventory per unit time, that is, $C_h = IC$, where C is the unit cost and I is the inventory carrying charge expressed as a percentage (%) value of the average inventory.

The aforementioned fundamental situations are shown in Fig. 21.2:

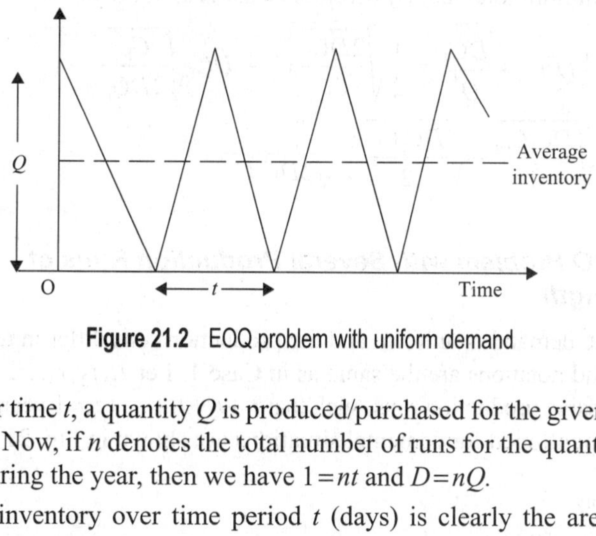

Figure 21.2 EOQ problem with uniform demand

Suppose after time t, a quantity Q is produced/purchased for the given time period, say one year. Now, if n denotes the total number of runs for the quantity produced/purchased during the year, then we have $1 = nt$ and $D = nQ$.

The total inventory over time period t (days) is clearly the area of the first triangle $\frac{1}{2}Qt$, the average amount is $\frac{\frac{1}{2}Qt}{t} = \frac{1}{2}Q$, and the annual inventory holding cost is $f(Q) = \frac{1}{2}QC_h$.

The annual costs associated with runs of size Q are given by

$$g(Q) = nC_1 = \frac{D}{Q}C_1$$

The minimum total cost occurs at the point where the ordering cost and the total inventory carrying costs are equal.

Hence we have,

$$f(Q) = g(Q) \Rightarrow \frac{1}{2}QC_h = \frac{D}{Q}C_1 \Rightarrow Q^2 = \frac{2DC_1}{C_h} \Rightarrow Q = \sqrt{\frac{2DC_1}{C_h}}$$

Hence, the optimum value of Q is $Q^* = \sqrt{\frac{2DC_1}{C_h}}$.

Characteristics The following are the characteristics of the fundamental EOQ problem:

1. The optimum number of orders placed per year is $n^* = \frac{D}{Q^*} = \sqrt{\frac{DC_h}{2C_1}}$
2. The optimum time length between orders is $t^* = \frac{T}{n^*} = T\sqrt{\frac{2C_1}{DC_h}}$

$t^* = \sqrt{\frac{2C_1}{DC_h}}$, (where T, the total time horizon, is one year)

3. The minimum total annual inventory cost is as follows:

$$\text{TC}^* = \frac{1}{2}Q^*C_h + \frac{DC_1}{Q^*} = \frac{1}{2}\sqrt{\frac{2DC_1}{C_h}}.C_h + DC_1\sqrt{\frac{C_h}{2DC_1}}$$

$$= \sqrt{\frac{DC_1C_h}{2}} + \sqrt{\frac{DC_1C_h}{2}} = \sqrt{2DC_1C_h}$$

Case 2—EOQ Problem with Several Production Runs of Unequal Length

In this case, the demand is uniform and the production runs differ in units; all other assumptions and notations are the same as in Case 1. Let $t_1, t_2, t_3, ..., t_n$ denote the times of successive production runs, such that $t_1+t_2+t_3+\ldots+t_n=1$ year.

The aforementioned fundamental situations are shown in Fig. 21.3:

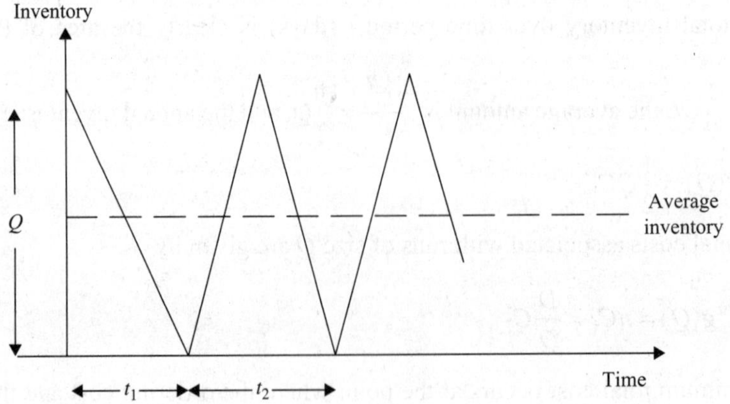

Figure 21.3 EOQ problem with several production runs of unequal length

The annual inventory holding cost is as follows:

$$f(Q)=\left(\frac{1}{2}Qt_1\right)C_h+\left(\frac{1}{2}Qt_2\right)C_h+\ldots+\left(\frac{1}{2}Qt_n\right)C_h$$

$$=\frac{1}{2}Q(t_1+t_2+\ldots+t_n)C_h=\frac{1}{2}QC_h$$

The set-up costs associated with runs of size Q are $g(Q)=nC_1=\frac{D}{Q}C_1$ $(\because nQ=D)$.

The total annual cost is $\text{TC}=f(Q)+g(Q)=\frac{1}{2}QC_h+\frac{D}{Q}C_1$

and $\frac{d}{dQ}(\text{TC})=\frac{C_h}{2}+\frac{DC_1}{(-Q^2)}=0$ for minima

$$\Rightarrow \frac{C_h}{2}=\frac{DC_1}{Q^2}\Rightarrow Q^2=\frac{2DC_1}{C_h}$$

$$\Rightarrow Q=\sqrt{\frac{2DC_1}{C_h}}$$

Now, check $\frac{d^2}{dQ^2}(\text{TC})>0,\ Q>0$

Hence, the optimum cost is $Q^*=\sqrt{\frac{2DC_1}{C_h}}$

and $\text{TC}^*=\frac{Q^*C_h}{2}+\frac{DC_1}{Q^*}=\frac{C_h}{2}\sqrt{\frac{2DC_1}{C_h}}+\frac{DC_1}{\sqrt{\frac{2DC_1}{C_h}}}=\sqrt{\frac{DC_1C_h}{2}}+\sqrt{\frac{DC_1C_h}{2}}$

$$=\sqrt{2DC_1C_h}$$

Example 21.1 Find the economic lot size, the associated total cost, and the duration between two orders. It is given that the set-up cost is ₹100, the daily holding cost per unit of inventory is 5 paise, and daily demand is approximately 30 units.

Solution Given that $D=30$ units, $C_1=₹100$, and $C_h=₹0.05$ per unit per day.

The economic lot size is $Q^*=\sqrt{\frac{2DC_1}{C_h}}=\sqrt{\frac{2\times30\times100}{0.05}}=100\sqrt{12}=100\times3.464=346.4$

Total cost $=\sqrt{2DC_1C_h}=\sqrt{2\times30\times100\times0.05}=\sqrt{300}=₹17.32$

and $t^*=\sqrt{\frac{2C_1}{DC_h}}=\sqrt{\frac{2\times100}{30\times0.05}}=10\sqrt{\frac{4}{3}}=11.549=11.55$ days.

Example 21.2 A car oil engine manufacturer purchases lubricants at the rate of ₹42 per piece from a vendor. The requirement of these lubricants is 1800 per year. What should be the order quantity per order, if the cost per placement of an order is ₹16 and inventory carrying charge per rupee per year is only 20 paise?

Solution Given that $D = 1800$ lubricants per year, $C_1 = ₹16$ per order, and $C_h = ₹42 \times ₹0.20 = ₹8.40$. We have

$$Q^* = \sqrt{\frac{2DC_1}{C_h}} = \sqrt{\frac{2 \times 1800 \times 16}{8.4}} = 82.8 \text{ or } 83 \text{ lubricants}$$

Example 21.3 A bike manufacturing company purchases 9000 parts of a machine for its annual requirements, ordering one month's usage at a time. Each part costs ₹20. The ordering cost per order is ₹15 and the carrying charges are 15% of the average inventory per year. You have been assigned to suggest a more economical purchasing policy for the company. What advice would you offer and how much would it save the company per year?

Solution Given that $D = 9000$ parts per year, $C_1 = ₹15$ per order, and $C_h = 15\%$ of the

average inventory per year $= \frac{20 \times 15}{100} = ₹3$ for each part per year.

$$Q^* = \sqrt{\frac{2DC_1}{C_h}} = \sqrt{\frac{2 \times 9000 \times 15}{3}} = \sqrt{30 \times 3000} = \sqrt{90000} = 300 \text{ units}$$

$$\text{and } t^* = \sqrt{\frac{2C_1}{DC_s}} = \sqrt{\frac{2 \times 15}{9000 \times 3}} = \frac{1}{30} \text{ years}$$

The minimum average yearly cost $(\text{TC}^*) = \sqrt{2DC_sC_1} = \sqrt{2 \times 9000 \times 15 \times 3} = ₹900$

If the company follows the policy of ordering every month, then the annual ordering cost would be $12 \times 15 = ₹180$

The lot size of inventory each month $= \frac{9000}{12} = 750$

The average inventory at any time $= \frac{Q}{2} = \frac{750}{2} = 375$

The storage cost at any time $= 375 \times 3 = ₹1125$

The total annual cost $= f(Q) + g(Q) = 1125 + 180 = 1305$

Thus, the company should purchase 300 parts at a time interval of 1/30 years, instead of ordering 750 parts each month. The net saving of the company is $= 1305 - 900 = ₹405$ per year.

Example 21.4 A manufacturer has to supply customers with 600 units of a product per year. Shortages are not allowed and the storage cost amounts to ₹0.60 per unit per year. The set-up cost per run is ₹80. Find the optimum run size and the minimum average yearly cost.

Solution Given that $D = 600$ unit per year, $C_1 = 80$ production runs, and $C_h = 0.60$ per unit per year.

$$Q^* = \sqrt{\frac{2DC_1}{C_h}} = \sqrt{\frac{2 \times 600 \times 80}{0.60}} = \sqrt{20 \times 80 \times 100} = \sqrt{1,60,000} = 400 \text{ units}$$

$$\text{and } t^* = \frac{Q^*}{D} = \frac{400}{600} = \frac{2}{3} \text{ years or 8 months.}$$

The minimum total inventory cost is as follows:

$$\text{TC}^* = \sqrt{2DC_1C_h} = \sqrt{2 \times 600 \times 80 \times 0.60} = ₹240$$

Case 3—EOQ Problem with Finite Replenishment or Production

All the assumptions are the same as in Case 1, except that of instantaneous replacement. This is because in many situations, the amount ordered is not delivered all at once, but at a finite rate (supply rate) per unit of time. This situation can arise in case of an order being filled by a machine having a finite production rate. Assume that each production run of length t consists of two parts, say t_1 and t_2, such that

1. the inventory builds up at a constant rate of $(k - r)$ units, per unit of time during t_1, $k > r$ and
2. there is no replenishment (production) during time t_2 and the inventory decreases at the rate of r per unit of time.

The aforementioned fundamental situations are shown in Fig. 21.4: The total order quantity Q is produced over a period t_1 which is defined by the production rate k. Hence, we have $Q = kt_1$ or $t_1 = Q/k$. During the production period t_1, the inventory increases at the rate of k and simultaneously decreases at the rate of r. Therefore, the inventory accumulates at the rate of $(k - r)$ units. Therefore, the maximum inventory level will be equal to $t_1(k - r)$.

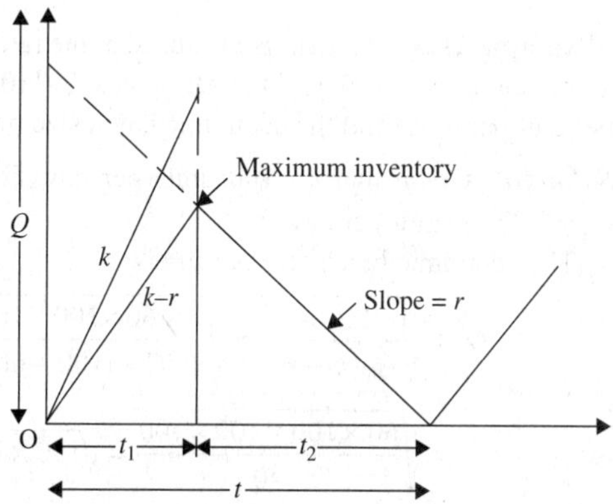

Figure 21.4 EOQ problem with finite replenishment

$$\text{Average inventory} = \frac{1}{2}t_1(K - r) = \frac{1}{2}Q\left(1 - \frac{r}{K}\right)$$

$$\text{Total annual holding cost, } f(Q) = \frac{1}{2}QC_h\left(1 - \frac{r}{K}\right)$$

$$\text{Annual ordering cost, } g(Q) = nC_1 = \frac{D}{Q}C_1$$

The minimum total cost occurs at the point where the annual ordering cost and the annual holding cost are equal. Hence, we have

$$f(Q) = g(Q) \Rightarrow \frac{1}{2}QC_h\left(1 - \frac{r}{k}\right) = \frac{D}{Q}C_1 \Rightarrow Q^2 = \frac{2DC_1}{C_h\left(1 - \frac{r}{k}\right)} \Rightarrow Q = \sqrt{\frac{2DC_1}{C_h\left(1 - \frac{r}{k}\right)}}$$

$$\text{Hence, the optimum value of } Q \text{ is } Q^* = \sqrt{\frac{2DC_1}{C_h\left(1 - \frac{r}{k}\right)}} = \sqrt{\frac{2DC_1}{C_h}\left(\frac{k}{k - r}\right)}$$

Characteristics The following are the characteristics of the EOQ problem with finite replenishment:

1. The optimum number of production runs per year, $n^* = \dfrac{D}{Q^*} = \sqrt{\dfrac{DC_h}{2C_1}\left(1 - \dfrac{r}{k}\right)}$
2. The optimum length of each lot's production run, $t_1^* = \dfrac{Q^*}{k} = \sqrt{\dfrac{2DC_1}{C_h k\,(k - r)}}$

3. The total minimum production inventory cost is as follows:

$$\text{TC}^* = \frac{C_1 D}{Q^*} + \frac{1}{2} Q^* \left(1 - \frac{r}{k}\right) C_h = C_1 D \sqrt{\frac{C_h (k-r)}{2DC_1 k}} + \frac{1}{2}\sqrt{\frac{2DC_1 k}{C_h (K-r)}} \left(1 - \frac{r}{K}\right) C_h$$

$$= \sqrt{\frac{DC_1 C_h (k-r)}{2k}} + \sqrt{\frac{DC_1 (k-r)}{2k}} = \sqrt{2DC_1 C_h \left(1 - \frac{r}{k}\right)}$$

Example 21.5 An item is produced at the rate of 100 units per day. The demand occurs at the rate of 60 per day, if the set-up cost is ₹100 per set-up and the holding cost is ₹0.02 per unit per day. Find the economic batch size per run.

Solution Given that $k = 100$ units per day, $D = 60$ per day, $C_1 = ₹100$ per set-up, and $C_h = 0.02$ per unit per day.

The economic batch size per run is

$$Q^* = \sqrt{\frac{2DC_1 k}{C_h (k-r)}} = \sqrt{\frac{2 \times 60 \times 100 \times 100}{0.02 \times (100 - 60)}}$$

$$= \sqrt{\frac{60 \times 100 \times 100 \times 100}{40}} = \sqrt{150 \times 100 \times 100} = 1224.5 \text{ units per day}$$

Example 21.6 An item is produced at the rate of 50 items per day. The demand occurs at the rate of 25 items per day if the set-up cost is ₹100 and holding cost is ₹0.01 per unit of item per day. Find the economic lot size for one run, assuming that shortages are not permitted. In addition, find the time of the cycle and the minimum total cost for one run.

Solution Given that $D = 25$, $C_h = ₹0.01$ per year, $C_1 = 100$ per set-up, $r = 25$ items per day, and $k = 50$ items per day.

We have $Q^* = \sqrt{\frac{2DC_1}{C_h} \frac{k}{(k-r)}} = \sqrt{\frac{2 \times 25 \times 100 \times 50}{0.01(50-25)}} = \sqrt{\frac{25 \times 100 \times 100 \times 100}{25}} = 1000$ items.

$$t^* = \frac{Q^*}{r} = \frac{1000}{25} = 40 \text{ days}$$

The total minimum product inventory cost is $\text{TC}^* = \sqrt{2DC_1 C_h \left(1 - \frac{r}{k}\right)}$

$$= \sqrt{2 \times 25 \times 100 \times 0.01 \left(1 - \frac{25}{50}\right)} = \sqrt{2 \times 25 \times \frac{25}{50}} = ₹5 \text{ per day}$$

Hence the total cost per run $= 5 \times 40 = ₹200$.

Example 21.7 A contractor has to supply 10,000 bearings per day to an automobile manufacturer. He finds that when he starts a production run, he can produce 25,000 bearings per day. The cost of holding a bearing in stock for one year is ₹2 and the set-up of a production run is ₹1800. How frequently should the production runs be made?

Solution Given that $D = 10{,}000$ per day and assuming 300 working days in a year then $D = 10{,}000 \times 300 = 30{,}00{,}000$ bearings per year, $C_h = ₹2$ per bearing per year, $C_1 = 18{,}000$ per production run, $r = 10{,}000$ per day, and $k = 25{,}000$ bearings per day

Now, we have $Q^* = \sqrt{\frac{2DC_1}{C_h} \cdot \frac{k}{(k-r)}}$

$$= \sqrt{\frac{2 \times 30{,}00{,}000 \times 1800}{2} \times \frac{25000}{25000 - 10000}}$$

$$= \sqrt{\frac{30{,}00{,}000 \times 1800 \times 25{,}000}{15{,}000}} = \sqrt{3{,}60{,}000 \times 25{,}000} = 94{,}868 \text{ bearings}$$

and $t_1^* = \frac{Q^*}{D} = \frac{94{,}868}{10{,}000} = 9.49$ days.

The length of the production cycle is $\frac{Q^*}{k} = \frac{94{,}868}{25{,}000} = 4$ days (approximate).

Hence, the production cycle starts at an interval of 9.49 days and the production continues for 4 days. Therefore, in each cycle, a batch of 94,868 bearings is produced.

21.8 DETERMINISTIC INVENTORY MODELS WITH SHORTAGES

This type of inventory problem becomes slightly more complicated as here, the company permits shortages or back orders to occur. In many situations, shortages are economically desirable. Permitting shortages allows the manufacturer or retailer to increase the cycle time, thereby spreading the set-up or the ordering cost over a longer time period. Allowing shortages may also be desirable when the unit value of the inventory, and therefore the inventory holding cost, is high. In the case of a back order, the customers place an order; if no stock is available, they simply wait until it becomes available.

Case 1—EOQ Problem with Instantaneous Production and Variable Order Cycle

In this case, all the assumptions are the same as mentioned in Section 21.7 (Case 1). The only difference is that here, the shortages are allowed. Let C_s be the shortage cost per unit of time per unit quantity. The aforementioned fundamental situations are shown in Fig. 21.5.

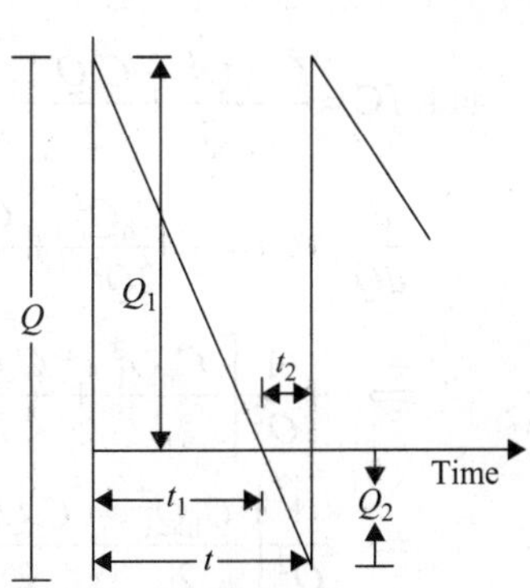

Figure 21.5 EOQ problem with instantaneous production

Total inventory over time period $t = \frac{1}{2}Q_1t_1$

Average inventory at any time period $t = \dfrac{\frac{1}{2}Q_1t_1}{t}$

Annual inventory holding cost at time t is

$$f(Q) = \frac{1}{2}\frac{Q_1t_1C_h}{t}$$

Similarly, total amount of shortage over time period $t = \frac{1}{2}Q_2t_2$

Annual shortage costs $= \dfrac{\frac{1}{2}Q_2t_2C_s}{t}$

Annual costs associated with run of size $Q = nC_1 = \frac{D}{Q}C_1$

Total annual cost, TC$=f(Q)+g(Q)+$Annual shortage costs

$$= \frac{1}{2}\frac{Q_1t_1C_h}{t} + \frac{1}{2}\frac{Q_2t_2C_s}{t} + \frac{D}{Q}C_1 = \frac{1}{2t}\left[Q_1t_1C_h + Q_2t_2C_s\right] + \frac{D}{Q}C_1 \quad (21.1)$$

Now, using the relationship for similar triangles, we have

$$\frac{t_1}{t} = \frac{Q_1}{Q} \Rightarrow t_1 = \frac{Q_1}{Q}t \text{ and } \frac{t_2}{t} = \frac{Q_2}{Q} \Rightarrow t_2 = \frac{Q_2}{Q}t \quad (21.2)$$

From Eqs (21.1) and (21.2), we have

$$\text{TC} = \frac{1}{2}\left[Q_1C_h\frac{Q_1}{Q} + Q_2C_s\frac{Q_2}{Q}\right] + \frac{D}{Q}C_1$$

$$= \frac{C_h}{2}\frac{Q_1^2}{Q} + \frac{C_s}{2}\frac{Q_2^2}{Q} + \frac{D}{Q}C_1 = \frac{C_h}{2}\frac{Q_1^2}{Q} + \frac{C_s}{2}\frac{(Q-Q_1^2)}{Q} + \frac{D}{Q}C_1$$

$$= \frac{C_h}{2}\frac{Q_1^2}{Q} + \frac{C_sQ^2 + C_sQ_1^2 - 2C_sQQ_1}{2Q} + \frac{D}{Q}C_1$$

Now $\frac{d}{dQ_1}(\text{TC}) = \frac{C_hQ_1}{Q} + \frac{0 + 2C_sQ_1 - 2C_sQ}{2Q} + 0 = 0$

$$\Rightarrow \frac{C_hQ_1}{Q} + C_s\frac{Q_1}{Q} - C_s = 0 \Rightarrow \frac{Q_1}{Q}(C_h + C_s) = C_s \Rightarrow Q_1 = \frac{C_sQ}{C_h + C_s}$$

and $\text{TC} = \frac{C_h}{2}\frac{Q_1^2}{Q} + \frac{C_sQ}{2} + \frac{C_sQ_1^2}{2Q} - C_sQ_1 + \frac{D}{Q}C_1$

$$\frac{d}{dQ}(\text{TC}) = -\frac{C_hQ_1^2}{2Q^2} + \frac{C_s}{2} - \frac{C_sQ_1^2}{2Q^2} - 0 - \frac{DC_1}{Q^2} = 0$$

$$\Rightarrow -\frac{1}{Q^2}\left[\frac{C_hQ_1^2}{2} + \frac{C_sQ_1^2}{2} + DC_1\right] + \frac{C_s}{2} = 0$$

$$\Rightarrow \frac{1}{Q^2}\left[\frac{C_hQ_1^2}{2} + \frac{C_sQ_1^2}{2} + DC_1\right] = \frac{C_s}{2}$$

$$\Rightarrow Q^2 = \frac{2}{C_s}\left[\frac{(C_h + C_s)}{2}Q_1^2 + DC_1\right] \Rightarrow Q = \sqrt{\frac{2DC_1 + C_hQ_1^2 + C_sQ_1^2}{C_s}}$$

$$\Rightarrow Q = \sqrt{\frac{2DC_1 + C_hQ_1^2}{C_s} + Q_1^2}$$

In addition, $\frac{\partial^2}{\partial Q_1^2}(\text{TC}) > 0,\ Q_1 > 0$ and $\frac{d^2}{dQ_2^2}(\text{TC}) > 0,\ Q_2 > 0$

Then $Q^* = \sqrt{\dfrac{2DC_1}{C_s} + \dfrac{(C_h + C_s)Q_1^{*2}}{C_s}}$ and $Q_1^* = \dfrac{C_s}{C_h + C_s}Q^*$

Now $Q^* = \sqrt{\dfrac{2DC_1}{C_s} + \dfrac{(C_h + C_s)}{C_s}\dfrac{C_s}{(C_h + C_s)}Q^{*2}}$

$$\Rightarrow Q^{*2} = \frac{2DC_1}{C_s} + \frac{C_s}{C_h + C_s}Q^{*2} \Rightarrow Q^{*2}\left[1 - \frac{C_s}{C_h + C_s}\right] = \frac{2DC_1}{C_s}$$

$$\Rightarrow Q^{*2}\left(\frac{C_h}{C_h + C_s}\right) = \frac{2DC_1}{C_s} \Rightarrow Q^* = \sqrt{\frac{2DC_1}{C_s}\left(\frac{C_h + C_s}{C_h}\right)}$$

We know that $Q_1^* = \dfrac{C_s}{C_h + C_s}Q^* = \dfrac{C_s}{C_h + C_s}\sqrt{\dfrac{2DC_1}{C_h}}\sqrt{\dfrac{C_h + C_s}{C_s}}$

$$= \sqrt{\frac{2DC_1}{C_h}}\sqrt{\frac{C_s}{C_h + C_s}}$$

Characteristics The characteristics of the EOQ problem with instantaneous production and variable order cycle are as follows:

1. The time between the receipt of orders is
$$t^* = \frac{Q^*}{D} = \frac{1}{D}\sqrt{\frac{2DC_1}{C_h}}\sqrt{\frac{C_h + C_s}{C_s}} = \sqrt{\frac{2C_1}{DC_h}\left(\frac{C_h + C_s}{C_s}\right)}$$
2. The total optimum inventory cost is $\text{TC}^* = \sqrt{2DC_1C_h}\sqrt{\dfrac{C_s}{C_h + C_s}}$
3. The reorder level is $Q^* - Q_2^* = Q^*\left(1 - \dfrac{C_s}{C_h + C_s}\right)$

Case 2—EOQ Problem with Instantaneous Production and Fixed Order Cycle

Consider the same assumptions as in the previous case and take t as fixed. Further assume that the items are supplied/produced at the rate of r units per unit of time during this fixed time period t.

Then, the optimum inventory level $Q_1^* = rt\left(\dfrac{C_s}{C_h + C_s}\right)$

and the total optimum inventory cost $\text{TC}^* = rt\left(\dfrac{C_hC_s}{C_h + C_s}\right)$.

Case 3—EOQ Problem with Finite Replenishment or Production

Consider the same assumptions as in the previous two cases. Further, if the rate of replenishment of the inventory is finite, say k units per unit of time, then the production cycle time is $t^* = \dfrac{Q^*}{r} = \sqrt{\dfrac{2C_1(C_h + C_s)}{C_hC_sr\left(1 - \dfrac{r}{k}\right)}}$.

The optimum inventory level is $Q_1^* = \sqrt{\frac{2C_1C_s r}{C_h(C_h + C_s)}\left(1 - \frac{r}{k}\right)}$

and the total optimum inventory cost is $TC^* = \sqrt{\frac{2C_hC_sC_1r}{(C_h + C_s)}\left(1 - \frac{r}{k}\right)}$.

Example 21.8 The demand of an item is uniform at a rate of 20 units per month. The fixed cost is ₹10, every time a production run is made. The production cost is ₹1 per item and the inventory carrying cost is ₹0.25 per item per month. If the shortage cost is ₹1.25 per item per month, how often should a production run be made and what size should it be?

Solution Given that $D = 20$ units per month, $C_1 = ₹10$ per production run, $C_h = ₹0.25$ per month, and $C_s = ₹1.25$ per month.

The optimum values of Q and t are

$$Q^* = \sqrt{\frac{2DC_1}{C_s}\left(\frac{C_h + C_s}{C_h}\right)}$$

$$= \sqrt{\frac{2 \times 20 \times 10 \times (0.25 + 1.25)}{1.25 \times 0.25}} = 43.8 = 44 \text{ items (approximate)}$$

and $t^* = \frac{Q^*}{r} = \frac{44}{20} = 2.2$ months.

Example 21.9 A commodity is to be supplied at a constant rate of 200 units per day. Supplies of any amount can be had at any required time, but each order costs ₹50. The cost of holding the commodity in the inventory is ₹2.00 per unit per day, whereas the delay in the supply of the item induces a penalty of ₹10.00 per unit per delay of one day. Find the optimal policy (q, t), where t is the reorder cycle period and q is the inventory level after the reorder. What would the best policy be if the penalty cost becomes ∞?

Solution Given that $D = 200$ units per day, $C_1 = ₹50$ per order, $C_h = ₹2$ per unit per day, and $C_s = ₹10$ per unit per day.

We have $Q^* = \sqrt{\frac{2DC_1(C_h + C_s)}{C_hC_s}} = \sqrt{\frac{2 \times 200 \times 50 \times (2 + 10)}{2 \times 10}} = \sqrt{1000 \times 12}$

$= \sqrt{12000} = 109.5$ units and $t^* = \frac{Q^*}{D} = \frac{109.5}{200} = \frac{1}{2}$ days

This shows that the optimum order quantity of 109.5 units must be supplied after 0.5 days. If the penalty cost becomes ∞, we have $Q^* = \sqrt{\frac{2DC_1}{C_h}} = \sqrt{\frac{2 \times 200 \times 50}{2}} = 100$ units

and $t^* = 100/200 = 1/2$ days.

Example 21.10 A contractor undertakes an order to supply diesel engines to a truck manufacturer at the rate of 25 per day. There is a clause in the contract penalizing him at ₹10 per engine per day for missing the scheduled delivery date. He finds that the cost of holding a complete engine in stock is ₹16 per month. His production process is such that each month, he starts a batch of engines through the shops, and all these engines are available for delivery any time after the end of the month. What should his inventory level be at the beginning of each month?

Solution Given that $C_s = ₹10$ per engine per day, $C_h = ₹16$ per engine per month = (16/30) per engine per day, $r = 25$ engines per day, and $t = 1$ month (30 days).

$$\text{The optimum values of } Q_1 \text{ is } Q_1^* = rt\left(\frac{C_s}{C_h + C_s}\right) = 25 \times 30\left(\frac{10}{\frac{16}{30} + 10}\right) = 712 \text{ engines}$$

Example 21.11 The demand for an item in a biscuit company is 18,000 units per year, and the company can produce the items at a rate of 3000 units per month. The cost of one set-up is ₹500 and the holding cost of one unit per month is ₹0.15. The shortage cost of one unit is ₹20 per month. Obtain the optimum production batch quantity.

Solution Given that $C_h = ₹0.15$ per unit per month, $C_s = ₹20$, $C_1 = ₹500$, $r = 18{,}000$ units per year or 1500 units per month, and $k = 3000$ units per month.

The optimum value of Q is as follows:

$$Q^* = \sqrt{\frac{2C_1(C_h + C_s)}{C_h C_s}\left(\frac{kr}{k - r}\right)}$$

$$= \sqrt{\frac{2 \times 500 \times (0.15 + 20)}{0.15 \times 20}\left(\frac{1{,}500 \times 3{,}000}{3{,}000 - 1{,}500}\right)} = 4488.89 \text{ units}$$

21.9 EOQ PROBLEM WITH PRICE BREAKS OR QUANTITY DISCOUNT

In inventory problems, sometimes the item cost is variable and depends upon the quantity manufactured/purchased. These conditions occur when discounts are offered while purchasing large quantities. These discounts are called *price break* in the inventory model. If we were to ignore the price discount factor, then $Q^* = \sqrt{\frac{2DC_1}{C_h}}$. However, if the price discount is considered, then the total cost per unit of the inventory system and items would be $\text{TC} = Dk_1 + \frac{D}{Q}C_1 + \frac{1}{2}Qk_1I$, where the cost of manufacturing/purchasing is k_1 per unit, C_1 is the cost per production run, and I denotes the holding cost per unit. Now, for the optimum order quantity we have $\frac{d}{dQ}(\text{TC}) = 0 \Rightarrow Q = \sqrt{\frac{2DC_1}{k_1I}}$, and $\frac{d^2}{dQ^2}(\text{TC}) > 0$ at $Q^* = \sqrt{\frac{2DC_1}{k_1I}}$.

Hence, the optimum value of Q will be $Q^* = \sqrt{\frac{2DC_1}{k_1I}}$

We shall now consider two cases of EOQ problems with single and n-price breaks.

Case 1—EOQ Problem with One Price Break

If there is only one price break, then the situation becomes as follows:

Range of quantity	Purchase cost per unit
$0 \le Q_1 < b_1$	k_{11}
$b_1 \le Q_2 < b_2$	k_{12}

where $k_{12} < k_{11}$.

The steps followed to solve an EOQ problem with one price break are as follows:

1. Find Q_2^* and compare it with b_1.
2. If $Q_2^* > b_1$, then the optimum order quantity will be Q_2^*, that is, $Q^* = Q_2^*$.
3. If $Q_2^* < b_1$, we cannot place an order at the reduced price k_{12}. Hence, we compare the total inventory cost for $Q^* = Q_1^*$ (for the price k_{11}) with $Q^* = b_1$.

$$\text{TC}(Q_1^*) = Dk_{11} + \frac{D}{Q_1^*}C_1 + \frac{1}{2}Q_1^* \times k_{11} \times I$$

$$\text{and TC}(b_1) = Dk_{12} + \frac{D}{b_1}C_1 \times \frac{1}{2}b_1 \times k_{12} \times I$$

If $\text{TC}(Q_1^*) > \text{TC}(b_1)$ then $Q^* = b_1$, otherwise $Q^* = Q_1^*$.

Case 2—EOQ Problem with n-price Breaks

If there are n price breaks, then the situation would be as follows:

Range of quantity	Purchase cost per unit
$0 \le Q_1 < b_1$	k_{11}
$b_1 \le Q_2 < b_2$	k_{12}
...	...
$Q_n \ge b_{n-1}$	k_{1n}

where $b_1, b_2, \ldots, b_{n-1}$ are those quantities that determine the price breaks.

The steps followed to solve the EOQ problem with n price breaks are as follows:

1. Find Q_n^* and compare it with b_{n-1}.
2. If $Q_n^* > b_{n-1}$, then the optimum order quantity will be Q_n^*, that is $Q^* = Q_n^*$.
3. If $Q_n^* < b_{n-1}$, obtain Q_{n-1}^*. If $Q_{n-1}^* \ge b_{n-1}$, then compare $\text{TC}(Q_{n-1}^*)$ with $\text{TC}(b_{n-1})$.
4. If $Q_{n-1}^* < b_{n-2}$ then obtain Q_{n-2}. If $Q_{n-2}^* \ge b_{n-3}$ then compare $\text{TC}(Q_{n-2}^*)$, $\text{TC}(b_{n-2})$,and $\text{TC}(b_{n-1})$.
5. If $Q_{n-2}^* < b_{n-2}$, obtain Q_{n-3}^*. If $Q_{n-3}^* \ge b_{n-4}$, then compare $\text{TC}(Q_{n-3}^*)$ with $\text{TC}(b_{n-3})$, $\text{TC}(b_{n-2})$, and $\text{TC}(b_{n-1})$.
6. Continue this way until $Q_{n-i}^* \ge b_{n-(i+1)}; 0 \le i \le n-1$ is achieved and then compare $\text{TC}(Q_{n-i}^*)$ with $\text{TC}(b_{n-i-2}), \ldots, \text{TC}(b_{n-1})$.

Example 21.12 Find the optimum order quantity of a product that has the following price breaks:

Range of quantity	Unit cost (₹)
$0 \le Q_1 < 700$	10
$700 \le Q_2$	8

The monthly demand for the product is 300 units, the cost of storage is 5% of the unit cost, and the cost of ordering is ₹500.

Solution Given that $C_1 = ₹500, D = 300$ units per month, $I = ₹0.05, k_{11} = ₹10$, and $k_{12} = ₹8$.

It is given that the highest discount available is ₹8. Therefore, we obtain Q_2^* by taking into consideration k_{12} as ₹8. Now, we have $Q_2^* = \sqrt{\dfrac{2DC_1}{k_{12}I}} = \sqrt{\dfrac{2 \times 300 \times 500}{8 \times 0.05}} = 866$ units.

Here, $Q_2^* > b_1$. Thus, the optimum purchase quantity is $Q^* = Q_2^* = 866$ units.

Example 21.13 Find the optimum order quantity of a product, that has the following price breaks:

Range of quantity	Purchasing cost per unit (₹)
$0 \le Q_1 < 100$	20
$100 \le Q_2 < 200$	18
$200 \le Q_3$	16

The monthly demand for the product is 400 units. The storage cost is 20% of the unit cost of the product and the cost of ordering is ₹25 per month.

Solution Given that $C_1 = ₹25$, $D = 400$ units per month, $I = ₹0.20$, $k_{11} = ₹20$, $k_{12} = ₹18$, and $k_{13} = ₹16$.

We have $Q_3^* = \sqrt{\dfrac{2DC_1}{k_{13}I}} = \sqrt{\dfrac{2 \times 400 \times 25}{16 \times 0.20}} = 79$ units.

Here, $Q_3^* < b_2$. Now, we have $Q_2^* = \sqrt{\dfrac{2DC_1}{k_{12}I}} = \sqrt{\dfrac{2 \times 400 \times 25}{18 \times 0.20}} = 75$ units.

Here also, $Q_2^* < b_1$. Now, we have $Q_1^* = \sqrt{\dfrac{2DC_1}{k_{11}I}} = \sqrt{\dfrac{2 \times 400 \times 25}{20 \times 0.20}} = 71$ units.

Now we obtain

$$\text{TC}(Q_1^*) = Dk_{11} + \frac{D}{Q_1^*}C_1 + \frac{1}{2}Q_1^* \times k_{11} \times I$$

$$= 400 \times 20 + \frac{400}{71} \times 25 + \frac{1}{2} \times 71 \times 20 \times 0.20 = ₹8283$$

$$\text{TC}(b_1) = Dk_{12} + \frac{D}{b_1}C_1 \times \frac{1}{2}b_1 \times k_{12} \times I$$

$$= 400 \times 18 + \frac{400}{100} \times 25 + \frac{1}{2} \times 100 \times 18 \times 0.20 = ₹7480$$

$$\text{and } TC(b_2) = Dk_{13} + \frac{D}{b_2}C_1 \times \frac{1}{2}b_2 \times k_{13} \times I$$

$$= 400 \times 16 + \frac{400}{200} \times 25 + \frac{1}{2} \times 200 \times 16 \times 0.20 = ₹6770$$

Thus, the optimum purchase quantity is $Q^* = b_1 = 100$ units, because $TC(b_2) < TC(b_1) < TC(Q_1^*)$.

21.10 PROBABILISTIC INVENTORY MODELS

We shall now consider a problem where the demand is probabilistic in nature. For such cases, the control variable is assumed to be the scheduling period, the order level, or both. To find the optimum order levels, we derive the minimum total expected cost rather than the actual cost involved.

21.10.1 Single Period Problem without Set-up Cost and Uniform Demand

To tackle such problems, the following assumptions and notations are used:

1. Let the demand be r units per time period.
2. The reorder time is fixed and known.
3. The set-up cost is not included in the total cost.
4. The production of the commodity is instantaneous.
5. The lead time is taken to be zero.
6. The holding cost per item per unit time is C_h.
7. The shortage cost per item per unit time is C_s.
8. Let Q be the inventory level.

We now proceed to find the optimal order level Q for cases when $r \leq Q$ or $r > Q$ at the beginning of each period. For both the cases, the inventory situation is shown in Figs 21.6(a) and (b).

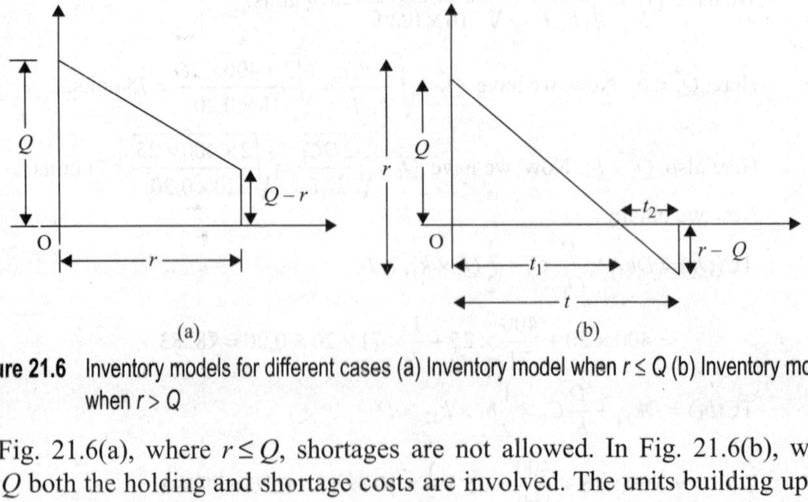

Figure 21.6 Inventory models for different cases (a) Inventory model when $r \leq Q$ (b) Inventory model when $r > Q$

In Fig. 21.6(a), where $r \leq Q$, shortages are not allowed. In Fig. 21.6(b), when $r > Q$ both the holding and shortage costs are involved. The units building up the inventory may be discrete or continuous units.

Case 1—Uniform Demand and Discrete Units

Let the demand for r units be estimated at a rate with the discrete probability $p(r)$, where $r = 1, 2, \ldots$

Hence, we have $\sum_{r=0}^{\infty} p(r) = 1$ and $p(r) \geq 0$

When $r \leq Q$, holding cost of carrying r units ($r \leq Q$) will be $\frac{1}{2}C_h(Q - r + Q)t$

The expected carrying cost is $\frac{1}{2}C_h t \sum_{r=0}^{Q} (2Q - r)p(r)$

When $r > Q$, the holding cost is $\frac{1}{2}Qt_1C_h$ and the shortage cost is $\frac{1}{2}(r - Q)\, t_1 t_2$

From Fig. 21.6(b), we have $\frac{t_1}{t} = \frac{Q}{r}$ and $\frac{t_2}{t} = \frac{r - Q}{r}$

The expected shortage cost is $\sum_{r=Q+1}^{\infty} \left[\frac{1}{2}QC_h \times \frac{Q}{r}t + \frac{1}{2}(r - Q)C_s \times \frac{r - Q}{r}t \right] p(r)$

The total expected cost (TC) associated with producing quantity Q is

$$\text{TC} = C_h t \sum_{r=0}^{Q} \left(Q - \frac{r}{2} \right) p(r) + \sum_{r=Q+1}^{\infty} t\left\{ C_h Q^2 + C_s (r - Q)^2 \right\} \frac{p(r)}{2r}.$$

The average cost or total expected cost for one period is given by TC(Q)

$$= C_h \sum_{r=0}^{Q} \left(Q - \frac{r}{2} \right) p(r) + \sum_{r=Q+1}^{\infty} t\left\{ C_h Q^2 + C_s (r - Q)^2 \right\} \frac{p(r)}{2r} \qquad (21.3)$$

To minimize TC(Q), the following conditions must be satisfied:

$$\text{TC}(Q + 1) > \text{TC}(Q) \text{ and } \text{TC}(Q - 1) > \text{TC}(Q)$$

That is, $\text{TC}(Q + 1) - \text{TC}(Q) > 0$ and $\text{TC}(Q) - \text{TC}(Q - 1) < 0$

Substituting ($Q + 1$) for Q in Eq. (21.3), we have

$$\text{TC}(Q + 1) = C_h \sum_{r=0}^{Q+1} \left(Q + 1 - \frac{r}{2} \right) p(r) + \sum_{r=Q+2}^{\infty} \left\{ C_h (Q + 1)^2 + C_s (r - Q - 1)^2 \right\} \frac{p(r)}{2r}$$

$$= \text{TC}(Q) + (C_h + C_s)F(Q) - C_s$$

where $F(Q) = \sum_{r=0}^{Q} p(r) + \left(Q + \frac{1}{2} \right) \sum_{r=Q+1}^{\infty} \frac{p(r)}{r}$

Similarly, we have $\text{TC}(Q - 1) = \text{TC}(Q) - (C_h + C_s)F(Q - 1) + C_s$

$$\therefore \left[\text{TC}(Q + 1) - \text{TC}(Q) \right] > 0 \Rightarrow (C_h + C_s)F(Q^*) - C_s > 0 \text{ or}$$

$$F(Q^*) > \frac{C_s}{C_h + C_s}$$

and $\left[TC(Q) - TC(Q-1)\right] < 0 \Rightarrow (C_h + C_s)F(Q^* - 1) - C_s < 0$ or

$$F(Q^*-1) < \frac{C_s}{C_h + C_s}$$

Hence $F(Q^*-1) < \dfrac{C_s}{C_h + C_s} < F(Q^*)$.

Case 2—Uniform Demand and Continuous Units

In the continuous case, replace $p(r)$ by $f(r)$ and summation by integration, where $\int_0^\infty f(r)dr = 1$ and $f(r) \geq 0$. Now proceed exactly in the same manner as in Case 1.

$$\int_0^Q \left(Q - \frac{r}{2}\right) f(r)dr \text{ and } \int_Q^\infty \{C_h Q^2 + C_s(r-Q)^2\} \frac{f(r)}{2r} dr$$

The average cost or total expected cost for one period is

$$TC(Q) = C_h \int_0^Q \left(Q - \frac{r}{2}\right) f(r)dr + C_h \int_Q^\infty Q^2 \frac{f(r)}{2r} dr + C_s \int_Q^\infty (r-Q)^2 \frac{f(r)}{2r} dr.$$

Now, $\dfrac{d}{dQ}[TC(Q)] = C_h \displaystyle\int_0^Q f(r)dr + C_h \int_Q^\infty \frac{Q}{r} f(r)dr - C_s \int_0^\infty \left(\frac{r-Q}{r}\right) f(r)dr$

$$= C_h \int_0^Q f(r)dr + C_h \int_Q^\infty \frac{Q}{r} f(r)dr - C_s \left[\int_0^\infty f(r)dr - \int_0^Q f(r)dr\right]$$

$$+ C_s \int_0^\infty \frac{Q}{r} f(r)dr$$

$$= (C_h + C_s)\left[\int_0^Q f(r)dr + \int_Q^\infty \frac{Q}{r} f(r)dr\right] - C_s (\text{since} \int_Q^\infty f(r)dr = 1)$$

For minima $\dfrac{d}{dQ}[TC(Q)] = 0 \Rightarrow F(Q^*) = \dfrac{C_s}{C_h + C_s}$

where $F(Q) = \displaystyle\int_0^Q f(r)dr + \int_Q^\infty \frac{Q}{r} f(r)dr$

and $\dfrac{d^2}{dQ^2} TC(Q)\Big|_{Q=Q^*} = (C_h + C_s)\displaystyle\int_Q^\infty \frac{f(r)}{r} dr > 0$

Thus, $F(Q^*) = \dfrac{C_s}{C_h + C_s}$ gives the condition for finding the optimum value of Q as Q^*.

21.10.2 Single Period Problems without Set-up Cost and Instantaneous Demand

Consider the same assumptions mentioned in Section 21.10.2, except that the demand is instantaneous, instead of being uniform. Here again, we consider the two cases $r \leq Q$ and $r > Q$.

The situation of this case is displayed in Figs 21.7(a) and (b).

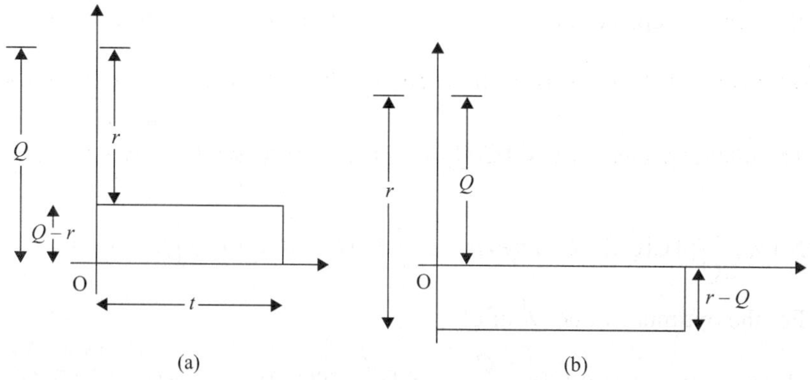

Figure 21.7 Inventory models with respect to shortages (a) Inventory model with no shortage when $r \le Q$ (b) Inventory model with shortage (no inventory) when $r > Q$

Case 1—Instantaneous Demand and Discrete Units

Suppose the demand for r units is estimated at a rate with discrete probability $p(r)$; $r = 1, 2, \ldots, n$. When $r \le Q$, the expected holding cost at the beginning of the period will be $C_h \sum_{r=0}^{Q} (Q - r)p(r)$.

When $r > Q$, the expected shortage cost at the beginning of the period will be $C_s \sum_{r=Q}^{\infty} (r - Q)p(r)$.

The total expected cost is $\text{TC}(Q) = C_h \sum_{r=0}^{Q} (Q - r)p(r) + C_s \sum_{r=Q+1}^{\infty} (r - Q)p(r)$.

For the optimum value of Q, we have $\text{TC}(Q+1) > \text{TC}(Q)$ and $\text{TC}(Q) < \text{TC}(Q-1)$.

Now, $\text{TC}(Q+1) = C_h \sum_{r=0}^{Q+1} (Q + 1 - r)p(r) + C_s \sum_{r=Q+2}^{\infty} (r - Q - 1)p(r)$

$$= \text{TC}(Q) + (C_h + C_s) \sum_{r=0}^{Q} p(r) - C_s$$

since $\sum_{r=Q+1}^{\infty} p(r) = \sum_{r=0}^{\infty} p(r) - \sum_{r=0}^{Q} p(r) = 1 - \sum_{r=0}^{Q} p(r)$.

Similarly, we have $\text{TC}(Q-1) = \text{TC}(Q) - (C_h + C_s) \sum_{r=0}^{Q-1} p(r) < C_s$

For the optimum value Q^*, we have $(C_h + C_s) \sum_{r=0}^{Q} p(r) > C_s$ and

$(C_h + C_s) \sum_{r=0}^{Q-1} p(r) < C_s$, that is, $F(Q^* - 1) < \dfrac{C_s}{C_h + C_s} < F(Q^*)$, where $F(Q^*) = \sum_{r=0}^{Q} p(r)$

Case 2—Instantaneous Demand and Continuous Units

In this case, replace $p(r)$ by $f(r)$ and summation by integration, where $\int_0^\infty f(r)dr = 1$ and $f(r) \geq 0$. Now, proceed exactly in the same manner as in Case 1.

The total expected cost is $\text{TC}(Q) = C_h \int_0^Q (Q-r)\, f(r)dr + C_s \int_Q^\infty (r-Q)\, f(r)dr$

Now, $\frac{d}{dQ}[\text{TC}(Q)] = C_h \int_O^Q f(r)dr - C_s \int_Q^\infty f(r)dr = (C_h + C_s)\int_0^Q f(r)dr - C_s$

For the optimum value Q^* of Q,

$$\frac{d}{dQ}\text{TC}(Q) = 0 \Rightarrow \int_0^Q f(r)dr = \frac{C_s}{C_h + C_s} \text{ and } \frac{d^2}{dQ^2}\text{TC}(Q)_{\text{at } Q=Q^*} = (C_h + C_s)\, \text{F}(Q^*) > 0$$

Hence, the $\text{F}(Q^*) = \frac{C_s}{C_h + C_s}$ gives the condition for finding the optimum value of Q.

Example 21.14 The probability distribution of the monthly sales of a certain item is as follows:

Monthly sales	0	1	2	3	4	5	6
Probability	0.02	0.05	0.30	0.27	0.20	0.10	0.06

The cost of holding inventory is ₹10 per unit per month. The current policy is to maintain a stock of four items at the beginning of each month. Assuming that the cost of shortage is proportional to both time and quantity, obtain the imputed cost of shortage of one item for one unit of time.

Solution Given that $C_h = ₹10$ per unit per month and $Q = 4$ units.
Since the demand is uniformly distributed over the month, we have

$$F(Q^* - 1) < \frac{C_s}{C_h + C_s} < F(Q^*)$$

$$\text{Now } F(Q^*-1) = F(4-1) = \sum_{r=0}^{4} p(r) + \left(4 - \frac{1}{2}\right)\sum_{r=4}^{6} \frac{p(r)}{r}$$

$$= 0.84 + \frac{7}{2}\left[\frac{0.20}{4} + \frac{0.10}{5} + \frac{0.06}{6}\right] = 0.92$$

Thus, the least value of C_s is given by $\frac{C_s}{C_h + C_s} = 0.92$ or $C_s = 0.92(10 + C_s) \Rightarrow C_s = ₹115$

Similarly, the greatest value of C_s is obtained by using the following equation: $\frac{C_s}{C_h + C_s}$

$$\frac{C_s}{C_h + C_s} = \sum_{r=0}^{4} p(r) + \left(4 + \frac{1}{2}\right)\sum_{r=5}^{6} \frac{p(r)}{r} = 0.84 + \frac{9}{2}\left[\frac{0.10}{5} + \frac{0.06}{6}\right] = 0.975$$

$$\Rightarrow C_s = 0.975(10 + C_s) \Rightarrow C_s = ₹390.$$

Hence, the imputed cost of shortage is $₹115 < C_s < ₹390$.

Example 21.15 A paper boy buys papers for 60 paise each and sells them for ₹1.40 each. He cannot return the unsold newspapers. The daily demand has the following distribution:

No. of customers	23	24	25	26	27	28	29	30	31	32
Probability	0.01	0.03	0.06	0.01	0.20	0.25	0.15	0.10	0.05	0.05

If each day's demand is independent of that of the previous day, how many papers should be ordered each day?

Solution Assume Q is the number of newspapers ordered per day and r is the demand for it, that is, the number of newspapers actually sold per day. Given that $C_h =$ ₹0.60 and $C_s = 1.40 - 0.60 =$ ₹0.80

We have, $C_s/(C_h + C_s) = 0.80/0.60 + 0.80 = 0.57$

The optimum solution is obtained by developing the cumulative probability distribution of daily demand as follows:

r	23	24	25	26	27	28	29	30	31	32
$p(r)$	0.01	0.03	0.06	0.10	0.20	0.25	0.15	0.10	0.05	0.05
$\sum_{r=23}^{Q}$	0.01	0.04	0.10	0.20	0.40	0.65	0.80	0.90	0.95	1.00

Here, $C_s/(C_h + C_s) = 0.57$ lies between 0.40 and 0.65 and the condition for optimality suggests that $Q^* = 28$, that is, the optimum number of newspapers to be ordered is 28.

Example 21.16 A baking company sells cake by weight (kg). It makes a profit of ₹0.50 per kg on every kg sold, the day it is baked. It disposes of all cakes that are not sold on the date they are baked, at a loss of 12 paise per kg. If the demand is known to be between 2000 and 3000 kg, calculate the optimum level of inventory if the demand is instantaneous.

Solution Given that $C_h =$ ₹0.12, $C_s =$ ₹0.50 and $f(r) = 1/1000$ for $2000 \le r \le 3000$.

For the optimum value Q^* of Q, we have $\int_0^Q f(r)dr = \frac{C_s}{C_h + C_s} \Rightarrow \int_{r=2000}^{Q} \frac{1}{1000} dr = \frac{C_s}{C_h + C_s}$

$$\Rightarrow \quad \frac{1}{1{,}000}(Q - 2000) = \frac{0.50}{0.12 + 0.50} \Rightarrow Q - 2000 = \frac{50}{62} \times 1000$$

$$\Rightarrow \quad Q - 2000 = 806.45 \Rightarrow Q = 2806.45 \text{ per kg}$$

Hence, the optimum level of inventory is $Q^* = Q = 2806.45$ per kg.

21.11 SOME IMPORTANT INVENTORY CONTROL TECHNIQUES

Control over inventories can be managed in several ways. One way, for example, is to keep a record of the receipts on the issue of inventory items every day. It is also called the perpetual inventory system. Another method is to find the EOQ of each item to be stocked, which we have discussed in Sections 21.8 and 21.9. There are several organizations consuming a variety of stored items. As all the items are not of equal importance, these items are classified into different categories. This theory, called the principle of selective control, is used to control the inventories. In this section, we shall discuss a few techniques of selective control.

ABC Analysis

Always better control (ABC) analysis is an important tool that aids the management in their efforts to achieve the best possible results. This technique comprises classifying the inventory items into three categories—A, B, and C—in the descending order of their annual consumption and the annual monetary value of each item. In ABC analysis, items of an inventory can be classified into three main categories based on their respective usage value.

Category* 'A' *items Costly items are classified as 'A' category. Items of this type involve a large investment, but may be fewer in number; for instance, 70% of the investment is used in 10% of the items in the inventory. Items of this category must be ordered frequently, but in smaller numbers. A periodic review policy should be followed to minimize the shortage percentage of such items. A well experienced team of people should control such items.

Category* 'B' *items Items having an average consumption value are classified as category 'B'. Nearly 25% of the inventory's items account for, say, 25% of the total investment.

Category* 'C' *items These are items of relatively lesser importance. For instance, nearly 80% of the inventory items account for only 5% of the total investment.

XYZ Analysis

The XYZ classification is based on the closing value of the items in storage. In the XYZ analysis, X stands for high inventory values and Z for low tie-up capital; Y is assigned to the items that lie between X and Z. This analysis helps in reducing the capital investment in high-value items and often leads to unlocking the blocked value.

VED Analysis

In VED analysis, V stands for vital items, E for essential items, and D for desirable items. This analysis is particularly essential for spare parts. Since the analysis is based on criticality, it helps in finding the critical spares that are difficult to procure.

FSND Analysis

In FSND analysis, F stands for fast-moving items, S for slow-moving items, N for non-moving items and D for dead items for which no future demand is foreseen.

JIT

JIT (Just-in-time production) system is also known as zero inventory. Here, the focus is on reducing inefficiency and unproductive time in the production process. This continuously improves the process and the quality of the product/service.

MRP

MRP stands for material requirement planning. In the MRP system, if the stock in not moving through the system, the production will still continue, to build inventory. Bills of materials, master production schedules, and inventory status of various items are mainly required for the efficient working of the MRP system.

RECAPITULATION

1. Inventory is made up of materials, commodities, and products, which are usually kept in stocks in order to be consumed or benefitted from when needed. The various types of inventory are movement inventories, buffer inventories, anticipation inventories, decoupling inventories, and lot size inventories.
2. The main objective of an inventory problem is to minimize the total (actual or expected) cost or to maximize the profit (actual or expected).
3. The main inventory costs are holding cost, set-up cost, ordering cost, and shortage cost.
4. The three main objectives of inventory management are improved customer service, reduced inventory investment, and increased productivity.
5. The inventory problems in which the demand is assumed to be fixed and completely predetermined are usually known as economic order quantity (EOQ) or lot-size problems.
6. The inventory problem becomes slightly more complicated when a company permits shortages or back orders to occur. However, in many situations, shortages are economically desirable.
7. Sometimes, the item cost is variable and depends upon the quantity manufactured or purchased. These conditions occur when discounts are offered on the purchase of large quantities. These discounts are called *price breaks* in the inventory model.
8. When the demand is probabilistic in nature, the control variable is assumed to be the scheduling period, the order level, or both.
9. ABC analysis is an important tool that aids the management in their efforts to achieve the best possible results.
10. In XYZ analysis, X stands for high inventory values, Z for low tie-up capital, and Y lies between X and Z.
11. In VED analysis, V stands for vital items, E for essential items, and D for desirable items.
12. In FSND analysis, F stands for fast-moving items, S for slow-moving items, N for non-moving items, and D for dead items, for which no future demand is foreseen.
13. JIT means just-in-time production. It is also known as zero inventory.
14. MRP stands for material requirement planning.

EXERCISES

Multiple-choice Questions

21.1 The rent for the stores where materials are kept falls under
(a) holding cost
(b) ordering cost
(c) set-up cost
(d) none of these

21.2 The lead time is the time
(a) between the receipt of the material and the usage of the material
(b) taken to place the order for the materials
(c) between placing the order and the receipt of materials
(d) none of these

21.3 In EOQ
(a) annual ordering cost = annual holding cost
(b) annual ordering cost = annual purchasing cost
(c) annual ordering cost = annual shortage cost
(d) none of these

21.4 Which of the following increases with the quantity ordered per order?
(a) Ordering cost (c) Shortage cost
(b) Holding cost (d) None of these

21.5 In the JIT system, the holding cost is assumed to be
(a) zero (c) equal to the purchasing cost
(b) equal to the ordering cost (d) none of these

21.6 The classification made on the basis of the weight of the material is known as
(a) ABC (b) XYZ (c) FSN (d) none of these

State True or False

21.1 One of the important objectives of inventory management is to employ the available capital efficiently so as to yield the maximum result.
21.2 As the order quantity decreases, it reduces the ordering cost.
21.3 Using inventory control techniques, we can reduce the material cost.
21.4 Shortage cost is incurred by having a demand for a product when the inventory is completely depleted.
21.5 If the unit cost increases, the optimal order quantity increases.
21.6 MRP stands for material requirement planning.

Fill in the Blanks

21.1 The stock of materials kept in the stores in the anticipation of future demand is known as ______.
21.2 Insurance charges of materials cost fall under the inventory ______ cost.
21.3 When the inventory reaches a specific level called ______, enough inventories are available to cover the expected demand during the lead time.
21.4 If the total investment in stock is limited, then the best order quantity for each item will be ______ the EOQ.
21.5 JIT stands for ______.
21.6 ______ analysis depends on the annual consumption value of the materials.

Review Questions

21.1 What do you mean by inventory?
21.2 What are the types of inventories?
21.3 Discuss the various costs of inventory.
21.4 Explain the EOQ fundamental model.
21.5 Define the following terms: demand, lead time, and reorder level.
21.6 Discuss the various factors affecting inventory control.
21.7 Write a short note on inventory management.
21.8 Explain the holding, shortage, and set-up costs.
21.9 Explain the deterministic inventory models without shortages.
21.10 Explain the deterministic inventory models with shortages.
21.11 Write a short note on the probabilistic inventory model.
21.12 Discuss some inventory control techniques.

Numerical Problems

21.1 Neon lights in an industrial park are replaced at the rate of 100 units per day. The park orders these neon lights periodically. It costs ₹100 to initiate a purchase order.

A neon light kept in storage is estimated to cost ₹0.02 per day. The lead time between placing and receiving an order is 12 days. Determine the optimum inventory policy for ordering the neon lights.

21.2 A certain item costs ₹235 per ton. The monthly requirement is 5 tons and each time the stock is replenished, there is a set-up cost of ₹1000. The holding cost has been estimated at 10% of the value of the stock per year. What is the optimal order quantity?

21.3 A shipbuilding company uses rivets at a constant rate of 20,000 rivets per year. The ordering cost is ₹30 per year. The rivets cost ₹1.50 per rivet. The holding cost of the rivets is estimated to be 12.5% of the unit cost per year. Find the EOQ.

21.4 An item is produced at the rate of 128 units per day. The annual demand is 6400 units. The set-up cost for each production run is ₹24 and the inventory holding cost is ₹3 per unit per year. There are 250 working days for production each year. Obtain an inventory policy for this item.

21.5 The storage of one item costs ₹1 per month and the set-up cost is ₹25 per run. If the production is instantaneous and the demand is 200 units per month, find the optimal size of the batch and the best time for the replenishment of the inventory.

21.6 A company uses 24,000 units of a raw material annually, which costs ₹1.25/unit. The order cost for each order is ₹22.5 and the carrying cost is 5.4% per year of the average inventory. Find the economic size and the total inventory cost (including the cost of material).

21.7 In a central grain store, it takes about 15 days to receive the stock after placing an order and 500 tons are dispatched to the neighboring markets every day. On an ad-hoc basis, the safety stock is assumed to be 10 days of stock. Calculate the reorder point.

21.8 Find the optimum order quantity of a product for which the costs are as follows:

Range of quantity	Purchase cost per unit (in ₹)
$0 \le Q_1 < 500$	10.00
$500 \le Q_2$	9.25

The monthly demand for the product is 200 units, the cost of storage is 2% of the unit cost, and the cost of ordering is ₹350.00.

21.9 Find the optimum order quantity of a product for which the costs are as follows:

Range of quantity	Purchasing cost per unit (₹)
$0 \le Q_1 < 500$	10.00
$500 \le Q_2 < 750$	9.25
$750 \le Q_3$	8.75

The monthly demand for the product is 200 units. The storage cost is 2% of the unit cost of the product and the cost of ordering is ₹350 per month.

21.10 A paper boy buys papers for ₹1.40 and sells them for ₹2.45 each. He cannot return the unsold newspapers. The daily demand has the following distribution:

Customers	25	26	27	28	29	30	31	32	33	34	35	36
Probability	0.03	0.05	0.05	0.10	0.15	0.15	0.12	0.10	0.10	0.07	0.06	0.02

If each day's demand is independent of the previous day's, how many papers should he order each day?

21.11 A TV dealer finds that the cost of holding a television in stock for a week is ₹20. Customers who cannot procure new televisions tend to go to other dealers immediately. The dealer estimates that for every customer who does not get immediate delivery, he loses ₹200 on an average. For one particular model of television, the probabilities for demand of 0, 1, 2, 3, 4, and 5 televisions in a week are 0.05, 0.10, 0.20, 0.30, 0.20, and 0.15 respectively. How many televisions should the dealer order per week? (Assume that there is no time lag between ordering and delivery).

21.12 A baking company sells cake by the kg. It makes a profit of ₹0.50 per kg on every cake sold, the day it is baked. It disposes of all the cakes that are not sold on the date they are baked at a loss of 12 paise per kg. If the demand is known to be between 2000 and 3000 kg, determine the optimum daily amount baked.

CONCEPTUAL PROBLEM

A company makes four orders a year for its inventory, with respect to each category, that is, A, B, and C, to get three months' supply of materials. The level of annual consumption for the three groups of items is as follows:

Items	Annual consumption (in ₹)	Number of orders
A	36,000	4
B	3600	4
C	360	4

(a) Calculate their average inventory (which is one-half of the order quantity) without using the ABC control technique.

(b) Using the ABC control technique, calculate the average inventory for making 12 orders but segregating items A, B, and C, according to their consumption value as 9, 2, and 1 (instead of 4, 4, and 4 as in the aforegiven table) orders respectively. Now do the same computations as in the first part of the problem and compare the inventory investment for the two cases.

Solution The computations for average inventory investment without using ABC analysis is as follows:

Items	Annual consumption (₹)	Number of orders	Average inventory investment (₹)
A	36,000	4	$36{,}000/4 \times 1/2 = 4500$
B	3600	4	$3600/4 \times 1/2 = 450$
C	360	4	$360/4 \times 1/2 = 45$
Total	39,960	12	4995

The average annual investment on inventories = ₹4995.

(Contd)

(*Contd*)

The computations made using ABC analysis are as follows:

Items	Annual consumption (₹)	Number of orders	Average inventory investment (₹)
A	36,000	9	$36{,}000/9 \times 1/2 = 2000$
B	3600	2	$3600/2 \times 1/2 = 900$
C	360	1	$360/1 \times 1/2 = 180$
Total	39,960	12	3080

The average annual inventory investment = ₹3080. Thus, the percentage reduction in the average investment $= \left(\dfrac{4995 - 3080}{4995}\right) \times 100 = 38.33\%$.

22 Classical Optimization Techniques

After studying this chapter, the reader will be able to

- understand single variable unconstrained optimization problems
- explain multi-variable unconstrained optimization problems
- comprehend multi-variable constrained optimization problems with equality constraints
- get acquainted with multi-variable constrained optimization problems with inequality constraints

22.1 INTRODUCTION

Classical optimization techniques are very useful in finding the optimal solution of problems that involve continuous and differentiable functions. Such techniques are analytical in nature and are used to obtain the maximum and minimum points of the objective function, under both unconstrained and constrained conditions. For equality constrained problems, Lagrange's multiplier method is used. Inequality constraints are converted to equality constraints and then, the Kuhn Tucker conditions are used to solve the problem using the Lagrange multiplier.

22.2 UNCONSTRAINED OPTIMIZATION PROBLEMS

In classical optimization, sometimes we come across problems where we have to find the maxima or minima of the functions of one or more variables with no restriction on the variable(s) involved in the problem. Such problems are known as unconstrained optimization problems. In this section, we shall discuss the necessary and sufficient conditions for the minima or maxima of single-variable and multi-variable unconstrained functions.

22.2.1 Single-variable Unconstrained Optimization Problems

An optimization problem with a single variable and without any restriction is called a single-variable unconstrained optimization problem. Let $f(x)$ be a continuous function of the single variable x defined in the interval $[a, b]$. Before finding the maxima and minima of $f(x)$, let us get familiar with the following terms:

Local maxima A function $f(x)$ with a single variable is said to have a local (relative) maxima at $x = x_0$, if $f(x_0) \geq f(x_0 + h)$, for all sufficiently small positive and negative values of h.

Local minima A function $f(x)$ with a single variable is said to have a local (relative) minima at $x = x_0$, if $f(x_0) \leq f(x_0 + h)$, for all sufficiently small positive and negative values of h.

Global maxima A function $f(x)$ with a single variable is said to have a global (absolute) maxima at $x = x_0$ in $[a, b]$, if $f(x_0) \geq f(x)$, for all x defined in the interval $[a, b]$.

Global minima A function $f(x)$ with a single variable is said to have a global (absolute) minima at $x = x_0$ in $[a, b]$, if $f(x_0) \leq f(x)$, for all x defined in the interval $[a, b]$.

Point of inflexion A function $f(x)$ is said to have a point of inflexion at $x = x_0$, if the curve represented by $f(x)$ changes its concavity at x_0.

In Fig. 22.1, points A, B, and C are the local maxima; P and Q are the local minima; C is the global maxima; Q is the global minima, and R is the point of inflexion.

22.2.2 Conditions for Local Maxima or Minima of Single-variable Function

The following are the conditions for the existence of local maxima or minima of a single-variable function:

Necessary condition Let the function $f(x)$ be defined in the interval $[a, b]$ and $f'(x_0)$ exist $(a < x_0 < b)$, then $f(x)$ has a local maxima or minima at $x = x_0$, if $f'(x_0) = 0$. In Fig. 22.2, we depict the connection that exists between the various points of the stationary points of a function $f(x)$.

Sufficient conditions Let $f'(x_0) = f''(x_0) = f'''(x_0) = \ldots = f^{(n-1)}(x_0)$. However, since $f^n(x_0) \neq 0$, then $f(x_0)$ has the following:

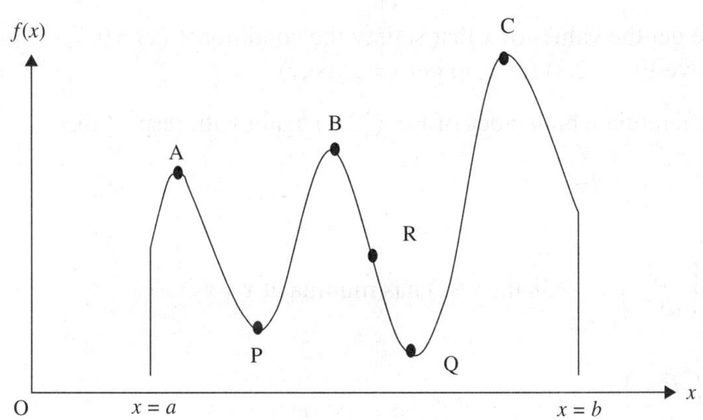

Figure 22.1 Local and global maxima or minima and point of inflexion

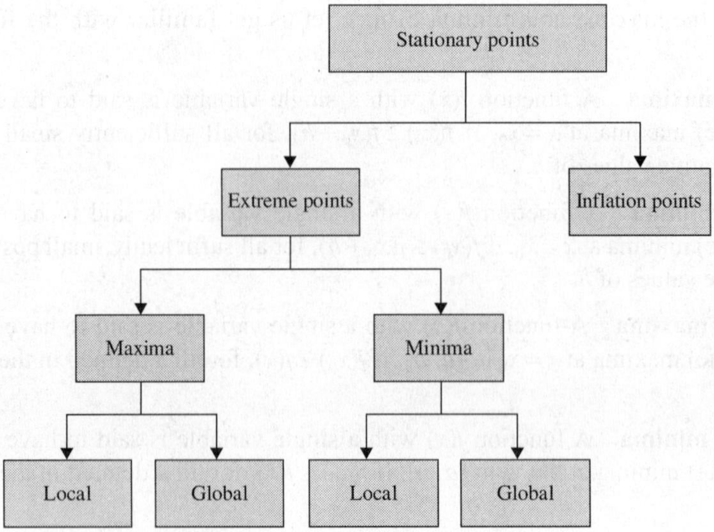

Figure 22.2 Connecting stationary points

1. A minimum value of $f(x)$, if $f^{(n)}(x_0) > 0$ and n is even
2. A maximum value of $f(x)$, if $f^{(n)}(x_0) < 0$ and n is even
3. Neither maxima nor minima, that is, point of inflexion if $f^{(n)}(x_0) \neq 0$ and n is odd

22.2.3 Procedure to Find Extreme Points of Functions of Single Variables

Let us consider a single variable function

$$y = f(x) \tag{22.1}$$

The following steps should be used to find the extreme points:

1. Differentiate both sides of Eq. (22.1) with respect to x to get

$$\frac{dy}{dx} = f'(x) \tag{22.2}$$

2. For extreme points, we have $\frac{dy}{dx} = 0 \Rightarrow f'(x) = 0$ (22.3)

   We get the values of x that satisfy the condition $f'(x) = 0$.
   Solve Eq. (22.3) for x, to get $x = x_0$ (say).

3. Differentiate both sides of Eq. (22.2) again with respect to x.

$$\frac{d^2y}{dx^2} = f''(x)$$

4. If $\left(\frac{d^2y}{dx^2}\right)_{x=x_0} > 0$, then $f(x)$ has minima at $x = x_0$.

5. If $\left(\frac{d^2y}{dx^2}\right)_{x=x_0} < 0$, then $f(x)$ has maxima at $x = x_0$.

6. If $\left(\dfrac{d^2y}{dx^2}\right)_{x=x_0} = 0$, then find $\dfrac{d^3y}{dx^3}$ and if we have $\left(\dfrac{d^3y}{dx^3}\right)_{x=x_0} \neq 0$ then there is neither a maxima nor a minima at $x = x_0$, that is, $x = x_0$ is the point of inflexion.
7. If $\left(\dfrac{d^3y}{dx^3}\right)_{x=x_0} = 0$, then find $\dfrac{d^4y}{dx^4}$ and repeat steps 4, 5, and 6.

Example 22.1 Find the maximum and minimum value of $y = 3x^5 - 5x^3$.

Solution It is given that

$$y = 3x^5 - 5x^3 \tag{22.4}$$

Differentiating both sides of Eq. (22.4) with respect to x, we get

$$\frac{dy}{dx} = 15x^4 - 15x^2 \tag{22.5}$$

For maxima and minima, we have

$$\frac{dy}{dx} = 0$$

$$\Rightarrow \quad 15x^4 - 15x^2 = 0$$

$$\Rightarrow \quad 15x^2(x-1)(x+1) = 0$$

$$x = 0, 1, -1$$

Differentiating both sides of Eq. (22.5) again with respect to x, we get

$$\frac{d^2y}{dx^2} = 60x^3 - 30x$$

At $x = 0$, $\dfrac{d^2y}{dx^2} = 0$, that is, x is a point of inflexion.

At $x = 1$, $\dfrac{d^2y}{dx^2} = 30 > 0$, that is, y is minimum at $x = 1$.

At $x = -1$, $\dfrac{d^2y}{dx^2} = -30 < 0$, that is, y is maximum at $x = -1$.

Example 22.2 Assume the following relationship for revenue and cost functions:

$$R(x) = 1000x - 2x^2 \text{ and } C(x) = x^3 - 59x^2 + 1315x + 5000$$

Find out at what level of output x (where x is measured in tons per week), is the profit maximum.

Solution The profit function is as follows:

$$\begin{aligned} P(x) &= R(x) - C(x) \\ &= 1000x - 2x^2 - x^3 + 59x^2 - 1315x - 5000 \\ &= -x^3 + 57x^2 - 315x - 5000 \end{aligned} \tag{22.6}$$

Differentiating both sides of Eq. (22.6) with respect to x, we get

$$\frac{dP}{dx} = -3x^2 + 114x - 315 \tag{22.7}$$

For maxima and minima, we have $\dfrac{dP}{dx} = 0$

$$\Rightarrow \quad -3x^2 + 114x - 315 = 0 \Rightarrow x = 3, 35$$

Differentiating both sides of Eq. (22.7) again with respect to x, we get

$$\frac{d^2P}{dx^2} = -6x + 114$$

At $x = 3$, $\frac{d^2P}{dx^2} = 96 > 0$, that is, P is minimum at $x = 3$.

At $x = 35$, $\frac{d^2P}{dx^2} = 96 < 0$, that is, P is maximum at $x = 35$.

Hence, the profit is maximum at $x = 35$ tons per week.

Example 22.3 In a submarine telegraph cable, the speed of signaling varies as $x^2 \log(1/x)$, where x is the ratio of the radius of the cube to that of the covering. Show that the greatest speed is attained when this ratio is $1 : \sqrt{e}$.

Solution Let u be the speed of signaling. Then

$$u = \lambda x^2 \log(1/x), \ \lambda \text{ is the constant of proportionally, } \lambda > 0, x \neq 0$$

$$u = -\lambda x^2 \log x \tag{22.8}$$

Differentiating both sides of Eq. (22.8) with respect to x, we get

$$\frac{du}{dx} = -2\lambda x \log x - \lambda x^2(1/x) = -\lambda[2x \log x + x] \tag{22.9}$$

For maxima and minima, we have $\frac{du}{dx} = 0$

$$\Rightarrow \quad -\lambda[2x \log x + x] = 0 \Rightarrow \log x = -1/2$$

$$\Rightarrow \quad x = e^{-1/2} = 1/\sqrt{e}$$

Differentiating both sides of Eq. (22.9) again with respect to x, we get

$$\frac{d^2u}{dx^2} = -\lambda[2x(1/x) + 2\log x] = -\lambda[2\log x + 3]$$

At $x = 1/\sqrt{e}$, $\frac{d^2u}{dx^2} = -2\lambda$ (negative), that is, u is maximum.

Hence, u is maximum, ratio for $x = 1 : \sqrt{e}$.

Example 22.4 Show that the height of a right circular cylinder of a given surface area (including its ends) and of maximum volume is equal to twice its radius.

Solution We know that $V = \pi r^2 h$ (22.10)

and $S = 2\pi rh + 2\pi r^2$ (let us assume a constant according to what is given)

$$\Rightarrow \quad 2\pi rh = 2k^2\pi - 2\pi r^2$$

$$\Rightarrow \quad h = \frac{k^2 - r^2}{r} \tag{22.11}$$

From Eqs (22.10) and (22.11), we have

$$V = \pi r(k^2 - r^2) \tag{22.12}$$

Differentiating both sides of Eq. (22.12) with respect to r, we get

$$\frac{dV}{dr} = \pi\left(k^2 - 3r^2\right) \tag{22.13}$$

For maxima and minima, we have $\frac{dV}{dr} = 0$

$$\Rightarrow \quad \pi\left(k^2 - 3r^2\right) = 0$$

$$\Rightarrow \quad r = \frac{k}{\sqrt{3}} \tag{22.14}$$

Differentiating both sides of Eq. (22.13) with respect to r, we get

$$\frac{d^2V}{dx^2} = -6\pi r \text{ (negative), that is, } V \text{ is maximum.}$$

Using Eqs (22.11) and (22.14), we have $hr = k^2 - r^2$

$$\Rightarrow h = 2r$$

Example 22.5 A DC generator has an integral resistance of R ohms and an open circuit voltage of V volts. Find the lead resistance r for which the power delivered by the generator is maximum.

Solution We know that the ohm's law is $V = i(R + r)$

$$\Rightarrow i = \frac{V}{R+r}$$

The power generated, $P = i^2 r = \dfrac{V^2 r}{(R+r)^2}$ (22.15)

Here V and R are constants.

Differentiating both sides of Eq. (22.15) with respect to r, we get

$$\frac{dP}{dr} = V^2\left[\frac{(R+r)^2 \cdot 1 - r \cdot 2(R+r)}{(R+r)^4}\right]$$

$$= V^2\left[\frac{(R+r) - 2r}{(R+r)^3}\right]$$

$$= V^2\left[\frac{R-r}{(R+r)^3}\right] \quad (22.16)$$

For maxima and minima, we have $\dfrac{dP}{dr} = 0$

$$\Rightarrow V^2\left[\frac{R-r}{(R+r)^3}\right] = 0 \Rightarrow R = r$$

Differentiating both sides of Eq. (22.16) again with respect to r, we get

$$\frac{d^2P}{dr^2} = V^2\left[\frac{(R+r)^3(-1) - (R-r)3(R+r)^2}{(R+r)^6}\right]$$

$$= V^2\left[\frac{-4R + 2r}{(R+r)^4}\right]$$

At $R = r$, $\dfrac{d^2P}{dr^2} = \dfrac{-V^2}{8R^3}$ (negative), that is, P is maximum for $r = R$.

Hence $P_{\max} = \dfrac{V^2}{4R}$.

Example 22.6 The efficiency of a screw jack is given by $\eta = \tan\alpha \cot(\alpha + \phi)$ where ϕ is constant. Prove that the efficiency is maximum at $\alpha = \dfrac{\pi}{4} - \dfrac{\phi}{2}$ and $\eta = \dfrac{1 - \sin\phi}{1 + \sin\phi}$.

Solution The efficiency of a screw jack is

$$\eta = \tan\alpha \cot(\alpha + \phi) \quad (22.17)$$

Differentiating both sides of Eq. (22.17) with respect to α, we get

$$\frac{d\eta}{d\alpha} = -\tan\alpha \operatorname{cosec}^2(\alpha+\phi) + \sec^2\alpha \cot(\alpha+\phi)$$

$$= \sec^2\alpha \cot(\alpha+\phi) - \tan\alpha \operatorname{cosec}^2(\alpha+\phi)$$

$$= \frac{1}{\cos^2\alpha \sin^2(\alpha+\phi)}[\sin(\alpha+\phi)\cos(\alpha+\phi) - \sin\alpha\cos\alpha] \qquad (22.18)$$

For maxima and minima, we have $\frac{d\eta}{d\alpha} = 0$

$$\Rightarrow \quad \frac{1}{\cos^2\alpha \sin^2(\alpha+\phi)}[\sin(\alpha+\phi)\cos(\alpha+\phi) - \sin\alpha\cos\alpha] = 0$$

$$\Rightarrow \quad \sin 2(\alpha+\phi) - \sin 2\alpha = 0$$

$$\Rightarrow \quad 2(\alpha+\phi) = \pi - 2\alpha$$

$$\Rightarrow \quad \alpha = \frac{\pi}{4} - \frac{\phi}{2}$$

From Eq. (22.18), we have $\cos^2\alpha \sin^2(\alpha+\phi)\frac{d\eta}{d\alpha} = [\sin(\alpha+\phi)\cos(\alpha+\phi) - \sin\alpha\cos\alpha]$

$$= \frac{1}{2}[\sin 2(\alpha+\phi) - \sin 2\alpha]$$

$$= \cos(2\alpha+\phi)\cos\phi \qquad (22.19)$$

Differentiating both sides of Eq. (22.19) again with respect to α, we get

$$\cos^2\alpha \sin^2(\alpha+\phi)\frac{d^2\eta}{d\alpha^2} + \frac{d}{d\alpha}\{\cos^2\alpha \sin^2(\alpha+\phi)\}\cdot\frac{d\eta}{d\alpha} = \frac{d}{d\alpha}[\cos(2\alpha+\phi)\cos\phi]$$

$$\cos^2\alpha \sin^2(\alpha+\phi)\frac{d^2\eta}{d\alpha^2} + \frac{d}{d\alpha}\{\cos^2\alpha \sin^2(\alpha+\phi)\}\cdot\frac{d\eta}{d\alpha} = -[\sin(2\alpha+\phi)\cos\phi]$$

At $\alpha = \frac{\pi}{4} - \frac{\phi}{2}$, $\frac{d\eta}{d\alpha} = 0$, then $\frac{d^2\eta}{d\alpha^2}$ is negative, that is, η is maximum.

At $\alpha = \frac{\pi}{4} - \frac{\phi}{2}$ the value of η by Eq. (22.17), we have

$$\eta = \frac{\sin\alpha\cos(\alpha+\phi)}{\sin(\alpha+\phi)\cos\alpha} = \frac{\sin(2\alpha+\phi) - \sin\phi}{\sin(2\alpha+\phi) + \sin\phi} = \frac{1-\sin\phi}{1+\sin\phi}$$

Example 22.7 A given quantity of metal is to be cast into a half cylinder, that is, with a rectangular base and semi-circular ends. Show that in order to have minimum surface area, the ratio of the height of the cylinder to the diameter of semi-circular ends is π:π+2.

Solution The volume of the half cylinder is given by this formula:

$$V = (1/2)\pi r^2 h \qquad (22.20)$$

where r and h are the radius and height of the half cylinder respectively.

Surface area of the rectangular base $= 2rh$
Surface area of the curved surface $= \pi rh$
Surface area of the two semi-circular ends $= \pi r^2$
The total surface area $S = 2rh + \pi rh + \pi r^2$

$$= rh(2+\pi) + \pi r^2 \qquad (22.21)$$

From equation Eq. (22.20), we have $h = \dfrac{2V}{\pi r^2}$

$$\text{Then } S = \frac{r \cdot 2V}{\pi r^2}(2+\pi) + \pi r^2$$

$$= \frac{2V}{\pi r}(2+\pi) + \pi r^2 \tag{22.22}$$

Differentiating both sides of Eq. (22.22) with respect to r, we get

$$\frac{dS}{dr} = 2\pi r - \frac{2V}{\pi r^2}(\pi + 2) \tag{22.23}$$

For maxima and minima, we have

$$\frac{dS}{dr} = 0 \Rightarrow 2\pi r - \frac{2V}{\pi r^2}(\pi+2) = 0$$

From Eq. (22.20), we have

$$2\pi r - h(\pi+2) = 0$$

$$\Rightarrow \frac{h}{2r} = \frac{\pi}{\pi+2} \tag{22.24}$$

Differentiating both sides of Eq. (22.23) again with respect to r, we get

$$\frac{d^2S}{dr^2} = 2\pi - \frac{4V}{\pi r^3}(\pi+2)$$

$= 6\pi$ (positive), that is, S is maximum.

Hence, surface area S is minimum at $\dfrac{h}{2r} = \dfrac{\pi}{\pi+2}$.

22.3 MULTIVARIABLE OPTIMIZATION PROBLEMS

An optimization problem with two or more variables without any restriction is called a multivariable unconstrained optimization problem. Whenever we deal with a function of more than one variable, we use partial derivatives, in order to differentiate them from total derivatives. For example, when f is a function of two variables x and y, then the total derivative of f is $df = \dfrac{\partial f}{\partial x}dx + \dfrac{\partial f}{\partial y}dy$, where $\dfrac{\partial f}{\partial x}$ and $\dfrac{\partial f}{\partial y}$ are the partial derivatives of f with respect to x and y respectively. Partial derivatives are used to find out the maxima and minima of the function of more than one variable. Here, we discuss the working rule to determine the extreme points of the functions of two variables and functions of n variables.

22.3.1 Working Rule to Find Extreme Points of Functions of Two Variables

Let $u = f(x_1, x_2)$ be a function of x_1 and x_2.

1. Differentiating $f(x_1, x_2)$ partially with respect to x_1 and x_2, we get

$$\frac{\partial f}{\partial x_1} \text{ and } \frac{\partial f}{\partial x_2}$$

2. For extreme points, we have $\frac{\partial f}{\partial x_1} = 0$ and $\frac{\partial f}{\partial x_2} = 0$

   Solving these equations, we get some points—(a_1, b_1), (a_2, b_2), …, etc.

3. Again differentiate the aforementioned equations partially to get the following derivations:

$$r = \frac{\partial^2 f}{\partial x_1^2},\ s = \frac{\partial^2 f}{\partial x_1 \partial x_2},\ \text{and } t = \frac{\partial^2 f}{\partial x_2^2}$$

4. At (a_1, b_1), calculate r and $rt - s^2$
   Case (a): If $rt - s^2 > 0$ and $r < 0$, then $f(x_1, x_2)$ is maximum at (a_1, b_1).
   Case (b): If $rt - s^2 > 0$ and $r > 0$, then $f(x_1, x_2)$ is minimum at (a_1, b_1).
   Case (c): If $rt - s^2 < 0$ then $f(x_1, x_2)$ has a saddle point at (a_1, b_1).

We extend the working rule to find the extreme points of the function of two variables to the function of n variables, as given in the following section.

22.3.2 Working Rule to Find Extreme Points of Functions of *n* Variables

Assume $u = f(x_1, x_2, x_3, \ldots, x_n)$ is a function of $x_1, x_2, x_3, \ldots, x_n$.

$$\text{Necessary conditions: } \frac{\partial f}{\partial x_1} = 0,\ \frac{\partial f}{\partial x_2} = 0,\ \ldots,\ \frac{\partial f}{\partial x_n} = 0$$

Sufficient conditions: The Hessian Matrix at P for n variables will be

$$H = \begin{bmatrix} \frac{\partial^2 f}{\partial x_1^2} & \frac{\partial^2 f}{\partial x_1 \partial x_2} & \cdots & \frac{\partial^2 f}{\partial x_1 \partial x_n} \\ \frac{\partial^2 f}{\partial x_2 \partial x_1} & \frac{\partial^2 f}{\partial x_2^2} & \cdots & \frac{\partial^2 f}{\partial x_2 \partial x_n} \\ \cdots & \cdots & \cdots & \cdots \\ \frac{\partial^2 f}{\partial x_n \partial x_1} & \frac{\partial^2 f}{\partial x_n \partial x_2} & \cdots & \frac{\partial^2 f}{\partial x_n^2} \end{bmatrix}$$

Its leading minors are defined as

$$H_1 = \left| \frac{\partial^2 f}{\partial x_1^2} \right|$$

$$H_2 = \begin{vmatrix} \frac{\partial^2 f}{\partial x_1^2} & \frac{\partial^2 f}{\partial x_1 \partial x_2} \\ \frac{\partial^2 f}{\partial x_2 \partial x_1} & \frac{\partial^2 f}{\partial x_2^2} \end{vmatrix}$$

$$H_3 = \begin{vmatrix} \dfrac{\partial^2 f}{\partial x_1^2} & \dfrac{\partial^2 f}{\partial x_1 \partial x_2} & \dfrac{\partial^2 f}{\partial x_1 \partial x_2} \\ \dfrac{\partial^2 f}{\partial x_2 \partial x_1} & \dfrac{\partial^2 f}{\partial x_2^2} & \dfrac{\partial^2 f}{\partial x_2 \partial x_3} \\ \dfrac{\partial^2 f}{\partial x_3 \partial x_1} & \dfrac{\partial^2 f}{\partial x_3 \partial x_2} & \dfrac{\partial^2 f}{\partial x_3^2} \end{vmatrix}$$

Hence, the following cases will arise:

Case (a): If H_1, H_2, H_3, … are positive (i.e., H is positive definite), then $f(x_1, x_2, x_3, \ldots, x_n)$ has its minimum at P.

Case (b): If H_1, H_2, H_3, … are alternately negative, positive, negative (i.e., H is negative definite), then $f(x_1, x_2, x_3, \ldots, x_n)$ has its maximum at P.

Case (c): H_1 and H_3, … are not of the same sign and $H_2 = 0$ (i.e., semi-definite or indefinite), then $f(x_1, x_2, x_3, \ldots, x_n)$ has a saddle point at P.

Example 22.8 Find the maximum and minimum value of u, where $u = \sin x \sin y \sin(x+y)$.

Solution The following equation is given:

$$u = \sin x \sin y \sin(x+y) \tag{22.25}$$

Differentiating Eq. (22.25) partially with respect to x and y on both sides, we get,

$$\frac{\partial u}{\partial x} = \sin y\,[\sin x \cos(x+y) + \cos x \sin(x+y)] \tag{22.26}$$

$$\text{and } \frac{\partial u}{\partial y} = \sin x[\sin y \cos(x+y) + \cos y \sin(x+y)] \tag{22.27}$$

For maxima and minima, we have $\dfrac{\partial u}{\partial x} = 0$ and $\dfrac{\partial u}{\partial y} = 0$

$$\sin y[\sin x \cos(x+y) + \cos x \sin(x+y)] = 0$$
$$\text{and } \sin x[\sin y \cos(x+\text{y}) + \cos y \sin(x+y)] = 0$$

Solving the expression, we get the following equations:

$$\tan(x+y) = -\tan x \tag{22.28}$$
$$\text{and } \tan(x+y) = -\tan y \tag{22.29}$$

Now, we have

$$\tan 2x = -\tan x = \tan(\pi - x)\ 2x = \pi - x$$
$$x = \pi/3 = y$$

In addition, $\sin y = 0 \Rightarrow y = 0$

and $\quad \sin x = 0 \Rightarrow x = 0$

Thus the stationary points are $(0, 0)$, $(\pi/3, \pi/3)$.

Differentiating Eqs (22.26) and (22.27) partially again, we get

$$r = \frac{\partial^2 u}{\partial x^2} = 2\sin y \cos(2x+y)$$

$$s = \frac{\partial^2 u}{\partial x \partial y} = \sin 2(x+y)$$

$$\text{and } t = \frac{\partial^2 u}{\partial y^2} = 2\sin x \cos(2y+x)$$

At (0, 0), we get $r = 0$, $s = 0$, $t = 0$.

$rt - s^2 = 0$, that is, u has a saddle point at (0, 0).

Now, at ($\pi/3$, $\pi/3$), we get $r = 2\sin(\pi/3)\cos\pi = -\sqrt{3}$

$$s = \sin(4\pi/3) = -\sin(\pi/3) = -\frac{\sqrt{3}}{2}$$

and $t = 2\sin(\pi/3)\sin\pi = -\sqrt{3}$

$$\Rightarrow \quad rt - s^2 = 9/4 \text{ is positive and } r < 0.$$

Hence, u is maximum at ($\pi/3$, $\pi/3$).

Example 22.9 Find the extreme points of the function $u = x^2 + 4y^2 + 4z^2 + 4xy + 4xz + 16yz$.

Solution Given that

$$u = x^2 + 4y^2 + 4z^2 + 4xy + 4xz + 16yz \tag{22.30}$$

Differentiating both sides of Eq. (22.30) partially with respect to x, y, and z respectively, we get

$$\frac{\partial u}{\partial x} = 2x + 2y + 2z$$

$$\frac{\partial u}{\partial y} = 8y + 4x + 16z$$

$$\text{and} \quad \frac{\partial u}{\partial z} = 8z + 4x + 16y$$

For extreme points, we have

$$\frac{\partial u}{\partial x} = 0, \quad \frac{\partial u}{\partial y} = 0, \text{ and } \frac{\partial u}{\partial z} = 0$$

$$\Rightarrow \quad 2x + 2y + 2z = 0 \text{ or } 2(x + y + z) = 0$$

$$8y + 4x + 16z = 0 \text{ or } 4(x + 2y + 4z) = 0$$

$$\text{and } 8z + 4x + 16y = 0 \text{ or } 4(x + 4y + 2z) = 0$$

Solving the expression, we get $x = 0$, $y = 0$, and $z = 0$.

Let the point P be (0, 0, 0).

Again differentiating partially, we get

$$\frac{\partial^2 u}{\partial x^2} = 2, \quad \frac{\partial^2 u}{\partial y^2} = 8, \quad \frac{\partial^2 u}{\partial z^2} = 8, \quad \frac{\partial^2 u}{\partial x \partial y} = 4, \quad \frac{\partial^2 u}{\partial y \partial x} = 4, \quad \frac{\partial^2 u}{\partial y \partial z} = 16,$$

$$\frac{\partial^2 u}{\partial z \partial y} = 16, \quad \frac{\partial^2 u}{\partial x \partial z} = 4, \quad \frac{\partial^2 u}{\partial z \partial x} = 4.$$

The Hessian matrix of $u(x, y, z)$ is $H = \begin{bmatrix} 2 & 4 & 4 \\ 4 & 8 & 16 \\ 4 & 16 & 8 \end{bmatrix}$

The leading minors of H are

$$H_1 = |2| = 2, \; H_2 = \begin{vmatrix} 2 & 4 \\ 4 & 8 \end{vmatrix} = 0, \text{ and } H_3 = \begin{vmatrix} 2 & 4 & 4 \\ 4 & 8 & 16 \\ 4 & 16 & 8 \end{vmatrix} < 0.$$

Here H_1 and H_3 are not of the same sign and $H_2 = 0$ (i.e., semi-definite). Hence, u has a saddle point at (0, 0, 0).

Example 22.10 Find the extreme points $f(x_1, x_2) = 20x_1 + 26x_2 + 4x_1x_2 - 4x_1^2 - 3x_2^2$.

Solution Given that

$$f(x_1, x_2) = 20x_1 + 26x_2 + 4x_1x_2 - 4x_1^2 - 3x_2^2 \tag{22.31}$$

Differentiating Eq. (22.31) partially with respect to x_1 and x_2 on both sides respectively, we get

$$\frac{\partial f}{\partial x_1} = 20 + 4x_2 - 8x_1 \tag{22.32}$$

$$\text{and } \frac{\partial f}{\partial x_2} = 26 + 4x_1 - 6x_2 \tag{22.33}$$

For extreme points, we have $\frac{\partial f}{\partial x_1} = 0$ and $\frac{\partial f}{\partial x_2} = 0$

$$\Rightarrow \quad 20 + 4x_2 - 8x_1 = 0 \text{ and } 26 + 4x_1 - 6x_2 = 0$$

Solving these, we get $x_1 = 7$ and $x_2 = 9$.

Again, differentiating Eq. (22.32) partially, we get

$$r = \frac{\partial^2 f}{\partial x_1^2} = -8, \; s = \frac{\partial^2 f}{\partial x_1 \partial x_2} = 4, \text{ and } t = \frac{\partial^2 f}{\partial x_2^2} = -6$$

We have $rt - s^2 = (-8)(-6) - (4)^2 = 32$

At (7, 9), $rt - s^2 > 0$ and $r < 0$, that is, f is maximum at (7, 9).

Example 22.11 Find the extreme points of the function $f(x_1, x_2) = -x_1^3 + 2x_2^3 + 3x_1^2 + 12x_2^2 + 24$. In addition, determine their nature.

Solution Given that

$$f(x_1, x_2) = -x_1^3 + 2x_2^3 + 3x_1^2 + 12x_2^2 + 24 \tag{22.34}$$

Differentiating Eq. (22.34) partially with respect to x_1 and x_2 on both sides, we get

$$\frac{\partial f}{\partial x_1} = 3x_1^2 + 6x_1 \tag{22.35}$$

$$\text{and } \frac{\partial f}{\partial x_2} = 6x_2^2 + 24x_2 \tag{22.36}$$

For extreme points, we have $\frac{\partial f}{\partial x_1} = 0$ and $\frac{\partial f}{\partial x_2} = 0$

$$\Rightarrow \quad 3x_1^2 + 6x_1 = 0 \text{ and } 6x_2^2 + 24x_2 = 0$$

Solving these, we get $x_1 = 0, -2, x_2 = 0, -4$.

Differentiating Eqs (22.35) and (22.36) partially again, we get

$$r = \frac{\partial^2 f}{\partial x_1^2} = 6(x_1 + 1)$$

$$s = \frac{\partial^2 f}{\partial x_1 \partial x_2} = 0$$

$$\text{and } t = \frac{\partial^2 f}{\partial x_2^2} = 12(x_2 + 2)$$

At (0, 0): $rt - s^2 = 72(x_1 + 1)(x_2 + 2) = 72 > 0$ and $r > 0$, that is, f is minimum at (0, 0).

At (0, –4): $rt - s^2 = 72(x_1 + 1)(x_2 + 2) = -144 < 0 \Rightarrow$ no extreme point, that is, f has a saddle point at (0, –4).

At (−2, 0): $rt - s^2 = 72(x_1 + 1)(x_2 + 2) = -144 < 0 \Rightarrow$ no extreme point, that is, f has a saddle point at (−2, 0).

At (−2, −4): $rt - s^2 = 72(x_1 + 1)(x_2 + 2) = 144 > 0$ and $r < 0$, that is, f is maximum at (−2, −4).

Example 22.12 Find the dimensions of a box of the largest volume that can be inscribed in a sphere of radius 3 m.

Solution Let the volume of the box be $V = 2x \cdot 2y \cdot 2z$

$$\Rightarrow \quad V = 8xyz \tag{22.37}$$

Changing the variables from cartesian to spherical polar coordinates as $x = r\sin\theta\cos\phi$, $y = r\sin\theta\sin\phi$, $z = r\cos\theta$, and using $r = 3$ (given).

$$V = 216\sin^2\theta\cos\theta\cos\phi\sin\phi$$
$$= 108\sin^2\theta\cos\theta\sin 2\phi \tag{22.38}$$

Differentiating Eq. (22.38) partially with respect to θ and ϕ on both the sides, we get

$$\frac{\partial V}{\partial \theta} = 108\sin 2\phi\left[\sin^2\theta(-\sin\theta) + 2\sin\theta\cos^2\theta\right]$$

$$= 108\sin 2\phi\sin\theta\left\{-\sin^2\theta + 2\cos^2\theta\right\} \tag{22.39}$$

$$\text{and} \quad \frac{\partial V}{\partial \phi} = 216\cos 2\phi\sin^2\theta\cos\theta \tag{22.40}$$

For maxima and minima, we have $\dfrac{\partial V}{\partial \theta} = 0$ and $\dfrac{\partial V}{\partial \phi} = 0$

$$\Rightarrow \quad 108\sin 2\phi\sin\theta\left\{-\sin^2\theta + 2\cos^2\theta\right\} = 0$$

That is, $\sin\theta = 0$ or $\tan\theta = \sqrt{2}$ or $\sin\phi = 0$.

That is, $\theta = 0$, $\tan^{-1}\sqrt{2}$, $\phi = 0$.

$$216\cos 2\phi\sin^2\theta\cos\theta = 0 \Rightarrow \theta = 0, \pi/4$$

Since $\theta = 0$ or $\phi = 0$ gives $V = 0$, we take $\theta = \tan^{-1}\sqrt{2}$ and $\phi = \pi/4$.

Differentiating Eq. (22.39) partially again, we get

$$\frac{\partial^2 V}{\partial \theta^2} = 108\sin 2\phi\left[\sin\theta(-4\cos\theta\sin\theta - 2\sin\theta\cos\theta) + \cos\theta(2\cos^2\theta - \sin^2\theta)\right]$$

At $(\tan^{-1}\sqrt{2}, \pi/4)$, $\dfrac{\partial^2 V}{\partial \theta^2} = 108\left[\dfrac{\sqrt{2}}{\sqrt{3}}\left(-6\times\dfrac{1}{\sqrt{3}}\times\dfrac{\sqrt{2}}{\sqrt{3}}\right) + \dfrac{1}{\sqrt{3}}\left(1\left(\dfrac{1}{\sqrt{3}}\right)^2 - \left(\dfrac{\sqrt{2}}{\sqrt{3}}\right)^2\right)\right]$

$$= -\frac{432}{\sqrt{3}}$$

$$\frac{\partial^2 V}{\partial \theta\,\partial \phi} = 216\cos 2\phi\left[\sin^2\theta(-\sin\theta) + \cos^2\theta(2\sin\theta)\right]$$

$$= 216\cos 2\phi\sin\theta\left[2\cos^2\theta - \sin^2\theta\right]$$

At $(\tan^{-1}\sqrt{2}, \pi/4)$, $\dfrac{\partial^2 V}{\partial \theta\,\partial \phi} = 0$

$$\frac{\partial^2 V}{\partial \phi^2} = -432\sin 2\phi\left[\sin^2\theta\cos\theta\right]$$

At $(\tan^{-1}\sqrt{2}, \pi/4)$, $\dfrac{\partial^2 V}{\partial \phi^2} = -432\times\left(\dfrac{\sqrt{2}}{\sqrt{3}}\right)^2\times\left(\dfrac{1}{\sqrt{3}}\right) = -\dfrac{864}{3\sqrt{3}}$

Now at $(\tan^{-1}\sqrt{2}, \pi/4)$, we have

$$\frac{\partial^2 V}{\partial \theta^2}\cdot\frac{\partial^2 V}{\partial \phi^2}-\left(\frac{\partial^2 V}{\partial \theta \partial \phi}\right)^2=\left(\frac{-432}{\sqrt{3}}\right)\left(\frac{-864}{3\sqrt{3}}\right)-0>0$$

That is, V is maximum.

The dimensions of the maximum inscribed box are

$$2x = 2r\sin\theta\cos\phi = 2(3)\cdot\frac{\sqrt{2}}{\sqrt{3}}\cdot\frac{1}{\sqrt{2}} = 2\sqrt{3}$$

$$2y = 2r\sin\theta\sin\phi = 2(3)\cdot\frac{\sqrt{2}}{\sqrt{3}}\cdot\frac{1}{\sqrt{2}} = 2\sqrt{3}$$

$$2z = 2r\cos\theta = 2(3)\cdot\frac{1}{\sqrt{3}} = 2\sqrt{3}$$

22.4 MULTIVARIABLE CONSTRAINED OPTIMIZATION PROBLEMS WITH EQUALITY CONSTRAINTS

Multivariable constrained optimization problems have great importance in engineering, science, and management. It is a fundamental technique used to solve engineering and economic problems with limitation of constraints. For example, in an inventory management problem, to minimize the inventory cost, many factors are considered, such as holding cost, ordering cost, deteriorating cost, purchasing cost, shortages cost, and partial backlogging cost. The optimization problem of a continuous and differentiable function subject to equality constraints is as follows:

Optimize (Max or Min) $Z = f(X)$

Subject to constraints $g_j(X) = 0; j = 1, 2, 3, \ldots, m$

where $X = \begin{pmatrix} x_1 \\ x_2 \\ \cdots \\ x_n \end{pmatrix}$

Here m is less than or equal to n. There are several methods for solving this type of problem. Here we discuss the following two methods:

1. Direct substitution method
2. Lagrange multipliers method

22.4.1 Direct Substitution Method

In this method, the value of any variable from the constraint set is put into the objective function. The problem reduces to an unconstrained optimization problem and can be solved using the unconstrained optimization method.

22.4.2 Lagrange Multipliers Method

Consider a general problem with n variables and m equality constraints:

Optimize $Z = f(X)$ (22.41)

$$\text{Subject to} \quad g_j(X) = 0; j = 1, 2, 3, \ldots, m \ (m < n) \tag{22.42}$$
$$X \geq 0$$

$$\text{where } X = \begin{pmatrix} x_1 \\ x_2 \\ \ldots \\ x_n \end{pmatrix}$$

Now we define a function

$$L(x_1, x_2, \ldots, x_n, \lambda_1, \lambda_2, \ldots, \lambda_m) = f(X) + \sum_{j=1}^{m} \lambda_j g_j(X) \tag{22.43}$$

Here, $\lambda_1, \lambda_2, \ldots, \lambda_m$ are known as Lagrange's undetermined multipliers.

The necessary conditions for L to have extreme points are

$$\frac{\partial L}{\partial x_i} = \frac{\partial f}{\partial x_i} + \sum_{j=1}^{m} \lambda_j \frac{\partial g_j}{\partial x_i} = 0 \tag{22.44}$$

$$\text{and } \frac{\partial L}{\partial \lambda_j} = 0; \begin{pmatrix} i = 1, 2, \ldots, n \\ j = 1, 2, \ldots, m \end{pmatrix} \tag{22.45}$$

Solving Eqs (22.44) and (22.45), we get

$$X^* = \begin{pmatrix} x_1^* \\ x_2^* \\ \ldots \\ x_n^* \end{pmatrix} \text{ and } \lambda^* = \begin{pmatrix} \lambda_1^* \\ \lambda_2^* \\ \ldots \\ \lambda_m^* \end{pmatrix}$$

$$\begin{bmatrix} L_{11} - k & L_{12} & \ldots & L_{1n} & g_{11} & g_{21} & \ldots & g_{m1} \\ L_{21} & L_{22} - k & \ldots & L_{2n} & g_{12} & g_{22} & \ldots & g_{m2} \\ \ldots & \ldots & \ldots & \ldots & \ldots & \ldots & \ldots & \ldots \\ L_{n1} & L_{n2} & \ldots & L_{nn} - k & g_{1n} & g_{2n} & \ldots & g_{mn} \\ g_{11} & g_{12} & \ldots & g_{1n} & 0 & 0 & \ldots & 0 \\ \ldots & \ldots & \ldots & \ldots & \ldots & \ldots & \ldots & \ldots \\ g_{m1} & g_{m2} & \ldots & g_{mn} & 0 & 0 & \ldots & 0 \end{bmatrix} = 0 \tag{22.46}$$

The sufficient condition for the function to have an extreme at point X^* is that the values of k obtained from Eq. 22.46 must be of the same sign. If all the Eigen values k are negative, then it is a maxima and if all the Eigen values k are positive, then it is a minima; however, if some Eigen values are zero or of a different sign, it is a saddle point. In Eq. (22.46), L_{ij} and g_{ij} are denoted by $\frac{\partial^2 L}{\partial x_i \partial x_j}$ and $\frac{\partial g_j}{\partial x_i}$ respectively.

22.5 MULTIVARIABLE CONSTRAINED OPTIMIZATION PROBLEMS WITH INEQUALITY CONSTRAINTS

Let us consider the following problem:

$$\text{Optimization (Max or Min) } Z = f(X) \tag{22.47}$$

$$\text{Subject to } \quad g_j(X) \le 0;\ j = 1, 2, 3, \ldots, m \tag{22.48}$$

$$\text{where } X = \begin{pmatrix} x_1 \\ x_2 \\ \ldots \\ x_n \end{pmatrix}$$

The inequality constraints in Eq. (22.48) can be converted into equality constraints by adding slack variables, such as $g_j(X) + y_j^2 = 0;\ j = 1, 2, 3, \ldots, m$

Now the problem becomes

$$\text{Optimize (Max or Min) } Z = f(X)$$

$$\text{Subject to } \quad G_j(X, Y) = g_j(X) + y_j^2 = 0;\ j = 1, 2, 3, \ldots, m$$

$$\text{where } X = \begin{pmatrix} x_1 \\ x_2 \\ \ldots \\ x_m \end{pmatrix} \text{ and } Y = \begin{pmatrix} y_1 \\ y_2 \\ \ldots \\ y_m \end{pmatrix}$$

Now, this can be solved by Lagrange's multiplier method.

Example 22.13 Find the optimum solution of the following constrained multivariable problem:

$$\text{Minimize } Z = x_1^2 + (x_2 + 1)^2 + (x_3 - 1)^2$$

$$\text{Subject to } \quad x_1 + 5x_2 - 3x_3 = 6$$

Solution Given that

$$Z = x_1^2 + (x_2 + 1)^2 + (x_3 - 1)^2 \tag{22.49}$$

$$\text{and } \quad x_1 + 5x_2 - 3x_3 = 6$$

$$\Rightarrow \quad x_3 = \frac{x_1 + 5x_2 - 6}{3} \tag{22.50}$$

Using Eqs (22.49) and (22.50), we get

$$Z = x_1^2 + (x_2 + 1)^2 + \frac{1}{9}(x_1 + 5x_2 - 9)^2 \tag{22.51}$$

Differentiating both sides of Eq. (22.51) partially with respect to x_1 and x_2 respectively, we get

$$\frac{\partial Z}{\partial x_1} = 2x_1 + \frac{2}{9}(x_1 + 5x_2 - 9)$$

$$\text{and } \quad \frac{\partial Z}{\partial x_2} = 2(x_2 + 1) + \frac{10}{9}(x_1 + 5x_2 - 9)$$

For maxima and minima, we have

$$\frac{\partial Z}{\partial x_1} = 0 \text{ and } \frac{\partial Z}{\partial x_2} = 0$$

$$2x_1 + \frac{2}{9}(x_1 + 5x_2 - 9) = 0 \text{ and } 2(x_2 + 1) + \frac{10}{9}(x_1 + 5x_2 - 9) = 0$$

Solving these, we get $x_1 = 2/5$ and $x_2 = 1$.

Differentiate again partially Eq. (22.52) and we get

$$r = \frac{\partial^2 Z}{\partial x_1^2} = 2 + \frac{2}{9} = \frac{20}{9}$$

$$t = \frac{\partial^2 Z}{\partial x_2^2} = 2 + \frac{50}{9} = \frac{68}{9}$$

$$\text{and } \frac{\partial^2 Z}{\partial x_1 x_2} = \frac{10}{9}$$

At (2/5, 1), $rt - s^2 = (20/9)\,(68/9) - (10/9)^2 = 1260 > 0$ and $r > 0$, that is, Z is minimum at (2/5, 1). The minimum value of Z is 28/5.

Example 22.14 Optimize $Z = x^2 + y^2 + z^2$ such that $4x + y^2 + 2z = 14$.

Solution It is given that

$$Z = x^2 + y^2 + z^2 \tag{22.53}$$

$$\text{and } 4x + y^2 + 2z = 14 \;\Rightarrow\; g(x, y, z) = 4x + y^2 + 2z - 14 = 0 \tag{22.54}$$

Constructing the Lagrangian function L:

$$L(x, y, z; \lambda) = x^2 + y^2 + z^2 + \lambda(4x + y^2 + 2z - 14) \tag{22.55}$$

The necessary conditions for extreme L are

$$\left.\begin{aligned} \frac{\partial L}{\partial x} &= 2x + 4\lambda = 0 \\ \frac{\partial L}{\partial y} &= 2y + 2y\lambda = 0 \\ \frac{\partial L}{\partial z} &= 2z + 2\lambda = 0 \\ \frac{\partial L}{\partial \lambda} &= 4x + y^2 + 2z - 14 = 0 \end{aligned}\right\} \tag{22.56}$$

Solving these we get $x = -2\lambda$, $z = -\lambda$. Now, two cases arise (a) if $y \neq 0$, then $\lambda = -1$. Using Eq. (22.56), we get $x = 2$, $z = 1$, and $y = \pm 2$, (b) If $y = 0$, then $\lambda = -1.4$. Using Eq. (22.56), $x = 2.8$, $z = 1.4$.

Here (2, 2, 1, −1), (2, −2, 1, −1) and $(-2\lambda, 0, -\lambda, \lambda)$ or (2.8, 0, 1.4, −1.4) are extreme points.

Differentiating Eq. (22.56) partially again, we get

$$\frac{\partial^2 L}{\partial x^2} = 2,\ \frac{\partial^2 L}{\partial x \partial y} = 0,\ \frac{\partial^2 L}{\partial x \partial z} = 0,\ \frac{\partial^2 L}{\partial y \partial x} = 0,\ \frac{\partial^2 L}{\partial y^2} = 2 + 2\lambda,\ \frac{\partial^2 L}{\partial y \partial z} = 2,\ \frac{\partial^2 L}{\partial z \partial x} = 0,$$

$$\frac{\partial^2 L}{\partial z \partial y} = 0,\ \frac{\partial^2 L}{\partial z^2} = 2,\ \frac{\partial g}{\partial x} = 4,\ \frac{\partial g}{\partial y} = 2y,\ \frac{\partial g}{\partial z} = 2.$$

The sufficient condition for the extreme point is

$$H = \begin{vmatrix} \dfrac{\partial^2 L}{\partial x^2} - k & \dfrac{\partial^2 L}{\partial x \partial y} & \dfrac{\partial^2 L}{\partial x \partial z} & \dfrac{\partial g}{\partial x} \\ \dfrac{\partial^2 L}{\partial yx \partial x} & \dfrac{\partial^2 L}{\partial y^2} - k & \dfrac{\partial^2 L}{\partial y \partial z} & \dfrac{\partial g}{\partial y} \\ \dfrac{\partial^2 L}{\partial z \partial x} & \dfrac{\partial^2 L}{\partial z \partial y} & \dfrac{\partial^2 L}{\partial z^2} - k & \dfrac{\partial g}{\partial z} \\ \dfrac{\partial g}{\partial x} & \dfrac{\partial g}{\partial y} & \dfrac{\partial g}{\partial z} & 0 \end{vmatrix} = 0$$

$$\text{That is, } H = \begin{vmatrix} 2-k & 0 & 0 & 4 \\ 0 & 2+2\lambda-k & 0 & 2y \\ 0 & 0 & 2-k & 2 \\ 4 & 2y & 2 & 0 \end{vmatrix} = 0$$

or $4(2-k)[-10+5k-10\lambda-2y^2+y^2k]=0$

At (2, 2,1,–1), from Eq. (22.54) we have $(2-k)(-10+5k+10-8+4k)=0 \Rightarrow k=2,\ 8/9$ that is, the values of k are positive, hence there is a minima.

At (2,–2, 1,–1), from Eq. (22.54) we have $(2-k)(-10+5k+10-8+4k)=0 \Rightarrow k=2,\ 8/9$ Again, the values of k are positive, hence there is a minima.

At (2.8, 0, 1.4, −1.4), from Eq. (22.54) we have $(2-k)(-10+5k+14-0+0)=0 \Rightarrow k=2,\ -4/5$. That is, the values of k are positive and negative (neither maxima nor minima), which means a saddle point.

Example 22.15 Minimize $f(x) = 1/2\ (x_1^2 + x_2^2 + x_3^2)$

Subject to $g_1(x) = x_1 - x_2 = 0$

and $g_2(x) = x_1 + x_2 + x_3 - 1 = 0$ by Lagrange's multiplier method.

Solution It is given that Minimize $f(x) = 1/2\ (x_1^2 + x_2^2 + x_3^2)$ (22.57)

Subject to $g_1(x) = x_1 - x_2 = 0$ (22.58)

and $g_2(x) = x_1 + x_2 + x_3 - 1 = 0$ (22.59)

Construct the Lagrangian function L is

$$L(x_1, x_2, x_3; \lambda_1, \lambda_2) = \frac{1}{2}(x_1^2 + x_2^2 + x_3^2) + \lambda_1(x_1 - x_2) + \lambda_2(x_1 + x_2 + x_3 - 1) \quad (22.60)$$

The necessary conditions for extreme of L are as follows:

$$\left.\begin{aligned} \frac{\partial L}{\partial x_1} &= x_1 + \lambda_1 + \lambda_2 = 0 \\ \frac{\partial L}{\partial x_2} &= x_2 - \lambda_1 + \lambda_2 = 0 \\ \frac{\partial L}{\partial x_3} &= x_3 + \lambda_2 = 0 \\ \frac{\partial L}{\partial \lambda_1} &= x_1 - x_2 = 0 \\ \frac{\partial L}{\partial \lambda} &= x_1 + x_2 + x_3 - 1 = 0 \end{aligned}\right\} \quad (22.61)$$

Solving these, we get

$$x_1 = x_2 = x_3 = \frac{1}{3};\ \lambda_1 = 0,\ \lambda_2 = -\frac{1}{3}$$

Differentiating Eq. (22.61) partially again, we get

$$\frac{\partial^2 L}{\partial x_1^2} = 1,\ \frac{\partial^2 L}{\partial x_1 \partial x_2} = 0,\ \frac{\partial^2 L}{\partial x_1 \partial x_3} = 0,\ \frac{\partial^2 L}{\partial x_2 \partial x_1} = 0,\ \frac{\partial^2 L}{\partial x_2^2} = 1,\ \frac{\partial^2 L}{\partial x_2 \partial x_3} = 0,\ \frac{\partial^2 L}{\partial x_3 \partial x_1} = 0,$$

$$\frac{\partial^2 L}{\partial x_3 \partial x_2} = 0,\ \frac{\partial^2 L}{\partial x_3^2} = 1,\ \frac{\partial g_1}{\partial x_1} = 1,\ \frac{\partial g_1}{\partial x_2} = -1,\ \frac{\partial g_1}{\partial x_3} = 0,\ \frac{\partial g_2}{\partial x_1} = 1,\ \frac{\partial g_2}{\partial x_2} = 1,\ \frac{\partial g_2}{\partial x_3} = 1.$$

The sufficient condition for extreme point is

$$H = \begin{vmatrix} 1-K & 0 & 0 & 1 & 1 \\ 0 & 1-K & 0 & -1 & 1 \\ 0 & 0 & 1-K & 0 & 1 \\ 1 & -1 & 0 & 0 & 0 \\ 1 & 1 & 1 & 0 & 0 \end{vmatrix} = 0$$

This implies we get $k = 1, 1, 1$. Here all the values of k are of the same sign and positive, that is, f is minimum at $\left(\frac{1}{3}, \frac{1}{3}, \frac{1}{3}; 0, -\frac{1}{3}\right)$.

The minimum value of $f = \frac{1}{2}\left(\frac{1}{9} + \frac{1}{9} + \frac{1}{9}\right) = \frac{1}{6}$.

RECAPITULATION

1. An optimization problem with no restriction on the decision variables is called an unconstrained optimization problem.
2. An optimization problem with a single variable without any restriction is called a single-variable unconstrained optimization problem.
3. An optimization problem with two or more than two variables and with no restriction on the decision variables is called a multivariable unconstrained optimization problem.
4. The sufficient conditions for the function $f(x_1, x_2)$ at the point (a, b) are as follows:
   (a) If $rt - s^2 > 0$ and $r < 0$ then $f(x_1, x_2)$ is maximum.
   (b) If $rt - s^2 > 0$ and $r > 0$ then $f(x_1, x_2)$ is minimum.
   (c) If $rt - s^2 < 0$ then $f(x_1, x_2)$ has a saddle point.
5. An optimization problem with a restriction on the decision variables is called a constrained optimization problem.
6. Lagrange's multiplier method is used to solve equality constrained problems.

EXERCISES

Multiple-choice Questions

22.1 A function is said to have its minimum value at the point $x = x_0$ if

(a) $f(x_0) > f(x_0 + h)\ \forall h$
(b) $f(x_0) < f(x_0 + h)\ \forall h$
(c) $f(x_0) = f(x_0 + h)\ \forall h$
(d) none of these

22.2 A maximization or minimization problem with no restriction on the decision variables is known as

(a) unconstrained optimization problem
(b) constrained optimization problem
(c) linear programming problem
(d) none of these

22.3 The sufficient conditions for the function $f(x_1, x_2)$ to have maxima at the point (a, b) are

(a) $rt - s^2 > 0$ and $r < 0$
(b) $rt - s^2 > 0$ and $r > 0$
(c) $rt - s^2 < 0$
(d) none of these

22.4 A Hessian matrix is positive definite, if all its leading principle minors are of order

(a) 1×1 are scalars
(b) 2×2 are scalars
(c) 1×1 are positives
(d) 2×2 are positives

State True or False

22.1 A function is said to have its maximum value at the point $x = x_0$, if $f(x_0) < f(x_0 + h)\ \forall h$.

22.2 An optimization problem with no restriction on the decision variables is known as a constrained optimization problem.

22.3 A Hessain matrix is negative definite, if all its leading principle minors are negative.

22.4 The Lagrange multiplier method is used when the constraints are given in equality sign.

Fill in the Blanks

22.1 An optimization problem with a single variable without any restriction is called a __________ optimization problem.

22.2 A function is said to have its __________ value at the point $x = x_0$, if $f(x_0) > f(x_0 + h)\ \forall h$.

22.3 A Hessain matrix is __________ if all its leading principle minors are negative, positive, negative.

22.4 Lagrange multiplier method is used when the constraints hold with __________ sign.

Review Questions

22.1 What is an unconstrained optimization problem?

22.2 What is the difference between constrained and unconstrained optimization problems?

22.3 Write the procedure to find the extreme points of a function of n variables.

22.4 Write a short note on Lagrange multiplier method.

Numerical Problems

22.1 The average cost per unit of manufacturing for Bharat food products is given by the following relationship $C(x) = 4000 - 16x + 0.2x$ where x is the quantity produced. Find the value of x that will minimize average cost per unit.

22.2 Prove that $\dfrac{x}{1 + x\tan x}$ is a maximum when $x = \cos x$.

22.3 Assuming that the petrol burnt (per hour) in driving a motorboat varies as the cube of its velocity, show that the most economical speed when going against a current of c km/h is $3c/2$ km/h.

22.4 A beam of length l is supported at one end. If w is the uniformly distributed load per unit length, the bending moment M at a distance x from the end is given by $M = (1/2)lx - (1/2)wx^2$. Find the value of x so that the bending moment is maximum.

22.5 A rectangular sheet of metal has four equal square portions removed at the corners and the sides are then turned up so as to form an open rectangular box. Show that when the volume contained in the box is maximum, the depth will be $1/6(a+b)-(a^2-ab+b^2)^{1/2}]$ where a and b are sides of the original dimensions of the rectangular sheet.

22.6 A person in a boat, who is 'a' miles from the nearest point on the beach, wishes to reach a point 'b' miles from that point along the sea shore as quickly as possible. The ratio of the person's rate of walking to his/her rate of rowing is sec a. Prove that he should land at a distance $b - a \cot a$ from the place to be reached.

22.7 Prove that the minimum radius vector of the curve $a^2/x^2 + b^2/y^2 = 1$ is of length $(a+b)$.

22.8 Find the maxima and minima of the function $xy(a-x-y)$.

22.9 Find the extreme points of the function $f(x_1, x_2) = x_1^3 + x_2^3 + 2x_1^2 + 4x_2^2 + 6$.

22.10 Find the extreme points of the function $f(x_1, x_2, x_3) = \sin x_1 \sin x_2 \sin x_3$, where x_1, x_2, and x_3 are the vertex angles of a triangle.

22.11 Find the stationary point of u, where $u = x^2 + y^2$.

22.12 Show that the minimum value of $u = xy(a^3/x) + a^3/y$ is $3a^2$.

22.13 Find the point(x_1, x_2, x_3) at which the function $f(x_1, x_2, x_3) = -x_1^2 - x_2^2 - x_3^2 + x_1x_2 + x_1 + 2x_3$ has optimum values.

22.14 Find the dimensions of a box of large volume, which can be inscribed in a sphere of radius a.

22.15 Use the Lagrange multiplier method to solve the following problem:

Minimize $f = 2x_1^2 + 2x_2^2 + x_3^2 + 2x_1x_2 - 8x_1 - 6x_2 - 4x_3 + 2x_1x_3 + 9$

Subject to $x_1 + x_2 + 2x_3 = 3$

22.16 Use the Lagrange multiplier method to solve the following problems:

(a) Minimize $f(x_1, x_2) = 3x_1^2 + x_2^2 + 2x_1x_2 + 6x_1 + 2x_2$

Subject to $2x_1 - x_2 = 4$

(b) Maximize $f(x_1, x_2) = 6x_1 + 8x_2 - x_1^2 - x_2^2$

Subject to $4x_1 + 3x_2 = 16$

and $3x_1 + 5x_2 = 15$

CONCEPTUAL PROBLEM

One of the most important concerns of a management is to decide when and how much to order or to manufacture so that the total cost associated with the inventory system is minimum. Effective inventory management enables an organization to meet or exceed customers' expectations of product availability while maximizing net profits or minimizing costs. Hold and manage inventory is essential for efficient and smooth running of any business organization. It is necessary for all kinds of organizations, be it a manufacturing industry, a five-star hotel, a printing press, or a hospital. Here we consider an inventory problem in which the objective is maximum profit. If a company minimizes the total inventory cost then the profit will be maximum. However, if the total inventory cost is greater than the sales revenue cost, then the company will run at a loss. Assume a company earns a profit P on some item, and is treated as a function of the units produced, say x, indicated by $P = 800x - 2x^2$. If the company's expenditure on interest, rent, and salary of the staff is ₹1 lakh, then the company would like to know whether it will have a loss or profit.

Solution It is given that

$$P = 800x - 2x^2 \tag{22.62}$$

(Contd)

(*Contd*)

Differentiating both sides of Eq. (22.62) with respect to x, we get

$$\frac{dP}{dx} = 800 - 4x \tag{22.63}$$

For maxima and minima, we have

$$\frac{dP}{dx} = 0 \Rightarrow 800 - 4x = 0 = x = 200$$

Differentiating both sides of Eq. (22.63) again with respect to x, we get

$$\frac{d^2P}{dx^2} = -4 \text{ (negative)}$$

Therefore, profit (P) is maximum for ($x = 200$).

The net profit = P – Expenditure

$= 800(200) - 2(200)^2 - 1,00,000 = -20,000$

Hence, the company will always run at a loss.

23 Non-linear Programming Problem I—Search Techniques

After studying this chapter, the reader will be able to understand the following:

For unconstrained non-linear programming problems:
- Fibonacci-search plan method
- Golden section method
- Univariate search technique
- Pattern search methods:
  - Hooke and Jeeves method
  - Powell's method
- Steepest descent method (Cauchy method)

For constrained problems:
- Complex method
- Zoutendijk method (Method of feasible directions)
- Transform method
- Penalty function and Penalty methods:
  - Interior penalty function method
  - Exterior penalty function method
- Rosen's gradient projection method

23.1 INTRODUCTION

In general, a mathematical programming problem involves determining the maximum or minimum of $f(X)$ restricted with constraints $g_i(X) \leq, =,$ or ≥ 0, where $f(X)$ and $g_i(X)$ are real-valued functions of $X = (x_1, x_2, \ldots, x_n)$ in n dimensional space E_n. In case some or all of the functions $f(X)$, $g_i(X)$, $i = 1, 2, \ldots, m$ are non-linear, then the mathematical programming problem is called a non-linear programming problem (NLPP). There is a large variety of NLPPs which can be categorized depending upon the nature of the objective function $f(X)$ and the constraints $g_i(X)$. Analytical methods have been devised to solve some of these NLPPs such as quadratic programming, separable programming, and geometric programming problems. However, there are many other NLPPs which do not yield solutions through the analytical methods. Our objective in this chapter is to discuss the methods to solve such problems using various techniques called *search techniques.* Such problems can be further divided into two categories: constrained NLPPs and unconstrained NLPPs.

In this chapter, the search methods to solve unconstrained and constrained NLPPs, for which analytical methods do not work, are discussed.

23.2 UNCONSTRAINED NON-LINEAR PROGRAMMING PROBLEM

The methods used to solve the unconstrained NLPPs are divided into two categories:

1. Direct search methods 2. Descent methods (or indirect search methods)

23.3 DIRECT SEARCH METHODS

In direct search methods, search is initiated from a fixed point in a direction through an optimum length to arrive at a new point and then the value of the function at the new point thus obtained is calculated. As derivatives of the function to be minimized are not required, these methods are suitable for functions that are not differentiable. Some of the direct search methods are as follows:

1. Fibonacci-search plan method (one-dimension)
2. Golden section method (one-dimension)
3. Univariate method
4. Pattern search methods
   (a) Hooke and Jeeves method (b) Powell's method

23.4 SEARCH TECHNIQUES IN ONE DIMENSION

We have already discussed classical techniques in Chapter 22, which are easily applicable if the expressions for an objective function and the constraints are simple. However, when an objective function and constraints are complicated, then classical analytical methods fail and we have to use the numerical methods. In most of these numerical techniques, the only requirement is that the function involved should be computable. Before discussing these methods, let us first discuss a few elementary definitions.

Unimodal function A function $f(x)$ with a unique optimal value (either a unique maxima or minima) in $[a, b]$ is called a unimodal function. Figures 23.1 and 23.2 depict the unimodal function in the interval $[a, b]$.
Let $y=f(x)$ be a unimodal function in $[a, b]$ and x_1, x_2 $(x_1<x_2)$ be two points in $[a, b]$. Only three outcomes, as shown in Figs 23.3(a), (b), and (c), are possible.

Interval of uncertainty Initially, the interval $[a, b]$ in which the optimum of the objective (response) function is needed is called the uncertainty interval and

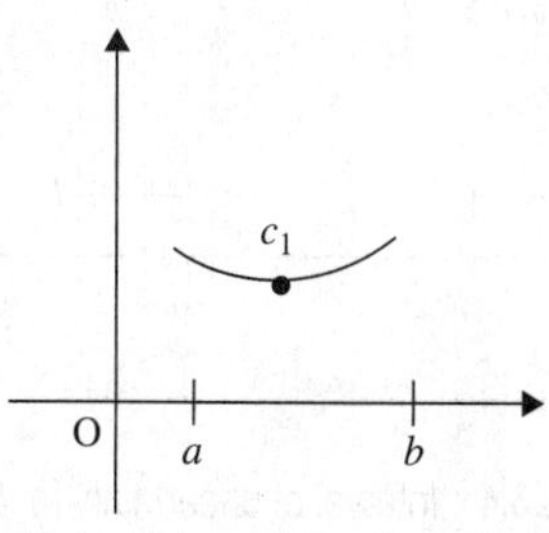

Figure 23.1 Unimodal function with global minima at c_1 in $[a, b]$

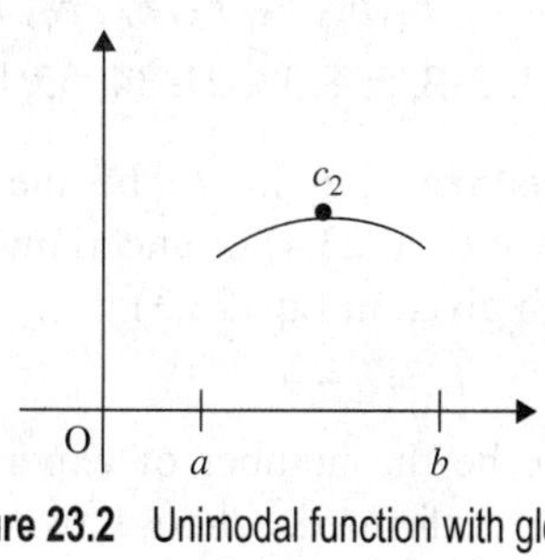

Figure 23.2 Unimodal function with global maxima at c_2 in $[a, b]$

after two experiments [finding $f(x_1)$ and $f(x_2)$], the uncertainty interval reduces to $(x_1, b]$, $[x_1, x_2]$, or $[a, x_2)$ (see Fig. 23.3).

Experiment The measure or calculation of the response (objective) function $f(x)$for any set of values x_i's in the uncertainty interval $[a, b]$ is called an experiment in the context of search techniques.

Measure of effectiveness Let L_0 be the initial uncertainty interval length and L_n be the uncertainty interval length after n experiments. Then the ratio $L_0/L_n = \alpha \leq 1$ is called the measure of effectiveness for comparing different search methods.

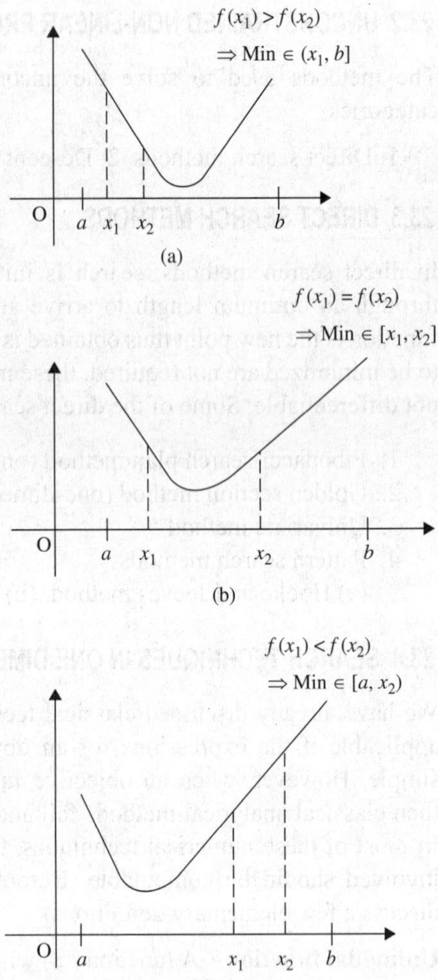

Figure 23.3 Graph displaying interval for minima (a) When $f(x_1)>f(x_2)$ (b) When $f(x_1)= f(x_2)$ (c) When $f(x_1)< f(x_2)$

23.4.1 Fibonacci Method of Search

In the Fibonacci method, the initial interval of uncertainty, say $[a, b]$, is given and the function to be optimized to a given degree of accuracy must be unimodal in the interval of uncertainty. Further, we use Fibonacci numbers. Hence, first we shall define the Fibonacci sequence. A Fibonacci sequence $\{F_n\}$ is defined as

$$F_n = F_{n-1} + F_{n-2} \qquad (23.1)$$

where n is an integer greater than $1 (n > 1)$ and

$$F_0 = F_1 = 1 \qquad (23.2)$$

Thus, the Fibonacci sequence is

$$F_0, F_1, F_2, F_3, F_4, F_5, F_6, F_7, F_8, F_9, F_{10}, F_{11}, F_{12} \ldots$$

or 1, 1, 2, 3, 5, 8, 13, 21, 34, 55, 89, 144, 233 ...

Procedure Let $[a, b]$ be the initial interval (Fig. 23.4) of uncertainty with length given in Eq. (23.3).

$$L_0 = b - a \qquad (23.3)$$

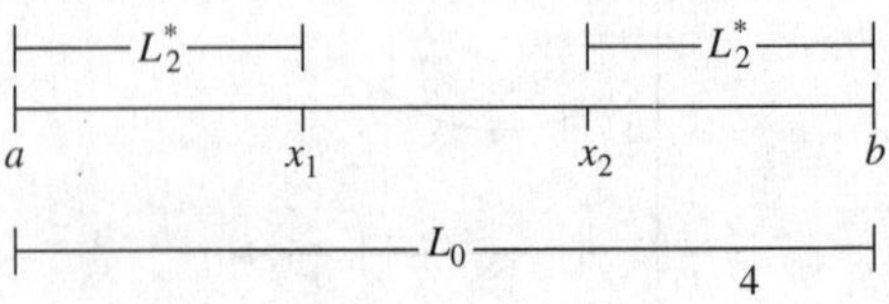

Figure 23.4 Interval of uncertainty $[a, b]$

and n be the number of experiments to be performed. Choose two points x_1 and x_2 in interval [a, b] of length L_0 (x_1, x_2) such that

$$x_1 = a + \frac{F_{n-2}}{F_n} L_0 = a + L_2^*, \text{ where } L_2^* = \frac{F_{n-2}}{F_n} L_0 \tag{23.4}$$

$$x_2 = b - \frac{F_{n-2}}{F_n} L_0 = b - L_2^* \tag{23.5}$$

$$\text{Thus, } b - x_2 = b - b + \frac{F_{n-2}}{F_n} L_0 \text{ (using Eq. 23.5)}$$

$$= x_1 - a \text{ (using Eq. 23.4)}$$

Here $x_1 - a = b - x_2$. It shows that x_1 and x_2 are symmetrically placed with respect to the end points, a and b, of interval of length L_0. From Eqs 23.3 and 23.4,

$$x_2 = b - \frac{F_{n-2}}{F_n} L_0 = a + L_0 - \frac{F_{n-2}}{F_n} L_0 = a + \left(\frac{F_n - F_{n-2}}{F_n}\right) L_0$$

$$= a + \frac{F_{n-1}}{F_n} L_0 \text{ (using Eq. 23.1)} \tag{23.6}$$

Let $y = f(x)$ be a unimodal function having minimum value in [a, b] with two points x_1 and x_2 in it, with the following conditions:

1. If $x_1 < x_2$ and $f(x_1) < f(x_2)$, then using the assumption of unimodal it can be concluded that the minimum value does not lie on the right of x_2 and hence reject interval $(x_2, b]$. The next uncertainty interval is $[a, x_2]$.
2. If $x_1 < x_2$ and $f(x_1) > f(x_2)$, discard $[a, x_1)$. The next uncertainty interval will then be $[x_1, b]$.

The length L_2 of this new uncertainty interval [for both cases (1) and (2)] will be

$$L_2 = L_0 - L_2^* = L_0 - \frac{F_{n-2}}{F_n} L_0 = L_0 \left(\frac{F_n - F_{n-2}}{F_n}\right) = \frac{F_{n-1}}{F_n} L_0 \tag{23.7}$$

For case (1), the interval of further search will be $[a, x_2)$ in which one observation is at x_1 (Fig. 23.5).

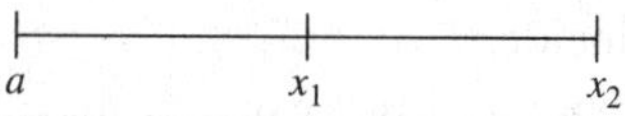

Figure 23.5 Interval of uncertainty $[a, x_2]$

For case (2), the interval of further search will be $(x_1, b]$ in which one observation is at x_2 (Fig. 23.6).

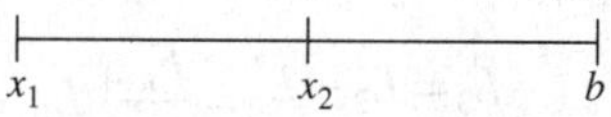

Figure 23.6 Interval of uncertainty $(x_1, b]$

The next iteration is on the interval of uncertainty $[a, x_2)$ of length L_2 (Fig 23.7) Let the interval $[a, x_2]$ be designated as $[a_1, b_1]$.

Now, choose x_3 and x_4 on the interval $[a, x_2]$ of length L_2 such that the conditions given in Eq. (23.8) are satisfied.

$$\left.\begin{aligned} x_3 &= a + \frac{F_{n-3}}{F_n} L_2 \\ \text{and } x_4 &= x_2 - \frac{F_{n-3}}{F_{n-1}} L_2 \end{aligned}\right\} \tag{23.8}$$

Now, $x_3 - a = \dfrac{F_{n-3}}{F_{n-1}} L_2 = x_2 - x_4$ (23.9)

Again, take $\dfrac{F_{n-3}}{F_{n-1}} L_2 = L_3^*$ (23.10)

The equality in Eq. (23.9) again implies that x_3 and x_4 selected according to Eq. (23.8) are symmetric with respect to the end points of L_2.

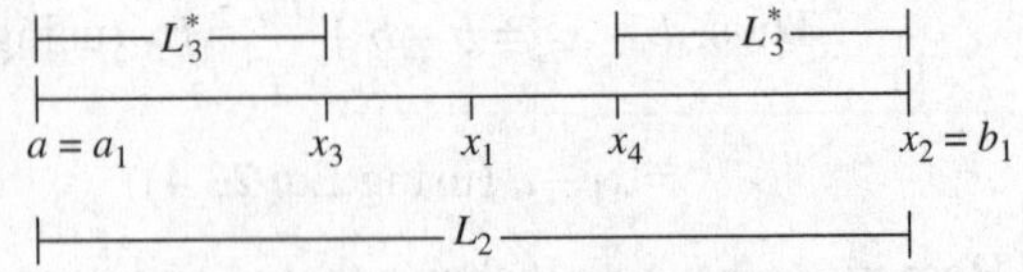

Figure 23.7 Interval of uncertainty $[a_1, b_1]$

Further, $x_2 - x_4 = \dfrac{F_{n-3}}{F_{n-1}} L_2 = \dfrac{F_{n-3}}{F_n} L_0$ (using Eqs 23.9 and 23.7) (23.11)

and $x_2 - x_1 = a + \dfrac{F_{n-1}}{F_n} L_0 - a - \dfrac{F_{n-2}}{F_n} L_0$ (using Eqs 23.4 and 23.6)

$$= \frac{F_{n-1} - F_{n-2}}{F_n} L_0 = \frac{F_{n-3}}{F_n} L_0 \quad (23.12)$$

Thus, $x_2 - x_1 = x_2 - x_4 = \dfrac{F_{n-3}}{F_n} L_0 = L_3^*$ (using Eqs 23.11 and 23.12)

This shows that x_4 coincides with x_1. This implies that in the second iteration and also in subsequent iterations, we need to evaluate the function value at only one new point (x_3 in this case).

Similarly, if for case (2) the new search interval $[x_1, b]$ is retained (Fig. 23.8), then we take $x_3 = x_1 + \dfrac{F_{n-3}}{F_n} L_2$ and $x_4 = b - \dfrac{F_{n-3}}{F_{n-1}} L_2$

Here $x_2 = x_3$, that is, x_2 and new point x_3 will coincide.

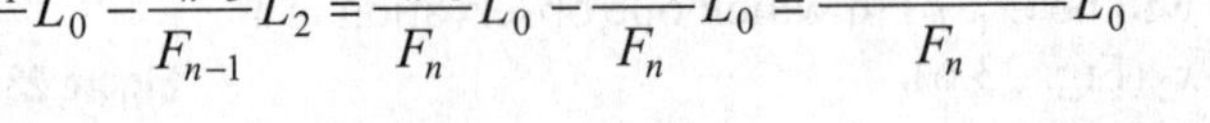

Figure 23.8 Interval of uncertainty $[x_1, b]$

The length of the uncertainty interval after the third experiment is given as follows:

$$L_3 = L_2 - L_3^* = \frac{F_{n-1}}{F_n} L_0 - \frac{F_{n-3}}{F_{n-1}} L_2 = \frac{F_{n-1}}{F_n} L_0 - \frac{F_{n-3}}{F_n} L_0 = \frac{F_{n-1} - F_{n-3}}{F_n} L_0$$

Thus, $L_3 = \dfrac{F_{n-2}}{F_n} L_0$ (23.13)

Now, repeat the whole process till the last two experiments are equidistant from the end points of the uncertainty interval $[a_n, b_n]$. Then, the optimum point x^* can be approximately taken as $x^* = \dfrac{a_n + b_n}{2}$.

Further, the distance of the eighth experiment from one end of the interval of uncertainty can be derived as follows:

$$L_j^* = \frac{F_{n-j}}{F_n} L_0, \quad (23.14)$$

The length of the uncertainty interval at this stage can be derived as given in Eq. (23.15)

$$L_j = \frac{F_{n-(j-1)}}{F_n} L_0, \; j = 2, 3, \ldots \tag{23.15}$$

Thus, $\dfrac{L_n}{L_0} = \dfrac{F_1}{F_n} = \dfrac{1}{F_n} = \alpha$ (say)

The ratio $L_n/L_0 = \alpha = 1/F_n$ permits the required number of experiments n that is necessary to achieve the desired accuracy for ascertaining the optimum of the given function. Let us find n when, as an example, the minimum of $x^2 - 3x$, $0 \le x \le 1.6$ is required within an interval of uncertainty equal to 0.25 L_0, where L_0 is the original interval of uncertainty. Then $\alpha = 1/F_n \le 0.25 \Rightarrow F_n \Rightarrow 1/0.25 = 4$ and hence $n = 4$.

Example 23.1 Find the minimum of $x^2 - 2x$, $0 \le x \le 1.5$, taking $n = 4$ using the Fibonacci method.

Solution Here $f(x) = x^2 - 2x$, $L_0 = 1.5 - 0 = 1.5$, and $n = 4$.

Clearly $f(x) = x^2 - 2x$ is unimodal in [0, 1.5].

First two experiments Let us take two points x_1 and x_2 at a distance $L_2^* = \dfrac{F_{n-2}}{F_n} L_0$ from the two points $a = 0$ and $b = 1.5$ where $x_1 = a + L_2^* = 0 + \dfrac{F_{n-2}}{F_n} L_0 = \dfrac{2}{5} \times 1.5 = 0.6$ (Fig. 23.9).

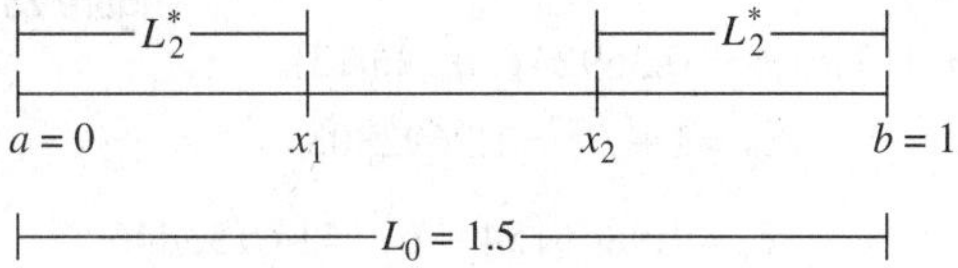

Figure 23.9 Interval of uncertainty of length $L_0 = 1.5$

Since x_1 and x_2 are symmetrically placed, x_2 will be at a distance 0.6 from $b = 1.5$, that is, $x_2 = 0.9$

$$f(x_1) = f(0.6) = -0.84 \text{ and } f(x_2) = f(0.9) = -0.99$$

$$\Rightarrow f(0.6) > f(0.9)$$

Thus, reject the interval [0, 0.6) as shown in Fig. 23.10; the new uncertainty interval is [0.6, 1.5].

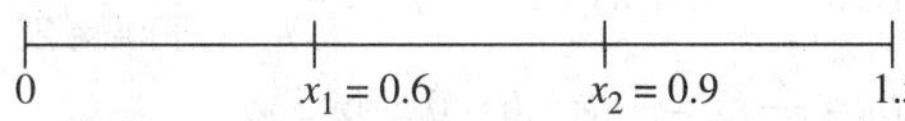

Figure 23.10 Interval of uncertainty [0, 1.5]

Third experiment Take two points $x_3(0.9)$ and x_4 in the new uncertainty interval [0.6, 1.5]. x_3 is the same as x_2, which is at a distance $L_3^* = 0.3$ from 0.6. Hence $x_4 = 1.2$, being symmetrical at a distance 0.3 from 1.5 (Fig. 23.11).

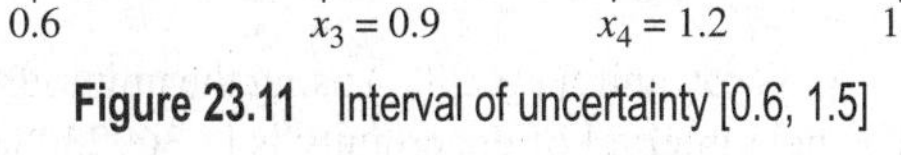

Figure 23.11 Interval of uncertainty [0.6, 1.5]

Thus, $x_3 = 0.9$ and $x_4 = 1.2$.

$$f(x_3) = f(0.9) = -0.99 \text{ and } f(x_4) = f(1.2) = -0.96 \Rightarrow f(x_4) > f(x_3)$$

Thus, reject (1.2, 1.5]. Hence, the next uncertainty interval is [0.6, 1.2].

Here 0.9 is the middle point of the uncertainty interval [0.6, 1.2]. Thus, we do not get a new point x_4 as $x_4 = x_3 = 0.9 = x^*$.

Thus, Min $f(x) = -0.99$ at $x = 0.9$. The correct minimum is $f(x) = -1$ at $x = 1$.

Example 23.2 Using the Fibonacci method, determine the minimum of $x^2 - 3x + 5$ in the interval [1, 2.6], taking $n = 6$.

Solution Here $f(x) = x^2 - 3x + 5$, $L_0 = 2.6 - 1 = 1.6$, initial interval [1, 2.6], and $n = 6$.

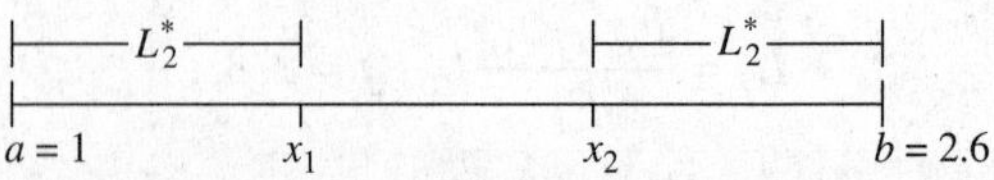

Figure 23.12 Interval of uncertainty [1, 2.6]

First two experiments Let us take two points x_1 and x_2 in [1, 2.6] at a distance $L_2^* = \frac{F_{n-2}}{Fn} L_0 = \frac{F_4}{F_6} L_0 = \frac{5}{13} \times 1.6 = 0.6153846$ from the end points 1 and 2.6 respectively (Fig. 23.12).

$$x_1 = a + L_2^* = 1.6153846 \text{ and } x_2 = b - L_2^* = 2.6 - 0.6153846 = 1.9846154$$

$$f(x_1) = 2.763314 \text{ and } f(x_2) = 2.984852$$

$$\Rightarrow f(x_1) < f(x_2),\ x_1 < x_2$$

Now, reject (1.9846154, 2.6] and the new uncertainty interval becomes [1, 1.9846154].

Third experiment Let x_3 and x_4 be two points in (1, 1.9846154] at a distance $L_3^* = \frac{F_{n-3}}{F_n} L_0 = \frac{F_3}{F_6} L_0$ $= 3/13 \times 1.6 = 0.36923$ (Fig. 23.13).

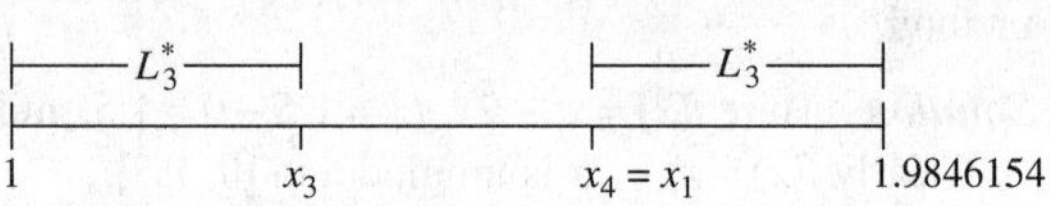

Figure 23.13 Interval of uncertainty [1, 1.9846154]

$$x_3 = 1 + L_3^* = 1.3692308$$

$$x_4 = 1.9846154 - L_3^* = 1.6153846$$

$$f(x_3) = 2.7671006,\ f(x_4) = 2.763314,\ x_3 < x_4, \text{ and } f(x_3) > f(x_4)$$

Now, reject [1, 1.3692308) and the new uncertainty interval becomes [1.3692308, 1.9846154] using the unimodality of the function.

Fourth experiment Let x_5 and x_6 be the two points in [1.3692308, 1.9846154] at a distance $L_4^* = \frac{F_{n-4}}{F_n}$ $L_0 = \frac{F_2}{F_6} L_0 = \frac{2}{13} \times 1.6 = 0.2461538$ (Fig. 23.14).

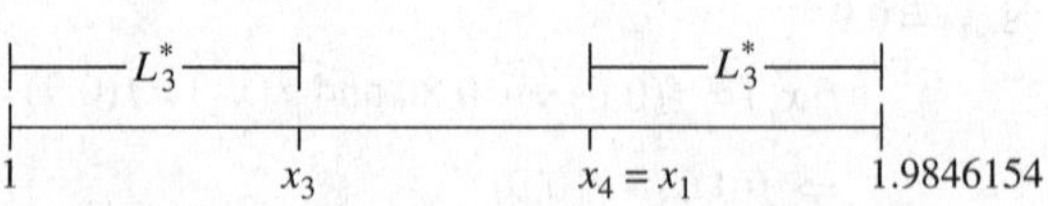

Figure 23.14 Interval of uncertainty (1.3692308, 1.9846154]

$$x_5 = 1.3692308 + L_4^* = 1.6153846$$

$$x_6 = 1.9846154 - L_4^* = 1.7384616$$

$$f(x_5) = 2.763314 \text{ and } f(x_6) = 2.8068639$$

As $x_5 < x_6$ and $f(x_5) < f(x_6)$ using the unimodality, reject interval [1.7384616, 1.9846154]. The next interval of uncertainty is [1.3692308, 1.7384616].

Fifth experiment Let x_7 and x_8 be the two points in [1.3692308, 1.7384616] at a distance $L_5^* = \frac{F_{n-5}}{F_n} L_0 = \frac{F_1}{F_6} L_0 =$ $\frac{1}{13} \times 1.6 = 0.1230769$ (Fig. 23.15).

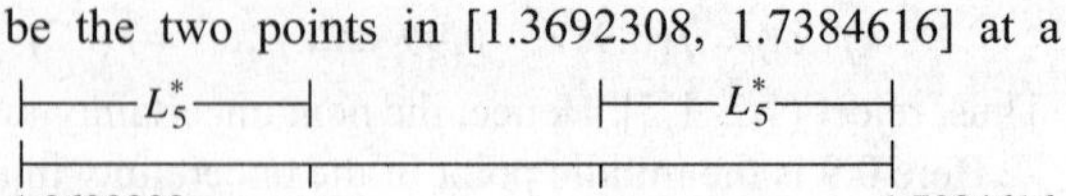

Figure 23.15 Interval of uncertainty (1.3692308, 1.7384616]

$$x_7 = 1.3692308 + L_5^* = 1.4923077 \text{ by symmetry}$$

$$x_8 = 1.7384616 - L_5^* = 1.6153846$$

$$\because f(x_7) = 2.750059 \text{ and } f(x_8) = 2.7633136$$

$\because x_7 < x_8$ and $f(x_7) < f(x_8)$, using unimodality, reject interval [1.6153846, 1.7384616]. The next new uncertainty interval is [1.3692308, 1.6153846].

Sixth experiment $L_6^* = \dfrac{F_{n-6}}{F_n} L_0 = \dfrac{F_0}{F_6} L_0 = \dfrac{1}{13} \times 1.6 = 0.1230769 = L_5^*$

Thus, there is no fresh point and hence the final uncertainty interval is [1.3692308, 1.6153846] whose middle point is derived as follows:

$$x^* = \frac{1.3692308 + 1.6153846}{2} = \frac{2.9846154}{2} = 1.4923077$$

Thus, Min $f(x) = 2.750059$ for $x = 1.4923077$.

23.4.2 Golden Section Method

The next search method is the golden section method wherein unlike the Fibonacci method, the total number of experiments to be conducted are unlimited. Let us assume that n is the total number of experiments to be conducted and is large enough to satisfy the limit.

$$\lim_{n\to\infty} \frac{F_{n-1}}{F_n} = \lim_{n\to\infty} \frac{F_{n-2}}{F_{n-1}} = \lim_{n\to\infty} \frac{F_{n-3}}{F_{n-2}} = \frac{1}{\lambda} \text{ (say)}$$

Then $L_2 = \lim_{n\to\infty} \dfrac{F_{n-1}}{F_n} L_0 = \dfrac{L_0}{\lambda}$, $L_3 = \lim_{n\to\infty} \dfrac{F_{n-2}}{F_n} L_0 = \lim_{n\to\infty} \dfrac{F_{n-2}}{F_{n-1}} \dfrac{F_{n-1}}{F_n} L_0 = \dfrac{L_0}{\lambda^2}$

Now, proceeding this way k times, we get $L_K = \dfrac{L_0}{\lambda^{K-1}} = \left(\dfrac{1}{\lambda}\right)^{K-1} L_0$

From the Fibonacci sequence $\{F_n\}$, we have

$F_n = F_{n-1} + F_{n-2}$, n is an integer greater than $1 (n > 1)$.

$$\therefore \frac{F_n}{F_{n-1}} = 1 + \frac{F_{n-2}}{F_{n-1}}$$

Taking the limit as $n \to \infty$, we have

$$\lim_{n\to\infty} \frac{F_n}{F_{n-1}} = \lim_{n\to\infty} \left(1 + \frac{F_{n-2}}{F_{n-1}}\right) \text{ or } \lambda = 1 + \frac{1}{\lambda},\ \lambda^2 - \lambda - 1 = 0,$$

$$\lambda = \frac{1+\sqrt{5}}{2} = 1.618033988. \therefore \frac{1}{\lambda} = 0.618$$

Working procedure Take two points in the initial uncertainty interval $[a, b]$ which are symmetrically placed from both the end points a and b at a distance $L_2^* = \lim_{n\to\infty} \dfrac{F_{n-2}}{F_n} L_0 = \lim_{n\to\infty} \dfrac{F_{n-2}}{F_{n-1}} \dfrac{F_{n-1}}{F_n} L_0 = \dfrac{L_0}{\lambda^2} = (0.618)^2 L_0$ where $L_0 = b - a =$ length of the initial uncertainty interval.

In the kth iteration, take $L_K^* = (0.618)^K L_0$. Now, repeat these iterations until the stage where the interval of uncertainty is as small as desired. The optimum points can be taken as the middle points of this final uncertainty interval.

Example 23.3 Use the golden section method to determine the maximum of $f(x) = x(5-x)$ given that $f(x)$ is a unimodal function in the interval [0, 8] wherein the maximum lies.

Solution Here $[a, b] = [0, 8]$ is the initial uncertainty interval and its length $L_0 = 8 - 0 = 8$.

First two experiments Let us take two points x_1 and x_2 between [0, 8] at a distance $L_2^* = (0.618)^2 L_0 = (0.618)^2 \times 8 = 3.055392$ from 0 and 8 (Fig. 23.16).

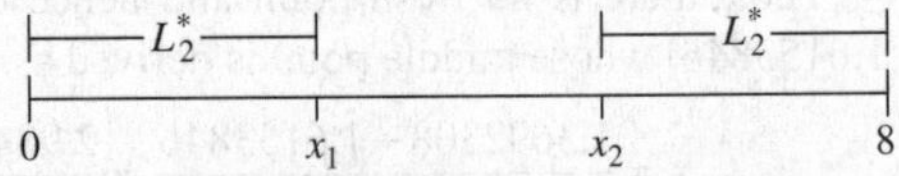

Figure 23.16 Interval of uncertainty [0, 8]

$x_1 = 0 + L_2^* = 3.055392, f(x_1) = 5.941975$

$x_2 = 8 - L_2^* = 8 - 3.055392 = 4.944,\ f(x_2) = 0.276864$

Thus, $x_1 < x_2$ and $f(x_1) > f(x_2)$

Therefore, by the unimodality assumption, we can infer that the maximum points do not lie on the right of x_2 and reject the interval (4.944, 8].

The next uncertainty interval is [0, 4.944] with length $L_2 = (0.628)^{2-1} L_0 \approx 4.944$.

Third experiment Take two points x_3 and x_4 in [0, 4.944] as a distance $L_3^* = (0.618)^3$ and $L_0 = (0.618)^3 \times 8 = 1.888$ (Fig. 23.17).

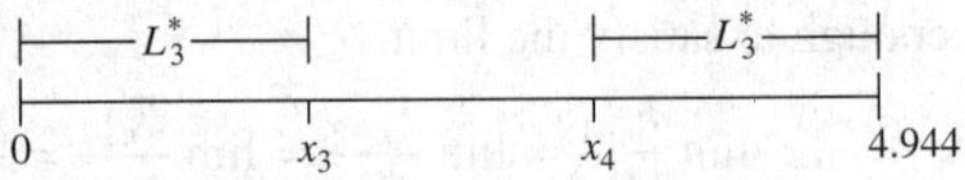

Figure 23.17 Interval of uncertainty [0, 4.944]

$x_3 = 0 + L_3^* = 1.888,\ f(x_3) = 5.875456$

$x_4 = 4.944 - L_3^* = 3.056,\ f(x_4) = 5.940864$

Thus, $x_3 < x_4$ and $f(x_3) < f(x_4)$.

Hence, by the unimodality assumption, the maximum points do not lie on the left of x_3. Therefore reject interval [0, 1.888). The next uncertainty interval is [1.888, 4.944].

Fourth experiment Take two points x_5 and x_6 in [1.888, 4.944] at a distance $L_4^* = (0.618)^4 L_0 = (1.1669)$ (Fig. 23.18).

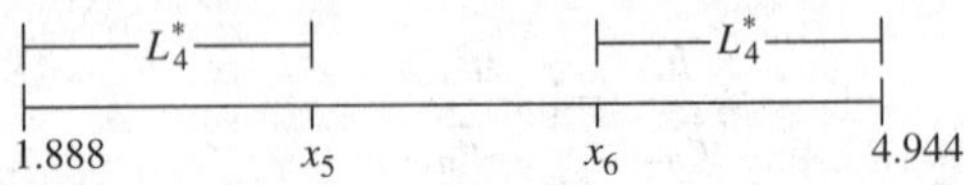

Figure 23.18 Interval of uncertainty [1.888, 4.944]

$x_5 = 1.888 + L_4^* = 3.055,\ f(x_5) = 5.941975$

$x_6 = 4.944 - L_4^* = 3.777,\ f(x_6) = 4.6190$

Thus, $x_5 < x_6$ and $f(x_5) > f(x_6)$

According to the unimodality assumption, the maximum points do not lie on the right of x_6 and therefore reject interval (3.777, 4.944]. Now, the next uncertainty interval is [1.888, 3.777].

Fifth experiment Take two points x_7 and x_8 in [1.888, 3.777] at a distance $L_5^* = (0.618)^5 L_0 = (0.721)$ from the two end points (Fig. 23.19).

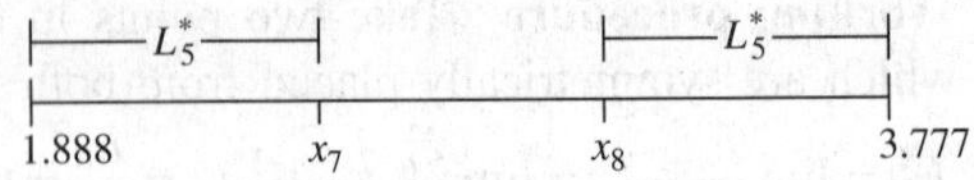

Figure 23.19 Interval of uncertainty [1.888, 3.777]

$x_7 = 1.888 + L_5^* = 2.609,\ f(x_7) = 6.238597$

$x_8 = 3.777 - L_5^* = 3.056,\ f(x_8) = 5.940864$

Thus, we get $x_7 < x_8$ and $f(x_7) > f(x_8)$

So by using the unimodality assumption, the maximum points do not lie on the right of x_8. Reject the interval (3.056, 3.777]. The next uncertainty interval is [1.888, 3.056].

After the fifth experiment, the uncertainty interval is sufficiently small (≈ 1.16) and the maximum points $x^* = \dfrac{1.888 + 3.056}{2} = 2.472$ and $f(x^*) = 6.249216$.

23.4.3 Univariate Method

Univariate method, as the name suggests, is a method in which the search is carried out in the axial directions covering all the directions one by one. This will complete one cycle of iterations. The process will be over when no further improvement in the value of the given function to be minimized is possible in any direction of a cycle. The following steps are involved in this method:

1. First, choose a fixed point X_1 (x_1, y_1) and a search direction $u_1 = (1, 0)^T$. Then, take a step of size λ_1 in this direction and obtain the optimum $\lambda_1 = \lambda_1^*$ for which $f(X_1 + \lambda_1^* u_1)$ is minimum. This gives a new point X_2 where $X_2 = X_1 + \lambda_1^* u_1$.
2. Repeat step 1 by taking the optimum step length λ_2^* in the direction $u_2 = (0, 1)^T$ from the point X_2 to arrive at the next point $X_3 = X_2 + \lambda_2^* u_2$.
3. Completion of steps 1 and 2 will form one cycle of iterations. Repeat the process for the completion of the next cycle of iterations until the variation in the value of function f is negligible.

Here, we shall consider functions in two variables x_1 and x_2 only, although the aforementioned procedural method can be extended for functions of more than two variables as well. Let us illustrate this method through Example 23.4.

Example 23.4 Minimize $f(X) = f(x_1, x_2) = 2x_1^2 + 3x_2^2 - x_1x_2$ using the univariate method taking $X_1 = \begin{pmatrix} 1 \\ 1 \end{pmatrix}$ as the starting point.

Solution We can minimize the given function using the following steps:

Cycle 1

1. Consider the unit vectors $u_1 = (1, 0)^T$ and $u_2 = (0, 1)^T$ in x_1 and x_2 (axial) directions respectively. Given: $X_1 = \begin{pmatrix} 1 \\ 1 \end{pmatrix}$, where $f(X_1) = 4$. We take a step size λ_1 in the x_1 direction to arrive at the value of X_2

$$X_2 = X_1 + \lambda_1 \begin{pmatrix} 1 \\ 0 \end{pmatrix} = \begin{pmatrix} 1 \\ 1 \end{pmatrix} + \begin{pmatrix} \lambda_1 \\ 0 \end{pmatrix} = \begin{pmatrix} 1 + \lambda_1 \\ 1 \end{pmatrix}$$

To obtain the optimum step size λ_1, we Min $f\begin{pmatrix} 1 + \lambda_1 \\ 1 \end{pmatrix} = 2(1 + \lambda_1)^2 + 3 - (1 + \lambda_1)$
where, $x_1 = 1 + \lambda_1$ and $x_2 = 1$

For minimum of f, $\dfrac{df}{d\lambda_1} = 4(1+\lambda_1) - 1 = 0 \Rightarrow \lambda_1 = -\dfrac{3}{4}$

and $\dfrac{d^2f}{d\lambda_1^2} = 4 > 0 \Rightarrow f$ is minimum for $\lambda_1 = -\dfrac{3}{4}$.

Therefore, new point $X_2 = \begin{pmatrix} 1+\lambda_1 \\ 1 \end{pmatrix} = \begin{pmatrix} 1/4 \\ 1 \end{pmatrix}$,

$$f(X_2) = 2\left(\frac{1}{4}\right)^2 + 3(1)^2 - \frac{1}{4}(1) = \frac{23}{8} = 2.875.$$

2. Now to obtain the optimum step size λ_2 in the x_2 direction,

Minimize $f[X_2 + \lambda_2(0,1)^T] = f\left(\begin{pmatrix} 1/4 \\ 1 \end{pmatrix} + \begin{pmatrix} 0 \\ \lambda_2 \end{pmatrix}\right) = f\begin{pmatrix} 1/4 \\ 1+\lambda_2 \end{pmatrix} = \dfrac{1}{8} + 3(1+\lambda_2)^2 - \dfrac{1}{4}(1+\lambda_2)$

For minimum of f, $\dfrac{df}{d\lambda_2} = 6(1+\lambda_2) - \dfrac{1}{4} = 0 \Rightarrow \lambda_2 = -\dfrac{23}{24}$

and $\dfrac{d^2f}{d\lambda_2^2} = 6 > 0 \Rightarrow f$ is minimum for $\lambda_2 = -\dfrac{23}{24}$.

New point $X_3 = X_2 + \lambda_2(0,\ 1)^T = \begin{pmatrix} 1/4 \\ 1+\lambda_2 \end{pmatrix} = \begin{pmatrix} 1/4 \\ 1/24 \end{pmatrix}$

$$f(X_3) = 2\left(\frac{1}{4}\right)^2 + 3\left(\frac{1}{24}\right)^2 - \left(\frac{1}{4}\right)\left(\frac{1}{24}\right) = 0.11979.$$

Thus, one cycle of iterations is completed.

Cycle 2

3. Now considering X_3 as the base point, to find optimum step size λ_3 in the x_1 direction,
Minimize $f[X_3 + \lambda_3(1,\ 0)^T]$

$$= f\left(\begin{pmatrix} 1/4 \\ 1/24 \end{pmatrix} + \begin{pmatrix} \lambda_3 \\ 0 \end{pmatrix}\right) = f\begin{pmatrix} 1/4+\lambda_3 \\ 1/24 \end{pmatrix} = 2\left(\frac{1}{4}+\lambda_3\right)^2 + 3\left(\frac{1}{24}\right)^2 - \frac{1}{24}\left(\frac{1}{4}+\lambda_3\right)$$

For minimum of f, $\dfrac{df}{d\lambda_3} = 4\left(\dfrac{1}{4}+\lambda_3\right) - \dfrac{1}{24} = 0 \Rightarrow \lambda_3 = -\dfrac{23}{96}$

$\dfrac{d^2f}{d\lambda_3^2} = 4 > 0 \Rightarrow f$ is minimum for $\lambda_3 = -\dfrac{23}{96}$.

$\therefore$ New point $X_4 = \begin{pmatrix} 1/4 + (-23/96) \\ 1/24 \end{pmatrix} = \begin{pmatrix} 1/96 \\ 1/24 \end{pmatrix}$,

$$f(X_4) = 2\left(\frac{1}{96}\right)^2 + 3\left(\frac{1}{24}\right)^2 - \left(\frac{1}{96}\right)\left(\frac{1}{24}\right) = \frac{23}{4608} = 0.00499.$$

4. Now considering X_4 as the base point, to find optimum step size λ_4 in the x_2 direction,
Minimize $f[X_4 + \lambda_4(0,\ 1)^T]$

$$= f\left(\begin{pmatrix} 1/96 \\ 1/24 \end{pmatrix} + \begin{pmatrix} 0 \\ \lambda_4 \end{pmatrix}\right) = f\begin{pmatrix} 1/96 \\ 1/24+\lambda_4 \end{pmatrix} = 2\left(\frac{1}{96}\right)^2 + 3\left(\frac{1}{24}+\lambda_4\right)^2 - \frac{1}{96}\left(\frac{1}{24}+\lambda_4\right)$$

For minimum of f, $\dfrac{df}{d\lambda_4} = 6\left(\dfrac{1}{24} + \lambda_4\right) - \dfrac{1}{96} = 0 \Rightarrow \lambda_4 = -\dfrac{23}{576}$.

$$\frac{d^2 f}{d\lambda_4^2} = 6 > 0 \Rightarrow f \text{ is minimum for } \lambda_4 = -\frac{23}{576}.$$

$$\therefore \text{ The new point } X_5 = \begin{pmatrix} 1/96 \\ (1/24) - (23/576) \end{pmatrix} = \begin{pmatrix} 1/96 \\ 1/576 \end{pmatrix},$$

$$f(X_5) = 2\left(\frac{1}{96}\right)^2 + 3\left(\frac{1}{576}\right)^2 - \left(\frac{1}{96}\right)\left(\frac{1}{576}\right) = \frac{33}{12 \times (96)^2} = 0.00020797.$$

Thus, minimum $f(X) = 0.00020797$, when $x_1 = 1/96$ and $x_2 = 1/576$.

Note From the aforementioned steps, we see that as the base point moves from (1, 1) towards (1/96, 1/576) $\approx$ (0, 0), $f(X)$ becomes smaller and smaller and tends to zero. Thus, Min $f(X) = 0$ at (0, 0).

23.4.4 Pattern Search Methods

In univariate method, which starts from a fixed point (the base point), search is made for the minimum of the given functions by moving in the axial directions (the directions of the axes) only. It can be observed that the rate of convergence in univariate method to arrive at the optimal point is very slow. To accelerate this rate of convergence, a number of algorithms have been devised. In the next subsection, the following algorithms are discussed:

1. Hooke and Jeeve's method 2. Powell's method

Hooke and Jeeve's Method

In this method, we start from a fixed point, say X_1. The step size in all the directions is considered to be constant (i.e., $\Delta x_1 = \Delta x_2 = \cdots = \Delta x_n$ = constant) and search is made in each direction u_i. u_i is the unit vector in the direction of the x_i-axis whose ith component is 1 and all other components are 0. The search is made first in the positive direction and then if necessary in the negative direction of each axis to arrive at a temporary base point. A temporary base point Y_{ki} obtained from X_k by slightly changing the ith component of X_k ($Y_{k0} = X_k$) is defined as follows:

$$Y_{ki} = \begin{cases} Y_{k,i-1} + \Delta x_i u_i & \text{if } f^+ = f(Y_{k,i-1} + \Delta x_i u_i) < f = f(Y_{k,i-1}) \\ Y_{k,i-1} - \Delta x_i u_i & \text{if } f^- = f(Y_{k,i-1} - \Delta x_i u_i) < f = f(Y_{k,i-1}) \\ Y_{k,i-1} & \text{if } f(Y_{k,i-1}) = f \le \min(f^+, f^-) \end{cases}$$

This process of finding a new base point is continued for $i = 1, 2, 3, \ldots, n$ until all directions are covered.

After the completion of one cycle of covering all the axial directions towards the final temporary base point of the cycle, say Y_{1n}, (n being the number of design variables in the problem), if Y_{1n} is different from $Y_{10} = X_1$, then a new base point is taken as $X_2 = Y_{1n}$ and a single step of the optimum size λ^* is taken in the direction $S = X_2 - X_1$. After this, the search along the axes is resumed and a second new pattern is generated to go ahead. This way, the pattern and the set of axial steps is repeated for some more iterations. Hence, this method is called the pattern search method. These steps are repeated until a desired accuracy is achieved or the change in the value of the function satisfies the given condition in the problem. We shall confine our discussion (see Example 23.5) by considering functions in two variables ($n = 2$), which can be extended for more than two variables easily.

Example 23.5 Minimize $f(X) = 2x_1^2 + x_2^2 + 2x_1x_2 + x_1 - x_2$ by Hooke's and Jeeve's method, taking the starting base point as $X_1 = (0, 0)$ and $\Delta x_1 = \Delta x_2 = 0.8$.

Solution We can minimize the given function using the following steps:

Cycle 1

1. Set $Y_{10} = X_1 = (0, 0)$, $f(X_1) = 0$. Now moving in direction u_1 from the base point X_1,

$$f^+(X_1 + \Delta x_1u_1) = f^+[(0, 0) + 0.8(1, 0)] = f^+(0.8, 0) = 2.08 \nless f(X_1)$$

$$f^-(X_1 - \Delta x_1u_1) = f^-(-0.8, 0) = 0.48 \nless f(X_1)$$

As f^+, f^- are not less than $f(X_1)$, movement in u_1 direction is not beneficial and is hence discarded. Further, $Y_{11} = X_1$ and $f(Y_{11}) = f(X_1) = 0$.

2. Resuming movement in u_2 direction,

$$f^+(X_1 + \Delta x_2u_2) = f^+[(0, 0) + 0.8(0, 1)] = f^+(0, 0.8) = -0.16 < f(X_1)$$

As $Y_{12} = X_1 + \Delta x_2u_2 = (0, 0.8)$ is different from X_1, the new base point is taken as $Y_{12} = X_2 = (0, 0.8)$ and $f(X_2) = -0.16$.

3. As movement has been made in both the axial directions u_1 and u_2, the third movement is to be made in the first pattern direction $S_{p1} = X_2 - X_1 = (0, 0.8) - (0, 0) = (0, 0.8)$ through step length λ from X_2 so that $f(X_2 + \lambda S_{p1})$ is minimum.

For minimum of $f[X_2 + \lambda(X_2 - X_1)] = f\{0, 0.8(1 + \lambda)\} = [0.8(1 + \lambda)]^2 - 0.8(1 + \lambda)$,

$$\frac{df}{d\lambda} = 2 \times 0.8(1 + \lambda) - 0.8 = 0 \Rightarrow \lambda = -3/8 \text{ and}$$

$$\frac{d^2f}{d\lambda^2} = 1.6 > 0 \Rightarrow f \text{ is minimum for } \lambda^* = -3/8.$$

Thus, we obtain the new base point $Y_{20} = X_3 = X_2 + \lambda^*(X_2 - X_1) = (0, 0.8) - (3/8)(0, 0.8) = (0, 0.5)$ and $f(X_3) = f(0, 0.5) = -0.25$. This completes the first cycle of iterations along the axial directions u_1 and u_2, and the first pattern direction S_{p1}.

Cycle 2

4. First, we move in the u_1 direction starting with base point $X_3 = (0, 0.5)$.

$$f^+(X_3 + \Delta x_1u_1) = f^+[(0, 0.5) + (0.8, 0)] = f^+(0.8, 0.5) = 2.63 \nless f(X_3)$$

$$f^-(X_3 - \Delta x_1u_1) = f^-(-0.8, 0.5) = -0.57 < f(X_3) = -0.25$$

Thus, the new temporary base point is $Y_{21} = X_3 - \Delta x_1u_1 = (-0.8, 0.5)$, $f(Y_{21}) = -0.57$.

5. Moving in the u_2 direction from the base point $Y_{21} = (-0.8, 0.5)$ to Y_{22}, we have

$$f^+(Y_{21} + \Delta x_2u_2) = f^+(-0.8, 1.3) = -1.21 < f(Y_{21})$$

Thus, the new base point $Y_{22} = X_4 = Y_{21} + \Delta x_2u_2 = (-0.8, 1.3)$ and $f(X_4) = -1.21$.

6. Now after exhausting both the axial directions in the second cycle of iterations, we move along the second pattern search direction $S_{p2} = X_4 - X_3$ starting from X_4 through optimal step length λ so that $f(X_4 + \lambda S_{p2}) = f[(-0.8, 1.3) + \lambda(-0.8, 0.8)] = f[-0.8(1 + \lambda), 1.3 + \lambda(0.8)]$ is minimum.

To obtain the minimum of $f(\lambda)$, $\frac{df}{d\lambda} = 1.28\lambda - 0.32 = 0 \Rightarrow \lambda = 0.25$

and $\frac{d^2 f}{d\lambda^2} = 1.28 > 0 \Rightarrow f$ is minimum for $\lambda = 0.25$

Thus, we get the point $X_5 = Y_{30} = X_4 + \lambda S_{p2} = (-1, 1.5)$ and $f(X_5) = -1.25$.

Cycle 3

7. Moving in the u_1 direction from $X_5 = (-1, 1.5)$.

$f^+(X_5 + \Delta x_1 u_1) = f^+[(-1, 1.5) + 0.8(1, 0)] = f^+(-0.2, 1.5) = 0.03 \nless f(X_5)$

$f^-(X_5 - \Delta x_1 u_1) = f^-[(-1, 1.5) - 0.8(1, 0)] = f^-(-1.8, 1.5) = 0.03 \nless f(X_5)$, so $Y_{31} = X_5$.

8. Making movement in u_2 direction from $Y_{31} = X_5$,

$f^+(X_5 + \Delta x_2 u_2) = f^+[(-1, 1.5) + 0.8(0, 1)] = f^+(-1, 2.3) = -0.61 \nless f(X_5)$

$f^-(X_5 - \Delta x_2 u_2) = f^-[(-1, 1.5) - 0.8(0, 1)] = f^-(-1, 0.7) = -0.61 \nless f(X_5)$

This shows that the movement from X_5 in any of the axial directions produces no change in the value of the function f at X_5. Thus, X_5 is the optimal solution. Therefore, Min $f = -1.25$, when $x_1 = -1$ and $x_2 = 1.5$.

Powell's Method

Let f be the function of n variables which is to be minimized and $u_1, u_2, \ldots, u_n$ be the axial directions. In Powell's method starting from a fixed point, search is carried out sequentially in the directions u_n; $u_1, u_2, \ldots, u_n$ in the first cycle, along $S_{p1}, u_2, u_3, \ldots, u_{n-1}, u_n, S_{p1}$ in the second cycle, and $S_{p2}, u_3, u_4, \ldots, u_{n-1}, u_n, S_{p1}, S_{p2}$ in the third cycle and so on until the minimum point is reached, where S_{pi}'s are the pattern search directions defined as $S_{pi} = X_i - X_{i-n}$ and n is the number of design (decision) variables. In particular taking $n = 2$, search is made in the directions u_2, u_1, u_2 and S_{p1}, u_2, S_{p1}, and so on. Thus, in the second cycle, one of the axial directions is replaced by the pattern search direction. We illustrate the method through examples.

Note In Powell's method, any of the unit vectors, u_1 or u_2 can be taken as first search direction and hence the cycle of iterations can also be in the order u_1; u_2, u_1, and S_{p1}; u_1, S_{p1}. Here u_2 is discarded in the second cycle of iteration. This method being the method of conjugate directions, converges in at most two cycles of iterations.

Example 23.6 Minimize $f(x_1, x_2) = 2x_1^2 + x_2^2 + 2x_1x_2 + x_1 - x_2$ using Powell's method, taking $X_1 = (0, 0)$ as the starting point.

Solution Given that $f = 2x_1^2 + x_2^2 + 2x_1x_2 + x_1 - x_2$ and base point $X_1 = (0, 0)$.

Cycle I In the first cycle of iteration, search will be made in the directions u_2, u_1, u_2. Take step of size λ_1 in the direction u_2 from X_1 to reach at X_2, such that

$$f(X_2) = f(X_1 + \lambda_1 u_2) = f\left[\begin{pmatrix}0\\0\end{pmatrix} + \lambda_1\begin{pmatrix}0\\1\end{pmatrix}\right] = f\begin{pmatrix}0\\\lambda_1\end{pmatrix} = \lambda_1^2 - \lambda_1 \text{ is minimum.}$$

For minimum, $\frac{df}{d\lambda_1} = 2\lambda_1 - 1 = 0 \Rightarrow \lambda_1 = \frac{1}{2}$ $\frac{d^2 f}{d\lambda_1^2} = 2 > 0 \Rightarrow f$ is minimum.

So $X_2 = X_1 + \lambda_1 u_2 = (0, 0) + (1/2)(0, 1) = (0, 1/2), f(X_2) = f(0, 1/2) = -0.25 < f(X_1)$.

Next moving from X_2 in the direction of $u_1 = (1, 0)$ through step length λ_2 arriving at point $X_3 = X_2 + \lambda_2 u_1$. We find λ_2 so that $f(X_3) = f(\lambda_2, 1/2) = 2\lambda_2^2 + 2\lambda_2 - 0.25$ is minimum.

For minimum f, $\dfrac{df}{d\lambda_2} = 4\lambda_2 + 2 = 0 \Rightarrow \lambda_2 = -1/2$

Thus, $\dfrac{d^2 f}{d\lambda_2^2} = 4 > 0 \Rightarrow f$ is minimum.

So $X_3 = X_2 + \lambda_2 u_1 = (0, 1/2) - 1/2(1, 0) = (-1/2, 1/2)$ and $f(X_3) = -0.75 < f(X_2) = -0.25$.

Now move from X_3 to X_4 taking step length of size λ_3 in the direction of $u_2 = (0, 1)$.

$$X_4 = X_3 + \lambda_3 (0, 1) = (-1/2, 1/2) + (0, \lambda_3) = (-1/2, 1/2 + \lambda_3).$$

Now, we determine λ_3 such that $f(X_4) = f(-1/2, 1/2 + \lambda_3) = \lambda_3^2 - \lambda_3 - 0.75$ is minimum.

For minimum f, $\dfrac{df}{d\lambda_3} = 2\lambda_3 - 1 = 0 \Rightarrow \lambda_3 = 1/2$

$\dfrac{d^2 f}{d\lambda_3^2} = 2 > 0 \Rightarrow f$ is minimum.

Thus, $X_4 = (-1/2, 1)$ and $f(X_4) = -1 < f(X_3) = -0.75$.

This completes one cycle of iterations.

Cycle 2 Here, $i = 4$

In the second cycle, the search will be done in the direction, S_{p1}, u_2, S_{p1}

We generate the pattern search direction S_{p1} as follows:

$$S_{p1} = X_i - X_{i-n} = X_4 - X_2 = (-1/2, 1) - (0, 1/2) = (-1/2, 1/2)$$

Now from X_4, we move to X_5 by taking step length λ_4 in the direction of $S_{p1} = (-1/2, 1/2)$.

$$X_5 = X_4 + \lambda_4 S_{p1} = (-1/2, 1) + \lambda_4(-1/2, 1/2) = [-1/2(1 + \lambda_4), 1 + 1/2\,\lambda_4]$$

We now determine λ_4 such that $f(X_5) = 0.25\lambda_4^2 - 0.5\lambda_4 - 1$ is minimum.

For minimum f, $\dfrac{df}{d\lambda_4} = 0.5\lambda_4 - 0.5 = 0 \Rightarrow \lambda_4 = 1$

$\dfrac{d^2 f}{d\lambda_4^2} = 0.5 > 0 \Rightarrow f$ is minimum, where $\lambda_4 = 1$.

$\therefore X_5 = X_4 + \lambda_4 S_{p1} = (-1/2, 1) + (-1/2)(1/2, 1/2) = (-1, 3/2)$ and $f(X_5) = -1.25 < f(X_4)$.

Now from X_5, we move to X_6 by taking step length λ_5 in the direction of u_2 (0, 1).

$$X_6 = X_5 + \lambda_5 u_2 = (-1, 3/2) + \lambda_5(0, 1) = (-1, 3/2 + \lambda_5).$$

We determine λ_5 such that $f(X_6) = f(-1, 3/2 + \lambda_5) = \lambda_5^2 - (5/4)$ is minimum.

For minimum f, $\dfrac{df}{d\lambda_5} = 2\lambda_5 = 0 \Rightarrow \lambda_5 = 0.$

$\therefore X_6 = (-1, 3/2) = X_5.$

($\lambda_5 = 0$) shows that f cannot be minimized in the direction of u_2 and there is no other direction to move. Thus, min $f = -1.25$, when $x_1 = -1, x_2 = 3/2$.

23.5 INDIRECT SEARCH METHODS

In direct search method, only function values are used; it is therefore suitable when functions are not differentiable. However if the gradient information is available, a gradient-based method may be more efficient. Indirect search methods use the derivatives along with determining the value of function at the search point. These methods, therefore, are also called 'gradient methods'.

23.5.1 Steepest Descent or Cauchy's Method

In this method, we start from an initial point called the base point and the movement is made in the steepest direction.

To solve a problem by the steepest descent method, we use the following steps:

1. Let X_1 be the starting point and $f(x_1, x_2)$ be the function to be minimized. A function decreases most rapidly in the negative direction of gradient. Calculate $f(X_1)$ and $S_1 = -\nabla f(X_1)$ (i.e., negative of gradient f at X_1).
2. Find the optimum step length λ_1 in this direction to arrive at the point $X_2 = X_1 + S_1\lambda_1$ so that $f(X_2) < f(X_1)$. Proceed this way until one of the following conditions is satisfied and terminate the process.

   (a) $\left|\dfrac{\partial f}{\partial x_i}\right| < \in$, $\in$ is a small positive number.

   (b) $|X_{i+1} - X_i| < \in$, that is, the change in design vectors in the consecutive iterations is small.

   (c) The relative change in the value of f at two consecutive steps is small, that is, $\left|\dfrac{f(X_{i+1}) - f(X_i)}{f(X_i)}\right| < \in$.

Example 23.7 Minimize $f(x_1, x_2) = 2x_1^2 + x_2^2 + 2x_1x_2 + x_1 - x_2$, starting from the point $X_1 = (0, 0)$ and using the steepest descent method, taking $\in = 0.01$.

Solution Given: $f(x_1, x_2) = 2x_1^2 + x_2^2 + 2x_1x_2 + x_1 - x_2$

$$\therefore \ \nabla f(X) = \left(\frac{\partial f}{\partial x_1}, \frac{\partial f}{\partial x_2}\right) = (4x_1 + 2x_2 + 1, 2x_2 + 2x_1 - 1)$$

Initial step: Given $X_1 = (0, 0)$ and hence $-\nabla f(X_1) = (-1, 1)$.

Iteration 1 We find λ_1 which minimizes

$$f[X_1 - \lambda_1 \nabla f(X_1)] = f[(0, 0) + \lambda_1(-1, 1)] = f(-\lambda_1, \lambda_1)$$
$$= 2\lambda_1^2 + \lambda_1^2 - 2\lambda_1^2 - \lambda_1 - \lambda_1 = \lambda_1^2 - 2\lambda_1$$

For minimum of f, $\dfrac{\partial f}{\partial \lambda_1} = 2\lambda_1 - 2 = 0 \Rightarrow \lambda_1 = 1$ and $\dfrac{\partial^2 f}{\partial \lambda_1^2} = 2 > 0$ (minimum).

Thus, the new point $X_2 = X_1 - \lambda_1 \nabla f(X_1) = (-1, 1)$.

$$\nabla f(X_2) = (-1, -1).$$

$$|\nabla f(X_2)| \not< \in \Rightarrow X_2 \text{ is not an optimal point.}$$

Iteration 2 Now we find λ_2, which minimizes

$$f[X_3] = f[X_2 - \lambda_2 \nabla f(X_2)] = f[(-1, 1) - \lambda_2(-1, -1)] = f(-1 + \lambda_2, 1 + \lambda_2)$$
$$\text{or } f(X_3) = 5\lambda_2^2 - 2\lambda_2 - 1$$

Now, we find λ_2, which minimizes $f(X_3)$ as a function of λ_2.

For minimum, $\dfrac{\partial f}{\partial \lambda_2} = 10\lambda_2 - 2 = 0 \Rightarrow \lambda_2 = 1/5$ and $\dfrac{\partial^2 f}{\partial \lambda_2^2} = 10 > 0$ (minimum)

$X_3 = X_2 - \lambda_2 \nabla f(X_2) = (-1 + (1/5), 1 + (1/5)) = (-0.8, 1.2)$.
$\nabla f(X_3) = (0.2, -0.2)$ and $|\nabla f(X_3)| \nless 0.01 \Rightarrow X_3$ is not an optimal point.

Iteration 3 Next we find λ_3, which minimizes

$$f[X_4] = f[X_3 - \lambda_3 \nabla f(X_3)] = f[(-0.8, 1.2) - \lambda_3(0.2, -0.2)]$$
$$= f(-0.8 - 0.2\lambda_3, 1.2 + 0.2\lambda_3)$$
$$\text{or } f(X_4) = 0.04\lambda_3^2 - 0.08\lambda_3 - 1.2$$

Now we find λ_3, which minimizes $f(X_3)$ as a function of λ_3.

For minimum, $\dfrac{\partial f}{\partial \lambda_3} = 0.08\lambda_3 - 0.08 = 0 \Rightarrow \lambda_3 = 1$ and $\dfrac{\partial^2 f}{\partial \lambda_3^2} = 0.08 > 0$ (minimum).

$X_4 = X_3 - \lambda_3 \nabla f(X_3) = (-1, 1.4)$ and $f(X_4) = -1.24$
$\nabla f(X_4) = (-0.2, -0.2)$ and $|\nabla f(X_4)| \nless 0.01 \Rightarrow X_4$ is not an optimal point.

Iteration 4 Further, we find λ_4 such that $f(X_5) = f[(X_4 - \lambda_4 \nabla f(X_4)]$ is minimum.
For $f(X_5) = f[X_4 - \lambda_4 \nabla f(X_4)] = 0.2\lambda_4^2 - 0.08\lambda_4 - 1.24$ minimum,

For minimum, $\dfrac{\partial f}{\partial \lambda_4} = 0.4\lambda_4 - 0.08 = 0 \Rightarrow \lambda_4 = 0.2$ and $\dfrac{\partial^2 f}{\partial \lambda_4^2} = 0.4 > 0$ (minimum)

$\therefore X_5 = X_4 - \lambda_4 \nabla f(X_4) = (-1, 1.4) - 0.2(-0.2, -0.2) = (-0.96, 1.44)$ and $\nabla f(X_4) = (0.04, -0.04)$ and $|\nabla f(X_5)| \nless 0.01 \Rightarrow X_5$ is not an optimal point.

Iteration 5 Minimize $f(X_6) = f(X_5 - \lambda_5 \nabla f(X_5) = 0.0016\lambda_5^2 - 0.0032\lambda_5 - 1.248$

$\dfrac{\partial f}{\partial \lambda_5} = 0.0032\lambda_5 - 0.0032 = 0 \Rightarrow \lambda_5 = 1$ and $\dfrac{\partial^2 f}{\partial \lambda_5^2} = 0.0032 > 0$ and hence minimum

$X_6 = X_5 - \lambda_5 \nabla f(X_5) = (-1, 1.48)$ and $\nabla f(X_6) = (-0.04, 0.04)$
$|\nabla f(X_6)| \nless 0.01 \Rightarrow X_6$ is not an optimal point.

Iteration 6 Minimize $f(X_7) = f(X_6) - \lambda_6 \nabla f(X_6) = 0.0016\lambda_6^2 - 1.2496$

$\dfrac{\partial f}{\partial \lambda_6} = 0.0032\lambda_6 = 0 \Rightarrow \lambda_6 = 0$ and $\dfrac{\partial^2 f}{\partial \lambda_6^2} = 0.0032 > 0 \Rightarrow f(X_7)$ is minimum.

With $X_7 = f(X_6) - \lambda_6 \nabla f(X_6) = (-1, 1.48)$
and $\lambda_6 = 0 \Rightarrow$ further improvement in f is not possible. Alternatively $X_6 = X_7 = (-1, 1.48)$
$\Rightarrow |X_7 - X_6| = (0, 0) < \in = (0.01, 0.01)$.
Thus, $\min f = f(X_6) = f(X_7) = -1.2496 \approx -1.25$, when $x_1 = -1$ and $x_2 = 1.48$.

23.6 CONSTRAINED NON-LINEAR PROGRAMMING PROBLEMS

To determine the solution of a constrained problem, we divide the solution method into two categories:

1. Direct methods 2. Indirect methods

23.7 DIRECT METHODS

In solving NLPP, if the function to be minimized is not differentiable but requires computational work, then direct methods are easy to use. Let us discuss two important direct methods:

1. Complex method
2. Zoutendijk method (Method of feasible direction)

23.7.1 Complex Method

Complex method deals with the constrained optimization problems of the following type:

$$\text{Minimize } Z = f(X) \tag{23.16}$$

$$\text{Subject to } g_i(X) \leq 0, j = 1, 2, 3, \ldots, m \tag{23.17}$$

$$X = (x_1, x_2, \ldots, x_n)^{\mathrm{T}} \text{ and } x_i^{(l)} \leq x_i \leq x_i^{(u)}, i = 1, 2, \ldots, n \tag{23.18}$$

$x_i^{(l)}$ = lower bound on x_i and $x_i^{(u)}$ = upper bound on x_i. The conditions in Eq. (23.18) are called the *side constraints*. For a minimization problem in n variables, if we consider k points where $k \leq n + 1$, then the figure formed on joining them is known as a complex. We shall consider the minimization problem in two variables. In two dimensions, it will have 4 ($n = 2$) vertices. In any dimension n, we will have $2n$ as an even number. Therefore when $n = 2$, k should be equal to 4, and hence take $k = 2n = 4$. These four points will form the vertices of the complex.

Working Procedure for Two Variables X_1 and X_2

The following steps are involved in solving problems comprising two variables:

1. In complex method, one point, X_1, is given and the remaining $(2n - 1) = 3$ ($n = 2$ here) points, X_2, X_3, and X_4 are obtained one at a time by using random members $r_{i,j}$, where $0 < r_{i,j} < 1$. Calculate $x_{i,j} = x_i^l + r_{i,j}[x_i^{(u)} - x_i^{(l)}]$, where $x_{i,j}$ = ith component of the point X_j, $i = 1, 2$ and $j = 2, 3, 4$. It is worth noticing that points X_2, X_3, and X_4 thus generated satisfy the side constraints in Eq. (23.18) but may not satisfy all the constraints in Eq. (23.17). In case, say, X_4, is not satisfying all constraints in Eq. (23.17), then a new point denoted by $X_4^{(1)}$ is obtained by moving X_4 halfway towards (in the direction of) the centroid $X_0 = (X_1 + X_2 + X_3)/3$ of the remaining points X_1, X_2, X_3, that is, $X_4^{(1)} = (X_0 + X_j)/2$. We check whether $X_4^{(1)}$ satisfies all constraints in Eq. (23.17) or not. If not, another new point $X_4^{(2)}$ is obtained by moving $X_4^{(1)}$ halfway towards the centroid X_0. We proceed this way until a feasible point of X_4 satisfying Eq. (23.17) is obtained. In this way, four feasible points X_1, X_2, X_3, and X_4 all satisfying conditions Eqs (23.17) and (23.18) are obtained, and they form the vertices of the starting complex.
2. Calculate $f(X_1)$, $f(X_2)$, $f(X_3)$, and $f(X_4)$ and mark the largest and the least value obtained. Let $f(X)$ be the largest at X_n and the smallest at X_l. Then the process of reflection is used to find a new point, $X_r = (1 + \alpha)X_0 - \alpha X_h$, where $\alpha \leq r$ and X_0 is the centroid of all vertices except X_h, that is, $X_0 = (X_1 + X_2 + X_3)/3$. Now, check the feasibility of X_r.

   (a) If X_r is feasible and $f(X_r) < f(X_h)$, then X_h is replaced by X_r and carry-on step 2.

   (b) If $f(X_r) \leq f(X_h)$, a new trial point X_r is found by taking a new value of α = (previous value of α)/2 and is tested further until the condition $f(X_r) < f(X_h)$ is satisfied and this way, the value of α will become smaller and smaller.

(c) If value of X_r does not satisfy $f(X_r) < f(X_h)$ in any way, neglect the whole reflection process and start a new reflection process by taking X_h, which gives the second largest value of the function. The procedure ends when the distance between any two vertices among X_1, X_2, X_3, and X_4 becomes smaller than the prescribed value of ε.

Example 23.8 Minimize $f(X) = f(x_1, x_2) = (x_1 - 1)^2 + (x_2 - 1.5)^2 - 0.25$,
Subject to $x_1 + x_2 \le 4$, $0 \le x_1 \le 2$ and $1 \le x_2 \le 3$ by the complex method with $X_1 = \begin{pmatrix} 0.7 \\ 1.1 \end{pmatrix}$.

Solution Here the constraint is $g(X) = x_1 + x_2 - 4 \le 0$ and the side constraints are $0 \le x_1 \le 2$ and $1 \le x_2 \le 3$.

Consider four vertices X_1, X_2, X_3, and X_4 of the complex with $X_1 = \begin{pmatrix} 0.7 \\ 1.1 \end{pmatrix}$.
Let us choose the random numbers

$$r_{1,2} = 0.4,\ r_{1,3} = 0.6,\ r_{1,4} = 0.8$$
$$r_{2,2} = 0.5,\ r_{2,3} = 0.7,\ r_{2,4} = 0.9$$

to determine the remaining three vertices X_2, X_3, and X_4 using the following formula:

$$x_{i,j} = x_i^{(l)} + r_{i,j}\,[x_i^{(u)} - x_i^{(l)}],\ i = 1, 2;\ j = 1, 2, 3, 4.$$

where $x_{i,j}$ is the ith component of vertex X_j.
Here $x_1^{(l)} = 0$, $x_1^{(u)} = 2$ and $x_2^{(l)} = 1$, $x_2^{(u)} = 3$.
Therefore,

$$\begin{aligned} x_{1,2} &= x_1^{(l)} + r_{1,2}\,[x_1^{(u)} - x_1^{(l)}] = 0.8 \\ x_{1,3} &= x_1^{(l)} + r_{1,3}\,[x_1^{(u)} - x_1^{(l)}] = 1.2 \\ x_{1,4} &= x_1^{(l)} + r_{1,4}\,[x_1^{(u)} - x_1^{(l)}] = 1.6 \\ x_{2,2} &= x_2^{(l)} + r_{2,2}\,[x_2^{(u)} - x_2^{(l)}] = 2.0 \\ x_{2,3} &= x_2^{(l)} + r_{2,3}\,[x_2^{(u)} - x_2^{(l)}] = 2.4 \\ x_{2,4} &= x_2^{(l)} + r_{2,4}\,[x_2^{(u)} - x_2^{(l)}] = 2.8 \end{aligned}$$

Thus, the first simplex consist of the vertices:

$$X_1 = \begin{pmatrix} 0.7 \\ 1.1 \end{pmatrix},\ X_2 = \begin{pmatrix} 0.8 \\ 2 \end{pmatrix},\ X_3 = \begin{pmatrix} 1.2 \\ 2.4 \end{pmatrix},\ \text{and } X_4 = \begin{pmatrix} 1.6 \\ 2.8 \end{pmatrix},$$

$\therefore\ g(X_1) = x_1 + x_2 = 0.7 + 1.1 = 1.8 \le 4$
Similarly $g(X_2) = 2.8 \le 4$, $g(X_3) = 3.6 \le 4$, and $g(X_4) = 4.8 > 4$.
$\therefore\ g(x)$ is not satisfied at vertex X_4 and hence it is replaced by some point in the feasible region.

X_0, the centroid of satisfying vertices is as follows:

$$X_0 = \frac{X_1 + X_2 + X_3}{3} = \begin{pmatrix} 0.9 \\ 1.83 \end{pmatrix}$$

Thus, the new X_4 vertex is $X_4 = X_4^{(1)} = \dfrac{X_0 + X_4}{2} = \begin{pmatrix} 1.25 \\ 2.315 \end{pmatrix}$ and $g(X_4^{(1)}) = 3.565 \le 4$.

Thus, $X_4^{(1)}$ lies in the feasible region and the initial complex has the vertices X_1, X_2, X_3, and $X_4 = X_4^{(1)}$ with

$$X_1 = \begin{pmatrix} 0.7 \\ 1.1 \end{pmatrix},\ X_2 = \begin{pmatrix} 0.8 \\ 2 \end{pmatrix},\ X_3 = \begin{pmatrix} 1.2 \\ 2.4 \end{pmatrix},\ \text{and } X_4 = X_4^{(1)} = \begin{pmatrix} 1.25 \\ 2.315 \end{pmatrix},$$

and $f(X_1) = 0$, $f(X_2) = 0.04$, $f(X_3) = 0.60$, and $f(X_4^{(1)}) = 0.4767$.

$f(X_3) = 0.60$ gives the maximum value and $f(X_1) = 0$ is the minimum value
So take $X_3 = X_h$ with $f(X_h) = 0.6$ and $X_1 = X_l$, $f(X_l) = 0$
The new centroid X_0 is obtained as follows:

$$X_0 = \frac{X_1 + X_2 + X_4^{(1)}}{3} = \begin{pmatrix} 0.917 \\ 1.805 \end{pmatrix} \text{ and } f(X_0) = -0.15.$$

$f(X_0) < f(X_h) \Rightarrow f(X)$ is decreasing from X_h $(=X_3)$ toward X_0. So let us find X_r using the reflection process as $X_r = (1 + \alpha)X_0 - \alpha X_h$

$$\text{Take } \alpha = 1, X_r = 2X_0 - X_h \begin{pmatrix} 0.634 \\ 1.21 \end{pmatrix} \text{ and } f(X_r) = -0.034944.$$

As X_r is feasible and $f(X_r) < f(X_h)$, we proceed further by considering the different values of α as follows:

1. Taking $\alpha = 0.1$, we have
$$X_r^{(1)} = 1.1X_0 - 0.1X_h = (1.1)\begin{pmatrix} 0.917 \\ 1.805 \end{pmatrix} - 0.1\begin{pmatrix} 1.2 \\ 2.4 \end{pmatrix} = \begin{pmatrix} 0.8887 \\ 1.7455 \end{pmatrix}$$
and $f(X_r^{(1)}) = -0.1773$.

2. Taking $\alpha = 0.2$, we have
$$X_r^{(2)} = 1.2X_0 - 0.2X_h = (1.2)\begin{pmatrix} 0.917 \\ 1.805 \end{pmatrix} - 0.2\begin{pmatrix} 1.2 \\ 2.4 \end{pmatrix} = \begin{pmatrix} 0.8604 \\ 1.686 \end{pmatrix}$$
and $f(X_r^{(2)}) = -0.1959$.

3. Taking $\alpha = 0.3$, we have $X_r^{(3)} = 1.3X_0 - 0.3X_h = (1.3)\begin{pmatrix} 0.917 \\ 1.805 \end{pmatrix} - 0.3\begin{pmatrix} 1.2 \\ 2.4 \end{pmatrix} = \begin{pmatrix} 0.8321 \\ 1.6265 \end{pmatrix}$
and $f(X_r^{(3)}) = -0.2058$.

4. Taking $\alpha = 0.4$, we have
$$X_r^{(4)} = 1.4X_0 - 0.4X_h = (1.4)\begin{pmatrix} 0.917 \\ 1.805 \end{pmatrix} - 0.4\begin{pmatrix} 1.2 \\ 2.4 \end{pmatrix} = \begin{pmatrix} 0.8038 \\ 1.567 \end{pmatrix}$$
and $f(X_r^{(4)}) = -0.2070$.

5. Taking $\alpha = 0.5$, we have
$$X_r^{(5)} = 1.5X_0 - 0.5X_h = (1.5)\begin{pmatrix} 0.917 \\ 1.805 \end{pmatrix} - 0.5\begin{pmatrix} 1.2 \\ 2.4 \end{pmatrix} = \begin{pmatrix} 0.7755 \\ 1.5075 \end{pmatrix}$$
and $f(X_r^{(5)}) = -0.1995$.

Since the values of f continue to decrease up to $X_r^{(4)}$, replace the vertex $X_h = X_3$ with the highest values by $X_r^{(4)}$ to obtain the new complex with vertices.

$$X_1 = \begin{pmatrix} 0.7 \\ 1.1 \end{pmatrix} \text{ and } f(X_1) = 0$$

$$X_2 = \begin{pmatrix} 0.8 \\ 1.1 \end{pmatrix} \text{ and } f(X_2) = 0.04$$

$$X_3 = X_r^{(4)} = \begin{pmatrix} 0.8038 \\ 1.567 \end{pmatrix} \text{ with } f(X_3) = f(X_r^{(4)}) = -0.207016$$

$$\text{and } X_4 = \begin{pmatrix} 1.25 \\ 2.315 \end{pmatrix} \text{ with } f(X_4) = f(X_4^{(1)}) = 0.476725$$

Thus, $f(X_4) = f(X_4^{(1)}) = 0.476725$ gives the maximum values and $f(X_3) = f(X_r^{(4)}) = -0.207016$ provides the minimum value.

$$\text{So } X_h = X_4 = \begin{pmatrix} 1.25 \\ 2.315 \end{pmatrix} \text{ with } f(X_4) = 0.476725$$

and $X_l = X_3 = X_r^{(4)} = \begin{pmatrix} 0.8038 \\ 1.567 \end{pmatrix}$ with $f(X_3) = f(X_l) = -0.207016$

The centroid $X_0 = (X_1 + X_2 + X_3)/3 = \begin{pmatrix} 0.7679 \\ 1.5557 \end{pmatrix}$ with $f(X_0) = -0.19303$

$|f(X_l) - f(X_0)| = 0.014.$

Thus, if the desired accuracy is $\varepsilon = 0.01$, then the aforementioned solution is accepted and therefore

Min $f(X) = -0.2070$ at X_3, that is, when $x_1 = 0.8038$ and $x_2 = 1.5557$.

23.7.2 Zoutendijk Method or Method of Feasible Direction

Consider the problem

Minimize $f(X)$ (23.19)

Subject to $g_j(X) \le 0, j = 1, 2, \ldots, m \ldots$ (23.20)

$X = (x_1, x_2, \ldots, x_m)^T$

In the method of feasible direction, choose the starting point X_i ($i = 1$ for starting point), which satisfies all the constraints in Eq. (23.20) and move to a better point X_{i+1} (the point where value of the function in Eq. (23.19) is lesser than that it has at X_i) as per the iterative formula given here:

$X_{i+1} = X_i + \lambda_i S_i$

where X_i = Starting point for the ith iteration

S_i = Direction in which to move

λ_i = Length of the step to be taken in the direction of S_i

X_{i+1} = Point obtained at the end of the ith iteration

Care should be taken to choose λ_i such that the new point X_{i+1} is in a feasible region. Further the search direction is found, such that

1. even a very small movement in that direction violates none of the constraints in Eq. (23.20) and
2. the value of the function $f(X)$ in Eq. (23.19) decreases.

Feasible direction A direction S which satisfies condition (1) is called feasible direction. One way to find feasible direction is given here: If at X_i, $\left\{\frac{d}{d\lambda}\left[g_j(X_i + \lambda_i S_i)\right]\right\}_{\lambda_i=0} = S_i^T \nabla g_j(X_i) \le 0$ then, the direction S_i is the feasible direction.

Usable feasible direction A direction S_i, which satisfies both the conditions given in (1) and (2) is called usable feasible direction. Further, a direction S_i will be a usable feasible direction if

(a) $\left\{\frac{d}{d\lambda}\left[f(X_i + \lambda_i S_i)\right]\right\}_{\lambda_i=0} = S_i^T \nabla f(X_i) \prec 0$ (23.21)

(b) $\left\{\frac{d}{d\lambda}\left[g_j(X_i + \lambda_i S_i)\right]\right\}_{\lambda_i=0} = S_i^T \nabla g_j(X_i) \le 0$ (23.22)

We can reduce the value of f by taking step lengths in such a direction S_i as described here. The process terminates at X_i if $\nabla f(X_i) = 0$. Zoutendijk method is a method of feasible directions.

Example 23.9 Using the method of feasible directions (Zoutendijk method)
Minimize $f(X) = x_1^2 + x_2^2 - 2x_1 - 3x_2 + 3$
Subject to $g_1(X) = x_1 + x_2 \le 4$ taking the starting point $X_1 = \begin{pmatrix} 0 \\ 0 \end{pmatrix}$.

Solution Given that $X_1 = \begin{pmatrix} 0 \\ 0 \end{pmatrix}$, $f(X) = x_1^2 + x_2^2 - 2x_1 - 3x_2 + 3$, $x_1 + x_2 \le 4$

Thus, $g_1(X) = x_1 + x_2 - 4$. Then, $f(X_1) = 3$ and $g(X_1) = -4 < 0$.

1. Here $g(X_1) < 0$ so search direction S_1 is $S_1 = -\nabla f(X_1) = -\begin{Bmatrix} \partial f/\partial x_1 \\ \partial f/\partial x_2 \end{Bmatrix}_{X_1=(0,\,0)} = \begin{pmatrix} 2 \\ 3 \end{pmatrix}$.

   After normalization, we have $S_1 = \begin{pmatrix} 2/3 \\ 1 \end{pmatrix}$.

To obtain a new point X_2, we take a step of length λ_1 in the direction of $-\nabla f(X_1)$ to arrive at $X_2 = X_1 + \lambda_1 S_1 = [(2/3)\lambda_1, \lambda_1]$

$$f(X_2) = f[(2/3)\lambda_1, \lambda_1] = (13/9)\lambda_1^2 - (13/3)\lambda_1 + 3$$

Now, determine λ_1 for minimum f, as follows:

$\frac{\partial f}{\partial \lambda_1} = \frac{26}{9}\lambda_1 - \frac{13}{3} = 0 \Rightarrow \lambda_1 = 1.5$ and $\frac{\partial^2 f}{\partial \lambda_1^2} = \frac{26}{9} \succ 0$. Thus f is minimum when $\lambda_1 = 1.5$ giving $X_2 = (0, 0) + 1.5(2/3, 1) = (1, 1.5)$ and $g(X_2) = -1.5 < 0$.

So, the new search direction S_2 is $S_2 = -\nabla f(X_2) = \begin{Bmatrix} \partial f/\partial x_1 \\ \partial f/\partial x_2 \end{Bmatrix}_{X_1=(1,\,1.5)} = \begin{pmatrix} 0 \\ 0 \end{pmatrix}$ which implies that there is no search direction available to obtain minimum f. Hence minimum $f = -0.25$ at $X = X_2 = (1, 1.5)$, that is, $x_1 = 1$ and $x_2 = 1.5$.

23.8 INDIRECT METHODS

While using the indirect method for solving an NLPP, the function to be minimized must be differentiable. Let us discuss the following three important indirect methods:

1. Transform techniques
2. Penalty function methods
3. Rosen's gradient projection method

23.8.1 Transform Techniques

If constraints $g_i(X)$ in a non-linear optimization problem have certain simple forms in independent variables, then it may be possible to make use of the transformation of these variables in such a way that constraints are automatically satisfied. Thus, it is possible to convert a constrained optimization problem into an unconstrained one using transformation techniques.

Some of the typical transformations of independent variables are as follows:

1. If the lower and upper bounds on x_i are given as follows:

$$l_i \le x_i \le u_i, \tag{23.23}$$

then these can be satisfied by transforming the variable x_i as follows:

$$x_i = l_i + (u_i - l_i)\sin^2 y_i \tag{23.24}$$

where y_i is the new variable, which can take any value

2. If $x_i \in (0,1)$, then one of the following transformations can be used:

$$\text{(a) } x_i = \sin^2 y_i \quad \text{(b) } x_i = \cos^2 y_i \quad \text{(c) } x_i = \frac{e^{y_i}}{e^{y_i} + e^{-y_i}} \quad \text{(d) } x_i = \frac{y_i^2}{1 + y_i^2} \tag{23.25}$$

3. If the design variable is restricted to assume only positive values, then one of the following transformations can be used:

$$\text{(a) } x_i = |y_i|, \text{ (b) } x_i = y_i^2, \text{ or (c) } x_i = e^{yi} \tag{23.26}$$

4. If $x_i \in (-1, 1)$, we can use

$$\text{(a) } x_i = \sin y_i, \text{ (b) } x_i = \cos y_i, \text{ or (c) } x_i = \frac{2y_i}{1 + y_i^2} \tag{23.27}$$

Note that, to use the transformed transformations, the constrained function $g_i(X)$ should be simple. In case, it is not possible to eliminate all the constraints by changing the variables, then it is better not to use the transformation method.

Example 23.10 A courier service does not accept rectangular packets of more than 42 cm length. If the maximum [length + 2(width + height)] is 72 cms, compute the maximum volume of the rectangular packet.

Solution Let x_1, x_2, and x_3 be the length, width, and height of the rectangular packets, respectively. Then the formulation of the problem is as follows:

$$\text{Max } f(X) = x_1 x_2 x_3 \tag{23.28}$$

$$\text{Subject to } \quad x_1 + 2x_2 + 2x_3 \le 72 \tag{23.29}$$

$$x_1 \le 42 \tag{23.30}$$

$$x_1, x_2, x_3 \ge 0 \tag{23.31}$$

Let us transform x_i's into y_i's by taking $y_1 = x_1$, $y_2 = x_2$, and $y_3 = x_1 + 2x_2 + 2x_3$

These imply that $x_3 = (1/2)(y_3 - y_1 - 2y_2)$ (23.32)

Thus, constraints in Eqs (23.29), (23.30), and (23.31) can be written as follows:

$$0 \le y_1 \le 42$$

$$0 \le y_2 \le 36$$

$$\text{and} \quad 0 \le y_3 \le 72 \tag{23.33}$$

The upper bound for y_i's in Eq. (23.33) can easily be obtained (say for y_2, taking $x_1 = x_3 = 0$ in Eq. (23.29), we get $2x_2 \le 72$. Thus, $x_2 = y_2 \le 36$).

The constrains in Eq. (23.33) are automatically satisfied if we define z_1, z_2, and z_3 as follows:

$$y_i = l_i + (u_i - l_i)\sin^2 z_i \tag{23.34}$$

Using Eqs (23.33) and (23.34), we have

$$y_1 = l_1 + (u_1 - l_1)\sin^2 z_1 = 42\sin^2 z_1 \tag{23.35}$$

$$y_2 = l_2 + (u_2 - l_2)\sin^2 z_2 = 36\sin^2 z_2 \tag{23.36}$$

$$y_3 = l_3 + (u_3 - l_3)\sin^2 z_3 = 72\sin^2 z_3 \tag{23.37}$$

So, the original problem reduces to the following:

$$\text{Max } f = (1/2)\, y_1 y_2\, (y_3 - y_1 - 2y_2)$$

$$= (1/2)\,(42\sin^2 z_1)\,(36\sin^2 z_2)(72\sin^2 z_3 - 42\sin^2 z_1 - 72\sin^2 z_2)$$

Subject to $0 \le \sin^2 z_i \le 1$, $i = 1, 2, 3$ (23.38)

For minimum f,

$$\frac{\partial f}{\partial z_1} = 0 \Rightarrow 1512 \sin z_1 \cos z_1 \sin^2 z_2 \left(\sin^2 z_3 - 7/6 \sin^2 z_1 - 72 \sin^2 z_2 \right) = 0 \quad (23.39)$$

$$\frac{\partial f}{\partial z_2} = 0 \Rightarrow 108864 \sin^2 z_1 \sin z_2 \cos z_2 \left(\sin^2 z_3 - 7/12 \sin^2 z_1 - 2 \sin^2 z_2 \right) = 0 \quad (23.40)$$

$$\frac{\partial f}{\partial z_3} = 0 \Rightarrow 108864 \sin^2 z_1 \sin^2 z_2 \sin z_3 \cos z_3 = 0 \quad (23.41)$$

From Eq. (23.40), we have

$\sin^2 z_1 = 0 \Rightarrow z_1 = 0 \Rightarrow y_1 = 0 \Rightarrow f = 0$, which is not acceptable.

$\sin^2 z_2 = 0 \Rightarrow z_2 = 0 \Rightarrow y_2 = 0 \Rightarrow f = 0$, which is not acceptable.

If $\sin z_3 = 0 \Rightarrow z_3 = 0 \Rightarrow y_3 = 0 \Rightarrow f$ is negative, which is not acceptable.

If $\cos z_3 = 0 \Rightarrow z_3 = \pi/2$ (23.42)

From Eq. (23.39), $\sin^2 z_3 - 7/6 \sin^2 z_1 - 72 \sin^2 z_2 = 0 \Rightarrow 7 \sin^2 z_1 + 6 \sin^2 z_2 = 6$ (23.43)

From Eq. (23.40), $\sin^2 z_3 - 7/12 \sin^2 z_1 - 2 \sin^2 z_2 = 0 \Rightarrow 7 \sin^2 z_1 + 24 \sin^2 z_2 = 12$ (23.44)

Subtract Eq. (23.43) from Eq. (23.44) $\Rightarrow 18 \sin^2 z_2 = 6 \Rightarrow \sin^2 z_2 = 1/3$.

From Eq. (23.43), $7 \sin^2 z_1 + 6(1/3) = 6$ $\sin^2 z_1 = 4/7$.

Hence, $y_1 = 42\ (4/7) = 24$, $y_2 = 36(1/3) = 12$, and $y_3 = 72$.

$\therefore x_1 = 24$, $x_2 = 12$, $x_3 = 12$, and Max $f = 3456\ \text{cm}^3$.

23.8.2 Penalty Function Methods

$$\left.\begin{aligned} &\text{Consider the problem Min } f(X) \\ &\text{Subject to } \quad g_j(X) \le 0,\ j = 1, 2, \ldots, m \\ &h_i(X) = 0,\ i = 1, 2, \ldots, p,\ X \in E_n \end{aligned}\right\} \quad (23.45)$$

Penalty method transforms problem Eq. (23.45) into a sequence of problems, each with no constraint. To solve (23.45), penalty method creates the effect of constraints by bringing in the modification in the objective function, as done in the Big-M method of artificial variable technique in the simplex method. Thus in order to solve the constrained optimization problem given in Eq. (23.45), we introduce an auxiliary unconstrained function

$$F(X, r_k) = f(X) + P(X, r_k) \quad (23.46)$$

where $P(X, r_k)$ is a function of constraints $g_j(X)$ and $h_i(X)$ and r_k is a positive parameter such that $\lim_{r_k \to 0} \min F(X, r_k) = \min f(X)$.

Formulation of Penalty Function

On the basis of the formulation of a penalty function, the penalty function method is divided into two categories:

1. The interior penalty method 2. The exterior penalty method

In the interior penalty method, the form of penalty function $P(X, r_k)$, which is a function of constraint functions $g_j(X)$ and $h_i(X)$ and a positive parameter r_k is

$$P(X, r_k) = -r_k \sum_{j=1}^{m} \frac{1}{g_j(X)} + \frac{1}{\sqrt{r_k}} \sum_{i=1}^{p} h_i^2(X) \tag{23.47}$$

In the interior penalty method, the minimum of the auxiliary function $F(X, r_k)$ as defined in Eq. (23.46) approaches the minimum of the objective function $f(X)$ from points inside the feasible region as r_k tends to zero.

In the exterior penalty method, we take the form of penalty function $P(X, r_k)$ as

$$P(X, r_k) = \frac{1}{r_k} \sum_{j=1}^{m} \left[\max\left(0, g_j(X)\right)\right]^2 + \frac{1}{r_k} \sum_{i=1}^{p} h_i^2(X) \tag{23.48}$$

and minimize the auxiliary function $F(X, r_k)$ for a sequence of decreasing values of r_k, that is, the minimum of $F(X, r_k)$ approaches to minimum of $f(X)$ from the points of infeasible region as r_k tends to be zero.

We illustrate the method through the following examples:

Example 23.11 Minimize $f(X) = f(x_1, x_2) = (x_1 + 2)^3 + 3x_2 + 1$

Subject to $x_1 \geq 2, x_2 \geq 0$

Solution The given problem can be rewritten as follows:

$f(X) = (x_1 + 2)^3 + 3x_2 + 1$

Subject to $2 - x_1 \leq 0, -x_2 \leq 0.$

Hence, the auxiliary unconstrained problem becomes

$$\text{Minimize } F(X, r_k) = (x_1 + 2)^3 + 3x_2 + 1 - r_k \left(\frac{1}{2 - x_1} - \frac{1}{x_2}\right).$$

For minimization of F,

$$\frac{\partial F}{\partial x_1} = 3(x_1 + 2)^2 - \frac{r_k}{(2 - x_1)^2} = 0 \Rightarrow \left(x_1^2 - 4\right)^2 = \frac{r_k}{3} \quad \text{or} \quad x_1 = \left(4 + \sqrt{\frac{r_k}{3}}\right)^{1/2}$$

$$\frac{\partial F}{\partial x_2} = 3 - \sqrt{\frac{r_k}{x_2^2}} \quad \Rightarrow \quad x_2 = \sqrt{\frac{r_k}{3}}$$

Thus, $x_1 = \left(4 + \sqrt{\frac{r_k}{3}}\right)^{1/2}$ and $x_2 = \sqrt{\frac{r_k}{3}}$ are the possible feasible values of x_1 and x_2.

Now as $r_k \to 0$, $x_1 = 2$ and $x_2 = 0$.

$$\lim_{r_k \to 0} \min F(X, r_k) = \min f(X) = 65, \text{ when } x_1 = 2 \text{ and } x_2 = 0.$$

The solution of the given problem tending to $x_1 = 2$ and $x_2 = 0$ can be seen from the following table as $r_k \to 0$ through four values 10, 0.1, 0.001, and 0.0001 given to r_k for $k = 1, 2, 3$, and 4.

k	r_k	x_1	x_2	$F(X, r_k)$	$f(X)$
1	10	2.414	10.826	117.556	92.478
2	0.1	2.045	0.183	70.508	67.739
3	0.001	2.005	0.018	65.529	65.274
4	0.0001	2.000	0.000	65.000	65.000

The aforegiven table shows how the values of the auxiliary function $F(X, r_k)$ approach the values of the given function $f(X)$ as r_k tends to zero through positive values.

Example 23.12 Minimize $f(X) = 2x$

Subject to $x \geq 3$, using the interior penalty method

Solution The given problem can be written as follows:

Min $f(X) = 2x$

Subject to $g_1(x) = 3 - x \leq 0$

The auxiliary unconstrained problem can be written as follows:

Minimize $F(X, r_k) = 2x - r_k\left(\dfrac{1}{3-x}\right)$ such that $\lim\limits_{r_k \to 0} \min F(X, r_k) = \min f(X)$

For minimization of F,

$$\frac{\partial F}{\partial x} = 2 - \frac{r_k}{(3-x)^2} = 0 \Rightarrow x = 3 + \sqrt{\frac{r_k}{2}},$$

$$\frac{\partial^2 F}{\partial x^2} = \frac{-2r_k}{(3-x)^3} > 0 \text{ at } x = 3 + \sqrt{\frac{r_k}{2}}$$

Now as $r_k \to 0$ at $x = 3$, $\lim\limits_{r_k \to 0} \min F(X, r_k) = \min f(X) = 6$, when $x = 3$.

Note that this is the problem in one variable x.

Example 23.13 Minimize $f(X) = x_1^2 + 2x_2^2$

Subject to $2x_1 + 5x_2 \leq 10$ using the exterior penalty method.

Solution The given problem can be written as follows:

$f(X) = x_1^2 + 2x_2^2$

Subject to $g_1(x) = 2x_1 + 5x_2 - 10 \leq 0$

The auxiliary function $F(X, r_k)$ in exterior penalty method is given as follows:

Minimize $F(X, r_k) = x_1^2 + 2x_2^2 + \dfrac{1}{r_k}[\text{Max}(0, 2x_1 + 5x_2 - 10)]^2$

For minimization of F, $(r = 1/r_k)$

$$\frac{\partial F}{\partial x} = 2x_1 + \frac{4}{r_k}(2x_1 + 5x_2 - 10) = 0 \Rightarrow (2 + 8r)x_1 + 20rx_2 - 40r = 0 \quad (23.49)$$

$$\frac{\partial F}{\partial x_2} = 4x_2 + 10r(2x_1 + 5x_2 - 10) = 0 \Rightarrow 20rx_1 + (4 + 50r)x_2 - 100r = 0 \quad (23.50)$$

From Eqs (23.49) and (23.50), we have

$$\frac{x_1}{160r} = \frac{x_2}{200r} = \frac{1}{8 + 132r} \Rightarrow x_1 = \frac{40r}{2 + 33r} = \frac{40}{2r_k + 33}$$

$$\text{and } x_2 = \frac{50r}{2 + 33r} = \frac{50}{2r_k + 33}$$

Now as $r_k \to 0$, $x_1 = \dfrac{40}{33}$ and $x_2 = \dfrac{50}{33}$

and minimum $f(X) = (40/33)^2 + 2(50/33)^2 = 200/33$.

23.9 ROSEN'S GRADIENT PROJECTION METHOD

Rosen's gradient projection method is an efficient method of solving constrained NLPPs. The method has been described by Rosen for general NLPPs where even the constraints can be non-linear along with the non-linear objective function. However, the method is more effectively used to solve problems wherein all the constraints are linear. To use this method, we consider a problem with linear constraints in Eq. 23.51.

$$\left.\begin{aligned} &\text{Minimize } f(X) \\ &\text{Subject to } g_j(X) = \sum_{i=1}^{n} a_{ij}x_i - b_j \le 0, j = 1, 2, ..., m \\ &\text{where } X = (x_1, x_2, ..., x_n) \end{aligned}\right\} \quad (23.51)$$

First, we provide an explanation for certain terms used in solving the non-linear problem in Eq. (23.51) using Rosen's gradient projection method:

Feasible point The set of all X such that $g_j(X) \le 0$ for all $j = 1, 2, \ldots, m$ is called the set of feasible solutions of Eq. (23.51) and is denoted as S_F. Thus, $S_F = \{X | g_j(X) \le 0, j = 1, 2, \ldots, m\}$. A point $X \in S_F$ is called a feasible point.

Active constraint A constraint $(g_j(X) \le 0)$ is called active at a feasible point X if $g_j(X) = 0$.

Projection matrix If $j_1, j_2, \ldots, j_p$ are the indices of active constraints at a feasible point X, then the $n \times p$ matrix of the gradients of these active constraints $N_p = [\nabla g_{j1}$ $\nabla g_{j2} \ldots \nabla g_{jp}]$, where $\nabla g_j(X) = \begin{Bmatrix} a_{1j} \\ a_{2j} \\ \vdots \\ a_{nj} \end{Bmatrix}$, $j = j_1, j_2, ..., j_p$ and the projection matrix P is defined as $P = I - N[N^{\mathrm{T}}N]^{-1}N^T$, where I is the identity matrix.

To solve an NLPP using Rosen's gradient projection method, we use the following steps:

Step 1 First find out the value $g_j(X_i) < 0$ at X_i for $i = 1$ and $j = 1, 2, \ldots, m$. Here two cases arise:

Case 1 If $g_j(X_i) < 0$ for $j = 1, 2, \ldots, m$, then find the normalized search direction $S_i = \dfrac{-\nabla f(X_i)}{\|\nabla f(X_i)\|}$.

Case 2 If $g_j(X_i) = 0$ for $j = j_1, j_2, \ldots, j_p$, find the normalized search direction $S_i = \dfrac{-P_i \nabla f(X_i)}{\|P_i \nabla f(X_i)\|}$.

where $P_1 = I - N_1[N_1^{\mathrm{T}}N_1]^{-1}N_1^{\mathrm{T}}$ and $N_p = [\nabla g_{j_1}(X_i) \nabla g_{j_2}(X_i) \ldots \nabla g_{j_p}(X_i)]$.

Step 2 Here again, two cases arise:

Case 1 If $S_i = 0$, compute the vector λ at X_i, $\lambda = -[N_p^{\mathrm{T}}N_P]^{-1}N_p^{\mathrm{T}}\nabla f(X_i)$.

If each component of λ is non-negative, then X_i is the optimum value and stop iterations.

If some of the components of λ are negative, then identify the component λ_q with the most negative value and form a new matrix N_p as $[\nabla g_{j_1}(X_i) \nabla g_{j_2}(X_i) \ldots$ $\nabla g_{j_{q-1}}(X_i) \nabla g_{j_{q+1}}(X_i) \ldots \nabla g_{j_p}(X_i)]$ and go to Step 1.

Case 2 If $S_j \neq 0$, then find the maximum step length λ_m where $\lambda_m = \min(\lambda_k)$, $\lambda_k > 0$ and k is any integer between 1 to m other than $j_1, j_2, \ldots, j_p$.

Calculate $\dfrac{df}{d\lambda}$ at $\lambda = \lambda_m$. If at $\lambda = \lambda_m$, $\dfrac{df}{d\lambda} = 0$ or negative, then take the step length $\lambda_i = \lambda_m$.

If at $\lambda = \lambda_m$, $\dfrac{df}{d\lambda}$ is positive, then find the minimum step length $\lambda = \lambda_i$ by substituting $\dfrac{df}{d\lambda} = 0$.

Now, the minimum point will be $X_{i+1} = X_i + \lambda_i S_i$

If $\lambda_i = \lambda_m$ or if $\lambda_m \leq \lambda_i^*$, some new constraints become active at X_{i+1} and hence generate the new matrix N_p to include the gradients of all the active constraints evaluated at X_{i+1}, and fix the new iteration number as $i = i + 1$ and go to Step 1 (Case 2).

If $\lambda_i = \lambda_i^*$ and $\lambda_i^* < \lambda_m$, no new constraint will be active at X_{i+1} and hence the matrix N_p remains unaltered. Fix the new iteration number as $i = i + 1$ and go to Step 1 (Case 1).

Example 23.14 Use the Rosen's gradient projection method to solve the following:

Minimize $f(X) = x_1^2 + x_2^2 - 2x_1 - 4x_2$

Subject to $g_1(x_1, x_2) = x_1 + 4x_2 - 5 \leq 0$

$g_2(x_1, x_2) = 2x_1 + 3x_2 - 6 \leq 0$

$g_3(x_1, x_2) = -x_1 \leq 0$

$g_4(x_1, x_2) = -x_2 \leq 0$

starting from the point $X = \begin{Bmatrix} 1 \\ 1 \end{Bmatrix}$.

Solution Here $g_1(1, 1) = 1 + 4 - 5 = 0$ for $j = 1$, so only the first constraint is active, $p = 1$, $j = 1$.

The normalized search direction $S_i = \dfrac{-P_i \nabla f(X_i)}{\|P_i \nabla f(X_i)\|}$

where $P_i = I - N_p[N_p^T N_p]^{-1} N_p^T$, and $N_p = [\nabla g_{j_1}(X_i)\, \nabla g_{j_2}(X_i) \ldots \nabla g_{j_p}(X_i)]$.

We have $N_1 = [\nabla g_1(X_1)] = \begin{bmatrix} 1 \\ 4 \end{bmatrix}$.

The projection matrix $P_1 = I - N_1[N_1^T N_1]^{-1} N_1^T$

$$= \begin{bmatrix} 1 & 0 \\ 0 & 1 \end{bmatrix} - \begin{bmatrix} 1 \\ 4 \end{bmatrix}\left[\begin{bmatrix} 1 \\ 4 \end{bmatrix}[1\ 4]\right]^{-1}[1\ 4]$$

$$= \begin{bmatrix} 1 & 0 \\ 0 & 1 \end{bmatrix} - \frac{1}{17}\begin{bmatrix} 1 \\ 4 \end{bmatrix}[1\ 4] = \begin{bmatrix} 1 & 0 \\ 0 & 1 \end{bmatrix} - \frac{1}{17}\begin{bmatrix} 1 & 4 \\ 4 & 16 \end{bmatrix}$$

$$= \frac{1}{17}\begin{bmatrix} 16 & -4 \\ -4 & 1 \end{bmatrix}$$

and $\nabla f(X_1) = \begin{Bmatrix} 2x_1 - 2 \\ 2x_2 - 4 \end{Bmatrix}_{(1,1)} = \begin{Bmatrix} 0 \\ -2 \end{Bmatrix}$

Now, we have
$$S_i = \frac{-P_i \nabla f(X_i)}{\|P_i \nabla f(X_i)\|} = \frac{-\dfrac{1}{17}\begin{bmatrix} 16 & -4 \\ -4 & 1 \end{bmatrix}\begin{bmatrix} 0 \\ -2 \end{bmatrix}}{\left\| -\dfrac{1}{17}\begin{bmatrix} 16 & -4 \\ -4 & 1 \end{bmatrix}\begin{bmatrix} 0 \\ -2 \end{bmatrix} \right\|} = \begin{Bmatrix} -0.9701 \\ 0.2425 \end{Bmatrix}.$$

Here $S_1 \neq 0$, so we find λ_m where $\lambda_m = \min(\lambda_k)$, $\lambda_k > 0$ and k is any one of the integers 2, 3, and 4, we set $X = \begin{Bmatrix} x_1 \\ x_2 \end{Bmatrix} = X_1 + \lambda S_1 = \begin{Bmatrix} 1 - 0.9701\lambda \\ 1 + 0.2425\lambda \end{Bmatrix}$

For $j = 2$, $g_2(X) = (2 - 1.9402\lambda) + (3 + 0.7275) - 6 = 0$, at $\lambda = \lambda_2 = -0.8245$.

For $j = 3$, $g_3(X) = -(1 - 0.9701\lambda) = 0$, at $\lambda = \lambda_3 = 1.03$.

For $j = 4$, $g_4(X) = -(1 + 0.2425\lambda) = 0$, at $\lambda = \lambda_4 = -4.124$.

Here λ_3 is the minimum positive value, so we take $\lambda_m = \lambda_3 = 1.03$.

Again, $f(X) = f(\lambda) = (1 - 0.9701\lambda)^2 + (1 + 0.2425\lambda)^2 - 2(1 - 0.9701\lambda) - 4(1 + 0.2425\lambda) = 0.9998\lambda^2 - 0.4850\lambda - 4$.

$$\frac{df}{d\lambda} = -1.9996\lambda - 0.4850$$

At $\lambda = \lambda_m$, $\dfrac{df}{d\lambda} = 1.9996(1.03) - 0.4850 = 1.5746 > 0$, so we compute minimum step length λ_1^* by putting $\dfrac{df}{d\lambda} = 0$. Thus, we get $\lambda_1 = \lambda_1^* = \dfrac{0.4850}{1.9996} = 0.2425$.

Now, the minimum point will be $X_2 = X_1 + \lambda_1 S_1 = \begin{Bmatrix} 1 \\ 1 \end{Bmatrix} + 0.2425 \begin{Bmatrix} -0.9701 \\ 0.2425 \end{Bmatrix} = \begin{Bmatrix} -0.764701 \\ 1.0588 \end{Bmatrix}$.

Here $\lambda_1 = \lambda_1^*$ and $\lambda_1^* < \lambda_m$. No new constraint will be active at X_2 and hence the matrix N_p remains unaltered. Now, fix the new iteration number as $i = 2$.

Since $g_2(X_2) = 0$, we set $p = 1$, $j_1 = 1$, the normalized search direction $S_2 = \dfrac{-P_2 \nabla f(X_2)}{\|P_2 \nabla f(X_2)\|}$.

Here $N_2 = \begin{bmatrix} 1 \\ 4 \end{bmatrix}$, $P_2 = \dfrac{1}{17}\begin{bmatrix} 16 & -4 \\ -4 & 1 \end{bmatrix}$,

$$\nabla f(X_2) = \begin{Bmatrix} 2x_1 - 2 \\ 2x_2 - 4 \end{Bmatrix}_{X_2} = \begin{Bmatrix} -0.4706 \\ -1.8824 \end{Bmatrix}$$

Now, we have $P_2 \nabla f(X_2) = \dfrac{1}{17}\begin{bmatrix} 16 & -4 \\ -4 & 1 \end{bmatrix}\begin{Bmatrix} -0.4706 \\ -1.8824 \end{Bmatrix} = \begin{Bmatrix} 0 \\ 0 \end{Bmatrix}$.

Since $S_2 = 0$, so we compute the step length λ at X_2,

$$\lambda = -[N_p^T N_p]^{-1} N_p^T \nabla f(X_i) = -\frac{1}{17}[1 \quad 4]\begin{Bmatrix} -0.4706 \\ -1.8824 \end{Bmatrix} = 0.4707 > 0.$$

Thus, the non-negative value of λ indicates that we have obtained the optimum value and therefore stop further iterations.

$$X_2 = \begin{Bmatrix} 0.7647 \\ 1.0588 \end{Bmatrix} \text{ and } f = -4.059.$$

RECAPITULATION

1. In direct search methods, derivatives of the function to be minimized are not needed; these methods are suitable for functions that are not differentiable.

(*Contd*)

(Contd)

2. A function $f(x)$ with a unique optimal value (either a unique maxima or minima) in $[a, b]$ is called unimodal function.
3. In Fibonacci method, the initial interval of uncertainty say $[a, b]$ is given. Further, the function to be optimized to a given degree of accuracy must be unimodal in the interval of uncertainty.
4. The golden section method differs from the Fibonacci method in that the total number of experiments to be conducted in the case of the former is unlimited.
5. In univariate method, we move in the axial directions covering all directions taking one direction at a time. This will complete one cycle of iterations. The process will be over when no further improvement in the value of the given function to be minimized is possible in any direction of a cycle. In this method, the rate of convergence is slow.
6. The pattern search method is used to accelerate the rate of convergence. Powell's method being the method of conjugate directions, converges in at the most two cycles of iterations.
7. Indirect search methods use derivatives to find the value of a function at the search point. These methods are also called the gradient methods.
8. In steepest descent method, we start from an initial point called the base point and movement is made in the steepest direction. This method is also called Cauchy's method.
9. In solving NLPP, the function to be minimized need not be differentiable; only computational work is needed. In such conditions, direct methods such as complex method and Zoutendijk method are convenient to use.
10. In using the technique of indirect methods for solving an NLPP, the function to be minimized must be differentiable. In such cases, indirect methods such as transform techniques and penalty function methods are used.
11. Rosen's gradient projection method is effectively used to solve the problems in which the constraints are all linear.

EXERCISES

Multiple-choice Questions

23.1 The method of feasible direction, in solving NLPPs using search technique is
(a) Powell's method
(b) Hooke's and Jeeve's method
(c) Zoutendijk method
(d) Univariate method

23.2 In search technique, a constrained NLPP is transformed into a sequence of problems, each with no constraint using the
(a) complex method
(b) penalty function method
(c) transform techniques
(d) none of these

23.3 Which of the following is an unconstrained search technique or method?
(a) Penalty function method
(b) Univariate method
(c) Transform technique
(d) None of these

23.4 The search technique that uses a gradient of the function to be minimized in solving an NLPP is
(a) transform technique
(b) complex method
(c) steepest descent method
(d) Zoutendijk method

State True or False

23.1 Only a few analytical methods are available to find the exact solution of an NLPP.
23.2 In indirect search methods, the function to be minimized should be differentiable.
23.3 For NLPPs where the function is not differentiable, we use the direct search methods.
23.4 The NLPPs not yielding any analytical solution can be solved using both the direct search as well as descent methods.
23.5 Indirect search methods use the derivatives while determining the value of a function at the search point.
23.6 Indirect search methods are used when the NLPP is constrained in an optimization problem.

Fill in the Blanks

23.1 For LPPs, which cannot be solved using analytical methods, ________ have been devised.
23.2 To solve an NLPP using descent search technique methods, the function to be minimized should be ________.
23.3 In univariate search, the convergence is ________.
23.4 In Hooke and Jeeve's method for an NLPP in two dimensions, after the movements have been made in both the axial directions, the third movement is made in the first ________ direction.
23.5 ________ method is the method of feasible direction.

Review Questions

23.1 Write a short note on the following methods:
(a) Fibonacci-search plan method
(b) Golden section method
(c) Complex method
23.2 What do you mean by feasible directions? Describe in detail the procedure used in the method of feasible direction.
23.3 What is the univariate search technique and how can it be applied to obtain the optimum solution?
23.4 Explain the Hooke and Jeeve's method.
23.5 What do you understand by Powell's method?
23.6 What is the direction of the steepest descent method and how can it be used to obtain the minimum of a non-linear function?
23.7 Write a short note on the penalty method.

Numerical Problems

23.1 Use the Fibonacci method to determine the maximum of $f(x) = x(5-x)$ given that $f(x)$ is a unimodal function in the interval [0, 8] wherein the maximum lies.
23.2 Minimize the function $f(x) = 0.65 - \dfrac{0.75}{1+x^2} - 0.65x\tan^{-1}(1/x)$ using the golden section method in the interval [0, 8] with $n = 6$.
23.3 Minimize $f(X) = f(x_1, x_2) = x_1^2 + x_2^2 - 2x_1 - 4x_2$ using the univariate method considering $X_1 = \begin{pmatrix} 0 \\ 0 \end{pmatrix}$ as the starting (base) point.
23.4 Minimize $f(X) = f(x_1, x_2) = x_1^2 + 3x_2^2 + x_1 - 17x_2 - x_1x_2$ using the univariate method taking $X_1 = \begin{pmatrix} 0 \\ 0 \end{pmatrix}$ as the starting (base) point.

23.5 Minimize $f(X) = f(x_1, x_2) = 2x_1^2 + x_2^2 + x_1 - x_2 + 2x_1x_2$ using the univariate method taking $X_1 = \begin{pmatrix} 0 \\ 0 \end{pmatrix}$ as the starting (base) point.

23.6 Minimize $f(X) = 3x_1^2 + x_2^2 - 2x_1x_2 - 4x_1 - 3x_2$ using the Hooke's and Jeeve's method, taking the starting base point as $X_1 = (0, 0)$ and $\Delta x_1 = \Delta x_2 = 1$.

23.7 Minimize $f(x_1, x_2) = 4x_1^2 + 3x_2^2 - 5x_1x_2 - 8x_1$ using the Powell's method taking $X_1 = (0, 0)$ as the starting point.

23.8 Minimize $f(x_1, x_2) = 2x_1^2 + x_2^2$ starting from the point $X_1 = (1, 2)$ and using the steepest descent method solve up to two iterations.

23.9 Minimize $f(X) = f(x_1, x_2) = (x_1 - 1)^2 + (x_2 - 2)^2$
Subject to $x_1 + x_2 \le 4$, $x_1 - x_2 \le 2$, $x_1 \ge 0$, and $x_2 \ge 0$ using the complex method with $X_1 = \begin{pmatrix} 0.5 \\ 1.5 \end{pmatrix}$.

23.10 Using the method of feasible directions of the Zoutendijk method
Minimize $f(X) = x_1^2 + x_2^2 - 4x_1 - 4x_2 + 8$
Subject to $g_1(X) = x_1 + 2x_2 - 4 \le 0$ taking the starting point $X_1 = \begin{pmatrix} 0 \\ 0 \end{pmatrix}$.

23.11 Find the dimensions of a rectangular prism-type box that has the largest volume when the sum of its length, width, and height is limited to a maximum value of 60 cm and its length is restricted to a maximum value of 36 cms.

23.12 Minimize $f(X) = x_1^2 + x_2^2 - 6x_1 - 8x_2 + 10$
Subject to $4x_1^2 + x_2^2 \le 16$, $3x_1 + 5x_2 \le 15$
using (a) interior penalty method and (b) exterior penalty method.

CONCEPTUAL PROBLEM

Inventory is a very critical aspect of successful management. When production exceeds sales then inventory increases. Thus, inventory is a consequence of a company's day-to-day activities. It generally represents a very significant proportion of the total assets. Inventories are the lifeblood of a business and are essential to running it efficiently and profitably. While conducting business with a company, customers will often not tolerate product unavailability or delays in delivery. The sales department often interacts regularly with a finished goods inventory controller to check if the required products are available in stock so that they can be used to fulfil a customer request. Here, we consider an inventory problem wherein a company faces a responsive price–volume relationship for its products, the lower a product's price, the greater is the sales quantity, even in the face of resultant price reductions by competitors. If the sales revenue does not vary proportionately with the price, reflect this phenomenon in the non-linear objective function of the price.

Solution Consider $s(r)$ to be the sales quantity as a function of the price r, say in the product mix problem. Thus, the associated sales revenue is $rs(r)$.

Suppose the sales quantity function is given as $s(r) = a - br$
If this is the demand equation for constants a and b, over the range of r, then the sales revenue component in the objective function is quadratic, denoted by $z = rs(r) = ar - br^2$ in the decision variables r.

If each unit costs c to produce (where r and c are in the same units) then the total profit P is given as follows:

$$\begin{aligned} P &= z - cs(r) \\ &= ar - br^2 - ac + bcr \\ &= (a + bc)r - ac - br^2 \end{aligned}$$

24 Non-linear Programming II—Quadratic and Separable

After studying this chapter, the reader will be able to

- understand the convex and concave functions
- explain the Kuhn–Tucker conditions
- solve the quadratic programming problem (QPP)
- discuss the separable programming problem (SPP)

24.1 INTRODUCTION

In this chapter, we shall discuss how the optimal solution to a non-linear programming problem (NLPP) is obtained using Kuhn–Tucker conditions. It should, however, be noted that there is no way to ascertain that the solution obtained is locally or globally optimum. To solve NLPPs, we use quadratic and separable programming for which specific computational algorithms have been designed. In a mathematical programming problem, when the objective function is quadratic and constraints are linear, we apply quadratic programming techniques. Separable programming like quadrate programming is a simplex-based method and is applicable if the objective function and the constraints are separable functions.

24.2 KUHN–TUCKER CONDITIONS

In this section, we shall discuss the Kuhn–Tucker conditions under certain restrictions, for identifying the stationary points of constrained non-linear optimization problems. Consider the following optimization problem:

$$\text{Optimization (Max or Min) } Z = f(X) \tag{24.1}$$

$$\text{Subject to } \quad g_j(X) \leq 0; j = 1, 2, 3, \ldots, m \tag{24.2}$$

Converting the inequality constraints into equality constraints by adding the slack variables s_j^2, we have

$$g_j(X) + s_j^2 = 0 \tag{24.3}$$

Now, we define a Lagrangian function

$$L(x_1, x_2, \ldots, x_n, \lambda_1, \lambda_2, \ldots, \lambda_m) = f(X) + \sum_{j=1}^{m} \lambda_j \left\{ g_j(X) + s_j^2 \right\} \tag{24.4}$$

The necessary Kuhn–Tucker conditions for extreme points are $\frac{\partial L}{\partial x_i} = 0$; $i = 1, 2, 3, \ldots, n$.

$$\Rightarrow \frac{\partial f}{\partial x_i} + \sum_{j=1}^{m} \lambda_j \frac{\partial g_j}{\partial x_i} = 0 \tag{24.5}$$

$$\frac{\partial L}{\partial \lambda_j} = 0$$

$$\Rightarrow g_j(X) + s_j^2 = 0; \quad j = 1, 2, 3, \ldots, m \tag{24.6}$$

and $\frac{\partial L}{\partial y_j} = 0$

$$\Rightarrow 2\lambda_j s_j = 0; \quad j = 1, 2, 3, \ldots, m \tag{24.7}$$

From Eqs (24.6) and (24.7), we have $\lambda_j g_j(X) = 0 \Rightarrow \lambda_j = 0$ or $g_j(X) = 0$

The following two cases may arise:

1. If $g_j(X) = 0$ at the optimum point, it is called an active constraint and we can determine the optimum solution.
2. If $\lambda_j = 0$ at the optimum point, it is called an inactive constraint.

Note If the given optimization problem is a minimization problem with constraints of the form $g_j(X) \geq 0$, then $\lambda_j \leq 0$; however if the problem is a maximization problem with constraints of the form $g_j(X) \leq 0$, then $\lambda_j \leq 0$.

Let us consider some maximization or minimization problems given in the following terms:

1. Maximize $Z = f(X)$

   Subject to $g_j(X) \leq 0; j = 1, 2, 3, \ldots, m$

   For the function $f(X)$ to have a maxima, we have

   $$\frac{\partial f}{\partial x_i} + \sum_{j=1}^{m} \lambda_j \frac{\partial g_j}{\partial x_i} = 0; i = 1, 2, 3, \ldots, m$$

   $\lambda_j g_j(X) = 0; j = 1, 2, 3, \ldots, m$ and $\lambda_j \leq 0$

2. Maximize $Z = f(X)$

   Subject to $g_j(X) \geq 0\, j = 1, 2, 3, \ldots, m$

   For the function $f(X)$ to have maxima, we have

   $$\frac{\partial f}{\partial x_i} + \sum_{j=1}^{m} \lambda_j \frac{\partial g_j}{\partial x_i} = 0; i = 1, 2, 3, \ldots, m$$

   $\lambda_j g_j(X) = 0; \; j = 1, 2, 3, \ldots, m$ and $\lambda_j \geq 0$

3. Minimize $Z = f(X)$

   Subject to $g_j(X) \leq 0; j = 1, 2, 3, \ldots, m$

   For the function $f(X)$ to have maxima, we have

   $$\frac{\partial f}{\partial x_i} + \sum_{j=1}^{m} \lambda_j \frac{\partial g_j}{\partial x_i} = 0; i = 1, 2, 3, \ldots, m$$

   $\lambda_j g_j(X) = 0; j = 1, 2, 3, \ldots, m$ and $\lambda_j \geq 0$

4. Minimize $Z = f(X)$
   Subject to $g_j(X) \geq 0; j = 1, 2, 3, \ldots, m$

For the function $f(X)$ to have maxima, we have

$$\frac{\partial f}{\partial x_i} + \sum_{j=1}^{m} \lambda_j \frac{\partial g_j}{\partial x_i} = 0; i = 1, 2, 3, \ldots, m$$

$$\lambda_j g_j(X) = 0; j = 1, 2, 3, \ldots, m \text{ and } \lambda_j \leq 0.$$

Example 24.1 Solve the following problem:

Minimize $f(X) = x_1^2 + x_2^2 + x_3^2$

Subject to
$$g_1(X) = 2x_1 + x_2 - 5 \leq 0$$
$$g_2(X) = x_1 + x_3 - 2 \leq 0$$
$$g_3(X) = 1 - x_1 \leq 0$$
$$g_4(X) = 2 - x_2 \leq 0$$
$$g_5(X) = -x_3 \leq 0$$

Solution It is given that

Minimize $f(X) = x_1^2 + x_2^2 + x_3^2$ (24.8)

Subject to
$$\left.\begin{aligned} g_1(X) &= 2x_1 + x_2 - 5 \leq 0 \\ g_2(X) &= x_1 + x_3 - 2 \leq 0 \\ g_3(X) &= 1 - x_1 \leq 0 \\ g_4(X) &= 2 - x_2 \leq 0 \\ g_5(X) &= -x_3 \leq 0 \end{aligned}\right\} \quad (24.9)$$

Now, we define a Lagrangian function by introducing a slack variable s_j^2,

$$L(x_1, x_2, x_3; \lambda_1, \lambda_2, \lambda_3, \lambda_4, \lambda_5; s_1, s_2, s_3, s_4, s_5) = f(X) + \sum_{j=1}^{5} \lambda_j \left[g_j(X) + s_j^2 \right] \quad (24.10)$$

From Eqs (24.8), (24.9), and (24.10), we have

$$L = x_1^2 + x_2^2 + x_3^2 + \lambda_1(2x_1 + x_2 - 5 + s_1^2) + \lambda_2(x_1 + x_3 - 2 + s_2^2) + \lambda_3(1 - x_1 + s_3^2) + \lambda_4(2 - x_2 + s_4^2) + \lambda_5(-x_3 + s_5^2) \quad (24.11)$$

The Kuhn–Tucker conditions necessary for the minimization of L (with $g_j(X) \leq 0$) are

$$\frac{\partial L}{\partial x_i} = 0; \quad i = 1, 2, 3$$

$$\lambda_j g_j = 0; \quad j = 1, 2, 3, 4, 5$$

$$\lambda_j \geq 0; \quad j = 1, 2, 3, 4, 5$$

Differentiate Eq. (24.11) partially to get

$$\left.\begin{aligned} \frac{\partial L}{\partial x_1} &= 2x_1 + 2\lambda_1 + \lambda_2 - \lambda_3 = 0 \\ \frac{\partial L}{\partial x_2} &= 2x_2 + \lambda_1 - \lambda_4 = 0 \\ \frac{\partial L}{\partial x_3} &= 2x_3 + \lambda_2 - \lambda_5 = 0 \end{aligned}\right\} \quad (24.12)$$

$$\left.\begin{aligned} \lambda_1(2x_1+x_2-5)&=0\\ \lambda_2(x_1+x_3-2)&=0\\ \lambda_3(1-x_1)&=0\\ \lambda_4(2-x_2)&=0\\ \lambda_5(-x_3)&=0 \end{aligned}\right\} \tag{24.13}$$

$$\lambda_j \geq 0;\ j=1,2,3,4,5 \tag{24.14}$$

Let $\lambda_3 \neq 0$ and $\lambda_4 \neq 0$. We have $x_1 = 1, x_2 = 2, x_3 = 0$, and $\lambda_5 = 0$.

If we take $x_1 = 1, x_2 = 2, x_3 = 0$, then Eq. (24.13) is satisfied. Now, from Eq. (24.13) we have $2x_1 + x_2 - 5 \neq 0$ and $x_1 + x_2 - 2 \neq 0$. Hence, $\lambda_1 = 0$ and $\lambda_2 = 0$.

From Eq. (24.12), we have

$$2-\lambda_3=0 \Rightarrow \lambda_3=2;\ 4-\lambda_4=0 \Rightarrow \lambda_4=4;\ \lambda_2-\lambda_5=0 \Rightarrow \lambda_2=\lambda_5$$

Hence, the optimum solution is $x_1 = 1, x_2 = 2, x_3 = 0, \lambda_1 = \lambda_2 = \lambda_5 = 0, \lambda_3 = 2$, and $\lambda_4 = 4$, and minimize $f = (1)^2 + (2)^2 + (0)^2 = 5$.

24.3 QUADRATIC PROGRAMMING

An NLPP in which the objective function is quadratic and the constraints are linear is known as a quadratic programming problem. Here, we shall discuss the following two methods to solve the QPPs: (a) Wolfe's modified simplex method and (b) Beale's method.

24.3.1 Wolfe's Modified Simplex Method

To solve a QPP using Wolfe's modified simplex method, we first use Kuhn–Tucker conditions to express the problem in such a form to apply the computational procedure based on the simplex method. The procedure to solve a QPP using Wolfe's modified simplex method is as follows:

Consider the following QPP:

$$\text{Max } Z = f(X) = \sum_{j=1}^{n} c_j x_j + \frac{1}{2}\sum_{j,k=1}^{n} x_j d_{jk} x_k$$

$$\sum_{j=1}^{n} a_{ij} x_j \leq b_i$$

$$x_j \geq 0,\ i=1,2,\ldots,m,\ j=1,2,\ldots,n$$

where $d_{jk} = d_{kj}$ for all j and k and where $b_i \geq 0$

1. If the given QPP is in the minimization form, then convert it into the maximization form.
2. Convert the inequality constraints into equations by introducing the slack variables s_i^2 in the ith constraint ($i = 1, 2, \ldots, m$) and the slack variables s_{m+j}^2 in the ith non-negativity constraints ($j = 1, 2, \ldots, n$).

3. Now construct the Lagrangian function

$$L(x, s, \lambda) = f(x) - \sum_{i=1}^{m} \lambda_i \left[\sum_{j=1}^{n} a_{ij} x_j - b_i + s_i^2 \right] - \sum_{j=1}^{n} \lambda_{m+j}(-x_j + s_{m+j}^2)$$

where $x = (x_1, x_2, ..., x_n)$, $s = (s_1, s_2, ..., s_{m+n})$, $\lambda = (\lambda_1, \lambda_2, \ldots, \lambda_{m+n})$

4. Differentiate $L(x, s, \lambda)$ partially with respect to the components of x, s, and λ and equate the first order partial derivatives equal to zero. Construct the Kuhn–Tucker conditions from the resulting equations.
5. Introduce the non-negative artificial variables A_j, $j = 1, 2, \ldots, n$ in the Kuhn–Tucker condition. We obtain the following equation:

$$c_j + \sum_{k=1}^{n} d_{jk} x_k - \sum_{i=1}^{m} \lambda_i a_{ij} + \lambda_{m+j} = 0 \quad \text{for } j = 1, 2, \ldots, n$$

6. Construct the objective function Max $Z = -A_1 - A_2 - \ldots - A_n$.
7. Find the initial basic feasible solution (BFS) to the following linear programming problem (LPP):

   Max $Z = -A_1 - A_2 - \ldots - A_n$ subject to the constraints:

$$\sum_{k=1}^{n} d_{jk} x_k - \sum_{i=1}^{m} \lambda_i a_{ij} + \lambda_{m+j} + A_j = -c_j$$

$$\sum_{j=1}^{n} a_{ij} x_j + x_{n+i} = b_i$$

$$A_j, \lambda_i, \lambda_{m+j}, x_j \geq 0$$

   where $x_{n+1} = s_i^2$, $i = 1, 2, ..., m$ and satisfying the complementary slackness conditions (restricted basis conditions)

$$\sum_{j=1}^{n} \lambda_{m+j} x_j + \sum_{j=1}^{m} x_{n+i} \lambda_i = 0.$$

8. Now use phase I of the artificial variable techniques (two-phase method) to determine an optimum solution to the LPP of step 7, which satisfies the complementary slackness conditions. This solution is optimum for the given QPP as well.

Example 24.2 Use Wolfe's modified simplex method to solve the following QPP:

Max $Z = 4x_1 + 6x_2 - 2x_1^2 - 2x_1x_2 - 2x_2^2$

Subject to $x_1 + 2x_1 \leq 2$

$x_1, x_2 \geq 0$

Solution The given problem is a maximization problem. Consider $x_1 \geq 0$ and $x_2 \geq 0$, also as the inequality constraints and convert the inequality constraints into equations by introducing slack variables s_1^2, s_2^2, and s_3^2 respectively. The modified problem is as follows:

Max $Z = 4x_1 + 6x_2 - 2x_1^2 - 2x_1x_2 - 2x_2^2$

$x_1 + 2x_2 + s_1^2 = 2$

$$-x_1 + s_2^2 = 0$$
$$-x_2 + s_3^2 = 0$$

The Lagrangian function is as follows: $L = L(x_1, x_2, s_1, s_2, s_3, \lambda_1, \lambda_2, \lambda_3)$

$= (4x_1 + 6x_2 - 2x_1^2 - 2x_1x_2 - 2x_2^2) - \lambda_1(x_1 + 2x_2 + s_1^2 - 2) - \lambda_2(-x_1 + s_2^2) - \lambda_3(-x_2 + s_3^2) = 0$

Differentiating partially with respect to $x_1, x_2, s_1, s_2, s_3, \lambda_1, \lambda_2, \lambda_3$ and equating to zero, we get

$$\frac{\partial L}{\partial x_1} = 4 - 4x_1 - 2x_2 - \lambda_1 + \lambda_2 = 0, \quad \frac{\partial L}{\partial x_2} = 6 - 2x_1 - 4x_2 - 2\lambda_1 + \lambda_3 = 0,$$

$$\frac{\partial L}{\partial s_1} = -2\lambda_1 s_1 = 0, \quad \frac{\partial L}{\partial s_2} = -2\lambda_2 s_2 = 0, \quad \frac{\partial L}{\partial s_3} = -2\lambda_3 s_3 = 0,$$

$$\frac{\partial L}{\partial \lambda_1} = x_1 + 2x_2 + s_1^2 - 2 = 0, \quad \frac{\partial L}{\partial \lambda_2} = -x_1 + s_2^2 = 0, \quad \frac{\partial L}{\partial \lambda_3} = -x_2 + s_3^2 = 0.$$

On simplification, we have

$$\left.\begin{aligned} &4x_1 + 2x_2 + \lambda_1 - \lambda_2 = 4 \\ &2x_1 + 4x_2 - 2\lambda_1 - \lambda_3 = 6 \\ &x_1 + 2x_2 + s_1^2 = 2 \end{aligned}\right\} \quad (24.14)$$

$$\left.\begin{aligned} &\lambda_1 s_1^2 + \lambda_2 s_2^2 + \lambda_3 s_3^2 = 0 \Rightarrow \lambda_1 s_1^2 + \lambda_2 x_1 + \lambda_3 x_2 = 0 \\ &x_1, x_2, s_1^2, \lambda_i \geq 0, \ i = 1, 2, 3 \end{aligned}\right\} \quad (24.15)$$

A solution x_j, $j = 1, 2$ of Eq. (24.14), satisfying Eq. (24.15), shall necessarily be an optimal one for maximizing L. To obtain the solution to the aforementioned simultaneous Eq. (24.14), we introduce the artificial variables A_1 and A_2 (both non-negative) in the first two constraints of Eq. (24.14) and construct the new objective function Max $Z = -A_1 - A_2$

Max $Z = -A_1 - A_2$

Subject to
$$4x_1 + 2x_2 + \lambda_1 - \lambda_2 + A_1 = 4$$
$$2x_1 + 4x_2 + 2\lambda_1 - \lambda_3 + A_2 = 6$$
$$x_1 + 2x_2 + x_3 = 2$$
$$x_1, x_2, x_3 \geq 0$$
$$A_1, A_2, \lambda_i \geq 0, \ i = 1, 2, 3$$

Here we have replaced s_1^2 by x_3 and satisfied the complementary slackness condition $\sum \lambda_i x_i = 0$. The optimum solution to the aforementioned LPP can now be obtained using the phase I of the artificial variable technique. Thus, the starting simplex table is as follows:

Cost →				0	0	0	0	0	0	−1	−1	Minimum ratio $\frac{X_B}{x_1}$
Variable →				x_1	x_2	x_3	λ_1	λ_2	λ_3	A_1	A_2	
Table no.	C_B	**Basic variable**	X_B	α^1	α^2	α^3	α^4	α^5	α^6	α^7	α^8	
1	−1	A_1	4	[4]	2	0	1	−1	0	1	0	1→
	−1	A_2	6	2	4	0	2	0	−1	0	1	3
	0	x_3	2	1	2	1	0	0	0	0	0	2
$c_j - Z_j$				6↑	6	0	3	−1	−1	0	0	

In the aforegiven table, we see that x_1 or x_2 are the variables with the most positive entry 6 in the $c_j - Z_j$ row. Let x_1 enter the basis and A_1 leave. The next simplex table is as follows:

Cost →				0	0	0	0	0	0	−1	Minimum ratio
Variable →				x_1	x_2	x_3	λ_1	λ_2	λ_3	A_2	$\frac{X_B}{x_2}$
Table no.	C_B	Basic variable	X_B	α^1	α^2	α^3	α^4	α^5	α^6	α^8	
2	0	x_1	1	1	1/2	0	1/4	−1/4	0	0	2
	−1	A_2	4	0	3	0	3/2	1/2	−1	1	4/3
	0	x_3	1	0	[3/2]	1	−1/4	1/4	0	0	2/3→
$c_j - Z_j$				0	3↑	0	3/2	1/2	−1	0	

In the aforegiven table, we see that x_2 is the variable with the most positive entry 3 in the $c_j - Z_j$ row and hence will enter the basis. x_3 will leave. The next simplex table is as follows:

Cost →				0	0	0	0	0	0	−1	Minimum ratio
Variable →				x_1	x_2	x_3	λ_1	λ_2	λ_3	A_2	$\frac{X_B}{\lambda_1}$
Table no.	C_B	Basic variable	X_B	α^1	α^2	α^3	α^4	α^5	α^6	α^8	
3	0	x_1	2/3	1	0	−1/3	1/3	−1/3	0	0	2
	−1	A_2	2	0	0	−2	[2]	0	−1	1	1→
	0	x_2	2/3	0	1	2/3	−1/6	1/6	0	0	...
$c_j - Z_j$				0	0	−2	2↑	0	−1	0	

In the aforegiven table, we observe that λ_1 is the variable with the most positive entry 2 in the $c_j - Z_j$ row and hence will enter the basis; A_2 will leave. The next simplex table is as follows:

Cost →				0	0	0	0	0	0	Minimum ratio
Variable →				x_1	x_2	x_3	λ_1	λ_2	λ_3	
Table no.	C_B	Basic variable	X_B	α^1	α^2	α^3	α^4	α^5	α^6	
4	0	x_1	1/3	1	0	0	0	−1/3	1/6	
	0	λ_1	1	0	0	−1	1	0	−1/6	
	0	x_2	5/6	0	1	1/2	0	1/6	−1/6	
$c_j - Z_j$				0	0	0	0	0	0	

In the aforegiven table 4, $c_j - Z_j \le 0$ for all j, and therefore the solution is optimal. Thus, the optimal solution is $x_1 = 1/3, x_2 = 5/6$, and

$$\text{Max } Z = 4x_1 + 6x_2 - 2x_1^2 - 2x_1x_2 - 2x_2^2$$

$$= 4\left(\frac{1}{3}\right) + 6\left(\frac{5}{6}\right) - 2\left(\frac{1}{3}\right)^2 - 2\left(\frac{1}{3}\right)\left(\frac{5}{6}\right) - 2\left(\frac{5}{6}\right)^2 = \frac{4}{3} + \frac{30}{6} - \frac{2}{9} - \frac{10}{18} - \frac{50}{36} = \frac{25}{6}.$$

Example 24.3 Use Wolfe's modified simplex method to solve the following QPP:

$$\text{Max } Z = 2x_1 + 3x_2 - 2x_1^2$$

Subject to $\quad x_1 + 4x_2 \le 4$

$$x_1 + x_2 \le 2$$
$$x_1, x_2 \ge 0$$

Solution The given problem is a maximization problem. Consider $x_1 \ge 0$ and $x_2 \ge 0$ as inequality constraints, and convert the inequality constraints into equations by introducing slack variables s_1^2, s_2^2, s_3^2, and s_4^2 respectively. The modified problem is as follows:

$$\text{Maximize } Z = 2x_1 + 3x_2 - 2x_1^2$$
$$\text{Subject to} \quad x_1 + 4x_2 + s_1^2 = 4$$
$$x_1 + x_2 + s_2^2 = 2$$
$$-x_1 + s_3^2 = 0$$
$$-x_2 + s_4^2 = 0$$

Construct the following Lagrangian function:

$$L = L(x_1, x_2, s_1, s_2, s_3, s_4, \lambda_1, \lambda_2, \lambda_3, \lambda_4)$$
$$= (2x_1 + 3x_2 - 2x_1^2) - \lambda_1(x_1 + 4x_2 + s_1^2 - 4) - \lambda_2(x_1 + x_2 + s_2^2 - 2)$$
$$-\lambda_3(-x_1 + s_3^2) - \lambda_4(-x_2 + s_4^2)$$

Differentiating partially with respect to x_1, x_2, s_1, s_2, s_3, s_3, λ_1, λ_2, λ_3, λ_4 and equating to zero, we have

$$\frac{\partial L}{\partial x_1} = 2 - 4x_1 - \lambda_1 - \lambda_2 + \lambda_3 = 0, \quad \frac{\partial L}{\partial x_2} = 3 - 4\lambda_1 - \lambda_2 + \lambda_4 = 0,$$

$$\frac{\partial L}{\partial s_1} = -2\lambda_1 s_1 = 0, \quad \frac{\partial L}{\partial s_2} = -2\lambda_2 s_2 = 0, \quad \frac{\partial L}{\partial s_3} = -2\lambda_3 s_3 = 0, \quad \frac{\partial L}{\partial s_4} = -2\lambda_4 s_4 = 0,$$

$$\frac{\partial L}{\partial \lambda_1} = x_1 + 4x_2 + s_1^2 - 4 = 0, \quad \frac{\partial L}{\partial \lambda_2} = x_1 + x_2 + s_2^2 - 2 = 0, \quad \frac{\partial L}{\partial \lambda_3} = -x_1 + s_3^2 = 0,$$

$$\frac{\partial L}{\partial \lambda_4} = -x_2 + s_4^2 = 0.$$

On simplification, we have

$$\left.\begin{aligned} &4x_1 + \lambda_1 + \lambda_2 - \lambda_3 = 2 \\ &4\lambda_1 + \lambda_2 - \lambda_4 = 3 \\ &x_1 + 4x_2 + s_1^2 = 4 \\ &x_1 + x_2 + s_2^2 = 2 \end{aligned}\right. \qquad (24.16)$$

$$\lambda_1 s_1^2 + \lambda_2 s_2^2 + x_1\lambda_3 + x_2\lambda_4 = 0 \qquad (24.17)$$

$$x_1, x_2, s_1^2, \lambda_i \ge 0, i = 1, 2, 3, 4$$

A solution x_j, $j = 1$, 2 of Eq. (24.16), satisfying Eq. (24.17), will necessarily be an optimal one for maximizing L. To obtain the solution to the simultaneous Eq. (24.16), we introduce the artificial variables A_1 and A_2 (both non-negative) in the first two constraints of Eq. (24.16) and construct the new objective function Max $Z = -A_1 - A_2$

Now, the problem is converted to the following form:

$$\text{Max } Z = -A_1 - A_2$$
$$\text{Subject to} \quad 4x_1 + \lambda_1 + \lambda_2 - \lambda_3 + A_1 = 2$$
$$4\lambda_1 + \lambda_2 - \lambda_4 + A_2 = 3$$
$$x_1 + 4x_2 + x_3 = 4$$
$$x_1 + x_2 + x_4 = 2$$
$$x_1, x_2, x_3, x_4, A_1, A_2, \lambda_i \ge 0, i = 1, 2, 3, 4$$

Here we have replaced s_1^2 by x_3 and s_2^2 by x_4 and satisfied the complementary slackness condition $\sum \lambda_i x_i = 0$. The optimum solution to the aforementioned LPP shall now be obtained using phase I of the artificial variable technique. Thus, the starting simplex table is as follows:

	Cost →			0	0	0	0	0	0	0	0	−1	−1	Minimum ratio
	Variable →			x_1	x_2	x_3	x_4	λ_1	λ_2	λ_3	λ_4	A_1	A_2	$\frac{X_B}{x_1}$
Table no.	C_B	Basic variable	X_B	α^1	α^2	α^3	α^4	α^5	α^6	α^7	α^8	α^9	α^{10}	
1	−1	A_1	2	$\boxed{4}$	0	0	0	1	1	−1	0	1	0	1/2→
	−1	A_2	3	0	0	0	0	4	1	0	−1	0	1	…
	0	x_3	4	1	4	1	0	0	0	0	0	0	0	4
	0	x_4	2	1	1	0	1	0	0	0	0	0	0	2
	$c_j - Z_j$			4↑	0	0	0	5	2	−1	−1	0	0	

In the aforegiven table 1, we see that λ_1 is the variable with the most positive entry 5 in the $c_j - Z_j$ row and hence, it should enter the basis. However, λ_1 will not enter the basis because x_3 is in the basis (complementary slackness conditions $\lambda_1 x_3 = 0$). The next most positive entry is 4 for the x_1 column in the $c_j - Z_j$ row, so x_1 enters the basis and A_1 leaves. The next simplex table is as follows:

	Cost →			0	0	0	0	0	0	0	0	−1	Minimum ratio
	Variable →			x_1	x_2	x_3	x_4	λ_1	λ_2	λ_3	λ_4	A_2	$\frac{X_B}{x_2}$
Table no.	C_B	Basic variable	X_B	α^1	α^2	α^3	α^4	α^5	α^6	α^7	α^8	α^{10}	
2	0	x_1	1/2	1	0	0	0	1/4	1/4	−1/4	0	0	…
	−1	A_2	3	0	0	0	0	4	1	0	−1	1	…
	0	x_3	7/2	0	$\boxed{4}$	1	0	−1/4	−1/4	1/4	0	0	7/8→
	0	x_4	3/2	0	1	0	1	−1/4	−1/4	1/4	0	0	3/2
	$c_j - Z_j$			0	0↑	0	0	4	1	0	−1	0	

In the aforegiven table 2, we observe that either λ_1 or λ_2 is the candidate to enter the basis. However, since x_3 and x_4 are still in the basis, they cannot enter the basis because of the complementary slackness conditions $\lambda_1 x_3 = 0$ and $\lambda_2 x_4 = 0$. Here, x_2 is the candidate to enter the basis because λ_4 is not in the basis (complementary slackness conditions $\lambda_4 x_2 = 0$). Now, use the entering and leaving variable rules to get the new simplex table as follows:

	Cost →			0	0	0	0	0	0	0	0	−1	Minimum ratio
	Variable →			x_1	x_2	x_3	x_4	λ_1	λ_2	λ_3	λ_4	A_2	$\frac{X_B}{\lambda_2}$
Table no.	C_B	Basic variable	X_B	α^1	α^2	α^3	α^4	α^5	α^6	α^7	α^8	α^{10}	
3	0	x_1	1/2	1	0	0	0	1/4	−1/4	−1/4	0	0	2
	−1	A_2	3	0	0	0	0	$\boxed{4}$	1	0	−1	1	3/8→
	0	x_2	7/8	0	1	1/4	0	−1/16	−1/16	1/16	0	0	…
	0	x_4	5/8	0	0	−1/4	1	−3/16	−3/16	3/16	0	0	…
	$c_j - Z_j$			0	0	0	0	4↑	1	0	−1	0	

In the aforegiven table 3 we see that either λ_1 or λ_2 is the candidate to enter the basis. However, x_4 is still in the basis; so λ_2 cannot enter the basis because of the complementary slackness condition $\lambda_2 x_4 = 0$. Hence, λ_1 enters the basis. Now use the entering and leaving variable rules to get the new simplex table as follows:

		Cost →		0	0	0	0	0	0	0	0	Minimum ratio
		Variable →		x_1	x_2	x_3	x_4	λ_1	λ_2	λ_3	λ_4	
Table no.	C_B	**Basic variable**	X_B	α^1	α^2	α^3	α^4	α^5	α^6	α^7	α^8	
4	0	x_1	5/16	1	0	0	0	0	3/16	−1/4	1/16	
	0	λ_2	3/4	0	0	0	0	1	1/4	0	−1/4	
	0	x_2	59/64	0	1	1/4	0	0	−3/64	1/16	−1/16	
	0	x_4	49/64	0	0	−1/4	1	0	−9/16	3/16	−3/64	
		$c_j - Z_j$		0	0	0	0	0	0	0	0	

In the aforegiven table 4, $c_j - Z_j \le 0$ for all j; therefore the solution is optimal.

Thus, the optimal solution is $x_1 = 5/16$, $x_2 = 59/64$, and Max $Z = 2x_1 + 3x_2 - 2x_1^2$

$$= 2\left(\frac{5}{16}\right) + 3\left(\frac{59}{64}\right) - 2\left(\frac{25}{256}\right) = \frac{160 + 708 - 50}{256} = \frac{818}{256} = 3.19$$

Example 24.4 Use Wolfe's modified simplex method to solve the following QPP:

$$\text{Max } Z = 6x_1 + 3x_2 - 4x_1x_2 - 2x_1^2 - 3x_2^2$$

$$\text{Subject to } \quad x_1 + x_2 \le 1$$

$$2x_1 + 3x_2 \le 4$$

$$x_1, x_2 \ge 0$$

Solution The given problem is a maximization problem. Consider $x_1 \ge 0$ and $x_2 \ge 0$ also as the inequality constraints and convert the inequality constraints into equations by introducing slack variables s_1^2, s_2^2, s_3^2, and s_4^2 respectively. The modified problem is as follows:

$$\text{Max } Z = 6x_1 + 3x_2 - 4x_1x_2 - 2x_1^2 - 3x_2^2$$

$$x_1 + x_2 + s_1^2 = 1$$

$$2x_1 + 3x_2 + s_2^2 = 4$$

$$-x_1 + s_3^2 = 0$$

$$-x_2 + s_4^2 = 0$$

Construct the following Lagrangian function:

$$L = L(x_1, x_2, s_1, s_2, s_3, s_4, \lambda_1, \lambda_2, \lambda_3, \lambda_4)$$

$$= (6x_1 + 3x_2 - 4x_1x_2 - 2x_1^2 - 3x_2^2) - \lambda_1(x_1 + x_2 + s_1^2 - 1) - \lambda_2(2x_1 + 3x_2 + s_2^2 - 4)$$

$$- \lambda_3(-x_1 + s_3^2) - \lambda_4(-x_2 + s_4^2)$$

Differentiating partially with respect to $x_1, x_2, s_1, s_2, s_3, s_4, \lambda_1, \lambda_2, \lambda_3, \lambda_4$ and equating to zero, we have

$$\frac{\partial L}{\partial x_1} = 6 - 4x_2 - 4x_1 - \lambda_1 - 2\lambda_2 + \lambda_3 = 0, \qquad \frac{\partial L}{\partial x_2} = 3 - 4x_1 - 6x_2 - \lambda_1 - 3\lambda_2 + \lambda_4 = 0,$$

$$\frac{\partial L}{\partial s_1} = -2\lambda_1 s_1 = 0, \qquad \frac{\partial L}{\partial s_2} = -2\lambda_2 s_2 = 0,$$

$$\frac{\partial L}{\partial s_3} = -2\lambda_3 s_3 = 0, \qquad \frac{\partial L}{\partial s_4} = -2\lambda_4 s_4 = 0,$$

$$\frac{\partial L}{\partial \lambda_1} = x_1 + x_2 + s_1^2 - 1 = 0, \qquad \frac{\partial L}{\partial \lambda_2} = 2x_1 + 3x_2 + s_2^2 - 4 = 0,$$

$$\frac{\partial L}{\partial \lambda_3} = -x_1 + s_3^2 = 0, \qquad \frac{\partial L}{\partial \lambda_4} = -x_2 + s_4^2 = 0.$$

On simplification, we have

$$\left.\begin{aligned} &4x_1 + 4x_2 + \lambda_1 + 2\lambda_2 - \lambda_3 = 6 \\ &4x_1 + 6x_2 + \lambda_1 + 3\lambda_2 - \lambda_4 = 3 \\ &x_1 + x_2 + s_1^2 = 1 \\ &2x_1 + 3x_2 + s_2^2 = 4 \end{aligned}\right\} \quad (24.18)$$

$$\left.\begin{aligned} &\lambda_1 s_1^2 + \lambda_2 s_2^2 + x_1\lambda_3 + x_2\lambda_4 = 0 \\ &x_1, x_2, s_1^2, s_2^2, \lambda_i \geq 0,\ i = 1, 2, 3, 4 \end{aligned}\right\} \quad (24.19)$$

A solution x_j, $j = 1, 2$ to Eq. (24.18) and satisfying (24.19) will necessarily be an optimal one for maximizing L. To determine the solution to the aforementioned simultaneous Eq. (24.18), we introduce the artificial variables A_1 and A_2 (both non-negative) in the first two constraints of Eq. (24.18) and construct the dummy objective function Max $Z = -A_1 - A_2$.

Now, the problem is converted to the following form:

$$\begin{aligned} &\text{Max } Z = -A_1 - A_2 \\ &\text{Subject to } \quad 4x_1 + 4x_2 + \lambda_1 + 2\lambda_2 - \lambda_3 + A_1 = 6 \\ &\qquad 4x_1 + 6x_2 + \lambda_1 + 3\lambda_2 - \lambda_4 + A_2 = 3 \\ &\qquad x_1 + x_2 + x_3 = 1 \\ &\qquad 2x_1 + 3x_2 + x_4 = 4 \\ &\qquad x_1, x_2, x_3, x_4 \geq 0 \\ &\qquad A_1, A_2, \lambda_i \geq 0,\ i = 1, 2, 3, 4 \end{aligned}$$

In this problem, we have replaced s_1^2 by x_3 and s_2^2 by x_4 and satisfied the complementary slackness condition $\sum \lambda_i x_i = 0$. The optimum solution to the aforementioned LPP will now be obtained by phase I of the artificial variable technique. Thus, the starting simplex table is as follows:

	Cost →			0	0	0	0	0	0	0	0	−1	−1	Minimum ratio X_B/x_1
	Variable →			x_1	x_2	x_3	x_4	λ_1	λ_2	λ_3	λ_4	A_1	A_2	
Table no.	C_B	**Basic variable**	X_B	α^1	α^2	α^3	α^4	α^5	α^6	α^7	α^8	α^9	α^{10}	
1	−1	A_1	6	4	4	0	0	1	2	−1	0	1	0	3/2
	−1	A_2	3	4	[6]	0	0	1	3	0	−1	0	1	1/2→
	0	x_3	1	1	1	1	0	0	0	0	0	0	0	1
	0	x_4	4	2	3	0	1	0	0	0	0	0	0	4/3
	$c_j - Z_j$			8	10↑	0	0	2	5	−1	−1	0	0	

In the aforegiven table 1, we see that x_2 is the variable with the most positive entry 10 in the $c_j - Z_j$ row and hence, will enter the basis. A_2 will leave. The next simplex table is as follows:

		Cost →		0	0	0	0	0	0	0	0	−1	Minimum ratio
		Variable →		x_1	x_2	x_3	x_4	λ_1	λ_2	λ_3	λ_4	A_1	$\frac{X_B}{x_1}$
Table no.	C_B	**Basic variable**	X_B	α^1	α^2	α^3	α^4	α^5	α^6	α^7	α^8	α^9	
2	−1	A_1	4	4/3	0	0	0	1/3	0	−1	2/3	1	3
	0	x_2	1/2	[2/3]	1	0	0	1/6	1/2	0	−1/6	0	3/4→
	0	x_3	1/2	1/3	0	1	0	−1/6	−1/2	0	1/6	0	3/2
	0	x_4	5/2	0	0	0	1	−1/2	−3/2	0	1/2	0	…
		$c_j - Z_j$		4/3↑	0	0	0	1/3	0	−1	2/3	0	

In the aforegiven table 2, we see that x_1 is the variable with the most positive entry 4/3 in the $c_j - Z_j$ row and hence, will enter the basis. x_2 will leave. The next simplex table is as follows:

		Cost →		0	0	0	0	0	0	0	0	−1	Minimum ratio
		Variable →		x_1	x_2	x_3	x_4	λ_1	λ_2	λ_3	λ_4	A_1	$\frac{X_B}{x_1}$
Table no.	C_B	**Basic variable**	X_B	α^1	α^2	α^3	α^4	α^5	α^6	α^7	α^8	α^9	
3	−1	A_1	3	0	−2	0	0	0	−1	−1	1	1	3
	0	x_1	3/4	1	3/2	0	0	1/4	3/4	0	−1/4	0	…
	0	x_3	1/4	0	−1/2	1	0	−1/4	−3/4	0	[1/4]	0	4/3→
	0	x_4	5/2	0	0	0	1	−1/2	−3/2	0	1/2	0	5
		$c_j - Z_j$		0	−2	0	0	0	−1	−1	1↑	0	

In the aforegiven table 3, we see that λ_4 is the variable with the most positive entry 1 in $c_j - Z_j$ row and hence, will enter the basis. x_3 will leave. The next simplex table is as follows:

		Cost →		0	0	0	0	0	0	0	0	−1	Minimum ratio
		Variable →		x_1	x_2	x_3	x_4	λ_1	λ_2	λ_3	λ_4	A_1	$\frac{X_B}{x_1}$
Table no.	C_B	**Basic variable**	X_B	α^1	α^2	α^3	α^4	α^5	α^6	α^7	α^8	α^9	
4	−1	A_1	6	4	4	0	0	[1]	2	−1	0	1	6→
	0	x_1	1	1	1	1	0	0	0	0	0	0	…
	0	λ_4	1	0	−2	4	0	−1	−3	0	1	0	…
	0	x_4	2	0	1	−2	1	0	0	0	0	0	…
		$c_j - Z_j$		0	4	0	0	1↑	2	−1	0	0	

In the aforegiven table 4, we see that either x_2 or λ_1 is the candidate to enter the basis. However λ_4 is still in the basis, so x_2 cannot enter the basis because of the complementary slackness condition $\lambda_2 x_4 = 0$. Hence, λ_1 enters the basis and A_1 leaves. Now use the entering and leaving variable rules to get the new simplex table as follows:

	Cost →			0	0	0	0	0	0	0	0	Minimum ratio
	Variable →			x_1	x_2	x_3	x_4	λ_1	λ_2	λ_3	λ_4	
Table no.	C_B	Basic variable	X_B	α^1	α^2	α^3	α^4	α^5	α^6	α^7	α^8	
5	0	λ_1	6	4	4	0	0	1	2	−1	0	
	0	x_1	1	1	1	1	0	0	0	0	0	
	0	λ_4	7	4	2	4	0	0	−1	−1	1	
	0	x_4	2	0	1	−2	1	0	0	0	0	
	$c_j - Z_j$			0	0	0	0	0	0	0	0	

In the aforegiven table 5, $c_j - Z_j \le 0$ for all j; therefore the solution is optimal. Thus, the optimal solution is $x_1 = 1, x_2 = 0$ and Minimum $Z = 6x_1 + 3x_2 - 4x_1x_2 - 2x_1^2 - 3x_2^2 = 6 + 0 - 0 - 2 - 0 = 4$.

24.3.2 Beale's Method

Consider the following QPP:

$$\text{Max } f(X) = c^T X + \frac{1}{2} X^T Q X$$

$$\text{Subject to} \quad AX \{\le, \ge, \text{or} =\} b \text{ and } X \ge 0$$

where $X \in R^n$, A is $m \times n$, b is $m \times 1$, c is $n \times 1$, and Q is an $n \times n$ symmetric matrix.

The procedure to solve a QPP using Beale's method is as follows:

Step 1 If the given QPP is in the minimization form, convert it into the maximization form.

Step 2 Set the given QPP in standard form using slack and surplus variables.

Step 3 Arbitrarily choose any m variables as the basic variables, so that the remaining $n - m$ variables become non-basic. The basic and non-basic variables are $X_B = (X_{B_1}, X_{B_2}, \ldots, X_{B_m})$ and $X_{NB} = (X_{NB_1}, X_{NB_2}, \ldots, X_{NB_{n-m}})$ respectively.

Step 4 Write each basic variable x_{B_i} in terms of non-basic variables X_{NB_i}'s (and u_i's, if any) using the given (as well as additional, if any) constraints.

Step 5 Write the objective function $f(X)$ in terms of non-basic X_{NB_i}'s (and u_i's, if any).

Step 6 Evaluate the partial derivatives of $f(X)$ formulated with respect to the non-basic variables at the point $X_{NB} = 0$ (and $u = 0$).

Here three cases arise:

Case 1 If $\left(\dfrac{\partial f(X)}{\partial x_{NB_k}}\right)_{\substack{X_{NB}=0\\u=0}} = 0$ for each $k = 1, 2, \ldots, n - m$

and $\left(\dfrac{\partial f(X)}{\partial u_i}\right)_{\substack{X_{NB}=0\\u=0}} = 0$ for each i

then the current basic solution is optimal. Go to Step 9.

Case 2 If $\left(\dfrac{\partial f(X)}{\partial x_{NB_k}}\right)_{\substack{X_{NB}=0\\u=0}} > 0$ for at least one k

choose the most positive one. The corresponding non-basic variables will enter the basis. Go to Step 7.

Case 3 If $\left(\dfrac{\partial f(X)}{\partial x_{NB_k}}\right)_{\substack{X_{NB}=0\\u=0}} = 0$ for each $k = 1, 2, \ldots, n - m$

and $\left(\dfrac{\partial f(X)}{\partial u_i}\right)_{\substack{X_{NB_k}=0\\u=0}} \neq 0$ for sum $i = r$

then introduce a new non-basic variables u_j defined by $u = \dfrac{1}{2}\left(\dfrac{\partial f}{\partial u_r}\right)$ and treat u_r as a basic variable (to be ignored later) and go to Step 4.

Step 7 Consider $X_{NB_i} = x_k$ to be the entering variable identified in Step 6 (Case 2). Find the minimum of the ratios, $\min\left[\dfrac{\alpha_{ho}}{|\alpha_{hk}|}, \dfrac{\gamma_{ko}}{|\gamma_{kk}|}\right]$ for all basic variables x_h where α_{ho} is the constant term and α_{hk} is the coefficient of x_k in the expression of basic variables x_h when expressed in terms of the non-basic ones, and γ_{ko} is the constant term, and γ_k is the coefficient of x_k in $\dfrac{\partial f}{\partial x_k}$.

Case 1 If the minimum ratio occurs for some $\dfrac{\alpha_{ho}}{|\alpha_{ho}|}$ the corresponding basic variables, x_h will leave the basis.

Case 2 If the minimum ratio occurs for some $\dfrac{\gamma_{ko}}{|\gamma_{kk}|}$, the exit criterion corresponds to a non-basic variables. In this case, introduce an additional non-basic variable, called a free variable, defined by

$$u_i = \frac{1}{2}\frac{\partial f}{\partial x_k} \quad (u_i \text{ is unrestricted})$$

This relation becomes an additional constraint equation.

Step 8 Now go to Step 4 and repeat the procedure until an optimal basic solution is determined.

Step 9 Determine the optimal values of X_B and $f(X)$ by setting $X_{NB} = 0$ in the expressions obtained in Steps 4 and 5.

Example 24.5 Use Beale's method to solve the following NLPP:

$$\begin{aligned} &\text{Max } Z = 2x_1 + 3x_2 - x_1^2 \\ &\text{Subject to } \quad x_1 + 2x_2 \le 4 \\ &\qquad\quad \text{and } \quad x_1, x_2 \ge 0 \end{aligned}$$

Solution The given QPP is as follows:

$$\text{Max } Z = 2x_1 + 3x_2 - x_1^2$$

Subject to $x_1 + 2x_2 \le 4$

and $x_1, x_2 \ge 0$

The given QPP can be written in the standard form using slack variable x_3 as follows:

$$\text{Max } Z = 2x_1 + 3x_2 - x_1^2$$

$$\text{Subject to } \quad x_1 + 2x_2 + x_3 = 4$$

$$\text{and } \quad x_1, x_2, x_3 \ge 0$$

Now, we have x_3 as the basis variable and x_1, x_2 as non-basis variables, that is, $X_B=(x_3)$, $X_{NB}=(x_1, x_2)$.

Then, we have $x_3 = 4 - x_1 - 2x_2$ and $f(X) = 2x_1 + 3x_2 - x_1^2$.

Differentiating f partially with respect to x_1 and x_2 on both sides, we get

$$\left(\frac{\partial f}{\partial x_1}\right)_{X_{NB}=0} = (2 - 2x_1)_{x_1,x_2=0} = 2$$

$$\text{and} \left(\frac{\partial f}{\partial x_2}\right)_{X_{NB}=0} = (3)_{x_1,x_2=0} = 3$$

Here, the most positive is x_2, so x_2 enters the basis.

$$\min\left[\frac{\alpha_{30}}{|\alpha_{32}|}, \frac{\gamma_{20}}{|\gamma_{22}|}\right] = \min\left[\frac{4}{|-2|}, \frac{3}{0}\right] = 2 \text{ and } x_3 \text{ leaves the basis.}$$

Now $X_B = (x_2)$, $X_{NB} = (x_1, x_3)$, then we have

$$x_2 = \frac{1}{2}(4 - x_1 - x_3) = 2 - \frac{1}{2}x_1 - \frac{1}{2}x_3$$

$$\text{and } f = 2x_1 + 3\left(2 - \frac{1}{2}x_1 - \frac{1}{2}x_3\right) - x_1^2 = 2x_1 + 6 - \frac{3}{2}x_1 - \frac{3}{2}x_3 - x_1^2$$

$$= 6 + \frac{1}{2}x_1 - x_1^2 - \frac{3}{2}x_3$$

Differentiating f partially with respect to x_1 and x_3 on both sides, we get

$$\left(\frac{\partial f}{\partial x_1}\right)_{X_{NB}=0} = \left(\frac{1}{2} - 2x_1\right)_{x_1,x_3=0} = \frac{1}{2}$$

$$\text{and} \left(\frac{\partial f}{\partial x_3}\right)_{X_{NB}=0} = \left(-\frac{3}{2}\right) = -\frac{3}{2}$$

Here x_1 enters the basis.

$\min\left[\frac{\alpha_{20}}{|\alpha_{21}|}, \frac{\gamma_{10}}{|\gamma_{11}|}\right] = \min\left[\frac{2}{|-1/2|}, \frac{1/2}{|-2|}\right] = \frac{1}{4}$ which corresponds to $\frac{\gamma_{10}}{|\gamma_{11}|}$ so x_2 does not leave the basis.

Now, we introduce a new non-basis variable u_1 such that

$$u_1 = \frac{1}{2}\frac{\partial f}{\partial x_1} = \frac{1}{2}\left(\frac{1}{2} - 2x_1\right) = \frac{1}{4} - x_1$$

$$X_B = (x_1, x_2),\ X_{NB} = (u_1, x_3)$$

$$x_1 = \frac{1}{4} - u_1$$

$$x_2 = 2 - \frac{1}{2}x_1 - \frac{1}{2}x_3 = 2 - \frac{1}{2}\left(\frac{1}{4} - u_1\right) - \frac{1}{2}x_3$$

$$= 2 - \frac{1}{8} + \frac{1}{2}u_1 - \frac{1}{2}x_3 = \frac{15}{8} + \frac{1}{2}u_1 - \frac{1}{2}x_3$$

and $f = 6 + x_1\left(\frac{1}{2} - x_1\right) - \frac{3}{2}x_3 = 6 + \left(\frac{1}{4} - u_1\right)\left(\frac{1}{2} - \frac{1}{4} + u_1\right) - \frac{3}{2}x_3$

$$= 6 + \left(\frac{1}{4} - u_1\right)\left(\frac{1}{4} + u_1\right) - \frac{3}{2}x_3 = 6 + \frac{1}{16} - u_1^2 - \frac{3}{2}x_3$$

$$= \frac{97}{16} - u_1^2 - \frac{3}{2}x_3$$

Differentiating f partially with respect to x_3 and u_1 on both sides, we get

$$\left(\frac{\partial f}{\partial x_3}\right)_{X_{NB}=0} = \left(-\frac{3}{2}\right)_{x_1, x_3=0} = -\frac{3}{2}$$

and $\left(\frac{\partial f}{\partial u_1}\right)_{X_{NB}=0} = (-2u_1)_{x_3, u_1} = 0$

This gives the optimal solution. Hence, the optimal solution is $x_1 = 1/4$, $x_2 = 15/8$, and Max Z = 97/16.

Example 24.6 Use Beale's method to solve the following NLPP:

Min $Z = 6 - 6x_1 + 2x_1^2 - 2x_1x_2 + 2x_2^2$

Subject to $x_1 + x_2 \leq 2$

and $x_1, x_2 \geq 0$

Solution The given QPP is as follows:

Min $Z = 6 - 6x_1 + 2x_1^2 - 2x_1x_2 + 2x_2^2$

Subject to $x_1 + x_2 \leq 2$

and $x_1, x_2 \geq 0$

Convert the minimization objective function to the maximization function to get

$$\text{Max } -Z = f(X) = -6 + 6x_1 - 2x_1^2 + 2x_1x_2 - 2x_2^2$$

The given QPP can be written in the standard form using slack variable x_3 as follows:

Min $Z = 6 - 6x_1 + 2x_1^2 - 2x_1x_2 + 2x_2^2$

Subject to $x_1 + x_2 + x_3 = 2$

and $x_1, x_2 \geq 0$

Arbitrarily, choosing x_3 as the basic variable, we have $X_B = (x_3)$ and $X_{NB} = (x_1, x_2)$. Expressing X_B and $f(x)$ in terms of X_{NB}, we have

$$x_3 = 2 - x_1 - x_2$$

and $f = -6 + 6x_1 - 2x_1^2 + 2x_1x_2 - 2x_2^2$

Differentiating f partially with respect to x_1 and x_2 on both sides, we get

$$\left(\frac{\partial f}{\partial x_1}\right)_{X_{NB}=0} = (6 - 4x_1 + 2x_2)_{\substack{x_1=0 \\ x_2=0}} = 6$$

$$\text{and } \left(\frac{\partial f}{\partial x_2}\right)_{X_{NB}=0} = (-2x_1 + 4x_2)_{\substack{x_1=0\\x_2=0}} = 0$$

Here, the most positive is x_1; so x_1 enters the basis.

$\min\left[\frac{\alpha_{30}}{|\alpha_{31}|}, \frac{\gamma_{10}}{|\gamma_{11}|}\right] = \min\left[\frac{2}{|-1|}, \frac{6}{|-4|}\right] = \frac{3}{2}$, which corresponds to $\frac{\gamma_{10}}{|\gamma_{11}|}$ so x_3 does not leave the basis.

Now, we introduce a new non-basic free variable u_1, defined by

$$u_1 = \frac{1}{2}\frac{\partial f}{\partial x_1} = 3 - 2x_1 + x_2$$

Now, $X_B = (x_3, x_1)$, $X_{NB} = (x_2, u_1)$

$$x_1 = \frac{3}{2} - \frac{1}{2}u_1 + \frac{1}{2}x_2$$

$$x_3 = 2 - x_1 - x_2 = 2 - \left(\frac{3}{2} - \frac{1}{2}u_1 + \frac{1}{2}x_2\right) - x_2 = \frac{1}{2} + \frac{1}{2}u_1 - \frac{3}{2}x_2$$

and $f = -6 + 6x_1 - 2x_1^2 + 2x_1x_2 - 2x_2^2 = -6 + 2x_1(3 - x_1 + x_2) - 2x_2^2$

$$= -6 + 2\left(\frac{3}{2} - \frac{1}{2}u_1 + \frac{1}{2}x_2\right)\left(3 - \frac{3}{2} + \frac{1}{2}u_1 - \frac{1}{2}x_2 + x_2\right) - 2x_2^2$$

$$= -6 + (3 - u_1 + x_2)\left(\frac{3}{2} + \frac{1}{2}u_1 + \frac{1}{2}x_2\right) - 2x_2^2 = -6 + \frac{1}{2}(3 + x_2 - u_1)(3 + x_2 + u_1) - 2x_2^2$$

$$= -6 + \frac{1}{2}\{(3 + x_2)^2\} - 2x_2^2 = -6 + \frac{1}{2}\{9 + x_2 - 6x_2 - u_1^2\} - 2x_2^2$$

$$= -6 + \frac{9}{2} + \frac{x_2^2}{2} - \frac{6x_2}{2} - \frac{u_1^2}{2} - 2x_2^2 = \frac{-3}{2} - \frac{u_1^2}{2} - 3x_2 - \frac{3}{2}x_2^2$$

Differentiating f partially with respect to x_2 and u_1 on both sides, we get

$$\left(\frac{\partial f}{\partial x_2}\right)_{X_{NB}=0} = (3 - 3x_2)_{\substack{x_1=0\\u_2=0}} = 3$$

$$\text{and } \left(\frac{\partial f}{\partial u_1}\right)_{X_{NB}=0} = (-u_1)_{\substack{x_2=0\\u_1=0}} = 0$$

Here x_2 enters the basis.

Now, $\text{Min}\left[\frac{\alpha_{30}}{|\alpha_{32}|}, \frac{\alpha_{10}}{|\alpha_{12}|}, \frac{\gamma_{20}}{|\gamma_{22}|}\right] = \text{Min}\left[\frac{1/2}{|-3/2|}, \frac{3/2}{|1/2|}, \frac{3}{|-3|}\right] = \frac{1}{3}$

This implies x_3 will leave the basis and the new variables are as follows: $X_B = (x_1, x_2)$, $X_{NB} = (u_1, x_3)$.

$$\text{Now, } x_2 = \frac{1}{3} + \frac{1}{3}u_1 - \frac{2}{3}x_3$$

$$x_1 = \frac{3}{2} - \frac{1}{2}u_1 + \frac{1}{2}x_2 = \frac{3}{2} - \frac{1}{2}u_1 + \frac{1}{2}\left(\frac{1}{3} + \frac{1}{3}u_1 - \frac{2}{3}x_3\right) = \frac{3}{2} - \frac{1}{2}u_1 + \frac{1}{6} + \frac{1}{6}u_1 - \frac{1}{3}x_3$$

$$= \frac{5}{3} - \frac{1}{3}u_1 + \frac{1}{3}x_3$$

$$\text{and } f = -\frac{3}{2} - \frac{u_1^2}{2} + \frac{3}{2}x_2(2-x_2)$$

$$= -\frac{3}{2} - \frac{u_1^2}{2} + \frac{3}{2}\left(\frac{1}{3} + \frac{1}{3}u_1 - \frac{2}{3}x_3\right)\left(2 - \frac{1}{3} - \frac{1}{3}u_1 + \frac{2}{3}x_3\right)$$

$$= -\frac{3}{2} - \frac{u_1^2}{2} + \frac{1}{6}(1 + u_1 - 2x_3)(5 - u_1 + 2x_3)$$

$$= -\frac{3}{2} - \frac{u_1^2}{2} + \frac{1}{6}(5 - u_1 + 2x_3 + 5u_1 - u_1^2 + 2x_3u_1 - 10x_3 + 2x_3u_1 - 4x_3^2)$$

$$= -\frac{3}{2} - \frac{u_1^2}{2} + \frac{5}{6} + \frac{4u_1}{6} - \frac{8}{6}x_3 - \frac{u_1^2}{6} + \frac{4}{6}x_3u_1 - \frac{4}{6}x_3^2$$

$$= -\frac{4}{6} - \frac{4}{6}u_1^2 + \frac{4}{6}u_1 - \frac{4}{3}x_3 - \frac{4}{6}x_3^2 + \frac{4}{6}x_3u_1$$

$$= -\frac{2}{3} + \frac{2}{3}u_1 - \frac{2}{3}u_1^2 - \frac{4}{3}x_3 + \frac{2}{3}x_3u_1 - \frac{2}{3}x_3^2$$

Differentiating f partially with respect to x_3 and u_1 on both sides, we get

$$\left(\frac{\partial f}{\partial x_3}\right)_{X_{NB}=0} = \left(-\frac{4}{3} - \frac{2}{3}u_1 - \frac{4}{3}x_3\right)_{\substack{x_3=0\\u_1=0}} = \frac{-4}{3}$$

$$\text{and } \left(\frac{\partial f}{\partial u_1}\right)_{X_{NB}=0} = \left(\frac{2}{3} - \frac{4}{3}u_1 + \frac{2}{3}x_3\right)_{\substack{x_3=0\\u_1=0}} = \frac{2}{3}$$

Since $\dfrac{\partial f}{\partial u_1} \neq 0$, this solution can be further improved.

Now, x_3 does not enter the basis. Thus, we introduce another non-basic free variable u_2 defined by $u_2 = \dfrac{1}{2}\dfrac{\partial f}{\partial u_1} = \dfrac{1}{3} - \dfrac{1}{3}u_1 + \dfrac{1}{3}x_3$

Now, $X_B = (x_1, x_2, u_1), X_{NB} = (x_3, u_2)$

$$u_1 = \frac{1}{2} - \frac{3}{2}u_2 + \frac{1}{2}x_3$$

$$x_2 = \frac{1}{3} + \frac{1}{3}u_1 - \frac{2}{3}x_3 = \frac{1}{3} + \frac{1}{3}\left(\frac{1}{2} - \frac{3}{2}u_2 + \frac{1}{2}x_3\right) - \frac{2}{3}x_3 = \frac{1}{2} - \frac{1}{2}u_2 - \frac{1}{2}x_3$$

$$x_1 = \frac{5}{3} - \frac{1}{3}u_1 - \frac{1}{3}x_3 = \frac{5}{3} - \frac{1}{3}\left(\frac{1}{2} - \frac{3}{2}u_2 + \frac{1}{2}x_3\right) - \frac{1}{3}x_3 = \frac{5}{3} - \frac{1}{6} + \frac{1}{2}u_2 - \frac{1}{6}x_3 - \frac{1}{3}x_3$$

$$= \frac{3}{2} + \frac{1}{2}u_2 - \frac{1}{2}x_3$$

$$f = -\frac{2}{3} + \frac{2}{3}u_1 - \frac{4}{3}x_3 + \frac{2}{3}x_3u_1 - \frac{2}{3}u_1^2 - \frac{2}{3}x_3^2$$

$$= -\frac{2}{3} + \frac{2}{3}u_1(1 + x_3 - u_1) - \frac{4}{3}x_3 - \frac{2}{3}x_3^2$$

$$= -\frac{2}{3} + \frac{2}{3}\left(\frac{1}{2} - \frac{3}{2}u_2 + \frac{1}{2}x_3\right)\left(1 + x_3 - \frac{1}{2} + \frac{3}{2}u_2 - \frac{1}{2}x_3\right) - \frac{4}{3}x_3 - \frac{2}{3}x_3^2$$

$$= -\frac{2}{3} + \frac{1}{6}(1 - 3u_2 + x_3)(2 + 2x_3 - 1 + 3u_2 - x_3) - \frac{4}{3}x_3 - \frac{2}{3}x_3^2$$

$$= -\frac{2}{3} + \frac{1}{6}(1 - 3u_2 + x_3)(1 + x_3 + 3u_2) - \frac{4}{3}x_3 - \frac{2}{3}x_3^2$$

$$= -\frac{2}{3} + \frac{1}{6}\left\{(1 + x_3)^2 - 9u_2^2\right\} - \frac{4}{3}x_3 - \frac{2}{3}x_3^2$$

$$= -\frac{2}{3} + \frac{1}{6}\left[1 + x_3^2 + 2x_3 - 9u_2^2\right] - \frac{4}{3}x_3 - \frac{2}{3}x_3^2$$

$$= -\frac{2}{3} + \frac{1}{6} + \frac{x_3^2}{6} + \frac{1}{3}x_3 - \frac{3}{2}u_2^2 - \frac{4}{3}x_3 - \frac{2}{3}x_3^2$$

$$f = -\frac{1}{2} - \frac{1}{2}x_3^2 - \frac{5}{3}x_3 - \frac{3}{2}u_2^2$$

Differentiating f partially with respect to x_3 and u_2 on both sides, we get

$$\left(\frac{\partial f}{\partial x_3}\right)_{X_{NB}=0} = \left(-x_3 - \frac{5}{3}\right) = -\frac{5}{3} \text{ and } \left(\frac{\partial f}{\partial u_2}\right)_{X_{NB}=0} = \left(-\frac{3}{2}u_2\right) = 0$$

This gives an optimal solution. Ignoring the free variable u_1, the optimal solution is $x_1 = \frac{3}{2}$, $x_2 = \frac{1}{2}$, and $Z = -f = -\left(-\frac{1}{2}\right) = \frac{1}{2}$.

24.4 SEPARABLE PROGRAMMING

A function of n variables $f(x_1, x_2, ..., x_n)$ is said to be *separable* if it can be written as the sum of n functions $f_1(x_1), f_2(x_2), ..., f_n(x_n)$, wherein each is a function of a single variable x_i, $i = 1, \ldots, n$. Thus, $f(x_1, ..., x_n)$ is called a separable function if

$$f(x_1, ..., x_n) = f_1(x_1) + f_2(x_2) + ... + f_n(x_n) \tag{24.20}$$

As a result, an SPP can be considered as a mathematical programming problem in the objective function and constraints. Thus, the general form of separable programming is as follows:

$$\begin{aligned}
\text{Max (or min)}\, Z &= f(X) = f_1(x_1) + f_2(x_2) + ... + f_n(x_n) \\
\text{Subject to}\quad & g_1^1(x_1) + g_2^1(x_2) + ... + g_n^1(x_n) \le = \ge b_1 \\
& g_1^2(x_1) + g_2^2(x_2) + ... + g_n^2(x_n) \le = \ge b_2 \\
& g_1^m(x_1) + g_2^m(x_2) + ... + g_n^m(x_n) \le = \ge b_m \\
& x_1, x_2, \ldots x_n \ge 0.
\end{aligned}$$

To use the simplex method for solving an SPP, first the functions involved in the problem are approximated by piecewise linear functions. For this, let us consider a function $f(x)$, which is continuous in the interval $[a_1, a_4]$ and $\hat{f}(x)$ be the function shown by dotted lines in Fig (24.1).

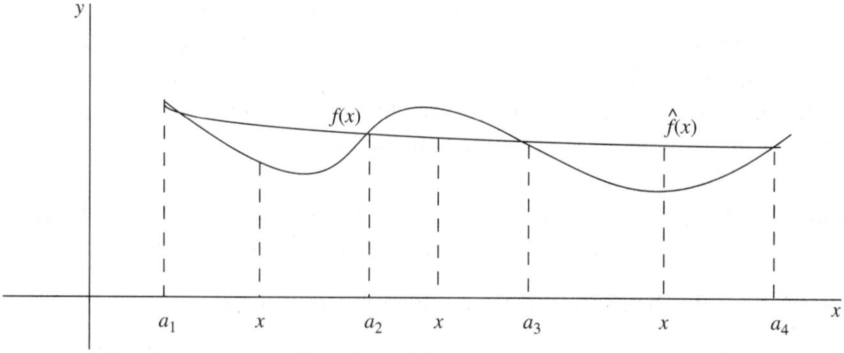

Figure 24.1 Approximation of function $f(x)$ by continuous line segments $\hat{f}(x)$

$\hat{f}(x)$ in $[a_1, a_2]$ is a straight line segment whose equation is as follows:

$$\hat{f}(x) - f(a_1) = \frac{f(a_2) - f(a_1)}{a_2 - a_1}(x - a_1),\ a_1 \le x \le a_2$$

$$\text{or } \hat{f}(x) = f(a_1) + \frac{f(a_2) - f(a_1)}{a_2 - a_1}(x - a_1),\ a_1 \le x \le a_2 \tag{24.21}$$

As $x \in [a_1, a_2]$, $x = \lambda_1 a_1 + (1 - \lambda_1)a_2$, (for same λ_1 such that $0 \le \lambda_1 \le 1$)

$$\text{or } x = \lambda_1 a_1 + \lambda_2 a_2, \lambda_1 + \lambda_2 = 1,\ \lambda_1, \lambda_2 \ge 0 \tag{24.22}$$

$$\text{or } x - a_1 = \lambda_2 (a_2 - a_1) \tag{24.23}$$

Now substituting Eq. (24.21) in Eq. (24.23), we get

$$\hat{f}(x) = \lambda_1 f(a_1) + \lambda_2 f(a_2) \tag{24.24}$$

Similarly for interval $[a_2, a_3]$, we have

$$\hat{f}(x) = \lambda_2 f(a_2) + \lambda_3 f(a_3),\ x \in [a_2, a_3] \tag{24.25}$$

$$\text{or } x = \lambda_2 a_2 + \lambda_3 a_3 \tag{24.26}$$

$$\hat{f}(x) = \lambda_3 f(a_3) + \lambda_4 f(a_4) \tag{24.27}$$

$$x = \lambda_3 a_3 + \lambda_4 a_4,\ (x \in [a_3, a_4]) \tag{24.28}$$

Combining Eq. (24.22) and Eqs (24.24) to (24.27), we have

$$\hat{f}(x) = \lambda_1 f(a_1) + \lambda_2 f(a_2) + \lambda_3 f(a_3) + \lambda_4 f(a_4) \tag{24.29}$$

$$x = \lambda_1 a_1 + \lambda_2 a_2 + \lambda_3 a_3 + \lambda_4 a_4, \lambda_1 + \lambda_2 + \lambda_3 + \lambda_4 = 1, \text{ and } \lambda_i\text{'s} \ge 0$$

$i = 1, 2, 3, 4$

The following are the two restrictions: (a) maximum of two λ_i's with rest equal to zero and (b) only adjacent λ_i's are allowed to be positive (24.29)
Conditions in Eq. (24.29) are called restricted basis conditions.

In the aforementioned derivation $\hat{f}(x)$ is the piecewise linear approximation of a non-linear function $f(x)$ in the interval $[a_1, a_4]$. λ_i's are called the weights and a_i's are called the breaking points of the ith interval. Thus to solve an NLPP, we shall denote the breaking points of the ith variable x_i by a_i^1, a_i^2 and the weight associated with the Kth breaking point of the ith variable by λ_i^K.

Notes 1. Separable programming gives the approximate solution of the problem.
2. Greater the number of breaking points, better will be the approximation.

Procedure In solving an SPP, we use the following steps:

Step 1 Identify the terms that are not linear in the objective function and the constraints.

Step 2 Find the range of the variable x_i wherein terms are not linear and divide it into integer breaking points (for simplified calculation) a_i^K's (kth breaking points) and associate weights λ_i^K's with these points.

Step 3 Prepare a table of calculated values of the functions at the breaking points.

Step 4 Write the given problem in the form of an approximated LPP.

Step 5 Solve the approximated LPP using the simplex method, using restricted basis condition till optimality.

Example 24.7 Using separable programming, solve the following NLPP.

$$\text{Max } Z = x_1 + x_2^4$$

$$\text{Subject to} \quad 3x_1 + 2x_2^2 \le 9 \qquad (24.30)$$

$$\text{and} \quad x_1, x_2 \ge 0$$

Solution Here $f_1(x_1) = x_1$, $f_2(x_2) = x_2^4$, $g_1^1(x_1) = 3x_1$, $g_2^2(x_2) = 2x_2^2$

Here $f_2(x_2)$ and $g_2^2(x_2)$ are not linear. From the constraint in Eq. (24.30), $x_2 < 3$. Let us linearize $f_2(x_2)$ and $g_2^2(x_2)$ in $[0, 3]$

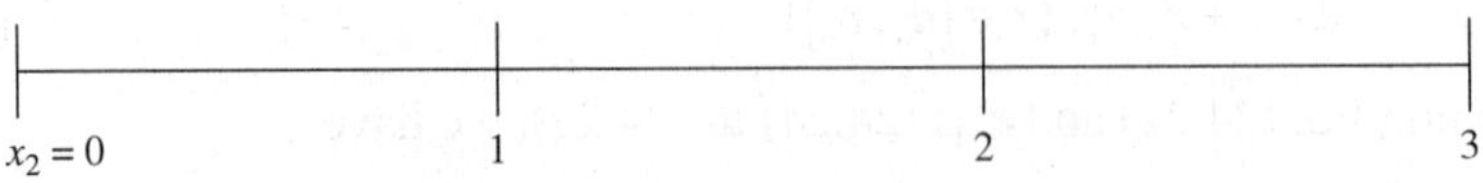

K	a_2^K	$f_2(a_2^K)$	$g_2^2(a_2^K)$
1	0	0	0
2	1	1	2
3	2	16	8
4	3	81	18

Thus, $f_2(x_2) \approx \hat{f}(x_2) = \lambda_2^1 f_2(a_2^1) + \lambda_2^2 f_2(a_2^2) + \lambda_2^3 f(a_2^3) + \lambda_2^4 f(a_2^4)$

$$= \lambda_2^2 + 16\lambda_2^3 + 81\lambda_2^4$$

and $g_2^1(x_2) \approx \hat{g}_2^1(x_2) = 2\lambda_2^2 + 8\lambda_2^3 + 18\lambda_2^4$

Thus, the given problem is approximated to the following LPP:

$$\text{Max } Z = f_1(x_1) + f_2(x_2) \approx x_1 + \lambda_2^2 + 16\lambda_2^3 + 81\lambda_2^4$$

$$\text{Subject to } \; 3x_1 + 2\lambda_2^2 + 8\lambda_2^3 + 18\lambda_2^4 \le 9$$

$$\lambda_2^1 + \lambda_2^2 + \lambda_2^3 + \lambda_2^4 = 1,\; x_1 \ge 0,\; \lambda_2^K \ge 0,\; K = 1, 2, 3, 4$$

and at the most two consecutive $\lambda_2^K > 0$ (restricted basis conditions).
The standard form of the aforementioned LPP is as follows:

$$\text{Max } Z = x_1 + \lambda_2^2 + 16\lambda_2^3 + 81\lambda_2^4$$

$$\text{Subject to } \; 3x_1 + 2\lambda_2^2 + 8\lambda_2^3 + 18\lambda_2^4 + x_3 = 9$$

$$\lambda_2^1 + \lambda_2^2 + \lambda_2^3 + \lambda_2^4 = 1$$

$x_1 \ge 0, \lambda_2^K \ge 0,\; K = 1, 2, 3, 4,\; x_3 \ge 0$ is a slack variable

		Cost →		1	1	16	81	0	0	Minimum ratio
		Variable →		x_1	λ_2^2	λ_2^3	λ_2^4	x_3	λ_2^1	$\frac{X_{B_i}}{\alpha_i^j}$
Table no.	C_B	**Basic variable**	X_B	α^1	α^2	α^3	α^4	α^5	α^6	
1	0	x_3	9	3	2	8	18	1	0	9/8
	0	$\leftarrow \lambda_2^1$	1	0	1	[1]	1	0	1	1/1 = 1 →
		$c_j - Z_j$		1	1	16↑	81	0	0	
2	0	$\leftarrow x_3$	1	3	−6	0	[10]	1	−8	1/10 →
	0	λ_2^3	1	0	1	1	1	0	1	1/1 = 1
		$c_j - Z_j$		1	−15	0	65↑	0	−16	
3	0	λ_2^4	1/10	3/10	−6/10	0	1	1/10	−8/10	
	0	λ_2^3	9/10	−3/10	16/10	1	0	−1/10	18/10	
		$c_j - Z_j$		−18.5	24	0	0	−65/10	36	

Now if λ_2^1 is the candidate to enter, then λ_2^3 will leave. However, λ_2^1 and λ_2^4 are not consecutive so λ_2^1 cannot enter. The next promising candidate to enter is λ_2^2. However, if λ_2^2 enters, then again λ_2^3 will have to leave, which is not possible as λ_2^2 and λ_2^4 are not consecutive. So table 3 gives the approximate optimal solution of the problem and is given by $\lambda_2^3 = 9/10$, $\lambda_2^4 = 1/10$, and Max $Z = 22.5$

$$\text{or } x_1 = 0,\; x_2^4 \approx \lambda_2^2 + 16\lambda_2^3 + 81\lambda_2^4 = 0 + \frac{144}{10} + \frac{81}{10} = \frac{225}{10} = 22.5.$$

Note If a variable in any separable programming is non-linear even at one place in the objective function or constraint(s), then that variable must be expressed in terms of λ_i^j's everywhere.

Example 24.8 Use separable programming to solve the following NLPP:

$$\text{Max } Z = x_1 + x_2^4$$

$$\text{Subject to } 3x_1^2 + 2x_2^2 \le 9,\ x_1,\ x_2 \ge 0$$

Solution Here $f_1(x_1) = x_1,\ f_2(x_2) = x_2^4,\ g_1^1(x_1) = 3x_1,\ g_2^1(x_2) = 2x_2^2$ (24.31)

From the constraint $3x_1^2 + 2x_2^2 \le 9$, we have $x_1 < 2$ and $x_2 < 3$.

Let $a_1^1 = 0,\ a_1^2 = 1,\ a_1^3 = 2,$ and $a_2^1 = 0,\ a_2^2 = 1,\ a_2^3 = 2,$ and $a_2^4 = 3$ be the breaking points for variables x_1 and x_2 respectively.

We calculate the values of f_i's and g_i's as per the following tables:

Table 24.1 Values of f_i and g_i with respect to a_1^K

K	a_1^K	$f_1(a_1^K)$	$g_1^1(a_1^K)$
1	0	0	0
2	1	1	3
3	2	2	12

Table 24.2 Values of f_i and g_i with respect to a_2^K

K	a_2^K	$f_2(a_2^K)$	$g_2^1(a_2^K)$
1	0	0	0
2	1	1	2
3	2	16	8
4	3	81	18

Using calculated values in Tables 24.1 and 24.2 and from Eq. (24.31), we have

$$f_1(x_1) \approx \lambda_1^1 \times 0 + \lambda_1^2 \times 1 + \lambda_1^3 \times 2 = \lambda_1^2 + 2\lambda_1^3$$

$$f_2(x_2) = \lambda_2^2 + 16\lambda_2^3 + 81\lambda_2^4$$

$$g_1^{\,1}(x_1) = 3\lambda_1^2 + 12\lambda_1^3$$

$$g_2^1(x_2) = 2\lambda_2^2 + 8\lambda_2^3 + 18\lambda_2^4$$

Thus, the given problem approximated in the form of an LPP is as follows:

$$\text{Max } Z = \lambda_1^2 + 2\lambda_1^3 + \lambda_2^2 + 16\lambda_2^3 + 81\lambda_2^4$$

$$\text{Subject to } 3\lambda_1^2 + 12\lambda_1^3 + 2\lambda_2^2 + 8\lambda_2^3 + 18\lambda_2^4 \le 9$$

$$\lambda_1^1 + \lambda_1^2 + \lambda_1^3 = 1$$

$$\lambda_2^1 + \lambda_2^2 + \lambda_2^3 + \lambda_2^4 = 1$$

and the restricted basis conditions are all λ_i^j's≥ 0; $i=1, 2; j=1, 2, 3, 4$ should be consecutive.

The standard form is as follows:

$$\text{Max } Z = \lambda_1^2 + 2\lambda_1^3 + \lambda_2^2 + 16\lambda_2^3 + 81\lambda_2^4$$

$$\text{Subject to } 3\lambda_1^2 + 12\lambda_1^3 + 2\lambda_2^2 + 8\lambda_2^3 + 18\lambda_2^4 = 9$$

$$\lambda_1^1 + \lambda_1^2 + \lambda_1^3 = 1$$

$$\lambda_2^1 + \lambda_2^2 + \lambda_2^3 + \lambda_2^4 = 1$$

Almost two λ_1^j's>0 and λ_2^j's>0 and they should be consecutive.

Cost→				1	2	1	16	81	0	0	0	Minimum ratio
Variable→				λ_2^1	λ_1^3	λ_2^2	λ_2^3	λ_2^4	x_3	λ_1^1	λ_2^1	$\frac{X_{B_i}}{\alpha_i^j}$
Table no.	C_B	**Basic variable**	X_B	α^1	α^2	α^3	α^4	α^5	α^6	α^7	α^8	
1	0	x_3	9	3	12	2	8	18	1	0	0	9/8
	0	λ_1^1	1	1	1	0	0	0	0	1	0	…
	0	$\leftarrow\lambda_2^1$	1	0	0	1	[1]	1	0	0	1	$1/1 = 1\rightarrow$
	$c_j - Z_j$			1	2	1	$16\uparrow$	81	0	0	0	–
2	0	$\leftarrow x_3$	1	3	12	–6	0	[10]	1	0	–8	$1/10\rightarrow$
	0	λ_1^1	1	1	1	0	0	0	0	1	0	…
	16	λ_2^3	1	0	0	1	1	1	0	0	1	$1/1 = 1$
	$c_j - Z_j$			1	2	–15	0	$65\uparrow$	0	0	–16	
3	81	λ_2^4	1/10	3/10	12/10	–6/10	0	1	1/10	0	–8/10	
	0	λ_1^1	1	1	1	0	0	0	0	1	0	
	16	λ_2^3	9/10	–3/10	–12/10	16/10	1	0	–1/10	0	18/10	
	$c_j - Z_j$			–37/2	–76	24	0	0	–13/2	0	36	

In table 1, λ_2^4 is the candidate to enter the basis, however it cannot enter because of the restriction in basis conditions. The next promising candidate to enter the basis is λ_2^3. Now when λ_2^3 enters the basis, λ_2^1 leaves. In table 2, λ_2^4 enters and x_3 leaves. In table 3, λ_2^1 and λ_2^2 are the candidates to enter but they cannot enter because of the restrictions in basic conditions in both cases.

Hence, the approximate optimal solution is $\lambda_1^1 = 1$, $\lambda_2^3 = 9/10$, $\lambda_2^4 = 1/10$.

Further, $x_1 = 0\lambda_1^1 + 1\lambda_1^2 + 2\lambda_1^3 = 0$

$$x_2 = 0\lambda_2^1 + 1\cdot\lambda_2^2 + 2\lambda_2^3 + 3\lambda_2^4 = 2\times\frac{9}{10} + 3\times\frac{1}{10} = \frac{18+3}{10} = 2.1$$

That is, $x_1 = 0$, $x_2 = 2.1$, and Max $Z = 45/2$

RECAPITULATION

1. A function $f(x)$ is said to be convex over a convex set S, if for any two point x_1 and x_2 in S $f[\lambda x_1 + (1-\lambda)x_2] \leq \lambda f(x_1) + (1-\lambda) f(x_2)$; $0 \leq \lambda \leq 1$.
2. A function $f(x)$ is said to be concave over a convex set S if for any two points x_1 and x_2 in S $f[\lambda x_1 + (1-\lambda)x_2] \geq \lambda f(x_1) + (1-\lambda) f(x_2)$; $0 \leq \lambda \leq 1$.
3. Kuhn–Tucker conditions are used to solve inequality constrained problems.
4. An NLPP in which the objective function is quadratic and constraints are linear is known as a QPP.
5. Separable programming is a mathematical programming wherein the functions or terms involved in an objective function and constraints are separable.

EXERCISES

Multiple-choice Questions

24.1 Consider the following problem:
Maximize $Z = f(X)$
Subject to $g_j(X) \le 0; j = 1, 2, 3, \ldots, m$
For the function $f(X)$ to have maxima, we have

$$\frac{\partial f}{\partial x_i} + \sum_{j=1}^{m} \lambda_j \frac{\partial g_j}{\partial x_i} = 0; i = 1, 2, 3, \ldots, m$$

$\lambda_j g_j(X) = 0; j = 1, 2, 3, \ldots, m$ and ________.

(a) $\lambda_j \le 0$
(b) $\lambda_j \ge 0$
(c) $\lambda_j > 0$
(d) none of these

24.2 A function $f(x)$ is said to be convex over a convex set S, if for any two points x_1 and x_2 in S $f[\lambda x_j + (1 - \lambda)x_2] \le \lambda f(x_1) + (1 - \lambda)f(x_2); 0 \le \lambda \le 1$

(a) $f[\lambda x_1 + (1 - \lambda)x_2] \le \lambda f(x_1) + (1 - \lambda) f(x_2); 0 \le \lambda \le 1$
(b) $f[\lambda x_1 + (1 - \lambda)x_2] \ge \lambda f(x_1) + (1 - \lambda) f(x_2); 0 \le \lambda \le 1$
(c) $f[\lambda x_1 + (1 - \lambda)x_2] \le \lambda f(x_1) + (1 - \lambda) f(x_2); \lambda \ge 0$
(d) None of these

24.3 While solving an NLPP, it is reduced (approximately) to an LPP if it is

(a) an SPP
(b) a QPP
(c) both (a) and (b)
(d) none of these

24.4 Every NLPP can be solved using

(a) an SPP
(b) a QPP
(c) both (a) and (b)
(d) none of these

24.5 An approximate (not exact) solution of an NLPP is obtained using

(a) an SPP
(b) a QPP
(c) both (a) and (b)
(d) none of these

State True or False

24.1 Consider the following optimization problem:
Minimize $Z = f(X)$
Subject to $g_j(X) \le 0; j = 1, 2, 3, \ldots, m$
For the function $f(X)$ to have maxima, we have,

$$\frac{\partial f}{\partial x_i} + \sum_{j=1}^{m} \lambda_j \frac{\partial g_j}{\partial x_i} = 0; i = 1, 2, 3, \ldots, m$$

$\lambda_j g_j(X) = 0; j = 1, 2, 3, \ldots, m$ and $\lambda_j \ge 0$.

24.2 A function $f(x)$ is said to be concave over a convex set S if for any two points x_1 and x_2 in S $f[\lambda x_1 + (1 - \lambda)x_2] \ge \lambda f(x_1) + (1 - \lambda)f(x_2); 0 \le \lambda \le 1$.

24.3 Every NLPP can be solved using a separable programming problem.

24.4 In a mathematical programming problem, when the objective function is quadratic and constraints are linear, we can use the quadratic programming techniques.

Fill in the Blanks

24.1 A function $f(x)$ is said to be convex over a convex set S, if for any two points x_1 and x_2 in S $f[\lambda x_1 + (1-\lambda)x_2]$ ________ $\lambda f(x_1) + (1-\lambda)f(x_2)$; $0 \le \lambda \le 1$.

24.2 Consider the following optimization problem:
Maximize $Z = f(X)$
Subject to $g_j(X) \ge 0; j = 1, 2, 3, \ldots, m$
For the function $f(X)$ to have maxima, we have

$$\frac{\partial f}{\partial x_i} + \sum_{j=1}^{m} \lambda_j \frac{\partial g_j}{\partial x_i} = 0;\ i = 1, 2, 3, \ldots, m$$

$\lambda_j g_j(X) = 0; j = 1, 2, 3, \ldots, m$ and λ_j ________ 0.

24.3 The separable programming problem cannot be used if the objective function is separable but constraints are ________ function of the decision variables.

24.4 Quadratic programming techniques provide an ________ solution to an NLPP.

Review Questions

24.1 Explain the convex and concave functions.
24.2 Explain Kuhn–Tucker conditions in detail.
24.3 What is quadratic programming? Explain Wolfe's method.
24.4 Write the procedure to solve a QPP using the Beale's method.
24.5 What do you mean by separable programming?
24.6 Explain the restricted basis conditions used for solving separable programming problems.

Numerical Problems

24.1 Use the Kuhn–Tucker conditions to solve the following problems:

(a) Max $f(x_1, x_2) = 8x_1^2 + 2x_2^2$
Subject to $x_1^2 + x_2^2 \le 9$
and $x_1 \ge 0$

(b) Max $f(x_1, x_2) = -x_1^2 - x_2^2 + x_1x_2 + 8x_1 + 4x_2$
Subject to $2x_1 + 3x_2 \le 24$
$-5x_1 + 12x_2 \le 24$
$x_2 \le 5$

24.2 State the Kuhn–Tucker conditions. Use them to solve
Min $f(x, y, z) = x^2 + y^2 + z^2 + 20x + 10y$
Subject to $x \ge 40$
$x + y \ge 80$
$x + y + z \ge 120$

24.3 Use Wolfe's modified simplex method to solve the following QPP:

(a) Max $Z = 2x_1 + x_2 - x_1^2$
Subject to $2x_1 + 3x_2 \le 6$
$2x_1 + x_2 \le 4$
and $x_1, x_2 \ge 0$

(b) Min $f = x_1^2 + 2x_2^2 - 2x_1x_2 - 6x_1$
Subject to $2x_1 - x_2 \le 16$
and $x_1, x_2 \ge 0$

24.4 Use Beale's method to solve the following QPP:

(a) Max $f = 4x_1 + 6x_2 - x_1^2 - 3x_2^2$
Subject to $x_1 + 2x_2 \le 4$
and $x_1, x_2 \ge 0$

(b) Max $f(X) = \frac{1}{4}(2x_3 - x_1) - \frac{1}{2}(x_1^2 + x_2^2 + x_3^2)$
Subject to $x_1 - x_2 + x_3 = 1$
and $x_1, x_2, x_3 \ge 0$

24.5 Use separable programming to solve the following QPP:

(a) Max $f(X) = 3x_1 + 2x_2$
Subject to $4x_1^2 + x_2^2 \leq 16$
and $x_1, x_2 \geq 0$

(b) Max $f(X) = 16 - 2(x_1 - 3)^2 - (x_2 - 7)^2$
Subject to $x_1^2 + x_2 \leq 16$
and $x_1, x_2 \geq 0$

CONCEPTUAL PROBLEM

Here is a case study of the price–production relationship of two products manufactured by a company. The company manufactures bulbs and tube lights. While bulbs with a quantity demand of $1500 - 5p_1$ carry a unit price of p_1, tube lights with a quantity demand of $3800 - 10p_2$ carry a unit price of p_2. The total cost functions for these two products are given by $200x_1 + 0.1x_1^2$ and $300x_2 + 0.1 - x_2^2$ respectively. The production takes place on two assembly lines. While bulbs are assembled on assembly line A, tube lights are assembled on assembly line B. Because of the limitations of the assembly line capacities, the daily production is limited to a maximum of 80 bulbs and 60 tube lights. The production of both types of products requires electronic components. The production of each of these sets requires five units and six units of electronic equipment components respectively. The electronic components are supplied by another manufacturer, and the supply is limited to 600 units per day. The company has 160 employees. While the production of one unit of bulb requires one man-day of labour, 2 man-days of labour are required for the production of one unit of tube lights. How many units of bulbs and tube lights should the company produce to maximize the total profit? Formulate the problem as an NLPP.

Solution Consider x_1 and x_2 as the quantities of bulbs and tube lights manufactured by the company respectively. Then, we have $x_1 = 1500 - 5p_1$ or $p_1 = 300 - 0.2x_1$ and $x_2 = 3800 - 10p_2$ or $p_2 = 380 - 0.1x_2$. It is also given that the total cost functions for these two products are $200x_1 + 0.1x_1^2$ and $300x_2 + 0.1x_2^2$ respectively. Then, we have $C_1 = 200x_1 + 0.1x_1^2$ and $C_2 = 300x_2 + 0.1x_2^2$. The revenue on bulbs and tube lights are p_1x_1 and p_2x_2 respectively. Thus, the total revenue is as follows:

$$R = p_1x_1 + p_2x_2 = (300 - 0.2x_1)x_1 + (380 - 0.1x_2)x_2 = 300x_1 - 0.2x_1^2 + 380x_2 - 0.1x_2^2$$

The total profit $Z = \text{Revenue} - \text{Total cost} = R - C_1 - C_2 = 300x_1 - 0.2x_1^2 + 380x_2 - 0.1x_2^2$

$$- 200x_1 - 0.1x_1^2 - 300x_2 - 0.1x_2^2$$
$$= 100x_1 - 0.3x_1^2 + 80x_2 - 0.2x_2^2$$

Since a maximum of 80 bulbs can be assembled on assembly line A and 60 tube lights on assembly line B per day, we have $x_1 \leq 80$ and $x_2 \leq 60$. The daily requirement of the electronic components is $5x_1 + 6x_2 \leq 600$. As the availability of employees is limited to 160 man-days, we have $x_1 + 2x_2 \leq 160$. Since the manufacturer cannot produce a negative number of units, we have $x_1, x_2 \geq 0$. Hence, the formulation of the NLPP is as follows:

Max $Z = 100x_1 - 0.3x_1^2 + 80x_2 - 0.2x_2^2$
Subject to $5x_1 + 6x_2 \leq 600$
$x_1 + 2x_2 \leq 160$
$x_1 \leq 80$
$x_2 \leq 60$
and $x_1, x_2 \geq 0$

Appendices

A. LINEAR ALGEBRA

The following terms are used frequently in linear algebra.

Linear combination Let $X_1, X_2, \ldots, X_n$ be n vectors from vector space E_n and $\alpha_1, \alpha_2, \ldots, \alpha_n$ be n scalars (real numbers). Then $X = \alpha_1 X_1 + \alpha_2 X_2 + \ldots + \alpha_n X_n$ is called a linear combination of vectors $X_1, X_2, \ldots, X_n$.

Linear dependence A set $\{X_1, X_2, \ldots, X_n\}$ of vectors in E_n is said to be linearly dependent, if there exists real numbers $\alpha_1, \alpha_2, \ldots, \alpha_n$, not all zero, such that

$$X = \alpha_1 X_1 + \alpha_2 X_2 + \ldots + \alpha_n X_n = 0.$$

Every superset of a linearly dependent set is linearly dependent.

Linear independence A set $\{X_1, X_2, \ldots, X_n\}$ of vectors in E_n is said to be linearly independent, if $X = \alpha_1 X_1 + \alpha_2 X_2 + \ldots + \alpha_n X_n = 0$ holds if and only if $\alpha_1 = \alpha_2 = \ldots = \alpha_n = 0$. Every subset of a linearly independent set is linearly independent.

An infinite subset S of E_n is said to be linearly independent, if every finite subset of S is linearly independent.

Span of a set The span of a set, $S \subset E_n$, is the set of all linear combinations of vectors in S. The span of S is denoted by $[S]$.

$$\text{Thus } [S] = \{\alpha_1 X_1 + \alpha_2 X_2 + \ldots + \alpha_k X_k / X_i \in S \text{ and } \alpha_i \in R, i = 1, 2, \ldots, k\}$$

Basis A subset B of E_n is called the basis of E_n if (a) B is linearly independent and (b) B spans E_n that is $[B] = E_n$.

Dimension The number of vectors in the basis of E_n is called the dimension of E_n. If the basis has a finite number of vectors, then the vector space is called *finite dimensional*, otherwise it is called *infinite dimensional*.

Note Basis of a vector space is not unique. A vector space can have an unlimited number of bases. However, the number of vectors in every basis of a vector space is the same.

Rank of matrix The number of linearly independent columns (rows) of a matrix is called the rank of the matrix.

B. MATRICES

The following are the basic terms used with respect to matrices.

Definition of matrix A rectangular array of numbers of the form

$$A = \begin{bmatrix} a_{11} & a_{12} & \dots & a_{1n} \\ a_{21} & a_{22} & \dots & a_{2n} \\ \dots & \dots & \dots & \dots \\ a_{m1} & a_{m2} & \dots & a_{mn} \end{bmatrix}$$ is called a matrix of order $m \times n$. We can also denote this matrix as $A = [a_{ij}]_{m\times n}$.

Notes
1. The horizontal lines are called rows and the vertical lines are called columns.
2. A $1\times n$ matrix is called a row matrix or row vector. For example A = [3 1 4 0].
3. An $m\times 1$ matrix is called a column matrix or a column vector. For example
$$A = \begin{bmatrix} 4 \\ 2 \\ 3 \end{bmatrix}.$$
4. If a matrix contains equal number of rows and columns, then it is called a square matrix.
5. A square matrix is called a unit matrix if all the elements along the principal diagonal are equal to one and the remaining elements are all zero. A unit matrix is also called an identity matrix and is denoted as I.

Sub-matrix of matrix If A is a matrix of order $m \times n$, then another matrix obtained by leaving some rows and columns of A is called the sub-matrix of A.

Determinant of matrix The determinant formed by the elements of a square matrix A is called the determinant of A and is denoted by $|A|$ or det. A.

If $|A| = 0$, then A is called a singular matrix and if $|A| \neq 0$, then A is called a non-singular matrix.

Transpose of matrix Let $A = [a_{ij}]_{m\times n}$, then a matrix of order $n \times m$ obtained by changing the rows of A into columns and columns of A into rows is called the transpose of A, and is denoted by A' or A^t or A^T.

Inverse of square matrix A square matrix A is said to be invertible if there exists another square matrix B of the same order such that $AB = I = BA$. Here B is called inverse of A or $A^{-1} = B$.

Note A matrix is invertible if and only if it is non-singular (i.e., $|A| \neq 0$).

Rank of matrix A non-negative number (integer) r is called the rank of a matrix A if
1. there is at least one square sub-matrix of order r, whose determinant is non-zero;
2. every square sub-matrix of order $r+1$ or more (if any) is such that its determinant of the sub-matrix is zero.

Elementary transformations of matrix An elementary transformation is an operation of any one of the following types:
1. The interchange of any two rows (columns)
2. The multiplication of the elements of a row (column) by a non-zero number
3. The addition to the elements of any row (column) the constant multiple of the corresponding elements of any other row (column)

C. CALCULUS

The following terms are important with regard to calculus.

Differentiability A function $f(x)$ is differentiable if the limit $\lim_{\Delta x \to 0} \frac{f(x+\Delta x)-f(x)}{\Delta x}$ exists.

Partial derivatives Let $u = f(x, y)$ be the function of the two variables x and y. The partial derivative of u with respect to x is denoted as $\partial u/\partial x$ or u_x, which is defined as $\frac{\partial f}{\partial x} = u_x = \lim_{\Delta x \to 0} \frac{f(x+\Delta x, y) - f(x,y)}{\Delta x}$, provided the limit exists, and the partial derivative of u with respect to y is $u_y = \lim_{\Delta y \to 0} \frac{f(x, y+\Delta y) - f(x,y)}{\Delta y}$.

Gradient vector If f is the function of three variables, then the gradient of f, denoted as grad f or ∇f (del f), is defined as grad $f = i\frac{\partial f}{\partial x} + j\frac{\partial f}{\partial y} + k\frac{\partial f}{\partial z}$. The gradient of f is always normal to the surface represented by $f(x, y, z) = 0$.

Directional derivative The directional derivative of $f(x, y)$ at the point P_0 in the direction of the unit vector $\vec{u}$ denoted $\left(D_{\vec{u}} f\right)_{P_0}$ is given as $\left(D_{\vec{u}} f\right)_{P_0} = (\nabla f)_{P_0} \cdot \vec{u}$.

The directional derivative of f is maximum in the direction of grad $f(\nabla f)$ and minimum in the direction of $(-\nabla f)$.

Taylor's theorem Let $f(X)$, $X=(x_1, x_2, \ldots, x_n)$ and its first $(m+1)$ partial derivatives be defined and continuous at a point $X_0 \in E_n$. Then $f(X) = f(X_0) +$

$$Df(X_0) + \frac{1}{2!}D^2 f(X_0) + \ldots + \frac{1}{m!}D^m f(X_0) + \frac{1}{(m+1)!}D^{m+1} f(c)$$

where $c = (X_0 + \theta h)$, $0 < \theta < 1$ and $h = X - X_0$.

The term $\frac{1}{(m+1)!}D^{m+1} f(c)$ is called the remainder after $(m+1)$ terms,

where $n = 1$, $X = x$, $X_0 = x_0$, and

$$D = \frac{d}{dx},\ Df(X_0) = f'(x_0)\,dx = h\,f'(x_0),\ D^2 f(x_0) = h^2 f''(x_0), \text{ and so on.}$$

When $n = 2$, $X = (x, y)$, $X_0 = (x_0, y_0)$, $D = h\frac{\partial}{\partial x} + k\frac{\partial}{\partial y}$ where $h = x - x_0$ and $k = y - y_0$

$$D^2 f = f_{xx}h^2 + 2f_{xy}hk + k^2 f_{yy} = \left(h\frac{\partial}{\partial x} + k\frac{\partial}{\partial y}\right)^2 f = (hf_x + kf_y)^2.$$

Convex and concave functions A function $f(x)$ is said to be convex over a convex set S, if for any two points x_1 and x_2 in S

$$f[\lambda x_1 + (1-\lambda)x_2] \le \lambda f(x_1) + (1-\lambda)f(x_2);\ 0 \le \lambda \le 1$$

A function $f(x)$ is said to be concave over a convex set S if for any two points x_1 and x_2 in S

$$f[\lambda x_1 + (1-\lambda)x_2] \geq \lambda f(x_1) + (1-\lambda)f(x_2);\ 0 \leq \lambda \leq 1$$

Notes
1. A function $f(x)$ is said to be convex, if the Hessian matrix $H(x)$ of the second partial derivatives is positive semi-definite.
2. A function $f(x)$ is said to be concave, if the Hessian matrix $H(x)$ of the second partial derivatives is negative semi-definite.

D. PROBABILITY

Some important definitions associated with probability are given here.

Mutually exclusive set of events Let $S_1, S_2, \ldots, S_n$ be n events in a sample space S. Then the set $\{S_1, S_2, \ldots, S_n\}$ is called mutually exclusive, if $S_i \cap S_j = \varphi$ (empty set) for $i \neq j$.

Exhaustive events Let $S_1, S_2, \ldots, S_n$ be n events in a sample space S, then $S_1, S_2, \ldots, S_n$ are called exhaustive events in S, if $\overset{n}{\underset{i=1}{U}} S_i = S$.

Conditional probability Let A and B be two events in a sample space S. The conditional probability of A, given that event B has occurred is denoted as $P(A/B)$ and is defined as $P\left(\frac{A}{B}\right) = \frac{P(A \cap B)}{P(B)}$.

If two events A and B are independent, then $P\left(\frac{A}{B}\right) = P(A)$ and $P\left(\frac{B}{A}\right) = P(B)$.

Hence $P\left(\frac{A}{B}\right) = \frac{P(A \cap B)}{P(B)}$. This implies that $P(A \cap B) = P(A)P(B)$.

Thus, two events are independent if the probability of their intersection equals the product of their respective probabilities.

Bayes' theorem If $A_1, A_2, \ldots, A_n$ are n mutually exclusive and exhaustive events in a sample space S and B is an arbitrary event in S, then $P\left(\frac{A_i}{B}\right) = \frac{P\left(\frac{B}{A_i}\right)P(A_i)}{\sum_{i=1}^{n} P\left(\frac{B}{A_i}\right)P(A_i)}$.

Random variables A random variable is a function from the sample space S to the set of real numbers R. Random variables are denoted by capital letters such as X, Y, and Z. Thus a random variable associated with each sample point (elementary event) in S is a unique real number $X(s)$. A random variable is also called a chance variable or a stochastic variable.

Example Experiment E: Toss a coin twice. Then sample space $S = \{$HH, HT, TH, TT$\}$ has four possible outcomes. Now, define a random variable X on S as $X =$ number of heads obtained. Then X can assume values 0, 1, or 2; that is, [X(TT) = 0, X(TH) = X(HT) = 1, and X(HH) = 2].

Continuous random variable A random variable X is said to be continuous, if it can assume any value within an interval, that is, all the values of a continuous scale. For example, the height of a group of individuals is a continuous random variable. Let X be a continuous random variable. A function $f(x)$ is called the probability density function (PDF) of X if (a) $f(x) \geq 0$; (b) $\int_{-\infty}^{\infty} f(x)dx = 1$. Further a function $F(x)$ defined as $F(x) = P(X \leq x) = \int_{-\infty}^{x} f(x)dx$ is called the cumulative distribution function or simply, the distribution function of X.

Further, $P(a \leq X \leq b) = \int_{a}^{b} f(x)dx = F(b) - F(a)$ and $\frac{d}{dx}F(x) = f(x)$.

Discrete random variable A random variable assuming isolated values is called a discrete random variable. As an example, the number of heads appearing in five tosses of a coin is a discrete random variable. If X represents the number of heads obtained in five tosses, then it assumes the values 0, 1, 2, 3, 4, or 5.

Let X be a discrete random variable assuming values $x_1, x_2, \ldots, x_n$ with probabilities $p_1, p_2, \ldots, p_n$ where $P(X = x_i) = p_i$, $i = 1, 2, \ldots, n$ and $p_1 + p_2 + \ldots + p_n = 1$.

Then, $\left.\begin{matrix} X : x_1, x_2, \ldots, x_n \\ P(X) : p_1, p_2, \ldots, p_n \end{matrix}\right\}$ is called the probability distribution (or point probability function) for X and it exhibits how a total probability of 1 is distributed over several values $x_1, x_2, \ldots, x_n$ of X.

The cumulative distribution function of X, denoted as $F(x)$ is defined as follows:

$F(x) = P(X \leq x) = \sum P(X \leq x_i) = \sum_{x_i < x} p_i$, where $P(X = x_i) = p_i$.

That is, $F(x)$ = sum of all those probabilities at x_i's with $x_i < x$.

Expected value and variance of random variable Expected (mean) value of a random variable is denoted as $E(X)$ or (μ) and is defined as follows:

1. $E(X) = \sum_{L=1}^{n} x_i p_i$, if X is discrete, assuming values $x_1, x_2, \ldots, x_n$ with probabilities $p_1, p_2, \ldots, p_n$ respectively.
2. $E(X) = \int_{-\infty}^{\infty} x f(x)dx$, if X is a continuous random variable with PDF $f(x)$.

Variance of a random variable X is denoted as $V(X)$ (or σ^2) and is defined as $V(X) = \sum_{i=1}^{n} p_i(x_i - \mu)$, if X is discrete and $V(X) = \int_{-\infty}^{\infty} (x - \mu)^2 f(x)dx$, if X is a continuous random variable.

That is, $V(X) = E(X^2) - [E(X)]^2$ (on simplification).

Binomial distribution Repeat experiment E, n times. Let A be an event such that $P(A) = p$ is constant in all repetitions of E. Now define a random variable X as X = number of times event A occurs during n repetitions of E.

The probability that X assumes value r is denoted as $P(X=r)$ and is given as

$$P(X=r)={}^nc_r p^r(1-p)^{n-r}, r=0, 1, 2, \ldots, n \qquad \text{D.1}$$

A random variable with the probability distribution given in Eq. D.1 is said to have binomial distribution with parameters n and p. If X is a binomial distribution, then $E(X) = np$ and $V(X) = np(1-p)$.

Poisson distribution A discrete random variable X having probability distribution $P(X = r) = \dfrac{e^{-\lambda}\lambda^r}{r!}$, $r = 0, 1, 2, 3, \ldots$ is called a Poisson distribution with parameter λ. This distribution is a limiting case of binomial distribution when the number of trials n is very large and the proportion of successes p is very small such that the product $np=\lambda$ is constant.

If X has a Poisson distribution, then $E(X) = V(X) = \lambda$.

Uniform distribution A random variable whose PDF $f(x)$ is of the type $f(x)= \dfrac{1}{b-a}$, $a \le x \le b$ is called uniform distribution. If a random variable X has uniform distribution over the interval $a \le x \le b$, then the expected value $E(X)$ and variance $V(X)$ of X are given as $\mu = E(X) = \dfrac{a+b}{2}$, $a \le x \le b$, and $\sigma^2 = V(X) = \dfrac{(b-a)^2}{12}$.

Exponential distribution A random variable X is said to have an exponential distribution with parameter λ if its PDF is of the type

$$f(x) = \lambda e^{-\lambda x}, x > 0, \lambda > 0, E(X) = \frac{1}{\lambda}, \text{ and } V(X) = \frac{1}{\lambda^2}$$

There is an important relationship between the exponential distribution and the Poisson distribution. If the Poisson distribution represents the number of failures per unit time, then the exponential distribution will describe the time between two successive failures. Similarly in the queuing system, if the number of customers arriving has Poisson distribution, then the inter-arrival time has exponential distribution.

Normal distribution A continuous random variable X is said to have normal distribution with parameters μ and σ if its density function $f(x)$ is of the form

$$f(x) = \frac{1}{\sigma\sqrt{2\pi}} e^{-\frac{1}{2}\{(x-\mu)/\sigma\}^2}, -\infty < x < \infty, -\infty < \mu < \infty, \sigma > 0.$$

with $E(X) = \mu$ and $V(X) = \sigma^2$.

It is usually denoted as $N(\mu, \sigma^2)$. If X is a normal random variable with mean μ and variance σ^2, then the random variable $Z = \dfrac{X-\mu}{\sigma}$ is said to have standard normal distribution and is denoted as $N(0, 1)$. The cumulative distribution function of Z, $F(z)$ is $F(z) = P(Z \le z) = \int_{-\infty}^{z} f(z)dz$ where $f(z) = \dfrac{1}{\sqrt{2\pi}} e^{-\frac{1}{2}z^2}$, $-\infty < z < \infty$.

The values of $F(z)$ have been tabulated in standard books on probability.

Functions of random variable The most important case arises when X is a continuous random variable with PDF $f(x)$ and H is a continuous function, then $Y = H(X)$ is a continuous random variable say, with PDF $g(y)$. To find $g(y)$, we have the following theorem:

Theorem Let X be a continuous random variable with PDF f, where $f(x) > 0$ for $a < x < b$. Assume $y = H(x)$ is a strictly monotonic (increasing or decreasing) function of x. Assuming that this function is differentiable and hence continuous for all x, the random variable Y defined as $Y = H(X)$ has a PDF given as

$$g(y) = f(x) \cdot \left|\frac{dx}{dy}\right|$$

where x is expressed in terms of y. If H is increasing, then g is non-zero, for those values of y satisfying $H(a) < y < H(b)$. If H decreases, then g is non-zero for those values of y satisfying $H(b) < y < H(a)$.

E. AREA UNDER THE STANDARD NORMAL DISTRIBUTION Z

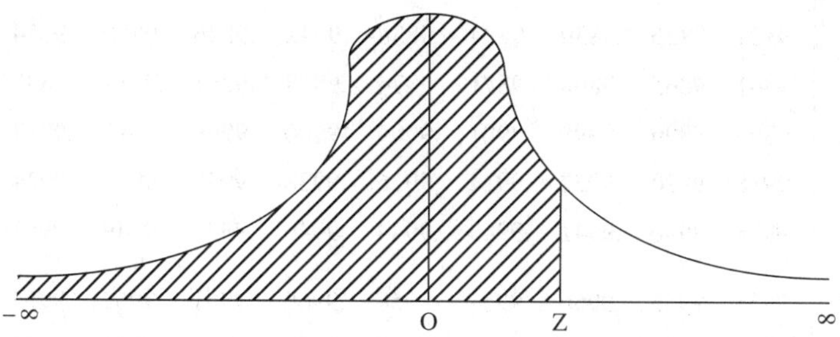

Shaded region $= P(Z \leq z) = \int_{-\infty}^{z} f(z)dz$

Z to first decimal	Second decimal									
	.00	.01	.02	.03	.04	.05	.06	.07	.08	.09
0.0	.5000	.5040	.5080	.5120	.5160	.5199	.5239	.5279	.5319	.5359
0.1	.5398	.5438	.5478	.5517	.5557	.5596	.5636	.5675	.5714	.5753
0.2	.5793	.5832	.5871	.5910	.5948	.5987	.6026	.6064	.6103	.6141
0.3	.6179	.6217	.6255	.6293	.6331	.6368	.6406	.6443	.6480	.6517
0.4	.6554	.6591	.6628	.6664	.6700	.6736	.6772	.6808	.6841	.6879
0.5	.6915	.6950	.6985	.9019	.7054	.7088	.7123	.7157	.7190	.7224
0.6	.7257	.7291	.7324	.7357	.7389	.7422	.7454	.7406	.7517	.7549
0.7	.7580	.7611	.7642	.7673	.7703	.7734	.7764	.7794	.7823	.7852
0.8	.7881	.7910	.7939	.7967	.7995	.8023	.8051	.8078	.8106	.8133
0.9	.8159	.8186	.8212	.8238	.8264	.8289	.8315	.8340	.8365	.8389
1.0	.8413	.8438	.8461	.8485	.8508	.8531	.8554	.8577	.8599	.8621

(Contd)

(*Contd*)

Z to first decimal	Second decimal									
	.00	.01	.02	.03	.04	.05	.06	.07	.08	.09
1.1	.8643	.8665	.8686	.8708	.8729	.8749	.8770	.8790	.8910	.8830
1.2	.8849	.8869	.8888	.8907	.8925	.8944	.8962	.8980	.8997	.9015
1.3	.9032	.9049	.9066	.9082	.9099	.9115	.9131	.9147	.9162	.9177
1.4	.9192	.9207	.9222	.9236	.9251	.9265	.9278	.9292	.9306	.9319
1.5	.9332	.9345	.9357	.9370	.9382	.9394	.9406	.9418	.9430	.9441
1.6	.9452	.9463	.9474	.9485	.9495	.9505	.9515	.9525	.9535	.9545
1.7	.9554	.9564	.9573	.9582	.9591	.9599	.9606	.9616	.9625	.9633
1.8	.9641	.9649	.9656	.9664	.9671	.9678	.9686	.9693	.9700	.9706
1.9	.9713	.9719	.9726	.9732	.9783	.9744	.9750	.9756	.9762	.9767
2.0	.9772	.9778	.9783	.9788	.9793	.9798	.9803	.9808	.9812	.9817
2.1	.9821	.9826	.9830	.9834	.9838	.9842	.9846	.9850	.9854	.9857
2.2	.9861	.9865	.9868	.9871	.9874	.9878	.9881	.9884	.9887	.9890
2.3	.9893	.9896	.9898	.9901	.9904	.9906	.9909	.9911	.9913	.9916
2.4	.9918	.9920	.9922	.9924	.9926	.9928	.9930	.9932	.9934	.9936
2.5	.9938	.9940	.9941	.9943	.9944	.9946	.9932	.9949	.9951	.9952
2.6	.9953	.9955	.9956	.9957	.9958	.9960	.9961	.9962	.9963	.9964
2.7	.9965	.9966	.9967	.9968	.9969	.9970	.9971	.9972	.9973	.9974
2.8	.9974	.9975	.9976	.9977	.9977	.9979	.9979	.9979	.9980	.9981
2.9	.9981	.9982	.9982	.9983	.9984	.9984	.9985	.9985	.9986	.9986
3.0	.9986	.9987	.9987	.9988	.9988	.9988	.9999	.9989	.9990	.9990

F. TABLE OF RANDOM NUMBERS

39 65 76 45 45	19 90 69 64 61	20 26 36 31 62	58 24 57 14 97	95 06 70 99 00
93 71 23 70 90	65 97 60 12 11	31 56 34 19 19	47 83 75 51 33	30 62 38 20 44
72 20 47 33 84	51 67 47 97 19	98 40 07 17 66	23 05 09 51 80	59 78 11 52 69
75 17 25 69 17	17 95 21 78 58	24 33 45 77 48	69 81 84 09 29	93 22 70 45 80
37 48 79 88 74	63 52 06 34 30	01 31 60 10 27	35 07 79 71 53	28 99 52 01 64
02 89 08 16 94	85 53 83 29 95	56 27 09 24 43	21 78 55 09 82	72 61 88 73 61
87 18 15 70 07	37 79 49 12 38	48 13 93 15 96	41 92 45 71 51	09 18 25 58 94
98 83 71 70 15	89 09 39 59 24	00 06 41 14 20	14 36 59 25 47	54 45 17 24 89
10 08 58 07 04	76 62 16 48 68	58 76 17 14 86	59 53 11 52 21	66 04 18 72 87
47 90 56 37 31	71 82 13 50 14	27 55 10 24 92	28 04 67 53 44	95 23 00 84 47

93 05 31 03 07	34 18 04 52 35	74 13 39 35 22	68 95 23 92 35	36 63 70 35 31
21 89 11 47 99	11 20 99 45 18	76 51 94 84 86	13 79 93 37 55	98 16 04 41 67
95 18 94 36 97	27 37 83 28 71	79 57 95 13 91	09 61 87 25 21	56 11 20 32 44
97 08 31 55 73	10 65 81 92 59	77 31 61 95 46	20 44 90 32 64	26 99 76 75 63
69 26 88 86 13	59 71 74 17 32	48 38 75 93 29	73 37 32 04 05	60 82 29 20 25
41 47 10 25 03	87 63 93 95 17	81 83 83 04 49	77 45 85 50 51	79 88 01 97 30
91 94 15 63 62	08 61 74 51 68	92 79 43 83 79	29 18 94 51 23	14 85 11 47 23
80 06 54 18 47	08 52 85 08 40	48 40 35 94 22	72 65 71 08 86	50 03 42 99 36
76 72 77 63 99	89 85 84 46 06	64 71 06 21 66	89 37 20 70 01	61 65 70 22 12
59 40 24 13 75	42 29 82 23 19	06 94 79 10 08	81 30 15 39 14	81 83 17 16 33
63 62 06 34 41	79 53 36 02 95	94 61 09 43 62	20 21 14 68 86	84 95 48 46 45
78 47 23 53 90	79 93 96 38 63	34 85 52 05 09	84 43 01 72 73	14 93 87 81 40
87 68 62 15 43	97 48 72 66 48	53 16 71 13 81	59 97 50 99 52	24 62 20 42 31
47 60 92 10 77	26 97 05 73 51	88 46 38 00 58	72 68 49 29 31	75 70 16 08 24
56 88 87 59 41	06 87 37 78 48	65 88 69 58 39	88 02 84 27 83	85 81 56 59 38
22 17 68 65 84	86 02 22 57 51	68 69 80 95 44	11 29 01 95 80	49 34 35 96 47
19 36 27 59 46	39 77 32 77 09	79 57 92 36 39	89 74 39 82 15	08 50 94 34 74
16 77 23 02 77	28 06 24 25 93	22 45 44 84 11	87 80 61 65 31	09 71 91 74 25
78 43 66 71 61	97 66 63 99 61	80 45 67 93 82	59 73 19 85 23	53 33 65 97 21
03 28 28 26 08	69 30 16 09 05	53 58 47 70 93	66 56 45 63 79	45 56 20 19 47
04 31 17 21 56	33 63 99 19 87	26 72 39 27 67	53 77 57 69 93	60 61 97 22 61
61 06 98 03 91	87 14 77 43 96	43 00 65 98 50	45 60 33 01 07	98 99 46 50 47
23 58 35 26 00	99 53 93 61 28	52 70 05 48 34	56 65 05 61 86	90 92 10 78 80
15 39 25 70 99	93 86 52 77 65	15 35 59 05 28	22 87 26 07 47	86 96 98 29 06
58 71 96 30 24	18 46 23 34 27	85 13 99 24 44	49 18 09 79 49	74 16 32 23 02
93 22 53 64 39	07 10 63 76 35	87 03 04 79 88	08 33 33 85 51	55 34 57 72 69
78 76 58 54 74	92 38 70 96 92	52 06 79 79 45	82 63 18 27 44	69 66 92 19 09
61 81 31 96 82	00 57 25 60 29	46 72 60 18 77	55 66 12 62 11	08 99 55 64 57
42 88 31 96 82	24 98 65 08 21	47 21 61 88 32	27 80 30 21 60	10 92 35 64 57
77 94 30 05 33	28 10 99 00 27	12 73 73 99 12	39 99 57 94 82	96 88 57 17 91

G. POISSON PROBABILITY DISTRIBUTION *X* TABLE

$$P(X = r) = \frac{\lambda^r e^{-\lambda}}{r!}, \lambda > 0, r = 0, 1, 2, \ldots$$

$r \rightarrow$ $\lambda \downarrow$	0	1	2	3	4	5	6	7	8	9	10	11	12	13	14	15
0.1	.9048	.0905	.0045	.0002	.0000	.0000	.0000	.0000	.0000	.0000	.0000	.0000	.0000	.0000	.0000	.0000
0.2	.8187	.1637	.0164	.0001	.0000	.0000	.0000	.0000	.0000	.0000	.0000	.0000	.0000	.0000	.0000	.0000
0.3	.7408	.2222	.0333	.0033	.0002	.0000	.0000	.0000	.0000	.0000	.0000	.0000	.0000	.0000	.0000	.0000
0.4	.6703	.2681	.0536	.0072	.0007	.0001	.0000	.0000	.0000	.0000	.0000	.0000	.0000	.0000	.0000	.0000
0.5	.6065	.3033	.0758	0.126	.0016	.0002	.0000	.0000	.0000	.0000	.0000	.0000	.0000	.0000	.0000	.0000
0.6	.5488	.3293	.0988	.0198	.0030	.0004	.0000	.0000	.0000	.0000	.0000	.0000	.0000	.0000	.0000	.0000
0.7	.4966	.3476	.1217	.0284	.0050	.0007	.0001	.0000	.0000	.0000	.0000	.0000	.0000	.0000	.0000	.0000
0.8	.4493	.3595	.1438	.0383	.0077	.0012	.0002	.0000	.0000	.0000	.0000	.0000	.0000	.0000	.0000	.0000
0.9	.4066	.3659	.1647	.0494	.0111	.0020	.0003	.0000	.0000	.0000	.0000	.0000	.0000	.0000	.0000	.0000
1.0	.3679	.3679	.1839	.0613	.0153	.0031	.0005	.0001	.0000	.0000	.0000	.0000	.0000	.0000	.0000	.0000
1.1	.3329	.3662	.2014	.0738	.0203	.0045	.0008	.0001	.0000	.0000	.0000	.0000	.0000	.0000	.0000	.0000
1.2	.3012	.3614	.2169	.0867	.0260	.0062	.0012	.0002	.0000	.0000	.0000	.0000	.0000	.0000	.0000	.0000
1.3	.2725	.3543	.2303	.0998	.0324	.0084	.0018	.0013	.0001	.0000	.0000	.0000	.0000	.0000	.0000	.0000
1.4	.2466	.3452	.2417	.1128	.0395	.0111	.0026	.0005	.0001	.0000	.0000	.0000	.0000	.0000	.0000	.0000
1.5	.2231	.3347	.2510	.1255	.0471	.0141	.0035	.0008	.0001	.0000	.0000	.0000	.0000	.0000	.0000	.0000
1.6	.2019	.3230	.2584	.1378	.0551	.0176	.0047	.0011	.0002	.0000	.0000	.0000	.0000	.0000	.0000	.0000
1.7	.1827	.3106	.2640	.1496	.0636	.0216	.0061	.0015	.0003	.0001	.0000	.0000	.0000	.0000	.0000	.0000
1.8	.1653	.2975	.2678	.1607	.0723	.0260	.0078	.0020	.0005	.0001	.0000	.0000	.0000	.0000	.0000	.0000
1.9	.1496	.2842	.2700	.1710	.0812	.0309	.0098	.0027	.0006	.0001	.0000	.0000	.0000	.0000	.0000	.0000
2.0	.1353	.2707	.2707	.1804	.0902	.0361	.0120	.0034	.0009	.0002	.0000	.0000	.0000	.0000	.0000	.0000

(Contd)

(*Contd*)

$r \rightarrow$ $\lambda \downarrow$	0	1	2	3	4	5	6	7	8	9	10	11	12	13	14	15
2.1	.1225	.2572	.2700	.1890	.0992	.0417	.0146	.0044	.0011	.0003	.0001	.0000	.0000	.0000	.0000	.0000
2.2	.1108	.2438	.2681	.1966	.1082	.0476	.0174	.0055	.0015	.0004	.0001	.0000	.0000	.0000	.0000	.0000
2.3	.1003	.2306	.2652	.2033	.1169	.0538	.0206	.0068	.0019	.0005	.0001	.0000	.0000	.0000	.0000	.0000
2.4	.0907	.2177	.2613	.2090	.1254	.0602	.0241	.0083	.0025	.0007	.0002	.0000	.0000	.0000	.0000	.0000
2.5	.0821	.2052	.2565	.2138	.1336	.0668	.0278	.0099	.0031	.0009	.0002	.0000	.0000	.0000	.0000	.0000
2.6	.0743	.1931	.2510	.2176	.1414	.0735	.0319	.0118	.0038	.0011	.0003	.0001	.0000	.0000	.0000	.0000
2.7	.0672	.1815	.2450	.2205	.1488	.0804	.0362	.0139	.0047	.0014	.0004	.0001	.0000	.0000	.0000	.0000
2.8	.0608	.1703	.2384	.2225	.1557	.0872	.0407	.0163	.0057	.0018	.0005	.0001	.0000	.0000	.0000	.0000
2.9	.0550	.1596	.2314	.2237	.1622	.0940	.0455	.0188	.0068	.0022	.0006	.0002	.0000	.0000	.0000	.0000
3.0	.0498	.1494	.2240	.2240	.1680	.1008	.0504	.0216	.0081	.0027	.0008	.0002	.0001	.0000	.0000	.0000
3.1	.0450	.1397	.2165	.2237	.1734	.1075	.0555	.0246	.0095	.0033	.0010	.0003	.0001	.0000	.0000	.0000
3.2	.0408	.1304	.2087	.3226	.1781	.1140	.0608	.0278	.0111	.0049	.0013	.0004	.0001	.0000	.0000	.0000
3.3	.0369	.1217	.2008	.2209	.1823	.1203	.0662	.0312	.0129	.0047	.0016	.0005	.0001	.0000	.0000	.0000
3.4	.0334	.1135	.1929	.2186	.1858	.1264	.0716	.0348	.0148	.0056	.0019	.0006	.0002	.0000	.0000	.0000
3.5	.0302	.1057	.1850	.2158	.1888	.1322	.0771	.0385	.0169	.0066	.0023	.0007	.0002	.0001	.0000	.0000
3.6	.0273	.0084	.1771	.2125	.1912	.1377	.0826	.0425	.0191	.0076	.0028	.0009	.0003	.0001	.0000	.0000
3.7	.0247	.0915	.1692	.2087	.1931	.1429	.0881	.0466	.0215	.0089	.0033	.0011	.0003	.0001	.0000	.0000
3.8	.0224	.0850	.1615	.2046	.1944	.1477	.0936	.0508	.0241	.0102	.0039	.0013	.0004	.0001	.0000	.0000

(*Contd*)

(*Contd*)

$r \rightarrow$ $\lambda \downarrow$	0	1	2	3	4	5	6	7	8	9	10	11	12	13	14	15
3.9	.0202	.0789	.1539	.2001	.1951	.1522	.0989	.0551	.0269	.0116	.0045	.0016	.0005	.0002	.0000	.0000
4.0	.0183	.0733	.1465	.1954	.1954	.1563	.1042	.0595	.0298	.0132	.0053	.0019	.0006	.0002	.0001	.0000
4.1	.0166	.0679	.1393	.1904	.1951	.1600	.1093	.0640	.0328	.0150	.0061	.0023	.0008	.0002	.0001	.0000
4.2	.0150	.0630	.1323	.1852	.1944	.1633	.1143	.0686	.0360	.0168	.0071	.0027	.0009	.0003	.0001	.0000
4.3	.0136	.0583	.1254	.1798	.1933	.1662	.1191	.0732	.0393	.0188	.0081	.0032	.0011	.0004	.0001	.0000
4.4	.0123	.0540	.1188	.1743	.1917	.1687	.1237	.0778	.0428	.0209	.0092	.0037	.0014	.0005	.0001	.0000
4.5	.0111	.0500	.1125	.1687	.1898	.1708	.1281	.0824	.0463	.0232	.0104	.0043	.0016	.0006	.0002	.0001
4.6	.0101	.0462	.1063	.1631	.1875	.1725	.1323	.0869	.0500	.0255	.0118	.0049	.0019	.0007	.0002	.0001
4.7	.0091	.0427	.1005	.1517	.1849	.1738	.1362	.0914	.0537	.0280	.0132	.0056	.0022	.0008	.0003	.0001
4.8	.0082	.0395	.0948	.1517	.1820	.1747	.1432	.0959	.0575	.0307	.0147	.0064	.0026	.0009	.0003	.0001
4.9	.0074	.0365	.0894	.1460	.1789	.1753	.1432	.1002	.0614	.0334	.0164	.0073	.0030	.0011	.0004	.0001
5.0	.0067	.0337	.0842	.1404	.1755	.1755	.1462	.1044	.0653	.0363	.0181	.0082	.0034	.0013	.0005	.0001

Answers to Exercises

CHAPTER-1

Multiple-choice Questions

1.1 (c)
1.2 (d)

State True or False

1.1 True
1.2 False
1.3 True

Fill in the Blanks

1.1 Second
1.2 1957
1.3 England

CHAPTER-2

Multiple-choice Questions

2.1 (a)
2.2 (a)
2.3 (a)
2.4 (a)

Numerical Problems

2.1 (a)

Curtain length	Number of curtains of different sizes in patterns												
	P_1	P_2	P_3	P_4	P_5	P_6	P_7	P_8	P_9	P_{10}	P_{11}	P_{12}	P_{13}
3	6	5	4	3	3	2	2	1	1	0	0	0	0
4	0	1	2	1	0	2	1	4	3	5	2	1	0
5	0	0	0	1	2	1	2	0	1	0	2	3	4
Trim loss	2	1	0	2	1	1	0	1	0	0	2	1	0

(b) Let x_i be the number of pieces cut in the pattern p_i, $i = 1, 2, 3, \ldots, 13$.

(c) Min $Z = 2x_1 + x_2 + 2x_4 + x_5 + x_6 + x_8 + 2x_{11} + x_{12}$

(d) $6x_1 + 5x_2 + 4x_3 + 3x_4 + 3x_5 + 2x_6 + 2x_7 + x_8 + x_9 \geq 1000$

$x_2 + 2x_3 + x_4 + 2x_6 + x_7 + 4x_8 + 3x_9 + 5x_{10} + 2x_{11} + x_{12} \geq 1500$

$x_4 + 2x_5 + x_6 + 2x_7 + x_9 + 2x_{11} + 3x_{12} + 4x_{13} \geq 3000$

(e) Min $Z = 2x_1 + x_2 + 2x_4 + x_5 + x_6 + x_8 + 2x_{11} + x_{12}$

Subject to $6x_1 + 5x_2 + 4x_3 + 3x_4 + 3x_5 + 2x_6 + 2x_7 + x_8 + x_9 \geq 1000$

$x_2 + 2x_3 + x_4 + 2x_6 + x_7 + 4x_8 + 3x_9 + 5x_{10} + 2x_{11} + x_{12} \geq 1500$

$x_4 + 2x_5 + x_6 + 2x_7 + x_9 + 2x_{11} + 3x_{12} + 4x_{13} \geq 3000$

and $x_1, x_2, x_3, x_4, x_5, x_6, x_7, x_8, x_9, x_{10}, x_{11}, x_{12}, x_{13} \geq 0$

2.2 Maximize $Z = 50x_1 + 55x_2$

Subject to $3x_1 + x_2 \leq 60$

$x_1 + x_2 \leq 50$

$2x_1 + 3x_2 \leq 80$

and $x_1 \geq 0, x_2 \geq 0$

2.3 (a) Let x_1 and x_2 be the respective tonnes of exterior and interior paints produced weekly.

(b) (iii)

(c) Max $Z = 6x_1 + 4x_2$

(d) Max $Z = 6x_1 + 4x_2$

Subject to $7x_1 + 5x_2 \le 35$

$2x_1 + 3x_2 \le 6$

$x_1 \le 3$

$x_1 - x_2 \le 2$

and $x_1, x_2 \ge 0$

2.4 Maximize $Z = 15x_1 + 10x_2$

Subject to $4x_1 + 3x_2 \le 100$

$2x_1 + x_2 \le 80$

and $x_1 \ge 0, x_2 \ge 0$

2.5 Maximize $Z = 15x_1 + 5x_2$

Subject to $2x_1 + x_2 \le 15$

$x_1 + x_2 \le 20$

and $x_1 \ge 0, x_2 \ge 0$

2.6 Maximize $Z = 2x_1 + 5x_2$

Subject to $x_1 + 2x_2 \le 500$

$x_1 + x_2 \le 400$

$x_2 \le 200$

and $x_1 \ge 0, x_2 \ge 0$

2.7 Maximize $Z = 5x_1 + 8x_2$

Subject to $3x_1 + x_2 \le 100$

$x_1 + 4x_2 \le 80$

and $x_1 \ge 0, x_2 \ge 0$

2.8 Maximize $Z = 10{,}000x_1 + 8{,}000x_2$

Subject to $x_1 + x_2 \le 50$

$3x_1 + x_2 \le 70$

$x_1 \le 30$

$x_2 \le 40$

and $x_1 \ge 0, x_2 \ge 0$

2.9 Maximize $Z = 5x_1 + 2x_2$

Subject to $3x_1 + x_2 \le 1200$

$x_1 + x_2 \le 1000$

$x_1 \le 600$

$x_2 \le 900$

and $x_1 \ge 0, x_2 \ge 0$

2.10 Maximize $Z = 1850x_1 + 2080x_2 + 1875x_3$

Subject to $x_1 + x_2 + x_3 \le 150$

$5x_1 + 6x_2 + 5x_3 \le 500$

and $x_1 \ge 0, x_2 \ge 0, x_3 \ge 0$

2.11 Maximize $Z = 5x_1 + 10x_2 + 15x_3$

Subject to $x_1 + 3x_2 + 3x_3 \le 12$

$2x_1 + 4x_2 + x_3 \le 20$

and $x_1 \ge 0, x_2 \ge 0, x_3 \ge 0$

2.12 Minimize $Z = 4x_1 + 5x_2$

Subject to $5x_1 + 3x_2 \ge 20$

$4x_1 + 4x_2 \ge 25$

$3x_1 + 6x_2 \ge 15$

and $x_1 \ge 0, x_2 \ge 0$

2.13 Minimize $Z = 72x_1 + 56x_2$

Subject to $4x_1 + 2x_2 \ge 25$

$x_1 \le 20$

$x_2 \le 15$

and $x_1 \ge 0, x_2 \ge 0$

2.14 Maximize $Z = \text{Min}\ [1/4(6x_1 + 5x_2 + 7x_3),$ $1/3(4x_1 + 85x_2 + 3x_3)] = p(\text{say})$

Subject to $6x_1 + 5x_2 - 7x_3 - 4p \ge 0$

$4x_1 + 8x_2 + 3x_3 - 3p \ge 0$

$7x_1 + 4x_2 + 2x_3 \le 100$

$5x_1 + 8x_2 + 7x_3 \le 200$

and $x_1 \ge 0, x_2 \ge 0, x_3 \ge 0$

2.15 Maximize $Z = 160x_1 + 240x_2$

Subject to $3x_1 + 2x_2 \le 210$

$2x_1 + 4x_2 \le 300$

$x_2 \le 65$

and $x_1 \ge 0, x_2 \ge 0$

2.16 Minimize $Z = 60x_1 + 21x_2$

Subject to $1000x_1 + 2000x_2 \ge 3000$

$25x_1 + 10x_2 \ge 100$

and $x_1 \ge 0, x_2 \ge 0$

CHAPTER-3

Multiple-choice Questions

3.1 (i) (a) (ii) (c) 3.2 (b) 3.3 (a) 3.4 (d)
3.5 (a) 3.6 (a) 3.7 (c) 3.8 (b)
3.9 (d) 3.10 (d) 3.11 (d) 3.12 (a)
3.13 (d) 3.14 (i) (c) (ii) (a) 3.15 (i) (a) (ii) (a)

State True or False

3.1 True 3.2 False 3.3 True 3.4 False

Fill in the Blanks

3.1 unbounded 3.2 optimum 3.3 feasible 3.4 two

Numerical Problems

3.1 (a) and (b) are solvable.
(c) Not solvable, as the number of variables is more than two.

3.4 (a) Yes (b) No (c) Yes.

3.5 (a) $x = 0$ (b) $x = 1$
(c) $x = 0$ and 1 (d) All boundary point of a circle $x^2 + y^2 = 1$

3.8 (a) $x_1 = 6, x_2 = 12$, Min $Z = 240$ (b) Unbounded solution
(c) No feasible solution (d) Infinite number of solutions
(e) $x_1 = 3, x_2 = 4$, Max $Z = 10$

3.9 $x_1 = 1000, x_2 = 500$, Max $Z = 5500$

3.10 Max $Z = 300x_1 + 100x_2$
Subject to $2x_1 + x_2 \leq 32$
$x_1 + x_2 \leq 24$
and $x_1, x_2 \geq 0$
Maximum profit is 4800 when $x_1 = 16$ and $x_2 = 0$.

3.11 $x_1 = 15, x_2 = 30$, Max $Z = 12,000$

3.12 $x_1 = 2, x_2 = 4$, Min $Z = 38$

CHAPTER-4

Multiple-choice Questions

4.1 (a) 4.2 (b) 4.3 (b) 4.4 (c) 4.5 (a)
4.6 (c) 4.7 (c) 4.8 (a) 4.9 (b) 4.10 (b)

State True or False

4.1 True 4.2 False 4.3 True 4.4 False 4.5 False
4.6 False 4.7 True 4.8 False 4.9 True 4.10 True
4.11 False 4.12 False 4.13 True 4.14 True 4.15 False

Fill in the Blanks

4.1 George Dantzig 4.2 slack, surplus
4.3 (–1) 4.4 Basic
4.5 Constraints 4.6 Zero

Numerical Problems

4.1 (a) $x_1 = 4, x_2 = 0$, and Max $Z = 8$ (b) $x_1 = 1, x_2 = 0$, and Min $Z = 1$

4.2 (a) $x_1 = 3, x_2 = 0$, and Max $Z = 21$ (b) $x_1 = 20/19, x_2 = 45/19$, and Max $Z = 235/19$
(c) $x_1 = 0, x_2 = 4$, and Max $Z = 28$ (d) $x_1 = 4, x_2 = 2$, and Max $Z = 10$
(e) $x_1 = 3, x_2 = 2$, and Max $Z = 17$ (f) $x_1 = 0, x_2 = 0, x_3 = 20$, and Max $Z = 40$
(g) $x_1 = 0, x_2 = 0, x_3 = 0$, and Min $Z = 0$ (h) $x_1 = 0, x_2 = 8/3, x_3 = 2$, and Max $Z = 64/3$
(i) $x_1 = 0, x_2 = 100, x_3 = 230$, and Max $Z = 1350$ (j) Unbounded solution
(k) $x_1 = 0, x_2 = 20$ or $x_1 = 75/4, x_2 = 25/2$, and Max $Z = 200$

4.3 (a) $x_1 = 15, x_2 = 20$, and Min $Z = 25$ (b) $x_1 = 4, x_2 = 1$, and Min $Z = 11$
(c) $x_1 = 5/2, x_2 = 0, x_3 = 0$, and Min $Z = 10$ (d) $x_1 = x_2 = x_3 = 5/2, x_4 = 0$, and Max $Z = 15$
(e) No feasible solution

4.4 (a) $x_1 = 1, x_2 = 5$, Min $Z = 13$ and actual requirements are 10, 12, and 21 units (b) $x_1 = 21/13, x_2 = 10/13$, and Min $Z = 31/13$
(c) $x_1 = 0, x_2 = 5$, and Max $Z = 40$ (d) No feasible solution

4.5 (a) $x_1 = 16/11, x_2 = -2/11$ (b) $x_1 = 2, x_2 = -1$ (c) $A^{-1} = \frac{1}{4}\begin{bmatrix} -2 & 2 \\ 3 & -1 \end{bmatrix}$

CHAPTER-5

Multiple-choice Questions

5.1 (a) 5.2 (c)

State True or False

5.1 (a) False (b) False (c) True (d) True (e) True

Fill in the Blanks

5.1 Simplex 5 2 less 5.3 revised simplex 5.4 large 5.5 row

Numerical Problems

5.1 (a) (2, 2) (b) (−2, −2, −9) (c) (i) (d) $x_1 = 4, x_2 = 6, x_3 = 0$, and Max $Z = 12$

5.2 (a) $x_1 = 3, x_2 = 0$, and Max $Z = 21$ (b) Unbounded solution
(c) $x_1 = 4, x_2 = 2$, and Max $Z = 10$ (d) $x_1 = 0, x_2 = 8/3, x_3 = 2$, and Max $Z = 64/3$
(e) $x_1 = 0, x_2 = 100, x_3 = 230$, and Max $Z = 1350$ (f) $x_1 = 0$, $x_2 = 20$ or $x_1 = 75/4$, $x_2 = 25/2$, and Max $Z = 200$
(g) $x_1 = 5/2, x_2 = 0, x_3 = 0$, and Min $Z = 10$ (h) $x_1 = x_2 = x_3 = 5/2, x_4 = 0$, and Max $Z = 15$
(i) No feasible solution (j) $x_1 = 0, x_2 = 5$, and Max $Z = 40$

CHAPTER-6

Multiple-choice Questions

6.1 (c) 6.2 (a) 6.3 (d) 6.4 (b)
6.5 (a) 6.6 (c) 6.7 (a) 6.8 (a)

State True or False

6.1 (a) True (b) False (c) False

6.2 (a) True (b) False (c) False
(d) False (e) False (f) False

Fill in the Blanks

6.1 unrestricted in sign
6.2 optimal
6.3 no restrictions
6.4 dual
6.5 Min $Z = b^T Y$, $A^T Y \geq C$, $Y \geq 0$
6.6 (a) minimum (b) $\leq$ (c) maximum

Numerical Problems

6.1 (a) Min $Z = 8y_1 + 10y_2 + 15y_3$
Subject to $2y_1 + 3y_3 \geq 3$
$3y_1 + 2y_2 + 2y_3 \geq 5$
$5y_2 + 4y_3 \geq 4$
and $y_1, y_2, y_3 \geq 0$

(b) Max $Z = 2y_1 + y_2$
Subject to $2y_1 + y_2 \leq 3$
$3y_1 + y_2 \leq 2$
and $y_1, y_2 \geq 0$

(c) Max $Z = 18y_1 + 8y_2 - 6y_3$
Subject to $3y_1 + y_2 - 2y_3 \leq 10$
$2y_1 + 3y_2 + y_3 \leq 20$
and $y_1, y_2, y_3 \geq 0$

(d) Max $Z = 3y_1 + 4y_2 + y_3 + 7y_4$
Subject to $8y_l + 3y_2 + 4y_3 + y_4 \leq 7$
$2y_1 + 6y_2 + y_3 + 5y_4 \leq 3$
$y_1 + 4y_2 + 5y_3 + 2y_4 \leq 8$
and $y_1, y_2, y_3, y_4 \geq 0$

(e) Max $Z = 2y_1 - 3y_2 - 5y_3$
Subject to $2y_1 - 3y_2 - y_3 \leq 2$
$3y_1 - y_2 - 4y_3 \leq 2$
$5y_1 - 4y_2 - 6y_3 \leq 4$
and $y_1, y_2, y_3 \geq 0$

(f) Min $Z = 6y_1 + 9y_2 + 10y_3$
Subject to $4y_1 + 2y_2 + y_3 \geq 3$
$-5y_1 + 3y_2 + y_3 \geq 1$
$-9y_1 + 4y_2 - 5y_3 \geq 4$
$y_1 - 5y_2 - 7y_3 \geq 1$
$-2y_1 + y_2 + 11y_3 \geq 9$
and $y_1, y_2, y_3 \geq 0$

(g) Max $Z = 7y_1 + 4y_2 - 10y_3 + 3y_4 + 2y_5$
Subject to $3y_1 + 6y_2 - 7y_3 + y_4 + 4y_5 \leq 3$
$5y_1 + y_2 + 2y_3 - 2y_4 + 7y_5 \leq -2$
$4y_1 + 3y_2 + y_3 + 5y_4 - 2y_5 \leq 4$
and $y_1, y_2, y_3, y_4, y_5 \geq 0$

(h) Min $Z = 6y_1 + 4y_2$
Subject to $4y_1 + y_2 \geq 2$
$3y_1 + 2y_2 \geq 3$
$y_1 + 5y_2 \geq 1$
y_1 and y_2 are unrestricted in sign.

(i) Min $Z = 6y_1 + 4y_2$
Subject to $3y_1 + 3y_2 \geq 1$
$2y_1 + y_2 \geq 3$
$y_1 \geq 0$ and y_2 is unrestricted in sign.

(j) Max $Z = 5y_1 - 3y_2 + 4y_3$
Subject to $y_1 - y_2 \leq 1$
$-3y_1 + 2y_2 + 2y_3 \leq 1$
$4y_1 - y_3 = 1$
$y_2, y_3 \geq 0$, y_1 are unrestricted in sign.

6.2 (a) $x_1 = x_2 = x_4 = 0$, $x_3 = 2000$, $x_5 = 500$, and Min $Z = 7000$
(b) $x_1 = 7$, $x_2 = 0$, $x_3 = 4$, $x_4 = 0$, and Min $Z = 7$
(c) $x_1 = 0$, $x_2 = 3$, $x_3 = 1$, and Min $Z = 36$
(d) $x_1 = 7$, $x_2 = 0$, $x_3 = 4$, $x_4 = 0$, and Max $Z = -7$

CHAPTER-7

Multiple-choice Questions

7.1 (a) 7.2 (b) 7.3 (a) 7.4 (d)
7.5 (b) 7.6 (b) 7.7 (b) 7.8 (c)

State True or False

7.1 True　　7.2 True　　7.3 True
7.4 True　　7.5 True　　7.6 False
7.7 True　　7.8 False　　7.9 True

Fill in the Blanks

7.1 feasibility　　7.2 both feasibility and optimality
7.3 both feasibility and optimality
7.4 unbounded and consequently the solution may become unbounded
7.5 empty　　7.6 linear

Review Questions

7.3 No

Numerical Problems

7.1 $c_1 \le 10/3$ and $c_2 \ge 9/2$
7.2 $x_1 = 4$, $x_2 = 3$, and Max $Z = 15$
7.3 $b_1 \ge 4$, $b_2 \le 16$, and $3/2 \le b_3 \le 10$
7.4 $x_1 = 0$, $x_2 = 5$, $x_3 = 0$, $x_4 = 0$, $x_5 = 4$, and Max $Z = 53$
7.5 (a) (i) $x_1 = 0, x_2 = 0, x_3 = 2$, and Max $Z = -2$　　(ii) $x_1 = 4/5$, $x_2 = 0$, $x_3 = 18/5$, and Max $Z = -14/5$
(b) Unbounded solution　　(c) No solution
(d) Unbounded solution　　(e) $x_1 = 2/3$, $x_2 = 4/3$, $x_3 = 2$, and Max $Z = -8/3$
(f) Unbounded solution　　(g) $c_2 \le -1/5$　　(h) $b_1 \ge 1$

7.6

λ	x_1	x_2	x_3	Z
$0 \le \lambda \le 1$	0	100	230	$1350 + 950\lambda$
$\lambda \ge 1$	0	0	230	$1150 + 1150\lambda$

7.7

λ	x_1	x_2	x_3	Z
$0 \le \lambda \le 10/3$	0	$5 - \lambda$	$30 + \lambda$	$160 + 3\lambda$
$10/3 \le \lambda \ge 30/7$	0	$(30 - 7\lambda)/4$	$30 + \lambda$	$165 + (3/2)\lambda$
$\lambda > 30/7$	No feasible solution exists.			

CHAPTER-8

Objective Questions

8.1 (b)　　8.2 (c)　　8.3 (b)　　8.4 (c)
8.5 (c)　　8.6 (b)　　8.7 (b)

State True or False

8.1 False　　8.2 True　　8.3 True　　8.4 True
8.5 False　　8.6 True　　8.7 True　　8.8 True

Fill in the Blanks

8.1 VAM
8.2 equal to
8.3 LP model
8.4 Simplex

8.5 (a) $\sum a_i = \sum b_j$
(b) $\sum a_i \neq \sum b_j$
(c) balanced transportation problem
(d) unbalanced transportation problem
(e) balance
(f) $m + n$
(g) $m + n - 1$
(h) minimization
(i) minimization
(j) $m + n$

Numerical Problems

8.1 (a) $x_{11} = 2, x_{21} = 1, x_{22} = 3, x_{23} = 2, x_{33} = 2, x_{43} = 5$
(b) $x_{11} = 3, x_{12} = 1, x_{22} = 2, x_{23} = 4, x_{24} = 2, x_{34} = 3, x_{35} = 6$

8.2 $x_{11} = 9, x_{12} = 8, x_{14} = 17, x_{23} = 15, x_{31} = 12, x_{42} = 17, x_{43} = 2$

8.3 (a) $x_{11} = 20, x_{13} = 10, x_{22} = 20, x_{23} = 20, x_{24} = 10, x_{32} = 20$
(b) $x_{11} = 20, x_{13} = 10, x_{22} = 20, x_{23} = 20, x_{24} = 10, x_{32} = 20$

8.4 (a) $x_{11} = 5, x_{12} = 2, x_{22} = 6, x_{23} = 3, x_{33} = 4, x_{34} = 14$. The minimum transportation cost is ₹1015.
(b) $x_{14} = 7, x_{21} = 2, x_{23} = 7, x_{31} = 3, x_{32} = 8, x_{34} = 7$. The minimum transportation cost is ₹814.
(c) $x_{11} = 5, x_{14} = 2, x_{23} = 7, x_{24} = 2, x_{32} = 8, x_{34} = 10$. The minimum transportation cost is ₹779.

8.5 (a) $x_{14} = 30, x_{21} = 5, x_{22} = 35, x_{31} = 17, x_{33} = 25, x_{34} = 11$. The minimum transportation cost is ₹2221.
(b) $x_{11} = 20, x_{13} = 10, x_{22} = 20, x_{23} = 20, x_{24} = 10, x_{32} = 20$. The minimum transportation cost is ₹180.

8.6 $x_{14} = 11, x_{21} = 6, x_{22} = 3, x_{24} = 4, x_{32} = 7, x_{33} = 12$. The minimum transportation cost is ₹710.

8.7 $x_{12} = 5, x_{13} = 9, x_{21} = 6, x_{23} = 6, x_{32} = 5$. The minimum transportation cost is ₹128.

8.8 $x_{11} = 5, x_{14} = 2, x_{22} = 2, x_{23} = 7, x_{32} = 6, x_{34} = 12$. The minimum transportation cost is ₹743.

8.9 (a) $x_{13} = 8, x_{24} = 4, x_{25} = 3, x_{31} = 5, x_{32} = 4, x_{41} = 1, x_{43} = 2, x_{52} = 4, x_{55} = 1$. The minimum transportation cost is ₹1102.
(b) $x_{12} = 2, x_{22} = 1, x_{23} = 4, x_{24} = 1, x_{31} = 3, x_{34} = 4$. The minimum transportation cost is ₹174.
(c) $x_{14} = 11, x_{21} = 6, x_{22} = 3, x_{24} = 4, x_{32} = 7, x_{33} = 12$. The minimum transportation cost is ₹796.
(d) $x_{11} = 6, x_{13} = 3, x_{15} = 4, x_{24} = 4, x_{25} = 2, x_{32} = 7, x_{34} = 1$. The minimum transportation cost is ₹163.

8.10 $x_{13} = 2, x_{22} = 1, x_{23} = 2, x_{31} = 4, x_{33} = 1$. The minimum transportation cost is ₹33.

8.11 (a) $x_{12} = 4, x_{14} = 5, x_{21} = 3, x_{22} = 2, x_{25} = 8, x_{32} = 1, x_{33} = 8$. The minimum transportation cost is ₹154.
(b) $x_{13} = 5, x_{22} = 3, x_{23} = 1, x_{26} = 2, x_{31} = 1, x_{32} = 1, x_{41} = 3, x_{44} = 2, x_{45} = 4$. The minimum transportation cost is ₹12.

8.12 $x_{12} = 100, x_{22} = 70, x_{25} = 80, x_{31} = 20, x_{33} = 50, x_{35} = 110, x_{41} = 70, x_{44} = 210$. The minimum transportation cost is ₹6600.

8.13 $x_{12} = 30, x_{14} = 40, x_{22} = 5, x_{23} = 50, x_{31} = 85, x_{34} = 5$. The minimum transportation cost is ₹1160.

8.14 $x_{12} = 5, x_{21} = 6, x_{22} = 1, x_{23} = 8, x_{31} = 9$. The minimum transportation cost is ₹91.

8.15 $x_{11} = 8, x_{12} = 7, x_{22} = 5, x_{24} = 6, x_{33} = 13$. The minimum transportation cost is ₹224.

8.16 $x_{12} = 1$, $x_{16} = 2$, $x_{21} = 3$, $x_{25} = 1$, $x_{32} = 2$, $x_{42} = 2$, $x_{43} = 4$, $x_{44} = 2$. The minimum transportation cost is ₹101.

8.17 $x_{12} = 6$, $x_{23} = 2$, $x_{24} = 6$, $x_{31} = 4$, $x_{33} = 6$. The minimum transportation cost is ₹28.

8.18 $x_{13} = 60$, $x_{21} = 50$, $x_{23} = 20$, $x_{32} = 80$. The minimum transportation cost is ₹750.

8.19 $x_{12} = 30$, $x_{14} = 40$, $x_{22} = 5$, $x_{23} = 50$, $x_{31} = 85$, $x_{34} = 5$. The minimum transportation cost is ₹1160.

8.20 (a) $x_{11}=5$, $x_{13}=15$, $x_{14}=20$, $x_{22}=30$, $x_{31}=15$, $x_{35}=5$, $x_{41}=10$. The minimum transportation cost is ₹210.

(b) $x_{11} = 4$, $x_{12} = 4$, $x_{24} = 8$, $x_{25} = 4$, $x_{33} = 6$, $x_{35} = 4$. The minimum transportation cost is ₹80.

8.21 $x_{14} = 160$, $x_{21} = 80$, $x_{22} = 10$, $x_{32} = 80$, $x_{33} = 110$. The minimum transportation cost is ₹17,050.

CHAPTER-9

Multiple-choice Questions

9.1 (a) 9.2 (b) 9.3 (b) 9.4 (b)

State True or False

9.1 True 9.2 False 9.3 False
9.4 True 9.5 True

Fill in the Blanks

9.1 Transportation problem
9.2 Hungarian algorithm
9.3 LP
9.4 (a) $2n$ (b) n (c) $n - 1$

Numerical Problems

9.1 1 → C, 2 → B, 3 → A, and the total computer minutes is 280.

9.2 (a) I → 3, II → 1, III → 2, IV → 4, and V → 5 or I → 3, II → 1, III → 4, IV → 2, and V → 5.

(b) A → II, B → III, C → IV, and D → 1, and the minimum cost is 20.

(c) A → *e*, B → *c*, C → *b*, D → *a*, and E → *d*, and the minimum cost is 570.

(d) A → IV, B → I, C → VI, D → III, E → II, and F → V or A → IV, B → VI, C → II, D → III, and E → V, F → I, in both the cases, the minimum cost is 142.

(e) A → I, B → IV, C → V, D → III, and E → II, and the minimum cost is 60.

(f) A → 3, B → 4, C → 2, and D → 1, and the minimum cost is 16.

(g) A → V, B → IV, C → I, D → III, and E → II, in both cases the minimum cost is 23.

(h) A → I, B → IV, C → III, D → II, and E → V or A → II, B → IV, C → III, D → V, and E → I, in both the cases, the minimum cost is 10.

(i) 1 → 11, 2 → 8, 3 → 7, 4 → 9, 5 → 10, and 6 → 12, and the minimum cost is 185.

(j) 1 → 11, 2 → 8, 3 → 7, 4 → 9, 5 → 10, and 6 → 12, and the minimum cost is 125.

(k) 1 → I, 2 → III, 3 → IV, 4 → II, and 5 → V, and the minimum cost is 39 miles.

(l) 1 → III, 2 → II, 3 → V, 4 → I, and 5 → IV.

9.3 $J_1 \rightarrow M_1$, $J_2 \rightarrow M_2$, $J_3 \rightarrow M_3$

9.4 (a) $(1 \rightarrow 102)^*, (2 \rightarrow 104), (3 \rightarrow 101), (4 \rightarrow 103)^*$ (b) $(1 \rightarrow 103)^*, (2 \rightarrow 104), (3 \rightarrow 101), (4 \rightarrow 102)^*$; minimum layover time is 210 quarter hours, that is 52 h 30 minutes.

9.5 $1 \to J_3$, $2 \to J_5$, $3 \to J_4$, $4 \to J_2$, and $5 \to J_1$ and the maximum profit is ₹50.
9.6 $A \to P_2$, $B \to P_4$, $C \to P_1$, and $D \to P_3$.
9.7 $1 \to C$, $2 \to A$, $3 \to D$, and $4 \to B$.
9.8 $A \to I$, $B \to II$, $C \to III$, and $D \to IV$ or $A \to I$, $B \to III$, $C \to II$, and $D \to IV$.
9.9 $A \to W$, $B \to X$, and $C \to Y$ and $A \to W$, $B \to Y$, and $C \to X$.
9.10 $P_1 \to J_3$, $P_2 \to J_2$, and $P_3 \to J_4$ or $P_1 \to J_4$, $P_2 \to J_2$, and $P_3 \to J_5$ and the minimum cost is 17.
9.11 $1 \to 4$, $2 \to 2$, $3 \to 5$, $4 \to 3$, and $5 \to 1$ or $1 \to 4$, $2 \to 1$, $3 \to 5$, $4 \to 3$, and $5 \to 2$ and the total minimum cost is 20, that is ₹2000.
9.12 $2 \to A$, $3 \to B$, $4 \to D$, and $6 \to C$ and the maximum profit is 28.
9.13 $1 \to D$, $2 \to B$, $3 \to C$, and $4 \to E$.
9.14 $A \to IV$, $B \to III$, $C \to II$, $D \to I$, and $E \to V$ or $A \to IV$, $B \to III$, $C \to II$, $D \to V$, and $E \to I$ or $A \to IV$, $B \to III$, $C \to V$, $D \to II$, and $E \to I$.
9.15 (a) $1 \to 3 \to 5 \to 2 \to 4 \to 1$ and the total minimum cost is 27.
(b) $A_1 \to A_2 \to A_3 \to A_4 \to A_5 \to A_1$ and the set-up cost is 15.
(c) $1 \to 5 \to 6 \to 3 \to 4 \to 2 \to 1$ and the total minimum cost is 103.
(d) $1 \to 3 \to 2 \to 5 \to 4 \to 1$ and the total minimum cost is 49.
(e) $A \to C \to D \to B \to E \to A$ and $A \to E \to B \to D \to C \to A$ and in both the cases, the total distance travelled = 30, that is 3000 miles.

CHAPTER-10

Multiple-choice Questions

10.1 (b) | 10.2 (b) | 10.3 (a)

State True or False

10.1 False | 10.2 True | 10.3 True

Fill in the Blanks

10.1 total elapsed time | 10.2 idle time | 10.3 $\sum_{j=1}^{n} M_{2j} + \sum_{j=1}^{n} I_{2j}$

Numerical Problems

10.1 (a) Optimal sequence = 3-1-5-6-2-4 and (b) elapsed time = 35 h.
10.2 The optimal sequence = 3-1-5-6-2-4 and the elapsed time = 42 h.
10.3 (a) Optimal sequence = 4-1-3-2-5-6, (b) elapsed time = 430 h, and (c) idle time for machine M_1 is 10 h and for M_2 is 60 h.
10.4 (a) Optimal sequence = 2-4-3-5-1, (b) elapsed time = 48 h, and (c) idle time is 12 h each for machines M_1 and M_2.
10.5 The optimal sequence = 1-6-4-2-3-5 and the elapsed time = 96 h.
10.6 (a) Optimal sequence = 1-3-9-2-8-6-4-7-5 or 1-3-9-8-2-6-4-7-5 and (b) elapsed time = 61 h.
10.7 (a) Optimal sequence = 6-4-1-7-2-3-5 and (b) elapsed time = 83 h.
10.8 The optimal sequence = 1-4-5-3-2-7-6 or 1-4-5-2-3-7-6 and the elapsed time = 67 h.
10.9 (a) Optimal sequence = 8-3-4-7-6-5-2-1 and (b) elapsed time = 6 h and 45 min.
10.10 The optimal sequence = 2-4-3-5-1 and elapsed time = 60 h
10.11 (a) Optimal sequence = 2-4-3-5-1, (b) elapsed time = 36 h, and (c) idle time for both machines = 7 h (3 h for M_1 and 4 h for M_2)

10.12 (a) Optimal sequence = 1-3-5-4-2, (b) elapsed time = 28 h, and (c) idle time for M_1 is 2 h and idle time for M_2 is 6 h.
10.13 The optimal sequence = 1-2-3-4 or 1-2-4-3 or 1-3-4-2 or 1-4-2-3, etc. and the elapsed time = 84 h.
10.14 The optimal sequence = 3-2-1-4-5 or 3-2-1-5-4 or 3-2-5-4-1 or 3-2-4-5-1 or 3-2-4-1-5- or 3-2-5-1-4 and the elapsed time = 44 h.
10.15 The optimal sequence = 3-2-4-1-5 and the elapsed time = 51 h.
10.16 (a) Optimal sequence = 2-1-4-6-5-3, (b) elapsed time = 209 h, and (c) idle time for machine M_1, M_2, and M_3 are 34, 159, and 44 h respectively.
10.17 The optimal sequence = 3-1-5-4-2 and elapsed time = 76 h.
10.18 (a) Optimal sequence = 2-5-4-3-1, (b) elapsed time = 63 h, and (c) idle time for machine M_1, M_2, and M_3 are 35, 35, and 7 h respectively.
10.19 (a) Optimal sequence = 2-5-4-3-1 or 5-4-3-2-1 or 5-2-4-3-1, (b) elapsed time = 40 h, and (c) idle time for machine M_1, M_2, and M_3 are 8, 25, and 12 h respectively.
10.20 The optimal sequence = 1-2-4-3.
10.21 The optimal sequence = 4-5-6-2-1-3 and the elapsed time = 53 h.
10.22 The optimal sequence = 4-1-2-3 and the elapsed time = 97 h.
10.23 The optimal sequence = 1-4-5-3-2 and the elapsed time = 42 h.
10.24 The optimal sequence = 1-2-3-4 or 1-2-4-3, etc. and the elapsed time = 61 h.
10.25 The total elapsed time = 150 h.
10.26 The minimum total elapsed time = 24 h.
10.27 The minimum total elapsed time = 20 h.
10.28 The minimum total elapsed time = 44 h.
10.29 The total elapsed time = 26 h.

CHAPTER-11

Multiple-choice Questions

11.1 (d) 11.2 (d) 11.3 (a)

State True or False

11.1 False 11.2 True 11.3 False 11.4 True

Fill in the Blanks

11.1 deterministic
11.2 probabilistic
11.3 node
11.4 latest completion, earliest starting

Numerical Problems

11.1

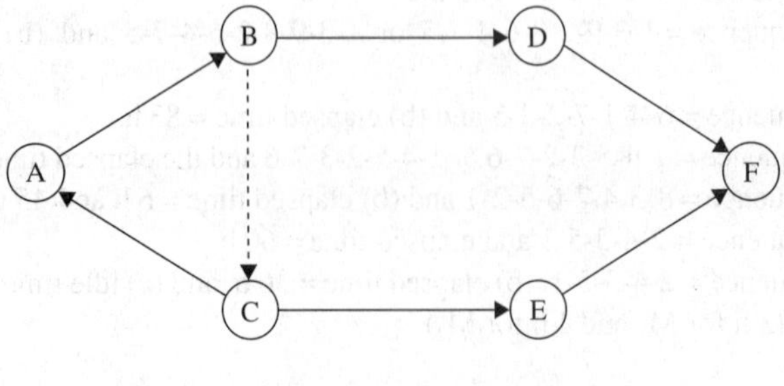

11.2

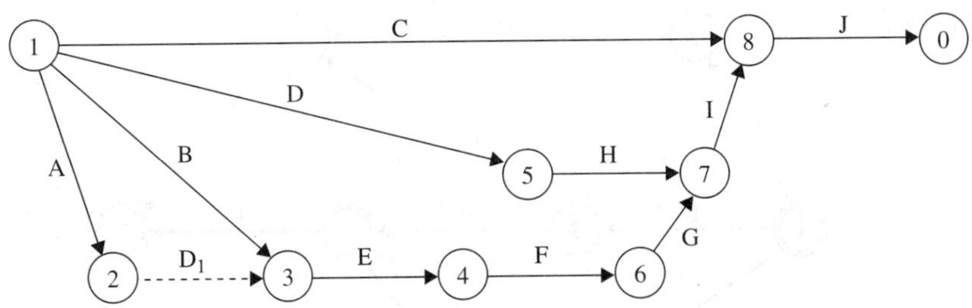

11.3

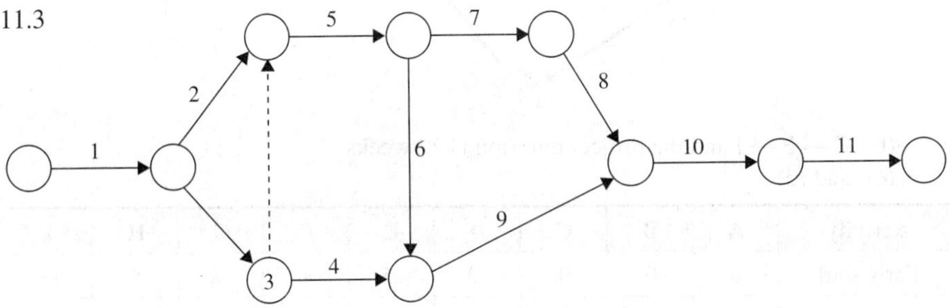

11.4

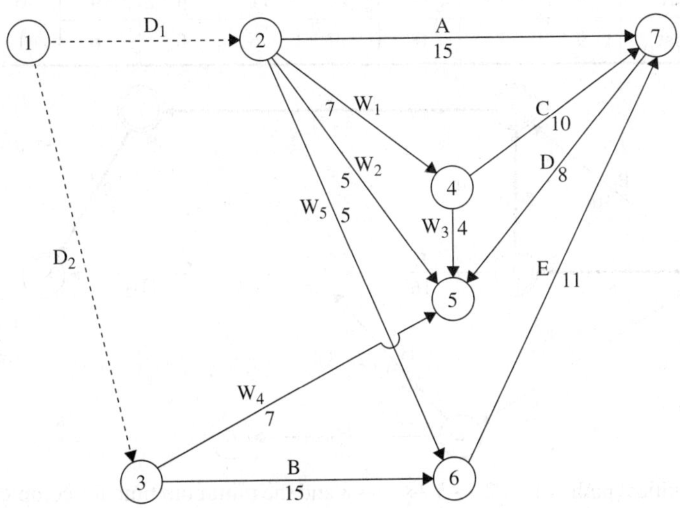

11.5

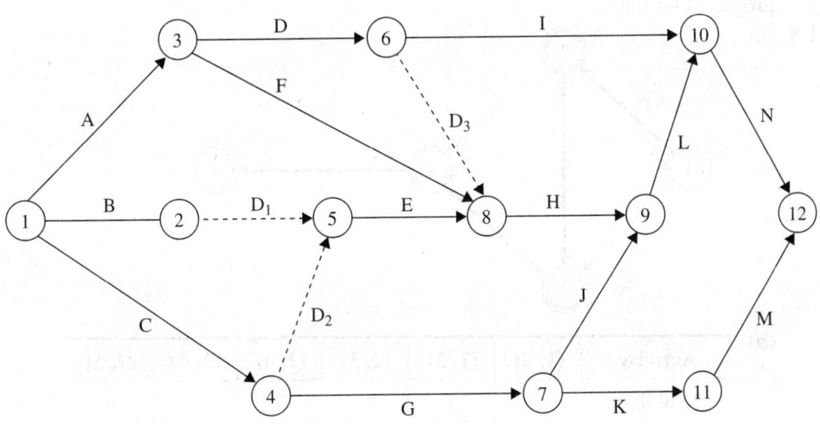

11.6 (a)

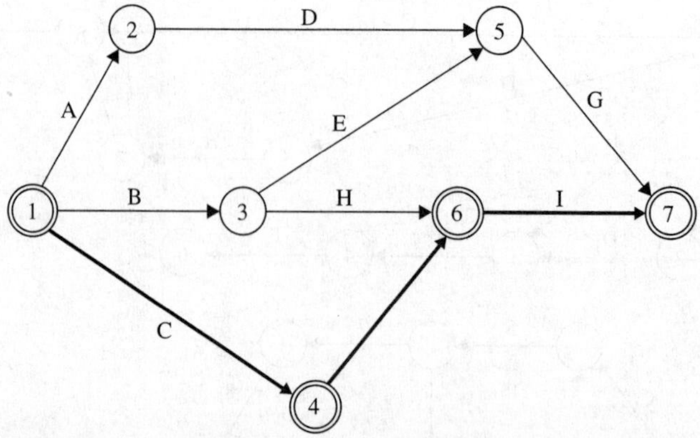

(b) C → F → I and the project duration is 22 weeks.

(c) and (d)

Activity	A	B	C	D	E	F	G	H	I
Early start	0	0	0	3	5	4	8	5	13
Late start	9	1	0	12	11	4	14	6	13
Total float	9	1	0	9	6	0	6	1	0

11.7

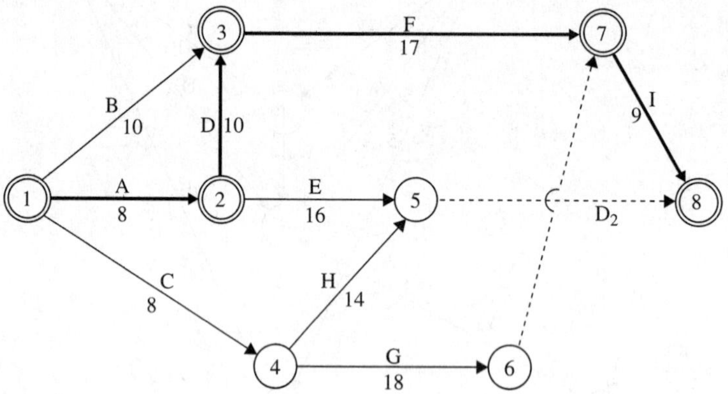

The critical path is 1 → 2 → 3 → 7 → 8 and the minimum time for completion of the project is 44 days.

11.8 (a)

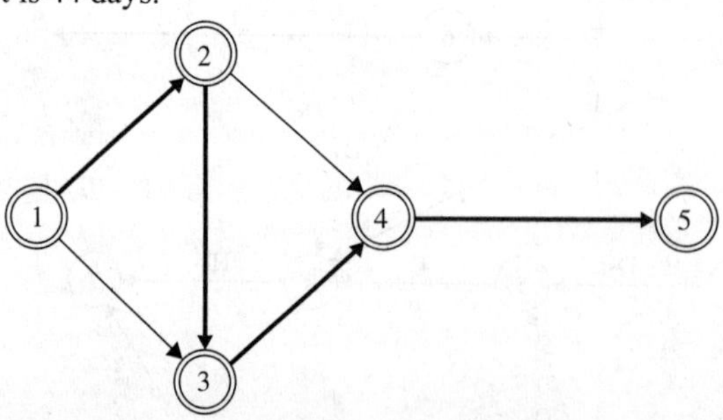

(b)

Activity	(1, 2)	(1, 3)	(2, 3)	(2, 4)	(3, 4)	(4, 5)
Total float	0	5	0	3	0	0
Free float	0	5	0	3	0	0

(c) The critical path is $1 \to 2 \to 3 \to 4 \to 5$ and the project duration is 45 days.

11.9

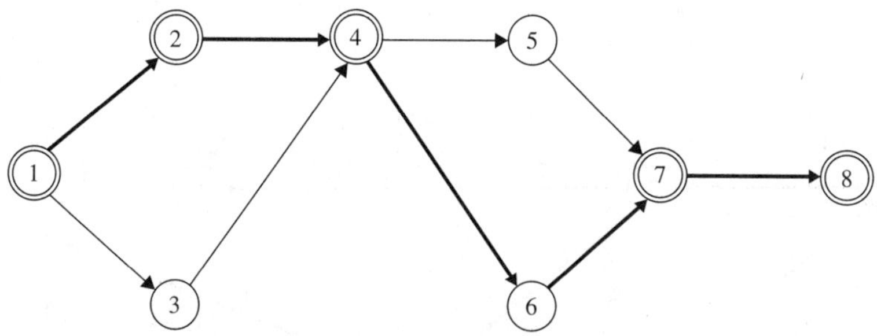

Event	1	2	3	4	5	6	7	8
Earliest	0	5	4	11	12	18	22	25
Latest	0	5	9	11	14	18	22	25

$1 \to 2 \to 4 \to 6 \to 7 \to 8$ (Critical path)

11.10 (a)

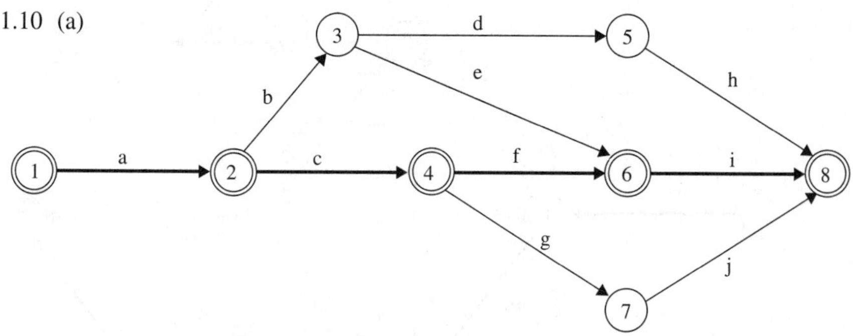

(b)

Activity	Earliest time		Latest time	
	Start	Finish	Start	Finish
1–2	0	2	0	2
2–3	2	5	5	8
3–4	2	7	2	7
3–5	5	9	8	12
3–6	5	6	12	13
4–6	7	13	7	13
4–7	7	9	14	16
5–8	9	17	12	20
6–8	13	20	13	20
7–8	9	13	16	20

(c) The critical activities are (1, 2), (2, 4), (4, 6), and (6, 8).

(d) 3 days

11.11 (a)

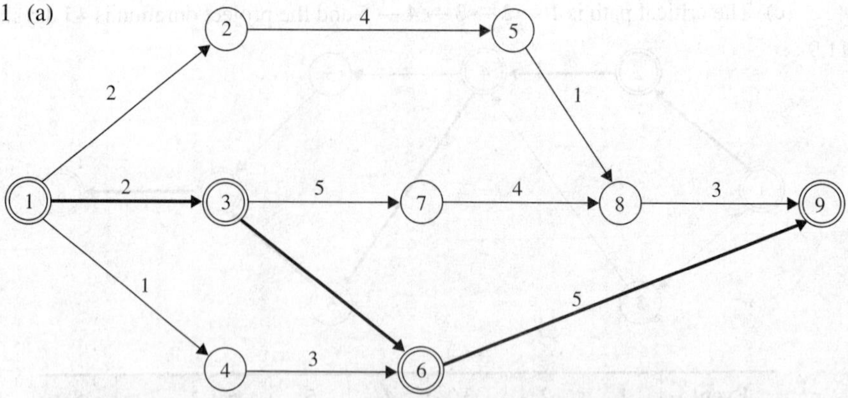

(b)

Activity	1–2	1–3	1–4	2–5	3–6	3–7	4–6	5–8	6–9	7–8	8–9
Total float	5	0	6	5	0	1	6	5	0	1	1

(c) The critical path is $1 \to 3 \to 6 \to 9$ and the project duration is 15 weeks.

11.12 (a)

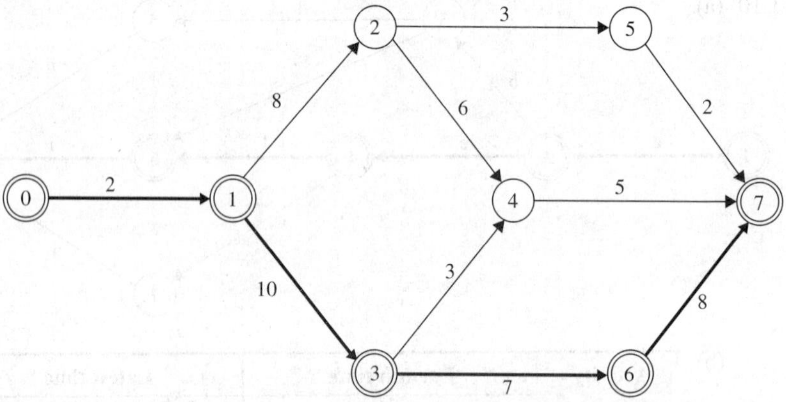

(b) The critical path is $0 \to 1 \to 3 \to 6 \to 7$ and the project duration is 27 days.

(c)

Activity	0–1	1–2	1–3	2–4	2–5	3–4	3–6	4–7	5–7	6–7
Total float	0	6	0	6	12	7	0	6	12	0
Free float	0	0	0	0	0	1	0	6	12	0
Independent float	0	0	0	0	0	1	0	0	0	0

11.13 (a) The critical path is $A \to C \to D \to F$ and the project duration is 12 weeks.

(b) and (c)

Activity	A	B	C	D	E	F
Earliest start time	0	2	2	6	0	2
Earliest finish time	2	5	6	12	2	10
Float (slack)	0	1	0	0	4	0

11.14

Activity	t_o	t_m	t_p	$T_e = \frac{t_o + 4t_m + t_p}{6}$	$\sigma_{ij} = \frac{t_p - t_o}{6}$	$V_{ij} = \sigma_{ij}^2 = \left(\frac{t_p - t_o}{6}\right)^2$
1	2	4	6	$\frac{2+4\times4+6}{6} = 4$	2/3	4/9
2	6	8	10	$\frac{6+4\times8+10}{6} = 8$	2/3	4/9
3	1	5	15	$\frac{1+4\times5+15}{6} = 6$	7/3	49/9
4	2	5	9	$\frac{2+4\times5+9}{6} = 5.166$	7/3	49/9
5	6	8	10	$\frac{6+4\times8+10}{6} = 8$	2/3	4/9
6	6	7	9	$\frac{6+4\times7+9}{6} = 7.166$	1/2	1/4

11.15 The critical path is $1 \to 2 \to 3 \to 5 \to 8$ and the project duration is 33 days.

11.16 (a)

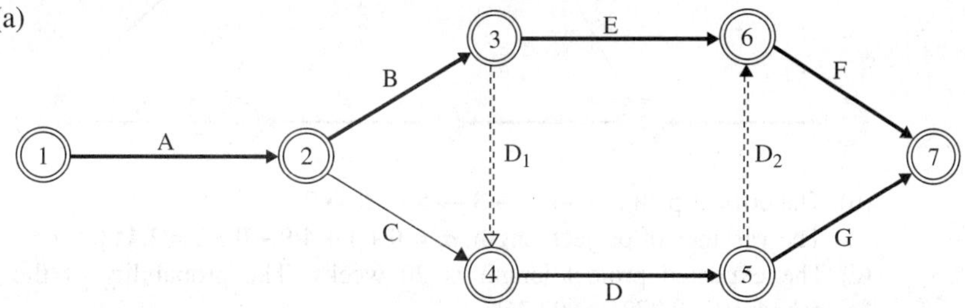

(b) The critical path is A → B → D → F; the expected project duration is 27 days. The next critical path is A → B → D → G; the expected project duration is 25 days.

(c) The project duration which has 95% probability of completion is

$$Z(1.64) = \frac{T_s - 27}{\sqrt{18}} \quad \text{or } T_s = 34 \text{ days}$$

11.17 (a)

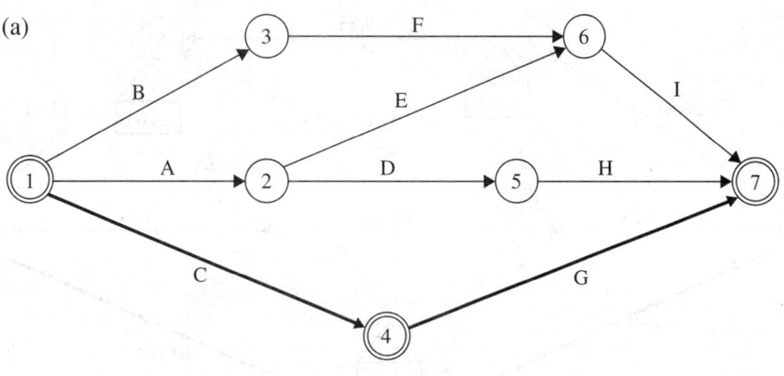

(b) The critical path is C → G, that is, $1 \to 4 \to 7$.

Task	T_e	$V_{ij} = \sigma_{ij}^2$	Earliest time		Latest time		Total float
			Start	Finish	Start	Finish	
A	7.8	0.69	0	7.8	9.0	16.8	9.0
B	20.0	0.44	0	20.0	9.0	29.8	9.8
C	33.0	5.34	0	33.0	0	33.0	0
D	18.0	0.44	7.8	25.8	16.8	34.8	9.0
E	20.0	2.78	0.8	27.8	1.8	38.8	11.0
F	9.0	1.00	20.0	29.0	29.8	38.8	9.8
G	9.8	0.69	33.0	42.8	33.0	42.8	0
H	8.0	0.11	25.8	33.8	34.8	42.8	9.0
I	4.0	0.11	29.0	33.8	38.8	42.8	9.8

The variance of the time schedule is 6.03.

(c) The probability of completing the project in 41.5 weeks is 0.30. The duration of the project having 95% chance of completion is 46.85 weeks.

11.18 (a)

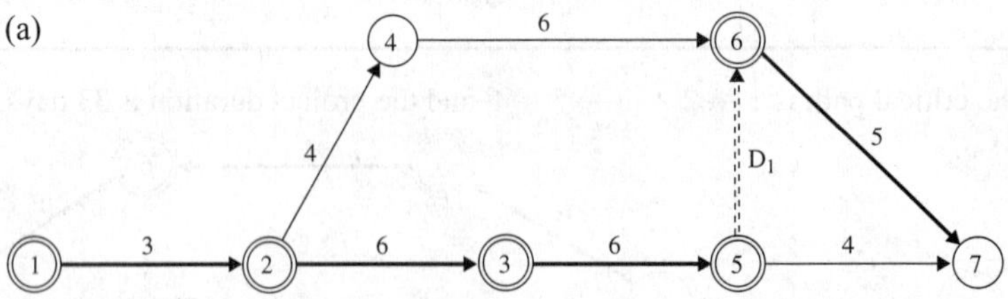

(b) The critical path is $1 \to 2 \to 3 \to 5 \to 6 \to 7$.
The variance of project length, $\sigma_t^2 = 0 + 1 + 4/9 + 0 + 1 = 2.444$.

(c) The expected project length is 20 weeks. The probability of the meeting schedule is 0.9726 or 97.26%.

11.19

Activity	1–2	2–3	2–4	3–5	4–6	5–6	5–7	6–7
T_e	3	6	4	6	6	0	4	5
$V_{ij} = \sigma_t^2$	0	1	4/9	4/9	4/9	1	1/9	1

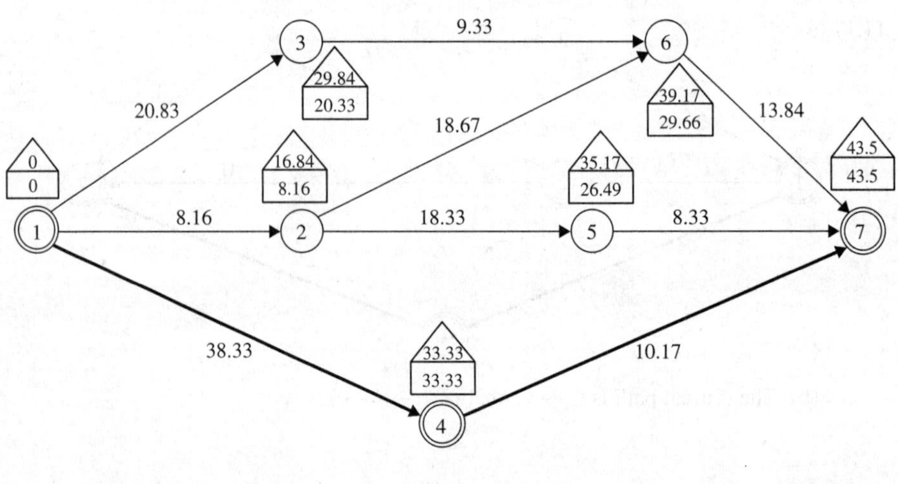

The critical path is 1 → 4 → 7. The project duration is 43.50.
The variance of the time schedule is 6.14.

11.20 (a)

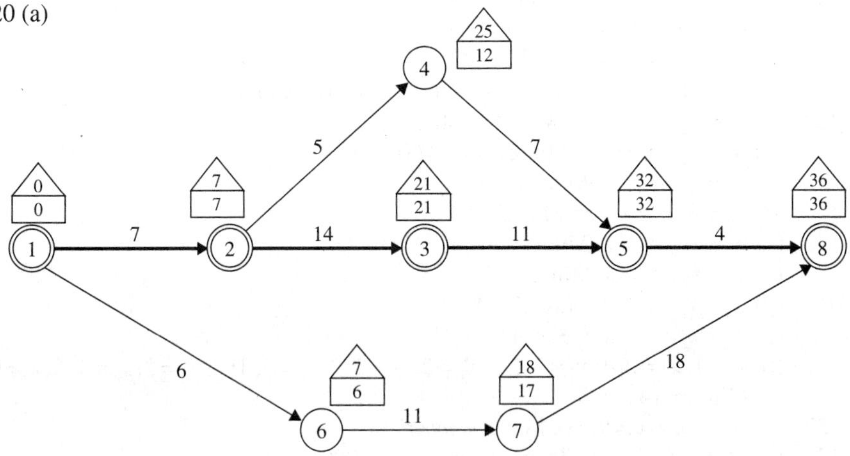

(b) The critical path is 1 → 2 → 3 → 5 → 8. The expected time duration is 36. The variance of the expected duration of the project is 25.

11.21 The optimum project duration is 17 days and the project cost is ₹15,000.

11.22 (a)

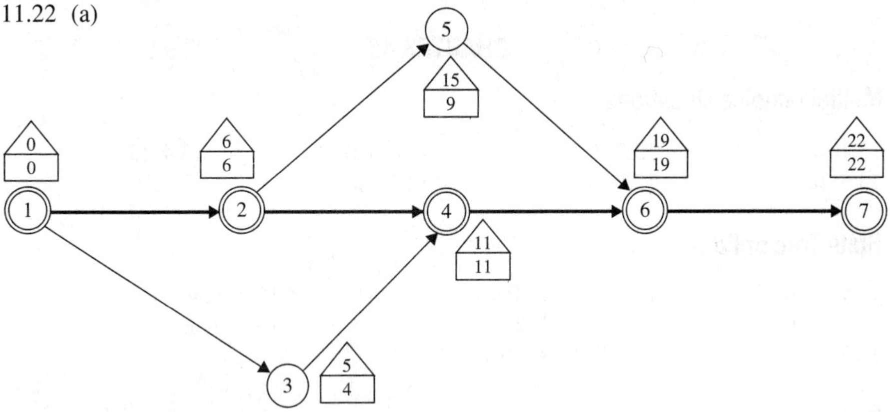

(b) The critical path is 1 → 2 → 4 → 6 → 7.

(c) The minimum total time is 13 days and the corresponding cost is ₹1145.

11.23 The optimum project duration is 12 days and the project cost is ₹17,800.

CHAPTER-12

Multiple-choice Questions

12.1 (c) 12.2 (b) 12.3 (a)

State True or False

12.1 True 12.2 True 12.3 False 12.4 False

Fill in the Blanks

12.1 dynamic programming 12.2 stages
12.3 *n* 12.4 input

Numerical Problems

12.1 $1 \rightarrow 3 \rightarrow 4 \rightarrow 8 \rightarrow 9$; 47 units

12.2 $1 \rightarrow 2 \rightarrow 3 \rightarrow 5 \rightarrow 7$; 21 units

12.3 $1 \rightarrow 3 \rightarrow 7 \rightarrow 11 \rightarrow 12$; ₹20

12.4 $1 \rightarrow 2 \rightarrow 6 \rightarrow 8 \rightarrow 10$ or $1 \rightarrow 4 \rightarrow 6 \rightarrow 8 \rightarrow 10$; 14 units

12.5 (a) $x_1 = x_2 = x_3 = 10/3$. Min $Z = 100/3$

(b) $x_1 = x_2 = x_3 = 1/3$. Min $Z = \log(1/3)$

(c) $x_1 = x_2 = x_3 = 3$. Min $Z = 36$

(d) $x_1 = 8, x_2 = x_3 = 0$. Max $Z = 64$

(e) $x_1 = x_2 = x_3 = 2$. Min $Z = 20$

(f) $x_1 = x_2 = x_3 = 3$. Min $Z = 27$

(g) $x_1 = x_2 = x_3 = 4$. Max $Z = 64$

(h) $x_1 = x_2 = x_3 = x_4 = 3$. Max $Z = 81$

(i) $x_1 = 1; x_2 = 2, 3; x_3 = 3, 2$ or $x_1 = 2; x_2 = 1, 3; x_3 = 3, 1$; or $x_1 = 3; x_2 = 1, 2; x_3 = 2, 1$. Min $Z = 14$

12.6 $x_1 = x_2 = \ldots = x_n = (c/n)$. Max $Z = (c/n)^n$

12.7 $x_1 = 2, x_2 = 0, x_3 = 0$. Maximum value is 62,000

12.8 First, second, third, and fourth companies to E_2, E_3, E_3, and E_3 type educational institutes respectively. Maximum return is ₹15 lakh

12.9 $x_1 = 2, x_2 = 5, x_3 = 4, x_4 = 4$. Maximum reliability is .1512

CHAPTER-13

Multiple-choice Questions

13.1 (a) 13.2 (c) 13.3 (d) 13.4 (a)

13.5 (d) 13.6 (a) 13.7 (d)

State True or False

13.1 True 13.2 True 13.3 False

13.4 True 13.5 True 13.6 True

13.7 True 13.8 False

Fill in the blanks

13.1 AILPP 13.2 MILPP 13.3 free

13.4 fathomed 13.5 ≤ 13.6 MILPP

Numerical Problems

13.1 (a) $x_1 = 0, x_2 = 2$, Max $Z = 6$ (b) $x_1 = 1, x_2 = 1$, Max $Z = 5$

(c) $x_1 = 4, x_2 = 4$, Max $Z = 48$ (d) $x_1 = 0, x_2 = 2$, Max $Z = 2$

13.2 (a) $x_1 = 3, x_2 = 5$, Min $Z = 27$ (b) $x_1 = 0, x_2 = 2$, Max $Z = 16$

(c) $x_1 = 3, x_2 = 4$, Max $Z = 18$ (d) $x_1 = 0, x_2 = 0, x_3 = 0$, Max $Z = 0$

13.3 Max $Z = 4x_1 + 7x_2 + 6x_3 + 5x_4 + 4x_5$

Subject to $5x_1 + 8x_2 + 3x_3 + 2x_4 + 7x_5 \leq 112$

$x_1 + 8x_2 + 6x_3 + 5x_4 + 4x_5 \leq 109$

$x_j \geq 0$ and are integers.

CHAPTER-14

Multiple-choice Questions

14.1 (a) 14.2 (b) 14.3 (b)
14.4 (c) 14.5 (b) 14.6 (d)

State True or False

14.1 False 14.2 True 14.3 False
14.4 False 14.5 True

Fill in the Blanks

14.1 $1/\lambda$ 14.2 both 14.3 arrival, departure

Numerical Problems

14.1 (a) 3 trucks (b) 0.06
(c) 1 h and 48 min (d) 2.25 trucks
14.2 14/3 h
14.3 (a) 50% (b) 12 min (c) 2.5 customers, 1 customer.
14.4 (a) 5/6 (b) 5 customers (c) 12 min
(d) 4.17 customers (e) 10 min
14.5 (a) 0.5 (b) 0.5 (c) 1 customer (d) 12 min
14.6 0.3.
14.7 0.166 customer and 0.5 customer
14.8 ₹100 per patient to ₹125 per patient
14.9 8 h
14.10 (a) 3.09 cars (b) 6.06 cars (c) 12.3 min (d) 30.3 cars per h.

CHAPTER-15

Multiple-choice Questions

15.1 (c) 15.2 (a) 15.3 (d) 15.4 (c)

State True or False

15.1 True 15.2 True 15.3 True 15.4 False

Fill in the Blanks

15.1 underachievement 15.2 constraint 15.3 LP 15.4 most positive

Numerical Problems

15.1 Min $Z = d_1^+ + d_2^- + d_3^-$
Subject to $x_1 + x_2 + d_1^- - d_1^+ = 80$
$x_1 + d_2^- = 8$
$x_2 + d_3^- = 10$
and $x_1, x_2, d_1^-, d_1^+, d_2^-, d_3^- \geq 0.$

15.2 Min $Z = p_1 d_1^- + p_2(4d_2^- + 3d_3^-) + p_3 d_1^+$
Subject to $x_1 + x_2 \leq 300$

$$2x_1+x_2+d_1^- - d_1^+=400$$
$$x_1+d_2^-=100$$
$$x_2+d_3^-=250$$
and $x_1, x_2, d_1^-, d_1^+, d_2^-, d_3^- \geq 0.$

15.3 Min $Z = p_1 d_1^- + p_2(d_2^+ + d_2^-) + p_3 d_3^-$

Subject to $8x_1 + 6x_2 + d_1^- - d_1^+ = 36$
$$x_1 - 2x_2 + d_2^- - d_2^+ = 0$$
$$2x_1 + 4x_2 + d_3^- = 20$$
$$3x_1 + 4x_2 + d_4^- = 24$$
and $x_1, x_2, d_1^-, d_1^+, d_2^-, d_2^+, d_3^-, d_4^- \geq 0.$

15.4 (a) $x_1 = 5/2,\ x_2 = 15,\ d_3^- = 5/2,\ x_3 = 5,\ d_1^+ = 200.$

(b) $x_1 = 10,\ x_2 = 0,\ d_2^- = 1000,\ d_3^- = 10.$

15.5 (a) $x_1 = 150,\ x_2 = 150,\ d_1^+ = 50.$

(b) $x_1 = 9/2,\ x_2 = 7,\ d_1^+ = 3/2,\ d_2^- = 7/2.$

CHAPTER-16

Multiple-choice Questions

16.1 (a) 16.2 (b) 16.3 (c) 16.4 (a) 16.5 (a) 16.6 (a)

State True or False

16.1 False 16.2 False 16.3 True 16.4 True
16.5 False 16.6 True 16.7 True 16.8 False
16.9 False 16.10 True

Fill in the Blanks

16.1 fixed 16.2 randomly 16.3 equal
16.4 fair 16.5 same 16.6 dominating

Numerical Problems

16.1 (a) Lower value = 6 and upper value = 6, saddle point exists
(b) Lower value = 0 and upper value = 2, saddle point does not exists
(c) Lower value = –5 and upper value = –5, saddle point exists
(d) Lower value = 4 and upper value = 5, saddle point does not exist
(e) Lower value = 7 and upper value = 7, saddle point exists.

16.2 $\begin{bmatrix} A_1 & A_2 \\ 1/5 & 4/5 \end{bmatrix}, \begin{bmatrix} B_1 & B_2 \\ 3/5 & 2/5 \end{bmatrix}$ and game value = 17/5

16.3 (a) $\begin{bmatrix} A_1 & A_2 \\ 2/5 & 3/5 \end{bmatrix}, \begin{bmatrix} B_1 & B_2 & B_3 & B_4 \\ 0 & 4/5 & 0 & 1/5 \end{bmatrix}$ and game value = 2/5

(b) $\begin{bmatrix} A_1 & A_2 & A_3 & A_4 & A_5 & A_6 \\ 0 & 3/5 & 0 & 2/5 & 0 & 0 \end{bmatrix}, \begin{bmatrix} B_1 & B_2 \\ 4/5 & 1/5 \end{bmatrix}$ and game value = 17/5

16.5 $\begin{bmatrix} A_1 & A_2 & A_3 & A_4 \\ 0 & 0 & 2/3 & 1/3 \end{bmatrix}, \begin{bmatrix} B_1 & B_2 & B_3 & B_4 \\ 0 & 0 & 2/3 & 1/3 \end{bmatrix}$ and game value = 8/3

16.6 $\begin{bmatrix} A_1 & A_2 & A_3 \\ 0 & 0 & 1 \end{bmatrix}, \begin{bmatrix} B_1 & B_2 & B_3 \\ 4/5 & 1/5 & 0 \end{bmatrix}$ and game value = 2

16.7 $\begin{bmatrix} A_1 & A_2 \\ 1/2 & 1/2 \end{bmatrix}, \begin{bmatrix} B_1 & B_2 \\ 1/2 & 1/2 \end{bmatrix}$ and game value = 0

16.8 (a) $\begin{bmatrix} A_1 & A_2 & A_3 \\ 6/13 & 3/13 & 4/13 \end{bmatrix}, \begin{bmatrix} B_1 & B_2 & B_3 \\ 6/13 & 4/13 & 3/13 \end{bmatrix}$

(b) $\begin{bmatrix} A_1 & A_2 \\ 1/2 & 1/2 \end{bmatrix}, \begin{bmatrix} B_1 & B_2 \\ 1/2 & 1/2 \end{bmatrix}$ and game value = 0

16.14 $\begin{bmatrix} A_1 & A_2 & A_3 \\ 2/3 & 1/3 & 0 \end{bmatrix}, \begin{bmatrix} B_1 & B_2 & B_3 \\ 0 & 1/2 & 1/2 \end{bmatrix}$ and game value = 1

CHAPTER-17

Multiple-choice Questions

17.1 (b) 17.2 (a) 17.3 (b) 17.4 (a) 17.5 (d)

State True or False

17.1 (a) False (b) True (c) False (d) True (e) True

Fill in the Blanks

17.1 risk 17.2 courses of action (strategies) 17.3 probabilistic
17.4 uncertainty 17.5 expected opportunity loss

Numerical Problems

17.1 (a) Demand (moderate, high), modernization (S_1), expansion (S_2)
(b) EPV (S_1) = ₹3.5 lakhs, EPV (S_2) = ₹7.5 lakhs
(c) EPVI = ₹6.5 lakhs
17.2 9 dozen, ₹168
17.4 3 per week, ₹4080
17.5 17 copies, ₹7.26
17.6 (a) A, ₹2400 (b) B, ₹6000 (c) C, ₹1000 (d) C, ₹10,600
17.7 660 litres, ₹833.33 17.9 S_1, 3.8, 1

CHAPTER-18

Multiple-choice Questions

18.1 (c) 18.2 (a) 18.3 (a)

State True or False

18.1 True 18.2 True 18.3 False

Fill in the Blanks

18.1 temporary, permanent 18.2 not belonging 18.3 labelled graph

Numerical Problems

18.1 (a) 7 units, 1→2→4
(b) 13 units, 1→3→4→6 or 1→3→5→6 or 1→2→5→6
(c) 9 units, 1→4→5→7
18.2 (a) [(3, 5), (1, 7), (2, 6), (5, 6), (4, 6), (6, 7)]
(b) [(1, 2), (2, 3), (4, 6), (5, 6), (2, 6)]
18.3 80 units 18.4 22 units

CHAPTER-19

Multiple-choice Questions

19.1 (b) 19.2 (c) 19.3 (c) 19.4 (b)

State True or False

19.1 False 19.2 True 19.3 True 19.4 True

Fill in the Blanks

19.1 less 19.2 gradually and suddenly 19.3 annual 19.4 $(1+r)^n$

Numerical Problems

19.1 after 6 years
19.2 3rd year, ₹5200
19.3 we should keep machine A for one year and replace it afterwards.
19.4 every fourth year
19.5 after the sixth year
19.6 after 6 years
19.7 30 years
19.8 group replacement after the second week

CHAPTER 20

Multiple-choice Questions

20.1 (b) 20.2 (c) 20.3 (b)

State True or False

20.1 True 20.2 False 20.3 True 20.4 True 20.5 True

Fill in the Blanks

20.1 Simulation
20.2 cumulative
20.3 design, conduct
20.4 seed
20.5 Random variate

Numerical Problems

20.3 u_2, u_3, u_4, u_5 is the requisite sequence of random numbers
20.4 −1.9, 0.125, −1.91, −0.065, −1.32
20.5 (a) 620 (b) 7 (c) 19
20.6 600, 6.4, and 14.5
20.8 22 breads per day
20.9 (i) 0.8 minutes (ii) 3 minutes

CHAPTER-21

Multiple-choice Questions

21.1 (a) 21.2 (c) 21.3 (a) 21.4 (b) 21.5 (a) 21.6 (c)

State True or False

21.1 True 21.2 False 21.3 True 21.4 True 21.5 False 21.6 True

Fill in the Blanks

21.1 inventory 21.2 holding 21.3 reorder level 21.4 less than
21.5 just-in-time production 21.6 ABC

Numerical Problems

21.1 Order 100 units whenever the inventory level drops to 200 units.
21.2 71.5 tons
21.3 1005
21.4 $Q^* = 358$ units, $t^* = 14$ working days, TC = ₹858.65 per year
21.5 $Q^* = 100$ units, $t^* = 15$ days
21.6 $Q^* = 4000$ units, TC = ₹30,270
21.7 12,500 tons 21.8 870 units 21.9 894 units
21.10 30 newspapers 21.11 4 TVs per week 21.12 between 500 and 600

CHAPTER-22

Multiple-choice Questions

22.1 (b) 22.2 (a) 22.3 (a) 22.4 (c)

State True or False

22.1 False 22.2 False 22.3 False 22.4 True

Fill in the Blanks

22.1 single variable unconstrained 22.2 maximum
22.3 negative definite 22.4 equality

Numerical Problems

22.1 At $x = 400$
22.4 $l^2/8w$
22.8 At (0, 0), (0, a), and (a, 0), u is neither the maxima nor minima. At (a/3, a/3) u is minimum if $a < 0$ and u is maximum if $a > 0$.
22.9 At (0, 0) f is minimum and at (–4/3, 8/3) f is maximum.
22.10 At (π/3, π/3, π/3) f is maximum.
22.11 Saddle point is at (0, 0).
22.13 Saddle point is P (1/2, 2/3, 4/3).
22.14 $8a^3/3\sqrt{3}$
22.15 (4/3, 7/9, 4/9)
22.16 (a) (1, –2) minimum $f = 5$ (b) (35/11, 12/11) maximum $f = 281$.

CHAPTER-23

Multiple-choice Questions

23.1 (c) 23.2 (b) 23.3 (b) 23.4 (d) 23.5 (c)

State True or False

23.1 False 23.2 True 23.3 True 23.4 False 23.5 True 23.6 True

Fill in the Blanks

23.1 search techniques
23.2 differentiable
23.3 slow
23.4 pattern
23.5 Zoutendijk

Numerical Problems

23.1 $x = 2.66667, f(x) = 6.22222$
23.2 $x = 0.4055, f(x) = -0.30658$
23.3 $x_1 = 1, x_2 = 2$
23.4 $x_1 = 1, x_2 = 3, f(X) = -25$
23.5 $x_1 = -1, x_2 = 1.5, f(X) = -1.25$
23.6 $x_1 = 1.75, x_2 = 3.2.5, f(X) = -8.375$
23.7 $x_1 = 48/23, x_2 = 40/23$
23.8 $f(X)= 1.0336$
23.9 $x_1 = 1, x_2 =2$
23.10 $x_1 = 1.6, x_2 = 1.2, f(X) = 0.8$
23.11 $x_1 = 20, x_2 = 20, x_3 = 20$, and Max $f = 8000$ cm^3
23.12 $x_1 = 3, x_2 = 4$

CHAPTER-24

Multiple-choice Questions

24.1 (a) 24.2 (a) 24.3 (a) 24.4 (d) 24.5 (c)

State True or False

24.1 True 24.2 True 24.3 False 24.4 True

Fill in the Blanks

24.1 $\leq$ 24.2 $\geq$ 24.3 non-separable 24.4 approximate

Numerical Problems

24.1 (a) $(2, \sqrt{5})$, max $f = 42$ (b) No solution
24.2 At (40, 40, 40), $f = 6000$
24.3 (a) $x_1 = 2/3, x_2 = 14/9$, Max $Z = 22/9$ (b) $x_1 = 0, x_2 = 1$, min $f = 2$
24.4 (a) $x_1 = 2, x_2 = 1$, max $f = 7$ (b) $x_1 = 1/8, x_2 = 0, x_3 = 7/8$, max $f = 1/64$
24.5 (a) $x_1 = 1, x_2 = 24/7$, max $f = 69/7$ (b) $x_1 = 3, x_2 = 7$, max $f = 16$

Index

Related Titles

Robotics: Fundamental Concepts and Analysis (9780195673913)

Ashitava Ghosal, Professor, Department of Mechanical Engineering, Indian Institute of Science, Bangalore

Robotics: Fundamental Concepts and Analysis is designed to fully meet the requirements of undergraduate students of mechanical engineering for a course in robotics. With the inclusion of some key advanced topics, the book also serves as an introductory text for postgraduate students specializing in robotics. The book introduces the science and engineering of robotics and provides in-depth coverage of mechanical manipulation.

Key Features

- Emphasizes the fundamental concepts and tools for analysis
- Reinforces theoretical concepts through numerous examples
- Provides insight into advanced topics such as flexible manipulators and mobile robots

Introduction to Mechatronics (9780195687811)

Appuu Kuttan K.K., Director, MANIT Bhopal

Introduction to Mechatronics is designed to serve as a text for undergraduate and postgraduate students of mechanical, electronics, and electrical engineering disciplines. It covers the fundamental principles of mechatronics and their applications. Beginning with an introduction to the subject, the book discusses at length the mechanical, electronics/electrical, and computer engineering aspects of mechatronics.

Key Features

- Covers system modelling, analysis, and application of microprocessors and microcontrollers in mechatronic systems
- Provides detailed review of sensors and actuators in mechatronic systems
- Includes application of mechatronic systems in autotronics, bionics, and avionics

Engineering Metrology and Measurements (9780198085492)

N.V. Raghavendra, Professor and Head, Department of Mechanical Engineering, National Institute of Engineering, Mysore

L. Krishnamurthy, Professor, Department of Mechanical Engineering, National Institute of Engineering, Mysore

Engineering Metrology and Measurements is a textbook designed for students of mechanical engineering and allied disciplines to help them understand the various shop floor measurement techniques and the basics of mechanical measurements. Beginning with a conventional introduction to the principles and standards of measurement, the book takes the reader through important topics of metrology such as limits, fits and tolerances, linear measurements, angular measurements, comparators, optical measurements, inspection, and quality control.

Key Features

- Covers both metrology and mechanical measurements in one volume to suit the needs of different syllabi
- Offers guidelines for the proper use and maintenance of important instruments such as vernier calliper, autocollimator, slip gauge, and pyrometer
- Includes a detailed laboratory manual containing 20 experiments with photographs of related equipment, wherever relevant, and comprehensive procedures, observation templates, and model characteristics

Other Related Titles

9780198079972 Ghoshdastidar: *Heat Transfer, 2/e*
9780198078876 Chattopadhyay: *Engineering Thermodynamics*
9780195699630 Ojha, Chandramouli, and Berndtsson: *Fluid Mechanics and Machinery*
9780198070771 Bhattacharyya: *Machine Drawing*
9780198086345 Subramanian: *Professional Ethics*